Ecologically Based Integrated Pest Management

Ecologically Based Integrated Pest Management

Edited by

Opender Koul

Insect Biopesticide Research Centre
Jalandhar, India

and

Gerrit W. Cuperus

1008 E. Franklin, Stillwater,
OK 74074, USA

www.cabi.org

CABI is a trading name of CAB International

CABI Head Office
Nosworthy Way
Wallingford
Oxfordshire OX10 8DE
UK
Tel: +44 (0)1491 832111
Fax: +44 (0)1491 833508
E-mail: cabi@cabi.org
Website: www.cabi.org

CABI North American Office
875 Massachusetts Avenue
7th Floor
Cambridge, MA 02139
USA
Tel: +1 617 395 4056
Fax: +1 617 354 6875
E-mail: cabi-nao@cabi.org

A catalogue record for this book is available from the British Library, London, UK.

A catalogue record for this book is available from the Library of Congress, Washington, DC, USA

ISBN-10: 1 84593 064 9
ISBN-13: 978 184593 064 6

Typeset by SPi, Pondicherry, India.
Printed and bound in the UK by Biddles Ltd, King's Lynn

Contents

About the Editors

Opender Koul, Fellow of the National Academy of Agricultural Sciences and the Indian Academy of Entomology, is an insect toxicologist/physiologist/chemical ecologist and currently the Director of the Insect Biopesticide Research Centre, Jalandhar, India. After obtaining his PhD in 1975 he joined the Regional Research Laboratory (CSIR), Jammu, and then became Senior Group Leader of Entomology at Malti-Chem Research Centre, Vadodara, India (1980–1988). He has been a visiting scientist at the University of Kanazawa, Japan (1985–1986), University of British Columbia, Canada (1988–1992) and Institute of Plant Protection, Poznan, Poland (2001). His extensive research experience concerns insect–plant interactions, spanning toxicological, physiological and agricultural aspects. Honoured with an Indian National Science Academy medal (INSA), the Kothari Scientific Research Institute award, KEC Science Society award and the Recognition award of National Academy of Agricultural Sciences of India for outstanding contribution in the field of Insect Toxicology/Physiology and Plant Protection, he has authored over 150 research papers and articles, and is the author/editor of the books *Insecticides of Natural Origin* (1997), *Phytochemical Biopesticides* (2001), *Microbial Biopesticides* (2002), *Predators and Parasitoids* (2003), *Biopesticides and Pest Management*, Volumes I and II (2003), *Neem: Today and in the New Millennium* (2004), *Integrated Pest Management: Potential, Constraints and Challenges* (2004), *Transgenic Crop Protection: Concepts and Strategies* (2004), *Insect Antifeedants* (2005), published by leading publishers globally. Dr Koul is on the panel of experts in many committees and leading international and national journals. He has also been an informal consultant to BOSTID, NRC of USA at ICIPE, Nairobi.

Gerrit W. Cuperus, was a Regent's Professor and Integrated Pest Management Coordinator at Oklahoma State University for more than 20 years. Dr Cuperus obtained his PhD in 1982, joined the Department of Entomology at Oklahoma State University and has since been involved in national IPM programmes of the USA aiming at the interdisciplinary focus to solve management issues. Dr Cuperus has chaired and served in different capacities in various state and national committees on food safety and pest management. He has made specific contributions in extension/research and has won distinguished service awards from USDA. His Research efforts focused on stored-product pest management have helped to build the Stored Product Research and Education Center (SPREC) at Oklahoma State University. He has authored more than 60 research papers and articles and is an editor of *Successful Implementation of IPM for Agriculture Crops* (1992), *Stored Product Management* (1995) and *Integrated Pest Management: Potential, Constraints and Challenges* (2004).

Contributors

Frank Arthur, *United States Department of Agriculture, Agricultural Research Service, Grain Marketing and Production Research Center, 1515 College Avenue, Manhattan, KS 66502, USA, E-mail: arthur@gmprc.ksu.edu*

Alberto T. Barrion, *International Rice Research Institute (IRRI), DAPO Box 7777, Metro Manila, Philippines*

Johann Baumgärtner, *International Centre of Insect Physiology and Ecology (ICIPE), Nairobi, Kenya; and Center for Analysis of Sustainable Agricultural Systems (CASAS), Kensington, USA, E-mail: j.baumgaertner@bluewin.ch*

R.R. Bellinder, *Department of Horticulture, Cornell University, Ithaca, NY 14850, USA*

Douglas D. Buhler, *College of Agriculture and Natural Resources, 286 Plant and Soil Science, East Lansing, MI 48824, USA*

Paul A. Burgener, *The University of Nebraska Panhandle Research & Extension Center, Scottsbluff, NE 69361, USA*

James F. Campbell, *United States Department of Agriculture, Agricultural Research Service, Grain Marketing and Production Research Center, 1515 College Avenue, Manhattan, KS 66502, USA, E-mail: campbell@gmprc.ksu.edu*

Yolanda Chen, *International Rice Research Institute (IRRI), DAPO Box 7777, Metro Manila, Philippines*

David A. Christian, *The University of Nebraska Panhandle Research & Extension Center, Scottsbluff, NE 69361, USA*

Gerrit W. Cuperus, *1008 E. Franklin, Stillwater, OK 74074, USA, E-mail: gcuperus1@cox.net*

Norman C. Elliott, *USDA–ARS, Plant Science and Water Conservation Laboratory, Stillwater, OK 74075, USA, E-mail: norman.elliott@ars.usda.gov*

Thomas W. Fuchs, *Extension IPM Coordinator, Texas Cooperative Extension Center, San Angelo, TX 76901, USA, E-mail: t-fuchs@tamu.edu*

Kristopher L. Giles, *Department of Entomology and Plant Pathology, Oklahoma State University, Stillwater, OK 74078, USA, E-mail: kgiles@okstate.edu*

Geoff M. Gurr, *Pest Biology and Management Group, Faculty of Rural Management, The University of Sydney, PO Box 833, Orange, New South Wales, 2800, Australia, E-mail: ggurr@orange.usyd.edu.au*

Robert G. Hartzler, *College of Agriculture, Department of Agronomy, Iowa State University, Ames, IA 50011, USA, E-mail: hartzler@iastate.edu*

Barry J. Jacobson, *Department of Plant Sciences and Plant Pathology, 119 AgBiosciences Facility, Montana State University, Bozeman, Montana, USA,* E-mail: *uplbj@montana.edu*

Gary C. Jahn, *International Rice Research Institute (IRRI), DAPO Box 7777, Metro Manila, Philippines,* E-mail: *G.Jahn@CGIAR.ORG*

Sean P. Keenan, *Department of Entomology and Plant Pathology, Oklahoma State University, Stillwater, OK 74078, USA,* E-mail: *keenans@okstate.edu*

Philip Kenkel, *Department of Agricultural Economics, Oklahoma State University, Stillwater, OK 74075, USA,* E-mail: *kenkel@okstate.edu*

Julia Kinane, *Risø National Laboratory, Plant Research Department, PO Box 49, 4000 Roskilde, Denmark,* E-mail: *julia.kinane@risoe.dk*

Opender Koul, *Insect Biopesticide Research Centre, 30 Parkash Nagar, Jalandhar 144 003, India,* E-mail: *koul@jla.vsnl.net.in*

Vibeke Langer, *Royal Veterinary and Agricultural University, Hojbakkegaards Alle 13, DK 2630 Taastrup, Denmark,* E-mail: *Vibeke.Langer@agsci.kvl.dk*

Leslie C. Lewis, *USDA–ARS Corn Insects and Crop Genetics Research Unit, Iowa State University, Genetics Laboratory, Ames, IA 50011, USA,* E-mail: *leslewis@iastate.edu*

James A. Litsinger, *1365 Jacobs Place, Dixon, CA 95620, USA,* E-mail: *jameslitsinger@yahoo.com*

R.G. Luttrell, *Department of Entomology, University of Arkansas, Fayetteville, AR 72701, USA,* E-mail: *luttrell@uark.edu*

Michael Lyngkjær, *Risø National Laboratory, Plant Research Department, PO Box 49, 4000 Roskilde, Denmark*

Phillip G. Mulder Jr, *Department of Entomology and Plant Pathology, Oklahoma State University, Stillwater, OK 74078, USA,* E-mail: *philmul@okstate.edu*

George W. Norton, *Department of Economics, Virginia Tech, Blacksburg, VA 24061, USA,* E-mail: *gnorton@vt.edu*

Achola O. Pala, *International Centre of Insect Physiology and Ecology (ICIPE), Nairobi, Kenya*

David R. Porter, *USDA–ARS, Plant Science and Water Conservation Laboratory, Stillwater, OK 74075, USA*

Peter W. Price, *Department of Biological Sciences, Northern Arizona University, Flagstaff, AZ 86011-5640, USA,* E-mail: *peter.price@nau.edu*

Tom A. Royer, *Department of Entomology and Plant Pathology, Oklahoma State University, Stillwater, OK 74078, USA,* E-mail: *rtom@okstate.edu*

A.M. Shelton, *Department of Entomology, Cornell University, New York State Agricultural Experiment Station, Geneva, NY 14456, USA,* E-mail: *ams5@cornell.edu*

Aaron T. Simmons, *Pest Biology and Management Group, Faculty of Rural Management, The University of Sydney, PO Box 833, Orange, New South Wales, 2800, Australia,* E-mail: *asimmons@orange.usyd.edu.au*

Michael W. Smith, *Department of Horticulture and Landscape Architecture, Oklahoma State University, Stillwater, OK 74078, USA,* E-mail: *mike.smith@okstate.edu*

Dale W. Spurgeon, *United States Department of Agriculture, Agricultural Research Service, Areawide Pest Management Research Unit, 2771 F and B Road, College Station, TX 77845, USA,* E-mail: *spurgeon@usda-apmru.tamu.edu*

Donald B. Thomas, *United States Department of Agriculture, Agricultural Research Service, Kika de la Garza Subtropical Agricultural Research Center, 2413 E. Hwy 83, Weslaco, TX 78596, USA,* E-mail: *dthomas@weslaco.ars.usda.gov*

Pasquale Trematerra, *Università degli Studi del Molise, Campobasso, Italy,* E-mail: *trema@unimol.it*

Mauricio Urrutia, *National Centre for Advanced Bio-protection Technologies, PO Box 84, Lincoln University, Canterbury, New Zealand,* E-mail: *urrutiam@lincoln.ac.nz*

Mark Wade, *National Centre for Advanced Bio-protection Technologies, PO Box 84, Lincoln University, Canterbury, New Zealand,* E-mail: *wadem@lincoln.ac.nz*

Steve D. Wratten, *National Centre for Advanced Bio-protection Technologies, PO Box 84, Lincoln University, Canterbury, New Zealand,* E-mail: *wrattens@lincoln.ac.nz*

Preface

_____

Much has been learned over the past decade about implementing effective IPM programmes in both developed and developing countries. While many pests (insects, weeds, diseases, etc.) are global, factors such as agroecological, cultural, economic and institutional differences dictate location-specific, participatory IPM research. However, in recent past more emphasis has been on ecologically based approaches and there is earnest need to implement them. IPM programmes that include use of natural, host-specific microbial agents have been found effective, for instance, in Indonesia, India, and elsewhere in substituting for chemical pesticides when means for their multiplication and dissemination are appropriately developed. A critical issue with many biocontrol tools is reducing barriers to their commercialization. Similarly, host plant resistance is a fundamental component in most IPM programmes. Fortunately, many breeding programmes in various research institutions and, in some cases, the private sector are producing material that can be integrated into IPM programmes. The possibility of materials being developed through genetic engineering enhances the potential for having host plant resistance as a key IPM strategy.

What is required in an ecologically based IPM (EBIPM) today is to look into the ecological concepts in relation to the incorporation of biotechnology wherever appropriate and analyse the policy, regulatory and socio-cultural factors influencing IPM adoption and impacts. Use of systems modelling along with a major effort to design and implement a technology transfer plan to achieve broad adoption of IPM practices and strategies is necessary. Standardized targets, indicators and benchmarks, especially those related to widespread adoption and impact of ecologically based IPM technologies and systems; need to be used as measures of programme accomplishments. For reaching the conclusive goals it is important to know what has been achieved in terms of EBIPM systems, so far and what needs to be done in the future. We have tried to compile the major aspects of EBIPM in this volume through 18 chapters emphasizing on the ecology/ecological theory, objectives of IPM programmes, economic aspects, tactics and examples of programme delivery. Although we have tried to concentrate on the issues due to limited resources available for IPM, the prospects are bright as discussed in Chapter 1. Examples of emerging technologies and issues include biotechnology, precision agriculture and agroecology. The rapidly increasing computer capacity globally should facilitate the use of systems approach deployment of improved IPM strategies and tactics with ecological concepts in continuous crop management systems. Before going into the details of systems it is imperative to know about the ecology of different pests and Chapters 2 and 3 have comprehensively dealt with these aspects using agriculture weeds and plant pathogens as the base examples. Chapters 4 and 5 discuss the concept of ecological theory emphasizing on the role of cover crops and intercropping using the ecological concepts. As the environment is one of the major components of EBIPM, Chapter 6 deals with the ecological effects of chemical control with an environmental perspective and subsequently social impacts have been comprehensively discussed in Chapter 7. Economics plays a major role in the success of ecologically based pest management programmes. A wide variety of economic analyses of pest management practices and policies have been conducted since the first assessment of economic thresholds more than 40 years ago. Many of the analyses have involved projections of profitability, risk, health and environmental effects, returns to research and implications for public policies affecting pest management decisions. Especially prevalent have been simple per acre budget analyses of IPM practices and analyses of factors influencing IPM adoption. Fewer analyses have addressed aggregate income and environmental/health impacts, and early dynamic modelling of crop–pest–predator interactions have been slow to develop into routine analyses. Dynamic analyses are especially important for assessing pesticide resistance implications of public policies. All these aspects have been discussed comprehensively in Chapters 8 and 9. Various tactics used in IPM programmes are very important vis-à-vis the ecological considerations and Chapters 10–13 discuss the concept in detail.

It is critical to understand that the adoption process is not a discrete, dichotomous event by which one moves from a non-adopter to an adopter by a single decision, but rather involves a process by which adoption occurs. One of the most basic reasons clientele adopt new technology is need, i.e. the grower recognizes a problem or need with which the new technology has potential to provide help. Therefore, programme delivery and dissemination studies have important role to play in the transfer of EBIPM-based technology and require in-depth study based on various crops (both preharvest and postharvest) with specific suggestions for dissemination and delivery. Chapters 14–18 contribute to these aspects. Although the candidate subject is vast, we have tried our best to bring forth state-of-the-art information available on EBIPM strategies through this comprehensive volume.

We received tremendous response and support from all the authors for preparing their chapters in tune with the theme of the book, for which we express our gratitude to them. We are also thankful to Tim Hardwick, CABI, for his patience and cooperation and help at various stages of preparation of this volume.

We hope the book will prove useful to all those interested in promoting the cause of IPM in formal and informal applications in both the developed and developing countries so that sustainability in agriculture system and the environmental protection for future generations is achieved.

Opender Koul
Gerrit W. Cuperus

1 Ecologically Based Integrated Pest Management: Present Concept and New Solutions

Opender Koul[1] and Gerrit W. Cuperus[2]

[1]*Insect Biopesticide Research Centre, 30 Parkash Nagar, Jalandhar 144 003, India;* [2]*1008 E. Franklin, Stillwater, OK 74074, USA*

Introduction

Our basic premise that integrated pest management (IPM) is essential to sustainability stems from our contention that insect pests, pathogenic microorganisms and weeds pose substantial threats to the yields and quality of agricultural commodities. Development of effective means for managing pests is essential to productivity and profitability of agriculture. Production systems that do not include effective pest regulation cannot sustain long-term profitability. We also have concern that the increasing difficulties in controlling these pests experienced over the last 50 years is a result of reliance on single control agents, particularly chemical pesticides. One of the first to voice these concerns was Rachel Carson, in her classic commentary on pesticides entitled *Silent Spring* (Carson, 1962). Since 1962, it has become increasingly apparent that employing chemical controls unilaterally will not provide safe and effective regulation of pests over the long term (Cuperus *et al.*, 1990; Zettler and Cuperus, 1990; Reed *et al.*, 1993). Problems ranging from pesticide resistance in target species (resulting in control failures) to environmental degradation and contamination of food products by pesticide residues have proven that reliance on single tactics seriously detracts from sustainability.

Given the situation, we need to implement a management system to deal with these pests. It is an established fact that IPM has its own potential, constraints and challenges (Koul *et al.*, 2004); however, IPM practices preventing damage to the environment is an essential component of sustainable agriculture. Therefore, one of the goals would be the deployment of ecologically based integrated pest management (EBIPM) practices (Kennedy and Sutton, 2000), the call for which was made in 1996 (Overton, 1996), along with the bio-intensive IPM (Benbrook *et al.*, 1996).

The idea was to shift the IPM paradigm from focusing on pest management strategies relying on pesticide management to a system approach relying primarily on biological knowledge of pests and on their interaction with the crops. This suggests that EBIPM programmes should represent 'a sustainable approach to manage pests combining biological, chemical, physical and cultural tools to ensure favorable economic, ecological and sociological consequences' (Kennedy and Sutton, 2000), i.e. a system based on the underlying knowledge of the managed ecosystem, including natural processes that suppress pest populations. These practices are integrated with biological control organisms and products, resistant host plants, cultural practices and narrow-spectrum pesticides. Our intention is not to influence the reader on a correct IPM definition but to bring clarity on the difference and similarity in definitions that were proposed by Overton (1996) and Benbrook *et al.* (1996). In fact, Royer *et al.* (1999) argued that the conceptual framework that defines IPM, as it was originally conceived, is ecologically based and flexible enough to accommodate any prevailing technologies and economic or social axioms required. Furthermore, it is the responsibility of IPM practitioners to shift their thinking and to design IPM programmes that reflect these important ecological and economic foundations. Emphasis on deployment of EBIPM systems has indeed increased greatly in recent years (Altieri, 1994; Barnett *et al.*, 1996; Boland and Kuykendall, 1998; Kennedy and Sutton, 2000; Altieri and Nicholls, 2003). It has been shown that IPM and integrated cropping systems can control pests effectively, which can contribute to the build-up of beneficial organisms and to the subsequent decrease in soil-borne pests. However, some key issues that need to be highlighted are as follows:

1. In EBIPM, programmes should emphasize an understanding of the ecological relationships between the host plant and the management practices like cultural control, biological control and host plant resistance.
2. Integration of management practices involves biological (e.g. parasites, predators and fungi), chemical (e.g. selective pesticides and pheromones), cultural (e.g. crop rotation, planting date and soil fertility) and physical aspects (e.g. tillage and aeration). These ecological factors regulate the system.
3. Sustainability implies durability over time. This is the true test for a management system.
4. EBIPM programmes should minimize economic, environmental and health risks (National Research Council, 1996).

Whether it is EBIPM or the classical approach of IPM, economics, environment and sociology are the basic forces that give shape to the system. Economic consider-

ations represent the cornerstone of a rational approach focusing on balancing inputs with returns to maximizing profits (see Chapter 8 for more details). It is obvious from the present scenario that comprehensive economic thresholds will integrate dynamic marketing strategies and economic values with variable control costs to make better choices for long-term economic returns (Cuperus *et al.*, 2000). The promotion of IPM must emphasize management of the surrounding ecological system at a cost that does not adversely affect quality. Farmers need to be persuaded to adopt IPM through economic analysis of the options, and then regulatory or environmental constraints may ultimately dictate their management levels.

Environment emphasizes the protection of land, water and other ecological components in an IPM system. In fact, environmental risks associated with pest management include detrimental effects to beneficial and non-target organisms, aquatic toxicity, avian toxicity, and have direct links with the ecological concerns through resource allocations (Benbrook *et al.*, 1996; Cuperus *et al.*, 1997). Therefore, any IPM strategy requires a balance between the economics of production and the environmental stewardship. To establish this balance, it is necessary to integrate long-term planning and implementation on both micro- and macroscales and more from an ecological perspective (Riha *et al.*, 1996).

Social issues present an equally important force (see Chapter 7) because they affect the acceptability of pesticides as a management tool and of new tools like biotechnology (see Chapter 13 for details). These issues encompass endangered species, safety of farm workers and food quality protection. Overall, IPM has to be a socially acceptable approach that would help in developing more sustainable, environmentally sound and economically viable systems (Buttel *et al.*, 1990). Looking at the basic forces vis-à-vis the development of IPM strategies, it is important to consider the consumer concerns before any IPM strategy with effective tools are finalized for any system. Lucerne and stored grains with interesting ecology, where pesticide load is reduced, worker safety has improved and the system is sustainable over time, are dealt with later in the discussion of two model systems.

Consumer Concerns

Health and well-being are highly valued in today's society, and safety of the food supply is of critical concern to consumers. The food system is consumer-driven with the purchaser significantly dictating which foods are produced and how food is produced, processed and distributed. Consumers want a safe, wholesome food supply that is produced without harming the environment, with no danger from contamination by microorganisms, naturally occurring toxins or other potentially hazardous chemicals that may be deliberately or inadvertently added into the food supply. Although scientists generally agree that microorganisms represent the biggest threat to food safety, the general public feels that pesticides, pesticide residues and biotech plants are the most critical issues facing consumers regarding food safety. At present, the European Union is not accepting biotechnology (IFIC, 2003).

National surveys show that more than 75% of consumers are very concerned with the safety of examining food supply because of pesticide residues. Consumers are also concerned about the genetic origin of the crop, though recent surveys indicate consumers are increasingly comfortable with biotechnology (IFIC, 2002; Hallman et al., 2003).

Research shows that the public is concerned about the environment and the overall stewardship and safety of the food system (Collins et al., 1992; Henneberry et al., 1999; IFIC, 2003). Consumer concerns are partially aggravated by:

1. Greatly increased use of biotech varieties;
2. Significantly increased pesticide use and other chemical inputs in the food system (Benbrook et al., 1996);
3. Perceptions that pesticide use will result in increased cancer rates;
4. Uncertainty about the impact of biotech plants on human health;
5. Increased analytical capability to detect minute agricultural chemical levels in food;
6. Public lack of exposure to and understanding of agricultural systems;
7. Lack of control over consumption of biotech plants and pesticide residues;
8. Lack of confidence in the food production system;
9. Great amount of publicity about health issues and agriculture.

Increasing the utilization of agricultural chemicals and biotech varieties has significantly improved the productivity, but with consumer concerns (Table 1.1). Lusk and Sullivan (2002) examined the public concern about biotechnology and found they were similar to their concerns over pesticides. Risk communication is critical for people working with agriculture (Benbrook et al., 1996).

A major concern deals with concerns regarding consumption of biotech plants in the food production, processing and distribution system. There is no control over consumption of biotech plants, and tremendous

Table 1.1. Percentage of public response to perceived risk of impact of pesticides on farm workers, wildlife, producers and the general public.[a] (From Shelton et al., 1997.)

Affected group	Perceived risk (percentage of response)			
	Great deal	Some	Little	No hazard
Farm workers	36	46	11	2
Wildlife	44	46	4	4
Farmer/rancher	31	46	16	2
General public	24	46	21	1

[a]Survey was conducted in Oklahoma City and two rural communities ($n = 400$).

growth of biotech crops has occurred during the past several years (Hallman *et al.*, 2003; PEW Initiative on Food and Biotechnology, 2003). This is, however, true for pesticide residue consumption as well as due to the high worker or farmer exposure (Cuperus *et al.*, 1990; Kenkel *et al.*, 1994a). Therefore, concerns exist with pesticide use, pesticide residues and food safety (Collins *et al.*, 1992; IFIC, 2003) and disappearance of traditional pests due to biotechnology (Jahn *et al.*, 2001). Pesticide resistance in pests has increased (Pedigo, 1999) along with the increase in control costs (Benbrook *et al.*, 1996), which has affected the profit margins. Increased vulnerability may now exist due to widespread planting of a single variety or single gene resistance, and environmental concerns continue to be important, along with food safety issues from pesticides and biotech plants (Cuperus *et al.*, 1991, 2004; Henneberry *et al.*, 1999; IFIC, 2003).

This public perception implies that there is overall bias in views of both pesticide usage and biotechnology. There is clear evidence from studies that while the European Union does not accept biotechnology (IFIC, 2003), others embrace biotech crops as the foundation for IPM (Fitt, 2000). These conflicting views exacerbate the consumers' concerns about the treatments applied to the crops. It is apparent that communication skills will be critical between the layers of production, processing, grocery and the consumer. Cuperus *et al.* (1996) examined how the levels interact and trust each other (Table 1.2). Clearly, a Hazard Analysis Critical Control Point (HACCP) approach seems appropriate (Cuperus *et al.*, 1991) where the food system is examined as a whole focusing on food safety. This approach is critical, especially if the genetic origin of crops with biotech varieties is important for domestic and international consumers.

Table 1.2. Food industry segments rating each other about the existing level of concerns about pesticide residues. (From Cuperus *et al.*, 1996.)

Queried audience	Percentage of response[a]				
	Very concerned	Moderately concerned	Somewhat concerned	No concern	Average rating
Grocers[b]					
Consumers	42.6	35.7	9.9	11.0	1.91
Grocers	53.4	25.4	10.4	10.7	1.78
Processors	46.7	28.7	12.4	12.1	1.90
Producers	43.2	31.7	13.7	11.3	1.93
Processors[c]					
Consumers	33.3	50.0	16.7	0	1.88
Grocers	11.1	50.0	38.8	0	2.28
Processors	44.4	44.4	11.1	0	1.67
Producers	33.3	44.4	22.2	0	1.83
Producers[d]					
Consumers	35.8	12.8	28.2	23.2	2.40
Grocers	17.8	10.2	28.2	43.6	2.97
Processors	7.6	12.8	28.2	51.4	3.24
Producers	30.7	17.9	25.6	25.6	2.43
Government					
Agencies	53.8	12.8	7.6	25.6	2.04
Consumers[e]	37.1	35.3	15.1	13.2	2.04

[a]Rating: very concerned = 1, no concern = 4.
[b]$n = 400$.
[c]$n = 35$.
[d]$n = 38$, [e]$n = 957$.

In contrast, scientific evidence suggests that there is little scientific evidence of any health problems that are due directly to biotechnology or agricultural chemical use (Fernandez-Cornejo and McBride, 2003a). Food and Drug Administration (FDA) reports indicate that no detectable residues were found in more than 65% of all samples and less than 1% of all samples contain illegal pesticide residues (Food and Drug Administration, 2001). Risk perception research indicates that the consumer concerns arise from factors other than scientific risks (Sandman, 1986; Cuperus et al., 1991) like:

- Control – consumers have no control on treatments applied to the production of food or whether the crops are biotechnology varieties.
- Scientific uncertainty about health effects.
- Risks unfairly distributed–producers benefit, consumers are at risk.
- Environmental concerns by the public – present production practices may threaten the environment.
- Political situations are seen as risky or safe.
- Value of chemical inputs is often misunderstood and benefits are poorly delivered to the public.
- Regulatory system is often misunderstood and not trusted by the public.
- Pesticide issues are unfamiliar to the public. Consumers perceive that pesticides are not effectively regulated.

Concerns pertaining to food safety, water quality, pesticide residues, farm worker safety and other perceived issues stem from the basic core of concern with environmental and human risk (Sandman, 1986). Surveys also indicated that consumer concerns about these complex issues stemmed from the above-mentioned issues and from other secondary issues (Table 1.2). Several authors indicate the manner the questions are asked can bias consumer surveys (Fernandez-Cornejo and McBride, 2003b; Hallman et al., 2003). Many consumer concerns were not about their personal safety but about the environment, wildlife, groundwater, and lack of confidence in the production and

regulatory systems. The regulatory system is a label-driven system through which the consumer receives the information about the product. Ecologically produced food is sold with ecolabel, which is the topic of discussion among the international organizations concerned with trade (Barham, 1997). Generally, a consumer looks at these labels for life cycle assessments and processing and production methods, and it is the basis for government regulation. Such regulation for environmental safety and health purposes is permitted under multilateral trade rules and does not cause problems for domestic use. As per General Agreement on Tariffs and Trade, the right of individual countries to set their standards within their boundaries has been established. From the consumer point of view, this has created non-tariff barriers in terms of competitiveness. Thus, at the consumer level, the issues are complex and emotional and must be addressed from scientific and educational aspects. Issues of risk are addressed from the same perspective regarding: food safety, environmental risks and ecological risk (Cuperus et al., 1991; Benbrook et al., 1996; Henneberry et al., 1999); therefore, such situations do influence the implementation of the EBIPM system.

Implementation

The above discussion implies that, although there are many issues, farmers are gaining faith in IPM and try to adopt IPM tactics and techniques; they are going to be the leaders in calling for new research and technology transfer programmes. In terms of implementing IPM, we have some potential examples of IPM with apple, potato and cotton growers. Various tactics used in EBIPM are crop-specific (Coble and Pedigo, 2003). However, to achieve IPM that is ecologically grounded, these tactics need to be seen as elements in an IPM continuum that farmers should implement based on information from a multitude of disciplines. The requirement is to focus on the implementation needs of the farmer, and that should be the goal of research and extension education. To be successful in implementing IPM, thinking only of pest control is not the answer. Implementation requires the partnership of economists, sociologists, ecologists, horticulturists, agronomists, agricultural engineers, geographic information specialists, soil scientists, food processors, crop consultants, pesticide applicators, regulatory agencies, computer scientists, consumers, public policy interest groups and again, most importantly, farmers (Coble and Pedigo, 2003).

Thus, from an implementation point of view, EBIPM should have the basic objectives of safety to humans and the environment, assured profitability for the farmers and long-term sustainability with a focus on host plant resistance, biological control and cultural IPM. By setting clear objectives, research and education programmes have a clear mission. The programmes can then deliver ecologically based approaches that include proper use of aeration, tillage, variety selection, biological control and planting dates (National Research Council, 1996). To implement and evaluate progress of EBIPM (Stark et al., 1990; Frisbie, 1994), agencies like Cooperative Extension are needed to help deliver programmes to the producers and agricultural industry (Stark et al., 1990; Frisbie, 1994; Coble and Pedigo, 2003). Although it is preferable to have one agency that develops and delivers information, it is critical to have the grower input (Lipke et al., 1987; Cuperus and Berberet, 1994; Cuperus et al., 2004).

How should IPM information be delivered? The source from where producers get IPM information is interesting, with the significant changes occurring from universities with focused cropping system efforts like stored grain (http://ipm.okstate.edu/ipm/stored_products/index.html), soybeans (http://www.ipm.uiuc.edu/fieldcrops/insects/soybean_aphids/index.html), cotton (ipmwww.ncsu.edu/cottonpickin), maize (http://www.ipm.iastate.edu/ipm/) and lucerne (Caddel et al., 2001; www.agr.okstate.edu/alfalfa) to private sector sources (Probst and Smolen, 2003; Cuperus et al., 2004) like the Certified Crop Advisors (CCA) could be significant delivery sources. According to Stark et al. (1990), newsletters (hard copy), email

(electronic), fax, etc. are key IPM information sources for growers. Also crop management associations, consultants, field tours, specific publications and videos and Internet-based computer delivery can substantially help the cause of EBIPM. In a small survey of IPM producers in several regions, computers were used by 40% in developed countries; 12.7% used decision-making software for pest control; and 31.1% used the Internet (Sorensen et al., 2000).

Constraints to implementation

- Lack of agency like the US Department of Agriculture (USDA) Extension can be a limitation. It is apparent that one agency needs to feel that it is their responsibility to develop and deliver IPM technology (Coble and Pedigo, 2003).
- Lack of research is clearly a constraint to IPM implementation (Cuperus and Berberet, 1994; Cuperus et al., 2004). Knowing what works and what does not and the consequence of action are critical in programme development.
- Ensuring that growers have a sound understanding of IPM is a critical first step (Cuperus et al., 1990, 2004).
- Lack of interdisciplinary programmes in a given commodity is a constraint that is often not discussed (Lipke et al., 1987).
- Perceived risks and real risks of IPM programme implementation are often constraints that are not addressed (Stark et al., 1990; Cuperus et al., 2004).
- Programme evaluation is commonly not addressed and needs to help focus and improve programmes (Rajotte et al., 1987; Stark et al., 1990).

EBIPM implementation and sustainable agriculture

Two of the most common phrases used over the last 20 years in relation to systems designed for production of food and fibre are 'sustainable agriculture' and 'integrated pest management'. These phrases appear almost invariably in publications that stress efficiency and profitability of production systems and, more emphatically, the necessity of protecting soil, water and the human food supply from contamination by agrochemicals. Our goal in this chapter is to support the concept that improving sustainability of production systems and implementation of EBIPM must be linked. Our basic premise is that employment of principles of IPM is essential to optimizing sustainability of agricultural systems, with a focus on all ecological aspects. We believe that future development and success of IPM are quite important to sustainability of agriculture for the coming centuries.

Many definitions have been proposed to describe sustainable agriculture and IPM, and we realize the necessity of presenting those that we intend to use in assessing the merits of our basic premise. We consider sustainable agriculture to be 'an agriculture that can evolve indefinitely towards greater human utility, greater efficiency of resource use and a balance with the environment that is favorable both to humans and to most other species' (Harwood, 1990). The following conditions (modified from Benbrook, 1990) must be satisfied for agricultural systems to be sustainable:

1. Soil resources must not be degraded through erosion, salination or contamination with toxic compounds (e.g. pesticides).
2. Water resources must be managed to meet the needs for irrigation and to prevent degradation with silt and toxic compounds.
3. The biological and ecological integrity of systems must be preserved through careful management of genetic resources (for both crops and livestock), nutrient cycles and pest species.
4. Production systems must be economically viable, returning an acceptable profit to farmers.
5. Social expectations must be satisfied, and food and fibre needs must be met in terms of quality and quantity of commodities available at reasonable prices to consumers.

These attributes give clear emphasis on two primary goals of sustainable systems: (i) these systems must be economically viable; and (ii) they must contribute to desirable environmental qualities over the long term.

The basic approach for pest regulation that we envision is consistent with these primary goals for sustainability as is evident in a recent accepted definition for IPM. This definition implies that IPM employs ecologically based management processes developed with an understanding of natural cycles and natural regulators of those species that compete with humans for resources in agricultural production systems (Cate and Hinkle, 1994). Successful IPM programmes, by this definition, are those that enhance profitability of agricultural enterprises and protect the environment for the indefinite future. Whether it involves pesticides, food safety or biotechnology, communication is critical for the producer, processor or consumer (Caldwell et al., 2000).

Emphasis on deployment of systems relatively committed to EBIPMs has increased greatly in recent years (Altieri, 1994; National Research Council, 1996; Boland and Kuykendall, 1998; Kennedy and Sutton, 2000) because of continued reliance on crop pesticides. Various studies in the USA and Europe have shown that IPM or integrated cropping systems can control pests effectively and can contribute to the build-up of beneficial organisms and suppression of soil-borne pests. This has helped in developing sustainable cropping systems. Furthermore, deployment of EBIPM can contribute to long-term restoration ecology by facilitating recovery of degraded lands (Dobson et al., 1997). To achieve the ecological goals of sustainable agriculture, crop and pest management practices need to be directed at: (i) suppressing incidence and intensity of a wide range of crop pests; (ii) increasing soil organic matter; and (iii) enhancing soil and crop health. In fact, biodiversity of soil microflora and microfauna is related to the rotation of primary crops with cover crops and green manure crops. Therefore, specific cropping systems as well as cover crops must mesh with the requirements of the primary crops and the farmer in a fully developed EBIPM system and will contribute significantly to the sustainability.

EBIPM implementation and food safety

With the concern of the European Union about genetically modified organisms (GMOs), identity preservation is important in our food system. Consumer concerns about genetic origin and pesticide residues are critical marketing issues. Communication among the different segments of the food system is critical (Cuperus *et al.*, 1991; Henneberry *et al.*, 1999; Phillips *et al.*, 1999; IFIC, 2003). Communication regarding GMOs shows that public reaction responds very much to the way the topic is presented (Hallman *et al.*, 2003). With any risk issue, the public response emphasizes the importance of risk communication and the bias that can occur with improper communication. Although food safety risks are often shown for horticultural crops, concerns are equally pronounced for field crops that are fed to dairy cattle (Ward *et al.*, 1995). Studies also emphasize that all these concerns are related (Cuperus *et al.*, 1991, 2004). Table 1.2 looks at the different levels of the food system and the perception of their concern about food safety. This may be critical if genetic characteristics are important.

Influence of Markets

Pest management is a critical part of the system. Losses from pests from just insects are 13%. Losses in postharvest often exceed 50% in developing countries (PEW Initiative on Food and Biotechnology, 2003; Tabashink *et al.*, 2003; Wolt *et al.*, 2003). Clearly, management of pests must be ecologically addressed to be economically sustainable.

Markets worldwide enable consumers (buyers) to register their preferences with the appropriate currency. Certainly, it is important to register the concerns of producers, processors and others along the marketing channel, and ultimately consumers regarding many aspects of food safety, worker safety

and environmental safety. Their interaction within the marketing system is highlighted in Table 1.2. Concerns and issues need to translate into sustainable production practices to respond to market demand.

EBIPM products can be more expensive to produce or can result in smaller yields. If they prove to be more expensive, consumers must consent to pay a premium price for them. However, consumer willingness to pay higher prices may not necessarily match their level of concern.

Biotechnology might result in lower costs of production and higher yields. Transgenic cultivars were first developed in the 1980s and they became available in the mid-1990s (PEW Initiative on Food and Biotechnology, 2003). They now dominate some systems like cotton, soybeans, and maize in the USA and worldwide. Consumers are very concerned over consuming these biotech varieties (Henneberry *et al.*, 1999; IFIC, 2003; Cuperus *et al.*, 2004) and this issue has not yet adequately addressed. Clearly, it would appear something would have to change when the majority of area is produced in varieties with consumers' concerns. With high losses in crop production and storage, a consumer educational programme is needed. With any concern, there is also an opportunity. However, again, if consumers are concerned about the biotechnology process or some aspects of safety regarding the end product, they may not be willing to purchase them, even if offered at a lower price. Consumer concerns may offset their apparent pocketbook savings. The issue of marketing EBIPM products is often discussed. Consumers have demonstrated a concern with management systems in production. That is particularly true in fresh produce but also with field crops (Collins *et al.*, 1992; Ward *et al.*, 1995; Heneberry *et al.*, 1999). Consumers desire a pesticide residue−free food produced in an ecologically sound fashion (Cuperus *et al.*, 1996).

Numerous studies have documented the opportunities for marketing ecologically based management approaches (Henneberry *et al.*, 1999). If the appropriate approach is made, a market premium may be available for ecologically based management systems.

Clearly, marketing will be the critical factor in acceptance of EBIPM.

Keys in Examples Related to Programme Delivery Information Source

A major application of weather data is key in decision making for current or future pest control activities. By coupling current weather data with the predictions based on models, 'forecasts' for crop development or pest activity are prepared. These forecasts often have great value in allowing farmers to conduct pest control activities in a more timely manner, as is often critical with applications of fungistatic compounds for limiting infections by pathogens. Weather parameters such as degree-day accumulations for insect development or relative humidity conditions will be increasingly important to improve decision making for pest control activities.

Major improvements in data collection for IPM programmes have occurred with the establishment of weather networks such as the Oklahoma Mesonet system. This system collects weather data from 117 sites in Oklahoma, which are used to develop comprehensive summaries of current and past conditions. A number of programmes that make important contributions to the IPM in the state are integrated with this system, e.g. calculations of degree-day accumulations for development of the alfalfa weevil, *Hypera postica* (Gyllenhal). These calculations are used in conjunction with current field-sampling data for decision making relative to the need for insecticide applications (Mulder and Berberet, 1993; Brock *et al.*, 1994). Increasingly, 'site-specific' weather data systems are being developed that will enhance decision making on farm-by-farm or field-by-field bases in the future.

Producers access numerous sources of IPM information (Cuperus *et al.*, 1990; Cuperus and Berberet, 1994). Most producers prefer traditional delivery methods (Certified Crop Advisor, 2003; Probst and

Smolen, 2003). Many times the local contact for fertilizers, pesticides and crop management are the critical information sources (http://www.agronomy.org/cca/). IPM can be the nucleus for interdisciplinary efforts (Lipke *et al.*, 1987; Stark *et al.*, 1990; Bolin, 2003). Research shows that to be an effective interdisciplinary effort, you need a core coordinating effort (Lipke *et al.*, 1987; Stark *et al.*, 1990; Cuperus *et al.*, 1991). This is the challenge of delivery of IPM throughout the world (Lipke *et al.*, 1987; Stark *et al.*, 1990; Jahn *et al.*, 2001; Bolin, 2003). IPM programmes can be the key coordinating focus in programmes.

Producers get most of their IPM information on this subject from their local fertilizer dealer (Table 1.3). About 17% indicated they get advice from the extension agent and the cooperative elevator, and about 11% go exclusively to the extension agent for this type of advice. This emphasizes that this is the source of information where they get the product.

Model Systems

We will use two IPM systems as examples for programme development and delivery: lucerne and stored grain. These systems are fairly perennial, and long-term management is required with short-term actions. These two systems were selected because

Table 1.3. Oklahoma producers' response to where they get advice on insects, weeds and diseases. (From Probst and Smolen, 2003.)

Information source	Number	Percentage
Cooperative elevator	41	47.1
Coop and extension	15	17.2
Extension	10	11.5
Local communication	7	8.1
Coop and commercial	4	4.6
Commercial applicator	1	1.2
Extension and commercial	2	2.3
Publications	1	1.2
Others	4	4.6

Survey of 87 producers.

they represent two contrasting systems seen in IPM delivery: (i) a well-known, well-researched system with an excellent extension programme (lucerne); and (ii) a relatively little-known system with the applied research and very little extension done after the programme started (stored grain). We will talk about how the social, economic and environmental goals were achieved. The ecology of both of these IPM systems was ignored and pesticides were used to try to overcome the lack of understanding. The systems represented two contrasting situations with the research and education fairly high in one system and very low in the other.

Case study of lucerne IPM

Lucerne is produced on 240,000 ha in Oklahoma and is sold for dairy, beef and horse consumption primarily in Oklahoma and Texas (Ward et al., 1995). Lucerne is a perennial crop that can live 5–6 years in Oklahoma. Several insects including alfalfa weevil and spotted alfalfa aphid, severe disease problems including the root rot diseases and severe weed problems are all annual problems of this crop. Economic analysis reveals that potential profit is directly related to stand life. The profitable years are after the costs of establishment are recovered (Caddel et al., 2001). In the 1980s, farmers were not adopting resistant varieties and were not effectively monitoring for weeds and insects, and often had difficulty making a profit because of input costs and impacts on yield (Miller, 1984; Cuperus et al., 2000). Producers often applied herbicides as stands age to maintain a full stand (Caddel et al., 2001). Herbicides were used on nearly 100% of acres of established stands and insecticides were applied 1.8 times/year (Stark et al., 1990). This is an interesting crop sold as a cash crop to dairymen in Texas and Oklahoma (Ward et al., 1995; Caddel et al., 2001). With this market, there also is the emphasis on being weed free. There was also a concern over pesticide residues because the hay is fed to dairy cattle (Ward et al., 1995).

Alfalfa weevil is a severe insect pest of lucerne in Oklahoma. Producers often have to spray more than once on first cutting depending on the product used. The estimates of damage and timing of spray applications are mostly based on visible damage (Table 1.4) and can greatly mislead growers on damage potential. In the early years of alfalfa weevil introduction, fields were white from severe alfalfa weevil defoliation. Producers often mistimed insecticide applications because they did not understand what was occurring in the field. Then they started making multiple applications to prevent damage, but profit was reduced. A more predictive system was desired for timely insecticide applications. A multivariate approach for pesticide spray is presently used, which effectively captures the large ecological variability of this key pest (Stark et al., 1992; Mulder and Berberet, 1993). Weevil damage potential is based on egg numbers, severity of winter, and degree-days in the spring. Significant progress has been made with the growers educated on the increasing importance of the degree-day method and the reduction in treatments based on visible damage (Table 1.5). The ecologically based approach has been shown to increase over time (Cuperus et al., 2000). This approach is based on the winter severity, egg numbers and spring heat units, and is made available every spring on

Table 1.4. Lucerne producers' ranking of factors used by producers to reduce alfalfa weevil. Oklahoma 1988 and 1997. (From Stark et al., 1990[a]; Mulder et al., 2003[b].)

Control strategy	1988		1997	
	N	%	n	%
Grazing	206	32.6	351	42.4
Early harvest	146	23.1	305	36.9
Fall harvest	92	14.6	140	16.9
Variety	89	14.1	195	23.5
Parasites	46	7.3	133	6.8
Predators	41	6.5	184	22.2
Fungal pathogens	12	1.9	17	2.05
Insecticides	–	–	633	75.3

[a]$n = 520$.
[b]$n = 827$.

Table 1.5. Producer ranking of factors when to spray for alfalfa weevil: 1988 compared with 1997. (From Stark *et al.*, 1990; Mulder *et al.*, 2003.)

Method	Total in 1988 (%)	Total in 1997 (%)
Visible damage	56	37
Degree-day method	36	39
Time of year	4	18
Applicator suggestions	2	4
Other producers' treatment	2	2

Day-degree method is the Oklahoma recommended method (Mulder and Berberet, 1993).

Table 1.6. Number of producers who indicated topics of information received and desired from OSU Cooperative Extension, 1988. (From Stark *et al.*, 1990.)

Topic	Information received	Information desired
Seedbed preparation	24	60
Insect control	162	253
Weed control	122	220
Storage	6	40
Marketing	11	98
Variety selection	108	198
Fertilizer	80	202
Harvest management	10	70
Feeding	8	–

the website http://alfalfa.okstate.edu/ (Stark *et al.*, 1993). This has improved with the availability of the data from the Oklahoma Mesonet system (http://agweather.mesonet. org/). Since the growers are involved with the development of this system, they readily understood and accepted the predictions.

In 1983, an interdisciplinary effort was started at Oklahoma State University called the Alfalfa Integrated Management (AIM) effort that worked closely with farmers and applicators focusing on the ecologically based approach (Stark *et al.*, 1990; Cuperus *et al.*, 2000). This comprised local and state extension and research staff that worked closely with local county lucerne growers associations (Miller, 1984; Stark *et al.*, 1990). When asked their needs for lucerne information, producers indicated the ecologically based information on host plant resistance, cultural management and weed and insect control (Table 1.6). This local ownership has been an important aspect of programme delivery.

The AIM team has successfully integrated diverse topics and has always tried to consider input costs and benefits. Example products of this include the Oklahoma Alfalfa Production Calendar (http://alfalfa.okstate. edu/alfa-cal.htm) on the World Wide Web, a project undertaken by the AIM group. The AIM team has written the *Alfalfa Production Handbook* and published it as a hard copy extension circular and as a web page: http:// alfalfa.okstate.edu (Caddel *et al.*, 2001).

Growers focused on grazing the lucerne to reduce weeds and insects (Stark *et al.*, 1990) (Table 1.3). When asked factors other than pesticides that reduced pest number, grazing and early harvest stood out because these were standard practices. Growers reported using several ecologically based, research-proven technologies (Caddel *et al.*, 2001). Table 1.6 demonstrates the progress lucerne growers made due to an intensive educational programme. Growers now receive alerts based on degree-days, weevil numbers and severity of winter instead of treatments based on visible damage.

Producer response to information sources of lucerne IPM information on insect management, weed management and variety selection (Table 1.7) and also producer response to sources of information on weed management (Table 1.8) were interesting including the comfort level that the producers felt with the subject matter and showed how critical neighbours can be in implementing EBIPM such as variety selection. Various accomplishments were:

- Producers have responded to the AIM programme very well. Insecticides are applied 1.2 times now (Mulder *et al.*, 2003) compared with 1.8 in 1988 (Stark *et al.*, 1990). The AIM team now has an early warning system in place that predicts the severity and timing of populations based on egg samples taken at key locations, temperature accumulations and winter severity (http://alfalfa.okstate.edu/).

- Producers have adopted recommended varieties, with nearly 50% using

Table 1.7. Producer response to information sources of lucerne IPM information on insect management, weed management and variety selection. (From Stark *et al.*, 1990.)

Source	Insect management	Weed management	Variety selection
Extension newsletter	29.6	24.4	32.1
Extension agents	18.2	21.3	13.7
Agricultural sales representatives	15.4	30.2	15.5
Neighbours	13.6	13.9	29.6
Farm magazines	12.7	13.0	–
Agricultural lenders	0.6	0.5	–
Others	9.8	3.4	5.8

recommended varieties and probably only 5% using common or unknown varieties. Before the AIM effort, the majority of producers were using unknown or common varieties (Cuperus *et al.*, 2004; Mulder, 2004, unpublished data). Growers using IPM reported a 50% insecticide reduction due to more precise timing (Mulder *et al.*, 2003). This satisfied the economic, social and environmental goals of EBIPM.

- The social and economic accomplishments are met through involvement of growers throughout the programme.
- Producers now look for research-based, ecologically based technology (Stark *et al.*, 1990).

Case study of stored product IPM

The stored product system was characterized by high losses from insects, diseases,

Table 1.8. Oklahoma producer response to sources of information on weed management. (From Mulder *et al.*, 2003.)

Sources of information	Producer response (%) 1988	Producer response (%) 1997
Cooperative extension	43	46
Agriculture sales representative	42	13
Farm publications	8	24
Local farmers	3	16
Others	4	1

high pesticide inputs, high worker exposure to pesticides and grain dust and high pest resistance to pesticides, and clearly was not sustainable. When this system was first addressed in the early 1980s there were no research or extension personnel focused on stored product IPM in Oklahoma. It was not known what species of insects and mould to look for or have any idea on pest population dynamics.

Worker exposure to pesticides is high in the stored wheat postharvest system (Cuperus *et al.*, 1990; Kenkel *et al.*, 1994b; Platt *et al.*, 1998). In the Southern US high plains area, the majority of grain is stored in commercial elevators due primarily to losses in on-farm storage (Cuperus *et al.*, 1990; Kenkel *et al.*, 1994a,b). Losses have been high due to pests including insects and moulds. Oklahoma grain elevator operators' and Oklahoma farmers' responses to where they get advice on stored grain management (Cuperus *et al.*, 1990; Kenkel *et al.*, 1994b) show that large differences occur where people get information depending upon their situation and perspective. Elevator operators primarily get IPM and pesticide information from industry representatives and industry literature. Producers primarily get their information from cooperative extension, industry and industry literature.

In the early years, few elevator operators looked to Oklahoma State University (OSU) for stored grain technology, while growers did so readily (Table 1.9). Both elevator operators and producers store grain but their systems were very different (Cuperus *et al.*, 1990; Kenkel *et al.*, 1994b). In 1989, an

Table 1.9. Elevator operators' response to ranking of reliability of information from sources (Oklahoma, 1991).

Information source	Ranking[a]
FGIS personnel	1.2
Newspapers	3.3
Chemical field man	1.9
Fumigation manual	1.9
Commercial fumigator	2.3
Commodity Credit Corporation/USDA	2.8
Trade magazine	2.6
Extension personnel	3.4
Chemical supplier	2.6
Other farmers/elevator operators	3.5

[a]Data are from surveys of 112 elevator operators. Ranking: 1 = most reliable, 7 = least reliable.

industry-sponsored stored grain IPM effort was started. An advisory committee of elevator operators was formed that stimulated good industry ownership. At the time, the industry interest was due to Oklahoma regulations that required continuing education units (CEUs), the desire for new information in IPM, grain management and equipment and numerous elevator manager suggestions like closed loop fumigation (CLF). There were many pesticide failures due to high pesticide resistance and poor application, and there were safety concerns. We started an industry advisory committee composed of 15 elevator operators, their executive director and the OSU staff involved. The committee still operates. Because of these items, they came and continued to come to the workshops. The group continually addressed issues they were asked to address and the group put items in front of them like preserved identity, CLF and respiratory protection. This, coupled with the change of fumigants from liquids to phosphine in the early 1980s, led to many fumigation failures from poor application and worker exposure.

The industry response from farmers and elevator operators has been the high usage of insecticides as protectants and fumigants (Cuperus et al., 1990). With this high pesticide usage, high pesticide resistance developed to malathion, diclorvos, phosphine and chlorpyrifos methyl (Zettler

and Cuperus, 1990). In some studies, insect numbers are found to be higher in the malathion-treated system than in the control bins (Reed et al., 1993). When the systems are ecologically examined, the population dynamics of insects and moulds, careful utilization of aeration to reduce pest risk and careful utilization of a one-time application of the grain fumigant phosphine allowed the system to operate efficiently and effectively with very limited pest losses (Cuperus and Krischik, 1995; Cuperus et al., 2004). A focus on the ecology, reduced costs, pest resistance and worker exposure allow this system to operate effectively (Pest Management Strategic Plan, 2004). This has resulted in annual elevator IPM and fumigation workshops with nearly 100% of the elevators in attendance.

Individual rankings were also interesting where they ranked the reliability of information sources (Table 1.10). Industry representatives and commercial fumigators were looked on as reliable, whereas the extension personnel were not regarded as reliable for this group.

The accomplishments of this effort were:

1. Every year, attendance at workshops represented nearly 100% of commercial elevators.
2. Fumigant use was reduced substantially comparing the level of 2.6/year in 1990 (Cuperus et al., 1990) with the present usage level of 1.3/year (Kenkel et al., 1994a). This was through proper use of sanitation and aeration focusing on the ecology of the system.
3. Many growers and elevator operators were encouraged not to use malathion (Reed et al., 1993). Usage of malathion has greatly reduced, thus reducing the inputs and consumer concerns. OSU produced a flier that was specifically targetted at producers that might use malathion (Cuperus et al., 2000).
4. The economic, social and environmental goals of EBIPM were met by the 50% reduction of fumigant use and the elimination of malathion use.

Professionals developed and published a national user manual for stored grain IPM (Cuperus and Krischik, 1995), which was made available online (http://pearl.

Table 1.10. Sources of IPM information in farm and commercial grain storage (Oklahoma, USA). (From Cuperus *et al.*, 1990.)

Source	Producer (%)	Commercial (%)
Company field man	12	48
Fumigation manual	10	57
Commercial fumigator	0	31
Commodity Credit Corporation/USDA	0	2
Federal Grain Inspection Service	0	3
Trade magazine	0	19
Extension personnel	54	24
Chemical supplier	25	47
Other farmers/elevator operators	1	19

Data are from surveys of 89 producers and 112 elevator operators.

agcomm.okstate.edu/ipm/wheat/index.html) with a local stored wheat manual (Phillips *et al.*, 2001). As a result of the stored product IPM efforts, the Stored Product Research and Education Centre was built (http://ipm. okstate.edu/ipm/sprec/index.htm).

Goals of EBIPM

We believe that EBIPM should have a focused road map based on the ecological concepts with specific priorities and strategic plans, which are designed with consumer needs and implementation aspects in mind. The specific goal should be to focus on decreasing the economic and environmental risks (e.g. where ecology has to play an important role that cannot be summarily ignored). Although we have tried to concentrate on the issues due to limited resources available for IPM, yet the prospects are bright. Examples of emerging technologies (as will be discussed in subsequent chapters of this volume) and

issues include biotechnology, precision agriculture and ecology. Rapidly increasing computer capacity should facilitate the global use of systems approach deployment of improved IPM strategies and tactics with ecological concepts in continuous crop management systems. To attain EBIPM as described earlier (National Research Council, 1996), the goal is to generate ecological knowledge base for the multitude of cropping systems, environments and pest complexes constituting global agriculture, which has long-term implications achievable through collaboration with farmers. For instance, polyculture using multiline cropping systems has decreased the disease incidence of several foliar pathogens. A mixture of disease-resistant and susceptible cultivars seems a promising method for controlling *Uromyces* bean rust on *Phaseolus* in Columbia (Panse *et al.*, 1997). Some studies in Germany and India based on multiline have enabled farmers to decrease fungicide applications (Wolfe, 1990), a specific goal for EBIPM implementation.

References

Altieri, M.A. (1994) *Biodiversity and Pest Management in Agroecosystems*. Food Products Press, New York.
Altieri, M.A. and Nicholls, C.I. (2003) Ecologically based pest management: a key pathway to achieving agroecosystem health. Available at: http://www.unicamp.br/fea/ortega/agroecol/ecpestma.htm.
Barham, E. (1997) What's in a name? Eco-labelling in the global food system. Available at: http://www. pmac.net/barham.htm.

Barnett, O.W., Backman, P. and van Alfen, N. (1996) Integrated pest management: an evolution to ecologically based practices. *Phytopathology News* 30(June), 87.

Benbrook, C.M. (1990) Society's stake in sustainable agriculture. In: Edwards, C.A., Lal, R., Madden, P., Miller, R.H. and House, G. (eds) *Sustainable Agricultural Systems*. St Lucie Press, Delray Beach, Florida, pp. 68–76.

Benbrook, C.M., Groth, E. III, Halloran, J.M., Hansen, M.K. and Marquardt, S. (1996) *Pest Management at the Crossroads*. Consumers Union, Yonkers, New York.

Boland, G.J. and Kuykendall, L.D. (1998) *Plant–Microbe Interactions and Biological Control*. Marcel Dekker, New York.

Bolin, P. (2003) Oklahoma integrated pest management. Available at: http://entoplp.okstate.edu/IPM/.

Brock, F.V., Crawford, K.C., Elliott, R.L., Cuperus, G.W., Stadler, S.J., Johnson, H.L. and Eilts, M.D. (1994) The Oklahoma Mesonet: a technical overview. *Journal of Atmospheric and Oceanic Technology* 12, 5–19.

Buttel, F.H., Gillespie, G.W. Jr and Power, A. (1990) Sociological aspects of agricultural sustainability in the United States: a New York case study. In: ? (eds) *Sustainable Agricultural Systems*. Soil and Water Conservation Society, Ankeny, Iowa, pp. 515–532.

Caddel, J., Stritzke, J., Berberet, R., Bolin, P., Huhnke, R., Johnson, G., Kizer, M., Lahman, D., Mulder, P., Waldner, D., Ward, C., Zhang, H. and Cuperus, G.W. (2001) Alfalfa production guide for the southern Great Plains. *Oklahoma State University Cooperative Extension Services Circular*, E-826.

Caldwell, A.M., Bantle-Stoner, K., Cuperus, G.W., Payton, M.E. and Berberet, R.C. (2000) Coverage of integrated pest management in major urban newspapers. *American Entomologist* 46, 46–60.

Carson, R. (1962) *Silent Spring*. Fawcett Crest, New York.

Cate, J.R. and Hinkle, M. (1994) *Integrated Pest Management: The Path of a Paradigm*. National Audubon Society, Washington, DC.

Certified Crop Advisor (2003) *American Society of Agronomy*. Available at: http://www.agronomy.org/cca/.

Coble, H.D. and Pedigo, L. (2003) *Integrated Pest Management: Current and Future Strategies*. Council for Agriculture Science and Technology, Task Force Report 140, 246 pp.

Collins, J.K., Cuperus, G.W., Cartwright, B.O., Stark, J.A. and Ebro, L.L. (1992) Consumer attitudes on production systems for fresh produce. *Journal of Sustainable Agriculture* 3, 456–466.

Cuperus, G.W. and Berberet, R.C. (1994) Training specialists in sampling. In: Pedigo, L.P. and Buntin, C. (eds) *Arthropod Sampling in Agricultural Systems*. CRC Press, Boca Raton, Florida, pp. 669–681.

Cuperus, G.W. and Krischik, V. (1995) Stored product management. *Oklahoma State University Cooperative Extension Services Circular*, E-912.

Cuperus, G.W., Noyes, R.T., Fargo, W.S., Clary, B.L., Arnold, D.C. and Anderson, K. (1990) Successful management of a high risk stored wheat system in Oklahoma. *American Entomologist* 36, 129–134.

Cuperus, G.W., Kendall, P., Frisbee, R., Hall, K., Bruhn, C., Deer, D., Woods, F., Branthaver, B., Weber, G., Poli, B., Buege, D., Linker, M., Andress, E., Wintersteen, W., Dost, F., Hegley, F., Aselhage, J., Stelltflug, L., Doe, R. and Murray, G. (1991) Integration of food safety and water quality concepts throughout the food production, processing and distribution educational programs using HACCP philosophies. *Oklahoma State University Cooperative Extension Services Circular*, E-903.

Cuperus, G.W., Owens, G., Criswell, J.T. and Henneberry, S. (1996) Food safety perceptions and practices: implications for extension. *American Entomologist* 42, 201–203.

Cuperus, G.W., Berberet, R. and Kenkel, P. (1997) *The Future of Integrated Pest Management*. University of Minnesota, St Paul, Minnesota.

Cuperus, G.W., Mulder, P.G. and Royer, T.A. (2000) Implementation of ecologically based IPM. In: Rechcigl, J.E. and Rechcigl, N.A. (eds) *Insect Pest Management*. Lewis Publishers, New York, pp. 171–204.

Cuperus, G.W., Berberet, R.C. and Noyes, R.T. (2004) The essential role of IPM in promoting sustainability of agricultural production systems for future generations. In: Koul, O., Dhaliwal, G.S. and Cuperus, G.W, (eds) *Integrated Pest Management: Potential, Constraints and Challenges*. CAB International, Wallingford, UK, pp. 265–280.

Dobson, A.P., Bradshaw, A.D. and Baker, A.J.M. (1997) Hopes for the future: restoration ecology and conservation biology. *Science* 277, 515–522.

Fernandez-Cornejo, J. and McBride, W.D. (2003a) Adoption of bioengineered crops. *Economic Research Service*, AER-810.

Fernandez-Cornejo, J. and McBride, W.D. (2003b) Genetically engineered crops for pest management in U.S. agriculture. *Economic Research Service*, AER-786.

Fitt, G.P. (2000) An Australian approach to IPM in cotton: integrating new technologies to minimize insecticide dependence. *Crop Protection* 19, 793–800.

Food and Drug Administration (2001) *Food and Drug Administration Pesticide Program: Residues in Food.* Food and Drug Administration, Washington, DC.

Frisbie, R.E. (1994) Integrated pest management. *Encyclopedia of Agricultural Sciences* 2, 569–578.

Hallman, W.K., Hebden, W.C., Aquino, H.L., Cuite, C.L. and Lang, J.T. (2003) Public perceptions of genetically modified foods: a national study of American knowledge and opinion. (Publication number 1003–004) Available at: http://www.foodpolicyinstitute.org/docs/reports/NationalStudy2003.pdf; http://www.foodpolicyinstitute.org/docs/reports/NationalStudy2003.pdf.

Harwood, R.R. (1990) A history of sustainable agriculture. In: Edwards, C.A., Lal, R., Madden, P., Miller, R.H. and House, G. (eds) *Sustainable Agricultural Systems*. St Lucie Press, Delray Beach, Florida, pp. 3–19.

Henneberry, S., Quiang, H. and Cuperus, G.W. (1999) Consumer food safety concerns and fresh produce consumption. *Journal of Food Products Marketing* 5, 83–94.

IFIC (2002) Agriculture and food production. Available at: http://www.ific.org//agriculture/index.cfm.

IFIC (2003) More US consumers see potential benefits to biotechnology. Available at: http://www.ogmdangers.org/enjeu/politique/sondage/IFIC_01_03.

Jahn, G.C., Sanchez, E.R. and Cox, P.G. (2001) The quest for connections: developing a research agenda for integrated pest management and nutrient management. *International Rice Research Institute Discussion Paper* 42, 16 pp.

Kenkel, P., Criswell, J.T., Cuperus, G.W., Noyes, R., Anderson, K. and Fargo, W.S. (1994a) Stored product integrated pest management. *Food Reviews International* 10, 177–193.

Kenkel, P., Criswell, J.T., Cuperus, G.W., Noyes, R.T., Anderson, K., Fargo, W.S., Shelton, K., Morrison, W.P. and Adam, B. (1994b) Current management practice and impact of pesticide loss in the hard red wheat post-harvest system. *Oklahoma State University Cooperative Extension Services Circular*, E-930.

Kennedy, G.G. and Sutton, T.B. (2000) *Emerging Technologies for Integrated Pest Management: Concepts, Research, and Implementation.* APS Press, St Paul, Minnesota.

Koul, O., Dhaliwal, G.S. and Cuperus, G.W. (eds) (2004) *Integrated Pest Management: Potential, Constraints and Challenges.* CAB International, Wallingford, UK.

Lipke, L.A., Ladewig, H.W. and Taylor-Powell, E. (1987) *National Assessment of Extension Efforts to Increase Farm Profitability Through Integrated Programs*. Texas Agricultural Extension Service, USDA, Texas.

Lusk, J.L. and Sullivan, P. (2002) Consumer acceptance of genetically modified foods. *Food Technology* 56, 32–37.

Miller, P.D. (1984) Oklahoma alfalfa pest management program annual report 1984. *Oklahoma State University Cooperative Extension Services Circular*, E-845.

Mulder, P. and Berberet, R. (1993) Scouting for alfalfa weevil in Oklahoma. *Oklahoma Cooperative Extension Service Current Report*, CR-7177.

Mulder, P., Seuhs, S.K., Criswell, J.T. and New, M.G. (2003) Measuring adoption of IPM practices in Oklahoma alfalfa – socioeconomic implications for the growers. *Oklahoma Cooperative Extension Services*, L-315.

National Research Council (1996) *Ecologically Based Pest Management*. National Academy Press, Washington, DC.

Overton, J. (1996) *Ecologically Based Pest Management: New Solutions for a New Century*. National Academy Press, Washington, DC.

Panse, A., Davis, H.C. and Fischbeck, G. (1997) Yield formation in mixtures of rust resistant and susceptible plants of common bean. *Journal of Agronomy and Crop Science Berlin* 178, 111–116.

Pedigo, L.P. (1999) *Entomology and Pest Management*. Prentice-Hall, Upper Saddle River, New Jersey.

Pest Management Strategic Plan (2004) Stored hard red winter wheat Oklahoma. Available at: http://ipm.okstate.edu/ipm/sprec/StoredWheat PMSP.pdf.

PEW Initiative on Food and Biotechnology (2003) Genetically modified crops in the United States. Available at: http://pewagbiotech.org/resources/factsheets/display.php3?FactsheetID=2.

Phillips, T.W., Berberet, R.C. and Cuperus, G.W. (1999) Post harvest integrated pest management. In: Francis, F.J. (ed.) *Encyclopedia of Food Science and Technology*, 2nd edn. John Wiley & Sons, New York, pp. 2690–2700.

Phillips, T.W., Noyes, R.T., Cuperus, G.W., Criswell, J.T. and Kenkel, P. (2001) Stored wheat management. *Oklahoma Cooperative Extension Services*, E-831.

Platt, R., Cuperus, G.W., Bonjour, E., Payton, M. and Pinkston, K. (1998) Integrated pest management perceptions and practices in grocery stores. *Journal of Stored Product Research* 34, 1–10.

Probst, T. and Smolen, M. (2003) Salt Fork Watershed agricultural producer survey results. Available at: http://bioen.okstate.edu/waterquality/projects_Programs/Salt_Fork/index.htm.

Rajotte, E.G., Kazmierczak, R.F., Norton, G,W., Lambur, M.T. and Allen, W.A. (1987) The national evaluation of Extension's integrated pest management (IPM) programs. Publ. No. 491-010. Virginia Cooperative Extension Service, Blacksburg, Virginia.

Reed, C., Pederson, J. and Cuperus, G.W. (1993) Efficacy of grain protectants against stored grain insects in wheat stored on farms. *Journal of Economic Entomology* 86, 1590–1598.

Riha, S., Levitan, L. and Hutoon, J. (1996) Environmental-impact assessment – the quest for a holistic picture. In: *Proceedings of the Third National IPM Symposium/Workshop*. USDA, Misc. Publication No. 1542, pp. 40–58.

Royer, T.A., Mulder, P.G. and Cuperus, G.W. (1999). Renaming (redefining) integrated pest management: fumble, pass, or play? *American Entomologist* 45, 136–139.

Sandman, P.M. (1986) *Explaining Environmental Risk*. United States Environmental Protection Agency, Office of Toxic Substances, Washington, DC.

Shelton, K., Cuperus, G., Smolen, M., Criswell, J.T., Koelsch, C. and Pinkston, K. (1997) The urban environment: Oklahoma attitudes and practices. *Oklahoma Cooperative Extension Service*, IPM-6.

Sorenson, A.A., Day, E. and Stewart, P. (2000) Factors affecting the adoption of new technologies. In: Kennedy, G.G. and Sutton, T.B. (eds) *Emerging Trends in IPM: Concepts, Research and Implementation*. APS Press, American Phytopathological Society, St Paul, Minnesota, pp. 12–31.

Stark, J.A., Cuperus, G.W., Ward, C., Huhnke, R., Stritzke, J. and Berberet, R. (1990) Alfalfa integrated management: a case study. *Oklahoma Cooperative Extension Service*, E-899.

Stark, J.A., Berberet, R.C. and Cuperus, G.W. (1992) A multivariate method for temporal predictions of alfalfa weevil, *Hypera postica* (Gyllenhal), larval populations exceeding the economic threshold in Oklahoma. *Environmental Entomology* 22, 305–310.

Stark, J.A., Berberet, R.C. and Cuperus, G.W. (1993) Mortality of overwintering eggs and larvae of the alfalfa weevil, *Hypera postica* (Gyllenhal) in Oklahoma. *Environmental Entomology* 23, 35–40.

Tabashink, B.E., Carriere, Y., Dinnehy, T.J., Moran, S., Sisterson, M., Roush, R.T., Shelton, A.M. and Zahao, J.-Z. (2003) Insect resistance to transgenic Bt crops: lessons learned from the laboratory and field. *Journal of Economic Entomology* 96, 1031–1038.

Ward, C.E., Huhnke, R.L. and Cuperus, G.W. (1995) Alfalfa hay preference of Oklahoma and Texas dairy producers. *Oklahoma Cooperative Extension Services*, E-936.

Wolfe, M.S. (1990) *Intra-crop Diversification, Disease, Yield, and Quality*. British Crop Protection Council, Farnham, UK.

Wolt, J.D., Peterson, R.K.D., Bystrak, P. and Meade, T. (2003) A screening level approach for nontarget insect risk assessment: transgenic Bt corn pollen and the monarch butterfly (*Lepidoptera: Danaidae*). *Environmental Entomology* 32, 237–246.

Zettler, L. and Cuperus, G.W. (1990) Pesticide resistance in *Tribolium* and *Rhyzopertha* in Oklahoma. *Journal of Economic Entomology* 83, 1677–1681.

2 Ecologically Based Management of Plant Diseases

Barry J. Jacobsen

Department of Plant Sciences and Plant Pathology, 119 AgBiosciences Facility, Montana State University, Bozeman, Montana, USA

Introduction

Management of plant diseases involves the integration of four basic control methods: (i) exclusion/avoidance; (ii) eradication/inoculum reduction; (iii) protection; and (iv) genetic resistance (Howard, 1996; Agrios, 2005). Typically, because a complex of diseases needs to be controlled, integration of specific tools from each of these four basic methods is required for disease management on a specific crop. This chapter focuses on ecologically based, non-chemical methods of disease management. Determination of which set of tools to use in integrated disease management is dependent on understanding of the pathogen life cycle, influence of both the abiotic and the biotic environment on both the host plant and the pathogen, disease loss–crop value relationships, presence or absence of the pathogen, survival of the pathogen, availability of effective cultural controls (rotation crops, tillage, etc.), disease-resistant cultivars, biological controls and other tools (such as effective chemical controls).

Exclusion/Avoidance

Several techniques are used to exclude the pathogen from the host plant or to avoid infection by pathogens. These techniques include the use of quarantines, inspection or pathogen-free certification programmes, propagation or cultivation under conditions that do not favour pathogen infection, eradication of alternate hosts and physical methods of excluding pathogens (Fry, 1982; Kahn, 1991; Agrios, 2005). Such methods are effective in areas where the pathogen is absent, where the conditions do not favour pathogen infection or spread and where the pathogen is primarily spread in association with propagation material, soil, water or plant debris. In many situations, these strategies are very economical since no other inputs are required. It should be remembered that some of the most damaging pests worldwide are those introduced to new ecosystems from other countries.

Quarantines and inspections are the tools implemented by national, state or local governments to prevent the introduction of pathogens not known to be present. Every country has quarantine laws to protect their agriculture and plants from pests (insects, weeds, plant pathogens, vertebrates, etc.) not known to be present in their country (Gram, 1960; Kahn, 1991; www.aphis.usda.gov; www.fao.org). When based on science and properly implemented, quarantines and inspections are highly effective and economical. There are numerous examples of regions where foreign plant pathogens have been introduced that have caused devastating losses (Young *et al.*, 1991). Examples are: (i) chestnut blight that devastated American chestnuts in the eastern and central USA; (ii) Dutch elm disease, white pine blister rust, downy mildew of grape in Europe; (iii) coffee rust in South America; (iv) soybean rust in Asia and South America; (v) lethal yellows of coconut palm in Africa; (vi) citrus canker in the USA and South America; (vii) soybean cyst nematode in the USA; and (ix) new mating types of the potato late blight fungus in the USA and Europe (Agrios, 2005). Each of these disasters could have been prevented had the disease cycle been understood and proper science-based implementation of quarantine and inspection procedures been initiated.

The Plant Quarantine Act of 1912 provides the legal framework for programmes operated by the US Department of Agriculture-Animal Plant Health Inspection Service (USDA/APHIS) in the USA. Similar statutes provide the legal framework in other countries. The Food and Agriculture Organization (FAO) of the United Nations provides the technical services and a forum to resolve quarantine disputes between countries. The websites listed in the suggested readings provide extensive information on FAO and USDA/APHIS programmes. USDA/APHIS has inspectors in other countries to pre-inspect plants and plant propagation material shipments to the USA and maintains inspections at all ports of entry. This has been an important programme in allowing the shipment of flower bulbs from Europe, where potato cyst nematode and other quarantined pests are commonly found. Pre-inspection may require certified soil tests for soil-borne pests, specific crop rotations, use of soil fumigants and other procedures. All foreign travellers experience questionnaires and luggage inspection for contraband plant materials, soils, etc. when they enter the USA as part of US quarantine programmes. In addition, USDA/APHIS maintains quarantine facilities where imported plant materials such as tree stocks are observed and tested for pathogens and insect pests for

varying periods before release (Waterworth, 1993). Such procedures and facilities are maintained by many countries worldwide. Although quarantines and inspections are highly effective, they unfortunately have been used as an artificial means to create trade barriers, and currently the World Trade Organization (WTO) requires a scientific basis for their invocation. Critical to proper use of quarantines is the knowledge of the distribution of the pathogen, disease risk in the protected environment, potential controls, potential pathways by which the restricted pathogen might be introduced, the accuracy of detection methods and the likelihood or risk if no action is taken (Young *et al.*, 1991).

Besides the use of quarantines and inspections, other exclusion methods include the production of certified disease-free planting materials and the use of methods to avoid or evade pathogens or their vectors (Kahn, 1991). The use of certified disease-free planting stock should be the foundation disease control method for any plant that is propagated vegetatively. Generally, any viral, viroid, mollicute, vascular bacterial or fungal pathogen present in the mother plant can be expected to be transferred by cutting rhizomes, tubers, buds, rootstocks, etc. (Holling, 1965; Fry, 1982). Disease-free certification programmes have resulted in significant reduction of disease impacts for crops such as potato, trees, small fruits, such as citrus and banana, and many ornamentals, such as roses, geranium, carnation and chrysanthemum. Pathogen-free certification programmes are typically operated by governmental agencies, often at the state or provincial level. These regulatory agency-based programmes focus on specific pathogens and typically have specific tolerances for individual pathogens. For example, potato certification programmes (Shepard and Claflin, 1975; Fry, 1982; Slack, 1993) generally have a zero tolerance for ring rot, potato cyst or root knot nematode; 0.1–5% allowance for certain viruses such as potato virus X, potato virus Y, leaf roll virus and spindle tuber virus; and 1–5% allowance for fungal diseases such as late blight or *Verticillium* wilt. These tolerances may vary based on the number of generations from nuclear or foundation stock status. Requirements to use zero tolerance foundation seed in the production of certified seed requires the use of nuclear stocks that are proven to be pathogen-free, field inspections, appropriate and specific serological and other detection tests for pathogens, storage inspections, winter grow outs and limited generation propagation from nuclear or foundation stock.

A key component of successful potato seed certification programmes is the requirement that only a limited number of generations can be propagated before growers are required to go back to nuclear stocks. This type of programme is often described as a 'flush out' programme. Once again, knowledge of the disease cycle, vector relationships and pathogen survival characteristics are critical to success. In general, it is difficult, if not impossible, to produce certified disease-free propagation material in regions where these same plants are grown commercially. Certified production should be isolated and highly effective pathogen management inputs must be used. A final factor to consider in such regulatory programmes is that politically independent personnel must operate them under laws. Such programmes widely adapted for vegetatively propagated plants may also be used with true seed and these are typically coupled with production in areas not favourable to pathogen infection.

Avoiding pathogens is typically done by growing the propagation material under environmental conditions that do not favour pathogen spread and infection. Examples are the production of bean or crucifer crop seeds in areas of low rainfall using no overhead irrigation by using furrow irrigation to prevent the spread of pathogens, such as bacteria and some fungi, by water splash. Integrating the use of certified disease-free foundation seed with avoidance of a favourable environment for pathogen dispersal allows the production of bean seed free of bacterial blight (*Pseudomonas syringae* pv. *phaseolicola*, *P. syringae* pv. *syringae* and *Xanthomonas campestris* pv. *phaseoli*) and anthracnose (*Colletotrichum*

lindemuthianum) pathogens in the arid areas of the western USA. A similar situation exists for the production of crucifer (cabbage, broccoli, cauliflower, etc.) seed free of black leg (*Plenodomus lingam*) and black rot (*X. campestris* pv. *campestris*) in the Skagit river valley of Washington state in the USA. Before the production of seed in these areas, these diseases have been devastating in wet environments; but these crops can be grown in wet environments with little or no loss if both disease-free seed and appropriate crop rotations are used (Walker, 1952; Fry, 1982). A similar situation exists for potato "seed" production where most "seed" is produced in northern latitudes or high elevations to avoid the presence of high, season-long populations of aphid vectors of potato viruses (Slack, 1993).

Testing or indexing seeds to assure freedom from seed-borne pathogens (bacterial, fungal, nematode and viral) is critical for many vegetable crops including crucifer crops, beans, peas, lettuce and tomato (Bridge, 1996; Agrios, 2005). Research has shown that one infected seed in 10,000 seeds can result in epiphytotics of some diseases (Walker, 1952; McGee, 1995). Selection of cereal seeds from field or heads without signs of smut diseases can be important for growers who do not use fungicide seed treatments.

Other avoidance strategies include removal of plantings as far as possible from overwintering hosts, locating new plantings upwind of older plantings, planting on "new" land not infested with soil-borne pathogens, planting in soils with pH that is unfavourable to disease development and growing plants under conditions that exclude vectors or that are unfavourable for disease development (Fry, 1982; Agrios, 2005). For example, avoiding poorly drained soils where "water mould", root rot pathogens such as *Pythium*, *Phytophthora* or *Aphanomyces*, are the problems is basic to controlling these diseases as is maintaining the relative humidity below 95% in greenhouses to control *Botrytis* grey mould (Dik and Wubben, 2004; Agrios, 2005). Planting high-vigour potential seeds under conditions that favour rapid germination, emergence and plant establishment is used routinely to reduce losses to damping-off diseases. Critical information for use of this technique is the germination and vigour potential of seeds plus knowledge of the critical temperature and moisture requirements for germination. The use of screens and positive-pressure greenhouses are often utilized to exclude virus vectors, and plastic or other mulches can be used to avoid fruit rots or water-splashed inoculum from soil (Fry, 1982; Jarvis, 1992; Agrios, 2005).

The use of spacing, trellises, staking or pruning can be used to change the microclimate such that foliar infection by fungi is reduced. A good example is the use of pruning and leaf removal to reduce *Botrytis* grey mould and other fungal bunch rot damage in grapes (English *et al.*, 1989). Anything that increases air movement and speeds drying of foliage and fruit will reduce damage from fungal pathogens that require wet surfaces for spore germination and infection (Elmer and Michailides, 2004). Fans are often used in greenhouse vegetable production for this reason (Jarvis, 1992). Staking of tomato plants and use of plastic mulches will reduce damage from rain-splashed pathogens such as *Septoria lycopersici* (*Septoria* leafspot) and *Colletotrichum coccodes* (anthracnose) (Agrios, 2005).

Storage problems can be avoided by storage under conditions that are unfavourable for growth of fungi and bacteria or their infection. For many fruits and vegetables, this may mean refrigerated storage where free moisture is avoided. This technique is commonly used to reduce losses from soft rot bacteria and fungi such as *Sclerotinia* and *Botrytis*. For these latter fungi, maintaining storage relative humidity below 90–95% is effective. Also, for fleshy fruits and vegetables, avoiding injuries to the epidermis that are common infection sites is critical. Artificial atmospheres with reduced oxygen and increased levels of carbon dioxide, sulphur dioxide or chlorine dioxide can be used to reduce decay caused by soft rot bacteria, *Fusarium*, *Botrytis* and *Pennicillium* species (Dennis, 1983; Agrios,

2005). Seed storage at moistures below the minimum for avoiding the growth of fungi such as *Aspergillus* and *Penicillium* is widely used in grain storage. In this situation, it is critical to understand the interaction between temperature, seed moisture and the oil composition of seeds. For example, at higher temperatures storage fungi can grow faster and produce heat and moisture by a metabolism that will allow more rapid growth of these fungi, and the given time can allow the growth of fungi and bacteria requiring seed moistures greater than 20–30%. Fungi such as *Aspergillus halophilicus* or *A. restrictus* can grow in seeds in equilibrium with 65–70% humidity (12–13% moisture for starchy cereal seeds, 5–6% for oilseeds such as sunflower, safflower or groundnuts); as these fungi grow, they can raise seed moistures such that rapidly growing fungi such as *A. glaucus, A. candidus* or *A. flavus* can rapidly deteriorate stored grains or seeds. Although starchy cereal seeds can safely be stored at 13–14% moisture, seeds with high oil contents need to be stored at less than 8–10% moisture for safe long-term storage. Short-term storage can be at higher moistures, especially when cold temperatures limit storage mould growth. This is typically used in temperate climates; however, as temperatures warm in the spring, moistures must be reduced below that in equilibrium with 60–65% humidity if long-term storage is anticipated. Another factor that must be considered is that weed seeds are often at higher moistures than grain at harvest and they can provide focal points for initial mould growth in storage. It is critical to understand the storability of a grain, or seed lot is determined not by the average moisture but by the moisture level of the wettest seeds in the storage mass. Aeration with cool dry air can be used to reduce the risk of storage mould and insect damage (Christensen and Meronuck, 1986; Mills, 1989).

Pathogens can be excluded from the host by the use of physical barriers to insect vectors such as screens or films formed by dodecyl alcohol emulsions or paraffinic oils that inhibit pathogen contact or penetration. Such films have been shown to be effective in controlling powdery mildews and several leaf diseases including Black Sigatoka on bananas. Paraffin oil films have also been shown to prevent transmission of stylet-borne viruses. Kaolin clay films have been shown to be effective in protecting apple shoots from infection by the fire blight and powdery mildew pathogens and have been shown to interfere with transmission of *Xylella fastidiosa*, the causal agent of Pierce's disease, by the glassy winged sharpshooter leafhopper. Although not a physical barrier *per se*, reflective mulches have been shown to reduce transmission of pathogens vectored by insects. It is thought that the reflective materials disorient the insect vectors such that they do not feed or alight on protected plants by reflected light. This effect lasts only until the plant canopy covers the reflective mulch (Agrios, 2005).

Eradication and Reduction of Pathogen Inoculum

Methods that reduce or eradicate pathogen inoculum include host eradication, removal of infected plant tissue, sanitation, crop rotation, heat treatment, use of antagonistic plants, trap crops, soil fumigation, disinfection/disinfestation of equipment, storage facilities, pots and many forms of biological control (Zadoks and Schein, 1979; Fry, 1982; Gnanamanickam, 2002; Agrios, 2005). Host plant eradication is often used for management of initial infestations of foreign pathogens and in many nursery situations. Eradication of citrus in areas of Florida where citrus canker has been introduced is a good example of both the effectiveness and limitations of this method of control (Schubert *et al.*, 2001). Eradication of host plants where pathogens survive between crops or alternate hosts has been very effective. Examples include the use of crop-free periods, control of volunteer plants and wild plant hosts near crop areas (Agrios, 2005). Eradication and destruction of elm trees with Dutch elm disease has been a critical part of managing this disease. Control of Johnson grass

for control of maize dwarf mosaic virus is an example of eliminating wild virus overwintering reservoirs. For pathogens that require two hosts to complete their life cycle, eradication of alternate hosts reduces both inoculum and genetic variability. The barberry eradication programme started in the 1930s in the USA has done much to control wheat stem rust and to reduce genetic variability of the rust so that stem rust resistance genes are effective over longer periods. Eradication of barberry has been a critical factor in control of wheat stem rust in the north central USA and northern Europe. Alternate host eradication is also effective for control of white pine blister rust, cedar-apple rust and oat crown rust. However, movement of gooseberry seeds by birds has greatly complicated the eradication programmes designed to control white pine blister rust (Agrios, 2005).

Pruning to remove infected host tissue has been effective for management of fire blight and many fungal canker diseases of woody plants; removal of dropped, mummified peach fruit is an important component of brown rot control programmes. Removal of pruning and dead plant material is important for control of fire blight and apple black rot in orchards and *Botrytis* grey mould in greenhouses. In greenhouse culture, the use of coverings that eliminate ultraviolet light <360 nm will prevent spore production by *Alternaria*, *Stemphylium* and *Botrytis* fungi, thereby preventing spread of these fungal pathogens (Agrios, 2005).

Crop rotation is a powerful disease control tool because most pathogens can attack only the members of single plant families and many cannot survive after the infected host tissue is decayed. Such pathogens are termed soil invaders since they cannot compete with saprophytic organisms for food, and they do not produce long-term survival structures. Pathogens that produce long-term survival structures or those that can compete with saprophytes or that have broad host ranges are called soil inhabitants and rotation is generally of little value for their control (Palti, 1981; Fry, 1982; Koenning *et al.*, 1993; Bridge, 1996; Agrios,

2005). However, rotation of certain crops or green manures may be helpful in controlling some pathogens in the soil-inhabitant class if hatching occurs spontaneously or due to the effect of antagonistic organisms or plant exudates on long-term survival propagules (Bridge, 1996; Davis *et al.*, 1996; Barker and Koenning, 1998; Whitehead, 1998; Mazzola and Mullinix, 2005; Wiggins and Kinkle, 2005). Examples include *Verticillium*, *Sclerotinia* and some cyst nematode species that are negatively affected by rhizobacteria associated with the roots of rotation crops or toxins such as cyanide or glycosinolates released from rotation crop residues. Most bacterial and fungal foliar pathogens are soil invaders, and rotations will provide excellent control if infected crop residues are decayed before a host plant is replanted (Sumner *et al.*, 1981).

The use of bare soil or flood fallows that last for several weeks has been shown to eradicate or substantially reduce populations of several fungal pathogens and nematodes. Flood fallows are most effective in the tropics where anaerobic conditions develop during the flood period (Stover, 1979; Fry, 1982; Bridge, 1996). In some areas, burning of crop residues is used to reduce or eliminate inoculum that normally survives on straw (Hardison, 1976). Burning of residues for disease control is common in grass seed production (Hardison, 1980; Pfender and Alderman, 2002) and in both malt barley and rice culture (Bridge, 1996; Mathre *et al.*, 1997; Cintas and Webster, 2001).

Heat is commonly used to eradicate or reduce pathogen inoculum in soils, bulbs, corms, nursery stocks and seeds (Agrios, 2005). Most plant pathogenic fungi, nematodes and bacteria are killed by 30-min exposure to temperatures in the 50–80°C range, with moist heat being more effective than dry heat. In warm, sunny environments, the use of clear polyethylene plastic mulches will allow soil temperatures to reach 50–55°C to a depth of more than 5–25 cm. This process is called soil solarization and is used widely and effectively in tropical and subtropical regions (Katan and DeVay, 1991; Martin, 2003). When the mulch is kept in place for 2–8 weeks and soil temperatures reach

50–55°C, most soil-borne bacteria, fungi and nematodes in this zone will be killed. There is also evidence that populations of organisms antagonistic to plant pathogens will increase, thus increasing the effectiveness of solarization. Incorporation of ammonia-generating manures before solarization can dramatically increase the efficacy of this procedure. In some cases, the use of organic amendments under plastic covers will result in low oxygen levels and significant pathogen control not attributable to the heat from solarization (Blok *et al.*, 2000). Solarization will result in reduction of many fungal and nematode pathogens in the upper 25 cm of soils, but it will neither provide control of these pathogens at greater depth nor is effective in most temperate climates because high enough soil temperatures are not maintained for sufficient periods. Pasteurization of soils with steam (80°C for 30 min) is commonly used in greenhouse and nursery situations. It is important that soil is not sterilized because this will kill antagonistic organisms and create a biological vacuum such that reintroduced pathogens will cause more damage than in non-sterilized soil. Higher temperatures can result in toxic levels of ammonia and soluble salts (Jarvis, 1992; Agrios, 2005). High temperatures reached in composting can achieve similar results to pasteurization, and there is evidence that pathogen-suppressive microbial communities are established and that antimicrobial compounds can be released from some compost substrates (Scheurell and Mahaffee, 2005; Veeken *et al.*, 2005). Evidence for this is most obvious in containerized nursery stock where use of composted tree bark has provided dramatic control of root rots caused by *Phytophthora, Pythium, Rhizoctonia* and *Thelaviopsis* species (Hoitink *et al.*, 1991; Scheuerell *et al.*, 2005).

Heat treatment of dormant plant parts and seeds is effective since most plant pathogenic fungi, nematodes and bacteria are killed at temperatures lower than is lethal to plant tissues. Generally, plant parts and seeds are soaked in hot water at 50–52°C for 20–30 min though lower temperatures and longer duration are used to free bulbs and rootstocks from nematode infestation. Hot water treatment is used in treating seeds, bulbs and rootstocks to eradicate pathogens. Examples include: crucifer seeds to eradicate *X. campestris* (black rot) and *P. lingam* (blackleg) infection, citrus rootstocks to eradicate the burrowing nematode, *Radolpholus similis*, and in bulbs for control of the bulb and stem nematode, *Ditylenchus dipsaci*. Care must be taken to control both temperature and time as either seeds or other plant materials can be damaged by excessive heat or by long duration of hot water soaks (Bridge, 1996; Agrios, 2005).

Heat is used to obtain meristems free of certain viruses since some viruses do not replicate at high temperatures. For example, potato plants grown for 4–6 weeks at 36°C will have meristems free of most common potato viruses. Virus-free planting stock can be multiplied from these meristems. Dry heat (72°C for 10 days) can also be used to free barley seed from *X. campestris* pv. *translucens* (black chaff) infection without germination damage (Agrios, 2005).

Disinfestation of pruning or potting tools, pots, storage containers and equipment to eradicate pathogens that may be present on cutting tools, in adhering soil, roots or slime from decayed roots, tubers or rhizomes is critical to prevent spread of pathogens or to prevent exposing disease-free plant materials to pathogens associated with the materials. Steam pasteurization is the most effective control though "soft" chemicals such as alcohol, sodium hypochlorite or quaternary ammonium compounds can be effective disinfectants. Frequent disinfection of pruning tools is critical in preventing spread of bacterial pathogens such as fire blight (*Erwinia amylovora*) in apple and pear. Disinfection of cutting knives in vegetatively propagated plants is critical to preventing spread of many diseases caused by bacterial pathogens, e.g. black stem rot (*X. campestris* pv. *pelargonii*) and bacterial wilt (*Ralstonia solanacearum*) of geranium and bacterial stem rot (*Erwinia chrysanthemi*) of chrysanthemum. Spread of sap-transmitted viruses by workers' hands can be reduced by washing hands with soap between plants. Tobacco users

in tobacco and greenhouse tomato culture (Agrios, 2005) had long used this technique to prevent spread of tobacco mosaic virus.

Use of anatgonistic plants or trap crops can be used to control several nematodes and parasitic plants such as striga (Bridge, 1996; Barker and Koenning, 1998; Berner and Williams, 1998; Whitehead, 1998). Members of the marigold family produce root exudates that are toxic to several nematode species. Green manures and organic soil amendments (oil cakes, other by-products, etc.) of several different plants have been shown to reduce populations of plant pathogenic nematodes and fungi in soils (Rodriquez-Kabana, 1986; Sikora, 1992). It is thought that glycosinolates released during breakdown of crucifer residues are antimicrobial. Oat green manures have been shown to reduce the severity of *Aphanomyces* black root rot of sugarbeets. Trap crops can be highly effective for control of cyst nematodes. For example, oilseed radish or mustard cultivars that are resistant to *Heterodera schachtii* can be used to control this nematode. To be effective, these trap crops need to be sown densely and must be in the field for 60–75 days while soil temperatures exceed 12°C, because the nematode does not hatch at lower temperatures (Wilson *et al.*, 1993; Krall *et al.*, 2000). Crotalaria is a trap crop for root knot nematode and black nightshade can be used as a trap crop for the potato golden nematode. Striga can be controlled using trap crops such as cotton or soybeans (Agrios, 2005). The trap crop produces hatching factors that stimulate egg or seed hatch and the nematode invades the trap crop root but cannot complete its life cycle, because feeding sites cannot be established or because the trap crop is killed before completion of the life cycle.

The use of crop borders has been used to reduce spread of aphid-transmitted viruses, particularly those that are non-persistent or stylet-borne. The aphids first alight in these crop border plants and the aphids "clean" their stylets in these crops before moving into sensitive crops such as potatoes, beans, pepper, tomato or squash (Agrios, 2005). Soybean crop borders 3–6 m in width will reduce transmission of potato virus Y to potatoes so long as there is no break between the soybean border and the potato crop (DiFonzo *et al.*, 1996). Breaks between the border crop and the crop to be protected will facilitate aphid landing in the crop to be protected. Maize is often used as a border crop to reduce transmission of cucumber mosaic virus to peppers or squash. This tall crop barrier also reduces injury from blowing soil that can provide infection courts for bacterial diseases such as bacterial spot of pepper or tomato and angular leafspot of cucurbits.

Biological control by antagonistic micro-organisms is perhaps the most common disease control phenomenon worldwide, and their action could be considered as eradication in some cases and protection in others. In some cases the pathogen is killed or deprived of nutrients and in other cases the plant is protected by induction of induced resistance or by depriving the pathogen of an infection niche (Jacobsen and Backman, 1993). However, the ability to manipulate these natural disease-suppressing organisms such that high levels of disease control can be routinely achieved has largely evaded plant pathologists. Although rarely if ever biological control agents (BCAs) eradicate pathogens, they can dramatically decrease pathogen populations in soils and on root, leaf, fruit or tuber surfaces. The development of disease-suppressive soils has been described in many areas (Hornby, 1983; Thurston, 1990, 1991; Cook and Veseth, 1991). Perhaps the classic situation is 'take-all decline', whereby the severity of this wheat disease declines from devastating levels to non-economic levels after 5–7 years of continuous cropping to wheat. This phenomenon is thought to be associated with build-up of *Pseudomonas rhizobacteria* that are antagonistic to the take-all fungus. Development of suppressive soils has been described for other pathosystems including: potato scab, *Phytophthora* root rot of avocado and papaya, *Fusarium* wilt of cucurbits, *Sclerotinia* white mould of lettuce and several genera of nematodes attacking a wide range of plants. The change from a disease-conducive soil to a suppressive soil is generally attributed to increased populations of antagonistic microorganisms (Agrios, 2005). In some situations, specific microbes have been identified and attempts to introduce or supplement populations of these microbes in soils or on leaf surfaces have met with mixed results. The failure to routinely create disease-suppressive conditions by the introduction of microbes is thought to be due to lack of understanding of the ecology of the soil, root or leaf surfaces, and how native and introduced organisms compete for niches and nutrients.

Because of the variable effects of the biological and physical environment, the vast majority of applied BCA research has focused on glasshouse or storage environments where the physical environment is more predictable and stable. The majority of research has focused on control of *Pythium* and *Rhizoctonia* in the soil and foliar or fruit pathogens such as *Botrytis* grey mould, powdery mildew and fruit storage moulds such as *Penicillium*, *Monilinia* and *Botrytis* (Paulitz and Bélanger, 2000; Bélanger and Labbé, 2002; Janisiewicz and Korston, 2002). Reasons for this focus include the relatively high values of vegetables, ornamentals and fruits, the lower cost to register a BCA relative to synthetic pesticides, shorter or absent re-entry periods for BCA compared with synthetic pesticides and, perhaps most important, the greater acceptance of consumers and regulatory agencies for the presence of BCAs on produce (Elad and Stewart, 2004).

There have been some successes with specific BCAs. Their mode of action can be classified as antibiosis, parasitism, niche occupation, competition and induced systemic resistance. In addition, some BCAs such as plant growth-promoting, rhizosphere-colonizing bacteria promote plant growth such that seedlings are in a highly susceptible state for a shorter period of time or that plants are more efficient in water or nutrient utilization (Kloepper, *et al.*, 2004). Most BCAs utilize more than one of these modes of action to affect pathogen control. *Trichoderma* species are perhaps the best example since all of the modes of action mentioned above have been implicated for this BCA (Harman *et al.*, 2004).

Antibiosis

Antibiosis is the most common mode of action explored and this likely reflects a chemical control equivalent mindset of many researchers. It should be noted that *in vitro* antibiosis activity does not often predict antibiosis *in vivo* and that many reports of antibiosis being responsible for disease control are based on *in vitro* and not on *in vivo* observations. Examples of BCAs employing antibiosis to control diseases include control of fire blight by *Pantoea agglomerans, Erwinia herbicola* Eh 318 (Stockwell *et al.*, 2002) and *Pseudomonas fluorescens* A 506 (Wilson and Lindow, 1993) and control of take-all of wheat by *P. fluorescens* 2–79 (Thomashow and Weller, 1988). AgraQuest Inc. has commercialized the production of antimicrobial gases by *Muscodor albus* for control of postharvest diseases of fruits and flowers caused by *Botrytis* and *Penicillium* fungi. These gases have also been shown to be effective in controlling soil-borne pathogens such as *Pythium, Aphanomyces, Rhizoctonia* and *Verticillium*. The *M. albus* fungus produces the antimicrobial gases as a by-product of growing on organic substrates and the composition of volatile organic compounds produced varies according to the substrate (Jacobsen *et al.*, 2004b).

Although both *Trichoderma harzianum* and *T. virens* have been shown to produce several antibiotic substances that affect several plant pathogenic fungi and directly parasitize hyphae of fungal pathogens, recent work with antibiotic- or parasitism-deficient mutants has shown that induced resistance is the primary mode of action (Harman *et al.*, 2004). Therefore, *Trichoderma* sp. will be discussed under the induced-resistance section and under other modes of action.

Parasitism

BCAs with parasitism as their mode of action have been developed for both foliar and soil-borne pathogens. Examples of parasitic BCAs used for control of soil-borne pathogens include *T. harzianum* and *T. (Gliocladium)* *virens*; these fungi are formulated as the commercial products, PlantShield, RootShield and T-22 (*T. harzianum*) and SoilGard (*T. virens*), for control of root diseases caused by fungi such as of *Rhizoctonia, Sclerotium, Pythium, Phytophthora* and *Fusarium* (McSpadden Gardener and Fravel, 2002). The fungus *Coniothyrium minitans*, formulated as the products Contans WG and Intercept WG, provides excellent control of *Sclerotinia* white mould by parasitizing the sclerotia (Aertsens and Michi, 2004). The fungus *Paecilomyces lilacinus*, formulated as the product BioAct WG, is a very effective egg parasite of several plant parasitic nematodes and provides economical control of root knot nematode (Kiewnick and Sikora, 2006). Nematode control by the obligate parasite, *Pasteuria penetrans*, has been remarkable where this bacterium can be established, but the inability to mass-produce this organism has limited practical use (Cetintas and Dickson, 2004). Control of powdery mildews by *Ampelomyces quisqualis* is perhaps the classic example of biological control of a foliar disease by direct parasitism. This parasite infects the hyphae, conidiophores, conidia and cleistothecia and forms pycnidia in or on its powdery mildew host (Sztejnberg *et al.*, 1989). This parasite is now formulated as AQ-10 (Ecogen Inc.). Lysis of pathogen hyphae by hydrolytic enzymes is often characteristic of mycoparasitism. This has been demonstrated for several *Trichoderma* species that control fungal pathogens. Other examples include: *Microsphaeropsis* sp. that attack the hyphae and pseudothecia of the apple scab pathogen, *Venturia inequalis,* such that this pathogen does not overwinter or it produces fewer ascospores (Carisse and Rolland, 2004) and *Penicillium purpurogenum* that controls *Monilinia* twig blight on peach (Larena and Melgarejo, 1996).

A unique example of parasitism is the use of hypovirulent strains of *Endothia parasitica* that can be used to control chestnut blight. These hypovirulent strains are infected with double-stranded RNA viruses and these viruses are transferred to healthy mycelia in cankers with the result that the host can cope with the now virus-infected fungus. It is critical to understand that a compatible hypovirulent strain must be used, and there are four or more compatibility types known in nature (MacDonald and Fulbright, 1991).

Although the mode of action might more appropriately be described as mycophagy than parasitism, control of grape powdery mildew (*Uncinula necator*) by the tydeid mite, *Orthotydeus lambi*, is of considerable interest. The number of mildew colonies per square centimetre and the proportion of leaf infected were reduced by >85% in the presence of the mite relative to non-release controls. This mite is considered responsible for preventing major outbreaks of powdery mildew on wild grape (*Vitis riparia*) and that this common and abundant member of leaf fauna on grape and other plants provides significant control of powdery mildew on both wild and cultivated grapes. Leaf morphology can be important in mite colonization and plants with acarodomatia (tufts of hairs or pits located in major leaf vein axes) have higher populations of mycophagous mites (English-Loeb *et al.*, 1999).

Niche occupation and competition for nutrients

These modes of action for BCAs are not mutually exclusive and will be treated together in this section since the BCA commonly competes successfully for an ecological niche in the rhizosphere, on the leaf, flower, fruit or other plant tissues by both competing for nutrients and infection niches. A classic case of niche occupation is the use of *Agrobacterium radiobacter* strains K-84 and K-1026 for control of crown gall. These strains are commercialized as Galltrol and Nogall and are widely used as bare rootstocks dips to prevent crown gall infection. This BCA works by occupying infection niches, i.e. fresh wound on roots. It successfully occupies these niches by production of a bacteriocin that inhibits *Agrobacterium tumefaciens* (Agrios, 2005). Other examples of niche occupation are the use of *Phleviopsis* (*Peniophora*) *gigantea* to inoculate fresh cut pine stumps to prevent infection with the root rot pathogen

Heterobasidium annosum (*Fomes annosus*) (Agrios, 2005) and control of fire blight with *P. fluorescens* A 506 (Wilson and Lindow, 1993). It has been demonstrated that *P. fluorescens* A 506 pre-emptively colonized pear pistils by competing for a limiting nutrient (Fe^{+++}), thus making this nutrient unavailable to *E. amylovora*. The apple and pear stigma is an iron-limited environment and Fe^{+++} is sequestered by *P. fluorescens* A 506 siderophores such that it is less available to *E. amylovora* and that the additional Fe^{+++} is required for *P. fluorescens* A 506 to produce an antibiotic that is toxic to *E. amylovora* (Temple *et al.*, 2004). The use of binucleate *Rhizoctonia* sp. for control of *Rhizoctonia* or avirulent *Fusarium oxysporum* isolates for control of *Fusarium* wilt diseases are other examples of niche occupation (Agrios, 2005).

Biological control by rhizosphere-colonizing fluorescent *Psuedomonas* sp. and other plant growth–promoting rhizobacteria (PGPR) has been shown to be in part due to sequestration of Fe^{+++} by siderophores produced by these BCAs. This sequestration of iron by the BCAs in an iron-limited environment makes it unavailable to a wide range of soil-borne pathogens including *Fusarium* sp., *Rhizoctonia solani*, *Thelaviopsis basicola* and soft rot *Erwinia*. PGPR also competes for nutrients in the rhizosphere that are required for pathogen spore germination (Weller, 1988). Competition for nutrients is also involved with yeasts that are used for control of apple storage pathogens. The yeasts *Cryptococcus laurentii* BSR-Y22 and *Sporobolomyces roseus* FS-43-238 provide control by outcompeting *Botrytis cinerea* for fructose and glucose in apple wounds. Competition for nitrates has also been identified as a factor in the mode of action for the yeast, *Candida guilliermondii*, in the biological control of *Penicillium expansum* rot of apple. Competition for space or nutrients is considered responsible for biological control of apple fruit rots caused by *B. cinerea* by the yeasts *Candida* sp., *C. laurentii*, *S. roseus* and *Candida oleophila*. Competition for nutrients resulted in reduced *B. cinerea* spore germination and slower germ tube growth, thereby reducing pathogen infection (Janisiewicz and Korston, 2002).

Biological control of *B. cinerea* on a wide range of plants with the BCA, *Ulocladium atrum*, is another form of competition whereby the BCA competes for colonization of necrotic tissue with *B. cinerea*. *B. cinerea* is an unusual pathogen in that it colonizes necrotic tissue prior to infection of healthy tissue. *U. atrum* is an excellent saprophyte and, because of pre-emptive colonization of necrotic tissues, it effectively reduces *Botrytis* colonization resulting in reduced sporulation by *B. cinerea*. Because *U. atrum* colonizes necrotic tissues at approximately one-half the rate of *B. cinerea* over a temperature range of 5–25°C, it is critical that *U. atrum* conidia be sufficiently dense and even in distribution if optimal biological control is to be achieved. The interaction between this BCA and *Botrytis* only takes place in necrotic tissue. This is illustrated by the observation that *U. atrum* is not effective in controlling *Botrytis elliptica* on lily because this *Botrytis* sp. can infect healthy leaf tissue (Kessel *et al.*, 2005).

Induced systemic resistance

Effective biological control of plant diseases by BCAs that induce systemic resistance has been described for a wide range of pathosystems including those that involve root- and leaf-infecting fungi and bacteria, viruses and root knot nematodes. These BCAs are applied to seeds, potting media or to the foliage (Kloepper *et al.*, 2004). This type of resistance makes an otherwise susceptible plant resistant to a wide array of subsequent pathogen attack. Elicitation of systemic resistance can be achieved by several basic types of stimuli including: necrotizing pathogens; chemicals; PGPRs; foliar-applied bacteria and fungi; bacterial, oomycete and fungal cell wall fractions; and compounds or microbes associated with composts or compost teas. Several systemic resistance pathways have been described: (i) systemic acquired resistance characterized by salicylic acid signalling and the production of pathogenesis-related (PR) proteins; (ii) systemic acquired resistance characterized by independence from salicylic acid signalling, production of active oxygen species independent of hypersensitive cell death and production of PR proteins; and (iii) induced systemic resistance in which signalling is due to jasmonates and ethylene and is independent of PR proteins. Upregulation of the *NPR1* gene may or may not occur depending on the inducer (Pieterse *et al.*, 1998; Dong, 2004; Kloepper *et al.*, 2004). Implicit to the implication of systemic induced resistance as a mode of action is that the BCA is applied spatially separate from the location where pathogen control occurs. This author feels that this mode of action is more common than uncommon and that most biocontrol workers have not explored this mode of action for BCAs. Typically, BCA induced resistance does not result in immunity but results in reduced levels of pathogen infection and spread. Use of BCA-induced resistance with moderate levels of host resistance often results in excellent control, whereas BCA-induced resistance alone typically results in 50–80% control (Jacobsen *et al.*, 2004a). Therefore, unless other disease control tools are used, results may be unsatisfactory under conditions that are highly favourable to disease development.

The phenomenon of cross protection is another form of systemic acquired resistance. This is where infection of a plant with a mild strain of a given virus protects the whole plant from infection with a more severe strain of the same virus. This technique has been used for control of tobacco mosaic virus in greenhouse tomatoes and for management of papaya ringspot virus in papaya. Although this can be an effective strategy for control of a single virus, the user must be cognizant of problems where other viruses can infect with the result that severe disease losses occur from infection by two viruses. Such an example would be the use of a mild strain of tobacco mosaic virus to protect greenhouse tomatoes followed by infection with potato virus X or other viruses with the result that severe losses occur. The use of this technique has also been limited by the availability of suitable mild virus strains (Agrios, 2005).

Protection

Direct protection of the plant is typically used for endemic pathogens. Protection is accomplished through application of chemical controls of the pathogen and pathogen vectors or by application of biological controls discussed earlier. Chemicals used for disease control include fungicides, bactericides, nematicides, insecticides, miticides and various oils of plant and petroleum origin. The products are applied as seed treatments for control of seed-borne pathogens and to protect against seed rots and damping-off diseases, as foliar sprays, soil treatments, pruning wound treatments and postharvest dips and sprays.

Fungicides are classified as protective-contact non-systemic or systemic and can be either organic or inorganic compounds. For protective-contact fungicides to be effective, they must be present as a barrier to prevent infection or kill fungal tissue they come in contact. Inorganic compounds include sulphur, bicarbonates of sodium, potassium and lithium, mono and dipotassium phosphates, silicon and various copper compounds (some are organic but the active molecule is the copper ion). These are all protective-contact non-systemic fungicides. Organic fungicides have all been developed since the 1930s and are of a wide variety of types with some being strictly protective-contact (captan, thiram, ethylene bis dithiocarbamate fungicides–maneb, zineb, mancozeb, ferbam and heterocyclic compounds–PCNB, chlorothalonil and biphenyl), some locally systemic (imazalil, triflorine) and others are partially or fully systemic (benzimidazoles, triazoles, oxanthiins, strobilurins, metalaxyl). The majority of systemic fungicides are absorbed through the foliage or roots and are translocated in the xylem to new growth; however, some are translocated downward in the plant while others are only locally systemic

fungicides and are translocated to a few cell layers or through the leaf (translaminar). Systemic fungicides can be curative or can eradicate infections only hours or days old. This ability to eradicate established infections is termed kickback activity and it is particularly valuable when using disease prediction models. Initially, most fungicides were effective against a broad range of fungi owing to multiple modes of action and as more environmentally friendly, lower non-target toxicity (humans and other organisms in the environment) fungicides have been developed they have generally been effective on a much narrower spectrum of fungi. Broad-spectrum fungicides typically have multiple modes of action, whereas systemic fungicides often have only a single mode of action. This single mode of action characteristic has allowed many fungal pathogens to develop resistance and resistance management strategies are critical considerations when using systemic fungicides. Fungicide resistance management strategies include application of mixtures of fungicides with different modes of action or alternate applications with different modes of action fungicides. Monitoring pathogen populations for resistance to fungicides or bactericides is critical to the stable use of chemical control tools (Lyr, 1995; Agrios, 2005).

Bactericides are used to control bacterial pathogens. Commonly used bactericides are various inorganic or organic copper compounds and streptomycin and tetracycline antibiotics. These products are used as seed treatments, protectants for cuttings, tubers or rhizomes and as sprays to protect foliage, stems or blossoms. Resistance of bacterial pathogens to antibiotics and copper is known (McManus et al., 2002).

Products used for control of pathogen vectors include insecticides, miticides and nematicides. Insecticides are most commonly used for control of aphid and leafhopper vectors of viruses and phytoplasmas and for control of beetles and other vectors of bacterial pathogens. In general, vector control is only efficient for pathogens where the vector must feed on the plant for 20 min or more to achieve transmission. For this reason, control of vectors of stylet-borne viruses that are transmitted almost instantaneously is not practical (Agrios, 2005).

Nematicides commonly used include organophosphates, carbamate compounds that also have insecticide action and organic compounds that are applied as fumigants. These include dichloropropene–dichloropropane, methyl bromide, metam sodium and metam potassium (Whitehead, 1998; Martin, 2003; Agrios, 2005).

Petroleum and plant oils are used to control several fungal diseases including black Sigatoka leafspot of banana and some powdery mildews and for control of stylet-borne viruses. The oil film on the leaf surface apparently interferes with virus transmission. Use of reflective mulches that disorient aphids and the use of mineral oils are used in some high-value vegetable crops and ornamental nurseries to control stylet-borne viruses such as cucumber mosaic virus and pepper mottle virus (Agrios, 2005).

Genetic Resistance

Genetic resistance to pathogens is one of the most efficient methods of control and it has been the focus of breeding programmes for all types of cultivated plants. Use of disease-resistant varieties has been the control of choice for pathogens such as vascular wilts and viruses for which alternative controls are not highly effective. With the selection of disease-resistant varieties, the grower can largely eliminate the costs for other disease controls. Resistance in non-cultivated plants develops during the evolutionary selection process when plant populations are exposed to pathogens. Plants are resistant to pathogens either because of specific resistance genes or because they are not hosts of the pathogen. Genetic resistance can be complete (e.g. immunity) or incomplete (e.g. something anywhere on the continuum from immunity to susceptibility), and can be conditioned by single genes that condition resistance to specific races of the pathogen (vertical resistance) or by multiple genes

that provide resistance to all pathogen races (horizontal resistance). Vertical resistance is commonly inherited as single dominant genes. In general, most crop plants have specific single genes for resistance to specific pathogens and also have some degree of horizontal resistance to some other pathogens. Because horizontal resistance is not complete, it is often used with reduced application or rates of fungicides (Schtienberg et al., 1994) or in combination with other control tools. Resistance can be expressed as a hypersensitive reaction where the plant responds to initial infection by killing infected cells and initiating a cascade of antimicrobial biochemical reactions, as physical features that allow the plant to escape infection or as factors that reduce the rate of disease development. Rate of disease development can be reduced by extending the period from infection until new infectious propagules are produced (the latent period) or by lesions producing fewer spores or allowing reduced pathogen reproduction.

Critical to understanding of vertical resistance is the gene-for-gene concept. This concept suggests that for each gene for resistance in the host plant there is a corresponding gene for virulence in the pathogen. This explains the development of new races that can attack previously resistant plants, and vertical resistance is considered unstable when dealing with pathogens with sexual stages and multiple infection cycles in a given year. Selection pressure for new races is highest for obligate pathogens since either they must develop new races or biotypes or they become extinct. Examples would be cereal rusts or powdery mildew pathogens. An example of an organism with sexual reproduction and multiple infection cycles would be potato late blight where both mating types occur and favourable conditions exist for infection and sporulation throughout the crop season. Vertical resistance has been highly stable for pathogens with a single disease cycle per year or where no sexual stage exists. Examples of highly stable vertical resistance would be *Fusarium* and *Verticillium* wilt resistance in tomato, cotton and cucurbits. To combat the development of new pathogen races, breeders either utilize several genes for resistance in a single cultivar (pyramiding resistance genes) or they utilize mixtures of cultivars each with different resistance genes (multilines) that would require a pathogen race to have multiple virulence genes or they utilize horizontal resistance. An alternative to using cultivars with only vertical or horizontal resistance is the use blends of resistant and susceptible cultivars in a single field. This low-technology technique has proven practical for control of a number of rice and wheat diseases (Cowger and Mundt, 2002). This has proven highly effective in rice culture and is thought to slow epidemics in small grains, because of spatial separation of susceptible individuals, and provides minimal pressure for the selection of new pathogen races (Fry, 1982; Agrios, 2005).

In general, horizontal resistance is difficult to work with because many genes are involved and resistance is not complete. Breeders do not have simple inheritance to work with, and scoring for resistance is difficult since it is usually characterized by a reduced level of disease. Horizontal resistance is often found in landraces of crops since these crops are grown year after year in the presence of a set of pathogens and only those with a high level of resistance remain after years of recurrent selection (Fry, 1982; Thurston, 1990, 1991; Agrios, 2005).

Wherever genetic uniformity exists, particularly over large geographic areas, it is likely that devastating epidemics will occur (National Academy of Sciences, 1972; Duvick, 1984). This is often the situation where growers plant a single genotype over large areas or breeders have used a single source of resistance or the host evolved separately from the pathogen. This latter situation occurred with American chestnut and the chestnut blight pathogen that evolved in Asia. An example of serious problems resulting from genetic uniformity is the epidemic of southern corn leaf blight that affected US maize in the 1970 and 1971 crop years (Ullstrup, 1972). This epidemic was directly linked to the use of Texas male sterile cytoplasm on more than 70 million

contiguous acres. Monitoring pathogens for the presence of new biotypes or races is critical to the effective use of disease resistance.

Resistance genes are generally found at the centre of origin of the host plant where evolution with the pathogen has provided selection pressure for development of resistant individuals. Examples of centres of origin are the Andes Mountains of South America for potato, Eurasia and the Middle East for wheat and barley and Central America for maize. Other sources of resistance genes are the landrace cultivars used by indigenous people for centuries or the close plant relatives. Finally, with modern molecular technology, identified resistance genes can be transferred with specificity between unrelated plants where traditional breeding techniques would be impossible (Agrios, 2005).

Breeding for disease resistance is perhaps the dominant disease management strategy for crops with low unit values such as small grains, maize, soybeans and rice where growers cannot afford the use of multiple fungicide sprays or use of other tactics such as long-term rotation. Breeders have developed cultivars with resistance to multiple pathogens. For example, wheat with resistance to one or more viruses, one or more rust and smut pathogens and *Septoria* or *Helminthosporium* leafspot is available. Maize, soybean and tomato varieties with multiple disease resistance are the rule. Vine crops with resistance to bacterial wilt, *Fusarium* wilt, cucumber mosaic, anthracnose, downy mildew and powdery mildew are available. Breeding for disease resistance is done by private companies, universities, government agencies and international agricultural research centres (IARCs) such as CYMMYT (wheat and maize), International Rice Research Institute (IRRI), CIAT (beans) and CIPP (potato) (CAST, 2003).

Reducing pathogen variability and selection for more virulent biotypes are the considerations in the effective long-term use of resistant varieties. Pathogen variability can be reduced by eliminating alternate hosts where pathogen sexual recombination occurs or by eliminating materials where pathogens survive between growing seasons.

Long-term control of wheat stem rust using resistant varieties in North America is directly related to eradication of *Berberis* sp. alternate hosts. Selection pressure can be reduced by using different resistance genes in subsequent years (gene rotation), deploying different genes over broad geographical areas (regional deployment of genes) and the use of varietal blends and multilines.

In recent years, host plant resistance has been improved using transgenic plants. The use of genetically modified crop plants have effectively controlled papaya ringspot virus in papaya, potato viruses and several cucurbit crop viruses in grower fields (Gonsalves, 1998; Beachy and Bendahmane, 1999; Pang *et al.*, 2000). In these examples, plants were transformed to express the virus coat protein and, by virtue of the cross protection phenomenon, the plants are immune to these viruses. In other situations, plants have been transformed with viral replicase genes, antisense nucleocapsid gene sequences, viral movement protein genes, genes for enhanced chitinase activity, glutamate decarboxylase and other genes (Agrios, 2005). In addition, modern molecular biology techniques have allowed breeders to incorporate resistance genes from plants where traditional breeding is precluded. The use of transgenic disease-resistant plants offers a tremendous opportunity to reduce disease losses but there are potential problems with unwanted gene transfer to non-transgenic plants and currently significant problems with consumer acceptance in some countries.

Integrated Disease Management

Because of the diversity of pathogens that attack nearly every plant, the plant pathologist must integrate all of the above strategies with loss economics, pathogen epidemiology, history of disease and other pertinent information in the development of disease management programmes. Key concepts are the elimination or reduction of primary inoculum, delay in the onset of disease, slowing of secondary disease cycles and increase of host plant resistance with the understanding of the economic constraints of producers. This latter point is critical since growers must be able to market what they grow and the crop must produce economically viable yields relative to the cost of production. For example, a resistant variety must be adapted to the requirements of the marketplace and low unit value crops will not support multiple pesticide sprays or other high-labour disease management inputs.

Understanding the critical environmental events that favour infection, inoculum dispersal, survival of primary inoculum and disease development are critical to developing effective disease management strategies. Epidemiological models have been developed for diseases such as apple scab, apple and pear fire blight, brown rot of stone fruits, early blight and late blight of potato and tomato, peanut leafspot, *Cercospora* leafspot of sugarbeets, rust diseases of wheat, corn bacterial wilt and leaf blights, downy mildew, powdery mildew, *Botrytis* grey mould of grape and *Septoria* blight of celery that allow growers to avoid spraying on a calendar basis. Growers using these decision support system programmes use environmental monitors (temperature, relative humidity, duration of leaf wetness, rainfall, etc.) or computer-generated weather prediction systems and disease loss development models to decide when and how to implement specific disease control practices. In many situations, pesticide use has been reduced by more than 50% through use of these programmes. These decision support system programmes are available as downloadable computer programmes, through extension services and other crop advisory services and through several private weather or disease prediction services (Russo, 1999; Wolf *et al.*, 2002; CAST, 2003; Agrios, 2005).

The complexity of integrated disease management strategies depends on whether the crop is perennial or annual, its value, the range of economically important diseases present in the production area and the options available for disease management (Cook and Veseth, 1991; Rowe, 1993; CAST, 2003). Finally, disease management strategies must be compatible with management of insect pests, weeds, soil erosion control practices and environmental goals.

References

Aertsens, F. and Michi, H. (2004) *Coniothyrium minitans*: a soil fungus against *Sclerotinia* in multiple crops. *Phytoma* 571, 33–35.

Agrios, G.N. (2005) *Plant Pathology*, 5th edn. Academic Press, New York.

Barker, K.R. and Koenning, S.R. (1998) Developing sustainable systems for nematode management. *Annual Review of Phytopathology* 76, 165–205.

Beachy, R.N. and Bendahmane, M. (1999) Genetic engineering in IPM: a case history for virus disease resistance. In: Kennedy, G.C. and Sutton, T.B. (eds) *Emerging Technologies for Integrated Pest Management: Concepts, Research and Implementation*. APS Press, St Paul, Minnesota, pp. 101–105.

Bélanger, R.R. and Labbé, C. (2002) Control of powdery mildews without chemicals: prophylactic and biological alternatives for horticultural crops. In: Bélanger, R.R., Bushnell, W.R., Dik, A.J. and Carver, T.L.W. (eds) *The Powdery Mildews: A Comprehensive Treatise*. APS Press, St Paul, Minnesota, pp. 256–267.

Berner, D.K. and Williams, O.A. (1998) Germination stimulation of *Striga gesnerioides* seeds by hosts and nonhosts. *Plant Disease* 82, 1242–1247.

Blok, W.J., Termorshuizen, A.J. and Bollen, G.J. (2000) Control of soilborne plant pathogens by incorporating fresh organic amendments following tarping. *Plant Disease* 90, 253–259.

Bridge, J. (1996) Nematode management in sustainable and subsistence agriculture. *Annual Review of Phytopathology* 34, 201–225.

Carisse, O. and Rolland, D. (2004) Effect of timing of application of the biological control agent *Microsphaeropsis ochracea* on the production and ejection pattern of ascospores by *Venturia inequalis*. *Phytopathology* 94, 1305–1314.

CAST (2003) *Integrated Pest Management: Current and Future Strategies*. Task Force Report No. 140, Council for Agricultural Science and Technology, Ames, Iowa.

Cetintas, R. and Dickson, D.W. (2004) Persistence and suppressiveness of *Pasteuria penetrans* to *Meloidogyne arenaria* race 1. *Journal of Nematology* 36, 540–549.

Christensen, C.M. and Meronuck, R.A. (1986) *Quality Maintenance in Stored Grains and Seeds*. University of Minnesota, Minneapolis, Minnesota.

Cintas, N.A. and Webster, R.K. (2001) Effects of rice straw management on *Sclerotium oryzae* inoculum, stem rot severity and yield of rice in California. *Plant Disease* 85, 1140–1144.

Cook, R.J. and Veseth, R.J. (1991) *Wheat Health Management*. APS Press, St Paul, Minnesota.

Cowger, C. and Mundt, C.C. (2002) Effects of wheat cultivar mixtures on epidemic progression of *Septoria tritici* blotch and pathogenicity of *Mycosphaerella graminicola*. *Phytopathology* 92, 617–623.

Davis, J.R., Husiman, O.C., Westermann, D.T., Hafez, S.L., Everson, D.O., Sorensen, L.H. and Schneider, A.T. (1996) Effects of green manures on *Verticillium* wilt of potato. *Phytopathology* 86, 444–453.

Dennis, C. (1983) *Post-harvest Pathology of Fruits and Vegetables*. Academic Press, New York.

DiFonzo, C.D., Ragsdale, D.W., Radcliffe, E.B., Gudmestad, N.C. and Secor, G.A. (1996) Crop borders reduce potato virus Y incidence in seed potato. *Annals of Applied Biology* 129, 289–302.

Dik, A.J. and Wubben, J.P. (2004) Epidemiology of *Botrytis cinerea* diseases in greenhouses. In: Elad, Y., Williamson, B., Tudzynski, P. and Delen, N. (eds) *Botrytis: Biology, Pathology and Control*. Kluwer Academic Publishers, Dordrecht, The Netherlands, pp. 319–333.

Dong, X. (2004) NPR1, all things considered. *Current Opinion in Plant Biology* 7, 547–552.

Duvick, D.N. (1984) Genetic diversity in major farm crops on the farm and in reserve. *Economic Botany* 38, 161–178.

Elad, Y. and Stewart, A. (2004) Microbial control of *Botrytis* sp. In: Elad, Y., Williamson, B., Tudzynski, P. and Delen, N. (eds) *Botrytis: Biology, Pathology and Control.* Kluwer Academic Publishers, Dordrecht, The Netherlands, pp. 223–241.

Elmer, P.A.G. and Michailides, T.J. (2004) Epidemiology of *Botrytis cinerea* in orchard and vine crops. In: Elad, Y., Williamson, B., Tudzynski, P. and Delen, N. (eds) *Botrytis: Biology, Pathology and Control.* Kluwer Academic Publishers, Dordrecht, The Netherlands, pp. 243–272.

English, J.T., Thomas, C.S., Marois, J.J. and Gubler, W.D. (1989) Microclimates of grapevine canopies associated with leaf removal and control of *Botrytis* bunch rot. *Phytopathology* 79, 395–401.

English-Loeb, G., Norton, A.P., Gadoury, D.M., Seem, R.C. and Wilcox, W.F. (1999) Control of powdery mildew in wild and cultivated grapes by a tydeid mite. *Biological Control* 14, 97–103.

Fry, W.E. (1982) *Principles of Plant Disease Management.* Academic Press, New York.

Gnanamanickam, S.S. (2002) *Biological Control of Crop Diseases.* Marcel Dekker, New York.

Gonsalves, D. (1998) Control of papaya ringspot virus in papaya: a case study. *Annual Review of Phytopathology* 36, 415–437.

Gram, E. (1960) Quarantines. In: Horsefall, J.G. and Diamond, A.E. (eds) *Plant Pathology: An Advanced Treatise.* Academic Press, New York, pp. 313–356.

Hardison, J.R. (1976) Fire and flame for plant disease control. *Annual Review of Phytopathology* 14, 355–379.

Hardison, J.R. (1980) Role of fire for plant disease control in grass seed production. *Plant Disease* 64, 641–645.

Harman, G.E., Howell, C.R., Viterbo, A., Chet, I. and Lorito, M. (2004) *Trichoderma* species: opportunistic, avirulent plant symbionts. *Nature Review of Microbiology* 2, 43–56.

Hoitink, H.A.J., Inbar, Y. and Boehm, M.J. (1991) Status of compost-amended potting mixes naturally suppressive to soil borne diseases of floricultural crops. *Plant Disease* 75, 869–873.

Holling, M. (1965) Disease control through virus free stock. *Annual Review of Phytopathology* 3, 367–396.

Hornby, D. (1983) Suppressive soils. *Annual Review of Phytopathology* 21, 65–85.

Howard, R.J. (1996) Cultural control of plant diseases: a historical perspective. *Canadian Journal of Plant Pathology* 18, 145–150.

Jacobsen, B.J. and Backman, P.A. (1993) Biological and cultural plant disease controls: alternatives and supplements to chemicals in IPM systems. *Plant Disease* 77, 311–315.

Jacobsen, B.J., Zidack, N.K. and Larson, B.J. (2004a) The role of *Bacillus*-based biological control agents in integrated pest management systems: plant diseases. *Phytopathology* 94, 1272–1275.

Jacobsen, B.J., Zidack, N.K., Strobel, G.A., Ezra, D., Grimme, E. and Stinson, A.M. (2004b) Mycofumigation with *Muscodor albus* for control of soil borne microorganisms. *IOBC WPRS Bulletin* 27, 103–113.

Janisiewicz, W.J. and Korston, L. (2002) Biological control of postharvest diseases of fruit. *Annual Review of Phytopathology* 40, 411–441.

Jarvis, W.R. (1992) *Managing Diseases of Greenhouse Crops.* APS Press, St Paul, Minnesota.

Kahn, R.P. (1991) Exclusion as a plant disease control strategy. *Annual Review of Phytopathology* 29, 219–246.

Katan, J. and DeVay, J.E. (1991) *Soil Solarization.* CRC Press, Boca Raton, Florida.

Kessel, G.J.T., Köhl, J., Powell, J.A., Rabbinge, R. and van der Werf, W. (2005) Modeling spatial characteristics in the biological control of fungi at leaf scale: competitive colonization by *Botrytis cinerea* and the saprophytic antagonist *Ulocladium atrum. Phytopathology* 95, 439–448.

Kiewnick, S. and Sikora, R.A. (2006) Biological control of the root knot nematode, *Meloidogyne incognita*, by *Paecilomyces lilacinus* strain 251. *Biological Control* 38, 179–187.

Kloepper, J.W., Ryu, C. and Zhang, S. (2004) Induced systemic resistance and promotion of plant growth by *Bacillus* sp. *Phytopathology* 94, 1259–1266.

Koenning, S.R., Schmitt, D.P. and Barker, K.R. (1993) Effects of cropping systems on population density of *Heterodera glycines* and soybeans yield. *Plant Disease* 77, 780–786.

Krall, J.M., Koch, D.W., Gray, F.A. and Nachtman, J.J. (2000) Cultural management of trap crops for control of sugarbeet cyst nematode. *Journal of Sugar Beet Research* 37, 27–43.

Larena, I. and Melgarejo, P. (1996) Biological control of *Monilinia laxa* and *Fusarium oxysporum* f.sp. *lycopersici* by a lytic enzyme-producing *Penicillium purpurogenum. Biological Control* 6, 361–367.

Lyr, H. (1995) *Modern Selective Fungicides: Properties, Applications, Mechanisms of Action*, 2nd edn. Gustav Fischer Verlag, Jena, Germany.

MacDonald, W.L. and Fulbright, D.W. (1991) Biological control of chestnut blight: use and limitations of transmissible hypovirulence. *Plant Disease* 75, 656–651.

Martin, F.N. (2003) Development of alternative strategies for management of soilborne pathogens currently controlled with methyl bromide. *Annual Review of Phytopathology* 41, 325–350.

Mathre, D.E., Kushnak, G.D., Martin, J.M., Grey, W.E. and Johnston, R.H. (1997) Effect of residue management on barley production in the presence of net blotch disease. *Journal of Production Agriculture* 10, 323–326.

Mazzola, M. and Mullinix, K. (2005) Comparative field efficacy of management strategies containing *Brassica napus* seed meal or green manure for the control of apple replant disease. *Plant Disease* 89, 1207–1213.

McGee, D.C. (1995) Epidemiological approach to disease management through seed technology. *Annual Review of Phytopathology* 33, 445–445.

McManus, P.S., Stockwell, V.O., Sundin, G.W. and Jones, A.L. (2002) Antibiotic use in plant agriculture. *Annual Review of Phytopathology* 40, 443–465.

McSpadden Gardener, B.B. and Fravel, R.D. (2002) Biological control of plant pathogens: research, commercialization, and application in the USA. Available online. *Plant Health Progress* doi:10.1094/PHP-2002-0510-01-RV.

Mills, J.T. (1989) Spoilage and heating of stored agricultural products: prevention and control. *Agriculture Canada Publication*, 1823E.

National Academy of Sciences (NAS) (1972) *Genetic Vulnerability of Major Crops.* National Academy Press, Washington, DC.

Palti, J. (1981) *Cultural Practices and Infectious Crop Diseases.* Springer-Verlag, Berlin.

Pang, S.-Z., Fuhjyh, J., Tricoli, D.M., Russel, P.F., Hu, J.S., Fuchs, M., Quemadu, H.D. and Gonsalves, D. (2000) Resistance to squash mosaic comovirus in transgenic squash plants expressing its coat protein genes. *Molecular Biology* 6, 87–93.

Paulitz, T.C. and Bélanger, R.R. (2000) Biological control in greenhouse systems. *Annual Review of Phytopathology* 39, 103–133.

Pfender, W.F. and Alderman, S.C. (2002) Evaluation of postharvest burning and fungicides to reduce the polyetic rate of increase of choke disease in orchard grass seed production. *Plant Disease* 87, 375–379.

Pieterse, C.M..J., van Wees, S.C.M., van Pelt, J.A., Knoester, M., Laan, R., Gerrits, H., Weisbeek, P.J. and van Loon, J.C. (1998) A novel signaling pathway controlling induced systemic resistance in *Arabidopsis. Plant Cell* 10, 1571–1580.

Rodriquez-Kabana, R. (1986) Organic and inorganic amendments to soil as nematode suppressants. *Journal of Nematology* 18, 129–135.

Rowe, R.C. (1993) *Potato Health Management.* APS Press, St Paul, Minnesota.

Russo, J.M. (1999) Weather forecasting for IPM. In: Kennedy, G.G. and Sutton, T.B. (eds) *Emerging Technologies for Integrated Pest Management: Concepts, Research and Implementation.* APS Press, St Paul, Minnesota, pp. 453–473.

Scheuerell, S.J. and Mahaffee, W.F. (2005) Microbial recolonization of compost after peak heating needed for the rapid development of damping-off suppression. *Compost Science and Utility* 13, 65–71.

Scheuerell, S.J., Sullivan, D.M. and Mahafee, W.F. (2005) Suppression of seedling damping-off caused by *Pythium ultimum, P. irregulare* and *Rhizoctonia solani* in container media amended with a diverse range of Pacific Northwest compost sources. *Phytopathology* 95, 306–315.

Schtienberg, D., Raposo, R., Bergeron, S.N., Legard, D.E., Dyer, A.T. and Fry, W.E. (1994) Incorporation of cultivar resistance in a reduced-sprays strategy to suppress early and late blights on potato. *Plant Disease* 78, 23–26.

Schubert, T.S., Rizvi, S.A., Sun, X., Gottward, T.R., Graham, J.H. and Dixon, W.N. (2001) Meeting the challenge of eradicating citrus canker in Florida again. *Plant Disease* 85, 340–356.

Shepard, J.E. and Claflin, L.E. (1975) Critical analyses of the principles of seed potato certification. *Annual Review of Phytopathology* 13, 271–293.

Sikora, R.A. (1992) Management of the antagonistic potential in agricultural ecosystems for the biological control of plant parasitic nematodes. *Annual Review of Phytopathology* 30, 245–270.

Slack, S.A. (1993) Seed certification and seed improvement programs. In: Rowe, R.C. (ed.) *Potato Health Management.* APS Press, St Paul, Minnesota, pp. 61–65.

Stockwell, V.O., Johnson, K.B., Sugar, D. and Loper, J.E. (2002) Antibiosis contributes to biological control of fire blight by *Pantoea agglomerans* strain Eh 252 in orchards. *Phytopathology* 92, 1202–1209.

Stover, R.H. (1979) Flooding of soil for disease control. In: Mulder, D. (ed.) *Soil Disinfestation.* Elsevier, Amsterdam.

Sumner, D.R. Doupnik, B. Jr and Boosalis, M.G. (1981) Effect of reduced tillage and multiple cropping on plant diseases. *Annual Review of Phytopathology* 19, 167–187.

Sztejnberg, A., Galper, S., Mazar, S. and Lisker, N. (1989) *Ampelomyces quisqualis* for biological and integrated control of powdery mildews in Israel. *Journal of Phytopathology* 124, 285–289.

Temple, T.N., Stockwell, V.O., Loper, J.E. and Johnson, K.B. (2004) Bioavailabilty of iron to *Pseudomonas fluorescens* A 506 on flowers of pear and apple. *Phytopathology* 94, 1286–1294.

Thomashow, L.S. and Weller, D.M. (1988) Role of phenazine antibiotic from *Pseudomonas fluorescens* in biological control of *Gaeumannomyces graminis tritici. Journal of Bacteriology* 170, 3499–3508.

Thurston, H.D. (1990) Plant disease management practices of traditional farmers. *Plant Disease* 74, 96–102.

Thurston, H.D. (1991) *Sustainable Practices for Plant Disease Management in Traditional Farming Systems.* Westview, Boulder, Colorado.

Ullstrup, A.J. (1972) The impacts of the southern corn leaf blight epidemics of 1970–1972. [*Helminthosporium turcicum*]. *Annual Review of Phytopathology* 10, 37–50.

Veeken, A.H.M., Blok, W.J., Curci, F., Coenen, G.C.M., Termorshuizen, A.J. and Hamelers, H.V.M. (2005) Improving quality of composted biowaste to enhance disease suppressiveness of compost-amended, peat-based potting mixes. *Soil Biology and Biochemistry* 37, 2131–2140.

Walker, J.C. (1952) *Diseases of Vegetable Crops.* McGraw-Hill, New York.

Waterworth, H.E. (1993) Processing foreign plant germplasm at the National Plant Germplasm Quarantine Center. *Plant Disease* 77, 854–860.

Weller, D.M. (1988) Biological control of soilborne plant pathogens in the rhizosphere with bacteria. *Annual Review of Phytopathology* 26, 379–407.

Whitehead, A.G. (1998) *Plant Nematode Control.* CAB International, Wallingford, UK.

Wiggins, B.E. and Kinkle, L.L. (2005) Green manures and crop sequences influence potato diseases and pathogen inhibitory activity of indigenous streptomycetes. *Phytopathology* 95, 178–185.

Wilson, M. and Lindow, S.E. (1993) Interactions between the biological control agent *Pseudomonas fluorescens* strain A506 and *Erwinia amylovora* in pear blossoms. *Phytopathology* 83, 117–123.

Wilson, R.G., Kerr, E.D. and Provance, P. (1993) Growth and development of oil-radish and yellow mustard in Nebraska. *Journal of Sugar Beet Research* 30, 159–167.

Wolf, P.F.J. and Verreet, J.A. (2002) An integrated pest management system in Germany for the control of fungal leaf diseases in sugar beet: the IPM sugar beet model. *Plant Disease* 86, 336–344.

Young, X.B., Dowler, W.M and Royer, M.H. (1991) Assessing the risk and potential impact of exotic plant disease. *Plant Disease* 75, 976–982.

Zadoks, J.C. and Schein, R.D. (1979) *Epidemiology and Plant Disease Management.* Oxford University Press, Oxford.

3 Ecological Management of Agricultural Weeds

Robert G. Hartzler[1] and Douglas D. Buhler[2]

[1]College of Agriculture, Department of Agronomy, Iowa State University, Ames, IA 50011, USA; [2]College of Agriculture and Natural Resources, 286 Plant and Soil Science, East Lansing, MI 48824, USA

Introduction

Weeds continue to be a constant threat to agricultural productivity despite decades of modern weed control practices aimed at their elimination. Oerke and Dehne (2004) reported that weeds had the highest potential to cause loss of the world's major crops, with animal pests (arthropods, rodents, birds, etc.) and pathogens posing less of a threat. Overall, weeds were estimated to have the potential to reduce yields by 32%, nearly twice as much as the other pest complexes combined. Actual losses due to weeds following application of management tactics were estimated at 9%. The development of herbicide-resistant weeds and weed population shifts continue to challenge the effectiveness of modern practices (Buhler *et al.*, 2000). The continuous adaptation of weed communities to control practices suggests the need for the development of management systems based on an understanding of the ecological principles governing the growth and development of weeds within agroecosystems.

The development of integrated pest management (IPM) systems has been a goal in agriculture for many years (Bottrell, 1979; Cate and Hinkle, 1994; Benbrook *et al.*, 1996). IPM systems have been implemented for many important insect pests and plant diseases. However, development of IPM systems for weeds has lagged behind other pest management disciplines. Because weeds are the major pests in most cropping systems, weed management must be a part of IPM for it to be successful.

Weed control refers to the actions used to eliminate an existing weed population. Weed management is more than control of existing weed problems and places greater emphasis on preventing weed reproduction, reducing weed emergence after crop planting and minimizing weed competition with the crop (Zimdahl, 1991; Buhler, 1996). Integrated weed management (IWM) emphasizes combinations of techniques and knowledge that consider the causes of weed problems rather than reacting to problems after they occur. The goal of IWM is to optimize crop production and grower profit through the concerted use of preventive tactics, scientific knowledge, management skills, monitoring procedures and efficient use of control practices.

Ecological weed management places a greater emphasis on understanding the interactions between the cropping system and the weed community, especially on reducing the negative impacts of weeds. This can be achieved by either creating a less favourable environment for weeds or enhancing the competitiveness of the crop.

Much of the focus in IWM has been on optimizing efficiency through integration of the different control tactics. This chapter focuses primarily on cultural practices that provide crops with an ecological advantage over weeds, thereby reducing the need for intervention to manage weeds.

Weedy Characteristics

For a plant to be a successful weed in agronomic crops, it must be able to survive and reproduce in spite of repeated attempts to remove it from the field. Baker (1974) developed a list of characteristics contributing to weediness including:

1. Ability to germinate under many environments;
2. Persistent seed;
3. Rapid growth;
4. Prolonged period of seed production;
5. Ability to reproduce through both self- and cross-pollination;
6. Cross-pollination not dependent upon specialized vectors;
7. High-seed production under favourable growing conditions;
8. Ability to produce seed under a wide variety of environmental conditions;
9. Adaptations for short- and long-distance dispersal of reproductive units;
10. Perennial species that have vegetative reproductive capacity and are not easily removed from the soil;
11. Increased competitiveness by allelochemicals or other mechanisms.

Thus it is clear that reproductive characteristics are very important attributes of weediness. Baker (1974) pointed out that no single species has all of the above characteristics, and that as the number of these characteristics possessed by a species increased, the weediness of a species would also increase.

Plant communities in agronomic fields continually change in response to the annual disturbances and cultural practices used to control weeds. Plant succession is a change in the composition of species occupying a habitat over time (Radosevich et al.,

1997). If an environment is stable, the rate of change slows until the climax vegetation for the habitat is reached. The presence of weeds in an agronomic field is a form of annual succession since weeds occupy niches previously vegetated but opened up by tillage or herbicide application. Given the repeated disturbance in agronomic fields, climax vegetation does not develop and community composition is unstable.

There are three types of habitats for which plants have evolved (Grime, 1977), and plant types have specific differing characteristics that favour their survival in different habitats. Most weeds are classified as ruderals – plants adapted for survival in areas of frequent disturbance and high resource availability. Characteristics of ruderals include an annual or short-lived perennial life cycle, a rapid relative growth rate and a tendency to devote a large proportion of their biomass to reproductive structures. Relatively few species possess the characteristics required to survive the specialized habitat of agricultural fields. Holm et al. (1997) compiled a list of approximately 100 species that are classified as the world's worst weeds.

The composition of a weed community in a specific area is influenced by many factors, but tillage and crop rotation usually have a larger effect than specific control tactics (Buhler et al., 2000). Annual weeds typically dominate cropping systems featuring intensive tillage and annual crops, whereas perennials are more prevalent in systems with minimal soil disturbance (Staniforth and Wiese, 1985). Weed communities are generally more diverse when crop rotations are more complex (Hume, 1982; Aldrich, 1984). The diversity of weeds decreased with increasing levels of tillage in an Ohio study (Cardina et al., 1991), and numerous studies have documented significant shifts among weed species as tillage was reduced or eliminated (Buhler, 1995). Changes in herbicide use in agricultural fields generally affect the relative abundance of individual species in a community, rather than determining the absence or presence of species (Mahn and Helmecke, 1979; Hume, 1987). Development of herbicide-resistant biotypes is an example of a herbicide-induced shift changing the relative abundance of weed species within a community.

The small number of weeds adapted to agronomic fields results in a relatively stable group of dominant weeds. Surveys of weed scientists in southern USA found that only one species that ranked in the top ten weeds of soybean (*Glycine max*) in 1974 was not included in the similar list compiled in 1995 (Webster and Coble, 1997). Changes in the ranking among the weed species were attributed to the introduction of new herbicides with greater efficacy on specific species. Occasionally, species of little economic importance increase in prevalence due to changes in crop production practices or movement into new regions. For example, adoption of no-tillage systems allows the survival of many winter annual and perennial weeds that are unable to tolerate intensive soil disturbance (Buhler, 1995).

Control Tactics

Regulatory control

Noxious weed laws have commonly been created to prevent or reduce the spread of weeds. In the USA, 32 of the contiguous states have noxious weed laws, whereas all 50 states have regulations pertaining to noxious weed seed in commercially sold crop seed (Skinner *et al.*, 2000). Most noxious weed laws target exotic species, and legislative action has the greatest likelihood for success when enacted during the early stages of invasion (Panetta and Scanlan, 1995). An extreme example of regulatory control is provided by witchweed (*Striga asiatica*) management in North and South Carolina. Witchweed was first detected in the region in 1957. Infested areas were quarantined, resources provided for education, detection and eradication, and affected areas dramatically reduced (Eplee, 1992).

Preventative control

The most basic of all weed control methods is prevention by using measures to forestall the introduction and spread of weeds. Possible source of preventable weed spread include contaminated crop seed; transport of plant parts and seeds on planting, tillage, harvest and processing equipment; livestock; manure and compost; irrigation and drainage water; and forage and feed grains (Walker, 1995). Movement of weeds may be particularly significant for crop seeds and animal feeds that are transported long distances.

Improved seed-cleaning techniques and seed certification programmes have reduced the spread of weeds in crop seed in many regions of the world. However, many growers still plant crop seeds that are contaminated with propagules of important weeds (Tonkin, 1982; Dewey and Whitesides, 1990; Rao and Moody, 1990). Manure can be a source of weed seed for farms with integrated crop and animal systems (Dastgheib, 1989; Mt. Pleasant and Schlather, 1994). Viability of weed seeds following ingestion by animals varies among species (Harmon and Keim, 1934) and manure-handling procedures (Rupende *et al.*, 1998). Although significant numbers of weed seed can be introduced in crop seeds and manure, their influence on weed management will vary depending on the magnitude of the existing weed seedbank of the field. In a survey of New York dairy farms, it was concluded that the number of viable weed seed returned to the soil in cow (*Bos taurus*) manure was insignificant compared with the existing seedbank for most farms (Mt Pleasant and Schlather, 1994). Weed seed introduction through crop seeds or manure is a greater concern for the introduction of new weed species, rather than maintenance of the seedbank of species previously established in the field. However, transport of manure from large animal operations is increasing, thereby enhancing the potential for introduction and spread of weed seed through manure applications.

Although the concept of prevention is quite simple, success and feasibility are determined by weed species, means of dissemination, farm size and the amount of effort expended. Preventive weed management programmes are most successful in situations where humans are the vectors or where they have direct control over the seed source (Walker, 1995). These programmes can be implemented through community action by the enactment and enforcement of laws and regulations (Day, 1972). Seed purity and noxious weed laws are examples of successful weed prevention programmes.

Cultural control

Crop rotation

Crop rotation is the practice of growing different crops in a recurring sequence on the same land, and is regarded as an important strategy for managing weeds and other pests. Before the introduction of synthetic herbicides, Leighty (1938) has described crop rotation as the most effective means of managing weeds. A survey of 253 Canadian farmers found that 83% reported weed control benefits when they included a forage species in their crop rotation (Entz et al., 1995). However, since the introduction of synthetic nitrogen fertilizers and pesticides in the middle of the 20th century, there has been a dramatic reduction in the diversity of crop rotations used in industrialized countries.

Weeds thrive in the presence of crops whose growth requirements and characteristics are similar to theirs (Liebman and Dyck, 1992). Because of this, rotations, which include diverse crops, should prevent any particular species from increasing to densities that create serious management difficulties. The effect of three rotations on weed seedbanks at two locations was evaluated after 35 years (Cardina et al., 2002). The seedbank following continuous maize (Zea mays) was dominated by a single species at each location, but there was a greater diversity of weed species in the 2- and 3-year rotations. In a review of crop rotation studies that reported on weed population dynamics, it was reported that weed densities were lower in 78% of the situations in crop rotations rather than in a monoculture (Liebman and Dyck, 1992). Diverse rotations may improve weed management by

employing crops with different planting and harvest dates, as well as different growth habits and crop residue characteristics, and by utilizing different tillage and weed management practices. This diversity provided by crop rotation challenged the weeds with a wide range of stresses and mortality risks, thereby reducing the opportunities for unchecked growth and reproduction (Liebman and Staver, 2001).

Although the weed management benefits of crop rotations are frequently described, it is difficult to separate the rotation effects from that of weed management tactics used within the rotations (Doucet et al., 1999; Lègére and Stevenson, 2002). In a study evaluating the effects of weed management and crop rotation on weed communities, weed management practices accounted for 38% of the variation in total weed density, whereas rotation accounted for only 5% of the variation (Légére and Stevenson, 2002). In this study, rotation also had a minor impact on weed diversity.

Crop rotations frequently benefit weed management when more effective control tactics are available for a particular weed in the different phases of the rotation (Ball, 1992; Blackshaw, 1994a; Unger et al., 1999; Streit et al., 2003). Downy brome (Bromus tectorum) increased nearly 50-fold in a continuous winter wheat (Triticum aestivum) cropping system over a 6-year period, whereas in a winter wheat – canola (Brassica rapa) – rotation downy rome increased only twofold in the same period (Blackshaw, 1994a). Including canola in the rotation provided the advantage of having a crop with a different life cycle than downy brome and allowed the use of herbicides, which are highly effective against downy brome.

Rotations may be detrimental to weed management compared with continuous cropping if less effective control tactics are available for one of the crop phases. The seedbank in a barley (Hordeum vulgare)–vetch (Vicia sativa) rotation was greater than in continuous barley due to lower competitiveness of vetch compared with barley (Dorado et al., 1999). The seedbank of summer annual weeds declined annually during 3 years of maize and soybean production (Buhler et al., 2001). However, the seedbank increased greater than 35-fold following oats (*Avena sativa*) underseeded with forage legumes, due to the lack of competitiveness of the first-year legumes following the harvest of the oats.

The effect of crop rotation on weed communities is influenced by the characteristics of the crops grown, cultural and tillage practices used in each crop and the effectiveness of the control tactics employed. The dominance of certain weeds in particular cropping systems is determined by interactions between the site characteristics, the crops produced and the forms of disturbances employed (Légére and Samson, 1999). When evaluating the contribution of crop rotation in managing weeds, the relative likelihood of survival of weeds in each phase of the rotation must be assessed.

Crop competitiveness

Enhancing the ability of crops to compete with weeds is an attractive component for ecological weed management (McWhorter and Barrentine, 1975; Pester et al., 1999). This can be accomplished by providing the best possible environment for crop growth combined with practices that reduce the density and vigour of the weeds. Practices such as narrow row spacing, increased plant density, appropriate time of planting and fertility management are capable of shifting the competitive balance to favour crops over weeds (Malik et al., 1993; Teasdale, 1995; Buhler and Gunsolus, 1996). Decreasing row spacing and increasing crop plant density has been shown to increase the competitiveness of many crops (Teasdale and Frank, 1983; Stoller et al., 1987; Malik et al., 1993; Teasdale, 1995). Teasdale (1995) found that growing maize in 38-cm-wide rows with increased density (compared with 76-cm-wide rows) improved weed control and reduced herbicide requirements. Malik et al. (1993) found differences in competitiveness among white bean (*Phaseolus vulgaris*) cultivars, and that decreasing row spacing reduced yield losses due to weeds. In an extensive review of the effects of cultural

practices on soybean–weed interactions, Stoller et al. (1987) concluded that soybean cultivar selection, row spacing, plant density, planting date, crop rotation, tillage and herbicides can all be used to maximize the ability of soybean to compete with weeds.

Differential weed competitive ability has been documented among commonly grown cultivars of several important crops (Callaway, 1992; Lanning et al., 1997; Pester et al., 1999; Bertholdsson, 2005). Characteristics commonly associated with crop competitiveness included rapid germination and root development, rapid early vegetative growth and vigour, rapid canopy closure and high leaf area index, profuse tillering or branching, increased leaf duration, greater plant height and allelopathy (Callaway, 1992; Pester et al., 1999). In rice (*Oryza sativa*), leaf area index and biomass production early in the growing season were the traits most closely associated with competitiveness against weeds (Garrity et al., 1992). In dry bean (*P. vulgaris*), leaf area index and leaf size accounted for 73% of the total variation in weed biomass production (Wortmann, 1993). The most consistent conclusion from many studies is that vigorous growth characteristics enhance competitiveness by reducing light quantity and quality beneath the crop canopy.

In addition to crop yield responses, enhanced crop competitiveness may reduce the reproductive capacity of weeds. Cheat (*Bromus secalinus*) produced more seed when grown with semi-dwarf than tall cultivars of winter wheat (Koscelny et al., 1990). Blackshaw (1994b) concluded that planting a more competitive cultivar of winter wheat resulted in higher wheat yield, less downy brome seeds and lower weed infestations in subsequent crops.

Planting date

Crop planting date may significantly affect weed management and crop–weed interactions by influencing emergence and growth of crops and weeds. Most weed species have distinct emergence patterns, which are relatively consistent from year to year (Stoller and Wax, 1973; Roberts and Potter,

1980; Hartzler and Buhler, 1999). Knowledge regarding emergence timing of weeds may allow planting of crops during periods of limited weed germination, no weed germination or when a crop has the competitive advantage over weeds (Ghersa and Holt, 1995).

Relatively small changes in crop planting dates can significantly impact densities of weed species whose peak emergence periods occur near crop planting (Weaver et al., 1988; Buhler and Gunsolus, 1996; Spandl et al., 1998). Gunsolus (1990) reported that a 20-day delay in soybean planting reduced the densities of giant foxtail (Setaria faberi), redroot pigweed (Amaranthus retroflexus) and common lambsquarter (Chenopodium album), which emerged with soybean by 76, 63 and 85%, respectively. In the absence of weed control, the later planting date yielded 1277 kg/ha more than the early planting date due to less competition between the crops and the weeds. Planting date manipulation is most effective in managing weeds for which the majority of emergence occurs near optimum crop planting dates.

Crop planting date may also alter the competitive balance between crops and weeds by influencing the relative emergence time and growth rates of the competing plants. A strong relationship between crop yield loss and relative leaf cover of weeds shortly after crop emergence has been reported (Ghersa and Martinez-Ghersa, 1991; Kropff and Spitters, 1991). Planting crops when soil and environmental conditions favour emergence and early-season growth of crops over weeds will provide the crop a competitive advantage over weeds. Ivyleaf morning glory (Ipomoea hederacea) and sicklepod (Cassia occidentalis) were found to be twice as competitive in soybean planted in early June compared to soybean planted in early May (Klingaman and Oliver, 1994). However, velvetleaf (Abutilon theophrasti) was more competitive in early-planted than late-planted soybean in the same region (Oliver, 1979).

The alteration of planting date is a common practice of farmers who produce agronomic crops without herbicides (Fernholz,

1990; Gunsolus 1990). Possible reduction in crop yield potential due to delayed planting and availability of highly effective herbicides limits the implementation of this tactic in many cropping systems. A basic understanding of the seedbank ecology of annual weeds and comparative growth of crops and weeds is essential to maximize the benefits associated with manipulating planting date to enhance weed management (Forcella et al., 1993; Ghersa and Holt, 1995). Although efforts have been made to provide farmers with the information needed to anticipate or predict weed emergence, these resources are generally underutilized (Buhler et al., 1997; Forcella, 1998). An improved ability to predict peak weed emergence would eliminate much of the risk of adjusting planting dates to improve weed management. Approaches to predicting weed emergence and the utility of these tools have been previously reviewed (Ghersa and Holt, 1995; Grundy, 2003).

Fertility management

Applications of nutrients, particularly nitrogen (N), phosphorous (P) and potassium (K), are commonly used to satisfy crop fertility needs and enhance yields. However, many situations exist in which weeds benefit from fertilizers as much as, or more than, crops (Hoveland et al., 1976, Teyker et al., 1991; Ampong-Nyarko and De Datta, 1993; DiTomaso, 1995; Blackshaw et al., 2003). Opportunities exist to enhance fertilizer and weed management efficiency by applying nutrients in a manner that favours the crop rather than the weeds.

Fertilizer management can be manipulated through application rate, source of nutrient, application timing and fertilizer placement. Optimum management practices will vary among crops, weeds and nutrients (DiTomaso, 1995; Jornsgard et al., 1996; Blackshaw et al., 2003). Common chickweed (Stellaria media) biomass and seed production in wheat increased with soil nitrogen supply, whereas the opposite response occurred in potato (Solanum tuberosum) (Van-Delden et al., 2002). Jornsgard et al. (1996) reported that increasing nitrogen fertilization rates tended to decrease weed biomass in spring barley and winter wheat, but that weed species responded differentially. Certain weeds had higher nitrogen optimal rates than the crops, whereas others had lower optimal rates. Reductions in nitrogen application rates in maize decreased the flexibility in timing of weed control tactics, thereby potentially increasing the need for more intensive weed management systems (Evans *et al.*, 2003). In acidic soils with low phosphorus availability, liming and phosphorus applications favoured the growth and yield of barley more than weed growth (Légère *et al.*, 1994).

Crops and weeds may have different temporal needs for nutrients, thus providing opportunities to apply fertilizers at times that favour utilization by the crop over weeds. Nitrogen fertilization of barley increased wild oat survival and fecundity, especially when applied at early tillering rather than at sowing (Scursoni and Benech Arnold, 2002). Delayed nitrogen application reduced wild oat (*Avena fatua*) growth rate by 25%. The most efficient time of nitrogen application to enhance the competitive ability of sugarbeet (*Beta vulgaris*) with wild mustard (*Brassica kaber*) and common lambsquarter varied between the two weed species (Paolini *et al.*, 1999).

Numerous studies have documented that band applications of fertilizers, either surface or injected, can favour crop growth over that of weeds when compared with broadcast applications (Kirkland and Beckie, 1998; Rasmussen, 2002; Petersen, 2003). Weed density and biomass was 20% to 40% less where nitrogen was applied in a band in spring wheat compared with a broadcast application (Kirkland and Beckie, 1998). Injection of fertilizers decreased the weed biomass compared with surface application by enhancing the crops' competitiveness with weeds rather than reducing the nutrient uptake by weeds (Blackshaw *et al.*, 2002; Rasmussen, 2002).

The form that a nutrient, particularly nitrogen, was applied may influence crop and weed growth differentially (Haynes and Goh, 1978; Teyker *et al.*, 1991; Davis and Liebman, 2001). Shoot biomass of redroot pigweed was reduced by 75% when nitrogen was supplied as ammonium rather than nitrate, whereas maize shoot growth was unaffected by nitrogen form (Teyker *et al.*, 1991). Giant foxtail did not show any preference to nitrate or ammonium nitrate (Salas *et al.*, 1997). Davis and Liebman (2001) reported that the relative growth rate of wild mustard was 12% slower when nitrogen was supplied to maize by organic rather than inorganic sources. The response of maize and common lambsquarter to the nitrogen source was inconsistent, with weed growth favoured by inorganic nitrogen compared with a leguminous source in only two of four experiments (Dyck and Liebman, 1995).

There is little question that soil fertilization can alter the competitive balance between crops and weeds, but specific methods to use fertility management as part of IWM have not been developed. The effect of weeds may be reduced by management strategies that maximize nutrient uptake by crops and minimize nutrient availability to weeds (DiTomaso, 1995). Another approach to increasing fertilizer use efficiency may be to enhance the mechanisms and kinetics of mineral uptake by crop plants. In almost all cases where nutrient concentrations in crops and associated weeds were compared, accumulation of nutrients in the weeds exceeded the levels in the corresponding crop (DiTomaso, 1995). Therefore, to maximize nutrient uptake and fertilizer use efficiency by crops in competition with weeds, we need to better understand the mechanisms of nutrient fluxes in the roots of weeds and crop plants.

Organic amendments and soil quality

Organic matter amendments have long been used to enhance soil fertility, improve soil structure and recycle waste products (Magdoff, 2001). Recently, the development of weed-suppressive soils through enhancement of soil quality has been reported (Gallant *et al.*, 1999; Kremer and Li, 2003). Organic matter amendments alter temporal patterns of nutrient availability, especially for nitrogen and phosphorus. Compared with pulsed application of synthetic fertilizers, composted manure (DeLuca and DeLuca, 1997) and fresh plant residues and manures (Gallandt *et al.*, 1998) release nutrients more slowly over a longer period. Because germination and early growth of many weed species are strongly dependent on soil nutrient concentrations (Karssen and Hilhorst, 1992; DiTomaso, 1995), shifts in the timing of nutrient availability may affect weed density, emergence timing and community composition.

Organic matter amendments are a source of non-nutrient compounds that may inhibit or promote growth (Valdrighi *et al.*, 1996; Ozores-Hampton *et al.*, 1999). They may also increase soil microbial biomass and activity, and change the incidence and severity of soil-borne diseases of weeds and crops. Conklin *et al.* (1998) reported that wild mustard seedlings grown in soil amended with compost and red clover (*Trifolium pratense*) residue were smaller and had a higher incidence and severity of *Pythium* infection than seedlings grown in soil receiving ammonium nitrate fertilizer; maize seedlings were unaffected by soil amendment treatments. Kremer and Li (2003) reported that weed rhizospheres from organic and integrated cropping systems had greater proportions of weed-suppressive bacteria than rhizospheres from conventional systems.

Management practices that increase the organic matter content of soils promote diverse microbial populations, enhance soil enzyme activity and promote the development of weed-suppressive microbial communities (Kennedy, 1999; Li and Kremer, 2000). Kennedy and Kremer (1996) suggested possible farming practices that create weed-suppressive soils in which the microbial community composition and activity are altered to deplete the weed seedbank, reduce probabilities of weed seedling establishment and reduce weed growth and competitive ability. They suggested that this might be accomplished by managing residue and microbial activity to increase seed decay potential within the residue zone. A review of arbuscular mycorrhizal fungi's effects on crop–weed interactions suggested that an improved understanding of the role of these organisms on weed communities could lead to new ecologically based weed management systems (Jordan *et al.*, 2000). Manipulation of soil to enhance weed suppression by the microbial community is similar to the conservation approach to biological control (Horwath *et al.*, 1998; Scheepens *et al.*, 2001).

Cover and smother crops

Cover crops and smother crops may enhance weed management by creating an environment that is less favourable for weeds either by using resources that would otherwise be available to weeds or by the production of allelochemicals (Teasdale, 1998). Cover crops are planted during periods when the field would otherwise remain fallow, whereas smother crops are grown during part of, or all through, the cropping season. The objective of these strategies is to replace an unmanageable weed population with a more easily managed crop.

The effectiveness of cover and smother crops in controlling weeds has been highly variable (Buhler *et al.*, 1998; Teasdale, 1998; Barberi and Mazzoncini, 2001). Annual legume cover crops planted in wheat stubble in a maize–winter wheat rotation affected winter annual weeds more consistently than summer annual and perennial weeds (Fisk *et al.*, 2001). Cover crops reduced both the density and biomass of winter annuals, but only affected biomass production of summer annuals and perennials. Barberi and Mazzoncini (2001) reported that cover crops only influenced weed composition in years when high levels of cover crop biomass were produced.

To improve the efficacy of cover and smother crops, a better understanding of the mechanisms by which cover crops change weed population dynamics is required. Teasdale and Mohler (2000) studied the effects of several mulch materials, including cover crop residue, on weed establishment. They concluded that physical impedance and light deprivation were the principle mechanisms controlling weed emergence rates through the cover crops. In addition to restricting weed emergence by physical effects, certain cover crop species produce allelopathic compounds that might suppress weed establishment (Creamer *et al.*, 1996; Duke *et al.*, 2002; Bhowmik, 2003).

The benefits of cover and smother crops may be enhanced by developing varieties with specific traits for this use. Opportunities may exist to enhance the allelopathic traits of cover crops, either through traditional breeding (Reberg-Horton *et al.*, 2001) or through molecular approaches (Duke *et al.*, 2002). Stanislaus and Cheng (2002) demonstrated the concept of genetically engineering a winter cover crop to self-destruct in response to an environmental cue, thereby eliminating the need to kill the cover crop with tillage or herbicides. Teasdale (1998) concluded: '[A]ttention should be focused on defining the impact of cover crops on important rate-defining steps in the life cycle of weeds. This knowledge will help characterize how to use cover crops most effectively to disrupt the succession of important weed species.'

Biological control

Biological weed control has been defined as the intentional use of natural enemies (insects, pathogens, herbivores, etc.) to suppress weed populations (Scheepens *et al.*, 2001). Three different approaches are recognized in biological weed control:

1. Inoculative or classical approach – the biocontrol agent is introduced in an area outside of its native range. The intent is that the agent will become permanently established and reduce densities of the target weed. This approach has typically targeted weeds introduced into new geographical areas.
2. Inundative or bioherbicide approach – a pathogen is applied at a high density to rapidly control the existing weed population. The biocontrol organism may need to be applied repeatedly to control new infestations.
3. Conservation approach – the intent is to manipulate the environment so that existing or supplemented biocontrol agents are more effective.

Several factors have limited the adoption of classical biological weed control in agronomic crops, including limited host spectrum, lack of consistency and slow or inadequate weed suppression (Kennedy and Kremer, 1996). A few bioherbicides have been successfully marketed, typically in systems where no conventional herbicides are available for a specific weed problem (Boyetchko, 1997). Collego™ is a commercial formulation of the fungus *Colletotrichum gloeosporioides* used to control northern jointvetch (*Aeschynomene virginica*) in rice and soybean (Bowers, 1986). Future success in this area lies in enhancing the effectiveness of pathogens through genetic manipulation or enhanced delivery systems (Hallett, 2005).

The conservation approach to biological control is a fundamental component of ecological weed management systems. The ability of a soil to suppress weeds is enhanced by adopting practices that enhance soil biological activity (Kennedy, 1999; Kremer and Li, 2003). Another form of biological control that influences weed densities within agronomic fields is seed predation (Marino *et al.*, 2005; Mauchline *et al.*, 2005). The relative importance of vertebrate and invertebrate predators varies among cropping systems and the activity of predators varies throughout the year (Westerman *et al.*, 2003; Mauchline *et al.*, 2005). Westerman *et al.* (2005) reported that velvetleaf seed losses due to predation were nearly twice as great in a 4-year rotation system than in a 2-year rotation. Mauchline *et al.* (2005) concluded that to fully capture benefits of weed seed predation the primary predators must be identified and production practices be implemented that maximize the activity of these predators at the time of seed shed.

Mechanical control

Mechanical removal or disruption of weeds was the primary tool used to manage weeds from the development of agriculture until the introduction of modern herbicides in the 1950s. Some form of soil disturbance remains a component of virtually all crop-ping systems. Mechanical weed control tactics can be divided into two broad strate-gies: (i) pre-plant tillage, which is used to prepare a seedbed and which eliminates existing weeds to provide the crop with an even start with weeds; and (ii) post-plant tillage, which is used to kill weeds that emerge after crop planting.

Weed communities are influenced by tillage through changes in the soil environ-ment, changes in seed distribution in the soil, effects on seed predators and effects on weed control practices (Brust and House, 1988; Buhler, 1995). Tillage for seedbed preparation can reduce densities of annual weed populations, especially if planting is delayed to allow weed seed germination prior to final seedbed preparation (Gunsolus, 1990; Buhler and Gunsolus, 1996). Pre-plant tillage may also reduce weed densi-ties by placing weed seeds at depths in the soil profile that are too deep for successful establishment, particularly for species with small seeds (Pareja et al., 1985; Buhler and Mester, 1991). Seed burial can be particu-larly beneficial in years following high seed rain due to weed control failures (Hartzler and Roth, 1993).

Post-plant tillage can also be part of IWM, especially during the production of annual crops (Gunsolus, 1990). Shallow tillage before crop emergence and cultiva-tion between rows after crop establishment are effective in removing annual weeds and inhibiting the growth of perennial species. Timely rotary hoeing reduces weed density up to 85% (Buhler and Gunsolus, 1996). When used in combination with rotary hoe-ing (Gunsolus, 1990; Buhler and Gunsolus, 1996) or supplemented with broadcast (Gebhardt, 1981; Steckel et al., 1990) or band applications of herbicides (Buhler et al., 1992, 1993; Wiltshire et al., 2003), cul-tivation between rows can provide effective weed control, reduce the quantity of herbi-cide applied and bring diversity to the weed management system.

Although post-plant tillage can be highly effective in managing weeds, its importance in recent times has dimin-ished due to less flexibility in timing and greater labour requirements than selective herbicides (Gunsolus, 1990). Opportunities exist to enhance the efficacy of mechanical control strategies through new designs of mechanical weed control tools (Kouwen-hoven, 1997; Fogelberg and Kritz, 1999), and the use of automatic guidance systems (Søgaard, 1998; Tillett et al., 2002; Wiltshire et al., 2003) as well as of decision tools to improve timing of operations (Oriade and Forcella, 1998).

Herbicides

Herbicides are the principle weed manage-ment tool in many cropping systems across the world. In 2001, herbicides accounted for 44% of all pesticides used globally (Keily et al., 2004). The adoption of herbicides has facilitated the development of no-till and other forms of reduced tillage, reduced labour requirements for weed management and decreased risks of weed control failures associated with adverse weather condi-tions (Kudsk and Streibig, 2003). The use of modern genetic techniques to transfer resistance to non-selective herbicides such as glyphosate and gluphosinate into several important agronomic crops has provided additional herbicide options. Fundamen-tally, these herbicide-resistant crops do not change weed control. They are simply an advance in technology that provides her-bicides with a broader spectrum of control and more flexibility in application time (Burnside, 1992; Owen, 2000).

Although herbicides provide growers with several advantages compared to other control tactics, they have decoupled weed management from cultural practices that once were an integral component of the procedure. This has resulted in simplifica-tion of weed management, facilitating the development of herbicide-resistant weeds and rapid shifts in weed populations.

The opportunity to enhance herbicide effectiveness through improved manage-ment or integration with other strategies is well documented (Zhang et al., 2000).

Techniques that have allowed successful reductions in herbicide rates include: (i) combining band applications of herbicides over the crop row with interrow cultivation (Rosales-Robles *et al.*, 1999); (ii) enhancing crop competitiveness through cultivar selection (Gealy *et al.*, 2003) or planting arrangement (Wait *et al.*, 1999); (iii) adjusting herbicide rate to target species (Knezevik *et al.*, 1998); (iv) using synergistic herbicide combinations (Scott *et al.*, 1998); (v) optimizing application timing (O'Sullivan and Bouw, 1997); and (vi) using appropriate tillage practices (Bostrom and Fogelfors, 1999). Although reductions in rates may increase the likelihood of control failures compared with full label rates, this risk can be managed by utilizing this approach in fields with low initial weed densities and appropriate integration with other management strategies (Zhang *et al.*, 2000; Bussan and Boerboom, 2001).

Thresholds and Weed Management

Economic thresholds are usually considered a critical component of IPM systems (Bottrell, 1979). Various types of thresholds have been described, most developed with the purpose of assisting farmers in making better weed control decisions (Cousens, 1987). Economic threshold is the most commonly described and is defined as the weed density at which the cost of control equals the value of the crop that would be lost due to interference if the weeds were left in the field. At weed densities below the economic threshold, it is recommended that weeds be left in the field because net returns would be higher than if they were controlled.

Although much effort has been devoted in developing economic thresholds, their acceptance by both weed scientists and producers has been limited. A survey of Illinois farmers found that only 9% used economic thresholds as a basis for weed control (Czapar *et al.*, 1995). Biological and agronomic limitations of economic thresholds have been reviewed (O'Donovan, 1996; Norris, 1999). Norris (1999) stated that the con-

cept of economic thresholds was initially developed for arthropod control and then adapted for weed control with little change. He argued that differences in the ecology and population biology between weeds and arthropods limit the transferability of threshold concepts between pest classes.

A primary argument against economic thresholds is the future effect of weed seed production by weeds below threshold densities. Subthreshold densities of velvetleaf resulted in a rapid increase in the weed seedbank and subsequent velvetleaf densities (Zanin and Sattin, 1988; Hartzler, 1996; Cardina and Norquay, 1997). Economic optimum thresholds (EOTs) differ from economic thresholds in that EOTs consider the long-term cost associated with seed production (Cousens, 1987). The EOTs for velvetleaf and common sunflower in soybean were calculated to be respectively 7.5-fold and 3.6-fold lower than the economic threshold (Bauer and Mortensen, 1992).

The instability of yield losses has been cited as another limitation of economic thresholds. The yield loss attributable to a specific weed infestation can vary by a factor of two or more depending upon the environment and crop production practices (Stoller *et al.*, 1987; Bauer *et al.*, 1991). The distribution of weeds within agricultural fields is rarely uniform; weeds are typically found in patches having a high relative density surrounded by areas with a few plants (Cardina *et al.*, 1996). Because the spatial pattern of weeds is not regular, the mean density alone is of little value in predicting yield losses. Assuming a regular distribution of weeds when predicting yield losses resulted in an overestimation of weed-related yield losses (Wiles *et al.*, 1992).

Auld and Tisdell (1987) stated that risk aversion by farmers and uncertainty of the yield-loss function limited the relevance of economic thresholds. Cousens (1987) concluded that subjectivity in decision making is acceptable as long as knowledge of competition, weed population dynamics and herbicide performance is used to guide management practices.

Site-specific Management

Recent advances in site-specific agriculture have created opportunities for enhancing the efficiency of weed management. Site-specific agriculture has been defined as 'an information and technology based agricultural management system to identify, analyze, and manage spatial and temporal variability within fields for optimum profitability, sustainability, and protection of the environment' (Robert et al., 1994). This concept has direct application to ecological weed management because of the spatial and temporal heterogeneity of weed populations across agricultural landscapes (Johnson et al., 1995; Cardina et al., 1996). Although weeds are not uniformly distributed across fields, most weed control practices are applied uniformly. The uniform application of herbicides over non-uniform weed populations has been identified as an important source of inefficiency in weed management (Cardina et al., 1997). Large portions of crop fields are often below the threshold weed density when the average field density is above the threshold, or vice versa (Cardina et al., 1995; Johnson et al., 1995). Cardina et al. (1996) found that with a hypothetical threshold of ten weed plants per square metre, about 40% of a field did not require treatment in one year, but 90% of the same field required treatment in another year.

Site-specific management requires affordable methods to sample and map weed populations (Wiles, 2005). Typically, sampling programmes developed for IWM are based on measuring the weed density and are insufficient for site-specific management. Weed maps can be generated using various methods of interpolation with spatially referenced weed density data (Dille et al., 2002). Currently, the expense of generating maps with sufficient detail to guide management decisions has limited site-specific weed management (Wiles, 2005). An improved understanding of the spatial stability of weed infestations may allow decisions to be based on historic weed distribution maps rather than generating new maps each season (Gerhards et al., 1997;

Colbach et al., 2000). Developing methods to combine growers' knowledge of weed distribution with spatially referenced data may facilitate the adoption of site-specific management (Wiles, 2005).

Site-specific management of weeds involves new concepts of weed biology and new technology. Principles of weed management and biology will need to be applied more precisely. Much of the effort in site-specific management has focused on developing the technology required to capitalize on the spatial distribution of weeds within fields to minimize external inputs (Woebbecke et al., 1993; Medlin et al., 2000). Aerial photography utilizing multispectral digital images is able to detect high densities (more than ten plants per square metre) of seedling-pitted morning glory and sicklepod in soybean (Medlin et al., 2000). 'Smart sprayers' utilize real-time sensors mounted to the tractor that detect the presence of weeds and control herbicide application (Tiam, 2002). Opportunities also exist to determine the ecological factors that drive weed patchiness within fields (Dieleman et al., 2000; Walter et al., 2002). The relationship between weed occurrence and soil properties is found to be field-specific, and soil characteristics are one of several factors that affected weed patchiness (Walter et al., 2002). Identifying associations between site characteristics and weed abundance may facilitate the development of practices that create less favourable environments for weeds.

Conclusions

Ecological weed management differs from traditional weed management in that the primary focus is on creating an environment unfavourable for weed establishment, growth and reproduction rather than on specific control tactics. Many of the components of an ecological management system are inextricably intertwined, thereby making it difficult to measure the individual contributions of specific elements of the system. A better understanding of the underlying mechanisms that influence the success or failure of weeds in agroecosystems will further the development and adoption of ecological weed management systems for agricultural crops.

The central challenge of developing ecological weed management systems is the integration of the options and tools that are available to make the cropping system unfavourable for weeds and to minimize the impact of any weeds that survive. No single weed management tactic has proven to be the 'magic bullet' to eliminate weed problems, and given the nature of weed communities, we should not expect one to appear in the near future. The best approach may be to integrate cropping system design

and knowledge of ecological processes with all available weed control strategies into a comprehensive weed management system.

The integration of ecological principles into weed management decision making presents a major challenge to weed science researchers and practitioners. Weed scientists must play a larger role in leading ecological research in agricultural systems. An expanded theory of applied ecology provides an excellent framework for expanded approaches to weed management because it allows for new and creative ways of meeting the challenge of managing weeds in ways that are environmentally and economically viable over the long term.

References

Aldrich, R.J. (1984) *Weed-Crop Ecology: Principles in Weed Management*. Breton, North Scituate, Massachusetts.

Ampong-Nyarko, K. and De Datta, S.K. (1993) Effects of nitrogen application on growth, nitrogen use efficiency and rice–weed interaction. *Weed Research* 33, 269–276.

Auld, B.A. and Tisdell, C.A. (1987) Economic thresholds and response to uncertainty in weed control. *Agricultural Systems* 25, 219–228.

Baker, H.G. (1974) The evolution of weeds. *Annual Review of Ecology and Systematics* 5, 1–24.

Ball, D.A. (1992) Weed seedbank response to tillage, herbicides, and crop rotation sequence. *Weed Science* 40, 654–659.

Barberi, P. and Mazzoncini, M. (2001) Changes in weed community composition as influenced by cover crop and management system in continuous corn. *Weed Science* 49, 491–499.

Bauer, T.A. and Mortensen D.A. (1992). A comparison of economic and economic optimum thresholds for two annual weeds in soybeans. *Weed Technology* 6, 228–235.

Bauer, T.A., Mortensen, D.A., Wicks, G.A., Hayden, T.A. and Martin, A.R. (1991) Environmental variability associated with economic thresholds for soybeans. *Weed Science* 39, 564–569.

Benbrook, C.M., Hoppin, P. and Liebman, M. (1996) New tools to measure reliance and use of herbicides and the adoption of integrated weed management. *Weed Science Society of America Abstracts* 36, 94–95.

Bertholdsson, N.O. (2005) Early vigour and allelopathy: two useful traits for enhanced barley and wheat competitiveness against weeds. *Weed Research* 45, 94–102.

Bhowmik, P.C. (2003) Challenges and opportunities in implementing allelopathy for natural weed management. *Crop Protection* 22, 661–671.

Blackshaw, R.E. (1994a) Rotation affects downy brome (*Bromus tectorum*) in winter wheat (*Triticum aestivum*). *Weed Technology* 8, 728–732.

Blackshaw, R.E. (1994b) Differential competitive ability of winter wheat cultivars against downy brome. *Agronomy Journal* 86, 649–654.

Blackshaw, R.E., Semach, G. and Janzen, H.H. (2002) Fertilizer application method affects nitrogen uptake in weeds and wheat. *Weed Science* 50, 634–641.

Blackshaw, R.E., Brandt, R.N., Janzen, H.H., Grant, C.A. and Derksen, D.A. (2003) Differential response of weed species to added nitrogen. *Weed Science* 51, 532–539.

Bostrom, U. and Fogelfors, H. (1999) Type and time of autumn tillage with and without herbicides at reduced rates in southern Sweden. 2. Weed flora and diversity. *Soil Tillage Research* 50, 283–293.

Bottrell, D.R. (1979) *Integrated Pest Management*. US Government Printing Office, Washington, DC.

Bowers, R.C. (1986) Commercialization of Collego™: an industrialist's view. *Weed Science* 34(Suppl. 1), 24–25.

Boyetchko, S.M. (1997) Principles of biological weed control with microherbicides. *HortScience* 32, 201–205.

Brust, G.E. and House, G.J. (1988) Weed seed destruction by arthropods and rodents in low-input soybean agroecosystems. *American Journal of Alternative Agriculture* 3, 19–25.

Buhler, D.D. (1995) Influence of tillage systems on weed population dynamics and management in corn and soybean in the central USA. *Crop Science* 35, 1247–1258.

Buhler, D.D. (1996) Development of alternative weed management strategies. *Journal of Productive Agriculture* 9, 501–505.

Buhler, D.D. and Gunsolus J.L. (1996) Effect of date of preplant tillage and planting on weed populations and mechanical weed control in soybean (*Glycine max*). *Weed Science* 44, 373–379.

Buhler, D.D. and Mester, T.C. (1991) Effect of tillage systems on the emergence depth of giant foxtail (*Setaria faberi*) and green foxtail (*Setaria viridis*). *Weed Science* 39, 200–203.

Buhler, D.D., Gunsolus, J.L. and Ralston, D.F. (1992) Integrated weed management techniques to reduce herbicide inputs in soybean. *Agronomy Journal* 84, 973–978.

Buhler, D.D., Gunsolus, J.L. and Ralston, D.F. (1993) Common cocklebur (*Xanthium strumarium*) control in soybean (*Glycine max*) with reduced rates of bentazon and cultivation. *Weed Science* 41, 447–453.

Buhler, D.D., Hartzler, R.G., Forcella, F. and Gunsolus J.L. (1997) Relative emergence sequence for weeds of corn and soybeans. *Iowa State University Extension Bulletin*, SA-11, Ames, Iowa.

Buhler, D.D., Kohler, K.A. and Foster, M.S. (1998) Spring-seeded smother plants for weed control in corn and soybean. *Journal of Soil and Water Conservation* 53, 272–275.

Buhler, D.D., Liebman, M. and Obrycki, J.J. (2000) Theoretical and practical challenges to an IPM approach to weed management. *Weed Science* 48, 274–280.

Buhler, D.D., Kohler, K.A. and Thompson, R.L. (2001) Weed seed bank dynamics during a five-year crop rotation. *Weed Technology* 15, 170–176.

Burnside, O.C. (1992) Rationale for developing herbicide-resistant crops. *Weed Technology* 6, 621–625.

Bussan, A.J. and Boerboom, C.M. (2001) Modeling the integrated management of giant foxtail in corn-soybean. *Weed Science* 49, 675–684.

Callaway, M.B. (1992) A compendium of crop varietal tolerance to weeds. *American Journal of Alternative Agriculture* 7, 169–180.

Cardina, J. and Norquay, H.M. (1997) Seed production and seedbank dynamics in subthreshold velvetleaf (*Abutilon theophrasti*) populations. *Weed Science* 45, 85–90.

Cardina, J., Regnier, E. and Harrison, K. (1991) Long-term tillage effects of seed banks in three Ohio soils. *Weed Science* 39, 186–194.

Cardina, J., Sparrow, D.H. and McCoy, E.L. (1995) Analysis of spatial distribution of common lambsquarters (*Chenopodium album*) in no-till soybean (*Glycine max*). *Weed Science* 43, 258–268.

Cardina, J., Sparrow, D.H. and McCoy, E.L. (1996) Spatial relationships between seedbank and seedling populations of common lambsquarters (*Chenopodium album*) and annual grasses. *Weed Science* 44, 298–308.

Cardina, J., Johnson, G.A. and Sparrow, D.H. (1997) The nature and consequences of weed spatial distribution. *Weed Science* 45, 364–373.

Cardina, J., Herms, C.P. and Doohan, D.J. (2002) Crop rotation and tillage system effects on weed seedbanks. *Weed Science* 50, 448–460.

Cate, J.R. and Hinkle, M.K. (1994) *Integrated Pest Management: The Path of a Paradigm*. National Audubon Society, Washington, DC.

Colbach, N., Forcella, F. and Johnson, G.A. (2000) Spatial and temporal stability of weed populations over five years. *Weed Science* 48, 366–377.

Conklin, A.E., Erich, M.S., Liebman, M. and Lambert, D.H. (1998) Disease incidence and growth of wild mustard seedlings in red clover and compost amended soil. *Agronomy Abstracts* 90, 279.

Cousens, R. (1987) Theory and reality of weed control thresholds. *Plant Protection Quarterly* 2, 13–20.

Creamer, N.G., Bennett, M.A., Stinner, B.R., Cardina, J. and Regnier, E.E. (1996) Mechanisms of weed suppression in cover crop-based production systems. *HortScience* 31, 410–413.

Czapar, G.F., Curry, M.P. and Gray, M.E. (1995) Survey of integrated pest management practices in central Illinois. *Journal of Production Agriculture* 8, 483–486.

Dastgheib, F. (1989) Relative importance of crop seed, manure and irrigation water as sources of weed infestation. *Weed Research* 29, 113–116.

Davis, A.S. and Liebman, M. (2001) Nitrogen source influences wild mustard growth and competitive effect on sweet corn. *Weed Science* 49, 558–566.

Day, B.E. (1972) *Pest Control: Strategies for the Future*. National Academy of Sciences, Washington, DC.

DeLuca, T.H. and DeLuca, D.K. (1997) Composting for feedlot manure management and soil quality. *Journal of Production Agriculture* 10, 235–241.

Dewey, S.A. and Whitesides, R.E. (1990) Weed seed analysis from four decades of Utah small grain drill-box surveys. *Proceedings of Western Society of Weed Science* 43, 69–70.

Dieleman, J.A., Mortensen, D.A, Buhler, D.D. and Ferguson, R.B. (2000) Identifying associations among site properties and weed species abundance. II. Hypothesis generation. *Weed Science* 48, 576–587.

Dille, J.A., Milner, M., Groeteke, J.J., Mortensen, D.A. and Williams, M.M. II (2002) How good is your weed map? A comparison of spatial interpolators. *Weed Science* 51, 44–55.

DiTomaso, J.M. (1995) Approaches for improving crop competitiveness through the manipulation of fertilization strategies. *Weed Science* 43, 491–497.

Dorado, J., Del Monte, J.P. and López-Fando D. (1999) Weed seedbank response to crop rotation and tillage in semiarid agroecosystems. *Weed Science* 47, 67–73.

Doucet, C., Weaver, S.E., Hamill, A.S. and Zhang, J. (1999) Separating the effects of crop rotation from weed management on weed density and diversity. *Weed Science* 47, 729–735.

Duke, S.O., Rimando, A.M., Baerson, S.R., Scheffler, B.E., Ota, E. and Belz, R.G. (2002) Strategies for the use of natural products for weed management. *Journal of Pesticide Science* 27, 298–306.

Dyck, E. and Liebman, M. (1995) Crop–weed interference as influenced by a leguminous or synthetic fertilizer nitrogen source. II. Rotation experiments with crimson clover, field corn, and lambsquarter. *Agricultural Ecosystem and Environment* 56, 109–120.

Entz, M.H., Bullied, W.J. and Katepa-Mupondwa, F. (1995) Rotational benefits of forage crops in Canadian prairie cropping systems. *Journal of Production Agriculture* 8, 521–529.

Eplee, R.E. (1992) Witchweed (*Striga asiatica*): an overview of management strategies in the USA. *Crop Protection* 11, 3–7.

Evans, S.P., Knezevic, S.Z., Lindquist, J.L., Shapiro, C.A. and Blankenship, E.E. (2003) Nitrogen application influences the critical period for weed control in corn. *Weed Science* 51, 408–417.

Fernholz, C. (1990) How I control weeds without herbicides. *The New Farm* March/April, 17–20.

Fisk, J.W., Hesterman, O.B., Shrestha, A., Kells, J.J., Harwood, R.R., Squire, J.M. and Sheaffer, C.C. (2001) Weed suppression by annual legume cover crops in no-tillage corn. *Agronomy Journal* 93, 319–325.

Fogelberg, F. and Kritz, G. (1999) Intra-row weeding with brushes on vertical axes: factors influencing in-row soil height. *Soil and Tillage Research* 50, 149–157.

Forcella, F. (1998) Real-time assessment of seed dormancy and seedling growth for weed management. *Seed Science Research* 8, 201–209.

Forcella, F., Eradat-Oskoui, K. and Wagner, S.W. (1993) Application of weed seedbank ecology to low-input crop management. *Ecological Applications* 3, 74–83.

Gallandt, E.R., Liebman, M., Corson, S., Porter, G.A. and Ullrich, S.D. (1998) Effects of pest and soil management systems on weed dynamics in potato. *Weed Science* 46, 238–248.

Gallandt, E.R., Liebman, M. and Huggins, D.R. (1999) Improving soil quality: implications for weed management. *Journal of Crop Production* 2, 95–121.

Garrity, D.P., Movillon, M. and Moody, K. (1992) Differential weed suppression ability in upland rice cultivars. *Agronomy Journal* 84, 586–591.

Gealy, D.R., Wailes, E.J., Estorninos, L.E. Jr and Chavez, R.S.C. (2003) Rice cultivar differences in suppression of barnyardgrass (*Echinochloa crusgalli*) and economics of reduced propanil rates. *Weed Science* 51, 601–609.

Gebhardt, M.R. (1981) Preemergence herbicides and cultivation for soybeans (*Glycine max*). *Weed Science* 29, 165–168.

Gerhards, R., Wyse-Pester, D.Y., Mortensen, D. and Johnson, G.A. (1997) Characterizing spatial stability of weed populations using interpolated maps. *Weed Science* 45, 108–119.

Ghersa, C.M. and Holt, J.S. (1995) Using phenology prediction in weed management: a review. *Weed Research* 35, 461–467.

Ghersa, C.M. and Martinez-Ghersa, M.A. (1991) A field method for predicting yield losses in maize caused by Johnson grass (*Sorghum halepense*). *Weed Technology* 5, 279–285.

Grime, J.P. (1977) Evidence for the existence of three primary strategies in plants and its relevance to ecological and evolutionary theory. *American Naturalist* 111, 1169–1194.

Grundy, A.C. (2003) Predicting weed emergence: a review of approaches and challenges. *Weed Research* 43, 1–11.

Gunsolus, J.L. (1990) Mechanical and cultural weed control in corn and soybeans. *American Journal of Alternative Agriculture* 5, 114–119.

Hallet, S.G. (2005) Where are the bioherbicides? *Weed Science* 53, 404–415.

Harmon, G.W. and Keim, F.D. (1934) The percentage and viability of weed seeds recovered in the feces of farm animals and their longevity when buried in manure. *Agronomy Journal* 26, 762–767.

Hartzler, R.G. (1996) Velvetleaf (*Abutilon theophrasti*) population dynamics following a single year's seed rain. *Weed Technology* 10, 581–586.

Hartzler, R.G. and Buhler, D.D. (1999) Emergence characteristics of four annual weed species. *Weed Science* 47, 578–584.

Hartzler, R.G. and Roth, G.W. (1993) Effect of prior year's weed control on herbicide effectiveness in corn (*Zea mays*). *Weed Technology* 7, 611–614.

Haynes, R.J. and Goh, K.M. (1978) Ammonium and nitrate nutrition of plants. *Biological Reviews* 53, 465–510.

Holm, L., Doll, J., Holm, E., Pancho, J. and Herberger, J. (1997) *World Weeds: Natural Histories and Distribution*. John Wiley & Sons, New York.

Horwath, W.R., Elliott, L.F. and Lynch, J.M. (1998) Influence of soil quality on the function of inhibitory rhizobacteria. *Letters Applied Microbiology* 26, 87–92.

Hoveland, C.S., Buchanan, G.A. and Harris, M.C. (1976) Response of weeds to soil phosphorus and potassium. *Weed Science* 24, 194–201.

Hume, L. (1982) The long-term effects of fertilizer application and three rotations on weed communities in wheat. *Canadian Journal of Plant Science* 62, 741–750.

Hume, L. (1987) Long-term effects of 2,4-D application on plants. 1. Effects on the weed community in a wheat crop. *Canadian Journal of Botany* 65, 2530–2536.

Johnson, G.A., Mortensen, D.A. and Martin, A.R. (1995) A simulation of herbicide use based on weed spatial distribution. *Weed Research* 35, 197–205.

Jordan, N.R., Zhang, J. and Huerd, S. (2000) Arbuscular-mycorrhizal fungi: potential roles in weed management. *Weed Research* 40, 397–410.

Jornsgard, B., Rasmussen, K., Hill, J. and Christiansen, J.L. (1996) Influence of nitrogen on competition between cereals and their natural weed populations. *Weed Research* 36, 461–470.

Karssen, C.M. and Hillhorst, H.W.M. (1992) Effect of chemical environment on seed germination. In: Fenner. M. (ed.) *Seeds: The Ecology of Regeneration in Plant Communities*. CAB International, Wallingford, UK, pp. 327–348.

Keily, T., Donaldson, D. and Grube, A. (2004) *Pesticides Industry Sales and Usage: 2000 and 2001 Market Estimates*. US Environmental Protection Agency, Washington, DC.

Kennedy, A.C. (1999) Soil microorganisms for weed management. *Journal of Crop Production* 2, 123–138.

Kennedy, A.C. and Kremer, R.J. (1996) Microorganisms in weed control strategies. *Journal of Production Agriculture* 9, 480–484.

Kirkland, K.J. and Beckie, H.J. (1998) Contribution of nitrogen fertilizer placement to weed management in spring wheat (*Triticum aestivum*). *Weed Technology* 12, 507–514.

Klingaman, T.E. and Oliver, L.E. (1994) Influence of cotton (*Gossypium hirsutum*) and soybean (*Glycine max*) planting date on weed interference. *Weed Science* 42, 61–65.

Knezevik, S.Z., Sikkema, P.H., Tardif, F., Hammill, A.S., Chandler, K. and Swanton, C.J. (1998) Biologically effective dose and selectivity of RPA 201772 for preemergence weed control in corn (*Zea mays*). *Weed Technology* 12, 670–676.

Koscelny, J.A., Peeper, T.F., Solie, J.B. and Solomon, S.G. (1990) Effect of wheat (*Triticum aestivum*) row spacing, seeding rate, and cultivar on yield loss from cheat (*Bromus secalinus*). *Weed Technology* 4, 487–492.

Kouwenhoven, J.K. (1997) Intra-row mechanical weed control: possibilities and problems. *Soil and Tillage Research* 41, 87–104.

Kremer, R.J. and Li, J. (2003) Developing weed-suppressive soils through improved soil quality management. *Soil and Tillage Research* 72, 193–202.

Kropff, M.J. and Spitters, C.F.T. (1991) A simple model of crop loss by weed competition from early observations on relative leaf area of the weeds. *Weed Research* 31, 97–105.

Kudsk, P. and Streibig, J.C. (2003) Herbicides: a two-edged sword. *Weed Research* 43, 90–102.

Lanning, S.P., Talbert, L.E., Matrin, J.M., Blake, T.K. and Bruckner, P.L. (1997) Genotype of wheat and barley affects light penetration and wild oat growth. *Agronomy Journal* 89, 100–103.

Légère, A. and Samson, N. (1999) Relative influence of crop rotation, tillage, and weed management on weed associations in spring barley cropping systems. *Weed Science* 47,112–122.

Légère, A. and Stevenson, E.C. (2002) Residual effects of crop rotation and weed management on wheat test crop and weeds. *Weed Science* 50, 101–111.

Légère, A., Simard, R.R. and Lapierre, C. (1994) Response of spring barley and weed communities to lime, phosphorus and tillage. *Canadian Journal of Plant Science* 74, 421–428.

Leighty, C.E. (1938) Crop rotation. In: *Soils and Men: Yearbook of Agriculture 1938*, USDA, Government Printing Office, Washington, DC, pp. 406–430.

Li, J. and Kremer, R.J. (2000) Rhizobacteria associated with weed seedlings in different cropping systems. *Weed Science* 48, 734–741.

Liebman, M. and Dyck, E. (1992) Crop rotation and intercropping strategies for weed management. *Ecological Applications* 3, 92–122.

Liebman, M. and Staver, C.P. (2001) Crop diversification for weed management. In: Liebman, M., Mohler, C.L. and Staver, C.P. (eds) *Ecological Management of Agricultural Weeds*. Cambridge University Press, Cambridge, UK, pp. 322–374.

Magdoff, F. (2001) Concept, components, and strategies of soil health in agroecosystems. *Journal of Nematology* 33, 169–172.

Mahn, E.G. and Helmecke, K. (1979) Effects of herbicide treatment on the structure and functioning of agro-ecosystems. II. Structural changes in the plant community after the application of herbicides over several years. *Agro-Ecosystems* 5, 159–179.

Malik, V.S., Swanton, C.J. and Michaels, T.E. (1993) Interaction of white bean (*Phaseolus vulgaris* L.) cultivars, row spacing, and seeding density with annual weeds. *Weed Science* 41, 62–68.

Marino, P.C., Westerman, P.R., Pinkert, C. and van der Werf, W. (2005) Influence of seed density and aggregation on post-dispersal weed seed predation in cereal fields. *Agricultural Ecosystem and Environment* 106, 17–25.

Mauchline, A.L., Watson, S.J., Brown, V.K. and Williams, R.J. (2005) Post-dispersal seed predation of non-target weeds in arable crops. *Weed Research* 45, 157–164.

McWhorter, C.G. and Barrentine, W.L. (1975) Cocklebur control in soybeans as affected by cultivars, seeding rates, and methods of weed control. *Weed Science* 23, 386–390.

Medlin, C.R., Shaw, D.R., Gerard, P.D. and LaMastrus, F.E. (2000) Using remote sensing to detect weed infestations in *Glycine max. Weed Science* 48, 393–398.

Mt. Pleasant, J. and Schlather, K.J. (1994) Incidence of weed seed in cow manure and its importance as a weed source for cropland. *Weed Technology* 8, 304–310.

Norris, R.F. (1999) Ecological implications of using thresholds for weed management. In: Buhler, D.D. (ed.) *Expanding the Context of Weed Management*. The Haworth Press, New York, pp. 31–58.

O'Donovan, J.T. (1996) Weed economic thresholds: useful agronomic tool or pipe dream? *Phytoprotection* 77, 13–28.

Oerke, E.C. and Dehne, H.W. (2004). Safeguarding production: losses in major crops and the role of crop protection. *Crop Protection* 23, 275–285.

Oliver, L.R. (1979) Influence of soybean (*Glycine max*) planting date on velvetleaf (*Abutilon theophrasti*) competition. *Weed Science* 27, 183–188.

Oriade, C.A. and Forcella, F. (1998) Maximizing efficacy and economics of mechanical weed control in row crops through forecasts of weed emergence. *Journal of Crop Production* 2, 189–205.

O'Sullivan, J. and Bouw, W.J. (1997) Effect of timing and adjuvants on the efficacy of reduced herbicide rates for sweet corn (*Zea mays*). *Weed Technology* 11, 720–724.

Owen, M.D.K. (2000) Current use of transgenic herbicide resistant soybean and corn in the USA. *Crop Protection* 19, 765–771.

Ozores-Hampton, M., Stoffella, P.J., Bewick, T.A., Cantliffe, D.J. and T.A. Obreza. (1999) Effect of age of composted MSW and biosolids on weed seed germination. *Compost Science Utilization* 7, 51–57.

Panetta, F.D. and Scanlan, J.C. (1995) Human involvement in the spread of noxious weeds: what plants should be declared and when should control be enforced? *Plant Protection Quarterly* 10, 69–74.

Pareja, M.R., Staniforth, D.W. and Pareja, G.P. (1985) Distribution of weed seed among soil structural units. *Weed Science* 33, 182–189.

Pester, T.A., Burnside, O.C. and Orf, J.H. (1999) Increasing crop competitiveness to weeds through crop breeding. *Journal of Crop Production* 3, 31–58.

Petersen, J. (2003) Weed: spring barley competition for applied nitrogen in pig slurry. *Weed Research* 43, 33–39.

Poalini, R., Principi, M., Froud-Williams, R.J., Del-Puglia S. and Biancardi, E. (1999) Competition between sugarbeet and *Sinapis arvensis* and *Chenopodium album*, as affected by timing of nitrogen fertilization. *Weed Research* 39, 425–440.

Rao, A.N. and Moody, K. (1990) Weed seed contamination in rice seed. *Seed Science and Technology* 18, 139–146.

Radosevich, S., Holt, J. and Ghersa, C. (1997) *Weed Ecology: Implications for Management*, 2nd edn. John Wiley & Sons, New York.

Rasmussen, K. (2002) Influence of liquid manure application method on weed control in spring cereals. *Weed Research* 42, 287–298.

Reberg-Horton, C., Creamer, N., Burton, J., Ranells, N. and Murphy, P. (2001) Breeding rye (*Secale cereale*) for increased allelopathy. *HortScience* 36, 561.

Robert, P.C., Rust, R.H. and Larson, W.E. (1994) Preface. In: Robert, P.C., Rust, R.H. and Larson, W.E. (eds) *Site-Specific Management for Agricultural Systems*. American Society of Agronomy, Madison, Wisconsin, pp. 2–8.

Roberts, H.A. and Potter, M.E. (1980) Emergence patterns of weed seedlings in relation to cultivation and rainfall. *Weed Research* 20, 377–386.

Rosales-Robles, E., Chandler, J.M., Senseman, S.A. and Prostko, E.P. (1999) Integrated johnsongrass (*Sorghum halepense*) management in field corn (*Zea mays*) with reduced rates of nicosulfuron and cultivation. *Weed Technology* 13, 367–373.

Rupende, E., Chivinge, O.A. and Mariga, I.K. (1998) Effect of storage time on weed seedling emergence and nutrient release in cattle manure. *Experimental Agriculture* 34, 277–285.

Salas, M.L., Hickman, M.V., Huber, D.M. and Schreiber, M.M. (1997) Influence of nitrate and ammonium nutrition on the growth of giant foxtail (*Setaria faberi*). *Weed Science* 45, 664–669.

Scheepens, P.C., Müller-Schärer, H. and Kempenaar, C. (2001) Opportunities for biological weed control in Europe. *BioControl* 46, 127–138.

Scott, R.C., Shaw D.R., O'Neal, W.B. and Klingaman, T.D. (1998) Spray adjuvant, formulation, and environmental effects on synergism from post-applied tank mixtures of SAN 582H with fluazifop-P, imazethapyr, and sethoxydim. *Weed Technology* 12, 463–469.

Scursoni, J.A. and Benech Arnold, R. (2002) Effect of nitrogen fertilization timing on the demographic processes of wild oat (*Avena fatua*) in barley (*Hordeum vulgare*). *Weed Science* 50, 616–621.

Skinner, K., Smith, L. and Rice, P. (2000) Using noxious weed lists to prioritize targets for developing weed management strategies. *Weed Science* 48, 640–644.

Søgaard, H.T. (1998) Automatic control of a finger weeder with respect to the harrowing intensity at varying soil structures. *Journal of Agricultural Engineering and Research* 70, 157–163.

Spandl, E., Durgan, B.R. and Forcella, F. (1998) Tillage and planting date influence foxtail (*Setaria* spp.) emergence in continuous spring wheat (*Triticum aestivum*). *Weed Technology* 12, 223–229.

Staniforth, D.W. and Wiese, A.F. (1985) Weed biology and its relationship to weed control in limited-tillage systems. In Wiese, A.F. (ed.) *Weed Control in Limited-tillage Systems*. Weed Science Society of America, Champaign, Illinois, pp. 15–25.

Stanislaus, M.A. and Cheng, C.L. (2002) Genetically engineered self-destruction: an alternative to herbicides for cover crop systems. *Weed Science* 50, 794–801.

Steckel, L.E., DeFelice, M.S. and Sims, B.D. (1990) Integrating reduced rates of postemergence herbicides and cultivation for broadleaf weed control in soybeans (*Glycine max*). *Weed Science* 38, 541–545.

Stoller, E.W. and Wax, L.M. (1973) Periodicity of germination and emergence of some annual weeds. *Weed Science* 21, 574–580.

Stoller, E.W., Harrison, S.K, Wax, L.M., Regnier, E.E. and Nafzinger, E.D. (1987) Weed interference in soybeans (*Glycine max*). *Reviews of Weed Science* 3, 155–181.

Streit, B., Rieger, S.B., Stamp, P. and Richner, W. (2003). Weed populations in winter wheat as affected by crop sequence, intensity of tillage and time of herbicide application in a cool and humid climate. *Weed Research* 43, 20–32.

Teasdale, J.R. (1995) Influence of narrow row/high population corn (*Zea mays*) on weed control and light transmission. *Weed Technology* 9, 113–118.

Teasdale, J.R. (1998) Cover crops, smother plants, and weed management. In: Hatfield, J.L., Buhler, D.D. and Stewart, B.A. (eds) *Integrated Weed and Soil Management*. Ann Arbor Press, Chelsea, Michigan, pp. 247–270.

Teasdale, J.R. and Frank, J.R. (1983) Effect of row spacing on weed competition with snap beans (*Phaseolus vulgaris*). *Weed Science* 31, 81–85.

Teasdale, J.R. and Mohler, C.L. (2000) The quantitative relationship between weed emergence and the physical properties of mulches. *Weed Science* 48, 385–392.

Teyker, R.H., Hoelzer, H.D. and Liebl, R.A. (1991) Maize and pigweed response to nitrogen supply and form. *Plant Soil* 135, 287–292.

Tiam, L. (2002) Development of a sensor-based precision agriculture system. *Computers and Electronics in Agriculture* 36, 133–149.

Tillett, N.D., Hague, T. and Miles, S.J. (2002) Inter-row vision guidance for mechanical weed control in sugar beet. *Computers and Electronics in Agriculture* 33, 163–177.

Tonkin, J.H.B. (1982) The presence of seed impurities in samples of cereal seed tested at the official seed testing station, Cambridge in the period of 1978–1981. *Aspects of Applied Biology* 1, 163–171.

Unger, P.W., Miller, S.D. and Jones, O.R. (1999) Weed seeds in long-term dryland tillage and cropping system plots. *Weed Research* 39, 213–223.

Valdrighi, M.M., Pera, A., Agnolucci, M., Frassinetti, S., Lunardi, D. and Vallini, G. (1996) Effects of compost-derived humic acids on vegetable biomass production and microbial growth within a plant (*Cichorium intybus*)–soil system: a comparative study. *Agricultural Ecosystem and Environment* 58, 133–144.

Van-Delden, A., Lotz, L.A.P., Bastiaans, L., Franke, A.C., Smid, H.G., Groeneveld, R.M.W. and Kropff, J.J. (2002) The influence of nitrogen supply on the ability of wheat and potato to suppress *Stellaria media* growth and reproduction. *Weed Research* 42, 429–445.

Wait, J.D., Johnson, W.G. and Massey, R.E. (1999) Weed management with reduced rates of glyphosate in no-till, narrow-row, glyphosate-resistant soybean (*Glycine max*). *Weed Technology* 13, 478–483.

Walker, R.H. (1995) Preventative weed management. In: Smith, A.E. (ed.) *Handbook of Weed Management Systems*. Marcel Dekker, New York, pp. 35–50.

Walter, A.M., Christensen, S. and Simmelsgaard, S.E. (2002) Spatial correlation between weed species densities and soil properties. *Weed Research* 42, 26–38.

Weaver, S.E., Tan, C.S. and Brain, P. (1988) Effect of temperature and soil moisture on time of emergence of tomatoes and four weed species. *Canadian Journal of Plant Science* 68, 877–876.

Webster, T.M. and Coble, H.D. (1997) Changes in the weed species composition of the southern United States: 1974 to 1995. *Weed Technology* 11, 308–317.

Westerman, P.R., Hoffman, A., Vet, L.E.M. and van der Werf, W. (2003) Relative importance of vertebrates and invertebrates in epigaeic weed seed predation in organic cereal fields. *Agricultural Ecosystem and Environment* 95, 417–425.

Westerman, P.R., Liebman, M., Menalled, F.D., Heggenstaller, A.H., Hartzler, R.G. and Dixon, P.M. (2005) Are many little hammers effective? – Velvetleaf (*Abutilon theophrasti*) population dynamics in two- and four-year crop rotation systems. *Weed Science* 53, 382–392.

Wiles, L.J. (2005) Sampling to make maps for site-specific weed management. *Weed Science* 53, 228–235.

Wiles, L.J., Wilkerson, G.G., Gold, H.J. and Coble, H.D. (1992) Modeling weed distribution for improved postemergence control decisions. *Weed Science* 40, 546–553.

Wiltshire, J.J.J., Tillett, N.D. and Hague, T. (2003) Agronomic evaluation of precise mechanical hoeing and chemical weed control in sugar beet. *Weed Research* 43, 236–244.

Woebbecke, D.M., Meyer, G.E., Von Bargen, K. and Mortensen, D.A. (1993) Plant species identification, size, and enumeration using machine vision techniques on near-binary images. In DeShazer, J.A. (ed.) *Proceedings of the SPIE Conference on Optics in Agriculture and Forestry Volume 1836*. SPIE, Boston, Massachusetts, pp. 208–219.

Wortmann, C.S. (1993) Contribution of bean morphological characteristics to weed suppression. *Agronomy Journal* 85, 840–843.

Zanin, G. and Sattin, M. (1988) Threshold level and seed production of velvetleaf (*Abutilon theophrasti*) in maize. *Weed Research* 28, 347–352.

Zhang, J., Weaver, S.E. and Hamill, A.S. (2000). Risks and reliability of using herbicides at below-labeled rates. *Weed Technology* 14, 106–115.

Zimdahl, R.L. (1991) *Weed Science: A Plea for Thought*. USDA Cooperative State Research Service, Washington, DC.

4 Role of Cover Crops in the Management of Arthropod Pests in Orchards

Michael W. Smith[1] and Phillip G. Mulder Jr[2]

[1]Department of Horticulture and Landscape Architecture, Oklahoma State University, Stillwater, OK 74078, USA; [2]Department of Entomology and Plant Pathology, Oklahoma State University, Stillwater, OK 74078, USA

Introduction

Agricultural practices are continually evolving, beginning with our prehistoric hunter–gatherer ancestry that gradually transitionaled into societies that depended upon farming a few staple crops. Current farming practices tend to concentrate single crop species into large acreages. These monoculture systems make mechanization economically feasible and facilitate intensive agronomic and horticultural practices that increase yield. However, another outcome of concentrating a single species with limited diversity and stabilizing its production from year to year is the increased number of problems associated with certain pests. For example, in the 1970s the widespread outbreak of southern corn leaf blight, *Bipolaris maydis* (Nisikado and Miyake) Shoemaker, was caused by extensive plantings of a single species with cultivars that had similar sources of disease resistance. Likewise, pecan weevil, *Curculio caryae* Horn, has become a major problem in certain areas of the USA. This has been largely due to improved horticultural practices stabilizing and increasing production, and the control of early-season fruit pests. In the natural state, pecan, *Carya illinoinensis* (Wangenh.) C. Koch, is a masting species, producing larger crops every other year with very large crops produced in about 9-year cycles (Wood, 1993). During seasons of low fruit production, early-season pests destroy most of the meagre crops rendering it unavailable for pecan weevil reproduction and food. Improved horticultural practices, development of cultivars with more regular bearing tendencies and management of early-season fruit pests have reduced irregular bearing, thus allowing pecan weevil to multiply rapidly and cause serious damage if left unchecked.

The challenges of increased pest pressure have been met in a variety of ways, usually by combining several strategies. Plant resistance to certain pathogens and arthropod pests has been fundamental for managing or minimizing several pest problems. Removal of infected or infested plant parts can reduce some pest problems. Orchard cover crops were common in the 1950s, particularly cool-season legumes that served as a nitrogen source for the orchard. Frequently, orchards were maintained weed-free during the summer to avoid water competition with the cultivated trees. Several commonly used cover crops have increased beneficial arthropod populations indirectly, thereby contributing to pest control. Agricultural chemicals have been a major component of pest management strategies in recent years. The first available insecticides were broad spectrum in action, eliminating much of the orchard arthropod community following application. Initially, excellent pest control was obtained by 'calendar applications' with broad-spectrum chemicals. However, in many instances continual use resulted in resistance of the target pest and serious outbreaks of secondary pests (Mizell, 1991), and in some cases excessive pesticide use compromised environmental quality. For instance, populations of oriental fruit moth, *Grapholita molesta* (Busck), developed resistance to organophosphorous and carbamate insecticides (Pree *et al.*, 1998; Usmani and Shearer, 2001). Similarly, resistance to pesticides has been identified in tarnished plant bug, *Lygus lineolaris* (Palisot de Beauvois) (Snodgrass, 1996), and pecan leaf scorch mite, *Eotetranychus hicoriae* (McGregor) (Boethel, 1981). The combination of managing pest resistance, avoiding secondary pest outbreaks and reducing environmental loading with pesticide has been addressed by the development and promotion of integrated pest management (IPM) approaches for many orchard crops. This in turn has led to an integrated orchard management system that combines horticultural and pest management considerations into tenable packages. Although integrated orchard management is in its infancy, usable systems are available and continue to evolve. We will discuss the role of cover crops in the management of arthropod pests as a component of integrated management systems for pecan, peach and apple.

There remains a need for broad-spectrum pesticides in today's agriculture production programmes. One example is in pecan management. During the later stages of budbreak, pecan is attacked by three species of *Phylloxera*: (i) *Phylloxera* spp. (Homoptera); (ii) pecan nut casebearer, *Acrobasis nuxvorella* Neunzig, Lepidoptera; and (iii) hickory shoot curculio, *Conotrachelus aratus* (Germar), Coleoptera. These diverse pests can be effectively controlled with a single, well-timed application of a broad-spectrum insecticide, with little apparent effect on subsequent beneficial arthropod populations. If only target-specific insecticides were available, a tank mix of three chemicals would be needed to control these pests, but would substantially increase costs.

Integrated Orchard Management

IPM was the initial phase in developing a logical approach for managing diseases, arthropods, nematodes and weeds in agricultural systems. This approach integrated (i) scouting programmes to determine economic thresholds before pesticide application; (ii) new or improved monitoring systems for key pests; (iii) selective pesticides to avoid widespread loss of beneficial arthropods; (iv) computer-based models to aid pest control decision making; (v) sanitation practices wherever practical; and (vi) use of cultivars that are resistant to certain pests. As research has progressed, IPM programmes have incorporated recommendations for specific cover crops that indirectly increase beneficial arthropod populations, and trap crops that either aid in monitoring key pests or in some instances are used as control sites for certain pests.

Adoption of IPM approaches has progressed at different rates depending primarily on the crop and the situation. It soon became apparent that management of pests could not be independent of the horticultural components required to produce large yields of high-quality fruit. A producer's chosen management system that incorporates a cover crop may affect the orchard's water use pattern, nutrition management, temperature stress and eventually soil structure. Economic considerations include those related to pest control, costs related to management of the cover crop, changes in costs of managing the trees, changes in harvest and postharvest handling systems and the effect of the new management system on fruit yield and quality. Cultivar recommendations must take into consideration attributes such as pest resistance or tolerance, yield potential, fruit quality, cold-hardiness, growth habit and several others, necessitating an integration of disciplines to evaluate candidate cultivars and make sound recommendations. Additionally, IPM requires the producer to know more about the target pests, their life cycles, how the crop is attacked, monitoring systems for pests and specificity of various pesticides. This requires a massive education programme to transform the agriculture industry from a 'calendar spray' mindset to an integrated management approach. A complete systems approach to orchard management can best be termed integrated orchard management (IOM) since it incorporates principles of IPM and horticulture.

Cover crops

One role of orchard cover crops in IOM is to produce an alternative food source (e.g. pollen or nectar) for beneficial arthropods. The cover crop may also attract certain phytophagous insects, e.g. aphids or thrips, that can serve as an alternative prey for some beneficial arthropods. Otherwise, highly mobile arthropods may abandon the orchard when prey populations are depleted and the development of less mobile immature arthropods will either be slowed or fail. Cover crops must be selected with care since the literature is replete with examples of non-crop vegetation hosting arthropod pests (e.g. Fye, 1980; Killian and Meyer, 1984; Atanassov et al., 2002). Another role of cover crops in IPM might be the suppression of undesirable vegetation (Rice, 1995).

Competition

Orchard floors are commonly maintained vegetation-free within the tree rows, with row middles maintained in sod. In arid climates, many orchard floors are maintained completely vegetation-free. Studies have demonstrated that live vegetation surrounding trees competes for water and nutrients, and some are allelopathic, reducing tree growth and yield (Bould and Jarrett, 1962; Patterson et al.,

1990; Goff et al., 1991; Wolf and Smith, 1999; Smith et al., 2001, 2002). Therefore, selection of cover crops must include evaluation of their potential benefits in pest management and their impact on orchard performance. Maintaining a vegetation-free area surrounding the tree will partially mitigate the negative impact of the cover crop on the tree. Also, cool-season ground covers tend to be less competitive with the trees, particularly for water. In some cases, a reduction in tree growth may be desirable. For instance, in peaches (*Prunus persica* L.), annual ryegrass (*Lolium multiflorum* Lam.) and deficit irrigation were used to decrease early spring growth, thus reducing the amount of pruning required without affecting yield (Huslig et al., 1993). After the initial spring growth flush, irrigation was increased and annual ryegrass was killed with a contact herbicide. The annual ryegrass residue suppressed weed growth and conserved soil moisture, reducing the need for additional herbicide intervention to avoid competition with the tree (Huslig et al., 1993).

Nutrition requirements of the orchard may change with certain cover crops. For instance, legume (Fabaceae) cover crops can be utilized to supply nitrogen (Table 4.1), thus partially or totally mitigating their establishment costs. Greatest nitrogen fixation by legumes occurs when other nitrogen sources are not available; therefore, fertilizing an orchard with nitrogen when using a legume ground cover reduces the benefit. Also, nitrogen fertilization tends to decrease the legume stand (Rogers et al., 1948). Legumes typically have a higher phosphorus requirement than trees, frequently necessitating supplemental phosphorous application to benefit legume growth with no benefit derived by the trees. Many of the non-legume ground covers compete efficiently with the trees for nitrogen, and can potentially create shortages of nitrogen in the trees unless nitrogen rates and application time are adjusted accordingly. Other crops can create different nutrition problems. For instance, lucerne (*Medicago sativa* L.) is known as a luxury consumer of potassium (Tisdale

Table 4.1. Apparent nitrogen supplied by legumes to pecan trees at three Oklahoma locations. (From Smith *et al.*, 1996c.)

| Location | Legume | Apparent N supplied by legume (kg/ha) | | |
		1992	1993	1994
Sapulpa	Crimson clover + hairy vetch	143	159	129
	Red clover + white clover	–	93	105
Beggs	Red clover	120	0	66
	White clover	115	0	47
	Red clover + white clover	132	30	83
Burneyville	Crimson clover + hairy vetch	156	101	141

and Nelson, 1966). On the other hand, pecan inefficiently absorbs potassium, but has a high potassium requirement for fruit development. Lucerne planted in a pecan orchard can create a potassium shortage that may require years to correct.

Water management is critical for consistent production of high-quality fruit. Many, but not all, commercial orchards in the USA are irrigated. Before irrigation was common, many orchards were maintained vegetation-free during the summer to avoid water competition, and then a winter cover crop was planted in the fall. Frequently, a legume or a mixture containing a legume, for its nitrogen value, was used as the winter cover crop.

Irrigation methods include surface and subsurface drip, microsprinkler, solid set and movable sprinkler, and flood irrigation. The type of irrigation, scarcity of water and tillage practices affect possible ground cover choices. Water quality for irrigation varies dramatically. Generally, with some exceptions, ground covers are more tolerant of poor-quality water than are trees.

Many orchards that use flood irrigation maintain vegetation-free floors. This is particularly true in arid areas where water availability is limited. Also, the orchard floor may be tilled prior to irrigation to increase water infiltration and repair dikes that direct water flow. There is less opportunity to effectively use summer cover crops in this situation than in others. Also, water limitations may be paramount, making summer cover crops less attractive in an integrated orchard management programme. However, winter cover crops are compatible with such systems.

A drip system delivers small quantities of water to a specific area, thereby increasing water use efficiency. Water usually does not reach the row middles in sufficient quantities to irrigate the cover crop. Therefore, if trees are drip-irrigated, cover crops are most applicable in areas with sufficient rainfall to support the cover crop. Where rainfall is limited, drought-tolerant cover crops must be chosen.

Environment modification

Orchard ground covers moderate orchard temperature. During the summer, ambient temperature in an orchard with a vegetative floor cover is cooler than if maintained vegetation-free (Dancer, 1964). This may improve fruit colour, which is attributed to anthocyanin pigment formation by decreasing respiration rates, thus allowing carbohydrates to accumulate in the fruit. Orchards with vegetative ground covers are more susceptible to damage from radiation frosts in the spring than those with vegetation-free floors (Hamer, 1975). Bare soil absorbs more radiation during the day than soil covered with vegetation. The stored heat is then radiated into the atmosphere at night, reducing the chance of frost injury in the spring. However, there are significant disadvantages of maintaining a bare soil. First, soil erosion from water or wind is greater. Second, orchard traffic may be impaired, particularly during wet weather when fungicides need to be applied in a timely manner. Third, frequent soil tillage tends to reduce soil organic matter, which will eventually affect the nutrient-holding capacity of the soil, its friability and its water intake and retention.

Barriers to Pest Management

One barrier for using some ground covers to attract beneficial arthropods is the marketing plan for the orchard. Producers of certain crops may market a portion of their crop or their entire crop as 'u-pick'. This typically requires a manicured orchard appearance for best marketing results. However, ground covers used to attract beneficial arthropods are infrequently mowed or at least maintained taller than desirable for foot traffic. The u-pick marketing plan may preclude typical cover crop management systems employed to attract Entomophaga.

Ground cover management in a commercially harvested orchard presents fewer problems than with u-pick harvest in most instances. An exception might be in orchards that require mechanical harvesters to sweep the crop from the ground, e.g. pecan, walnut, macadamia and others. Pecans are typically harvested by dislodging the fruit from the tree with a shaker, and then sweeping up the nuts with a harvester. The harvester performs best with a short ground cover that can withstand substantial wear during the harvest process. Also, dirt clods that may be present from recent soil tillage to establish annual cover crops will substantially increase processing time and costs to prepare the crop for the market. Marketing strategy and harvest methods must be considered in the IOM programme.

Fruit quality standards present another barrier to certain aspects of pest management. For instance, there are no tolerances for 'worms' in sour cherry (*P. cerasus* L.) fruit (USDA, 1941); therefore, prophylactic pesticide applications are necessary to produce pest-free cherries. The calendar-based pesticide applications needed to maintain pest-free cherries reduce the possibility of controlling other pests with integrated programmes. In several

other orchard crops, tolerance to insect damage is quite low. This has led to more emphasis on controlling foliage-feeding pests or pests that do not appear in the final product. However, new technologies may reduce this problem. For example, a team is working on technology to detect and remove nuts with pecan weevil, *C. caryae* (Horn), larvae before they are processed. This should lead to greater pest tolerance in the raw product for processing, and allow scientists and producers to make full use of integrated programmes. Similar technology for other crops would greatly benefit IOM programme development.

Pecan

Pecans offer unique opportunities for integrated management programmes. Pecan production can be divided into native and cultivar systems. Pecan trees are a dominant climax vegetation type in the riparian river bottomland from central Texas east to the Mississippi River. Commercial native pecan production extends from northern Missouri southward into Mexico. These native orchards are developed by removing all woody vegetation, except pecan, then thinning the pecan trees to about 30–34 ft^2 of cross-sectional trunk area per acre (Hinrichs, 1961; McCraw, 2002). Thus, each tree in the orchard is a unique genotype, which is reflected in its pest resistance, yield potential, fruit size, fruit quality and other characteristics. Nearly all native pecans are marketed to processors, rather than directly to the public. The smaller nut size, multiple shapes and sizes and lower kernel percentage of native pecans compared with cultivars make these less attractive for direct marketing. The genetic diversity of native orchards imparts greater pest resistance and a marketing strategy that tolerates more kernel damage from insects than direct marketing. In addition this increased pest resistance facilitates an integrated approach to pest management with less reliance on pesticides and lower input costs.

Cultivar orchards typically consist of two to four cultivars, and therefore have less genetic diversity than a native pecan orchard. There are two pecan-growing regions with unique problems and opportunities. The first is the humid region, from central Texas eastward to the Atlantic Ocean. This encompasses the native pecan region and the area east of the Mississippi River where pecans are not native. Pecans in the humid region are often grown without irrigation, although many cultivar orchards are irrigated. Arthropod pests and diseases in the humid region include all that are endemic on pecan. The second region is the arid west, from west Texas to the Pacific Ocean. Pecan is not native to this region. Low rainfall in this region eliminates fungal disease problems, but necessitates that trees be irrigated. Most pecan pests, except aphids, have been excluded from the arid west, substantially reducing the burden of arthropod pest management.

Cultivar orchards in the humid region can be subdivided into two management systems: (i) low-input and (ii) high-input. Low-input orchards rely on cultivar resistance for disease control, are typically not irrigated, include cultivars with smaller fruit size, minimize pesticide application and tolerate a greater incidence of insect damage than high-input orchards. The marketing strategy of low-input orchards is targeted on processors. In contrast, high-input orchards may combine direct market, wholesale gift pack market and the processor market, the first two commanding a greater price and demanding higher fruit quality. High-input orchards rely more on pesticides than do low-input orchards, but cover crops can be useful in high-input orchards as well. Yield potential and input costs are lower for low-input orchards than for high-input orchards.

Use of recently developed pesticides that are target-specific and avoidance of broad-spectrum pesticides improve the effectiveness of cover crops for increasing beneficial arthropods in the orchard. Cover crops have the greatest impact on aphid – *Monellia caryella* (Fitch), *Monelliopis pecanis* Bissell and *Melanocallis caryaefoliae* (Davis) – management in pecan. These aphids are foliage-feeding pests that can rapidly develop

resistance to insecticides (Dutcher and Htay, 1985). Other disadvantages of frequent broad-spectrum pesticide applications to control aphids and other pests are outbreaks of secondary pests (Ball, 1981; Mizell, 1991) and reductions in beneficial arthropod populations that act as a natural control of pecan pests (Dutcher, 1983; Dutcher and Payne, 1983; Tedders, 1983; Mizell, 1991). The pecan aphid complex is attacked by predators and parasitoids (Shepard, 1973; Tedders, 1978; Flores, 1981; Edelson, 1982; Watterson and Stone, 1982; Liao *et al.*, 1984, 1985; Edelson and Estes, 1987). Bugg *et al.* (1991b) proposed a cover crop management system for pecans to enhance beneficial arthropods in the orchard after testing several warm-season and cool-season cover crops (Bugg and Dutcher, 1989; Bugg *et al.*, 1990, 1991a), and later reviewed the literature for managing arthropod pests in orchards with cover crops (Bugg and Waddington, 1994).

Warm-season cover crops that showed promise for increasing aphidophagous insects and other Entomophaga in Georgia pecan orchards were: American jointvetch, *Aeschynomene americana* L.; cowpea, *Vigna unguiculata* spp. *unguiculata* (L.) Walpers; sesbania, *Sesbania exaltata* (Rafinesque-Schmaltz) Cory; hairy indigo, *Indigofera hirsuta* L.; and hybrid sudan, *Sorghum bicolor* (L.) Moench (Bugg and Dutcher, 1989). These ground covers hosted alternative prey aphids that attracted or increased the abundance of various lady beetles (Coleoptera: Coccinellidae), syrphid flies (Diptera: Syrphidae) and various entomophagous wasps (Hymenoptera). Of the cover crops showing promise, sesbania appeared to be the best source of alternative prey aphids and a reservoir for aphidophaga. Unfortunately, sesbania stands decreased substantially during the growing season due to drought, shade and other causes. Showy partridge pea, *Cassia fasciculate* Michaux, was a source of extra floral nectar in the orchard that was attractive to entomophagous wasps. Later work showed that sesbania supported high densities of bandedwinged whitefly, *Trialeurodes abutilonea* (Haldeman) and cowpea aphid, *Aphis craccivora* Koch that were colonized by coccinellids (Bugg and Dutcher, 1993). However, they found no evidence that the sesbania cover crop influenced coccinellid density in the pecan canopies. Sesbania and cowpea harboured stink bug (Hemiptera: Pentatomidae) (R.L. Bugg and J.D. Dutcher, unpublished data, cited in Bugg and Waddington, 1994), a kernel feeding pest of pecan, making these cover crops unsuitable for pecan orchards.

Crape myrtle, *Lagerstroemia indica* L., is a promising woody perennial host for alternative prey of entomophagous species. The peak density of Crapemyrtle aphid, *Tinocallis kahawaluokalani* (Kirkaldy), preceded those of the pecan aphid complex (*M. caryella, M. pecanis* and *M. caryaefoliae*) (Mizell and Schiffhauer, 1987). Increased abundance of Coccinellidae, Syrphidae, Chrysopidae and Anthocoridae either coincided with, or occurred just after, *T. kahawaluokalani* peaks. Generalist predators that attack *T. kahawaluokalani* are effective predators of the pecan aphid complex; however, the relationship among predators increases when associated with *T. kahawaluokalani*, and abundance of pecan aphids or beneficial arthropods in the tree canopy was not determined.

In Georgia, 18 cool-season cover crop regimes were evaluated for their associated insect complexes (Bugg *et al.*, 1990). 'Cahaba White' and 'Vantage' vetches (*Vicia sativa* L. × *V. cordata* Wulf) harboured large populations of tarnished plant bug, *L. lineolaris* (Palisot de Beauvois), and other *Lygus* spp. that are pecan pests, making them unsuitable as cover crops. Hairy vetch, *Vicia villosa* Roth., and some other legumes also harboured *Lygus* spp., but in substantially lower densities than the two hybrid vetches. Nymphal stages of *Lygus* spp. were observed feeding at the floral nectaries of hybrid vetches, making these cover crops more attractive than the other species tested. Cover crops that attracted the largest densities of beneficial arthropods were those that harboured large populations of aphids and thrips (Thysanoptera: Thripidae). As the cover crops matured and died, prey became scarce and beneficial arthropods

dispersed (Fig. 4.1). Therefore, cover crop combinations with progressive maturations were recommended that included crimson clover (*Trifolium vesiculosum* Savi), hairy vetch, blue lupin (*Lupinus angustifolius* L.) and mustard (*Brassica hirta* Monech).

Nine cool-season annual and perennial legumes were evaluated in Oklahoma as pecan orchard ground covers to supply nitrogen and increase beneficial arthropod densities (Smith *et al.*, 1994). Density of coccinellids was positively correlated with that of aphids, but not that of spiders (Araneida). Several other beneficial arthropods were observed in the legumes; however, densities were low. Two annual legumes ('Dixie' crimson clover and hairy vetch) and two perennial legumes ('Kenland' red clover, *Trifolium pretense* L., and 'Louisiana S-1' white clover, *Trifolium repens* L.) were chosen for additional evaluation based on their abundant beneficial reservoirs and nitrogen-supplying characteristics.

Crimson clover and hairy vetch cover crops were compared with a grass sod, primarily bermudagrass, *Cynodon dactylon* (L.) Pers., in Oklahoma and Georgia pecan orchards (Rice *et al.*, 1998). Two native pecan orchards were used in Oklahoma and one cultivar orchard in Georgia. In Oklahoma, crimson clover and hairy vetch were seeded together over the entire orchard floor, and in Georgia row middles were alternated with clover and vetch. Coccinellids were abundant on legume ground covers, but coccinellid density in the tree canopy was rarely affected by ground cover treatment. There were substantial differences in the abundance of coccinellid species collected from the legumes and the trees. In Oklahoma, *Olla v-nigrum* (Mulsant) and *Cycloneda munda* (Say) were the primary species in the trees, while *Hippodamia convergens* Guerin, *Coccinella septempunctata* L. and *Coleomegilla maculata lengi* Timberlake were dominant species in the legumes (Fig. 4.2). In Georgia, *Harmonia axyridis* (Pallas) was the dominant species in the trees, and *C. septempunctata* in the legumes (Fig. 4.3). The study was conducted before *H. axyridis* became well established in Oklahoma.

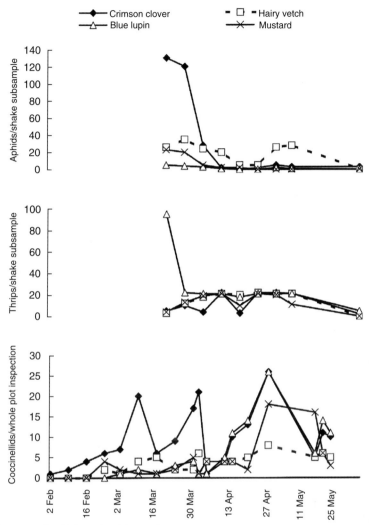

Fig. 4.1. Numbers of aphids, thrips (alternative prey) and coccinellids (predators) on four cool-season cover crops. (From Bugg *et al.*, 1990.)

Subsequent observations indicated that *H. axyridis* would be included as a major arboreal species. Habitat preference among Coccinellidae was obvious in this study, illustrating that a large concentration of beneficial arthropods associated with the ground cover may have little impact on densities in the tree canopy. Pecan aphids at the two Oklahoma sites were usually not affected by cover crop treatment, but at the Georgia site, early-season aphid densities were frequently less with a legume cover crop than with a grass sod (Rice *et al.*, 1998).

In Georgia, although coccinellid density in the trees was generally not affected by the cover crop, the general increase in aphid predators and parasitoids apparently reduced pecan aphids.

In Oklahoma, arthropods were collected from pecan canopies with grass or legume cover crops by applying esfenvalerate (a fast-action pyrethroid) to the canopy, and collecting the arthropods on plastic sheets stretched below the canopy (Smith *et al.*, 1996a). Tree canopies were sampled during spring, midsummer and fall. Beneficial

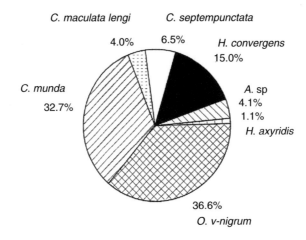

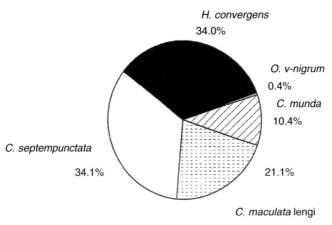

Fig. 4.2. Adult coccinellid distribution in the pecan canopy and the legume ground cover in an Oklahoma pecan orchard. (From Rice *et al.*, 1998.)

arthropods that were collected represented eight orders of Insecta, three orders of Arachnida and one order of Chilopoda. Ground cover treatment had little effect on the density or type of beneficial arthropods present in the pecan canopy, except that the densities of green lacewing, *Chrysoperla rufilabris* Burmeister, were greater during July from pecans with a legume ground cover than with a grass ground cover. Similarly, ground cover type had little influence on phytophagous arthropods in the canopy, suggesting that the ground covers tested

were unlikely to increase phytophagous pecan pests (Smith *et al.*, 1996b).

Stahmann Farms in the arid west near Las Cruces, New Mexico, implemented a successful IPM programme for aphids using native vegetation, releases of predators and a reduction in the application of broad-spectrum insecticides in a 4000 acre orchard (LaRock and Ellington, 1996). Predators released included green lacewing (*C. rufilabris*) and convergent ladybeetles (*H. convergens* and *H. axyridis*). Before the management change, frequent mowing

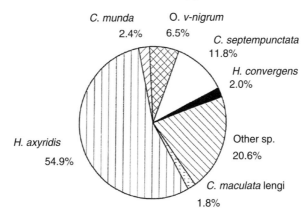

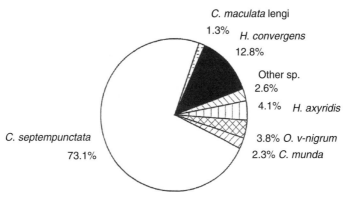

Fig. 4.3. Adult coccinellids distribution in the pecan canopy and the legume ground cover in a Georgia pecan orchard. (From Rice *et al.*, 1998.)

typically controlled the orchard ground cover. However, when the IPM programme was implemented indigenous vegetation was allowed to grow to about 2 ft tall before mowing, thus providing a nectar and pollen source, alternative prey and a refuge for beneficial arthropods. Over 7 years, pesticide costs were dramatically reduced (Fig. 4.4), aphids were maintained below economic thresholds in all but 1 year, and yield and nut quality remained high, except during 1 year. The year that nut quality and yield were reduced was attributed to a severe black aphid outbreak that could have been managed with pesticide intervention. Costs associated with purchase and release of

beneficial insects were substantially lower than that previously spent on insecticide applications (Fig. 4.4).

It has been difficult to demonstrate that a certain ground cover will increase beneficial arthropods in the pecan canopy and result in aphid control. However, it is clear that application of broad-spectrum pesticides eliminates beneficial arthropods that act as natural aphid control, thus inciting aphid outbreaks (Dutcher, 1983; Dutcher and Payne, 1983; Tedders, 1983; Mizell, 1991). It is also clear that aggressive aphid management with insecticides quickly results in resistance (Dutcher and Htay, 1985). Reliance on biological

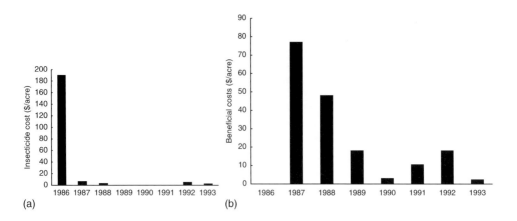

Fig. 4.4. Costs of (a) insecticides and (b) beneficial insects before biological aphid control was implemented (1986) and after implementation (1987–1993). (From LaRock and Ellington, 1996.)

aphid control that may be enhanced with certain cover crops is proving to be the most realistic approach. Also, judicious monitoring is required as biological control occasionally fails (LaRock and Ellington, 1996), necessitating pesticide intervention.

Apple

Apple aphid, *Aphis pomi* DeGeer, prefers young foliage, and if left undisturbed by natural enemies or pesticides, aphid densities can increase rapidly and remain high until shoot growth ceases (Carroll and Hoyt, 1984). Dense *A. pomi* populations deposit honeydew, which supports sooty mould and may cause fruit russet. Once new growth stops, *A. pomi* densities typically decrease, with a few colonies surviving on water sprouts, suckers and late-growing shoots. If new growth occurs in the fall, resurgence in *A. pomi* may occur. However, growth of young apple trees, *Malus domestica* Borkh., or heavily fertilized mature trees may continue throughout most of the summer, allowing *A. pomi* to continue increasing. Apple aphids have developed resistance or tolerance to many of the commonly used pesticides.

There are several predators and a few parasitoids that attack *A. pomi* (Evenhuis,

1965; Asgari, 1966; Holdsworth, 1970; Niemczyk *et al.*, 1973; Remaudiere *et al.*, 1973; Adams and Prokopy, 1980; Tracewski *et al.*, 1984). Carroll and Hoyt (1984) observed in Washington that spring *A. pomi* populations were governed by the density of over-wintering eggs, which depended on the late-fall *A. pomi* density and the amount of parasitism of young aphids by *Lysiphlebus testaceipes* (Cresson) and *Praon* sp., and predation by *Coccinella transversoguttata* Richardsoni Brown. Parasitism appeared more important than predation in controlling early spring *A. pomi* densities. However, in May and early June there was little aphid parasitism, and predator densities were usually too low to adequately control *A. pomi*. By late June, *A. pomi* populations grew rapidly with few adult predators laying eggs. Ground faunal predators, e.g. Forficulidae, Nabidae, Lygaeidae and Phalangidae, helped to slow the aphid increase during this period, but were usually inadequate. Thus, there was about a month-long gap from early June to mid-July when biological control of apple aphid was ineffective. Similarly, Adams and Prokopy (1980) reported that the cecidomyiid, *Aphidoletes aphidimyza* (Rondani), a major *A. pomi* predator in Massachusetts, did not appear until mid-June. Typically, the other predators of *A. pomi* dispersed in the spring to other habitats with more prey or preferred prey and did not return until *A. pomi* densities were high (Carroll and Hoyt, 1984). Predators returning later in the season included Coccinellidae, Chrysopidae, Miridae, Syrphidae and Chamaemyiidae. The habitats with alternative prey for these general predators were peach orchards, riparian trees and shrubs, some urban plants and certain weeds. Including a host in, or next to, the orchard that could provide an alternative to prey might mitigate the 'predator gap' for biological control of *A. pomi*.

Haley and Hogue (1990) found a lower seasonal density of *A. pomi* using a white clover and grass cover crop than with fall rye (*Secale cereale* L.) or grass-covered alleys with weed-free strips or woven plastic mats surrounding the trees during the first year of a study, but not the second. However, trees with the clover and grass ground cover had lower predator densities than trees with the other ground covers, and the only predator commonly found in both the trees and ground cover was the syrphid fly. Tree growth and leaf nitrogen concentration were lower in the white clover and grass cover crop than in the other treatments. Since *A. pomi* prefers young foliage, and aphid densities normally decline as the foliage matures (Carroll and Hoyt, 1984), the reduction in *A. pomi* density was probably caused by the negative impact of the cover crop on the tree rather than the cover crop's effect on predatory arthropods.

In a series of studies, Altieri and Schmidt (1985) tested cover crops of bell bean (*Vicia faba* L.), lana vetch (*Vicia dasycarpa* Ten.) plus assorted weeds, white clover, strawberry clover (*Trifolium fragiferum* L.) and subterranean clover (*Trifolium subterraneum* L.), to increase beneficial arthropods. The legume cover crops that remained in bloom throughout the season, e.g. white clover and strawberry clover, sustained the highest arthropod densities of the cover crops tested. These cover crops typically harboured large numbers of prey and their flowers furnished pollen and nectar that were attractive

or served as an alternate food source for certain species. Orchards with cover crops typically had less colonization and lower densities of aphids, leafhoppers, codling moths (*Cydia pomonella* L.) and other herbivores. Cover crops also enhanced the removal rate of prey artificially placed in the tree. However, a high predator density on the cover crop did not necessarily translate into higher predator densities in the trees.

In West Virginia, adjacent apple orchards were either managed conventionally or using a reduced spray programme combined with cover crops selected to attract beneficial arthropods (Brown and Glenn, 1999). The cover crops were alternating strips of dill (*Anethum graveolens* L.), buckwheat (*Fagopyrum esculentum* Moench), dwarf sorghum (*S. bicolor* L.) and rape (*Brassica napus* L.) in the tree row with tall fescue (*Festuca arundinacea* Schreb.) in the row middles. The conventional orchard had tall fescue in the row middles combined with vegetation-free strips surrounding the trees. Spirea aphid, (*Aphis spiraecola* Patch) and leafhoppers, (*Typhlocyba pomaria* McAtee), *Edwarsiana rosae* (L.) and *Empoasca fabae* (Harris) were more abundant in the ground cover orchard than the conventionally managed orchard. Fruit damage was similar in both orchards. Yield was reduced by the ground covers, making this system unacceptable compared to conventional management.

Flowering weeds provide a rich food source for adult parasites, affecting the amount of parasitism of host Lepidoptera. In Ontario, 15 abandoned apple orchards were ranked as rich, average or poor in spring-flowering weeds (Leius, 1967). Most of the flowering weeds were present in all the orchards; only their abundance was different. Parasitism of American tent caterpillar, *Malacosoma americanum* (Fabricius), pupae in the orchards with abundant wild flowers was 18 times greater than that found in orchards with less wild flowers (Table 4.2). Parasitism of tent caterpillar eggs and codling moth larvae were about four and five times greater, respectively, when orchards were rich in wild flowers.

Table 4.2. The relationship of wild flower abundance in abandoned Ontario apple orchards to parasitism of tent caterpillar eggs and pupae, and codling moth larvae. (From Leius, 1967.)

Wild flower abundance	Parasitism (%)		
	Tent caterpillar eggs	Tent caterpillar pupae	Codling moth larvae
Rich	4	65	34
Average	2	20	18
Poor	1	4	7

Peach

A major problem encountered in peach, *P. persica* (L.) Batch, orchards is the peach tree short life (PTSL). Individual trees in an orchard decline and die, frequently at the beginning of the fourth year after planting. The severity of the problem varies among geographic regions, but it exists wherever peaches are grown in the USA. A complex of factors is responsible for PTSL, but one of the contributing biotic problems is nematode. Difficulty in controlling nematodes has increased since effective post-plant nematicides are no longer available and methyl bromide, the most commonly used pre-plant nematicide, has been removed from the market. Consequently, alternative methods of nematode management are being sought.

Another common problem in peaches is fruit damage caused by the feeding of tarnished plant bug, *L. lineolaris* Palisot de Beauvois, and stink bugs (Hemiptera: Pentatomidae). These insects feed on young fruit, causing necrotic areas in the exocarp and mesocarp that fail to develop. The undamaged areas of the fruit continue to grow, leaving sunken areas in the fruit where the insects fed, causing a fruit deformity called catfacing. The development of pesticide resistance by tarnished plant bug (Snodgrass, 1996) has exacerbated the problem.

X-disease is believed to be caused by a mycoplasma (Nasu *et al.*, 1970) that is transmitted from wild hosts to peach by leafhoppers. Tomato ringspot virus (TmRSV) is found in several weedy hosts and is transmitted to peach, apple and several other important fruit crops by the nematodes *Xiphinema americanum* Cobb (Teliz *et al.*, 1966) and *X. rivesi* Dalmasso (Forer *et al.*, 1981). Selection and management of orchard cover crops should avoid species that attract leafhoppers and weed species that serve as hosts to the ringspot viruses. Cover crops that were particularly attractive to leafhoppers in one study were red clover and mixed weed ground covers with many Rosaceae spp., whereas bare ground or orchard grass, *Dactylis glomerata* L., was less attractive (McClure *et al.*, 1982). In one study, weed species with high frequencies of TmRSV infection were dandelion (*Taraxacum officinale* Weber), sheep sorrel (*Rumex acetosella* L.) and chickweed (*Stellaria* spp.) (Powell *et al.*, 1984).

Wheat, *Triticum aestivum* L., was ineffective in suppressing ring nematode, *Criconemella xenoplax* (Raski) Luc and Raski, in 4-year-old peach trees (Nyczepir *et al.*, 1998). In addition, wheat reduced growth of newly planted peach trees. A test conducted in North and South Carolina found that among several cover crops tested, nimblewill, *Muhlenbergia schreberi* J.F. Gmelin, suppressed ring nematode populations. In addition, this short-stature perennial grass tolerated drought, grew well in partial shade and did not harbour two-spotted spider mite, *Tetranychus urticae* Koch, or fruit-feeding Hemiptera (Meyer *et al.*, 1992).

A survey of North Carolina peach orchards found that winter weeds frequently found in the orchards attracted tarnished plant bugs and stink bugs (Killian and Meyer, 1984). Catfacing damage of the fruit was especially severe if the winter annuals reached the bloom stage before mowing. Apparently, the highest densities of Hemiptera were associated with flowering of winter weeds, and if mowed during this period, the Hemiptera migrated into peach trees and fed on the young fruit. Similarly, another study in North Carolina reported that weedy orchards had more fruit catfacing than other orchards (Meagher and Meyer, 1990).

Since weedy orchard growth is attractive to tarnished plant bugs and stink bugs, cover crops that are unappealing to these pests are desirable. Such cover crops were tested in New Jersey for the control of tarnished plant bugs and stink bugs, combined with reduced organophosphorous and carbamate insecticide use and mating disruption with pheromones, to decrease fruit damage from the oriental fruit moth, G. molesta Busck (Atanassov et al., 2002). Ground covers tested were hard fescue, Festuca longifolia Thuill, or tall fescue compared with conventional orchards with naturalized weedy vegetation between rows. The primary broadleaf weeds in the conventional orchards were: (i) sweet white clover, Melilotus alba Medik.; (ii) shepherdspurse, Capsella bursa-pastoris L.; (iii) dandelion, T. officinale; (iv) mayweed, Anthemis cotula L.; and (v) redroot pigweed, Amaranthus retroflexus L. The ground cover in the conventional orchards was maintained with periodic mowing or disking. The orchards with fescue were mowed as needed, and dicot weeds were treated with the selective herbicide 2,4-D in the fall, a treatment that would eliminate winter annual or biennial and perennial dicot weeds. Vegetation-free strips were maintained on both sides of the trees with appropriate herbicides. During the first year of the study the fescue orchard had 2.3 times fewer tarnished plant bugs and stink bugs and two times less Hemiptera fruit damage. During the second year, there were 4.9 times fewer Hemiptera insects in the fescue orchards, but catfacing was similar in both orchard management systems. Mating disruption of G. molesta controlled fruit damage from this pest as well as, or better

than, pest management programmes used in the conventionally managed orchards.

Peach growers in New Jersey observed that damage from oriental fruit moth was more severe in clean cultivated orchards than orchards that had substantial weed populations. These observations were confirmed by Pepper and Driggers (1934), who surveyed four of the most common weed species found in the orchards: (i) ragweed, Ambrosia artemisiaefolia L.; (ii) smartweed, Polygonum sp.; (iii) lambsquarter, Chenopodium album Aellen; and (iv) goldenrod, Solidago sp. From a stem borer (Lepidoptera) that fed on ragweed they reared the following parasites of oriental fruit moth: (i) Glypta rufiscutellaris Cress.; (ii) Pristomerus ocellatus Cush.; (iii) Cremastus minor Cush.; and (iv) Macrocentrus ancylivorus Rohr.. From the Lepidoptera larvae found in smartweed they reared M. ancylivorus and Macrocentrus delicatus Cress., parasitoids found in oriental fruit moth larvae. These results suggest that certain weed species that host Lepidoptera larvae with parasitoids that attack the oriental fruit moth may aid in the control of this introduced fruit pest. However, the weed species chosen to support alternative Lepidoptera hosts must not support tarnished plant bugs or stink bugs, as these pests are difficult to control and cause substantial fruit loss from catfacing. Cover crops chosen should also be undesirable to leafhoppers that transmit X-disease and weed species that are hosts for TmRSV. The cover crops, which meet these criteria and show the greatest promise in certain areas of the USA are nimblewill and both hard and tall fescue.

Conclusions

The development of integrated orchard management systems is in its infancy. This chapter highlights some of the challenges and successes in the development of such systems. Pest problems and horticultural requirements are different among crops and geographic regions, necessitating that crop and regional orchard management systems be developed and refined in several locations. This will require integration of several disciplines to achieve useful results. In addition, extensive educational programmes will be required to integrate these systems into existing agricultural management.

The infrastructure exists in the universities and the Agricultural Research Service to achieve these goals, but the expertise and funding for such research and extension programmes are being eroded. The problems and challenges in agriculture continually change; therefore, adequate personnel and resources are required to meet the new demands placed on agriculture producers. Agriculture today is very different than it was a generation ago, and will be substantially different in the next generation. To meet the challenges that agriculture will face in the future, we must regain the focus of the Land-Grant system and develop a plan for identifying and refining the priorities that new challenges will present in a worldwide marketplace.

References

Adams, R.G. and Prokopy, R.J. (1980) *Aphidoletes aphidimyza* (Rondani) (Diptera: Cecidomyiidae): an effective predator of the apple aphid (Homoptera: Aphididae) in Massachusetts. *Protection Ecology* 2, 27–39.

Altieri, M.A. and Schmidt, L.L. (1985) Cover crop manipulation in northern California orchards and vineyards: effects on arthropod communities. *Biological Agriculture and Horticulture* 3, 1–24.

Asgari, A. (1966) Untersuchungen über die im Raum Strutgard Hohenheim als wichstigste Prädatoren der grünen Apfelbalattlaus (*Aphidula pomi* DeG) auftretenden Arthropoden. *Zeitschrift für Angewandte Zoologie* 53, 35–93.

Atanassov, A., Shearer, P.W., Hamilton, G. and Polk, D. (2002) Development and implementation of a reduced risk peach arthropod management programme in New Jersey. *Journal of Economic Entomology* 95, 803–812.

Ball, J.C. (1981) Pesticide-induced differences in relative abundance of mites on pecans. *Journal of Economic Entomology* 74, 425–427.

Boethel, D.J. (1981) Resistance in the pecan leaf scorch mite, *Eotetranychus hicoriae* (McGregor) (Acari: Tetranychidae): implications in pecan pest management. *Miscellaneous Publication of Entomological Society of America* 12, 31–44.

Bould, C. and Jarrett, R.M. (1962) The effect of cover crops and NPK fertilizers on growth, crop yield and leaf nutrient status of young desert apple trees. *Journal of Horticulture Science* 37, 58–82.

Brown, M.W. and Glenn, D.M. (1999) Ground cover plants and selective insecticides as pest management tools in apple orchards. *Journal of Economic Entomology* 92, 899–905.

Bugg, R.L. and Dutcher, J.D. (1989) Warm-season cover crops for pecan orchards: horticultural and entomological implications. *Biological Agriculture and Horticulture* 6, 123–148.

Bugg, R.L. and Dutcher, J.D. (1993) *Sesbania exaltata* (Rafinesque-Schmaltz) Cory (Fabaceae) as a warm-season cover crop in pecan orchards: effects on aphidophagous *Coccinellidae* and pecan aphids. *Biological Agriculture and Horticulture* 9, 215–229.

Bugg, R.L. and Waddington, C. (1994) Using cover crops to manage arthropod pests of orchards: a review. *Agriculture Ecosystems and Environment* 50, 11–28.

Bugg, R.L., Phatak, S.C. and Dutcher, J.D. (1990) Insects associated with cool-season cover crops in southern Georgia: implications for control in truck-farm and pecan agroecosystems. *Biological Agriculture and Horticulture* 7, 17–45.

Bugg, R.L., Dutcher, J.D. and McNeil, P.J. (1991a) Cool-season cover crops in the pecan orchard understory: effects on *Coccinellidae* (Coleoptera) and pecan aphids (Homoptera: Aphididae). *Biological Control* 1, 8–15.

Bugg, R.L., Sarrantonio, M., Dutcher, J.D. and Phatak, S.C. (1991b) Understory cover crops in pecan orchards: possible management systems. *American Journal of Alternative Agriculture* 6, 50–62.

Carroll, D.P. and Hoyt, S.C. (1984) Natural enemies and their effects on apple aphid, *Aphis pomi* DeGeer (Homoptera: Aphididae), colonies on young apple trees in central Washington. *Environmental Entomology* 13, 469–481.

Dancer, J. (1964) The influence of soil moisture and temperature on growth of apple trees. *Horticulture Research* 4, 3–13.

Dutcher, J.D. (1983) Carbaryl and aphid resurgence in pecan orchards. *Journal of Georgia Entomological Society* 18, 492–495.

Dutcher, J.D. and Htay, U.T. (1985) Resurgence and insecticide resistance problems in pecan aphid management. In: Neel, W.W., Tedders, W.L. and Dutcher, J.D. (eds) *Aphids and Phylloxeras of Pecan*. Special Publication 38. Georgia Agricultural Experiment Station, Athens, Georgia, pp. 17–29.

Dutcher, J.D. and Payne, J.A. (1983) Impact assessment of carbaryl, dimethoate, and dialifor on foliar and nut pests of pecan orchards. *Journal of Georgia Entomological Society* 18, 495–507.

Edelson, J.V. (1982) Seasonal abundance, distribution and factors affecting population dynamics of yellow pecan aphids (*Monellia caryella* and *Monelliopis nigropunctata*). PhD thesis, Department of Entomology, Auburn University, Auburn, Alabama.

Edelson, J.V. and Estes, P.M. (1987) Seasonal abundance and distribution of predators and parasites associated with *Monelliopis pecanis* Bissell and *Monellia caryella* (Fitch) (Homoptera: Aphidae). *Journal of Entomological Science* 22, 336–347.

Evenhuis, H.H. (1965) Host specificity in the parasites and hypersites of apple aphids. In: Hodek, I. (ed.) *Ecology of Aphidophagous Insects*. Academia, Prague, Czech Republic, pp. 39–40.

Flores, R.J.F. (1981) A comparison of field population dynamics in relation to developmental time and fecundity under laboratory conditions of the foliage aphid species complex on pecan. PhD thesis, Department of Entomology, Texas A & M University, College Station, Texas.

Forer, L.B., Hill, N. and Powell, C.A. (1981) *Xiphinema riversi*, a new tomato ringspot virus vector. *Phytopathology* 7(Abstr.), 874.

Fye, R.E. (1980) Weed sources of Lygus bugs in the Yakima Valley and Columbia Basin in Washington. *Journal of Economic Entomology* 73, 469–473.

Goff, W.D., Patterson, M.G. and West, M.S. (1991) Orchard floor management practices influence elemental concentrations in young pecan trees. *HortScience* 26, 1379–1381.

Haley, S. and Hogue, E.J. (1990) Ground cover influence on apple aphid, *Aphis pomi* DeGeer (*Homoptera: Aphididae*), and its predators in a young apple orchard. *Crop Protection* 9, 225–230.

Hamer, P.J.C. (1975) Physics of frost. In: Pereira, H.C. (ed.) *Climate and the Orchard*. Research Review, No. 5., East Malling Research Station, Kent, England.

Hinrichs, H.A. (1961) The relationship of native pecan tree spacing to yield. *Oklahoma Agriculture Experimental Station Bulletin* B-574, 1–11.

Holdsworth, R.P. (1970) Aphids and aphid enemies effect of integrated control in an Ohio apple orchard. *Journal of Economic Entomology* 63, 530–535.

Huslig, S.M., Smith, M.W. and Brusewitz, G.H. (1993) Irrigation schedules and annual ryegrass as a ground cover to conserve water and control peach tree growth. *HortScience* 28, 908–913.

Killian, J.C. and Meyer, J.R. (1984) Effect of orchard weed management on catfacing damage to peaches in North Carolina. *Journal of Economic Entomology* 77, 1596–1600.

LaRock, D.R. and Ellington, J.J. (1996) An integrated pest management approach, emphasizing biological control, for pecan aphids. *Southwestern Entomologist* 21, 153–166.

Leius, K. (1967) Influence of wild flowers on parasitism of tent caterpillar and codling moth. *Canadian Entomologist* 99, 444–446.

Liao, H.T., Harris, M.K., Gilstrap, F.E., Dean, D.A., Agnew, C.W., Michels, G.J. and Mansour, F. (1984) Natural enemies and other factors affecting seasonal abundance of the blackmargined aphid on pecans. *Southwestern Entomologist* 9, 404–420.

Liao, H.T., Harris, M.K., Gilstrap, F.E. and Mansour, F. (1985) Impact of natural enemies on the blackmargined pecan aphid, *Monellia caryella* (Homoptera: Aphidae). *Environmental Entomology* 14, 122–126.

McClure, M.S., Andreadis, T.G. and Lacy, G.H. (1982) Manipulating orchard ground cover to reduce invasion by leafhopper vectors of peach X-disease. *Journal of Economic Entomology* 75, 64–68.

McCraw, D. (2002) Improving native pecan groves. *Oklahoma Cooperative Extension Services*, F-6208. (www.osuextra.com).

Meagher, R.L. Jr and Meyer, J.R. (1990) Effects of ground cover management on certain abiotic and biotic interactions in peach orchard ecosystems. *Crop Protection* 9, 65–72.

Meyer, J.R., Zehr, E.I., Meagher, R.L. Jr and Salvo, S.K. (1992) Survival and growth of peach trees and pest populations in orchard plots managed with experimental ground covers. *Agriculture Ecosystems and Environment* 41, 353–363.

Mizell, R.F. III (1991) Pesticides and beneficial insects: applications of current knowledge and future needs. In: Wood, B.W. and Payne, J.A. (eds) *Pecan Husbandry: Challenges and Opportunities.* First National Pecan Workshop Proceedings. ARS-96, Agriculture Research Service, US Department of Agriculture, Springfield, Virginia, pp. 47–54.

Mizell, R.F. III and Schiffhauer, D.E. (1987) Seasonal abundance of the crapemyrtle aphid, *Sarucallis kahawaluokalani*, in relation to the pecan aphids, *Monellia caryella* and *Monelliopis pecanis* and their common predators. *Entomophaga* 32, 511–520.

Nasu, S., Jensen, D.D. and Richardson, J. (1970) Electron microscopy of mycoplasma like bodies associated with insect and plant hosts of peach western X-disease. *Virology* 41, 583–595.

Niemczyk, E., Olsyak, E.R., Miszczak, M. and Bakowski, G. (1973) Effectiveness of some predacious insects in the control of phytophagous mites and aphids on apple trees. Annual Report 1974, Research Institute Pomology, Poland.

Nyczepir, A.P., Bertran, P.F., Parker, M.L., Meyer, J.R. and Zehr, E.I. (1998) Interplanting wheat is not an effective postplant management tactic for *Criconemella xenoplax* in peach production. *Plant Disease* 82, 573–577.

Patterson, M.G., Wehtje, G. and Goff, W.D. (1990) Effects of weed control and irrigation on the growth of young pecans. *Weed Technology* 4, 892–894.

Pepper, B.B. and Driggers, B.F. (1934) Non-economic insects as intermediate hosts of parasites of the oriental fruit moth. *Annals Entomological Society of America* 27, 593–598.

Powell, C.A., Forer, L.B., Stouffer, R.F., Cummins, J.N., Gonsalves, D., Rosenberger, D.A., Hoffman, J. and Lister, R.M. (1984) Orchard weeds as hosts of tomato ringspot and tobacco ringspot viruses. *Plant Disease* 68, 242–244.

Pree, D.J., Whitty, K.J. and Van Driel, L. (1998) Resistance to insecticides in oriental fruit moth populations (*Grapholita molesta*) from Niagara Peninsula of Ontario. *Canadian Entomologist* 130, 245–256.

Remaudiere, G., Iperti, G., Leclant, F., Lyon, J.P. and Michel, M.F. (1973) Biologie et écologie des aphids et leurs ennemis naturels: application à la lutte intégrée en vergers. *Entomophaga Memoirs* 6, 1–35.

Rice, E.L. (1995) *Biological Control of Weeds and Plant Diseases: Advances in Applied Allelopathy.* University of Oklahoma Press, Norman.

Rice, N.R., Smith, M.W., Eikenbary, R.D., Arnold, D., Tedders, W.L., Wood, B., Landgraf, B.S., Taylor, G.G. and Barlow, G.E. (1998) Assessment of legume and nonlegume ground covers on Coleoptera: Coccinellidae density for low-input pecan management. *American Journal of Alternative Agriculture* 13, 111–123.

Rogers, W.S., Raptopoulos, T.H. and Greenham, D.W.P. (1948) Cover crops for fruit plantations. V. Effect of form and time of application of nitrogen on orchard swards. *Journal of Horticulture Science* 24, 271–283.

Shepard, M. (1973) Seasonal incidence and vertical distribution of aphids and beneficial arthropods on pecan trees. *Proceedings Southeastern Pecan Growers' Association* 66, 153–159.

Smith, M.W., Eikenbary, R.D., Arnold, D.C., Landgraf, B.S., Taylor, G.G., Barlow, G.E., Carroll, B.L., Cheary, B.S., Rice, N.R. and Knight, R. (1994) Screening cool-season legume cover crops for pecan orchards. *American Journal of Alternative Agriculture* 9, 127–134.

Smith, M.W., Arnold, D.C., Eikenbary, R.D., Rice, N.R., Shiferaw, A., Cheary, B.S. and Carroll, B.L. (1996a) Influence of ground cover on beneficial arthropods in pecan. *Biological Control* 6, 164–176.

Smith, M.W., Arnold, D.C., Eikenbary, R.D., Rice, N.R., Shiferaw, A., Cheary, B.S. and Carroll, B.L. (1996b) Influence of ground cover on phytophagous and saprophagous arthropods in pecan trees. *Southwestern Entomologist* 21, 303–316.

Smith, M.W., Shiferaw, A. and Rice, N.R. (1996c) Legume cover crops as a nitrogen source for pecan. *Journal of Plant Nutrition* 19, 1117–1130.

Smith, M.W., Wolf, M.E., Cheary, B.S. and Carroll, B.L. (2001) Allelopathy of bermudagrass, tall fescue, redroot pigweed and cutleaf evening primrose on pecan. *HortScience* 36, 1047–1048.

Smith, M.W., Cheary, B.S. and Carroll, B.L. (2002) Fescue sod suppresses young pecan tree growth. *HortScience* 37, 1045–1048.

Snodgrass, G.L. (1996) Insecticide resistance in field populations of the tarnished plant bug (Heteroptera: Miridae) in cotton in the Mississippi Delta. *Journal of Economic Entomology* 89, 783–790.

Tedders, W.L. (1978) Important biological and morphological characteristics of the foliar-feeding aphids of pecan. Technical Bulletin 1579, Agricultural Research Service, USDA, Washington, DC.

Tedders, W.L. (1983) Insect management in deciduous orchard ecosystems: habitat manipulation. *Environmental Management* 7, 29–34.

Teliz, D., Lownsberry, B.F. and Grogran, R.G. (1966) Transmission of tomato ringspot virus by *Xiphinema americanum*. *Phytopathology* 56(Abstr.), 151.

Tisdale, S.L. and Nelson, W.L. (1966) *Soil Fertility and Fertilizers*. Macmillan, New York.

Tracewski, K.T., Johnson, P.C. and Eaton, A.T. (1984) Relative densities of predaceous Diptera (Cecidomyiidae, Chamaemyiidae, Syrphidae) and their aphid prey in New Hampshire, USA, apple orchards. *Protection Ecology* 6, 199–207.

USDA (1941) United States standards of grades of red sour cherries for manufacture. US Department of Agriculture, Agriculture Marketing Service, Fruit and Vegetable Division.

Usmani, K.A. and Shearer, P.W. (2001) Susceptibility of male Oriental fruit moth (*Lepidoptera: Tortricidae*) populations in New Jersey apple orchards to azinphosmethyl. *Journal of Economic Entomology* 94, 233–239.

Watterson, G.P. and Stone, J.D. (1982) Parasites of blackmargined aphids and their effects on aphid populations in far-west Texas. *Environmental Entomology* 11, 667–669.

Wolf, M.E. and Smith, M.W. (1999) Cutleaf evening primrose and Palmer amaranth reduce growth of nonbearing pecan trees. *HortScience* 34, 1082–1084.

Wood, B.W. (1993) Production characteristics of the United States pecan industry. *Journal of American Society of Horticulture Science* 118, 538–545.

5 Intercropping for Pest Management: The Ecological Concept

Vibeke Langer[1], Julia Kinane[2] and Michael Lyngkjær[2]

[1]Royal Veterinary and Agricultural University, Hojbakkegaards Alle 13, DK 2630 Taastrup, Denmark; [2]Risø National Laboratory, Plant Research Department, PO Box 49, 4000 Roskilde, Denmark

Introduction

Intercropping is a special case of multiple cropping, which is 'the intensification and diversification of cropping in time and space dimensions' (Francis, 1986). The aim of this chapter is to present and discuss the ecological basis of the effects of intercropping on pests in agricultural crops, i.e. plant pathogens (diseases) and herbivorous arthropods (mainly insects and mites). The basis for our current understanding of the ecological mechanisms is presented in many excellent works (Altieri and Letourneau, 1982; Altieri, 1983, 1999; Andow, 1983, 1991; Vandermeer, 1989; Hooks and Johnson, 2003), all of which focus on arthropod responses to increased vegetational diversity. The aim of this chapter is to add to these existing reviews and expand the discussion of the ecological principles to embrace the crop diseases also.

Although intercrops are often associated with pest control it should be acknowledged that they are very rarely used solely for their pest-reducing effects. Farmers make decisions on whether or not to use intercrops based on a variety of motives, all of which change continuously with social and environmental production conditions (García-Barrios, 2003), and most of which are more important than pest

management, e.g. income stability, resource exploitation and soil erosion (Vandermeer et al., 1998). This implies that the evaluation of intercrops as cropping systems cannot – and should not – be based solely on their performance regarding pest control, and that the overall success criteria used in an evaluation of systems are highly context-dependent and change quickly with farming opportunities and constraints.

The prevalence of intercrops

Intercropping systems are found throughout the world, with most diversity in the tropical parts, especially in regions with high pressure on land suitable for agriculture. Many of the world's large crops, e.g. maize, sorghum, rice, bean, pigeon pea and other legumes as well as oats, are more often found as components in intercrops (Rao, 1986). Many different combinations have developed, each of them adapted over time to local conditions and needs. Estimates of cropped land devoted to intercrops vary from below 20% in India to more than 90% in Malawi (Vandermeer, 1989).

Comprehensive lists of intercropping combinations may be found in Vandermeer et al. (1998) and some important multiple cropping systems in Asia, tropical Africa, America and temperate zones are presented in Francis (1986). With the exception of mixtures of cereals and legumes (e.g. oats, barley and vetch), which were commonly grown in many places in Europe for forage, and local traditional mixtures, e.g. rye–wheat mixed crops for bread in Slovenia (F. Bavec et al., 2004, Slovenia, personal communication), mixtures of crops on field level have not been used widely in practical agriculture in temperate zones during the last 50 years. Although concerns about agricultural sustainability have been voiced since the 1980s, suggesting that low-technology alternatives to pesticides such as intercropping be reassessed for use in developed countries (Horwith, 1985), mainstream agronomic research has focused on monocrops. But the ban of protein sources, e.g. bone meal from ruminant diets, may result in an increased interest in readopting the growing of intercrops in temperate regions to boost protein levels in forage, and some authors now use the expression 'a tropical technology for temperate regions' (Anil et al., 1998). Organic farms in industrialized countries use mixed fodder crops consisting of legumes and cereals more often than conventional farms, e.g. mixtures to maturity or silage cover 7% of the total organic crop area on Danish organic farms (Plant Directorate, 2003). However, a general adoption of mixed cropping systems on organic farms has not taken place. Mixed species cropping is only likely to be widely adopted by farmers, organic or conventional, in temperate zones once the benefits have been fully evaluated and shown to outweigh any increased management and labour input necessary for the more complex cropping system (Rämert et al., 2002).

Research in intercropping

During the past decades, research in intercropping has emanated both from the widely used practice of mixing crops in tropical small-scale agriculture and from the science of ecology. Growing a mixture of crops in the same field has been the topic of an intensive research effort and has produced an extensive body of empirical data on agronomic and socio-economic performance of diversified systems. Although monocropping has been widely propagated within agricultural research in many tropical areas, the persistent refusal of farmers to abandon the system of mixed cropping has resulted in a considerable effort in the field (Van Rheenen et al., 1981; García-Barrios, 2003). The main aim of this line of research is the testing and development of systems within specific climatic and socio-economic contexts, focusing on farmer needs for problem solving and system improvement. A range of criteria for system performance are used for evaluation, as not only economic yield, but yield stability, labour distribution, nutrient conservation, exploitation of labour resources and risk minimization also determine the potential for farmers (Parkhurst and Francis, 1986). Intercrops and other high-diversity systems are seen as 'an important (and maybe the most available) ingredient in the endeavour for a more sustainable agriculture in the tropics' (Garcia-Barrios, 2003).

In temperate areas, research in intercropping has been the concern of plant ecologists and entomologists, and the long-debated question of the link between species diversity and system stability and resilience in natural systems has been taken up by agroecologists working in agricultural systems. Their fascination has been the idea of 'copying nature' by diversifying systems and thus gaining some of the benefits ascribed to natural systems, e.g. good internal regulation of herbivores and efficient nutrient conservation. Ecologists have become increasingly focused on the application of basic ecological theory in applied fields, and since human manipulation is a fundamental characteristic of agroecosystems, much of the empirical data on the effects of vegetational diversification on populations have been generated in agricultural fields with an often narrow focus on arthropod herbivores, predators or parasitoids (Bommarco and Banks, 2003).

In this way, two almost separate research directions have evolved. One is the applied intercropping research well founded on empirical data and taking the starting point in traditional small-scale farmers' systems. This approach focuses on the practical design and development of intercropping systems, but rarely applies a theoretical framework to subtract general laws and explanations. The other direction is the theoretically based research on specific ecological mechanisms in diversified systems, in which solid empirical data have been scarce and which has thus been highly speculative in many cases. In the last two decades the two directions have approached each other, as the shared goal of designing and developing cropping systems has been acknowledged. It has been realized that understanding and explaining phenomena in diversified systems, as well as solving the agronomic problems of the mixed cropping systems, are necessary in order to fully exploit the potentials of these systems in sustainable agriculture. However, even researchers who have worked to develop predictive models for intercropping systems characterize the role of crop diversity for ecosystem function as enigmatic (Vandermeer *et al.*, 1998). They claim that associated (unplanned) biodiversity, which plays a much larger role in multispecies than in single species cropping systems, is largely overlooked by farming systems analysts (Vandermeer, 1989). In addition, the fact that much research on intercrops has been done within single disciplines like entomology, pathology, agronomy or economy is a challenge to make results useful in the context of IPM.

Types of intercropping: spatial and temporal patterns

Multiple cropping systems are complex systems in which the vegetational diversity, planned by the farmer, is high. Here we focus on intercropping, defined as the growing of two (or more) genetically different crops simultaneously on the same field. Crops are not necessarily grown at exactly the same time and their harvest times may be quite different, but they are usually 'simultaneous'

for a significant part of their growing period, distinguishing intercropping from 'relay' cropping in which growing periods only briefly overlap. The spatial arrangement can be separate crop rows, crops mixed within a row or crops mixed irregularly within the field, i.e. without rows. It would make sense to distinguish between various spatial arrangements in the parts dealing with dispersal of pests, but it is difficult to adhere to in reviewing the literature. For simplification in this chapter 'intercropping' is therefore used in a general sense without implying specific spatial arrangement. The component crops often belong to different plant families and differ in most characteristics, but they may also differ in characteristics only relevant for certain organisms, e.g. multiline cultivars, which are mixtures of lines bred for phenotypic uniformity of agronomic traits but with different reactions to a pathogen, or cultivar mixtures, which are mixtures of agronomically compatible cultivars with no additional breeding.

When measuring the advantage of intercrops over sole crops it is important to see how they differ from the crops with which they are compared. In additive intercrops not only the diversity of crops but also the total plant density is higher. Here, an 'additional crop' is added as extra plants to the initial monocrop. In replacement intercrops, plants of the additional crop(s) substitute plants of the initial crop. This way of describing intercrops in relation to monocrops of one or more of the component crops is mainly used in research and development to clarify which cropping systems are compared for their performance. Intercrop performance measures have been thoroughly discussed by Vandermeer (1989) and García-Barrios (2003).

In this chapter we include as intercrops the crops undersown with living mulches, e.g. clover, because this approach has been extensively tested in high-value crops. Especially in temperate vegetable systems where farmers struggle with many pests, undersowing with living mulches has gained attention among researchers as a manageable cropping system with potential benefits in regard to both nutrient and pest management (Rämert, 1995;

Theunissen *et al.*, 1995; Langer, 1996a, b; Finch and Kienegger, 1997; Theunissen, 1997; Brandsæter *et al.*, 1998; Theunissen and Schelling, 2000; Baumann *et al.*, 2002). Other ways of purposely increasing vegetation diversity on field scale will not be included here: the creation of non-crop borders around fields (but see Bigger and Chaney, 1998) or allowing the presence of a diverse weed flora (for reviews of crop, weed and insect interactions, see Altieri and Liebman, 1986, and Norris and Kogan, 2000). However, seen 'from the point of view' of the organisms in the system, these systems differ only little from intercrops, and many of the interactions discussed here will be similar.

Intercrops as a way of imitating natural systems

Natural ecosystems provide a reference point for understanding the ecological mechanisms regulating pests. Compared with the large areas of genetically homogeneous crop genotypes found in modern agriculture, most natural ecosystems have a higher plant diversity and are usually less prone to rapid and severe disease epidemics because host genotypes are more diverse and are distributed in smaller areas (Mundt, 1990). Diversification of host resistance therefore may be seen as a way to imitate natural systems, in order to achieve successful and durable management of crop pathogens by genetic means.

Also the lower herbivore numbers observed in practical agriculture and field experiments with diversified agricultural systems have been interpreted as a proof of a more stable system due to the more 'nature like' diverse vegetation.

Which diversity?

Searching for general rules governing the response of individual organisms reveals that the responses to vegetational texture of diseases, herbivores and arthropods on higher trophic levels are highly variable

and not very predictable. The response may not be linked to plant diversity per se, but is determined by specific interactions between the plant community, its herbivores, diseases and other organisms, many of which are not (and will probably never be) fully understood. A review of biodiversity manipulation experiments concludes that most studies show some effects of increased biodiversity on ecosystem processes, and also that sustained ecosystem functioning on a short term may require only a proportion of local species richness (Schwartz *et al.*, 2000). However, the debate on this issue has not been settled, and Bengtsson (1998) expresses a general opinion on the issue: in spite of his belief that the kinds of diversity, i.e. which species, are more important than diversity per se for ecosystem functions, he points out that high diversity in any system should be strived for in order to secure a pool of species, the presence of which may be crucial later in the system's 'life' or under changed environmental conditions, but which may not be important for ecosystem functions now. This apparent importance of the species composition for a well-functioning ecosystem emphasizes the need to take the starting point in farmers' knowledge and long-term experience with different combinations of component crops when developing diverse cropping systems (Morales and Perfecto, 2000; Altieri, 2004).

The Intercrop as an Environment for Plants and Pests

Arthropod and disease occurrence in cropping systems is determined by multidimensional interactions between the host plant, the environment, arthropods, pathogens and other organisms present in the system. The two main differences between a sole crop and an intercrop seen from the herbivore or the pathogen's point of view are: (i) host plant quality and (ii) crop environment. In the following sections we will review these two with emphasis on the parameters that are known to affect arthropods and plant pathogens in crops.

Intercrop effects on host plant quality

The various interactions between plants in intercrops have strong impacts on many parameters in host plant quality, e.g. nutrient status, plant morphology and secondary compound content. The nature of these interplant interactions depends on the characteristics of components crops, overall plant density and spatial distribution of crop plants, and has been reviewed comprehensively in Trenbath (1986), Vandermeer (1989) and García-Barrios (2003).

Nutrient and water status

The nutrient and water status of a plant affects its quality as a host plant: its growth patterns, morphology, anatomy and chemical composition have an effect on its interaction with pests. The competition for resources that takes place in an intercrop and that determines host plant access to nutrients and water depends on a range of factors including root growth rate, root density, root length and time and rate of uptake of the component crops. The considerable effects of intra- and interspecific competition, however, may be partly offset by other mechanisms working in the intercrop in both time and space. Crops may have their peak demands for nutrients at different stages of growth, which decreases competition, and intercrops are often more efficient in resource extraction, so that more nutrients are taken up in total by plants in an intercrop than in a sole crop (Willey, 1979). This has also been demonstrated for water where Reddy and Willey (1981) found a mixture of millet and groundnut using 406 mm water compared with 303 mm in the sole crop millet and 368 mm in the sole groundnut. This increased uptake is probably caused by root systems of different crop species utilizing different parts of the soil profile, which minimizes the degree of competition. In cereal–legumes mixtures also plant roots seem to exploit different soil layers. During early stages of crop development (i.e. before root nodules are fully developed in the legume) cereal nitrogen status depends largely on root characteristics of the component crop. When grown with bean or pea, both of which have shallow root systems, barley has a higher competitive ability for obtaining inorganic nitrogen than the legumes because its root grows faster and deeper (Jensen, 1996; Hauggaard-Nielsen *et al.*, 2001). Thus, in barley–bean and barley–pea mixtures nitrogen is taken up from different parts of the soil profile. However, in a lupin–barley mixture, because lupin has a deep taproot with a potential to grow deeper than barley roots (Hamblin and Hamblin, 1985) it competes with barley for soil inorganic nitrogen from similar regions of the soil profile.

Further influencing nutrient status of component crops is the release of nutrients from one component crop as a result of senescence of plant parts, which are then available to the other crop. Such transfer of nutrients between species has been observed in mixtures containing legumes, where nitrogen is deposited in the rhizosphere during crop growth as a result of continuous turnover of legume roots and nodules (Walker *et al.*, 2003) and is taken up by the legume itself or the neighbouring plant. In glasshouse intercropping experiments, Giller *et al.* (1991) estimated that up to 15% in the N_2-fixing plants was transferred to the intercropped maize, and similar transfers of 19% were found between pea and barley (Jensen, 1996). This rhizodeposition is a major source of nitrogen, carbon and other nutrients for the soil and its inhabitants (Walker *et al.*, 2003), and thus has significant effects on the density and activity of microorganisms in the rhizosphere and on the turnover and availability of nutrients to plants in the root zone (Hogh-Jensen and Schjoerring, 2000, 2001).

As many crop plants are obligately mycorrhizal, a large and diverse mycorrhizal population may offset competition effects in mixed crops, mediating nutrient transfer between plants of different species, as shown for transfer of nitrogen from soybean to maize mediated through arbuscular mycorrhizae (Van Kessel *et al.*, 1985). While it is thought that the absolute quantity of nitrogen or phosphorus transferred between living plants is insignificant (Johansen and Jensen, 1996), arbuscular mycorrhizae may enhance the transfer of nutrients from dead to living plants (Johansen and Jensen, 1996). Thus, in an intercrop with crops differing in growth durations, one could assume that there would be increased nutrient transfer after one crop is harvested (Jensen, 1996). On the other hand, including non-mycorrhizal plants like crucifer, beet or spinach may reduce fungus populations beneficial for other component crops, calling for more conscious use of current knowledge when designing intercrops (Picone, 2003).

In several studies the observed benefit of a mycorrhizal association has been shown to be an enhanced protection against pathogens: *Fusarium* spp. in soybean (Zambolim and Schenck, 1983), *Phytophthora nicotianae* var. *parasitica* in tomato (Cordier *et al.*, 1998), *Verticillium* spp. in lucerne (Hwang *et al.*, 1992) and cotton (Liu, 1995) and *Rhizoctonia solani* in potato (Yao *et al.*, 2002). Disease reduction due to mycorrhizae may be the result of either improved nutritional status of the host plant and thus the ability to resist the invader or the presence of one fungus impeding infection by others. Picone (2003) mentions studies of polycultures and weedy crops in which diversity-mediated changes in the fungus community impede soil pathogenic negative effects. The advantage in a maize–solanum intercrop was lost in a mycorrhizae-free environment, which led Vandermeer (1989) to suggest that what is interpreted as reduced competition between component crops brought about by the crops mining different parts of the soil column may actually be due to a transfer of nutrients from one crop species to another through mycorrhizal connections.

Plant architecture

The architecture and conformation of a plant, e.g. leaf shape, plant colour, branching and canopy pattern, can vary considerably depending on the interactions with neighbouring plants. Gold (1994) observed that intercropping cassava with cowpea

influenced cassava growth and conformation, with taller plants and larger internode lengths in the intercrop. Similar effects were observed by Fondong *et al.* (2002), who found cassava plants intercropped with maize or cowpea to have less foliage than cassava in a sole crop. Intercropping broccoli with clover resulted in lower broccoli plants with less leaves and side shoots than in a sole crop, although nutrient content as well as final weight and size of harvested heads were not affected (Hooks and Johnson, 2001).

Secondary compounds and induced resistance

The level and composition of secondary compounds in the component crops are influenced by growing conditions, e.g. water stress and other competition effects found in intercrops (Dale, 1988; Holtzer *et al.*, 1988). Secondary compounds have a broad range of functions in plants, including being attractants to arthropod pests and defensive agents against both crop pathogens and herbivores. A wide range of secondary compounds are known to be effective inducers of resistance and are used by the plant in its response to the presence of a potential pathogen. When exposed to a potential pathogen, the plant's secondary compounds in the form of elicitors trigger a cascade of resistance responses, thus inducing resistance, which can be active against several disease organisms. In an intercrop, due to the variety of host materials present, the pathogen population is likely to be more variable compared with that in a sole crop. Given that non-host pathogens have the potential to induce resistance to host pathogens, intercropping might increase the level of induced resistance.

Most of the reports on induced resistance involving plant pathogens have focused on induced resistance on the attacked plant, but there are indications that elicitors of induced resistance are volatile and move between plants. This means that following the attack of one plant, volatile elicitors are released inducing resistance in neighbouring plants. Investigations of such plant-to-plant communication have predominantly been focused on insect-induced resistance. For example, partial defoliation of alder (*Alnus glutinosa*) resulted in induced resistance to herbivory by the alder leaf beetle, *Agelastica alni* (Linnaeus), in the defoliated trees as well as in their non-specific neighbours. The effects waned with distance from the defoliated tree and with time since defoliation (Dolch and Tscharntke, 2000). In another example, native wild tobacco plants that grew next to damaged sagebush (*Artemisia tridentate*) plants had reduced levels of insect damage and higher levels of the defensive enzyme polyphenol oxidase compared with control plants next to undamaged sagebrush plants (Karban *et al.*, 2000, 2003). This type of mechanism may influence insect damage in mixed crops in subtle ways. This has been demonstrated in a mixture of barley varieties, where aphid acceptance of barley plants of one variety seemed to be affected by the mere presence of another variety. The response was seen both in the laboratory and in the field, and it was suggested that volatiles from one barley cultivar affect aphid acceptance of another barley cultivar (Ninkovic *et al.*, 2002). The same type of plant–plant interactions was suggested as an explanation of why more ladybird beetles were found in plots with a mixture of barley, *Cirsium* and *Elytrigia* than in pure barley, even in the absence of aphids or nectar resources. Furthermore, ladybirds were attracted to barley plots where barley plants had been exposed to volatiles from the two weeds, suggesting some kind of plant–plant communication (Ninkovic and Pettersson, 2003). Although the mechanisms are not known there are a number of indications that many types of plant–plant interactions affect host plant quality for pests in crop mixtures.

Intercrop effects on microclimatic conditions

The microclimatic conditions in an intercrop differ from that in a pure stand of the component crops, providing less or more favourable conditions for the system's specific herbivorous arthropods or pathogens, and thus potentially causing differences in the incidence and severity of attack. Data on microclimate in intercrops are sparse and mainly related to plant diseases, but are crucial for an understanding of the impact of microclimatic factors on disease development in intercrops and thus for the development of disease-forecasting systems and simulation models directed towards the evaluation of management methods.

In a maize–bean intercrop both Fininsa (2001) and Boudreau (1993) found small but statistically significant differences in microclimate between the sole crop and intercropping systems. Generally, temperature and wind velocity were lower in the intercrop than in the sole crop, whereas relative humidity was higher. Boudreau (1993) measured the leaf temperature on 5 days during the growing season and found a reduction of leaf temperature by an average of 0.2°C in the intercrop. In a more detailed experiment, Fininsa (2001) monitored microclimatic temperature in sole and three bean–maize intercropping systems (e.g. row intercropping, mixed intercropping within row and broadcast intercropping). He found that the temperature was consistently more stable in the intercropping systems, i.e. cooler during the day (3°C) and warmer during the evening and night (0.9°C). In the same experiments, Boudreau (1993) recorded a 1.8% increase in relative humidity in the intercrop, and Fininsa (2001) found that the relative humidity increased during the day (8%) and decreased during the evening and night (2.6%). Similar effects on relative humidity have been observed by Potts (1990) in a potato–dwarf bean intercrop in central Africa, where early morning dew remained longer on the intercropped potatoes than on the sole-cropped potatoes. In tomato–marigold and tomato–pigweed intercrops the same reductions in temperature were seen, but here maximum relative humidity was also reduced, possibly due to higher leaf transpiration in tomato in the intercrop (Gómez-Rodríguez *et al.*, 2003).

Wind velocity and quality (turbulence) may be different in an intercrop system compared with a monocrop, leading to changes in spore dispersal. In an additive intercrop with bean/maize ratios of 4:1 and 2:1, wind velocities just above the canopy averaged 63% and 47% of the velocity in the monocrop, respectively (Boudreau, 1993). Within the canopy, wind velocity could be further reduced and thus the spread of spores by the wind (Van Rheenen *et al.*, 1981). Reduced wind velocity removes fewer spores from lesions and lowers spore impaction efficiency (Boudreau, 1993), which may influence their adherence to leaves. Armstrong and Mitchell (1988) speculated that raindrops (which may contain propagules) can be 'converted' into larger, slower drops falling during times of reduced wind velocity, thus influencing propagules landing on the leaves.

The influence of the environmental differences on disease incidence between mono- and associated crops mentioned above is complex, largely unexplored and vital for understanding disease epidemiology within an intercrop. In their review on the interactions between weeds, arthropod pests and natural enemies, Norris and Kogan (2000) identified similar microclimatic effect of the presence of additional plants, e.g. weeds: less daily temperature fluctuations and protection from winds.

Intercropping Effects on Different Stages in Pest Biology: 'The Mechanisms'

One of the fundamental differences between arthropods and plant diseases of agricultural crops is that diseases passively 'enter' the crop, but many arthropod species exhibit active host finding and acceptance, i.e. they 'find' and 'lose' host plants. Many recently published studies report the positive effects of manipulating vegetation texture with the goal of reducing herbivore and disease loads. Why is this not enough – why is it also necessary to look for explanations? Firstly, the effects of intercropping seem to change with small changes in component plants, pest species and seasons, indicating that the predictability of these systems may be low in some cases. Secondly, the nature

of the mechanisms at work determines to what extent results generated in experiments with a traditional plot design with closely spaced small plots with different treatments, e.g. monocropping and intercropping, may be expected to be found in full field scale where only one treatment is present. This is discussed further in the next section. Therefore, identifying the mechanisms behind the effects of plant diversity on herbivores is crucial to their practical application.

Compared with the large body of literature on arthropod response to diversified cropping systems, the literature on diseases is considerably limited (see selected references in Table 5.4). Most available reports on diseases focus predominantly on the observed effect of intercropping on disease incidence, whereas mechanisms that contribute to intercrop–disease interactions are rarely investigated. As both arthropod pests and diseases exhibit distinguishable and similar stages in their relation with their host plant, we have chosen to review existing knowledge within the same framework (Table 5.1). As the two groups also differ substantially in some aspects, we will treat the first two phases for arthropod pests and diseases separately and the third phase together.

Intercrop effects on initial disease inoculation and dispersal

The primary factor determining initial inoculation of a disease in a crop is the dispersal mechanism of the pathogen: by wind, water or insects. Wind may carry propagules to quite long distances, and spores can move laterally and vertically in the air. The distance of movement is related to the wind speed, and size and shape of the spore. Typically, fungi disperse in a log reduction with distance from the source, meaning that concentrations of spores are highest at the source and 99% have been deposited on the ground within 100 m. Raindrops are also responsible for transferring pathogenic propagules from one surface to another. The impact of the raindrop carries the spore with it, and the effective distance of dispersal is small but significant

Table 5.1. Phases in the interaction between crop plants and their associated pests and diseases, which may be affected by the presence of non-host plants.

Phase	Diseases	Arthropod pests
Entering crop	Initial inoculation Dispersal in field	Immigration into field
Establishment in crop	Infection: germination, penetration of host plant	Landing on, examination of and acceptance of host plant
Surviving and reproducing	Survival, growth and reproduction in cropin crop	

in some plant pathogens, where splash may increase the spread of the pathogen within the crop extremely rapidly during suitable weather conditions. Many insects carry spores, and as they have the habit of visiting similar locations (usually with a relatively small area), the probability of spores being taken to suitable environments is high within that area. Consequently, the majority of the propagules are spread over short distances and long-range spread is limited for many pathogens. Therefore, any factor that may reduce short-distance dispersal may have a reducing effect on the disease development of the crop. A frequently proposed mechanism of disease reduction in intercrop systems is that a species mixture reduces the amount of inoculum reaching the host. This could be achieved by: (i) the presence of non-host plants physically impeding inoculum dispersal, either above or below the ground; (ii) spatial arrangement and density of susceptible hosts; and (iii) interactions (competition) between the host and non-host(s) that could alter the host in such a way as to affect dispersal.

Physical impediment of inoculum dispersal

Most pathogens are host-specific to a particular plant species, genus or family. Therefore, in an intercrop the non-host crop may act as a physical barrier (either above or below ground), limiting inoculum deposition between host plants in the crop mixture. Above the ground, the barrier effect could be due to direct interference of propagule movement by the non-host plants,

e.g. when spores impact on a non-host leaf they are 'trapped' (Boudreau and Mundt, 1997). Soleimani *et al.* (1996) examined the effect of a clover understorey in wheat on the splash dispersal of *Pseudocercosporella herpotrichoides* and found that in spite of the same level of spores at a height of 12 cm above the clover canopy, 50% fewer spores were found at ground level in the intercrop than in the sole crop. The difference was attributed to the physical presence of the clover understorey. Similarly, Bannon and Cooke (1998) attempted to quantify the effects of a white clover understorey in wheat on the splash dispersal of *Septoria tritici* pycnidiospores. Their growth chamber experiments showed that both horizontal and vertical spore movements at the wheat stem base were impeded by the presence of clover, which also reduced the number of spores at heights above that of the clover, leading to a reduction in lesions per flag leaf in the intercrop compared with the sole crop. This implies that the clover understorey acted like a sieve or a physical barrier to the *S. tritici* pycnidiospores. Exploiting row intercrops as physical barriers in tomato intercropped with maize or toria, Kumar and Sugha (2000) found *Septoria* leaf spot on tomato to be reduced only when grown in double rows and not in single rows, indicating that two rows of maize or toria provided an excellent barrier and reduced disease progress. In a walnut–olive intercrop, interference of olive with inoculum dispersal was cited as a mechanism involved for an observed 80% reduction in walnut anthracnose (*Gnomonia leptostyla*) incidence (Boudreau and Mundt,

1992). Tall non-host crops may protect the host by forcing wind-borne pathogens to pass over the top of the host crop, particularly if the tall crop is planted around the edges of small plots, as is often practised for plots of breeder's seed (Potts, 1990).

When considering root pathogens that are transferred between plants below the ground by root-to-root contact, the conventional wisdom that crops with different rooting characteristics should be grown together in order to exploit nutrient resources may have to be carefully balanced with the need for disease control. Autrique and Potts (1987) reported that when potatoes were intercropped with bean, the incidence of bacterial wilt was lower than in a potato–maize intercrop. The authors attributed the difference to beans having a shallow root system similar to that of potatoes, and thus being more effective in preventing root-to-root contact than the deeper-rooted maize. Similarly, Michel *et al.* (1997) found that the incidence of tomato bacterial wilt (*Pseudomonas solanacearum*) was reduced when tomato was intercropped with cowpea, but not when intercropped with soybean or welsh onion. Since the soil population of *P. solanacearum* was not reduced in the intercrop treatments, it appears that the cowpea roots acted as a physical barrier within the tomato roots, reducing root-to-root spread of the pathogen.

Spatial arrangement and density of susceptible hosts

Given that the majority of plant pathogenic propagules spread over short distances, with only limited long-range spread, increasing the distance between a source of inoculum (focal point) and a susceptible host will reduce the likelihood of propagules reaching host tissue. The dilution of inoculum that occurs due to increased distance between plants of the same genotype appears to be an important mechanism in disease reduction in intercrops. Especially when a pathogen is spread through root-to-root transmission, increasing the distance between host plants may be effective by

reducing or delaying interplant contact and thus disease spread. In experiments with *P. solanacearum*, the causal agent of bacterial wilt in potato, Autrique and Potts (1987) demonstrated that increasing the within-row spacing in potatoes from 30 to 90 cm reduced plants wilting from bacterial wilt, from 14% to 7%. The distance effect may be magnified by the presence of a non-host crop, which provides a physical impediment to inoculum, i.e. a barrier between crop plants of the same genotype. According to Burdon and Chilvers (1982), it is generally accepted that increasing plant density will increase disease levels. Indeed, some studies confirm this, such as those of Burdon and Chilvers (1976, 1977, 1982) in which varying amounts of host plants were grown in replacement crop combinations, i.e. total plant density was kept constant but host plant density decreased as non-hosts were added to the mixture. However, reviewing the published material in this area indicates that field results are highly variable and can vary enormously between sites and years (Buiel *et al.*, 1989; Pfleeger *et al.*, 1999; Garrett *et al.*, 2001). In some cases disease levels decreased with increasing host density, e.g. the severity of barley powdery mildew (Dinoor and Eshed, 1984) and rust (*Puccinia graminis f.* sp. *tritici*) on the two wheat cultivars tested (Dill-Macky and Roelfs, 2000). Thus, to truly isolate the effect of plant density on crop mixture efficiency for disease control, one needs to include all sole crops and all mixtures at a range of densities, a procedure that is often difficult to accomplish in the field. In one study for variety of mixtures (Fitt and McCartney, 1986), the effect of wheat cultivar mixtures on yellow rust severity was greatest at an intermediate planting density in 2 years, despite the fact that rust severity in pure stands increased with planting density in 1 year and decreased with planting density in the other year of the study. These variable results may indicate that host plant density is not the best parameter to correlate with disease severity. It might be more accurate to relate plant biomass with disease.

Interactions between the host and non-host plants

What is often ignored is that intraplant interactions, including competition between the host and non-host crop, may alter the host by affecting the dispersal of pathogenic propagules. In a bean–maize intercrop Boudreau and Mundt (1992) investigated whether (i) interaction/competition between bean–maize crops or (ii) the physical presence of maize affected the dispersal of the bean rust pathogen (*Uromyces appendiculatus*). They isolated these two factors by assessing the dispersal in bean plots from which maize had recently been cut out (i.e. interaction without physical presence), as well as in bean plots with artificially supported cut maize stalks (physical presence without interaction). The slopes of the primary dispersal gradients from a focal inoculum source were not significantly affected by the physical presence of maize. Instead by 50 days after planting, the slopes consistently became steeper solely due to interaction (competition) of the beans with maize (Fig. 5.1). In other words, the presence of maize had little effect on the dispersal of spores whereas the interaction between beans and maize reduced it. They also investigated how the fate of spores released from a source was affected by these factors by estimating the fraction of spores retained (deposited) vs. removed (carried up and away from the canopy by turbulent air currents) in these plots. Also this parameter, which affects the level of disease, was determined by intraplant interactions and not by the presence of maize. Therefore, in this system, not only is the physical barrier of the maize crop of little importance, but the competition within the intercrop is primarily responsible for any disease reductions associated with dispersal in bean–maize intercrops. This has implications for efforts to improve bean yields in bean–maize intercrops by fertilization, as this will alter interaction/competition between beans and maize, which may in turn counteract the reduced spore dispersal.

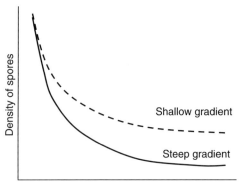

Fig. 5.1. Dispersal gradients: steep dispersal gradient – most spores are dispersed within close range of the focal point (source of inoculum); shallow dispersal gradient – spores are dispersed over a wider range from the focal point.

Intercrop effects on arthropod pest immigration

Immigration and emigration of herbivores into and out of a crop are about finding and losing host plants. Most herbivores are able to move moderate to large distances when searching for host plants, and although immigration into the crop has long been considered one of the key mechanisms by which herbivores respond to vegetation diversity, Kareiva (1985) demonstrated the great importance of emigration in an experiment with flea beetles from host plant patches of different size. Emigration has also been suggested as a key factor in natural enemy aggregation in diverse habitats, e.g. for parasitoids (Coll and Bottrell, 1996). In a number of studies reduced immigration is proposed as the cause of lower populations in crop mixtures, in the absence of data to support other explanations (Hooks and Johnson, 2003). Distinguishing the effects of intercropping on immigration from later stages of host plant acceptance and arthropod survival requires that arthropods entering a habitat with a specific vegetation texture must be assessed separately from those colonizing host plants in the habitat, i.e. the number of individuals attracted to the habitat 'as a whole' should be distinguished from the number arrested on host plants. If the numbers of herbivores landing on both host and non-host plants are totalled, as has been done for female cabbage root fly, *Delia radicum* Linnaeus, in controlled experiments (Coaker, 1980; Kostal and Finch, 1994) and for onion thrips in leek (Den Belder *et al.*, 2000), this gives an estimate of immigration. In both cases the females landed with the same frequency in a 'habitat' without clover and with clover, indicating that olfactory disruption, i.e. clover repelling flies from entering the habitat, does not play a role. If conclusions about immigration to host plant patches with different vegetation texture are based solely on herbivores landing on host plants, as done for flea beetles by Garcia and Altieri (1992), distinguishing between immigration and later stages in host finding is difficult.

Intercrop effects on disease establishment: infection

When a pathogen lands on a host plant, it must germinate and try to infect the plant, either through openings in the plant (e.g. stomata or wounds) or by penetrating the plant surface by physical or chemical means.

Microclimate and infection

While most plant pathogens are able to germinate and penetrate the host plant over a range of environmental conditions, each specific pathogen has particular requirements for maximum germination and infection. Therefore, any differences in microclimate in the intercrop compared with a sole crop (reduced temperature and wind velocity but increased relative humidity and leaf wetness) can affect the infection efficiency, which is usually measured in the field as disease incidence. Disease response to these differences in the microclimate are variable, being less favourable for some diseases, e.g. *Xanthomonas campestris* pv. *phaseoli*, the causal agent of common bacterial blight on bean, and *U. appendiculatus*, the causal agent of rust on bean. Boudreau and Mundt (1992) directly evaluated *U. appendiculatus* infection and subsequent fungal development

in bean when intercropped with maize by simultaneously inoculating bean plants in sole crops and intercrops and comparing latent periods and disease incidence after 14 days. The latent period was not affected by intercropping, whereas incidence was dramatically reduced. The proposed explanation was that the cooling of the bean leaves in the evening was delayed when a maize canopy was present, thus decreasing time in which dew, essential for *U. appendiculatus* infection, is present on the bean leaves (Imhoff *et al.*, 1981). However, high humidity and leaf wetness favour many pathogens that infect more readily in moist conditions, e.g. angular leaf spot (*Phaeoisariopsis griseola*) or white mould (*Sclerotinia*) whose levels are higher in intercrops of bean and maize than in sole crops (Van Rheenen *et al.*, 1981; Fininsa, 1996, 2001; Boudreau and Mundt, 1997). Similarly, Raymundo and Alcazar (1982) observed an increased incidence of late blight, *Phytophthora infestans*, in a potato–sweet potato intercrop and attributed this to an increase in humidity. Reduction in early blight disease on tomato when intercropped with pigweed was also due to the alteration of microclimatic conditions around the canopy, particularly by reducing the number of hours per day with relative humidity of more than 92%, thus diminishing conidial germination and formation of infection structures (Gómez-Rodríguez *et al.*, 2003).

Induced resistance

Defence responses in the host may be induced by many different types of inducers of such resistance including microorganisms (e.g. fungi and bacteria), certain chemicals, plant extracts and even ultraviolet light, and induced resistance is actually widespread in nature as plants are constantly bombarded with inducers. It could be argued that in an intercrop system there is a greater variety of microorganisms, and thus a greater likelihood of induced resistance occurs than in a sole crop system. Altered disease levels in intercrops have thus been attributed to an induction of defence responses in the host plants through challenges by avirulent races of the pathogens, especially for diseases caused by complex pathogen races, as well as by non-pathogens (Dean and Kuc, 1985). This induced/acquired resistance can persist from one to many weeks, and some reports describe season-long protection after activation of induced resistance. Induced resistance to bean rust (*U. appendiculatus*) from inoculation with sunflower rust spores (*Puccinia helianthi*) or maize rust spores (a *P. sorghi–P. polysora* mix) has been demonstrated in the laboratory (Allen *et al.*, 1983). Also, plant exposure to pollen can either increase or decrease infection, depending on the pathogen and host: pollen enhances infection and fungal growth by *Botrytis cinerea* on faba beans and by *Colletotrichum lindemuthianum* on cowpea, but reduces cowpea yellow mosaic infection (Chin and Wolfe, 1984; Kessmann *et al.*, 1994; Kuc, 1995). In simulation models, induced resistance is predicted to be an important factor in altering disease levels in intercrops, a factor that is nearly impossible to test in the field. Chin and Wolfe (1984) attempted to isolate plant density, induced resistance and interference components of mildew reduction in barley cultivar mixtures and found important density effects early in the season and significant induced resistance only later.

Allelopathy

Allelopathy in the broadest sense encompasses all types of inhibitory or sti-mulatory interactions among plants and between plants and microorganisms. In this context the question is whether intercropping including a plant species antagonistic to phytopathogens is efficient for disease control. Marigold (*Tagetes erecta*) is resistant to a number of herbivores and pathogens, and has fungicidal, nematicidal and insecticidal properties due to the presence of thiophenes in its tissues (Morallo-Rejesus and Decena, 1982; Montes and García, 1997). Early blight disease (*Alternaria solani*) severity on tomato was compared with pigweed or marigold in intercrops, and while both intercrops showed a reduction in the disease compared with a monocrop of tomato, the extent of the reduction was greater when tomato was intercropped with marigold (Gómez-Rodríguez *et al.*, 2003). The authors found a clear allelopathic effect of marigold leaves on *A. solani* germination and germling development on tomato leaves, which they suggested was due to volatile thiophenes liberated by the marigold. They speculated that the additional reduction in early blight on tomato when intercropped with marigold was primarily due to these allelopathic effects.

Intercrop effects on arthropod pest establishment

For most herbivores, landing on, examining and accepting, host plants are determined by individual combinations of interacting olfactory (smell), gustatory (taste) and visual (sight) clues, which may be weighted differently in each of the three stages. When exposed to different host plants, specialist herbivores discriminate between a broad range of host plant quality parameters, e.g. plant species, plant varieties and plant vigour. All the important changes in morphology or physiology, observed in plants exposed to competition, may affect crop plant attractiveness of herbivores through changes in these clues. These changes are seen regardless of the type of competition: intraspecific competition in dense sole crop stands or interspecific competition in mixed cultures. As an easily measurable parameter, plant size and age have been linked to attractiveness for several herbivores: ovi-positing *D. radicum* and *Pieris rapae* (Linnaeus) as well as flea beetles respond to plant age and plant size in a non-linear way, both preferring the larger among younger plants to both old and smaller plants, and avoiding stressed or stunted plants (Jones, 1977; Kareiva, 1982; Kostal and Finch, 1994; Langer, 1996a, b). However, the leaf reflectance pattern, chemical contents and leaf surface characteristics also change with age and environmental conditions, indicating that the search for one or few single host plant properties as responsible for herbivore response is not likely to succeed.

This discrimination between food plants, practised by herbivores, seems to be a labile trait in many species, depending largely on the opportunities for discrimination that individual movement provides (Kareiva, 1982). Flea beetles, with their limited mobility, respond to different host patch quality only when patches are close, whereas *P. rapae*, which are more mobile, respond even when patches are far apart. Aphids alighting on non-host plants become increasingly apt to accept host plants of lower quality (Klingauf, 1987a), and *P. rapae* and *D. radicum* females become less 'choosy' as to host plant quality with increasing egg load and thus the increasing need to locate a suitable plant for oviposition. To make the predictability even lower, recent studies on *P. rapae* have shown that individuals within a population may exhibit variable responses when presented with less acceptable host plants in a no-choice situation, some individuals not ovipositing at all, some with delayed and reduced egg laying and some individuals fully accepting the low-quality plants (Hopkins and Loon, 2001).

When moving in diverse vegetation, many specialist herbivores alight on non-host plants by mistake. Most often they leave non-host plants more frequently than hosts, with the risk of leaving the crop area during this movement (Kareiva, 1986). Cabbage aphids, flea beetles and cabbage root fly land on non-host plants by mistake, and all three species respond to this mistake by increased activity. Apterous cabbage aphids increase their rate of locomotion, while (winged) alates after having landed on a non-host set off on longer flights (Klingauf, 1987a). In cabbage root fly, landing on a non-host plant interrupts the female in the spiral flights necessary for oviposition (Fig. 5.2) (Kostal and Finch, 1994), and it eventually may leave the plant (Coaker, 1980), whereas flea beetles increase their general activity and more frequently emigrate out of the crop, having visited non-host plants (Garcia and Altieri, 1992).

Asman *et al.* (2001) compared the effects of intercropping leek and brassica with clover on two specialist moths in leek and brassica, respectively. Their work, which included laboratory, cage and field studies, is an example of an attempt to predict intercropping control effects by understanding the effects of non-host plants on specific parts of the plant–insect relationship, i.e. on oviposition and emigration. Host plant quality was kept constant during the experiment by using uniform potted plants. In choice experiments, oviposition by leek moth was not affected, whereas the oviposition by diamond back moth was shown to be reduced by the presence of clover only when the clover covered the host plant, but not with a low clover cover. Emigration from plants due to encounters with non-host plants was not larger in the intercrop. Similarly, Den Belder *et al.* (2000) studied settling (finding and accepting) of adult onion thrips, *Thrips tabaci* Linderman, in experiments with leek monocrop and leek undersown with clover. Potted plants were placed in the field to control host plant quality. Measurable plant growth parameters were kept uniform, but the fact that removal of the clover from pots just before introduction in the field still resulted in much lower thrips abundance indicates that plants were influenced by the clover in a subtle way, but enough to affect their attractiveness. Not only abundance was affected, but thrips on leek plants, which had grown with clover in pots without detectable reductions in plant size, also exhibited decreased feeding and increased restlessness.

Intercrop effects on survival and reproduction of diseases and arthropod pests

Host plant quality is one of the main determinants of both herbivore survival and reproduction and disease survival and reproduction. Nitrogen and water status resulting from fertilizer application, drought or competition have been the main host plant characteristics examined and, although detailed studies identify single important factors, the interaction with other plant characteristics, including the disease and the

(a)

(b)

Fig. 5.2. Behaviour of 100 female cabbage root flies in a no-choice situation after landing on a brassica plant growing on a background of bare soil (a) or grass (b). Host plant quality is kept uniform by using potted plants placed in the soil. Total numbers of movements are shown in arrows, and circled numbers indicate the positions of the females at the end of the observation period. After landing on the cabbage plant, most females carry out 'spiral flights'. In bare soil (a) most frequently spiral flights and with females landing again on the cabbage plan (342 times) whereas with a grass background (b) females attempting spiral flights often get diverted and land on the grass (337 times). Only 7% of the females eventually lay eggs on the cabbage plant in a grass background, whereas 36% of the females lay eggs on the cabbage plant in bare soil. (Modified from Kostal and Finch, 1994.)

herbivores themselves, makes it difficult to draw simple conclusions.

Herbivores can increase survival by adapting to different plant qualities within the plant by locating themselves on parts of optimal quality (e.g. young leaves) (Hoy and Shelton, 1987); others compensate for a low nitrogen food source by increased feeding (Dale, 1988). Whether a herbivore has one of these capabilities will determine whether it is adversely affected by lower host plant quality that is inflicted by the competition often seen in an intercrop.

Similarly, once a pathogen has gained entry into a plant, its survival in the plant depends on the plant resistance response and the nutritional status of the host plant. Besides the natural defences of a plant, induced resistance (which may be more prevalent in an intercrop system as described above) could reduce development of a pathogen. Survival and reproduction of pathogens could be greatly influenced by allelochemicals released by one crop, which may be toxic to arthropod pests or pathogens of the other crop(s). When tomato plants were intercropped with Chinese chive plants, bacterial wilt occurrence on tomato was delayed and suppressed. Furthermore, *P. solanacearum* populations decreased faster in soil with the intercrop compared with the sole crop (Yu, 1999), indicating that root exudates of Chinese chive were inhibitory to *P. solanacearum*.

When evaluating survival and repro-duction of herbivores in systems with non-host plants present, not only the survival and reproduction of herbivores that have accepted the host plant, but also the survival of herbivores that have landed by mistake on non-hosts or found in the surroundings of the host plant should be considered. In many instances specialist herbivores find themselves outside their host plant, and their chance of survival may differ between bare soil and non-host plants and between different non-hosts. In a mixed broccoli–legume system, Garcia and Altieri (1992) observed that flea beetles spent considerable time entangled in the branches of the leguminous non-host plants, indicating that high complexity in non-host plant

architecture may contribute to herbivore mortality. Similarly, *D. radicum* females landing on bare soil easily locate or relocate cabbage plants, whereas females having landed on surrounding grass have great difficulty in relocating the host plant (Fig. 5.2) (Kostal and Finch, 1994). Likewise, large numbers of *Brevicoryne brassicae* (Linnaeus) drop off plants, sitting on old leaves that are going to be shed (Raworth, 1984), and the success of relocating a host plant may be affected by the presence of a vegetation cover. A reduced survival on non-host plants may be a 'lasting' effect on herbivores, even in full field scale.

Intercropping effects on natural enemies of herbivores

From a pest management point of view, a system manipulation like intercropping aims to enhance natural enemy impact on important herbivores, regardless of whether this effect is produced by increased enemy density, activity or efficiency. Predators and parasitoids respond to vegetation texture in ways that are no less complicated to interpret and predict than herbivore response, and like herbivores their response in full field scale is determined by specific life traits of individual natural enemy species.

One of the patterns that may help predict how natural enemies are affected is the different responses we intuitively expect from a generalist and a specialist enemy (Sheenan, 1986). For generalist predators, a mixture of plants offers a more buffered microclimate and more abundant food resources than in monocultures, and many studies have shown higher total density/ activity as well as higher species diversity of generalist predators in mixed systems than in monocultures (Theunissen *et al.*, 1992). Judging from studies of abundance and species richness in different crop types (Booij and Noorlander, 1992; Booij, 1994) and with different weed densities (Purvis and Curry, 1984; Bosch, 1987), especially the presence of an early plant cover like in the cabbage–clover system, determines total density and species richness. However, different groups and species respond in an individual manner to both plant cover (Purvis and Curry, 1984; Powell *et al.*, 1985; Bosch, 1987; Sunderland and Samu, 2000) and grass or straw mulch (Hellqvist, 1996), and several of the species considered important polyphagous predators are found more abundantly in monocultures.

Higher abundance of generalist enemies may or may not result in increased enemy impact on crop herbivores, depending on food specificity and searching behaviour. In some studies of predation in different vegetation textures, higher predator abundance leads to higher predation on the specific herbivore (Speight and Lawton, 1976), but not in others (Cárcamo and Spence, 1994). The lack of correlation between predator abundance and predation may be partly caused by an alternative prey present on non-crop plants in high-diversity systems that may act as a 'diversionary prey' seen from a pest management point of view, actually leading to lowered enemy impact on the pest in focus. The importance of the availability of alternative food for generalist predator impact on a pest is illustrated by the studies of the cabbage root fly egg loss due to predation by generalist predators, which in early studies were found to be 50% per day and totalling up to 90% (Hughes, 1957; Freuler, 1975). Questioning this substantial impact of predation on egg survival, Finch and Skinner (1988) later argued that such large predation losses were found only in fields under intensive insecticide pressure and thus with few alternative food items available for the predators, and they estimated much lower egg mortality due to predators under less insecticide pressure.

Specialist enemies, including many parasitoids, are often affected both directly and indirectly by the presence of non-host plants (Sheenan, 1986). Work on adult parasitoids in diverse systems of vegetation has focused on *Hymenoptera* and *Diptera*, two groups that depend on available food sources such as nectar and pollen for adult parasitoids and are often found more abundantly in diverse and thus resource-rich habitats (Altieri and Liebman, 1986). The link between abundance and observed parasitism of the host arthropod, however, is complex. Many specialist parasitoids

use the host plant of their herbivore host in host finding, why parasitism may be negatively affected by the presence of non-host plants, just as specialist herbivores may be disturbed in their host finding. Langer (1996b) showed that intercropping reduced the risk of cabbage root fly (*D. radicum*) pupae being parasitized by one of its parasitoids, the staphylinid beetle, *Aleochara bilineata* Gyllenhal. Staphylinids are generally enhanced in diversified habitats (Purvis and Curry, 1984) and the lower parasitism illustrates that relating natural enemy abundance to enemy impact is difficult. Parasitism by another parasitoid, *T. rapae*, was not affected by intercropping. The two different patterns of parasitism also bring attention to the point that field-observed parasitism in experiments with varying vegetation texture is not only the result of physical effects of plant cover on parasitoid behaviour, but is also influenced by differences in host spatial pattern caused by vegetation texture, i.e. by host patch size, and density dependence of parasitism (Hassell and Goodfray, 1992). Similarly, Perfecto and Vet (2003) found in laboratory experiments that mixing Brussels sprouts with potatoes affected two parasitoids of *P. rapae*, one more specialized than the other, differently. The least specialized parasitoid was less efficient in the intercrop than in the monocrop, although efficiency increased with experience. Contrary to this, the more specialized parasitoid was more efficient in the intercrop, and here experience had no impact. The authors made the point that since natural enemies use clues from the first trophic level, the plants, less specialized parasitoids were diverted by damaged non-host plants – in this case, potato plants damaged by Colorado beetle.

In addition to the volatiles produced by host plants, plants emitting volatiles attractive to parasitoids have also been searched. Gohole *et al.* (2003) investigated whether the foraging activity of a pupal parasitoid on *Chilo partellus* (Swinhoe), a pest of maize, was enhanced by intercropping. The non-host plant tested was molasses grass, which is said to produce volatiles attractive to some parasitoids. In this case volatiles from molasses grass appeared to be repellent to the parasitoid, interfering with its attraction to host plant's odours, especially from plants infested by its host.

Designing Experiments on Pests in Intercrops

As outlined in the previous part of the chapter, diseases and arthropod pests are affected by intercrops at many points in their life cycles. Ideally, when comparing systems with different component/vegetational texture, experimental design should be suited to elucidate the mechanisms governing pest response, regardless of the phase, i.e. establishment, reproduction or survival, that is affected. The ideal experimental design, e.g. plot size, will often differ depending on the mechanism that is in focus. However, intercrop trials are often carried out to assess the overall performance of the cropping system. The aim of such experiments is to develop optimum spatial pattern, timing and variety choice. Assessments of pests are frequently included but are usually not the main goal. This means that many intercrop research experiments are designed to satisfy criteria other than the ones relevant for explaining pest response to intercrops.

Abundance and effects

When investigating the role of natural enemies in intercrops, many studies report on abundance of natural enemies rather than on their impact on arthropod pest species (Coaker and Williams, 1963; Girma *et al.*, 2000). However, we are not mainly interested in the population of natural enemies, but in what they do. Establishing this link between density or density/activity measures of natural enemies and the impact on a specific herbivore (e.g. between numbers of trapped ground beetles and their predation) is difficult for several reasons. Firstly, many predators are generalists, i.e. they eat a broad range of arthropods, and although feeding experiments in the laboratory can reveal the

potential of different predator species eating specific herbivore species or stages (Coaker and Williams, 1963; Finch and Elliott, 1991), estimation of actual predation under different field conditions is still not very reliable and demands detailed tests of field-collected predators. Similarly, other characteristics of cropping systems, e.g. the larger complexity of vegetation in an intercrop compared with a sole crop, could lead to larger overall abundance and diversity of arthropods, and thus to reduced generalist predator impact on the herbivore. For specialized predators and parasitoids the availability of alternative food prey or hosts is less important. Secondly, most predators and parasitoids are mobile and move within our experiments in ways that may make interpretations of enemy abundance difficult.

Extrapolating from plot experiments

When examining single species response to system manipulation, the aim is not only to understand the mechanisms governing the observed response in the concrete case, but also to predict their outcome under other conditions, ultimately making it possible to design agricultural systems that satisfy the goal of minimizing pest problems. Such system manipulations are often examined in plot experiments where the treatments consist of different vegetation textures. The big challenge here is to interpret the observed pest response in terms of their implications for full-field situations.

Experiments with herbivores

Many plot experiments examining specialist herbivore response to vegetational diversification have aimed to test whether the observed results could be attributed to the 'resource concentration' hypothesis (later separated in 'disruptive-crop hypothesis' and 'trap-crop hypothesis' by Vandermeer, 1989) or the 'enemies' hypothesis (Root, 1973). This does not solve the main problem that arises when the tested systems are to be implemented: to predict their performance in full field scale, i.e. whether the arthropod, be it a pest or a beneficial, will respond similarly when the cropping system is in full field scale as observed in plot experiments?

Kareiva (1983, 1986) raised the basic problem of traditional plot experiments. He argued that only the identification of features in the biology of individual species that make some arthropods sensitive and others insensitive to cropping patterns permits us to draw general conclusions from the many already performed crop-pattern experiments. The problems connected with presenting herbivores with a 'menu of cropping arrangements' were also pointed out by others (Bergelson and Kareiva, 1987; Rämert and Ekbom, 1996), and the importance of arthropod mobility for experimental design was stressed.

Misinterpretation of observed arthropod response in plot experiments has been aggravated by commonly used terminology linked to host plant finding in specialist herbivores. Local aggregation of herbivores in plot experiments is often described as a 'preference' for certain plants or for habitats of particular vegetation texture, and such plants or habitats are then called 'selected'. When extrapolating from observations in plot experiments, this terminology implies a perception that the individual cropping pattern has an inherent quality independent of the surroundings, and thus an anticipation that this 'herbivore-reducing ability' will stay with the system in full field scale. Miller and Strickler (1984) stress that although requirements may change, host finding most often consist of a series of 'take it or leave it' situations, and that herbivores rarely consider alternatives when responding to a plant.

Whether local aggregation (e.g. lower herbivore attack) observed in plot experiments with a variety of vegetation textures can be used for predicting full-scale response depends on what mechanisms are affected by vegetation texture. If the mechanisms governing host finding and acceptance (i.e. number of colonists in the plot) are affected by vegetation texture, response in one plot is likely to be influenced by vegetation textures in adjacent plots. If texture affects

the mechanisms determining survival and reproduction of the individuals established on the plants, the chances of refinding the response observed in plots even in a no-choice full-field situation are larger (Stanton, 1983). In Table 5.2 this classification of mechanisms is demonstrated for cabbage herbivores, of which many behavioural responses to both host and non-host plants are known. The different effects of non-host plants on host finding and survival in plot experiments are accompanied by comments on the conditions for refinding the results in full field scale.

As seen in Table 5.2, aggregations of arthropod pests observed in monocultures in plot experiments will probably not be found in full field scale, if it is a result of host plant discrimination and increased activity and emigration from mixed cultures, as seen in *D. radicum*, flea beetles, *P. rapae* and cabbage aphids (Jones, 1977; Klingauf, 1987a; Garcia and Altieri, 1992; Kostal and Finch, 1994). Often no-choice experiments, i.e. situations in which the herbivore has access only to one plant or habitat, reveal that herbivore discrimination between host plants changes with time, i.e. many herbivores gradually become less choosy. This has implications for the no-choice situation in which a vegetation texture, e.g. an intercrop, after having shown herbivore-reducing potential in a plot experiment is then tested in field scale. In this case many herbivores will become less discriminating in host finding and acceptance with time: both the ovipositing *P. rapae* and *D. radicum* females become less discriminating towards oviposition plants with increasing egg load (Jones, 1977; Kostal, 1993), and similarly, starved aphids become less choosy with time (Klingauf, 1987b). A combination of choice and no-choice experiments therefore offers valuable information on changes in host plant discrimination in the herbivore over time and may thus help explain the results generated in field experiments.

Contrasting with this, aggregations seen in plot experiments, which are caused by differences in survival or reproduction due to low host plant quality, will often persist in full field scale. This includes situations

Table 5.2. Examples from cabbage herbivores of direct and indirect effects of non-host plant presence on specialist herbivore host finding and survival in plot experiments, with comments on the probability of obtaining these effects in full field scale also (see text for further explanation).

Herbivore life stage	Direct non-host effects through	Indirect non-host effects through	Possible effect on herbivore	'Pest-reducing effect' refound in full field scale?
(1) Immigration	(1a) Olfactory disruption: masking host plant odours		Less immigrating individuals	
(2) Landing	(2a) Visual disruption: less plant–soil contrast (2b) Olfactory disruption	(2d) Plant competition: host plant colour, size/age, water status host plant surface characteristics	Less landings on host plant Landings on non-host	Yes, if with starvation or egg load
(2) Examining host plant	(2c) Physical disruption: landing on non-hosts interrupts accumulation of stimuli	host plant nutrient and secondary compound content	Increased activity Increased emigration Less eggs laid	Yes, if landing and staying on non-host plants increases mortality through vulnerability to enemies or to energy expenditure
(2) Accepting host plant			Less feeding damage	
(3) Surviving and reproducing in habitat	(3a) Vegetational texture: natural enemy abundance natural enemy effectiveness increased vulnerability when on non-hosts	(3b) *Plant competition*: host plant nutrient contents	Reduced survival Decreased fecundity Increased or decreased predation or parasitism	Yes, if herbivore is not able to compensate by moving or by increased feeding Yes, if natural enemy response does not result from local aggregation in attractive habitats and Altered enemy abundance results in altered impact on pest

in which increased herbivore movement, caused either by non-host plant presence or reduced host plant quality in a mixed culture, makes the herbivore more available to predation. In that case a full field scale effect may also be anticipated.

The above-mentioned limitations of experiments in which mobile arthropods may exhibit a 'choice' were documented by Bommarco and Banks (2003), who reviewed the design of studies of 25 predators and 41 herbivores in which arthropod abundance was compared between a treatment of higher plant diversity and a control. They distinguished between studies employing small plots (<16 m²), medium plots (16–256 m²) and large plots (>256 m²). The average 'herbivore-reducing effect' observed in diverse systems was largest in experiments using small plots, less evident when medium-size plots were used and non-existing in experiments with large plots. This result challenges the interpretation of the results of many previous studies, and Bommarco and Banks (2003) characterize the generally accepted fact of herbivore declines associated with increased vegetation diversity to be misleading.

Plant pathogens

When studying intercropping effects on disease, one of the important questions is whether a delay in the development of an epidemic observed in a small field trial will also be seen if you greatly increase the area of intercropping? Many ecological processes are strongly influenced by spatial scale, causing a major dilemma for experimental biologists, as large-scale field experiments are often prohibitively expensive. However, experimental procedure and the nature of pathogen dispersal can cause substantial underestimation of the impact of increased diversity on disease in small-scale experimental plots (Wolfe 1985; Garrett and Mundt, 1999).

Large amounts of outside inoculum can overwhelm any effect from a mixed species crop. All field experiments include a control (i.e. sole crop), which generally has a greater severity of disease, and produce greater amounts of inoculum. As the effect of the intercrop is ultimately to reduce inoculum, interference from other plots, e.g. large amounts of inoculum from the control plots, can obliterate the 'intercrop' effect. Reducing the size of the sole crop plots compared with the intercrop plots would reduce the influence on the intercrops, and guard or barrier rows may reduce interference between plots. However useful these changes may be in reducing between-plot interference, appreciation of problems of scale in experimental trials is crucial in evaluating the potential efficacy of intercropping to disease. Once results of interest are obtained in experimental plots, trials must be scaled up from the typical plot size of 20–30 m². The potential of participatory research, where farmers' fields function as experimental plots, was beautifully demonstrated by Zhu *et al.* (2000) following several reports that varietal diversification reduced disease severity in experimental plots. They persuaded thousands of rice farmers in a large area in China (initially 812 ha) to grow particular mixtures of rice varieties. Unlike standard experimental fields, control plots of sole crops were small compared with the total area of mixtures planted by farmers in the surrounding area, thus reducing the potential impact of spore dispersal from the more heavily infected monocultures to the mixture plots (Wolfe, 1985; Garrett and Mundt, 1999). The level of rice blast (*Magnaporthe grisea*) was highly decreased in this area, with blast severity in the panicle at 20% in the monoculture and 1% in the mixed rice population. From these results, unlike data for arthropods (Bommarco and Banks, 2003), intercrop efficacy on plant diseases could be greater in production-scale situations than in small-scale experimental plots (Table 5.3).

It should be noted that the problems of identifying the basic mechanisms responsible and thus the implications for observed herbivore and disease responses in plot experiments, discussed above, are not restricted to work dealing with intercropping. It is a problem that is shared by workers in crop resistance to arthropod pests and diseases, as the observed dif-

Table 5.3. Examples of non-host plant presence on pathogen dispersal and subsequent survival in plot experiments, with comments on the probability of obtaining these effects in full field scale.

Stage of life cycle for pathogen	Non-host effects through	Possible negative effect on pathogen	'Pathogen-reducing effect' refound in full field scale?
Initial inoculation dispersal	Physically impeding inoculum dispersal	Less pathogenic propagules land on host	Yes, if the amount of incoming inoculum is not excessively high
	Increasing distance between susceptible hosts		
	Interactions/competition between the host and non-host		Expected that the effect would be multiplied under larger field scale
Infection: germination and penetration of host plant	Altered microclimate Allelopathy Induced resistance by non-host pathogens Nutritional status of plant	Reduced germination of pathogen Reduced penetration efficiency of pathogen	Yes, however, differences in climatic conditions at particular times in the growing season could alter competition between the component crops, thus altering the effect of the non-host on the pathogen
Survival and reproduction	Induced resistance Nutritional status of plant	Reduced pathogen development and reproduction	Yes, provided induced resistance and plant nutritional status are not altered when plants are stressed, e.g. drought

ferences in field trials with natural infestations of pests are difficult to interpret, unless the mechanisms causing the response are clarified (Stoner and Shelton, 1988a, b; Eigenbrode *et al.*, 1991).

Predicting pest response to intercropping

The problems connected with understanding and predicting pest response in different spatial scales cannot be solved solely by the accumulation of more detailed knowledge of pest response to the manipulation of agricultural systems. Spatially, specific simulation models that use existing knowledge about arthropod behaviour and disease epidemiology as parameters may be a way of organizing our detailed knowledge, and are suitable for testing whether our perception of these parameters is valid under field conditions (Leonard, 1969; Jones, 1977; Kareiva, 1982; Boudreau and Mundt, 1994).

Predicting herbivore response using models

Mechanistic descriptions of, for instance, herbivore movement, can be seen as a way of organizing the 'rules of movement' of a specific species in a formal frame. Such an approach will facilitate understanding and predictions of consequences of movement, not only where the original observations are made but also in a variety of ecological settings (Kareiva, 1982; Morris and Kareiva, 1991). As an example, Kareiva (1982) developed a model to organize and test observations made of flea beetles moving between host plant patches of different distance and quality. Movement of individuals is described by the probability of an individual performing a variety of actions: moving to the next host patch, remaining in the host patch, immigration to a patch from outside the system, emigration out of the system and dying. With these simple rules of movement the distribution of flea beetles in patches of varying distances could be satisfactorily described. Another example is the model created by Corbett and Plant (1993), in which natural enemy movement in response to system diversification was described using the different dispersal rates of natural enemies in diversified vegetation strips and in the field. In this case different spatial scenarios were created, and the predictions of the outcome of different scenarios demonstrated how insects with different dispersal rates respond differently to a range of patterns of vegetation.

However, the development of models and the call for sophisticated understanding of the agroecosystem as a basis for successful IPM and as a promising path to improved design of intercropping systems is being challenged. In some cases the observed response is a combined result of mechanisms working in opposite directions (Coll and Bottrell, 1994) and models may not adequately capture such balances. Way and van Emden (2000) characterize the contribution of models aimed at providing novel insights and guidelines for IPM research as meagre. They point out that, at best, models can be a shortcut in suggesting outcomes for experimental testing, but they cannot be a substitute for empiricism. They claim that 'models cannot predict the unpredictable, whereas experiments can reveal ... unexpected synergism or compatibility of control methods'. This is even more important in complex systems like intercrops.

Predicting disease response

Before intercropping can be recommended as a means of disease control, we should be able to predict the net effect of combining a particular host crop with other crops, at a certain time and location, on the dynamics of a specific pathogen. Detailed studies specific to intercropping in particular pathosystems must be carried out to allow us to forecast the effect of the intercrop on arthropod pests and plant pathogens. The anticipation is that diseases characterized by the same general traits will respond in the same pattern: the inherent characteristics

of the individual disease, generation time, method of dispersal, the stage of the crop at which the epidemic occurs, and whether 'internal' or 'external' inoculum play a larger role.

The effect of generation time was seen in the models designed by Leonard (1969) to predict outcomes in species mixtures of highly susceptible and immune plants. In an equal mixture of susceptible and immune plants, Leonard predicted that the disease level on the susceptible plants in the mixture relative to the pure stand of susceptible plants will be 0.5 in the first generation of secondary spread, 0.25 in the second generation and 0.125 in the third generation. Thus, pathogens that go through many life cycles on a crop during a growing season are more likely to be influenced by the intercrop.

The stage of the growing season at which the epidemic occurs is also crucial. If the epidemic begins early in the season, the disease progress will be delayed in the intercrop. Later in the growing season the carrying capacity of the host will be reached earlier in the sole crop than the intercrop (Fig. 5.3) (Wolfe and Barrett, 1980). While the final impact of disease on both crops may be similar, the delay in disease progress in the intercrop means more green tissue is available for photosynthesis, and hence possibly increased yield. If the epidemic starts during the middle of the growing season, the plants may be harvested before maximum carrying capacity of the host is reached. Thus, the difference in disease is large (Fig. 5.4a). If the epidemic begins very late in the growing season, especially if the pathogen has a relatively long generation time, by the time the plants are mature and begin to senesce, the pathogens have gone through few life cycles, reducing the difference between the pure stand and the mixture (Fig. 5.4b).

The level of initial infection affects the rate of disease development in an intercrop, and large amounts of external inoculum greatly reduce the efficiency of disease control in the mixture (Fig. 5.5). Mundt and Leonard (1985) found that the effect of host diversity on oat rust was greater when epidemics

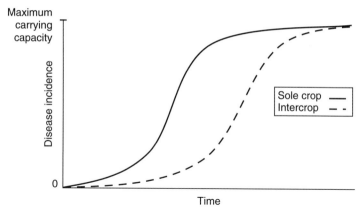

Fig. 5.3. Disease progress on a sole crop or intercrop over time. Disease levels in the intercrop and the sole crop to initially diverge over time, and then to converge as the host's carrying capacity is approached.

began with a limited number of foci instead of at sites neighbouring heavily infected oat fields, providing a continuously large source of incoming inoculum. Similarly, Garrett and Mundt (2000) compared the effects of potato genotype mixtures on late blight in France and Ecuador. In France host-diversity affected the disease, whereas in Ecuador host-

diversity effect was found only in one of three sites. In France, epidemics developed from obvious loci, whereas in Ecuador, disease was widespread throughout experimental plots from early on in epidemics.

The influence of epidemic intensity on intercrop response will thus depend on the cause(s) of disease increase. If high

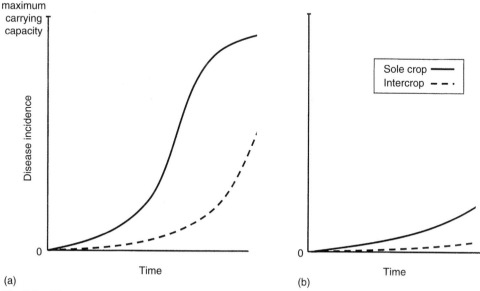

(a)   (b)

Fig. 5.4. Disease progress on a sole crop or intercrop over time if the pathogen arrives in mid-growing season (a) or very late in the season. (b) If the epidemic starts in the mid-growing season, the plants may have senesced before maximum carrying capacity if the host is reached. If the epidemic begins very late in the season, by the time the plants are mature and begin to senesce, the pathogen may only have gone through few life cycles, so the difference between the pure stand and the mixture is small.

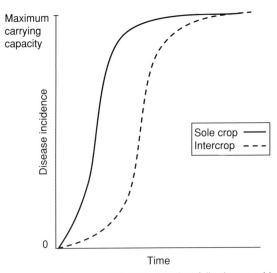

Fig. 5.5. Disease progress on a sole crop or intercrop over time following very high levels of initial infection. Disease levels in the intercrop and the sole crop reach the host's carrying capacity quickly after initial inoculation.

disease incidence in pure stands is due to a fast approach to carrying capacity or large amounts of outside inoculum, mixtures may be less effective than in less severe epidemics. If, on the other hand, high disease incidence in pure stands is driven by the number of pathogen generations, mixtures may be more effective in severe than in less severe epidemics.

Modelling disease response

Computer modelling has been a valuable component of the ecologists' efforts to understand complex communities and might also be useful in intercrop systems. Computer programs developed by plant pathologists include epidemics in particular crop systems, e.g. in variety mixtures or multilines (Leonard, 1969). Boudreau and Mundt (1994) modified such models and used input variables from their own field data to study epidemics in intercrops. They were able to evaluate the relative and combined roles of the various mechanisms of intercrop influence on disease that they had studied in a bean–maize intercrop – bean rust system. These computer simulations generated disease progress curves, which in many ways recreated the

sort of variability observed in field studies generally. Partitioning disease reductions that occurred in the first season into component mechanisms revealed that dispersal alteration due to maize was responsible for only about 19% of the reduction in the area under the disease progress curve (AUDPC); the remainder was due to reductions in post-deposition infection efficiency. In the field data on which these were based, the reduction in infection efficiency was probably due to a microclimatic effect and occurred late in the season; the simulation suggested that such late-season effects could still have a profound influence on disease severity. These simulations were the first attempt in applying computer modelling to an intercrop situation. They are consistent with the variable results observed in the field and point out some reasons for it, e.g. sensitivity of intercrop effects to the rate of disease progress, importance of late-season microclimatic effects and spore removal processes.

However encouraging these simulations are at evaluating epidemic development and pathogen evolution in crop mixtures, current simulations and models do not describe events in the field accurately enough to allow disease prediction in intercrops. Lack of empirical studies from well-designed field experiments continues to limit our understanding of the effect of mixtures on disease development. It is necessary to combine the results from several different studies such as meta-analysis. Combining the results from different studies may also make it possible to compare the effects, e.g. geographic region, climatic conditions and pathogen species. Further development could result in models that not only allow us to understand how epidemics are conditioned by species diversity in natural and agricultural ecosystems, but also provide predictive power for efficacy of intercropping a particular crop system under specific conditions.

Intercropping and IPM

Depending on the prevalent farming systems in the region, intercropping can be seen either as a pest management practice (i.e. one of many elements in IPM) or as a fundamental part of a cropping system in which IPM is to be practised. In temperate areas intercropping is often seen as a potential pest management method that has to be adapted to the other farming practices in the system, whereas in the tropics, crops are usually grown in mixed cropping systems (Way and van Emden, 2000), and in such instances intercropping is a condition for IPM and not a cultural pest management method that may or may not be included. Thus, pest management in traditional agriculture is often a built-in process in the overall crop production system rather than a separate well-defined activity (Abate *et al.*, 2000).

To explore whether a deeper understanding of the multitude of inter-actions in cropping systems leads to better prospects for action, Way and van Emden (2000) examined the research and development approaches that have succeeded, by offering examples of large-scale ecological studies never contributing to the advancement of practical control. They point out that even in a one-crop system like irrigated rice the potential number of trophic links is 9000 among

645 taxa. They conclude that we cannot possibly demand a sophisticated under-standing of the ecosystem involved when working with intercrops. Thus, searching for the fundamental rules, which make us capable of predicting responses and designing systems, is probably not a suitable strategy to develop the use of intercrops to their full potential in an IPM perspective. Norris and Kogan (2000) on the basis of several hundred papers on weed–arthropod interactions, which are similar to interactions in intercrops, conclude that the search for general rules is not worthwhile. Most interactions are species-specific and thus change with the species involved. They agree with Kareiva (1983) in that the variety of responses dissolves any hope that easy generalizations can be made, and with Andow (1988) in that it is not possible to predict whether a particular arthropod will have a higher or lower population in a certain vegetation texture, without running the experiment.

In areas where key constraints are poor soil fertility or uncertain rainfall, farmers employ a risk strategy including growing intercrops as insurance. Under these conditions, Way and van Emden (2000) argue that IPM practices may be considered complicated and costly if based on imported methods adopted from industrialized agriculture. They propose a concept of IPM that has as its fundamental pillars the integration of cultural practices like early sowing, the removal of infested residues, with intercropping as a promising research direction.

Their view of a need to build on farmer knowledge is shared by Morales (2002). In a survey of articles published in *Journal of Economic Entomology* in 1999–2001, Morales found that only 1% has intercropping as the main focus compared with almost 50% on synthetic or microbiological pesticides and 22% on resistance, half of which discuss resistance obtained by genetic modification. She criticizes IPM projects in developing countries for promoting management practices that farmers are unfamiliar with. IPM promoters should, take advantage of traditional farmers' familiarity with preventive pest management and their experience and willingness to test new polycultures and give special priority to these developmental paths in their programmes.

Local development of intercrop systems

Given that much of the fluctuation and low productivity in developing countries are due to the direct or indirect impact of biotic stresses, any factor that influences disease and insect pressure must be considered in crop protection. Intercropping has been shown repeatedly to significantly reduce pest levels (Table 5.4). While this reduction is sometimes small, in areas where disease and insect incidence are high, any reduction will have a positive effect on yield.

As shown above, the details of the application of intercropping under specific field conditions cannot be applied as a rule of thumb, but must be considered case by

Table 5.4. Disease management by intercropping.

Host Non-host	Pathogen	Effect on disease: − reduction 0 no effect + increase	Reference
Alfalfa Cocksfoot grass	Alfalfa mosaic virus	−	Gibbs and Harrison (1976)
Barley Grain legumes (pea, bean or lupins)	Pyrenophora teres	−	Kinane and Lyngkjær (2002)
	Leaf rust Puccinia hordei	−	Kinane and Lyngkjær (2002)
	Blumeria graminis f. sp. Hordei	−	Kinane and Lyngkjær (2002)
Barley Wheat	B. graminis f. sp. Hordei	−	Burdon and Chilvers, (1977)
Barley Oats	B. graminis f. sp. Hordei	−	Vilich-Meller (1992)
Phaseolus bean Groundnut	Groundnut rosette disease	−	Farrell (1976)
Bean Maize	P. griseola	−, 0, +	Moreno (1977); Van Rheenen et al. (1981); Msuku and Edje (1982); Sengooba (1990); Boudreau (1993)
	U. appendiculatus	−, 0	Soria et al. (1975); Monteiro et al. (1981); Van Rheenen et al. (1981); Moreno and Mora (1984); Boudreau and Mundt (1992)
	Pseudomonas phaseolicola	−	Van Rheenen et al. (1981)
	Bean common mosaic virus	−	Van Rheenen et al. (1981)
	C. lindemuthianum	−	Van Rheenen et al. (1981)
	Xanthomonas phaseoli	−	Van Rheenen et al. (1981)
	Elsinoe phaseoli	−	Van Rheenen et al. (1981)
	Phoma exigua var diversispora	−	Van Rheenen et al. (1981)
	Erysiphe polygoni	−	Van Rheenen et al. (1981)
	Sclerotium sclerotiorum	+	Van Rheenen et al. (1981)
Cassava Bean	Oidium manihotis	+	Soria et al. (1975)

Table 5.4. *Continued*

Host Non-host	Pathogen	Effect on disease: − reduction 0 no effect + increase	Reference
Cassava Cowpea	Cassava mosaic disease	−, 0	Fondong *et al.* (2002)
	Xanthomonas axonopodis pv. *manihotis* (*X. campestris* pv. *manihotis*)	−	Zinsou *et al.* (2004)
Cassava Maize	African cassava mosaic virus	−, 0, +	Fargette and Fauquet (1988)
	Sphaceloma sp.	−	Larios and Moreno (1977)
	Glomerella cingulata	−	Larios and Moreno (1977)
	Cassava mosaic viral disease	−, 0	Fondong *et al.* (2002)
Cassava Plantain	Severe mosaic virus	−	Gámez and Moreno (1983)
Cassava Sorghum	*X. axonopodis* pv. *manihotis* (*X. campestris* pv. *manihotis*)	−	Zinsou *et al.* (2004)
Chilli Maize or Brinjal	Chilli veinal mottle virus	−	Hussein and Samad (1993)
Cowpea Maize	*Ascochyta phaseolorum*	−	Larios and Moreno (1977)
	Cowpea mosaic virus	−	Larios and Moreno (1977)
	Chlorotic cowpea mosaic virus	−	Larios and Moreno (1977)
Cowpea Sorghum	Virus (CaMV and CMV)	+	Edema *et al.* (1997)
	Cercospora canescens	+	Edema *et al.* (1997)
	C. lindemuthianum	−	Edema *et al.* (1997)
	Sphaceloma	−	Edema *et al.* (1997)
	Synchytrium dolichi	−	Edema *et al.* (1997)
	Dactuliophora leaf spot	−	Edema *et al.* (1997)
	Erysiphe polygoni	−	Edema *et al.* (1997)
Cucumber Tomato	Tomato leaf curl viral disease	−	Al-Musa (1982)
Groundnut Beans	Rosette disease	−	Thresh (1982)
Maize Cowpea or Green gram	*Pratylenchus brachyurus*	−, 0, +	Egunjobi (1984)
Maize Bean	Corn Stunt spiroplasma	−	Power (1987)
Oats Barley	*Drechslera avenae*	−	Vilich-Meller (1992)
Pea Barley	*Ascochyta pisi*	−	Kinane and Lyngkjær (2002)
Groundnut Pigeon pea	*Cercospora arachidicola, Phaeoisariopsis personata*	−	Ghewande *et al.* (1993)
Groundnut Bean	Peanut rosette virus	−	Farrell (1976)

Continued

Table 5.4. *Continued*

Host Non-host	Pathogen	Effect on disease: – reduction 0 no effect + increase	Reference
Plantain Cassava	*Mycosphaerella fijiensis*	–	Emebiri and Obiefuna (1992)
Radish Rice or trefoil	Radish mosaic virus	–	Palti (1981)
Potato Sorghum or maize	*P. infectans*	–	Raymundo and Alcazar (1982)
Potato Sweet potato		+	Raymundo and Alcazar (1982)
Rice Maize and Casava	*Cochliobolus miyabeanus*	+	Kass (1978)
Soybean Sorghum	Soybean mosaic virus	–	Bottenberg and Irwin (1992)
Sorghum Maize	*Colletotrichum sublineolum*	–	Ngugi *et al.* (2001)
Sugarbeet Barley or mustard or other crops	Aphid-borne viruses	–	Palti (1981)
Tomato Maize or Toria (*Brassica campestris*)	*Septoria lycopersici*	–	Kumar and Sugha (2000)
Tomato Bean	*S. lycopersici*	0	Kumar and Sugha (2000)
Tomato Cowpea	*Pseudomonas solanacearum*	–	Michel *et al.* (1997)
Tomato Soybean/ welsh onion	*P. solanacearum*	0	Michel *et al.* (1997)
Tomato Marigold/pigweed	*A. solani*	–	Gómez-Rodríguez *et al.* (2003)
Oats Barley	*Drechslera avenae*	–	Vilich-Meller (1992)
Rye Wheat	*Rhyncosporium secalis*	–	Vilich-Meller (1992)
Wheat Clover	*S. tritici*	–, +	Bannon and Cooke (1998)
Wheat *Medicago lupulina*	*Gaeumannomyces graminis* var. *tritici*	–	Lennartsson (1988)
Wheat Rye	*Puccinia recondite f.*	–	Vilich-Meller (1992)

case, for each crop and pest. The challenge for future sustainability lies in developing cropping systems and increasing yield levels through the adaptation of traditional systems and the inclusion of new suitable techniques. This is a process of considerable complexity and social significance. Therefore, it is crucial to have an efficient advisory service that can understand the facts available on intercropping and is familiar enough with local conditions to be able to critically review their relevance for local farmers.

References

Abate, T., Huis, A. and van Ampofo, J.K.O. (2000) Pest management strategies in traditional agriculture: an African perspective. *Annual Review of Entomology* 45, 631–660.

Allen, S.J., Brown, J.F. and Kochman, J.K. (1983) Effects of leaf age, host growth stage, leaf injury, and pollen on the infection of sunflower by *Alternaria helianthi. Phytopathology* 73, 896–898.

Al-Musa, A. (1982) Incidence, economic importance and control of tomato yellow leaf curl in Jordan. *Plant Disease* 66, 561–563.

Altieri, M.A. (1983) Vegetational designs for insect-habitat management. *Environmental Management* 7, 3–7.

Altieri, M.A. (1999) The ecological role of biodiversity in agroecosystems. *Agriculture Ecosystems and Environment* 74, 19–31.

Altieri, M.A. (2004) Linking ecologists and traditional farmers in the search for sustainable agriculture. *Frontiers in Ecology and the Environment* 2, 35–42.

Altieri, M.A. and Letourneau, D.K. (1982) Vegetation management and biological control in agroecosystems. *Crop Protection* 1, 405–430.

Altieri, M.A. and Liebman, M. (1986) Insect, weed and plant disease management in multiple cropping systems. In: Francis, C.A. (ed.) *Multiple Cropping Systems.* Macmillan, New York, pp. 183–218.

Andow, D. (1983) Effect of agricultural diversity on insect populations. In: Lockeretz, W. (ed.) *Environmentally Sound Agriculture.* Praeger, New York, pp. 91–115.

Andow, D. (1988) Management of weeds for insect manipulation in agroecosystems. In: Altieri, M.A. and Liebman, M. (eds) *Weed Management in Agroecosystems: Ecological Approaches.* CRC Press, Boca Raton, Florida, pp. 265–301.

Andow, D. (1991) Vegetational diversity and arthropod population response. *Annual Review of Entomology* 36, 561–586.

Anil, L., Park, J., Phipps, R.H. and Miller, F.A. (1998) Temperate intercropping of cereals for forage: a review of the potential for growth and utilization with particular reference to the UK. *Grass and Forage Science* 53, 301–317.

Armstrong, C.L. and Mitchell, J.K. (1988) Plant canopy characteristics and processes, which affect transformation of rainfall properties. *Transactions of the ASAE* 31, 1400–1409.

Asman, K., Ekbom, B. and Rämert, B. (2001) Effect of intercropping on oviposition and emigration behavior of the leek moth (*Lepidoptera: Acrolepiidae*) and the diamondback moth (*Lepidoptera: Plutellidae*). *Environmental Entomology* 30, 288–294.

Autrique, A. and Potts, M.J. (1987) The influence of mixed cropping on the control of potato bacterial wilt (*Pseudomonas solanacearum*). *Annals of Applied Biology* 111, 125–133.

Bannon, F.J. and Cooke, B.M. (1998) Studies on dispersal of *Septoria tritici* pycnidiospores in wheat–clover intercrops. *Plant Pathology* 47, 49–56.

Baumann, D.T., Bastiaans, L. and Kropff, M.J. (2002) Intercropping system optimisation for yield, quality, and weed suppression combining mechanistic and descriptive models. *Agronomy Journal* 94, 734–742.

Bengtsson, J. (1998) Which species? What kind of diversity? Which ecosystem function? Some problems in studies of relations between biodiversity and ecosystem function. *Applied Soil Ecology* 10, 91–199.

Bergelson, J. and Kareiva, P. (1987) Barriers to movement and the response of herbivores to alternative cropping patterns. *Oecologia* 71, 457–460.

Bigger, D.S. and Chaney, W.E. (1998) Effects of *Iberis umbellata* (*Brassicaceae*) on insect pests of cabbage and on potential biological control agents. *Environmental Entomology* 27, 161–167.

Bommarco, R. and Banks, J.E. (2003) Scale as modifier in vegetation diversity experiments: effects on herbivores and predators. *Oikos* 102, 440–448.

Booij, C.J.H. (1994) Diversity patterns in carabid assemblages in relation to crop and farming systems. In: Desender, K. (ed.) *Carabid Beetles: Ecology and Evolution.* Kluwer Academic Publishers, The Netherlands, pp. 425–431.

Booij, C.J.H. and Noorlander, J. (1992) Farming systems and insect predators. *Agriculture, Ecosystems and Environment* 40, 125–135.

Bosch, J. (1987) Der Einfluss einiger dominanter Ackerunkräuter auf Nutz- und Schadarthropoden in einem Zuckerrübenfeld. *Zeitschrift für Pflanzenkrankheiten und Pflanzenschutz* 94, 398–408.

Bottenberg, H. and Irwin, M.E. (1992) Canopy structure in soybean monocultures and soybean sorghum mixtures – impact on aphid (*Homoptera: Aphididae*) landing rates. *Environmental Entomology* 21, 542–548.

Boudreau, M.A. (1993) Effect of intercropping beans with maize on the severity of angular leaf-spot of beans in Kenya. *Plant Pathology* 42, 16–25.

Boudreau, M.A. and Mundt, C.C. (1992) Mechanisms of alteration in bean rust epidemiology due to intercropping with maize. *Phytopathology* 82, 1051–1060.

Boudreau, M.A. and Mundt, C.C. (1994) Mechanisms of alteration in bean rust development due to intercropping, in computer-simulated epidemics. *Ecological Applications* 4, 729–740.

Boudreau, M.A. and Mundt, C.C. (1997) Ecological approaches to disease control. In: Rechcigl, J.E. and Rechcigl, N.A. (eds) *Environmentally Safe Approaches to Crop Disease Control*. CRC press, Boca Raton, Florida, pp. 33–62.

Brandsæter, L.O., Netland, J. and Meadow, R. (1998) Yields, weeds, pests and soil nitrogen in a white cabbage – living mulch system. *Biological Agriculture and Horticulture* 16, 291–309.

Buiel, A.A.M., Verhaar, M.A., Vandenbosch, F., Hoogkamer, W. and Zadoks, J.C. (1989) Effect of cultivar mixtures on the wave velocity of expanding yellow stripe rust foci in winter-wheat. *Netherlands Journal of Agricultural Science* 37, 75–78.

Burdon, J.J. and Chilvers, G.A. (1976) Controlled environment experiments on epidemics of barley mildew in different density host stands. *Oecologia* 26, 61–72.

Burdon, J.J. and Chilvers, G.A. (1977) Controlled environment experiments on epidemic rates of barley mildew in different mixtures of barley and wheat. *Oecologia* 28, 41–146.

Burdon, J.J. and Chilvers, G.A. (1982) Host density as a factor in plant disease ecology. *Annual Review of Phytopathology* 20, 143–166.

Cárcamo, H.A. and Spence, J.R. (1994) Crop type effects on the activity and distribution of ground beetles (*Coleoptera: Carabidae*). *Environmental Entomology* 23, 684–692.

Chin, K.M. and Wolfe, M.S. (1984) The spread of *Erysiphe graminis* f. sp. *hordei* in mixtures of barley varieties. *Plant Pathology* 33, 89–100.

Coaker, T.H. (1980) Insect pest management in brassica crops by intercropping. *IOBC/WPRS Bulletin* 3, 117–125.

Coaker, T.H. and Williams, D.A. (1963) The importance of some carabidae and staphylinidae as predators of the cabbage root fly, *Erioischia brassicae* (Bouché). *Entomologia Experimentalis et Applicata* 6, 156–164.

Coll, M. and Bottrell, D.G. (1994) Effects of nonhost plants on an insect herbivore in diverse habitats. *Ecology* 75, 723–731.

Coll, M. and Bottrell, D.G. (1996) Movement of an insect parasitoid in simple and diverse plant assemblages. *Ecological Entomology* 21, 141–149.

Corbett, A. and Plant, R.E. (1993) Role of movement in the response of natural enemies to agroecosystem diversification. *Environmental Entomology* 22, 519–531.

Cordier, C., Pozo, M.J., Barea, J.M., Gianinazzi, S. and Gianinazzi-Pearson, V. (1998) Cell defense responses associated with localized and systemic resistance to *Phytophthora parasitica* induced in tomato by an arbuscular mycorrhizal fungus. *Molecular Plant–microbe Interactions* 11, 1017–1028.

Dale, D. (1988) Plant-mediated effects of soil mineral stresses on insects. In: Heinrichs, E.A. (ed.) *Plant Stress – Insect Interactions*. John Wiley & Sons, New York, pp. 35–110.

Dean, R.A. and Kuc, J. (1985) Induced systemic protection in plants. *Trends in Biotechnology* 3, 125–129.

Den Belder, E., Elderson, J. and Vereijken, P.F.G. (2000) Effects of undersown clover on host-plant selections by *Thrips tabaci* adults in leek. *Entomologia Experimentalis et Applicata* 94, 173–182.

Dill-Macky, R. and Roelfs, A.P. (2000) The effect of stand density on the development of *Puccinia graminis* f. sp. *tritici* in barley. *Plant Disease* 84, 29–34.

Dinoor, A. and Eshed, N. (1984) The role and importance of pathogens in natural plant-communities. *Annual Review of Phytopathology* 22, 443–466.

Dolch, R. and Tscharntke, T. (2000) Defoliation of alders (*Alnus glutinosa*) affects herbivory by leaf beetles on undamaged neighbours. *Oecologia* 125, 504–511.

Edema, R., Adipala, E. and Florini, D.A. (1997) Influence of season and cropping system on occurrence of cowpea diseases in Uganda. *Plant Disease* 81, 465–468.

Egunjobi, O.A. (1984) Effects of intercropping maize with grain legumes and fertilizer treatment on populations of *Pratylenchus brachyurus* Godfrey (Nematoda) and on the yield of maize (*Zea mays* L.). *Protection Ecology* 6, 153–167.

Eigenbrode, S.D., Stoner, K.A., Shelton, A.M. and Kain, W.C. (1991) Characteristics of glossy leaf waxes associated with resistance to diamondback moth (*Lepidoptera: Plutellidae*) in *Brassica oleracea*. *Journal of Economic Entomology* 84, 1609–1618.

Emebiri, L.C. and Obiefuna, J.C. (1992) Effects of leaf removal and intercropping on the incidence and severity of black Sigatoka disease at the establishment phase of plantains (*Musa* spp. AAB). *Agriculture, Ecosystems and Environment* 39, 213–219.

Fargette, D. and Fauquet, C. (1988) A preliminary study on the influence of intercropping maize and cassava on the spread of African cassava mosaic virus by whiteflies. *Aspects of Applied Biology* 17, 195.

Farrell, J.A.K. (1976) Effects of intersowing with beans on the spread of groundnut rosette virus by *Aphis craccivora* Koch (*Hemiptera: Aphididae*) in Malawi. *Bulletin of Entomological Research* 66, 331–333.

Finch, S. and Elliott, M.S. (1991) Predation of cabbage root fly eggs by *Carabidae*. *IOBC/WPRS Bulletin* XV/4, 176–183.

Finch, S. and Kienegger, M. (1997) A behavioural study to help clarify how undersowing with clover affects host-plant selection by pest insects of brassica crops. *Entomologia Experimentalis et Applicata* 84, 165–172.

Finch, S. and Skinner, G. (1988) Mortality of the immature stages of the cabbage root fly. In: Cavalloro, R. and Pelerents, C. (eds) *Progress on Pest Management in Field Vegetables. Proceedings of the CEC/IOBS Experts' Group Meeting*. Rennes, pp. 45–48.

Fininsa, C. (1996) Effect of intercropping bean with maize on bean common bacterial blight and rust diseases. *International Journal of Pest Management* 42, 51–54.

Fininsa, C. (2001) Epidemiology of bean common bacterial blight and maize rust in intercropping. PhD thesis, Swedish University of Agricultural Sciences, Uppsala, Sweden.

Fitt, B.D.L and McCartney, H.A. (1986) Spore dispersal in relation to epidemic models. In: Leonard, K.J. and Fry, W.E. (eds) *Plant Disease Epidemiology*, Vol. 1. Macmillan, New York, pp. 311–345.

Fondong, V.N., Thresh, J.M. and Zok, S. (2002) Spatial and temporal spread of cassava mosaic virus disease in cassava grown alone and when intercropped with maize and/or cowpea. *Journal of Phytopathology–Phytopathologische Zeitschrift* 150, 365–374.

Francis, C.A. (1986) Introduction: distribution and importance of multiple cropping. In: Francis, C.A. (ed.) *Multiple Cropping Systems*. Macmillan, New York, pp. 1–19.

Freuler, J. (1975) Quantitative Erfassung der Populationsbewegungen von verschiedenen hylemya-Arten, inbesondere *Hylemya brassicae* Bouché (*Diptera: Anthomyiidae*), während des ganzen Jahres. *Mitteilungen der Schweizerischen Entomologischen Gesellschaft* 48, 341–355.

Gámez, R. and Moreno, R.A. (1983) Epidemiology of beetle-borne viruses of grain legumes in Central America. In: Plumb, R.T. and Thresh, J.M. (eds) *Plant Virus Epidemiology*. Blackwell Scientific Publications, Oxford, pp. 103–113.

Garcia, M.A. and Altieri, M.A. (1992) Explaining differences in flea beetle *Phyllotreta cruciferae* densities in simple and mixed broccoli cropping systems as a function of individual behaviour. *Entomologia Experimentalis et Applicata* 6, 201–209.

García-Barrios, L. (2003) Plant–plant interactions in tropical agriculture, In: Vandermeer, J.H. (ed.) *Tropical Agroecosystems*. CRC Press, Boca Raton, Florida, pp. 12–58.

Garrett, K.A. and Mundt, C.C. (1999) Epidemiology in mixed host populations. *Phytopathology* 89, 984–990.

Garrett, K.A. and Mundt, C.C. (2000) Host diversity can reduce potato late blight severity for focal and general patterns of primary inoculum. *Phytopathology* 90, 1307–1312.

Garrett, K.A., Nelson, R.J., Mundt, C.C., Chacon, G., Jaramillo, R.E. and Forbes, G.A. (2001) The effects of host diversity and other management components on epidemics of potato late blight in the humid highland tropics. *Phytopathology* 91, 993–1000.

Ghewande, M.P., Desai, S., Narayan, P. and Ingle, A.P. (1993) Integrated management of foliar diseases of groundnut (*Arachis hypogaea* L.) in India. *International Journal of Pest Management* 39, 375–378.

Gibbs, A.J. and Harrison, B.D. (1976) *Plant Virology: the Principles*. Arnold, London.

Giller, K.E., Ormesher, J. and Awah, F.M. (1991) Nitrogen transfer from *Phaseolus* bean to intercropped maize measured using N-15-enrichment and N-15-isotope dilution methods. *Soil Biology and Biochemistry* 23, 339–346.

Girma, H., Rao, M.R. and Sithanantham, S. (2000) Insect pests and beneficial arthropods population under different hedgerow intercropping systems in semiarid Kenya. *Agroforestry Systems* 50, 279–292.

Gohole, L.S., Overholt, W.A., Khan, Z.R. and Vet, L.E.M. (2003) Role of volatiles emitted by host and non-host plants in the foraging behaviour of *Dentichasmias busseolae*, a pupal parasitoid of the spotted stemborer, *Chilo partellus*. *Entomologia Experimentalis et Applicata* 107, 1–9.

Gold, C.S. (1994) The effects of cropping systems on cassava whiteflies in Colombia: implications for control of African Cassava mosaic virus disease. *African Crop Science Journal* 2, 423–436.

Gómez-Rodríguez, O., Zavaleta-Mejia, E., Gonzalez-Hernandez, V.A., Livera-Munoz, M. and Cardenas-Soriano, E. (2003) Allelopathy and microclimatic modification of intercropping with marigold on tomato early blight disease development. *Field Crops Research* 83, 27–34.

Hamblin, A.P. and Hamblin, J. (1985) Root characteristics of some temperate legume species and varieties on deep free-draining Entisols. *Australian Journal of Agricultural Research* 36, 63–72.

Hassell, M.P. and Goodfray, H.C.J. (1992) The population biology of insect parasitoids. In: Crawley, M.J. (ed.) *Natural Enemies.* Blackwell, The Netherlands, pp. 265–292.

Hauggaard-Nielsen, H., Ambus, P. and Jensen, E.S. (2001) Interspecific competition, N use and interference with weeds in pea–barley intercropping. *Field Crops Research* 70, 101–109.

Hellqvist, S. (1996) Mulching with grass-clippings in cauliflower: effects on yield and brassica root flies *Delia* spp. *International Journal of Pest management* 42, 39–46.

Hogh-Jensen, H. and Schjoerring, J.K. (2000) Below-ground nitrogen transfer between different grassland species: direct quantification by N-15 leaf feeding compared with indirect dilution of soil N-15. *Plant and Soil* 227, 171–183.

Hogh-Jensen, H. and Schjoerring, J.K. (2001) Rhizodeposition of nitrogen by red clover, white clover and ryegrass leys. *Soil Biology and Biochemistry* 33, 439–448.

Holtzer, T.O., Archer, T.L. and Norman, J.M. (1988) Host plant suitability in relation to water stress. In: Heinrichs, E.A. (ed.) *Plant Stress – Insect Interactions.* John Wiley & Sons, New York, pp. 111–138.

Hooks, C.R.R. and Johnson, M.W. (2001) Broccoli growth parameters and level of head infestations in simple and mixed plantings: impact of increased flora diversification. *Annals of Applied Biology* 138, 269–280.

Hooks, C.R.R. and Johnson, M.W. (2003) Impact of agricultural diversification on the insect community of cruciferous crops. *Crop Protection* 22, 223–238.

Hopkins, R.J. and van Loon, J.A. (2001) The effect of host acceptability on oviposition and egg accumulation by the small white butterfly, *Pieris rapae. Physiological Entomology* 26, 149–157.

Horwith, B. (1985) A role for intercropping in modern agriculture. *Bioscience* 35, 286–291.

Hoy, C.W. and Shelton, A.M. (1987) Feeding response of *Artogeia rapae* (*Lepidoptera: Pieridae*) and *Trchoplusia ni* (*Lepidoptera: Noctuidae*) to cabbage leaf age. *Environmental Entomology* 16, 680–682.

Hughes, R.D. (1957) The natural mortality of *Erioischia brassicae* (Bouché) (*Diptera: Anthomyiidae*) during the egg stage of the first generation. *Journal of Animal Ecology* 28, 231–241.

Hussein, M.Y. and Samad, A.N. (1993) Intercropping chilli with maize on brinjal to suppress populations of *Aphis gossypii* Glov., and transmission of chili viruses. *International Journal of Pest Management* 39, 216–222.

Hwang, S.F., Chang, K.F. and Chakravarty, P. (1992) Effects of vesicular–arbuscular mycorrhizal fungi on the development of *Verticillium* and *Fusarium* wilts of alfalfa. *Plant Disease* 76, 239–243.

Imhoff, M.W., Main, C.E., Leonard, K.J. (1981) Effect of temperature, dew period, and age of leaves, spores, and source pustules on germination of bean rust urediospores. *Phytopathology* 71, 577–583.

Jensen, E.S. (1996) Grain yield, symbiotic N-2 fixation and interspecific competition for inorganic N in pea–barley intercrops. *Plant and Soil* 182, 25–38.

Johansen, A. and Jensen, E.S. (1996) Transfer of N and P from intact or decomposing roots of pea to barley interconnected by an arbuscular mycorrhizal fungus. *Soil Biology and Biochemistry* 28, 73–81.

Jones, R.E. (1977) Movement patterns and egg distribution in cabbage butterflies. *Journal of Animal Ecology* 46, 195–212.

Karban, R., Baldwin, I.T., Baxter, K.J., Laue, G. and Felton, G.W. (2000) Communication between plants: induced resistance in wild tobacco plants following clipping of neighboring sagebrush. *Oecologia* 125, 66–71.

Karban, R., Maron, J., Felton, G.W., Ervin, G. and Eichenseer, H. (2003) Herbivore damage to sagebrush induces resistance in wild tobacco: evidence for eavesdropping between plants. *Oikos* 100, 325–332.

Kareiva, P. (1982) Experimental and mathematical analysis of herbivore movement: quantifying the influence of plant spacing and quality on foraging discrimination. *Ecological Monographs* 52, 261–282.

Kareiva, P. (1983) Influence of vegetation texture on herbivore populations: resource concentration and herbivore movement. In: Denno, R.F. and McClure, M.S. (eds) *Variable Plants and Herbivores in Natural and Managed Systems.* Academic Press, New York, pp. 259–289.

Kareiva, P. (1985) Finding and losing host plants by *Phyllotreta*: patch size and surrounding habitat. *Ecology* 66, 1809–1816.

Kareiva, P. (1986) Trivial movement and foraging by crop colonizers. In: Kogan, M. (ed.) *Ecological Theory and Integrated Pest Management Practice*. John Wiley & Sons, New York, pp. 59–82.

Kass, D.C.L. (1978) Polyculture cropping systems: review and analysis. *Cornell International Agriculture Bulletin* 32. Cornell University, Ithaca, New York, pp. 1–69.

Kessmann, H., Staub, T., Hofmann, C., Maetzke, T., Herzog, J., Ward, E., Uknes, S. and Ryals, J. (1994) Induction of systemic acquired disease resistance in plants by chemicals. *Annual Review of Phytopathology* 32, 439–459.

Kinane, J.T. and Lyngkjær, M. (2002) Effect of barley–legume intercrop on disease frequency in an organic farming system. *Plant Protection Science* 38, 227–231.

Klingauf, F.A. (1987a) Host plant finding and acceptance. In: Minks, A.K. and Harrewijn, P. (eds) *Aphids: Their Biology, Natural Enemies and Control*. Elsevier, London, pp. 209–223.

Klingauf, F.A. (1987b) Feeding, adaptation and excretion. In: Minks, A.K. and Harrewijn, P. (eds) *Aphids: Their Biology, Natural Enemies and Control*. Elsevier, London, pp. 225–253.

Kostal, V. (1993) Oogenesis and oviposition in the cabbage root fly, *Delia radicum* (Diptera: Anthomyiidae), influenced by food quality, mating and host-plant availability. *European Journal of Entomology* 90, 137–147.

Kostal, V. and Finch, S. (1994) Influence of background on host-plant selection and subsequent oviposition by the cabbage root fly, *Delia radicum*. *Entomologia Experimentalis et Applicata* 70, 109–118.

Kuc, J. (1995) Phytoalexins, stress metabolism, and disease resistance in plants. *Annual Review of Phytopathology* 33, 275–297.

Kumar, S. and Sugha, S.K. (2000) Role of cultural practices in the management of *Septoria* leaf spot of tomato. *Indian Phytopathology* 53, 105–106.

Langer, V. (1996a) Within-field diversification. A study of crop yield, insect pests and natural enemies in a cabbage–clover system. PhD thesis, The Royal Veterinary and Agricultural University, Copenhagen.

Langer, V. (1996b) Insect–crop interactions in a diversified cropping system: parasitism by *Aleochara bilineata* and *Trybliographa rapae* of the cabbage root fly, *Delia radicum*, on cabbage in the presence of white clover. *Entomologia Experimentalis et Applicata* 80, 365–374.

Larios, J.F. and Moreno, R.A. (1977) Epidemiologia de algunas enfermedades foliares de la yucca en diferentes sistemas de cultivo. II. Roya y muete descendente. *Turrialba* 27, 151–156.

Lennartsson, M. (1988) Effects of organic soil amendments and mixed species cropping on take-all of wheat. In: Allen, P. and Van Dusen, D. (eds) *Proceedings of 6th International IFOAM Scientific Conference*. University of California, Santa Cruz, California, pp. 575–580.

Leonard, K.J. (1969) Factors affecting rates of stem rust increase in mixed plantings of susceptible and resistant oat varieties. *Phytopathology* 18, 1845–1850.

Liu, R.J. (1995) Effect of vesicular arbuscular mycorrhizal fungi on *Verticillium* wilt of cotton. *Mycorrhiza* 5, 293–297.

Michel, V.V., Wang, J.F., Midmore, D.J. and Hartman, G.L. (1997) Effects of intercropping and soil amendment with urea and calcium oxide on the incidence of bacterial wilt of tomato and survival of soil-borne *Pseudomonas solanacearum* in Taiwan. *Plant Pathology* 46, 600–610.

Miller, J.R. and Strickler, K.L. (1984) Finding and accepting host plants. In: Bell, W.J. and Cardé, R.T. (eds) *Chemical Ecology of Insects*. Chapman & Hall, New York, pp. 127–157.

Monteiro, A.A.T., Vieira, C. and da Silva, C.C. (1981) Comportamentode de cultivares de feijao (*Phaseolus vulgaris* L.) na zona da mata de minas gerais. *Revista Ceres* 28, 588–606.

Montes, B.R. and García, L.R. (1997) Efecto de extractos vegetales en la germinación de esporas y en los niveles de dano de *Alternaria solani* en tomate. *Fitopatología* 32, 52–57.

Morales, H. (2002) Pest management in traditional tropical agroecosystems: lessons for pest prevention research and extension. *Integrated Pest Management Reviews* 7, 145–163.

Morales, H. and Perfecto, I. (2000) Traditional knowledge and pest management in the Guatemalan highlands. *Agriculture and Human Values* 17, 49–63.

Morallo-Rejesus, B. and Decena, A. (1982). The activity, isolation and purification of the insecticidal principles from Tages. *Philippine Journal of Crop Science* 7, 31–36.

Moreno, R.A. (1977) Efecto de diferentes sistemas de cultivo sobre la severidad de la mancha angular del frijol (*Phaseolus vulgaris* L.) causada por *Isariopsis griseola* Sacc. *Agronomica Costarricense* 1, 39–42.

Moreno, R.A. and Mora, L.E. (1984) Cropping pattern and soil management influence on plant disease: II. Bean rust epidemiology. *Turrialba* 34, 41–45.

Morris, W.F. and Kareiva, P. (1991) How insect herbivores find suitable host plants: the interplay between random and non-random movement. In: Bernays, E. (ed.) *Insect–Plant Interactions*, Vol. 3. CRC Press, Boca Raton, Florida, pp. 175–208.

Msuku, W.A.B. and Edje, O.T. (1982) Effect of mixed cropping with maize and bean on bean diseases. *Annual Report of the Bean Improvement Co-operative* 26, 16.

Mundt, C.C. (1990) Disease dynamics in agroecosystems. In: Carroll, C.R., Vandermeer, J.H. and Rossett, P.M. (eds) *Agroecology*. McGraw-Hill, New York.

Mundt, C.C. and Leonard, K.J. (1985) Effect of host genotype unit area on epidemic development of crown rust following focal and general inoculations of mixtures of immune and susceptible oat plants. *Phytopathology* 75, 1141–1145.

Ngugi, H.K., King, S.B., Holt, J. and Julian, A.M. (2001) Simultaneous temporal progress of sorghum anthracnose and leaf blight in crop mixtures with disparate patterns. *Phytopathology* 91, 720–729.

Ninkovic, V. and Pettersson, J. (2003) Searching behaviour of the sevenspotted ladybird, *Coccinella septempunctata* – effects of plant–plant odour interaction. *Oikos* 100, 65–70.

Ninkovic, V., Olsson, U. and Pettersson, J. (2002) Mixing barley cultivars affects aphid host plant acceptance in field experiments. *Entomologia Experimentalis et Applicata* 102, 177–182.

Norris, R.F. and Kogan, M. (2000) Interactions between weeds, arthropod pests, and their natural enemies in managed ecosystems. *Weed Science* 48, 94–158.

Palti, J. (1981) *Cultural Practices and Infectious Crop Diseases*. Springer-Verlag, Berlin, pp. 15–42.

Parkhurst, A.M. and Francis, C.A. (1986) Research methods for multiple cropping. In: Francis, C.A. (ed.) *Multiple Cropping Systems*. Macmillan, New York, pp. 285–316.

Perfecto, I. and Vet, L.E.M. (2003) Effect of a nonhost plant on the location behavior of two parasitoids: the tritrophic system of *Cotesia* (*Hymenoptera: Braconidae*), *Pieris rapae* (*Lepidoptera: Pieridae*), and *Brassica oleraceae. Environmental Entomology* 32, 163–174.

Pfleeger, T.G., da Luz, M.A. and Mundt, C.C. (1999) Lack of a synergistic interaction between ozone and wheat leaf rust in wheat swards. *Environmental and Experimental Botany* 41, 195–207.

Picone, C. (2003) Managing mycorrhizae for sustainable agriculture in the tropics. In: Vandermeer, J.H. (ed.) *Tropical Agroecosystems*. CRC Press, Boca Raton, Florida, pp. 95–132.

Plant Directorate (2003) Statistik om økologiske bedrifter 2003 – Autorisation og produktion. Danish Plant Directorate, Lyngby, Denmark.

Potts, M.J. (1990) Influence of intercropping in warm climates on pests and diseases of potato, with special reference to their control. *Field Crops Research* 25, 133–144.

Powell, W., Dean, G.J. and Dewar, A. (1985) The influence of weeds on polyphagous arthropod predators in winter wheat. *Crop Protection* 4, 298–312.

Power, A.G. (1987) Plant community diversity, herbivore movement, and an insect-transmitted disease of maize. *Ecology* 68, 1658–1669.

Purvis, G. and Curry, J.P. (1984) The influence of weeds and farmyard manure on the activity of carabidae and other ground-dwelling arthropods in a sugar beet crop. *Journal of Applied Ecology* 21, 271–283.

Rämert, B. (1995) Intercropping as a pest management strategy against carrot fly (*Psila rosae*). PhD dissertation, Swedish University of Agricultural Sciences, Uppsala, Sweden.

Rämert, B. and Ekbom, B. (1996) Intercropping as a management strategy against carrot rust fly (*Diptera*: *Psilidae*): a test of enemies and resource concentration hypotheses. *Environmental Entomology* 25, 1092–1100.

Rämert, B., Lennartsson, M. and Davies, G. (2002) The use of mixed species cropping to manage pests and diseases – theory and practice. In: Powell *et al.* (eds) *UK Organic Research 2002: Proceedings of the COR Conference*. Aberystwyth, UK, pp. 207–210.

Rao, M.R. (1986) Cereals in multiple cropping. In: Francis, C.A. (ed.) *Multiple Cropping Systems*. Macmillan, New York, pp. 96–132.

Raworth, D.A. (1984) Population dynamics of the cabbage aphid, *Brevicoryne brassicae* (*Homoptera*: *Aphididae*) at Vancouer, British Columbia. V. A simulation model. *Canadian Entomologist* 116, 895–911.

Raymundo, S.A. and Alcazar, J. (1982) Development and spread of late blight, *Phytophthora infestans* (Mont) Debary in sole and intercrop potato plots. *Phytopathology* 72, 962.

Reddy, M.S. and Willey, R.W. (1981) Growth and resource use in an intercrop of pearl millet/groundnut. *Field Crops Research* 4, 13–24.

Root, R.B., (1973) Organization of a plant–arthropod association in simple and diverse habitat: the fauna of collards (*Brassica oleracea*). *Ecological Monographs* 43, 95–124.

Schwartz, M.W., Brigham, C.A., Hoeksema, J.D., Lyons, K.G., Mills, M.H. and van Mantgem, P.J. (2000) Linking biodiversity to ecosystem function: implications for conservation ecology. *Oecologia* 122, 297–305.

Sengooba, T. (1990) Comparison of disease development in beans in pure stand and maize intercrop. *Annual Report of the Bean Improvement Co-operative* 33, 57–58.

Sheenan, W. (1986) Response by specialist and generalist natural enemies to agroecosystem diversification: a selective review. *Environmental Entomology* 15, 456–461.

Soleimani, M.J., Deadman, M.L. and McCartney, H.A. (1996) Splash dispersal of *Pseudocercosporella herpotrichoides* spores in wheat monocrop and wheat–clover bicrop canopies from simulated rain. *Plant Pathology* 45, 1065–1070.

Soria, J., Bazan, R., Pinchinat, A.M., Paez, G., Mateo, N., Moreno, R., Fargas, J. and Forsythe, W. (1975) Investigación sobre sistemas de producción agríola para el pequeno agricultor del trópico. *Turrialba* 25, 283–293.

Speight, M.R. and Lawton, J.H. (1976) The influence of weed cover on the mortality imposed on artificial prey by predatory ground beetles in cereal fields. *Oecologia* 23, 211–233.

Stanton, M.L. (1983) Spatial patterns in the plant community and their effects upon insect search. In: Ahmad, S. (ed.) *Herbivorous Insects: Host-seeking Behavior and Mechanisms*. Academic Press, New York, pp. 125–157.

Stoner, K.A. and Shelton, A.M. (1988a) Effect of planting date and timing of growth stages on damage to cabbage by onion thrips (*Thysanoptera*: *Thripidae*). *Journal of Economic Entomology* 81, 1186–1189.

Stoner, K.A. and Shelton, A.M. (1988b) Influence of variety on abundance and within plant distribution of onion thrips (*Thysanoptera*: *Thripidae*) on cabbage. *Journal of Economic Entomology* 81, 1190–1195.

Sunderland, K. and Samu, F. (2000) Effects of agricultural diversification on the abundance, distribution, and pest control potential of spiders: a review. *Entomologia Experimentalis et Applicata* 95, 1–13.

Theunissen, J. (1997) Application of intercropping in organic agriculture. *Biological Agriculture and Horticulture* 15, 251–259.

Theunissen, J. and Schelling, G. (2000) Undersowing carrots with clover: suppression of carrot rust fly (*Psila rosae*) and cavity spot (*Pythium* sp.) infestation. *Biological Agriculture and Horticulture* 18, 67–76.

Theunissen, J., Booij, C.J.H., Schelling, G. and Noorlander, J. (1992) Intercropping clovers in white cabbage. *IOBC/WPRS Bulletin* 192, 104–114.

Theunissen, J., Booij, C.J.H. and Lotz, L.A.P. (1995) Effects of intercropping white cabbage with clovers on pest infestation and yield. *Entomologia Experimentalis et Applicata* 74, 7–16.

Thresh, J.M. (1982) Cropping practices and virus spread. *Annual Review of Phytopathology* 20, 193–218.

Trenbath, B.R. (1986) Resource use by intercrops. In: Francis, C.A. (ed.) *Multiple Cropping Systems*. Macmillan, New York, pp. 57–81.

Vandermeer, J. (1989) *The Ecology of Intercropping*. Cambridge University Press, Cambridge.

Vandermeer, J., van Noordwijk, M., Anderson, J., Ong, C. and Perfecto, I. (1998) Global change and multi-species agroecosystems: concepts and issues. *Agriculture Ecosystems and Environment* 67, 1–22.

Van Kessel, C., Singleton, P.W. and Hoben, H.J. (1985) Enhanced N-transfer from soybean to maize by vesicular arbuscular mycorrhizal (VAM) fungi. *Plant Physiology* 79, 562–563.

Van Rheenen, H.A., Hasselbach, O.E. and Muigai, S.G.S. (1981) The effect of growing beans together with maize on the incidence bean diseases and pests. *Netherlands Journal of Plant Pathology* 87, 193–199.

Vilich-Meller, V. (1992) Mixed cropping of cereals to suppress plant-diseases and omit pesticide applications. *Biological Agriculture and Horticulture* 8, 299–308.

Walker, T.S., Bais, H.P., Grotewold, E. and Vivanco, J.M. (2003) Root exudation and rhizosphere biology. *Plant Physiology* 132, 44–51.

Way, M.J. and van Emden, H.F. (2000) Integrated pest management in practice – pathways towards successful application. *Crop Protection* 19, 81–103.

Willey, R.W. (1979) Intercropping – its importance and research needs. Part 1. Competition and yield advantages. *Field Crop Abstracts* 32, 1–10.

Wolfe, M.S. (1985) The current status and prospects of multiline cultivars and variety mixtures for disease resistance. *Annual Review of Phytopathology* 23, 251–273.

Wolfe, M.S. and Barrett, J.A. (1980) Can we lead the pathogen astray? *Plant Disease* 64, 148–155.

Yao, M.K., Tweddell, R.J. and Desilets, H. (2002) Effect of two vesicular–arbuscular mycorrhizal fungi on the growth of micropropagated potato plantlets and on the extent of disease caused by *Rhizoctonia solani. Mycorrhiza* 12, 235–242.

Yu, J.Q. (1999) Allelopathic suppression of *Pseudomonas solanacearum* infection of tomato (*Lycopersicon esculentum*) in a tomato–Chinese chive (*Allium tuberosum*) intercropping system. *Journal of Chemical Ecology* 25, 2409–2417.

Zambolim, L. and Schenck, N.C. (1983) Reduction of the effects of pathogenic, root-infecting fungi on soybean by the mycorrhizal fungus, *Glomus mosseae. Phytopathology* 73, 1402–1405.

Zhu, Y.Y., Chen, H.R., Fan, J.H., Wang, Y.Y., Li, Y., Chen, J.B., Fan, J.X., Yang, S.S., Hu, L.P., Leung, H., Mew, T.W., Teng, P.S., Wang, Z.H. and Mundt, C.C. (2000) Genetic diversity and disease control in rice. *Nature* 406, 718–722.

Zinsou, V., Wydra, K., Ahohuendo, B. and Hau, B. (2004) Effect of soil amendments, intercropping and planting time in combination on the severity of cassava bacterial blight and yield in two ecozones of West Africa. *Plant Pathology* 53, 585–595.

6 Ecological Effects of Chemical Control Practices: The Environmental Perspective

R.G. Luttrell

Department of Entomology, University of Arkansas, Fayetteville, AR 72701, USA

Introduction

To fully appreciate the evolution of chemical pest control and powerful new opportunities to manage pests with reduced environmental impact, one must consider the relatively short history of synthetic chemical controls, observed environmental impacts and evolving pest management systems. Chemicals derived from technologies post–First World War gave mankind unprecedented manipulative power over pest populations. For the first time in world history, a little more than 50 years ago, man had curative tools to fight agricultural pests and vectors of human disease. Chemical weed control provided an effective substitute for manual weed control and further advanced mechanized production systems in the latter half of the 20th century. This revolutionized agriculture and impacted social structure through reduced dependence on human labour. Increased agricultural productivity associated with unprecedented levels of pest control, improved crop and animal genetics, availability of fertilizers and mechanized production systems allowed food supplies to increase and standard of living to improve in many regions of the world. This green revolution was driven by a wide array of different chemical technologies. Broad management concepts

soon evolved to deal with both the positive results and the negative impacts of rapid and wide-scale adoption of these amazingly powerful new technologies. The evolution of integrated pest management (IPM) principles is a classic example of moderation and the need to balance short-term benefits and long-term costs of pest control technologies. Benefits from the use of organic pesticides were obvious and highly desired. Unfortunately, overreliance and injudicious use led to problems including lack of profitability, resistant pest populations, disruptions of ecological balance and elevation of nonpests to pest status (Metcalf and Luckmann, 1994). Long-term impacts on the physical environment illustrated expanding scales of ecological impact and an awareness that chemical pest control could have widespread impact on the environment. These broader environmental impacts forced agriculturalists to consider environmental and ecological impacts of pest control practices. A need to moderate the use of chemical controls became acutely obvious.

Classical concepts of pest management and integrated control (Stern *et al.*, 1959; Pedigo, 1996) have emerged as reasonable approaches to this challenge of balancing economic benefits and environmental costs (Horn, 1988; Dent, 1991). Ideas of moderation typified by articulation of economic injury and economic threshold (Stern *et al.*, 1959) required new information about pests, their ecology and assessments of the relative value of using these powerful new chemical control tools. This expanded agricultural sciences and created linkages among biology, economics and sociology that continue to evolve even today. Basic principles of pest management and their close conceptual tie to the impacts of indiscriminate use of chemical control on the environment can be found in many introductory pest management texts (Fry, 1982; Horn, 1988; Camphill and Madden, 1990; Dent, 1991; Pimentel, 1991; Strange, 1993; Metcalf and Luckmann, 1994; Agrios, 1996; Anderson, 1996; Aldrich and Kremer, 1997; Ruberson, 1998; Ross and Lembi, 1999; Zimdahl, 1999; Monaco *et al.*, 2002).

The evolution of pest management concepts and the need for innovative approaches to balance costs and benefits of chemical pest control are by no means complete (Benbrook *et al.*, 1996; Koul *et al.*, 2004). Worldwide changes in agricultural productivity and social infrastructures of agriculture continue to evolve, as do approaches to balance positive and negative aspects of pest control (Benbrook, 2003; CAST, 2003). Exponential growth of human populations and the critical need for new food sources will have dramatic impacts on the world's environment over the next 50 years (Pimentel *et al.*, 1992; Tilman *et al.*, 2001). Tilman *et al.* (2001) suggested that this may be the most critical time in human history and that man's approach to the challenge of feeding the world will have dramatic impacts on the environment and the future of human society.

Actual and perceived benefits of chemical control were enormous in the 20th century, and continue to this day, providing technological opportunities for the future (Pimentel *et al.*, 1992; Koul *et al.*, 2004). Just as the benefits of chemical use in the 20th century had to be balanced by moderation and a consideration of long-term environmental costs, those of the 21st century will face complex and evolving environmental issues (Reynolds, 1997; Yudelman *et al.*, 1998; Tilman *et al.*, 2001; Koul *et al.*, 2004). Less environmentally disruptive, yet highly active, pesticides and rapid worldwide

expansion of genetically modified crops are recent examples of important industry-based technologies (Larson, 1998; NAS, 2000; Holm and Baron, 2002; FAO, 2004; Ware and Whitacre, 2004a, b).

Moderating unintended non-target effects of pest control technologies is still a struggle for human society, but one that seems more manageable given the lessons of the 20th century (Whitten, 1992). IPM remains the most robust approach to efficiently moderate ecological impacts of pesticides on the environment (Cuperus *et al.*, 1997; Kogan, 1998; Koul *et al.*, 2004). Application of these ecologically based principles to vastly different socio-economic regions of the world requires sophisticated management systems for a wide array of different social infrastructures (Mengech *et al.*, 1995; Maredia, 2003; Koul *et al.*, 2004). Predictions for unprecedented human population growth in developing countries and the growing need for increased food complicate worldwide approaches to balance short- and long-term impacts of chemical control practices (Tilman *et al.*, 2001). Future management systems must include a range of temporal (short- and long-term) and spatial (field-by-field or farm, and community, region, nation or world) dimensions in addition to the typical field-by-field surveillance and reaction of traditional IPM. Evolving technologies and more sophisticated management approaches provide opportunities to accelerate the scope of managed chemical pest control, especially when related to environmental issues. A review by Stark and Banks (2003) describes recent efforts to quantify population-level impacts of pesticides on arthropods. From a broad environmental perspective, this is evidence of expanding management approaches and capacity.

The literature on chemical control and the environment is voluminous and beyond the capacity of a single review of chemical control practices. An abundance of literature exists within most pest-related scientific disciplines. There are also important examples of similar management problems with pharmaceuticals and various industrial contaminants of the environment. In agriculture and pest control, academic degree programmes in entomology, plant pathology, weed science and pest management flourished during the latter half of the 20th century because of societal recognition of a need for safer methods of pest control (Bosso, 1987; Horn, 1988; Dent, 1991; Pimentel and Lehman, 1993). Increased flow of technology and management capacity to developing countries remains a critical issue, and perhaps a new challenge for academic programmes. Need for efficient integration of biology, ecology and socio-economics is often cited as a critical issue in contemporary policy decisions. Worldwide approaches to chemical pest control are just now maturing to a point that critical environmental and social issues identified in industrialized nations in the later 20th century can be more broadly considered in developing countries. Extension education programmes and non-government pest management advisers have greatly expanded capacities to assist growers and society with critical management issues and practical decision making (Dent, 1991; Fitzner, 2002). International projects in pest control and maturing management systems in many developing countries are now addressing some of the adverse environmental impacts of pest control (Craig and Camisani-Calzolari, 2001; Karlsson, 2004; Davis, 2005). The US Agency for International Development (USAID) has a long history of economic aid and educational outreach in developing countries. The agency recently announced $34 million in collaborative research support for IPM and sustainable agricultural research (http://www.usaid.gov, http://usinfo.stategov, http://www.ag.vt.edu/ipmcrsp/index.asp). These funds will provide new support for large cooperative projects administered through Virginia Tech University. The projects are designed to link the expertise of land grant universities to people in developing countries. Ecologically based pest management is a stated emphasis of this new initiative, and the proposed programmes intend to advance pest management and land use methods that enhance productivity, increase food security and preserve natural resources. While other aspects of pest management impact on the capacity of these programmes, a continuous need for environmentally benign pest management options fuels worldwide invest-

ment in chemical control and advanced pest management systems (Pimentel and Lehman, 1993; Tilman et al., 2001).

An accelerating need for affordable, wholesome food (Pimentel, 1997; Tilman et al., 2001) complicates efforts to protect the environment and manage pests. But recent advances in technology, especially pest control options derived from biotechnologies and expanded management information systems, offer opportunities to balance pest control needs with minimal environmental impact (Cassman et al., 2003). Concepts of site-specific applications (West et al., 2003) and detailed tracking of pesticide-use patterns (Epstein and Bassein, 2003) are some examples. This is not to suggest that the process will be simple or that other, perhaps currently unrecognized, environmental issues will develop. Certainly, contemporary discussions of pest resistance (Mota-Sanchez et al., 2002; Zalom et al., 2005), food safety (Merrill, 1997; WHO, 1998; Ragsdale, 2000; Reese and Watson, 2000; Shelton et al., 2002) and gene containment (Toenniessen, 2003; Pilson and Prendeville, 2004) depict evolving management issues related to new pest control technologies.

Here we review the current state of environmental impacts of chemical control and consider opportunities to further expand IPM and moderate environmental risks in a worldwide environment of accelerating population growth and increasing demand for food. Examples tend to emphasize insect control on agricultural crops in the USA, as that is our research area, but our intent is a broader conceptual and practical consideration of chemical control as it relates to environmental impacts and day-to-day management decisions across different pest disciplines and worldwide environments. Many outstanding reviews on these subjects are available in the literature. Serious students are encouraged to index these studies through linkages to the more extensive literature cited in our referenced reviews. Electronic sources of information are rapidly expanding worldwide access to information about chemical pest control (Van Dyk, 2000) (Table 6.1).

An emphasis of this collective text is international variability in IPM. Considerable

Table 6.1.   Useful websites on pesticides, pesticide-use patterns and environmental information.

Center for Environmental Health Sciences (University of California–Berkeley) http://ist-socrates.berkely.
    edu/mutugen/center.pubs.html
Consortium for International Crop Protection, IPMnet News: http:www.ipmnet.org
Cornell University, Pesticide Management Application Program: http://pmep.cce.cornell.edu
ECOTOX (USA Environmental Protection Agency): http://www.epa.gov/ecotox/
EPA Office of Pesticide Programs: http://www.epa.gov/pesticides/
EXTOXNET (EXtension TOXicology NETwork) – University of California, Davis, Oregon State University,
    Michigan State University, Cornell University, University of Idaho: http://ace.orst.edu:80/info/extoxnet/
    ghindex.html
FRAC (Fungicide Resistance Action Committee): www.frac.info/frac_body.html
HRAC (Herbicide Resistance Action Committee): www.plantprotection.org/HRAC/
Integrated Pest Management Collaborative Research Support Program (USAID, Virginia Tech University):
    http://www.ag.vt.edu/ipmcrsp/index.asp
International Survey of Herbicide Resistant Weeds: http:www.weedscience.org/in.asp
IRAC (Insecticide Resistance Action Committee): www.irac-online.org
Michigan State University Database of Arthropod Resistance to Pesticides: www.pesticideresistance.org
National Pesticide Information Center: www.npic.orst.edu
National Water Quality Assessment Program, USA Geological Society: http://ca.water.usgs.gov/pnsp/
PAN International (Pesticide Action Network International): http:www.pan-international.org/
PAN Africa (Pesticide Action Network Africa): http:www.pan-africa.sn
PAN Latin America (Pesticide Action Network Latin America): http://www.pan-international.org/
    latinamericaen.html
PAN Asia and Pacific (Pesticide Action Network Asia and Pacific): http://www.pan-panap.net/
PAN NA (Pesticide Action Network North America): http://www.panna.org
PAN UK (Pesticide Action Network United Kingdom): http://www.pan-uk.org/
Pesticide Education Resources (University of Nebraska, Lincoln): http://pested.uni.edu
Pesticide National Synthesis Project, US Geological Survey's Water Quality Assessment Program
    (NAWQA): http://ca.water.usgs.gov/pnsp/index.html
Radcliffe's IPM World Textbook (University of Minnesota): http://ipmworld.umn.edu/
USAID: www.usaid.gov, usinfo.stategov, www.ag.vt.edu/ipmcrsp/index.asp/
World Resources Institute: www.wri.org

All websites were active by 1 March, 2005.

difference exists in the rate of development of environmental criteria for pest management decisions between industrialized and developing countries, although the general model of management maturity and the trend of evolving pest management activities from simple individual pest control to complex optimization of management decisions within a context of broad ecosystem impact are similar. The time span of infrastructure evolution differs between industrialized and developing countries, but may change with enhanced worldwide communication and dispersal of technical information (Ausher *et al.*, 1996). Developing countries may be able to adopt environmentally preferred technologies more efficiently and maintain management-intensive IPM programmes, assuming access to new technologies and management information. Capital-intense systems in the industrialized nations may

be becoming less suitable for management approaches that require routine data collection and reactionary responses. General views of contemporary pest management are considered by others in the collective work of this text and include information about pest management deployment worldwide.

In this chapter, we have attempted a broad examination of the environmental impacts of chemical pest control and associated management options across the major pest groups, especially insects, plant pathogens and weeds. Worldwide variabilities in conceptual and practical approaches to pest control are important considerations. We have approached this primarily as a contrast of the methods used in more industrialized countries to those used in more diversified and agriculturally based systems in developing countries. We do not intend to oversimplify the vast complexity of different issues and approaches that vary significantly within developing and industrialized systems. Excellent reviews of evolving pest management systems in developing countries include those by Kadir and Barlow (1993), Campanhola *et al.* (1995), Mengech *et al.* (1995), Raheja (1995), Zethner (1995), Ausher *et al.* (1996), Yudelman *et al.* (1998), Maredia *et al.* (2000), Yanez *et al.* (2002) and Toenniessen (2003).

General Background of Environmental Issues and Chemical Control

Environmental problems resulting from unilateral reliance on highly effective chemicals were first recognized by ineffective control of targeted pest populations and the emergence of new pests released from natural controls (Horn 1988; Metcalf and Luckmann, 1994). However, within decades of the introduction of organic pesticides scientists began to observe long-term cumulative effects on the physical environment and non-target organisms (e.g. cancer, mutagenicity, biomagnifications and residues in water, soil and air). Societal reaction in the USA collimated in the historic environmental action of the late 1960s and early 1970s, popularized by the seminal publication of Rachel Carson's (1962) *Silent Spring* and the formation of the USA Environmental Protection Agency (USA EPA) (Table 6.2). This was also a time of mandated emphasis for formalized research and education in IPM (Horn, 1988; Cate and Hinkle, 1993; Fitzner, 2002). A concept of moderating pesticide use through a 'treat only when needed' philosophy became the predominant conceptual approach to balancing the benefits and costs of chemical pest control. Stern's articulation of economic injury, economic threshold and integrated control based on sound ecological principles of pest control (Stern *et al.*, 1959) became internationally recognized benchmarks for moderation strategies of pesticide use and founding principles of modern pest management concepts (Kogan, 1998; Koul *et al.*, 2004).

Large multidiscipline research and extension projects were initiated in the USA in the 1970s (Horn, 1988; Fitzner, 2002). International efforts to moderate pesticide use, especially the more toxic and persistent chemicals, and transfer alternative technologies to developing countries were expanded (Dent, 1991; Kadir and Barlow, 1993; Ausher *et al.*, 1996; Maredia *et al.*, 2000; Maredia, 2003; Koul *et al.*, 2004). In the 35 years since the creation of the USA EPA, tremendous environmental gains have been made and the agency has become a central link in the development, registration and availability of chemical control technologies. Regulatory agencies focused on pesticide use and environmental protection exist around the world, especially in the industrialized nations (Table 6.2), but regulatory oversight is still lacking in some of the developing countries (PAN UK, 1996; WHO, 1996; Huan and Ang, 2000). Environmental concerns often lag behind the more immediate economic problems of pest control and the economic agenda of profitable agricultural export. Differences in pest problems and socio-economic status still exist and require different approaches to pesticide regulation in different regions. For example, the valued use of DDT as a persistent and non-acutely toxic option for control of malaria vectors is still a topic of intense discussion (Wandiga, 2001; Tren and Bate, 2004). New pesticides with more specific modes of action and reduced spectrums of activity (Grafton-Cardwell *et al.*, 2005) may be more suited for large-scale agricultural systems utilizing a vast arsenal of chemical control options. Small, resource-limited farmers in developing countries may be forced to use older products, especially those with broader spectrums of activity. Similarly, use of more environmentally acceptable pesticides linked to new biotechnologies may be less accessible in some regions of the world. Recent trends in development of transgenic cottons expressing endotoxin proteins of *Bacillus thuringiensis* (Bt-cotton) (Altman and Watanabe, 1995; Ismael *et al.*, 2002; James, 2002; Huang *et al.*, 2003) indicate that the technology is beginning to reach small farmers in developing countries who

Table 6.2. Evolution of environmental regulations and technologies: an overview of selected historical events in chemical pest control over the past two centuries.

Pre-1800s

Pesticides – Numerous reports of botanical and non-selective use of inorganic pesticides, oils, protective
structures, sulphur, mercury, arsencial compounds
Regulations/developments
Agricultural revolution in Europe
Historical use of pest controls based on pest biology and evolved experience

1800–1900

Pesticides – Increased reports of botanical and inorganic pesticides including arsenic, rotenone, nicotine,
soap, turpentine, sulphur, mercuric chloride, pyrethrum, quassia, whale-oil, derris, lime-sulphur, paris
green, petroleum emulsions, hydrogen cyanide, Bordeaux mixture, fish oil, carbolineum, lead arsenate,
dinitrophenol, copper sulphate citronella
Regulations/developments
World population is 1.65 billion (1900)
Bordeaux mixture controls powdery mildew in France
DDT synthesized, insecticidal activity unknown
First texts solely devoted to pest control include descriptions of chemical, biological, cultural and varietal
approaches to pest control
Grape phylloxera control in Europe with grafted resistant rootstock
Successful biological control of cottony cushion scale with imported natural enemies

1900–1930

Pesticides – Reported use of lime-sulphur, lubricating oil, nicotine sulphate, derris, zinc arsenate, calcium
cyanide, cryolite, selenium compounds, ethylene dichloride, pyrethrum, ethyl oxide, nicotine tannate,
calcium arsenate, organomercurials for seed treatments, activity of *Bacillus thuringiensis* against
insects discovered
Regulations/developments
American Phytopathological Society formed
American Society of Agronomy formed
Entomological Society of America formed
First aerial application of pesticides in the USA
Resistance to lime-sulphur reported in San Jose scale (1914)
USA Federal Food, Drug and Cosmetic Act (Pure Food Law)
USA Federal Insecticide Act

1930–1940

Pesticides – Thiram (first organic sulphur fungicide), methyl bromide, ethylene, acetylene, nicotine ben-
tonite, pentachlorophenol, TEPP, *B. thuringiensis*, DNOC, DDT
Regulations/developments
Mueller discovers insecticidal properties of DDT
USA Federal Pure Food Law amended to include pesticides and establish tolerances in food (Food, Drug
and Cosmetic Act)
Commercial *B. thuringiensis* products used in France

1940–1950

Pesticides – BHC, 2,4-D, DDT (in the USA), zineb, 2,4,5-T, ammonium sulphonate, chlordane, propham,
parathion, toxaphene, aldrin, dieldrin, dicarboximide fungicides, allethrin and malathion
Regulations/developments
World population is 2.52 billion (1950)
First insect resistance to DDT (housefly in Sweden 1946)
Lindane imported into Kenya
USA Federal Insecticide, Fungicide and Rodenticide Act (FIFRA)
USA Federal Drug Administration established 100:1 safety factor for pesticide residue tolerances in food

Table 6.2. *Continued*

1950–1960

Pesticides – First carbamate insecticides: carbaryl, diazionon and azinphosmethyl; first animal systemic organophosphates: DEET, carbaryl, gibberellic acid, atrazine, paraquat, triflurin, malathion, maneb and diazinon discovered

Regulations/developments

World population is 3 billion (1960)

DDT, dieldrin and toxaphene imported into Kenya

First reported observations of herbicide resistance

First commercial *B. thuringiensis* formulations used in the USA

India Prevention of Food Adulteration Act

Korean Pesticide Management Law

USA FIFRA amended to include all pesticides

USA Food Additives Amendment to FDCA prohibited carcinogens (Delaney Clause)

USA Food and Drug Administration (FDA) embargoed cranberries for pesticide residues

USA Miller Amendment to FDCA limited pesticide residues and required pre-testing of pesticides

1960–1970

Pesticides – Mancozeb, chlorophacinone, slow-release household fumigants, thiabendazole, aldicarb, carboxin, chlordimeform, benomyl, tetramethrin, resmethrin, bioresmethrin, *B. thuringiensis* labels expanded, trifluralin introduced, methomyl

Regulations/developments

World population is 3.7 billion (1970)

DDT use restricted in some US states

FAO/WHO Codex Alimentarius Commission (food protection standards)

India Insecticide Act

Rachel Carson publishes *Silent Spring*

United States Department of Agriculture (USDA) endorses policy of preferred non-persistent pesticides

USA Endangered Species Act

USA Federal Aviation Regulations regulated aerial applications

USA National Environmental Policy Act (NEPA)

1970–1980

Pesticides – Glyphosate, encapsulated methyl parathion, permethrin, Heliothis NPV, insecticide-impregnated tape for household use, methoprene, several photostable pyrethroid insecticides, triallate, ficam, bentazon and benzamide

Regulations/developments

B. thuringiensis activity against flies and beetles recognized from different subspecies

Canada Pest Control Products Act

Cancellation of aldrin and dieldrin use patterns in the USA

CGIAR Consultative Group on International Agricultural Research formed

China Environmental Protection Law

DDT use restricted in Canada

Gossyplure, an insect pheromone, registered for pest control use

Pakistan Agricultural Pesticide Rules

Pakistan Agricultural Pesticides Ordinances

Philippines Fertilizer and Pesticide Authority

Stockholm Conference on the Human Environment

Thailand Pesticide registration

UK Food and Environmental Protection Act

USA Environmental Protection Agency created

USA EPA cancels or suspends use of DDT, aldrin and dieldrin (except as temiticides) and most mercury compounds, e.g. DBCP, Mirex, 2,4,5-T and silvex

USA EPA concludes first RPAR of pesticide (chlorobenzilate)

USA EPA issues Rebuttable Presumption Against Registration (RPAR) for strychnine, endrin, Kepone, 1080 and BHC

USA EPA sets standards for worker re-entry

Continued

Table 6.2. *Continued*

USA Federal Environmental Pesticide Control Act
USA Federal Pesticide Act
USA FIFRA amended to improve pesticide registration process
USA Toxic Substances Control Act
Wide-scale cancellation of all non-medical uses of DDT

1980–1990
Pesticides – Benzoylureas, bromoxynil, dicamba, metribuzin, diclofop-methyl, metalaxyl, clofentazine, first
    group 2 ALS/AHAS inhibitor herbicides, metolachlor
Regulations/developments
World population is 5 billion (1987)
Bt transgenic plants produced
Canada adopts new Environmental Protection Act
Canada removes alachlor from market
Center for International Environmental Law
Denmark and Sweden make IPM national policy
FAO International Code of Conduct on the Distribution and Use of Pesticides
Germany makes IPM national policy
Indonesia observes major success with IPM in rice systems
Indonesia Pesticide Control Committee established
North American Commission for Environmental Cooperation
Pesticide Action Network formed from International Organization of Consumer Unions
Pesticide Action Network North America formed
Pesticide plant accident in Bhopal, India kills thousands
Remaining uses of DDT cancelled in Canada
Resistance first reported to *Bacillus thuringiensis* in *Plodia interpunctella*
Resistance to Bt endotoxin expressed in transgenic plants reported
UK Control of Pesticide Regulations
UK Food and Environmental Protection Act
USA Comprehensive Environmental Response, Compensation and Liability Act (Superfund)
USA controversial removal of daminozide from apple market
USA EPA cancels use of EDB, endrin, toxaphene, dinoseb, chlordane and heptachlor
USA EPA Endangered Species Protection Program (ESPP)
USA FIFRA Amendment established timetable for reregistration of existing pesticides
USA Occupational and Safety Housing Authority (OSHA) requires employers to provide Material Safety
    Data Sheets (MSDS) to employee
World Bank Update of Operational Manual (Operational Policies: Pest Management)
World Commission on Environmental Development

1990–2000
Pesticides – Imidacloprid and other chloronicotinyls, spinosad, pyridaben, fipronil, emamectin benzo-
    ate, thiamethoxam, triazamate, indoxacarb, dimethomorph, cymoxanil, actiguard, kresoxim-methyl,
    azoxystrobin, imizathepyr, nicosulphuron, methysulfuron-methyl, tribenuron methyl, fenoxaprop-ethyl,
    tralkoxydim, clodinaflop-proparlgyl, myclobutanil, flusilazole, clopyralid, glufosinate ammonium, her-
    bicide-tolerant crops glyphosate, glufosinate, imizathepyr, propamocarb, dimethomorph, cymoxanil;
    stobilurin, phenylpyrroles, anilinopyrimidines, phenoxyquinoline fungicides; herbicide-resistant crops
    include bromoxynil-cotton, sethoxydim-maize, glufosinate-canola, glufosinate-maize, glyphosate-soy-
    bean, glyphosate-canola, glyphosate-cotton, glyphosate-maize, imidazolinone-maize, imidazolinone-
    canola, sulphonylurea-soybean, triazine-canola, Bt-cotton, Bt-maize, Bt-potato, Bt-tomato, Bt-tobacco
    and Bt-canola
Regulations/developments
World population is 5.3 billion (1990)
Australia National Industrial Chemicals Notification and Assessment Scheme
Australia Pesticides and Veterinary Medicines Authority
International Organization for Standardization introduces ISO 14000 standard for Environmental Manage-
    ment Studies

Table 6.2. *Continued*

IOMC (Inter-organization Program for the Sound Management of Chemicals) including UNEP, ILO, FAW, WHO, UNIDO and OECD

Laos instituted a pesticide regulation to control imports

Myanmar Pesticide Law

Myanmar Pesticide Registration Board

The Netherlands makes IPM implicit in Crop Protection Plan

North American Free Trade Agreement

OECD (Organization for Economic Cooperation and Development) Pesticide Program

Thailand Hazardous Substances Act

UK Plant Protection Regulation

United Nations Conference on Environment and Development, Rio de Janeiro

USA Food Quality Protection Act

USA Food, Agriculture, Conservation and Trade Act (Farm Bill) amended FIFRA

Vietnam Pesticide Registration Committee

WHO estimates that 95% of member countries have some legislation for pesticide registration, 13% have no laws for vector control, 16% have no laws to govern pest control business, 28% have no laws governing household use of pesticides and 22% have not adopted WHO/FAO guidelines for public health pesticide use

2000–2005

Pesticides – Broxoxynil-canola, glufosinate-cotton, imidazolinone-wheat and imidazolinone-rice
Regulations/developments

World population is 6 billion (2000)

Consortium on Persistent Organic Pollutants set goals to eliminate a dozen of world's most dangerous pesticides

FAO/UN Revised International Code for Distribution and Use of Pesticide revised

Intergovernmental Forum on Chemical Safety

US Geological Survey maps pesticide-use patterns in USA watersheds

Organophosphorous insecticide use restricted in USA

Abstracted from Crop Life International: http://www.croplife.org, Global Herbicide Directory: http://www.agranova.co.uk/ herbhist.htm, PAN UK: http://www.pan-uk.org/pestnews/pn29pll.htm

can greatly benefit from reduced reliance on dangerous foliar sprays.

In spite of the differences in worldwide pest control needs and regulatory capacities, significant progress has been made in removing many of the most dangerous and highly persistent chemical controls of the mid-1900s (Table 6.3). Major international efforts (by FAO and WHO) have resulted in the International Code of Conduct on the Distribution and Use of Pesticides (FAO, 2003) that provides standards of conduct for governments and the pesticide industry on the import, handling and use of pesticides. These codes have become an important benchmark for international commerce and finance. An effective international effort was the development of the Prior Import Consent (PIC) policy (Table 6.2), which required

exporting countries to provide essential information about the toxicity and handling of toxic chemicals. International networks of information like Pesticide Action Network (PAN) (Table 6.1) provide technical information and links to vast databases relevant to pesticides and environmental impacts of pesticides. Specific agendas of many other environmental groups focus on the worldwide removal and banning of pesticides. A group of 18 target pesticides labelled 'the dirty dozen' became a focal point of these environmental social agendas in 2000. They include aldicarb, aldrin, chlordane, chlordimeform, DBEP, DDT, dieldrin, EDB, endrin, heptachlor, hexachloridebenzene, lindane, paraquat, methyl parathion, pentachlorophenol, toxaphene and 2,4,5-T. Many countries and international groups have systematically eliminated or banned these and other equally dangerous pesticides, which is a strong indication of the growing relevance of environmental concerns on pest management systems worldwide, and the importance of societal investment in regulatory activities (Table 6.3). Tait (2000) and Karlsson (2004) discuss the impact of these international agendas on differing needs of developing countries and associated availabilities of new technology. Table 6.3 includes a brief summary of current registrations for a few of these pesticides and comparative registrations for emerging, more environment-friendly replacement chemistries.

Table 6.2 provides an insight into the historical development of regulatory agencies and the evolving concepts of ecologically based environmental pest management. This is only a selected list of historical developments intended to illustrate trends in worldwide reaction to environmental problems related to chemical pest control. Many other worldwide and country-specific events are relevant, including those in non-English literature. The electronic databases referenced in Table 6.1 provide additional information, and interested students can explore information for specific countries or continents. For example, Sweden was the first nation to have a formal EPA. Pimentel (1997) recognized Sweden's efforts to balance pesticide use and protect the environment. He indicated that Sweden had reduced pesticide dependence by 68% without reducing crop yields or quality, and that pesticide poisonings had been reduced by 77%.

IPM and concepts of moderating pesticide use through sound and ecologically based pest management are adopted and accepted worldwide with generalized terminology, and communicate a range of different ideas and concepts to different sections of society (Dent, 1991; Cuperus et al., 1997; Kogan, 1998; Koul et al., 2004). Many examples of successful deployment of IPM with resulting reductions in pesticide use are reported in the literature (Metcalf and Luckmann, 1994; Ruberson, 1998). Recent dramatic examples are those in developing countries (Mengech et al., 1995; Maredia, 2003; Koul et al., 2004). The general model

of evolving management from sole focus on pest control to broader concerns for the environment is similar in industrialized and developing countries (Table 6.2). Evolution of these more holistic management systems continues with additional and complex detail being continually added to the decision process (Dent, 1991; Cuperus et al., 1997; Kogan, 1998; Koul et al., 2004), even in the more highly evolved systems of industrialized nations. As indicated above, an important and critical step in this evolving process is the creation of public sector regulatory groups that register and monitor pesticide use, consider long-term environmental impacts and protect the public from adverse non-target effects of pesticide misuse. Certainly, the evolution of the USA EPA is an example of the dynamic nature of this process and the potential impact of regulatory action on chemical control practices. State and local agencies also work within the broader national agenda to hone environmental policies for local needs and values. The California system of regulating and tracking pesticide use is among the most advanced in the world (Epstein and Bassein, 2003). In many ways the USA EPA and other sophisticated regulatory agencies are successful in prioritizing environmental concerns for chemical control practices. Without this regulatory framework, it would be difficult to further sophisticate the environmental use of pesticides around the world. The history of pesticide regulations and subsequent technological developments is given in Table 6.2, which provides a perspective of the importance of public sector interest, especially when viewed over the entire 35–40 years of public regulatory agencies providing oversight to pesticide use. Recent advances in pest control technologies developed by large multinational companies within the framework of stringent environmental oversight, e.g. transgenic crops and novel, more selective pesticides, exemplify man's capacity to deal with dynamic pest problems and a range of different environmental concerns. This process is still evolving and the complexity of changing socio-economic variables and evolving technologies defies simple description.

Table 6.3. Status of worldwide registration of selected pesticides. (Abstracted from PAN Pesticide Database: http://www.pesticideinfo.org/Index.html, February 2005.)

DDT

Banned or restricted in 56 countries. Illegal to import in 102 countries. Limited use, primarily for vector control, in Bhutan, India, Nepal, Philippines, Vietnam, Ethiopia, Guinea, Kenya, Madagascar, Mauritius, Sudan, Tanzania, Boliva, Brazil, Mexico and Venezuela. No reported legal use in Europe, the Middle East, or North America.

2,4,5-T

Banned or restricted in 33 countries. Illegal to import in 69 countries. Limited use, primarily for road brush control in Tanzania, Japan, European Union, Brazil and Suriname.

Paraquat

Banned or restricted in 10 countries. No import restrictions. Banned or restricted in Togo, Indonesia, Austria, Denmark, Finland, Germany, Hungary, Slovenia, Sweden and Kuwait. Registered uses in North America and Europe.

Methyl parathion

Banned or restricted in 19 countries. Illegal to import in 43 countries. Registered uses, usually for agricultural pests, in Algeria, Republic of Congo, Madagascar, South Africa, Sudan, Australia, China, India, New Zealand, Pakistan, Philippines, Thailand, Cyprus, European Union, German, Hungary, the Netherlands, Switzerland, Turkey, Brazil, Colombia, Costa Rica, Suriname and the USA.

EDB (ethylene dibromide)

Banned or restricted in 36 countries. Illegal to import in 91 countries. Limited use acceptable in South Africa, Tanzania, Zimbabwe, Cook Islands, Fiji, India, Japan, New Zealand, Barbados and Brazil. No legal use in North America.

Dieldrin

Banned or restricted in 51 countries. Illegal to import in 99 countries. Limited use, primarily for termites or vector control, in Ethiopia, Libya, Sudan, Tanzania, Uganda, Zambia, Zimbabwe, India, Napa, Sri Lanka, Brazil and Venezuela. No legal use in North America or Europe.

Glyphosate

Not banned or restricted in any country. No import restrictions. Listed as registered for use in 25 countries (Burkina Faso, Cameroon, Cape Verde, Chad, Gambia, Guinea-Bissau, Madagascar, Mali, Mauritania, Niger, Senegal, South Africa, Tanzania, Uganda, Australia, New Zealand, Philippines, Denmark, Germany, Hungary, the Netherlands, Portugal, the UK, Canada and the USA).

Spinosad

Not banned or restricted in any country. No import restrictions. Listed as registered for use in seven countries (South Africa, Australia, New Zealand, Philippines, Denmark, Canada and the USA)

Carbaryl

Banned or restricted in four countries (Angola, Austria, Germany and Sweden). No import restrictions. Listed as registered in 12 countries (Madagascar, South Africa, Tanzania, Australia, New Zealand, India, Philippines, Hungary, Portugal, the UK, Canada and the USA).

Aldicarb

Banned or restricted in 13 countries. No import restrictions. Listed as registered for use in 11 countries (South Africa, Sudan, Australia, India, Germany, Hungary, the Netherlands, Portugal, the UK, Argentina and the USA).

An extensive list of recent regulatory actions on most chemical pesticides can be obtained at the PAN website (http://www.pesticideinfo.org/Index.html).

As chemical controls have positive and negative impacts, society must consider the benefits and costs of regulation (Table 6.2). This is a very complex undertaking, especially when one considers the differing socio-economic characteristics of worldwide populations. Considerable debate is ongoing on the relative importance of environmental preservation, expanded safe and wholesome food supplies, and development of new technologies through the scientific capacities of private industry. Contemporary discussions about resistance management and mandated refugia are examples (James, 2002; Benbrook, 2003), as are the continued need for DDT and other persistent chemicals for vector control in developing countries (Krattiger, 1997; Hemingway and Rauson, 2000; Tren and Bate, 2004). The maturity of these debates evidenced by worldwide dialogue and various well-organized and focused advocacy groups is illustrative of the increased relevance of environmental issues on chemical pest control. Cate and Hinkle (1993) describe the importance of environmental lobbyists and special interest groups on governmental policies, pesticide regulations and research agendas in the USA. Tait (2000, 2001) presents examples of differing social concerns and practical policy perspectives on the relative importance of ecological issues to pest control practices.

Regulatory action is an important component of evolving chemical pest controls, especially as related to environmental impacts and deployment of new technologies, but it is not the only factor governing environmental use of chemical pest controls. Scientific principles and ecological realities have been communicated to the world through the umbrella concept of IPM. IPM has become official national policy in many nations (Table 6.2), and major international finance structures like the World Bank often require established IPM practices for participation in various programmes (Tait, 2000, 2001; Karlsson, 2004; Davis, 2005). The recent investment of USAID in programmes to expand ecologically based pest management in developing countries (http://www.usaid.gov, http://www.ag.vt.edu/ipmcrsp/) is an example of the worldwide importance of pest management and sustainable agriculture. Creating a sustainable and viable food supply with minimum adverse environmental impacts of pest control and other agricultural practices is a difficult and globally recognized problem. Concepts of IPM remain the most viable paradigm for moderating the use of unsustainable pest control technologies, even in the current environment of exciting new transgenic crops that provide season-long, constitutive expression of insecticidal proteins and encourage wide-scale spraying of herbicides on herbicide-resistant crops. To couch our discussion in a broader framework and provide benchmarks for a consideration of evolving management practices, we briefly review the historical costs and benefits of chemical pest control.

Benefits and costs of chemical pest control

Pesticides account for about 4% of total US farm production costs, resulting in 3- to 4- fold yield increases, i.e. protection from losses (NAS, 2000). In addition, pesticide use increases farming efficiency and availability of fruits and vegetables, contributes to the supply of low-cost food and fibre for consumers, improves food quality and soil conservation, decreases losses in transport and storage and ensures a stable and predictable food supply. Indeed, the agricultural benefits of chemical pest control are huge. Benefits associated with vector control and elimination of human disease are not considered specifically in this review, but may be of even greater importance to human society.

Agricultural benefits from chemical pest control are largely associated with increased food and fibre production. This is extremely important worldwide as human civilization copes with expanding human populations and increasing demands for food. World populations increased 3.7-fold in the 20th century (Tilman et al., 2001), with projected populations at 7.5 billion by 2020 and 9–10 billion by 2050. In 1991, daily worldwide population growth was estimated at about 250,000 (Pimentel, 1991).

Pimentel (1997) indicated that about 2 million people were malnourished and per capita resources of food, wealth and agricultural potential were highly variable among world populations. He indicated that about 0.6 ha of cropland was available per capita in the USA compared with 0.3 ha of cropland per capita worldwide. Per capita worth increased 4.6-fold in the 20th century. It is expected to increase 2.4-fold over current values by 2050 (Tilman *et al.*, 2001). In the USA, agricultural productivity grew 230% from 1943 to 1986 (Szmedra, 1991). Similar increases will be needed worldwide to address increasing demands for food.

Pimentel (1997) estimated that pests destroy 37% of all potential crops in the USA, even with pesticide use and current IPM programmes. About 13% is due to insects, 12% to plant diseases and 12% to weeds. Worldwide losses due to pests are similar, but postharvest losses may be higher (20% or more) than those in the USA (9%) due to lack of storage facilities, warmer climates and inefficient transportation and processing capacity. Pesticide use generally results in substantial economic returns at the farm level. Pimentel's (1997) estimates suggest $4 returns for each $1 invested in pest control. Total investment in pesticide purchases was about $10 billion in the USA in 1997, and about $40 billion worldwide. About 34% of total pesticide use was in the USA, and about 45% was in Europe. Worldwide use of pesticide was estimated at 3 billion kg in 1997 (Pimentel, 1997), which included 1500 active ingredients (Yanez *et al.*, 2002).

While the benefits of pesticide use are enormous, they are not without real costs to society and the environment. Pimentel *et al.* (1992) estimated that indirect impacts of pesticide use on the environment and economy in the USA might cost more than $8 billion per year, an amount that is about 50% of the total estimated economic returns from pesticide use (~$16 billion per year). Considering the significant economic and social costs of pesticide use, Pimentel (1991) reviewed options for a more efficient system and postulated that it was possible to maintain current per capita levels of food supplies until 2011 with careful management of land, water, energy and human resources. He also suggested that the economic costs of pesticide use to the USA economy could be reduced through moderation and more judicious use of pesticides. He indicated that it was technologically feasible to reduce pesticide use in the USA by 35–50% without reducing crop yields. The suggested reduction in pesticide use was associated with more efficient application and timing of application, and not the elimination of chemical controls directed at economically damaging pest populations. Estimating the economic impacts of pesticide use on human health and the societal investments in health care is extremely difficult. Kolcum *et al.* (2003) provide a detailed outline of the chronic health effects of pesticides. Pimentel and Lehman (1993) considered the health and environmental problems of pesticide use relative to increased agricultural productivity and raised various ethical, economic, environmental and health issues related to pesticides and evolving pest management practices. Edwards (1993) reviewed the impacts of pesticides on the environment including acute effects on living organisms, indirect effects on living organisms, broader effects on the ecosystem and contamination of the environment. The work included a detailed bibliography to many of the classical works on environmental effects of pesticide use, and considered various technical approaches to minimizing environmental effects through legislative and political activities. Improved pesticides with lower mammalian toxicities, improved pesticide application methods, biological control, IPM and sustainable agriculture were considered the major technological options for reduced environmental effects. Legislative and political activities included pesticide registration and government control, political activities and the growth of private interest groups interested in environmental conservation and protection. All were considered essential components in protecting the environment and moderating the use, or misuse, of pesticides around the world.

Pesticides are acute poisons. PAN UK (1996) indicated that China produced and used 8 million knapsack sprayers in 1995, many for the application of acutely toxic insecticides. Insecticides accounted for 74% of China's total pesticide use in 1995.

Huan and Ang (2000) described problems with unsafe use of pesticides in Vietnam. Eddleston *et al.* (2002) called for an international list of less toxic pesticides to reduce human poisoning in the developing world. Yanez *et al.* (2002) presented an overview of human health issues in developing countries associated with chemical pest control. Fleming *et al.* (2003), Kolcum *et al.* (2003) and Alavanja *et al.* (2004) provided perspectives on chronic human health effects in industrialized nations.

Concerns for food safety and pure water have major influence on evolving pest management programmes worldwide. FDA (2002) gives an example of the type of public information available on pesticide residues and the breadth of pesticide residue monitoring in the USA. Similar programmes exist in other countries and the linkage between FDA and other regulatory groups is briefly considered in the 2002 FDA report. In 2002, more than 4600 samples were analysed from food shipments into the USA from 100 countries. An additional 2100 samples were analysed from domestic food supplies. No residues were found in 72% of the domestic samples and 83% of the import samples. Less than 1% of the domestic samples and about 4% of the imported samples had pesticide residues at violative levels (FDA, 2002).

Patterns of chemical pesticide use

Agriculture uses 85% of all pesticides worldwide, but the relative proportions of different types of pesticides vary widely with crop structure and socio-economic status. Growing markets of pesticides are being observed in non-agricultural sectors of the industrialized nations. Donaldson *et al.* (2002) provide a detailed summary of pesticide-use patterns in USA markets in 1998 and 1999 with reference to worldwide trade. Of the total worldwide expenditures for pesticides ($33,593 million), 44% was for herbicides, 27% for insecticides, 20% for fungicides and 9% for other pesticide groups. The USA market accounted for 43%, 33%, 14% and 26%, of the worldwide markets for herbicides, insecticides, fungicides and other pesticides, respectively. About one-third of all worldwide expenditure for pesticide was in the USA. Similar trends in data for active ingredients were evident. Total pesticide use in the world in 1999 was estimated at 2578 million kg of active ingredients. About 36% was herbicide, 25% insecticide, 10% fungicide and 29% was other pesticide groups. About 4% of the total USA farm production expenditure was for pesticides, and about 80% of all herbicide use in the USA was for agriculture. Industrial and governmental uses accounted for 10%, and home and garden use accounted for 10%. Similarly, 74% of all US insecticide use was for agriculture, 15% for industrial and government use and 11% was for home and garden. Agriculture used 82% of the total fungicides in the USA, 17% was for industrial and government use and 13% for home and garden markets. Of the 25 most common active ingredients used in 1999, 14 were herbicides, 3 insecticides and 2 were fungicides. The remaining groups were fumigants and plant growth regulators. Atrazine and glyphosate were the most commonly used herbicides. Among the active ingredients used, malathion, chlorpyrifos and tebufos were the most commonly used insecticides. Chlorothalonil and mancozeb were the most commonly used fungicides. In the home and garden markets, six of the most common active ingredients were herbicides, with 2,4-D and glyphosate being the most common. The remaining four most common pesticides were the organophosphorous and carbamate insecticides: diazinon, chlorpyrifos, carbaryl and malathion. Illustrative of evolving regulatory actions, all non-agricultural uses of diazinon were cancelled in the USA in 2004 – an effort consistent with the 1996 USA Food Quality Protection Act to reduce the risks of pesticide exposure to children.

Organophosphorous insecticides are among the most acutely toxic pesticides used, although they are generally less persistent and do not exhibit the accumulated build-ups typical of DDT and chlorinated hydrocarbons. Targeted reduction of organophosphorous insecticide is a result of the 1996 USA Food Quality Protection Act (*California Agriculture*, January/

February 2005, Vol. 59), and threatens some agricultural production systems that rely on these broad-spectrum insecticides. Amounts of organophosphorous insecticides used in the USA were rather stable during the 1990s, and accounted for 72% of all insecticide use. Among the active ingredients, malathion and chlorpyrifos were the most commonly used insecticides in the USA in the 1990s. Of 104 million households in the USA, 77 million reported some use of pesticide. The most commonly used pesticide in the household environment was insecticide (58 million households). Donaldson *et al.* (2002) indicated that there were 18 major basic producers of pesticide in the USA and several hundred major formulators. There were also more than 33,000 commercial pest control firms, 800,000 private certified applicators and 380,000 commercially certified applicators.

Szmedra (1991) indicated that agricultural productivity in the USA grew 230% from 1947 to 1986. He attributed this increase in productivity to the expanded use of agricultural chemicals that grew eightfold during the period and increased mechanization. The linkage between chemical control and expanded mechanization is an important point that cautions against simplistic categorization of pesticide impacts, especially as related to the use of fossil fuels. Current benefits of herbicide-resistant crops are closely linked to conservative tillage and reduced use of mechanized weed control (Madsen and Sandoe, 2005). There are probably many other social and technological factors intertwined in the evolution of chemical control and perceived costs and benefits. Szmedra (1991) reported that US farmers spent about $4.8 billion on chemical pesticides in 1987. This accounted for 28% of total worldwide expenditures. Of this, US expenditures for herbicides, insecticides and fungicides were 42%, 19% and 15%, respectively. In 1974, US expenditures on pesticides accounted for 34% of total worldwide expenditures. Most of the major field crops in the USA were treated with herbicide as early as the late 1980s.

Hilstrom (2002) indicated that approximately 7000 chemicals are registered for use in Canada. In 2000–2001, 22 minor-use registrations were approved in Canada, 18 for food use and 4 for non-food use. During the same period, 1200 minor uses were approved in the USA, with 500 for food and 700 for non-food use. Hilstrom (2002) suggested that the Canadian Pest Management Regulatory Agency was slower to register new products than similar regulatory groups in the USA and Australia, which is an indication of more stringent environmental regulations in Canada. Further search of information indicated that 6000–7000 pesticide products are registered for use in Canada, and that 50 million kg of herbicides, insecticides and fungicides is used each year. About 91% of the total pesticide use in Canada is for agriculture. Canadian farmers account for only 3% of the worldwide total pesticide use.

Epstein and Bassein (2003) studied the use patterns of foliar-applied pesticides in California, especially fungicides and insecticides applied to vegetables, fruits and nuts from 1993 to 2000. Their intent was to explore options to reduce pesticide use and risks. They reported that there was no obvious trend in decreased use of materials for plant disease control in the 1990s. The more toxic organophosphorous insecticides used to control disease vectors were being readily replaced with pyrethroid insecticides when possible, and only limited use of microbial biocontrol agents was observed. They further suggested that major advantages often reported for replacing calendar spray programmes with environmentally driven programmes assume incorrectly that a majority of farmers are using calendar spray programmes.

The practice of IPM in tropical and subtropical Africa from 1970 to 1990 was reviewed by Zethner (1995). Farming practices in tropical Africa are typically associated with small landholdings of one to a few hectares. Farmers' knowledge of new productive practices has increased over the last two decades, including that related to pest management, and farmers have long adopted sound cultural practices that are consistent with IPM programmes, especially the use of optimum seeding rates and managed time of crop production. The use of pesticides was reportedly growing faster in Africa than in other regions of the world because of the immediate direct benefits from pest control. Less than half of the African countries appear to have legislation to regulate pesticide usage and imports. Most African countries comply with the Food and Agriculture Organization (FAO) Codes of Conduct (FAO, 2003), and there is certainly a rapid expansion of IPM and growing concern for environmental problems (Zethner, 1995; Tilman *et al.*, 2001; Wheeler, 2002). There was a noted lack of trained personnel, especially as related to teaching and expanding IPM concepts, and governmental infrastructures to regulate pesticides and encourage environmentally safe approaches to pest control.

Yanez *et al.* (2002) considered the pesticide problems faced by a wide range of developing countries and reported that 1500 active ingredients were registered as pesticides worldwide including 900 inert ingredients and 50,000 or more commercial products. About 85% of all pesticide use in the world is for agriculture and almost all countries use major pesticides. However, the relative amount of each type of pesticide varies from country to country. For example, 75% of the pesticides used in Malaysia is herbicides, and insecticides account for only 13%. In the Philippines, insecticides account for 55% of total pesticide use, and fungicides account for 20%. India represents about 2% of the total worldwide pesticide market. About 75% of the pesticides used in India is insecticides according to Yanez *et al.* (2002).

Raheja (1995) considered the collective status of IPM programmes in South and South–east Asia, which represents the world's most densely populated region. Raheja (1995) reported an increasing dependence on chemical pesticides, but a good growth in IPM practices during the 1990s. Ten countries in Asia have pesticide regulation laws, and IPM is official policy in China, India, Indonesia, Malaysia, Korea, Pakistan, Philippines and Thailand with government-supported pest surveillance and forecasting programmes. There are problems with pesticide subsidies and lack of governmental oversight for envi-

ronmental impacts and human safety. Crop practices in Asia were described as being dominated by an increasing dependence on chemical pesticides for agriculture, public health and home use. Calendar spraying and large-scale aerial spraying are widespread practices. In 1990, Asia's share of the worldwide pesticide market was 27%. An estimated 46% of the total was insecticides including DDT, BHC, methyl parathion, malathion and endosulfan. Rice received 40% of the total pesticide. Fruits and vegetables received 33%. In Indonesia, a 50% reduction in pesticide use was observed in 2 years (1986–1988) because of the adoption of IPM practices. Ten countries in Asia had legislation in place to regulate the production, import, export and handling of pesticides in 1990. There was some concern that prevailing agriculture policy hindered IPM. Problems cited were indirect pesticide subsidies, lack of government regulations and credit packages for farmers that included the purchase of pesticides. The emphasis on exports and increased production often stimulated excessive pesticide use.

Evolution of the agrochemical industry

The story of pesticide use in the USA has been an intriguing example of interaction between public and private interests and the continuing need to foster efficient cooperation and communications. Century-old, multinational companies traditionally founded on scientific discovery, chemistry and a range of innovative chemical technologies continue to evolve, merge and reorganize. Products from these science-based industries remain as overriding influences on man's ability to deal with pests. Holm and Baron (2002) provide an elaborate review of the evolution of chemical control and conclude that there is little doubt that pesticides have contributed to increased global agricultural productivity just as they have in the USA. They report a fourfold increase in the yield of maize in the USA from 1920 to 1990, threefold increases in yields of soybean, wheat and cotton, and an amazing 13-fold increase in the total agricultural output per farmer over the 70-year time frame. Osteen and Padgitt (2002) reported that a technological revolution in the USA increased productivity of agriculture by 2.5-fold between 1984 and 1994. Synthetic organic pesticides were cited as major contributors to this increased productivity, along with mechanization and other factors. Pesticide use continued to grow steadily in the USA between 1940 and the early 1980s when it reached market saturation (Holm and Baron, 2002). Total pesticide use increased at a slower rate through the 1990s. Trends in recent pesticide use are evolving as the impact of genetically altered crops continues. There has been some concern for increased pesticide use on transgenic crops (Wolfenbarger and Phifer, 2000; Benbrook, 2003), especially with increased reliance on herbicide-resistant crops since 1998 (Benbrook, 2003), but these new biotechnologies seem to offer tremendous potential to reduce inputs of pesticides into the environment when considered in a broader context (Wolfenbarger and Phifer, 2000; Altieri, 2001; Duke, 2005; Gianessi, 2005). Moving this new technology to developing countries with serious needs for safer pesticides, but little or no economic capital, remains a major social issue of evolving worldwide IPM (Altman and Watanabe, 1995; Craig and Camisani-Calzolari, 2001; Davis, 2005). The needs of the developing countries may be quite different from those of the industrialized nations where highly mechanized, capital-intensive agriculture is the norm. Conversely, the more traditional moderation approaches to chemical pest control that require sampling, and detailed management may become relatively more sustainable in the developing countries with smaller production units and more labour, assuming a fundamental base of essential technology and educational background.

The modern crop protection chemical industry evolved from companies formed hundreds of years ago. Giegy Company began in 1758, DuPont in 1852, Bayer in 1836, Eli Lilly and Company in 1876, Dow Chemical in 1887, Monsanto in 1901, Cyanamid in 1907 and Dow AgroSciences in 1907. A detailed chronology of the changing crop protection industry has been prepared by Appleby (2004) and can be found on the Oregon State University website (www.cropandsoil.oregonstate.edu). Holm and Baron (2002) describe the continuing evolution of the industry and the constant reorganization of major companies. Mergers since 2000 include BASF and American Cyanamid, Bayer Crop Science and Aventis Crop Science, Dow AgroScience and Rohm and Haas Ag. Chem., Valent BioScience Corporation and Abbott Ag. Specialties, as well as Syngenta, Novartis and Zeneca. Tait (2000) described the linkage between traditional chemistry of pesticides and the pharmaceutical industry. He also reviewed perceptions by the pharmaceutical groups regarding regulatory issues and new biotechnology products in agricultural markets. The suggestion was that this could impact future investments in new agricultural technologies. Total global sales of the top ten companies in 2000 (Syngenta, Aventis, Monsanto, BASF, Dupont, Bayer, Dow, Makhteshim-Agan, Sumitomo and FMC) were $29.7 billion, almost 75% of total worldwide sales. The world's main countries producing and exporting pesticide are France, Germany, Spain, the Netherlands, Australia, Canada and the USA. The fastest-growing markets in 2000 were Brazil, Germany, Spain, France, the Netherlands, Australia and Canada. These mega-, multinational corporations produce most of the new active ingredients and generally support international efforts for the safe and efficient deployment of chemical control materials. International pesticide resistance action committees (Table 6.1) are examples of worldwide stewardship of contemporary pesticides.

Local formulators and distributors also have a major influence on pesticide use at the farm level. Worldwide, generic products have become an important aspect of marketing pesticides and managing environmental impacts. Significant problems with substandard products, unlabelled and nonregistered pesticides, and fraudulent materials spiked with less expensive but more dangerous chemicals remain an international concern. PAN UK (1996) suggested that as much as 30% of China's total pesticide use

(70,000 t) might be of inferior quality. Huan and Ang (2000) indicated similar problems in Vietnam. Recent work in India to develop simple test kits to confirm pesticide potency is an example of efforts to address these problems (CFC, 2000). Gross profit margins of 20–30% for off-patent products vs. 50–60% for patented products influence marketing strategies and encourage distribution of older, sometimes more toxic, modes of action. Holm and Baron (2002) indicate that generic products may account for more than 70% of China's pesticide market, 60% of India's market, 50% of Korea's market and 40% of Taiwan's market.

Holm and Baron (2002) also provide a detailed account of marketing strategies and the transition from discovery chemistry to designer chemicals and seed-delivered technologies. Seed companies have become an integral component of all traditional pesticide companies. This has tremendous potential for delivery of pest control technologies to small farmers in developing countries (Altman and Watanabe, 1995), although it creates different infrastructure needs (Davis, 2005) including effective regulation and protection of proprietary rights, effective diffusion and marketing mechanisms, adequate investment and a strong international network. Holm and Baron (2002) also report a consistent increase in the number of new traditional pesticides, especially herbicides, introduced since the 1950s. Duke (2005) and Gianessi (2005) describe a possible concern that wide-scale adoption of the herbicide-resistant crops may reduce incentives for the industry to develop more site-specific, selective herbicides. Industry focus on specific modes of action and overall more active materials at reduced use rates has greatly impacted on environmental hazards and reduced much of the environmental risk associated with older pesticides used at exponentially higher rates (Casida and Quistad, 1998; Tomizawa and Casida, 2003). Effective use of these more specific chemicals often requires more sophisticated management systems. Knight et al. (1997), Larson (1998), Ragsdale (1999), Ware and Whitacre (2004a, b),

Whitacre and Ware (2004) and Duke (2005) review recent advances in chemical control technologies. New technologies and use patterns will require new educational efforts and programmes. The recent USAID investment in IPM programmes will be an important educational bridge to these new technologies in many systems.

The latest technologies, e.g. the products of biotechnology, are being rapidly adopted and are revolutionizing agriculture. In 1999, less than 5 years from the first commercial introduction of genetically engineered cotton, 40 million ha of cropland were planted worldwide with herbicide-resistant and insect-resistant crops. About 72% of the total was crops grown in the USA. Argentina and Canada accounted for 27% of the genetically altered crops in 1999. The remaining 1% was grown on limited acreages in China, Australia, South Africa, Mexico, Spain, Portugal, France, Romania and Ukraine. Since 1999, acreages of these genetically modified crops have greatly expanded around the world, primarily in commercialized high-input systems, but opportunities to reduce pesticide input and improve the health and economic status of poor farmers in developing countries are historically significant (Yudelman et al., 1998; Madsen and Sandoe, 2005). Numerous reviews of expanding genetically modified crops and a range of different management issues are available including Mazur and Falco (1989), Holt et al. (1993), Altman and Watanabe (1995), Wolfenbarger and Phifer (2000), Altieri (2001), James (2002), Wolf (2002), Benbrook (2003), Eaglesham et al. (2003), Waibel et al. (2003) and Pilson and Prendeville (2004).

Regulation of Chemical Pest Controls

We have repeatedly referenced the importance of regulatory activities in this chapter, not to overstate the importance of regulations or understate the importance of technology development, but rather to emphasize the central importance of societal oversight for environmental impacts of chemical control. The story of pesticide use in the USA is illustrative of patterns of technology development and management, and an intriguing example of interaction between public and private interests and continuing needs to foster efficient cooperation and communications. The general model of evolving pest management systems from the sole focus on pest control to broader concerns for the environment has been repeated around the world. One of the first actions in evolving systems of pest control is the creation of laws and the formation of regulatory agencies to limit unsafe use of pesticides. The impacts of these regulatory actions are not only restrictions on chemical controls but also encouraging efforts to expand alternative pest control options. The search for alternative pest management technologies and more diversified pest control alternatives continues today with major international efforts focused on biological control, host plant resistance and various cultural practices.

The regulatory impact cannot be minimized. It affects the scope of new science by setting boundaries for acceptable use of pesticides and articulated human society effects. Struggles between industry groups developing and marketing new pesticides and the regulatory restrictions of environmental agencies around the world continue to be the central topics of evolving pest management approaches. Significant gains have been made in protecting the environment by the use of more environmentally safe pesticides, especially within the last decade. A range of regulatory activities, associated social debate and strategic research have influenced this over the last 50 years. Many environmentally dangerous materials have been methodically and systematically removed from public use (PAN website, Table 6.1), not only in the USA but also in other developed nations. We are now seeing new technologies that have been designed not only for exceptional pest control but also for minimal impact on the environment (James *et al.*, 1993; Knight *et al.*, 1997; Duke, 2005; Grafton-Cardwell, 2005).

Edwards (1993) indicated that registration requirements for pesticides had become increasingly more stringent in industrialized countries and that the number of adverse environmental effects was accordingly reduced. He indicated that considerable work had been done, especially by the FAO of the United Nations, to limit the movement of pesticide-contaminated food across international borders (Lupien, 2000). The Organization for Economic Cooperation and Development (OECD) and the European Economic Community (EEC) were cited as organizations that had developed detailed protocols for managing the impact of chemical control agents on the environment. Results of a survey of pesticide regulations around the world can be found in OECD (2001). Edwards (1993) concluded that the effects of pesticides on the environment could be categorized into three main areas: effects on wildlife, effects on soil and water and effects on humans. Major effects on wildlife were problems with birds and other organisms at higher tropic levels than targeted pest species, impacts on fish and aquatic crustaceans, reductions in bees and other pollinators, emergence of new pest species due to ecological disruptions of the chemical control practices and broad-scale impacts on biodiversity of plants and animals. Pesticide-contaminated soil and water is considered a growing and critical issue in many developed countries like the USA. Edwards (1993) indicated that all of these ultimately affect the quality of human life. He concluded that much progress had been made in minimizing the impacts of pesticides in recent years, especially in developed countries, and encouraged more progressive exploration of alternatives to minimize the ecological impacts of chemical control practices.

Pimentel and Lehman (1993) suggested that a loss of public confidence in state and federal regulatory agencies and agricultural research institutions was related to major environmental and public health problems associated with pesticides. In the same text, Lehman (1993) and Perkins and Holochuck (1993) considered ethical issues impacting on pesticide use and minimizing environmental impacts of pesticides. Interested readers should refer to this excellent review of the social dilemma in balancing agricultural productivity and environmental impacts. We briefly consider some additional socio-economic influences here.

CAST (2003) provides a comprehensive analysis of issues impacting on pest control, including the scientific, environmental and political contexts in which IPM has developed over the last two decades. This report examined the environment as a series of distinct but interlinked ecosystems and included considerations of evolving management tools. The authors identified seven key issues that should be considered in formulating future IPM strategies: (i) impact of biotechnology on agriculture; (ii) genetic diversity and pest adaptability; (iii) ecology-based management systems; (iv) increased understanding of microflora/fauna in the environment; (v) training and technology transfer; (vi) government policies and regulations; and (vii) a need for continuous assessment of strategies. Issues identified as being essential in assessing the consequences of pesticide use included: health impacts on farm workers; consumer and the general public; lethal, chronic effects on other non-target biota; direct or indirect effects on natural and agroecosystems; calculation of air, soil and water pollution; and costs vs. benefits to producers and to society for decreasing pesticide use.

Socio-economic influences

Concepts of integrated control, economic injury and economic threshold have been widely considered, sophisticated and debated around the world (Kogan, 1998; Koul et al., 2004). High-level management systems have been proposed and refined from the simple concepts articulated by Stern et al. (1959). Yet the simple concepts of moderation communicated so effectively by Stern et al. (1959) remain the unquestionable foundation for IPM and modern pest control activities. They are relevant to all pest control decisions, especially to chemical control practices as they were forged in an environment of pesticide misuse and an awakening to the potential adverse impacts of this misuse on the environment. Elaborate reviews of pest management systems are cited throughout this chapter. Kogan and Jepson (2005) provide a conceptual linkage between pest management and broader ecological theory.

Environmental management systems beyond those historically linked with pest management may be useful models, given the increasing complexities of rapid technological change, expanding information and worldwide concerns for a range of different environmental problems. Numerous recent reviews of policy and management concerns for other environment-linked technologies are available (Miller and Holzman, 1994; Daily and Ehrlich, 1996; Naylor, 1996; Nash and Ehrenfeld, 1997; Sagar, 2000; Brooks, 2001; Kates, 2001).

Chemical pest control decisions are influenced by biological, ecological and socio-economic factors. Osteen and Padgitt (2002) indicated that the pesticide use responds to economic factors like markets and farm programmes, but pesticide regulatory process influences aggregate quantity by encouraging the types of new pesticides developed and eliminating others from the market through the regulatory process. They also indicated some support for the argument that farm programmes in the USA encourage more pesticide use than is necessary, but suggested that changes in farm programme legislation since 1977 had reduced incentives for pesticide use. Encouraging IPM has become a policy tool that supports reduction in the undesirable environmental aspects of pesticide use, and at the same time a policy tool to promote cost-effectiveness and profitable production. Sometimes these objects compete with each other. Osteen and Padgitt (2002) suggested that society generally views pesticides in a negative context because it values the benefits of efficient production less than the environmental costs associated with food safety, water quality, water safety and wildlife mortality. Changing societal attitudes and perceptions is the basis for most of the current policy on IPM and pesticide regulations in industrialized nations. Recent legislation on food safety is an example of socio-economic influence on regulatory policy in the USA. Other value systems operate in other nations, especially in developing countries.

Risk assessment is the central driving force in regulatory decisions and managing multiple stakeholder interests on most public issues. According to Reed (2002) there are rather extensive databases to support estimates of risks and benefits for pesticides compared with other environmental contaminants, but more information is needed on the sensitivity of children as well as aggregate and cumulative exposures. Reed (2002) also indicated that risk assessment is never a closed book and that changing technologies, improved monitoring capacities and more specific active ingredients would always alter pesticide-use patterns over time. Societal oversight, continual monitoring of potential impacts and reassessments of risks and benefits are essential if society is to benefit from newer technologies and protect itself from unexpected environmental problems.

Buttel (2003) examined the societal costs of agricultural production in the USA and recognized that some of the key production resources (seeds, tubers, soil, manures and rainwater) were renewable and potentially sustainable. He also recognized that other aspects of agriculture more closely resembled an extractive industry similar to mining and were unsustainable. He further considered long-term costs of agricultural practices on human health, environmental quality and the welfare and social livelihoods of humans, as well as the tremendous importance of agricultural production to human society. He recommended six precautionary solutions to these long-term societal costs:

1. efforts to discourage farming practices that tend to increase off-site movement of pesticides and fertilizer;
2. movement towards a consensus that agricultural production should be subjected to environmental standards of non-agricultural businesses;
3. increased US investments in wetland restoration;
4. policy initiatives that link multifunctionality policies to farm programme payments that include environmental protection, flood control, ecosystem maintenance and rural development;
5. reduction in government policies that internalize the costs of food and fibre production; and
6. increase in broad-based research on agroecology.

Socio-economic influences on chemical control practices in developing countries are quite different from those of the industrialized world. Toenniessen (2003) recently examined opportunities and challenges for small farmers in developing countries during the biotechnology era. More than 80% of the world's population lives in developing countries. Toenniessen (2003) recognized substantial reductions in poverty through increased food supplies over the last 50 years, but cited numerous examples of the continuing struggle to feed a rapidly expanding human population: 800 million still consume less than 2000 calories per day; 127 million pre-school children suffer from vitamin deficiencies that cause blindness and premature death; 24,000 people die from hunger each day (75% are children); 400 million childbearing women have iron deficiencies and are anaemic; 75% of one billion people live in rural poverty, earning less than $1 per day. International programmes of agriculture and investments in public research during the latter half of the 20th century resulted in major yield increases between 1962 and 2002: 200% for wheat, 117% for rice, 150% for maize, 57% for sorghum, 77% for potato and 43% for cassava. More than 400 public breeding programmes in 100 countries had released 8000 modern varieties of 11 major crops. Despite these gains, billions of small-scale farmers still rely on subsistence agriculture as a source of livelihood using traditional farming methods on depleted soils. Shrinking plots of land, scarce or polluted water, and inefficient agricultural markets are major problems. Toenniessen's (2003) plea for new technologies for these small rural farmers in developing countries, specifically for seed-borne pest control technologies, provides a stark contrast to the proprietary and environmental concerns of the industrialized world. He argued that billions of small farmers have the most to gain by efficient use of the newer, safer technologies, and worldwide benefits from reduced use of environment-damaging pesticides would be significant. Proprietary rights, accelerating registration costs and public acceptance of these new technologies in industrialized countries were seen as social challenges. Maredia *et al.* (2000) indicated that major global treaties, e.g. General Agreement on Tariffs and Trade (GATT), Trade-related Aspects of Intellectual Property Rights (TRIPs), World Trade Organization (WTO), International Union for Protection of New Varieties of Plants (UPOV) and Convention on Biological Diversity (CBD), require that member countries respect the intellectual property rights of other member countries. The USA and Europe have highly developed offices to handle trade agreements, intellectual property, copyright issues and patenting. These capabilities are lacking for many developing countries.

Roberts *et al.* (2003) describe an unexpected impact of a shift in pesticide use in Sri Lanka that resulted in an initial decline in pesticide deaths, but a subsequent increase when more dangerous replacement pesticides were deployed. Socio-economic values are not stable or uniformly applicable across worldwide populations. Understanding this variability is extremely important in refining global pesticide management strategies. A consideration of environmental impacts of chemical control practices from an international perspective demands an appreciation of the contrasting values and capacities of industrialized and developing nations.

Calzolari, 2001; Karlsson, 2004), but complex linkages among ecological, economic, environmental and social influences have been generally communicated and adopted through the IPM paradigm worldwide (Mengech *et al.*, 1995; Maredia, 2003). Successful environmental pest management requires more than mere continuous recycling of technologies and regulatory controls. Social variables, including those associated with education and different value systems, are critical links in environmental pest management on a global scale (Mengech *et al.*, 1995; Maredia, 2003; Koul *et al.*, 2004). Technologies derived from genetic engineering and more target site-specific chemicals are changing traditional pest management approaches, but the importance of economic, environmental and social linkages are still as important as they were in the mid-1900s. Sustainable management approaches that require routine sampling and data selection may be challenged by increased reliance on seed-borne technologies expected to provide season-long pest control. We examine here several evolving factors that potentially contribute to a more balanced use of chemical control practices and a more stable, sustainable agricultural environment in the 21st century.

Evolving Impact of Environmental Issues on Pest Management Decisions

Meaningful progress has been made over the last three decades in minimizing the environmental impacts of chemical pest control, much through worldwide adoption and acceptance of IPM (Mengech *et al.*, 1995; Kogan, 1998; Maredia, 2003) and an expanded general appreciation of the ecological and social scales involved (Horn, 1988; Kogan and Jepson, 2005). More work remains, especially in developing countries that lack governmental infrastructures to balance short-term economic benefits and long-term environmental impacts (WHO, 1996; Craig and Camisani-

Technical information, communication and education

Concerns for technology transfer and capacity of emerging pest management systems were important worldwide agendas in the early 1990s (Bosso, 1987; Tait, 2000; Craig and Camisani-Calzolari, 2001). A more educated and technically savvy pest manager was envisioned as one of the most critical needs of the time (Dent, 1991). Kadir and Barlow (1993) compiled papers of an international conference on 'Pest Management and the Environment in 2000' in 1991. The objective of the conference held more than 10 years ago was to increase awareness of the dangers, issues, trends and possible solutions in pest management.

Major challenges envisioned were integrating cost-effectiveness, sustainability and environmental friendliness for a range of different pest management options. Barriers to success in developing countries included a long list of institutional, attitudinal, technological and economic problems, especially those associated with technical capacity and education. Although much work remains and effective pest management requires a continuous educational support system, capacity to disperse vast amounts of technical information have dramatically improved over the last decade.

The 21st century is truly the age of information, and public access to technical information is expanding rapidly. Issues related to pesticides are especially suitable for this expanded exchange of information as education and communication of technical information are critical aspects of decisions to apply pesticides at the farm, corporate and governmental levels. An example of the growing amount of information available on pesticides and their potential impacts on the environment is a website (http://www.pesticideinfo.org) maintained by PAN. Table 6.1 lists several other websites used extensively to prepare this chapter and of almost endless list of electronic information sources is now available. The PAN site includes links to a wide range of databases, libraries and reference topics associated with pesticide chemistry, human health effects, medical and general toxicology information, pesticide incidents, pesticide use maps and statistics, economics of pesticide use, environmental effects and resistance, laws, regulations and treaties, national pesticide registration procedures in different countries, links to educational materials from the pesticide industry, links to various environmental organizations and pesticide activist groups, and various pesticide newsletters, reports and economic information. Similarly, the USA EPA, the Extension Toxicology Network (EXTOXNET) and the Consortium for International Crop Management (CIPM) websites (Table 6.1) link to other worldwide electronic resources. Assuming local pest managers, extension works and various government and non-government

advisers have access to the World Wide Web, technical information about pests, pesticides and a wide range of environmental concerns are readily accessible. Future IPM programmes must provide the essential educational background for proper integration and use of this information (Fitzner, 2002). Unfortunately, resource and education limitations will continue to be a problem in many developing countries (Koul et al., 2004), but the potential for worldwide dissemination of information is evolving at an astonishing rate. Perhaps, a focus of future efforts should be an emphasis on education and worldwide deployment of environmental and technical savvy managers.

In the early 1990s Whitten (1992) suggested that contemporary pest protection systems were marred by misapplication of technologies in both industrialized and developing countries, and that policies and efforts devoted to chemical pest control were dominated by needs to overcome the adverse impacts resulting from misapplication. He suggested that the overriding challenge in the future would not be the creation of new pest management options alone, but continued integration and application of available options in a sound economic, political and social context. This is an interesting idea given the recent development and rapid acceptance of transgenic crops that constitutively express insecticidal proteins season-long. Whitten (1992) further indicated that the world needs highly qualified technologists to communicate understandable technologies to policymakers. Balancing long-term environmental policies and effectively deploying new technologies require knowledgeable pest managers and communicators. It is more than utilization and cycling of new technologies. Whitten (1992) advocated new forms of informal education, delivery of relevant information to the masses, policies that benefit producers and consumers, and control of population growth. He was particularly interested in effective communication to governmental officials, regulatory agencies and other policymakers. He envisioned that tools of molecular biology would impact future pest management systems through refined identification, quantification and genetic monitoring of pest populations, and he correctly recognized the central roles of education and communication in capturing the benefits of this process.

Pest management systems in the USA (Fitzner, 2002) have emphasized practical development and expanded technology transfer in recent years, especially during the last decade. Needs for continuous training exist in the more industrialized nations just as they do in developing countries, and levels of understanding are never static, especially in technology-rich sciences like chemical pest control. Most licensed agricultural consultants and certified pesticide applicators are required to meet continuing education units. This serves as a useful vehicle to transfer new information and creative management approaches. This chapter discusses a useful example of the power of networking technical information. Infrastructures of communication exist on different scales: within local communities, across states or regions, within individual nations or countries, and internationally. Altman and Watanabe (1995), BCPC (1997), Lupien (2000), Maredia et al. (2000), Tait (2000) and FAO (2003, 2004) discuss different aspects of these information networks relative to pest control programmes in developing countries. The electronic resources listed in Table 6.1 are examples of expanding global networks of information.

Improved educational programmes focused on pest controls are being observed in developing countries (Mengech et al., 1995; Maredia, 2003; Koul et al., 2004). Plant clinics designed to educate farmers about pest biology, identification and appropriate controls have dramatically expanded over the last two decades. Ausher et al. (1996) surveyed the activities of plant clinics in Africa, Asia, Latin America and southern Europe. They indicated that considerable progress had been made in the physical creation of plant clinics in developing countries, and that they were generally staffed with well-trained personnel. They did find some inadequacies and inconsistencies that need to be addressed to fully provide plant protection services to their clientele, especially adequate facilities

and modern equipment. Numerous international organizations, many networked through the United Nations and the World Bank, target education and technology flow to poor farmers in developing countries. It is often a difficult and long-term effort, but one that is critical for improved pest control and environmentally safe use of chemical pesticide. Basic efforts initiated 50 years ago in response to adverse environmental impacts need to be continued and expanded on the basis of more recent information and growing technical capacity.

Education and technical information have long been essential to effective deployment of ecologically based pest management. Accurate pest identification, reasonable estimates of potential damage and determination of appropriate controls have given a 'tool box' of alternatives (CAST, 2004) that are basic components of reactionary pest control, regardless of the class of pest organism or the commodity involved. Many pest problems require a more preventative approach, and decisions to treat or not treat may be based on historical information rather than real-time samples. All approaches require some level of basic information about the pest, the crop or the commodity involved, and short- and long-term consequences of a range of different pest control options (Horn, 1988; Dent, 1991; Metcalf and Luckmann, 1994). This mandates that pest control decision makers possess fundamental understanding of the biological system involved and have access to dynamic information about changing benefits and costs. Achieving these fundamental levels of knowledge is a monumental effort given the number and diversity of potential decision makers involved and the rapid development of new information (Dent, 1991; Koul et al., 2004). Success in establishing these basic, fundamental knowledge bases is closely aligned with progress towards more ecologically based pest management (Mengech et al., 1995; Fitzner, 2002; Maredia, 2003; Koul et al., 2004). It is also an effort that consumes much of the resources for IPM. IPM is largely a human-based investment, specifically for the pest control decision makers and associated data collectors who use sample data and benchmark knowledge to determine the need to treat or not treat. In our review of the literature, we did not obtain an estimate of the magnitude of human capital involved in worldwide chemical control operations, but it would be huge if one considered all government personnel, all industry representatives and all special interest groups involved in the range of activities from technology synthesis to actual field deployment.

Large multinational agencies and agricultural chemical firms invest heavily in technology transfer, education and communication. Newer, more specific management tools often require more knowledge, and every new product or management tool has a cost of education. Futuyama (2000) reported that costs of developing a new pesticide might be more than $40 million and might take 8–10 years. This does not include market development or ongoing technology refinement after a product is introduced. Current estimates may be twice that, given more stringent regulatory requirements. BCPC (1997) reviews the continuing innovation associated with deployment of new technologies. Public institutions typically provide some level of technical education, but it varies greatly for new technologies among the socio-economic structures of worldwide populations. A range of different sources in both industrialized and developing countries finance growing numbers of private consultants, farm advisers and non-governmental advisers. More elaborate consideration of these information sources can be found in other chapters of this book and in the general pest management texts referenced. It is sufficient for this discussion to recognize the critical importance of fundamental knowledge and its influence on deployment of environmentally safe chemical controls in a range of different production systems. More refined management approaches often require more investment in human capital, information collection and data processing. Globalization of agriculture (BCPC, 1997) and increased costs of pest control (Freshwater and Wichenko, 1986; Pimentel et al., 1992; Naylor, 1996; Osteen and Padgitt, 2002; Buttel, 2003) heighten the need to address these expanding communication and educational needs.

Improved management models and decision aids

Cuperus et al. (1997) envisioned a broad application of IPM concepts and alternatives beyond traditional agriculture to urban environments and the public, including the food industry, the home and public facilities. IPM in public schools is now a major component of national IPM programmes in the USA. Cuperus et al. (1997) cited increased desires for healthy environments and safe foods as movements supporting the broader focus of IPM. The USA Food Quality Protection Act of 1996 (Table 6.2) required increased safeguards for infants and children, mandatory review of 'high risk' pesticides, increased information to consumers and accelerated regulations to reduce risks of pesticide exposure to humans. Considering these more stringent requirements, Cuperus et al. (1997) considered a range of new opportunities to create a safer environment and address the growing concern for pesticide residues and safety. These included precision agriculture and geographic information systems, more emphasis on food safety, more elaborate decision support software and genetically engineered plants. All were seen as opportunities to expand the base of traditional IPM applications, generally reserved for agronomic pests. Obviously, these changing social values require more complex decision models that consider a wide range of criteria.

Sutton (1996) also indicated that plant disease management on deciduous fruit trees in the USA was moving progressively towards a policy of less use of fungicide, much in response to the Food Quality Protection Act. He advocated a return to disease management programmes that relied less on pesticides. Interest in reduced reliance on fungicides was driven by overall public concern for the use of fungicides and loss of economically viable fungicides through pathogen-resistant populations and more stringent regulatory procedures (Ragsdale, 2000). Less reliance on protective fungicides would require more detailed management.

The literature contains reference to many multidecision models designed to integrate multiple factors in decisions to treat or not treat a pest problem. Levitan *et al.* (1995) and Levitan (1997) describe various approaches for incorporating risk of adverse environmental effects into decision models. HERB and SWC (Wilkerson *et al.*, 1991; Shaw *et al.*, 1998) are examples of sophisticated decision tools for weed control in soybean. Gossym/Comax and COTMAN are examples of relating management decisions to cotton crop development (McKinion *et al.*, 1989; Danforth and O'Leary, 1998). Higley and Peterson (1996) provide an excellent description of environmental risk and relationships to pest management decisions, especially as it relates to formulating an environmental economic injury level (EIL). van der Werf and Zimmer (1998) provide another approach to incorporate environmental risk of pesticide use in the decision-making process: the use of a fuzzy expert system. Walker *et al.* (1997) give an example of a pesticide-rating system for monitoring agricultural inputs in New Zealand's fruit industry. Dushoff *et al.* (1994) review efforts to incorporate a range of different environmental factors into pest management decisions.

Kovach *et al.* (1992) and Penrose *et al.* (1994) describe systems for incorporating environmental costs into pesticide spray decisions. The Kovach *et al.* (1992) model is widely cited as a practical method of considering environmental costs in pesticide decisions. This model utilizes an environmental impact quotient (EIQ) to incorporate available information on dermal toxicity, chronic toxicity, fish toxicity, leaching potential, surface loss potential, bird toxicity, soil half-life, bee toxicity, beneficial arthropod toxicity, plant surface half-life, applicator exposure and picker exposure. Additional literature on the subject of pesticide risk indicators for management decisions includes Reus *et al.* (2002) and Barnard (2000). NAS (2000) also summarized various risk models related to the evaluation of pest control strategies and associated environmental and health costs. Risk-assessment approaches were also included for worker safety costs, food safety costs and other exposure costs.

Even though numerous examples of more complex decision aids and support systems abound in the literature, pest control decisions at the farm level are often based on simple assessments of probable crop damage by the farmer or crop adviser. This sometimes encourages unnecessary sprays because farmers lack confidence in making decisions to 'not treat'. Benbrook *et al.* (1996) indicated that more than $1 billion was spent in the USA in regulation and enforcement programmes for pesticide safety, mostly to reduce unwise use of dangerous pesticides. They indicated that the large growth in regulatory bureaucracy in federal and state programmes rested on an assumption that science can accurately define the risk that pesticides will pose, and that regulators can insure that risks will never exceed benefits. Certainly this is a challenging accusation and one that is difficult to address without specified expectations, but it does raise an interesting question about our abilities to sophisticate chemical control decisions from an environmental perspective. They suggested that it may not be possible to define the risks. Benbrook *et al.* (1996) advocated a stronger focus on alternatives, specifically alternative cropping practices, physical barriers, bio-intensive IPM and use of low-risk pesticides, rather than moderated use of the more risky pesticides through increased regulation.

Whitten (1992) indicated that recent scientific and technical advances in information technology and molecular biology would shape possibilities for future pest management through improved modelling, expert forecast systems, genetically engineered crop plants and improved natural enemies. He suggested that the distinguishing feature of pest control for decades to come would be the integration of cultural, chemical, biological and physical approaches to pest management combined with genetically superior plants and animals, but as discussed earlier there is a fundamental need for educated pest managers to operate and deploy these more sophisticated management tools.

Evolving regulations, political debate, socio-economic variables and food safety

The importance of regulatory agencies has been referenced numerous times in this document, primarily from the perspective of societal protection. Given the evolving societal values, we should consider the dynamic nature of pesticide regulations and their broader impact on technology and innovative management. Generally, pesticides must be registered or approved by a government agency before they can be marketed, but the requirements of registration and stringency of enforcement vary from country to country (Willson, 1996). Most pesticides are marketed, exported and imported worldwide. The International Code of Conduct on the Distribution and Use of Pesticides adopted by the FAO of the United Nations (FAO, 2003) is considered the international guide for reasonable international movement and trade of pesticides. The code was amended in 2002 (Table 6.2) and is a subject of ongoing discussion and debate in the United Nations. In industrialized countries, especially in the USA and Europe, there are strong environmental groups that aggressively work to limit pesticide use and their impact on the environment. These groups are often at odds with farmers and various agricultural commodity groups. Considering the importance of this debate, Clarren (2003) reported that 32 pesticide products were registered in the USA in 1939. Now there are more than 20,000, and USA farmers use an estimated 0.5 billion kg annually. She indicated that the agricultural chemical industry in the USA is a big business with a lot of political clout and that they contributed $1.6 million to political campaigns in 2001–2002. She also reported that the average farm worker in the USA made $8750 in 1999–2000.

Benbrook et al. (1996) highlighted the perceived failures of chemical control technologies and advocated an emphasis on bio-intensive IPM that promotes biological control and alternative pest control practices. Adverse effects of continued reliance on chemical control mentioned were poisoning of farm workers and children, accumulation of pesticide in wildlife, decimation of eagles and other birds, birth defects, cancer, endocrine disruption and drinking water contamination. These are all critical points of concern for the USA public regarding pesticide regulatory policies and funding of agricultural research. Lehman (1993) carefully evaluated the policies and societal attitudes for pesticide use in the USA relative to Carson's (1962) Silent Spring and her references to Robert Frost's poem The Road Not Taken. This analogy of crossroads and imminent critical decisions has been a consistent theme in the debate of policies regulating pesticide use and its impact on the environment. Lehman (1993) concluded that the dramatic increase in pesticide use from the 1940s to the 1980s had stabilized in the USA by the 1980s and that it was likely to be further reduced in the future, suggesting that it was indeed possible to back up and take a different road. He indicated that the basic idea of IPM was that of mitigation and that it reflected an awareness of human limitations. He further suggested that the need to regulate pesticide use was a moral virtue and conflicting attitudes that reflected complete control over the natural world were arrogant. In describing the ethical issues and social environment around the pesticide–environment debate, he described conflicting ideas. The objective of the new road is less clear. Cheap and abundant food is a great benefit to people; illness and concern for possible illness are harmful.

Pimentel and Lehman (1993) concluded that increased pesticide use in the USA had stimulated agricultural productivity, but had also caused serious health and environmental problems. They discussed concerns for the influence of the agricultural chemical industry on governmental regulatory and research agencies and suggested that serious ethical issues remained with regard to the role of public scientists and priorities for research and public welfare and the environment. van den Bosch's (1978) *Pesticide Conspiracy* may have relevance to present-day policies based strongly on moderated data requirements and industry-funded research. Wheeler (2002) recently examined a breadth of issues related to agricultural pesticides and the environment, and conceptually considered environmental impacts relative to the 'state of the art' of pest management. The projected impact of the USA Food Quality Protection Act of 1996 was discussed in detail, as were evolving IPM policies, economic issues and risk assessment, environmental fate of pesticides and evolution of new technologies for delivering pesticides. It is a valuable reference for understanding broad socio-political factors influencing current USA policies on pesticide and the environment.

Food safety is the most recent emphasis of pesticide regulation. Merrill (1997) reviewed the history of food safety concerns and changes in legislation in the USA that govern food safety. He indicated that a growing advocate group believes that contemporary threats to food safety are more serious than those of the 1990s. Ragsdale (2000) described new regulatory decision criteria related to the USA Food Quality Protection Act including endocrine disruption, need for additional safety factors for children, aggregate exposure and cumulative exposure to pesticides with similar modes of action. She indicated that these new regulatory activities would have a direct effect on the availability of many fungicides registered in the USA, particularly on fruits and vegetables, and that loss of the use of these fungicides would contribute to increased costs of impacted foods for producers and consumers. She further indicated that these regulations would raise complex issues regarding the structure of agriculture in the USA and import/export relationships with other countries. Relationships to worldwide export opportunities could have impacts on pesticide use strategies of other countries. Delaplane (2000) summarized many of the issues related to the growing concern for food safety from the perspective of an agricultural scientist in the USA. A recent issue of *California Agriculture* (January/February 2005, Vol. 59) is devoted entirely to these issues.

The importance of pesticide regulation and its potential impact on social and economic issues, including the environment, are topics of intense contemporary debate far beyond the scope of this review, but we hope that readers are gaining an appreciation for the complexity of socio-economic values and resulting pesticide regulations that set boundaries for acceptable pesticide use. Interested students should refer to a growing collection of materials on the history and evolving issues linked to this discussion (Carson, 1962; Whorton, 1974; van den Bosch, 1978; NAS, 1980; Dunlap, 1982; Dunnett, 1983; Freshwater and Wichenko, 1986; Bosso, 1987; Marco *et al.*, 1987; NAS, 1987; Ragsdale and Kuhr, 1987; Salter, 1988; Cook, 1996; Wargo, 1996).

New concepts, new ideas and new technologies

Whitten (1992) considered lessons from the 20th century in suggesting new directions for pest control in the 21st century. Great advances in man's capacity to deal with pests were highlighted, as were problems of pesticide resistance and environmental failure. He indicated that several generations of pesticides had come and gone, and lessons were learned. Although pesticides had provided amazing control of many pests in the 20th century, there were examples where pesticides had failed to provide acceptable control, especially for the key pests of rice and vegetables. Pesticide-induced pests were also considered a characteristic feature of non-sustainable agriculture in many parts of the world.

The USA EPA registered more than 20 reduced-risk pesticides in the mid- to late 1990s, including: the herbicides alpha-metolachlor, cadre, carfentrazone, diflufenzopyr, imazamox and s-dimethenamid; the insecticides bifenazate, difulbenzuron, DPX-MP062, hexaflumuron, pyretrozine, pyriproxyfen, spinosad and tebufenozide; and the fungicides azoxystrobin, cyprodinil, fenhexamid, fludioxonil, mefenoxam and trifloxystrobin. These reduced-risk pesticides have lessened impacts on human health through lower toxicities than alternatives, lower potentials for groundwater contamination, lower pest resistance potential, increased efficacy and a high compatibility with IPM programmes (NAS, 2000).

Benbrook et al. (1996) provided a perspective on the origins of current USA pest management policies and wondered why people do not use IPM more and why chemical control was still so popular. They suggested that the complexity of management, the cost of multidisciplinary research requiring many years and the comparative inexpensive nature of chemical pesticides were the primary reasons. They indicated that we were at a crossroads where American agriculture and urban households had wholeheartedly embraced the use of chemical control, and there was a growing need for transition towards a bio-intensive IPM that would predominate regulatory policies and use patterns by the year 2020. They indicated that first step should be the creation of an infrastructure that can diagnose the nature and source of pest problems and rely on a range of preventative tactics of biological control. They advocated new and improved biopesticides and expanded access to IPM information and services. They suggested that some public institutions have policies that can slow the process of new bio-intensive technologies and indicated that research priorities in the public and private sectors were supporting and expanding chemical control practices. They suggested that priorities are required to be reversed. They especially advocated increased federal funding for pest management alternatives and suggested that a system should be created to reward IPM innovation, not to promote more pesticide use. Pimentel (1997) suggested that pesticide use could be reduced by 50% without resulting in crop loss by efficient deployment of existing IPM capacity, but he also recognized the tremendous economic benefits of pesticide use. A recent issue of California Agricul-ture (January/February 2005, Vol. 59) summarizes a wide range of issues associated with pesticide costs and benefits including options to effectively manage pests without the organophosphorous pesticides. The use of pesticides has been defended and provides a perspective of unnecessary losses associated with a hypothetical banning of pesticides (Knutson et al., 1990; Delaplane, 2000).

Larson (1998) and Ware and Whitacre (2004b) provide excellent reviews of new pesticide chemistry. Ware and Whitacre (2004b) and Whitacre and Ware (2004) provide summaries of their recent text and links to electronic public information on insecticides and herbicides. NAS (2000) considers the future role of pesticides in the USA and provides a balanced review of emerging issues, including environmental concerns. Knight et al. (1997) provide an industry perspective on the potential loss of traditional fungicides through more restrictive registration. They review the numerous new classes of fungicides developed in the 1990s and suggest priorities for new fungicides with low toxicity to humans and wildlife, low environmental impact and low residues in food. They envisioned more managed systems with an emphasis on IPM and more environmentally safe fungicides. Grafton-Cardwell et al. (2005), Tomizawa and Casida (2003) and Duke (2005) provide recent reviews of progress and concerns with new insecticides and herbicides.

Biotechnology and transgenic crops

Transgenic crops, specifically those genetically engineered to express insecticidal proteins and herbicide-resistant crops that can be oversprayed with broad-spectrum herbicides to kill a wide range of different weeds, have changed the scope of agriculture in the last 10 years. The amount of information and the changing science associated with these new agricultural tools are enormous. We can only reference a few works and focus specifically on the impacts on chemical control practices. Table 6.2 provides a brief index of historical developments over the last decade.

Duke (1995, 1998) provided early reviews and projections of the use of herbicide-resistant crops, and Duke (2005) provides an assessment of herbicide-resistant crops over the last 10 years. Mazur and Falco (1989) and Holt et al. (1993) also provide early descriptions of the weed control technology, and Owen and Zelaya (2005) consider transgenic crops in a recent review of weed resistance to herbicides. General reviews in the entomological literature include Tabashnik (1994), Gould (1998) and Shelton et al. (2002). A website at Colorado State University (Table 6.1) provides links to many of the contemporary works on transgenic crops. The complexity of regulatory issues and socio-economic considerations was considered in other sections.

Acreages of the transgenic crops are increasing dramatically, and perceived benefits of reduced pesticide use and superior pest control are accelerating interest in even more acreages, especially in the developing countries struggling with marginal pest control (Ismael et al., 2002; FAO, 2004). Owen and Zelaya (2005) report that 52 million ha of land are currently planted with genetically modified crops, 41 million ha are planted with herbicide-resistant crops and 33 million ha are planted with herbicide-resistant soybean. Pest adaptation to these powerful pest controls is perhaps the most discussed biological limitation. Legal and socio-economic hurdles of technology flow to developing countries, and the perceived environmental dangers of genetically altered food and gene containment are reviewed elsewhere. Pesticide-use patterns associated with transgenic crops were discussed briefly in the previous section. Almost all reports indicate reduced use of foliar insecticides with planting of Bt crops, although shifts in pest species and emergence of new pest problems, e.g. *hemipteran* pests previously controlled indirectly with *lepidopteran*-targeted sprays that are no longer necessary, are increasing the use of insecticides in some systems. The net benefit, however, has been dramatic and consistent reductions in insecticide use. This is most obvious in the cotton system where previous routine sprays for *lepidopteran* pests are no longer necessary.

Cotton production requires large inputs of insecticide. Meaningful reduction of insecticide use in cotton has a major impact on total worldwide use of insecticides.

Table 6.2 provides a list of the herbicide-resistant crops and associated herbicide-resistant traits currently registered for use in worldwide systems. The crops include canola, cotton, flax, maize, rice, soybean and sugarbeet. Countries with legal uses of the crops include Argentina, Australia, Brazil, Canada, Japan, Mexico, South Africa, Uruguay and the USA. Benbrook (2003) reported decreased use of herbicides when the herbicide-resistant crops were first introduced, but a general increase in pesticide use over time. He associated this increase to higher use rates and attributed the higher rates to difficult-to-control weeds and possible evolving weed resistance problems. Duke (2005) and Madsen and Sandoe (2005) also considered these increases and suggested that the recent estimates may overstate pesticide use. They cautioned against conclusions that would detract from the technology, indicating that the total environmental impacts of reduced use of fossil fuels and conservation tillage should be considered. An important marketing consideration is expanded marketing of generic products containing glyphosate. Duke (2005) suggested that new herbicide-resistant crops would continue to be developed, but that the rate of commercial deployment would likely be less than that of the last decade. He expressed concern about the impact of the transgenic crops on traditional herbicide discovery and continued corporate investment in more specific, selective chemistries.

Wolfenbarger and Phifer (2000) reviewed much of the early debate about the risks and benefits of genetically engineered plants and suggested that key ecological experiments were lacking. Carpenter (2001) suggested that Wolfenbarder and Phifer (2000) understated the potential benefits of these genetically engineered plants to reduce pesticide use. Krattiger (1997) reviewed the potentials of Bt crops to impact on pest control problems in developing countries. He indicated that the technology offers tremendous potential to improve crop production in an environmentally safe manner, but expressed concerns for long-term sustainability and potential pest resistance. He advocated consideration of alternative strategies if resistance to Bt crops developed. James (2004) provides a recent assessment of the global status of commercialized biotech crops.

The topics discussed by Eaglesham *et al.* (2003) included the dramatic impact of biotechnology products on agriculture and the influence of these products on a changing system of agricultural production. Issues included sustainability, environment and production, including discussion of a therapeutic intervention (e.g. threshold model) vs. restructuring the system, impact on the rural landscape and regulation of genetically modified foods. Opportunities and applications of biotechnology in developing countries were highlighted using an example of insect-resistant maize for Africa. Broader conceptual discussions dealt with philosophical perplexities and ethical enigmas of USA regulatory procedures.

Resistance and resistance management

Concepts of pest management emerged in the 1950s and 1960s as a need to manipulate uncontrollable pest densities. Resistance management similarly emerged as a philosophy in the 1980s (BOA, 1986) as a need to manipulate resistance genes and preserve susceptible alleles in pest population. Counteradaptation of pests to pesticides is a major tenet of the pest sciences and a vivid reminder of the adaptive powers of Mother Nature. Arthropod resistance has been a consistent problem for entomologists since the wide deployment of DDT and other organic insecticides in the 1950s and 1960s (Metcalf and Luckmann, 1994). Mota-Sanchez *et al.* (2002) reviewed the literature associated with arthropod resistance and reported that prior to 1980 there were 1613 reported cases (species × compound) of arthropod resistance to different pesticide classes. Prior to 1990, this had increased to 2325 reported cases, and prior to 2000 some 2585 cases of arthropod resis-

tance to pesticides were recorded. About 49% of the reported cases were species resistant to organophosphates. Resistance to the organochlorines accounted for 33% of the reported cases of resistance. Pyrethroids and carbamates accounted for 9% and 8% of the arthropod resistance cases, respectively. Reports of resistance to a wide range of other pesticides, including the newer insect growth regulators and other selected chemistries, account for less than 1% of the reported cases of arthropod resistance. Resistance to these newer modes of action and various microbials like *B. thuringiensis* continues to caution against complacency regarding the adaptive capacity of arthropods. The Database of Arthropods Resistant to Pesticides at Michigan State University (www.pesticideresistance.org) keeps a detailed list of all suspected and reported shifts in arthropod susceptibilities to insecticides. The current database includes more than 2500 cases of reported resistance in arthropod species to one or more pesticides. In 2000, the database included reports of resistance to 310 different compounds. BOA (1986), Roush and Tabashnik (1990), Hoy (1998) and Futuyama (2000) provide excellent benchmarks in the literature for arthropod resistance to pesticide. More than 500 species are considered resistant including important agricultural pests and disease vectors. Continued use of DDT (Hemingway and Rauson, 2000; Tren and Bate, 2004) is often within the context of managing resistance. Futuyama (2000) estimated that arthropod resistance costs the USA about $118 million per year in additional insecticides. Zalom *et al.* (2005) recognized the importance of managing resistance to the newer insecticide groups. Knight and Norton (1989) provide an economic assessment of the problem of pesticide resistance in arthropods.

Plant pathologists have long battled resistance to various fungicides. Widespread resistance to benomyl in the 1970s provided a conceptual framework for much of the resistance management programmes for insecticides and herbicides during the 1980s and 1990s. Damicone (1999) provides an excellent overview of fungicide resistance and applied approaches to deal with fungicides at the farm level.

Emerging herbicide resistance is one of the most critical areas of concern in modern agriculture (Llewellyn *et al.*, 2001; Owen and Zelaya, 2005). The Weed Science Society of America (www.weedscience.org) recognizes 295 resistant weed biotypes from 177 species (106 dicots and 71 monocots) in more than 270,000 production fields around the world. Recent resistance to glyphosate in several weed species is raising concern for the large-scale monocultures of herbicide-resistant crops. Holt *et al.* (1993) and Prather *et al.* (2000) provide historical reviewing of herbicide-resistant weeds and evolving management approaches. Herbicide-resistant weeds were first detected in the 1950s, but became very numerous in the last decade with wide-scale use of specific modes of action, especially the ASL inhibitors introduced in the 1980s.

Holt *et al.* (1993) and Duke (2005) recognize weed susceptibility as a resource to be managed. Industry clearly understands the impact of resistance on their technologies and the need to effectively manage resistance. Worldwide access to the Insecticide Resistance Action Committee, the Herbicide Resistance Action Committee and the Fungicide Resistance Action Committee is possible through the websites listed in Table 6.1. These are truly global committees with representatives from all major industry groups and key academic cooperators. Similarly, the *Resistance Management Newsletter* (Table 6.1) is a forum for worldwide discussion and an alert of evolving resistance problems. CAST (2004) summarizes emerging issues in resistance management across the different disciplines and indicates that managing resistance will become increasingly important in IPM programmes worldwide.

Application and handling of chemical control agents

Hall (1991) and Pimentel and Levitan (1991) indicate that many researchers have concluded that less than 1–2% of pesticide applied actually arrives at the desired target. This off-target pesticide contaminates land, water and air. They also indicate that the amount of pesticide released into the environment has risen by 1900% from 1930 to 1980. Improved application procedures and technologies coupled with effective education and transfer of these technologies to field management of pest problems offers opportunities to reduce overall pesticide use and minimize many non-target impacts (Pimentel, 1997), but much of the pesticide worldwide is still applied through century-old application equipment.

Problems with poor application of pesticides are probably worse in developing countries where many of the more toxic materials are applied by hand or by knapsack sprayers, but drift and movement of pesticides from the point of atomization to targeted deposition is a continuing problem with all application methods, including tractors and airplanes. PAN UK (1996) reported that China manufactured 8 million knapsack sprayers in 1995, the most widely used application equipment in China. Others (PAN UK, 1996; WHO, 1996; Eddleston *et al.*, 2002) report serious problems with unintentional and intentional poisoning of farmers and farm workers in developing countries. Improved designs of equipment use of modern technologies for accurate dosing and distribution including global position systems for monitoring swath widths and more efficient atomization technologies have increased the opportunities to more efficiently deliver chemical controls to target areas (Friedrich, 2000). Unfortunately, this requires more education and transfer of information to field-level applicators, and as discussed previously education remains a challenge at both on-farm and governmental levels.

Standards for handling and applying pesticides are common around the world. Both industry and public sector groups recognize the enormity of this educational challenge and are working to enhance the distribution of information. A current example in the USA is an Environmental Stewardship website maintained by Syngenta Crop Protection (http://www.syngentacroppro-tecton-us.com/environ/) that includes links to many research and educational resources designed to understand and minimize spray drift. Links include the EPA, a Spray Drift Task Force, the National Coalition on Drift Management, equipment manufacturers, university and USDA research programmes, the Association of American Pesticide Control Officials, Pesticide Applicator Associations, the American Society for Agricultural Engineering and other useful resources of information. In a summary of the current status of application technologies and impacts on worldwide use of pesticides, Friedrich (2000) suggested that technical developments in application equipment and methods have progressed with the evolution of new pesticides and modern products, but unfortunately these improved technologies and methods had not been adequately transferred to field practice around the world. He further indicated that this was a serious issue in developing countries where unsafe application of extremely toxic pesticides results in many cases of human poisoning each year. So the Agricultural Engineering Service of FAO established a programme to increase awareness of application problems. Specific actions included targeted communications about the problems at the government level, initiation of practical training programmes for farmers and operators and improved equipment quality through the introduction of standards and regular tests. Friedrich (2000) advocated the development of FAO standards for application equipment that focused on safety and efficiency. He suggested that the guidelines should address minimum standards for the basic requirements for pesticide application equipment and that more comprehensive FAO standards should be developed that would serve as guidelines for the code of conduct and review of proposals.

Future Opportunities

Just as various medicines and antibiotics remain critical tools for managing human health, pesticides remain necessary foundations for efficient pest management. Moderation continues as a central paradigm for both (NAS, 2000; Wenzel and Edmond,

2000; Hooten and Levy, 2001; CAST, 2004). Adverse effects of pesticide overuse on the environment are widely recognized, as are the various social and economic influences that tend to encourage short-term benefits from excessive pesticide use. Balancing these short- and long-term benefits remains a struggle in developing countries lacking strong societal influence and a regulatory infrastructure capable of limiting ineffective and dangerous chemicals. Continued focus on educational and individual benefits associated with appropriate pesticide selection, timing and application can have dramatic local impacts, just as they did in the evolution of IPM practices in many industrialized countries during the 20th century.

The conceptual appeal of economic injury and integrated control models (Stern et al., 1959) still offers strong benchmarks for moderating pesticide use in all countries. Regardless of the scale, conceptual vs. practical or societal vs individual, these simple models of moderation encourage cost-effective management of pests and limited disruption on the environment. This is not to suggest that the conceptual or practical approaches to pest management cannot be improved. Evolving chemical control practices in industrialized countries are driven by increased public awareness of potential problems, societal values for a safe food source and different perspectives on needed approaches to protect the environment. Regulatory policies and constraints on pesticide use are generally increasing worldwide. More environment-friendly chemical controls and transgenic technologies are rewriting some of the traditional concepts about reactive pest management and decisions to trigger chemical control. Issues like resistance management, environmental ownership, property rights and food safety are accelerating the conceptual context of pest management decisions and challenging traditional thinking about practical application of chemical control at the farm level. Dramatic growth of worldwide human populations and limited natural resources will accelerate the consideration of new conceptual approaches to pest control over the next century. Bridging the variable socio-economic values and the diversity of changing pest control needs will challenge global strategies for pest management, but chemical control will remain an important part of this increasingly complex scenario.

Technology is not stable, and tremendous opportunities always exist for human capacity to design and develop better tools. Understanding the social and biological complexity of the managed system is the key. Knowledge of chemical pest controls and recognition of wide-scale impacts on the environment will continue to evolve, but the basic lessons of the 20th century will be important benchmarks (Whitten, 1992). Creating economic incentives for environmentally safe chemical control technologies and practices is critical, but dramatically different socio-economic environments around the world include a diversity of different value systems. The most advantageous systems for ecological concerns in industrialized nations may be quite different from the short-term perspectives of developing nations seeking relief from severe pest problems. Pesticide chemistries developed over the last decade (Ware and Whitacre, 2004a, b; Whitacre and Ware, 2004) are a good example of the potential progress that can be made in defining long-term environmental concerns. Continued use of unsafe and ecologically disruptive, but affordable, chemicals in developing countries (PAN website, Table 6.1) is an evidence of worldwide inequities and the strong linkage between available resources and utilization of chemical control technologies. Increased communication and widely available technical information should enhance the flow of new environmentally advanced pest control to developing nations. Policy and socio-economic hurdles to this technology flow will be more aggressively addressed as issues of world hunger and environmental preservation heighten with expanding human population (Tilman et al., 2001).

Sagar (2000) considered capacity development in different worldwide regions and developing countries as related to environmental management. He suggested that policies governing environmental management in developing countries were similar to those in industrialized countries, but that implementation varies with infrastructures. Advances in the capacity to deal with complex environmental issues that transcend social and geographic boundaries will require a systematic effort by governments and multinational organizations to organize worldwide agendas. Progress has been made in the last two decades to strengthen worldwide information about highly toxic and persistent chemicals. Nash and Ehrenfeld (1997) described codes of environmental management practice that emerged as tools of broad environmental policies in the 1980s. Although aspects of these codes have some application to pesticide regulation and associated environmental impacts, they more generally describe environmental contaminants like industrial hazardous waste. Closer examination of these broader concepts of environmental management may provide insight into options for managing the impacts of chemical pest control on the environment. A consortium of interest groups including industry and public sector representatives typically develops these codes. Nash and Ehrenfeld (1997) recommend a range of new management practices including initiation of environmental management systems, public environmental reporting and community advisory panels.

Certainly, food safety and mandated standards for a safe and wholesome food supply have dramatic effects on chemical control practices. The 1996 Food Quality Protection Act in the USA is an example with broad and global influence on emerging pest management systems. A recent issue of California Agriculture (January/February 2005, Vol. 59) is devoted to a discussion of the Food Quality Protection Act and evolving pest management systems in California. Woteki and Kineman (2003) recently considered challenges to reducing food-borne illness and reported that regulations and laws comprising the food safety system frequently lag behind current scientific knowledge of the risks posed by food-borne pathogens. They suggested that future changes to food safety should better relate risk to specific pathogens in the food supply and include costs and benefits of mitigation strategies. Pesticide residues may follow a similar pattern. Direct and indirect effects of pesticide exposure to humans are a continuing struggle in managing the use of pesticides. In developing countries, education and access to new and safer controls seem to be of utmost importance. In industrialized countries, especially those in North America and Europe, standards for reduced exposure to pesticides and societal expectations for a safe environment have dramatically increased over the last three decades and will continue to evolve as long as adverse impacts on human health are measured (Fleming *et al.*, 2003). Alavanja *et al.* (2004) evaluated current procedures to limit human exposure to pesticides in the USA. They suggested that epidemiological studies indicate that current exposures to pesticides are associated with risks to human health despite the pre-market animal testing that is required for registration of a commercial pesticide. They recommended improved epidemiological studies and a better understanding of how toxicology data relate to human health risks and how public health policymakers should more accurately judge this information. Others (Delaplane, 2000) argue that the food supply is very safe and a broader perspective of risks vs. benefits should be considered (Ragsdale, 1999). Never the less, the concern for safe food reflects the evolving maturity of environmental concerns for chemical control practices.

NAS (2004) recently released a report ('Pesticides in the Diets of Infants and Children') that provides a summary of the current knowledge and regulations associated with pesticides in the diets of children. An increased awareness of the potential harmful effects of pesticides on children and cumulative exposures are fuelling mandates for better risk models and improved detection systems to quantitatively describe exposure and set tolerances. The Food Quality Protection Act is a direct result of this growing societal concern.

Perhaps the most specific and direct view of the future importance of pesticides and associated management systems in the USA is a recent report of the National Academy of Sciences (NAS, 2000). This committee of the National Research Council reviewed the current status of chemical pesticides and identified the most promising opportunities to reduce risks and increase benefits of pesticide use. The scope of the review was broad and included pesticide use in all sectors of the USA production system including processing, storage and transportation of field crops, fruits, vegetables, ornamentals, fibre and forest products, livestock and aquaculture. Weeds, pathogens, vertebrate and invertebrate organisms that attack crops, livestock and urban systems were considered. The temporal scope of the review was limited to the next 10–20 years because the influences of technological innovations beyond 20 years were difficult to project, yet ultimately influential on the capacity of the management systems. The executive summary for this report can be found on the National Academy Press website (http://www.nap.edu). The report suggested that there was no justification for completely abandoning chemical control. They recommended maintaining a diversity of tools for maximizing flexibility, precision and stability of pest management. They also advocated that a concerted effort be made to increase the competitiveness of alternatives to chemical pesticides through expanded research and policy. This was seen as a necessary prerequisite for diversifying the pest management 'toolbox' in an era of rapid economic and ecological change. The report further suggested that investments in research by the public sector should emphasize those areas of pest management that are now not being (and historically have never been) undertaken by private industry. Government policies should be adapted to foster innovation and reward risk reduction in private industry and agriculture, and the public sector has a unique role in supporting research on minor use cropping systems that lack environmentally acceptable alternatives. Perhaps the most important recommendation was that the public sector must provide quality education to ensure well-informed decision making in both the private and public sectors. Regional pest management programmes in the USA have recently emphasized technology transfer

and the role of extension education. Similarly, the new USAID programmes place
emphasis on technology transfer and farmer
education.

To achieve optimum production standards with minimum negative impacts on
the environment, the USA must further organize integrated management systems that
utilize combinations of pest control technologies in ecologically balanced programmes.
These programmes must be based on more
fundamental and applied science, and the
National Academy committee suggested that
appropriate research efforts be expanded.
Fitzner (2002) also reviewed the history of
IPM policy in the USA and recognized a
dichotomy of federal IPM policy as it relates
to the need to balance production efficiencies
and protect the environment. He argued that
great strides had been made through IPM and
that while some stakeholders become impa-
tient and desire more dramatic and measurable reductions in pesticide use, a continued
investment in research and education efforts
was essential to prevent future crises. Some
problems require a dedicated long-term commitment of resources.

NAS (2000) discussed a wide range
of economic and regulatory changes that
impact the future of pest management.
These included globalization of world food
markets, industrialization of agriculture
and food processing, issues of decentral-
ization and privatization (e.g. technology
transfer, royalties, protection against patent
violation and buyer's rights), privatization
of extension services and consulting, phase
out of commodity programmes and emergence of a knowledge-based economy, with
knowledge and information being dominant
factors in production, and as such major
sources of wealth. Examples discussed were
organic food marketing and ecolabelling
initiatives.

Yudelman (1998) indicated that the
supply of food would have to increase by
70% in developing countries so that the
6.5 billion people living in those countries
would be food-secure in 2020. IPM is seen
as an important component of the process,
but formulating strategies for improving pest
management is difficult because of inaccu-

rate information on benefits and costs of
pest control. The paper acknowledged new,
safer technologies being developed by international corporations, but the concern for
regulation, distribution and manufacturing
of the older, more toxic pesticides were considered more important because of lack of
environmental infrastructures in the developing countries. IPM is an obvious component of the solution. Genetically engineered
plants seem to hold tremendous potential,
but a concern for distribution of these private corporation technologies to the poorer,
developing countries was also expressed.
Other options discussed were increased
investments of the larger, more advanced
countries in biotechnology applications in
the developing countries, and a 'leapfrog'
approach to transfer of technology by providing incentives to corporations with the
technology. In a review of food production
evolution over the last 50 years, FAO (2004)
indicated that the development of new and
safe biotechnologies offered tremendous
potential for sustainability of agricultural
productivity. The report also recognized that
technology alone could not solve the current and future problems of the poor. More
consideration of socio-economic impacts,
food safety and environmental issues were
encouraged.

Whitten (1992) considered social issues
and policies in developing countries that
impact pest management and the environment. He considered growing human
populations as the underlying problem of
managing pests and protecting the environment. In his review, he referenced high-
input, high-output agriculture in Europe
that included taxpayer subsidized production, elaborate storage capacity and strong
export industries. He applauded Dutch
efforts to create national plans for reducing
pesticide use, and referenced the national
USA IPM plan of the Clinton administration. He suggested that 46% reduction in
pesticide use observed in the USA from
1976 to 1982 was a model of technological
potential. He further considered the impact
of national policies on pest control potential and resulting environmental impacts
referencing his personal experiences in

the USSR and evolving national policies in China. He recognized that external forces can dictate the use of pesticides in some countries and suggested that scientists need to communicate beyond their technical and socio-economic spheres. He also advocated more global data and sharing of data on pest management and pesticides. Major industry groups are a world resource of new technology, and Whitten (1992) suggested that they needed to take a hard look at sales policies that promote unethical use of pesticides with resulting adverse impacts on the environment. He further suggested that the worldwide community would need to raise its knowledge of realistic environmental expectations or risk losing major products through fear and ignorance.

Tilman *et al.* (2001) suggested that the next 50 years would be the final period of rapid agricultural expansion. During this period, the demand for food for a 50% larger and wealthier world popula-

tion will influence global ecology. Based on estimates of past impacts, they projected that 10^9 ha of natural ecosystem will be converted to agriculture with two to threefold increases in pesticide use and fertilizers. This is expected to cause unprecedented ecosystem simplification and species extinctions. They advocate significant and immediate regulatory, technological and policy changes to address these expected impacts of agricultural expansion. Pest control will be an important component of this challenge, especially chemical control technologies. As observed in the 20th century, moderating approaches to chemical use will be most sustainable and effective if developed in ecologically sound pest management systems. These systems may be quite different from those currently conceived, but new concepts and new management capacities will evolve with new technology and new challenges.

References

Agrios, G.N. (1996) *Plant Pathology*, 4th edn. Academic Press, New York.

Alavanja, M.C.R., Hoppin, J.A. and Kamel, F. (2004) Health effects of chronic pesticide exposure: cancer and neurotoxicity. *Annual Review of Public Health* 25, 155–197.

Aldrich, R.J. and Kremer, R.J. (1997) *Principles in Weed Management*, 2nd edn. Iowa State Press, Iowa.

Altieri, M.A. (2001) The ecological impacts of agricultural biotechnology. American Institute of Biological Sciences. Available at: www.actionbioscience.org

Altman, D. and Watanabe, K. (1995) *Plant Biotechnology Transfer to Developing Countries*. Academic Press, Texas.

Anderson, W.P. (1996) *Weed Science: Principles and Application*, 3rd edn. West Publishing, St Paul, Minnesota.

Appleby, A.P. (2004) Herbicide company genealogy. Crop Science Department, Oregon State University. Available at: http://cropandsoil.oregonstate.edu/herbgnl/tree.html

Ausher, R., Ben-Ze'ev, I.S. and Black, R. (1996) The role of plant clinics in plant disease diagnosis and education in developing countries. *Annual Review of Phytopathology* 34, 51–66.

Barnard, C. (2000) *OECD Survey of National Pesticide Risk Indicators*, 1999–2000/USA. Economic Research Service, Washington, DC.

BCPC (1997) *Future Research, Development and Technology Needs for UK Crop Protection*. British Crop Protection Council Report, August 1997, Farnham, Surrey, UK.

Benbrook, C.M. (2003) Impacts of genetically engineered crops on pesticide use in the United States: the first eight years. BioTech InfoNet, Technical Paper No. 6, November 2003. Available at: http://www.biotech-info.net/technicalpaper6.html

Benbrook, C.M., Groth, E. III, Halloran, J.M., Hansen, M.K. and Markuardt, S. (1996) *Pest Management at the Crossroads*. Consumers Union, Yonkers, New York.

BOA (1986) *Pesticide Resistance: Strategies and Tactics for Management*. National Research Council, Board on Agriculture, National Academy Press, Washington, DC.

Bosso, J.C. (1987) *Pesticides and Politics: The Life Cycle of a Public Issue*. University of Pittsburgh Press, Pittsburgh.

Brooks, H. (2001) Autonomous science and socially responsive science: a search for resolution. *Annual Review of Energy and the Environment* 26, 29–48.

Buttel, F.H. (2003) Internalizing the societal costs of agricultural production. *Plant Physiology* 133, 1656–1665.

California Agriculture (2005) Beyond organophosphates. *California Agriculture* 59. Division of Agriculture and Natural Resources, University of California.

Campanhola, C., José de Moraes, G. and de Sá', L.A.N. (1995) Review of IPM in South America. In: Mengech, A.N., Saxena, K.N. and Gopalan, H.N.B. (eds) *Integrated Pest Management in the Tropics: Current Status and Future Prospects*. John Wiley & Sons, New York, pp. 121–152.

Camphill, C.L. and Madden, L.V. (1990) *Introduction to Plant Disease Epidemiology*. John Wiley & Sons, New York.

Carpenter, J.E. (2001) *Case Studies in Benefits and Costs of Agricultural Biotechnology: Roundup Ready Soybean and Bt Field Corn*. National Center for Food and Agricultural Policy (NCFAP), Washington, DC.

Carson, R. (1962) *Silent Spring*. The Riverside Press, Cambridge.

Casida, J.E. and Quistad, G.B. (1998) Golden age of insecticide research: past, present, or future? *Annual Review of Entomology* 43, 1–16.

Cassman, K.G., Dobermann, A., Walters, D.T. and Yang, H. (2003) Meeting cereal demand while protecting natural resources and improving environmental quality. *Annual Review of Environment and Resources* 28, 315–358.

CAST (2003) *Integrated Pest Management: Current and Future Strategies*. Council for Agricultural Science and Technology, Washington, DC.

CAST (2004) *Management of Pest Resistance: Strategies Using Crop Management, Biotechnology, and Pesticides*. Council for Agricultural Science and Technology, Washington, DC.

Cate, J.R. and Hinkle, M.K. (1993) *Integrated Pest Management: The Path of a Paradigm*. National Audubon Society Special Report, Washington, DC.

CFC (2000) *Sustainable Control of the Cotton Bollworm,* Helicoverpa armigera, *in Small-scale Cotton Production Systems*. Common Fund for Commodities, Appraisal Report, Second Account, Amsterdam.

Clarren, R. (2003) Fields of Poison. *Brief Fact Articles, Nation* 277, 23–25.

Cook, R.J. (1996) Assuring the safe use of microbial biocontrol agents: a need for policy based on real rather than perceived risks. *Canadian Journal of Plant Pathology* 18, 439–445.

Craig, I.K. and Camisani-Calzolari, F.R. (2001) Technology transfer in developing countries 2000 – automation in infrastructure creation. *Proceedings of the International Federation Automation Conference, Pretoria, South Africa*. Elsevier, USA.

Cuperus, G., Berberet, R. and Kenkel, P. (1997) The future of integrated pest management. Available at: http://ipmworld.umn.edu

Daily, G.C. and Ehrlich, P.R. (1996) Global change and human susceptibility to disease. *Annual Review of Energy and Environment* 21, 125–144.

Damicone, J. (1999) Fungicide resistance management. Oklahoma Cooperative Extension Service, OSU Extension Facts, F-7662. Available at: http://pearl.agcomm.okstate.edu/plantdiseases/f-7663.html

Danforth, D.M. and O'Leary, P.F. (1998) *COTMAN Export System 5.0*. University of Arkansas., Agriculture Experimental Station, Arkansas.

Davis, K.E. (2005) Regulation of technology transfer to developing countries: the relevance of institutional capacity. *Law and Policy* 27, 6–32.

Delaplane, K.S. (2000) Pesticide usage in the United States; history, benefits, risks, and trends. *University of Georgia Cooperative Extension Service*, Bulletin 1121, University of Georgia, Athens, USA, 8.

Dent, D. (1991) *Insect Pest Management*. CAB International, Wallingford, UK.

Donaldson, D., Kiely, T. and Grube, A. (2002) *Pesticide Industry Sales and Usage: 1998 and 1999 Market Estimates*. US Environmental Protection Agency, Washington, DC.

Duke, S.O. (1995) *Herbicide-resistant Crops: Agricultural, Economic, Environmental, Regulatory, and Technological Aspects*. Lewis Publishers, Chelsea, Michigan.

Duke, S.O. (1998) Herbicide-resistant crops: their influence on weed science. *Journal of Weed Science and Technology* 43, 94–100.

Duke, S.O. (2005) Taking stock of herbicide-resistant crops ten years after introduction. *Pest Management Science* 61, 211–218.

Dunlap, T.R. (1982) *DDT: Scientists, Citizens, and Public Policy*. Princeton University Press, Princeton, New Jersey.

Dunnett, E. (1983) Regulation of pesticides and risk–benefit analysis: can it help. *Canadian Farm Economics* 18, 3–6.

Dushoff, J., Caldwell, B. and Mohler, C.L. (1994) Evaluating the environmental effects of pesticides: a critique of the environmental impact quotient. *American Entomologist* 40, 180–184.

Eaglesham, A., Ristow, S. and Hardy, R.W.F. (2003) *Biotechnology: Science and Society at a Crossroad*. National Agricultural Biotechnology Council, Report 15, Ithaca, New York.

Eddleston, M., Karalliedde, L., Buckley, N., Fernando, R., Hutchinson, G., Isbister, G., Konradsen, F., Murray, D., Piola, J.C., Senanayake, N., Sheriff, R., Singh, S., Siwach, S.B. and Smit, L. (2002) Pesticide poisoning in the developing world – a minimum pesticide list. *Lancet* 360, 1163–1167.

Edwards, C.A. (1993) The impact of pesticides on the environment. In: Pimentel, D. and Lehman, H. (eds) *The Pesticide Question: Environment, Economics and Ethic*. Chapman & Hall, New York, pp. 13–46.

Epstein, L. and Bassein, S. (2003) Patterns of pesticide use in California and the implications for strategies for reduction of pesticides. *Annual Review of Phytopathology* 41, 351–375.

FAO (2003) *International Code of Conduct on the Distribution and Use of Pesticides*. Food and Agriculture Organization of the United Nations, Plant Protection Services, Pesticide Management Unit, Rome. Available at: http://www.fao.org/waicent/faoinfo/agricult/agp/agpp/pesticid/code/pm_code.html

FAO (2004) *Agriculture Biotechnology: Meeting the Needs of the Poor?* Food and Agriculture Organization of the United Nations, Rome.

FDA (2002) *Residue Monitoring 2002*. Food and Drug Administration, Pesticide Program, Washington, DC.

Fitzner, M.S. (2002) The decades of federal integrated pest management policy. In: Wheeler, W.B. (ed.) *Pesticides in Agriculture and the Environment*. Marcel Dekker, New York, pp. 1–24.

Fleming, L.E., Gomez-Marin, O., Zheng, D., Ma, F. and Lee, D. (2003) National health interview survey mortality among US farmers and pesticide applicators. *American Journal of Industrial Medicine* 43, 227–233.

Freshwater, D. and Wichenko, G. (1986) Coming to grips with pesticide regulation. *Canadian Farm Economics* 20, 29–37.

Friedrich, F. (2000) *Agricultural Pesticide Application: Concepts for Improvement*. Food and Agriculture Organization of the United Nations (FAO), Agricultural Engineering Branch, Rome, Italy. Available at: http://www.fao.org/ag/ags/ages/pae.html

Fry, W.E. (1982) *Principles of Plant Disease Management*. Academic Press, New York.

Futuyama, D.J. (2000) *Insect Pests: Resistance and Management*. Evolution, Science and Society. Available at: http://evonet.sdsc.edu/evoscisociety/insect_pests.html

Gianessi, L.P. (2005) Economic and herbicide use impacts of glyphosate-resistant crops. *Pest Management Science* 61, 241–245.

Gould, F. (1998) Sustainability of transgenic insecticidal cultivars: integrating pest genetics and ecology. *Annual Review of Entomology* 43, 701–729.

Grafton-Cardwell, E.E., Godfrey, L.D., Chaney, W.E. and Bently, W.J. (2005) Various novel insecticides are less toxic to humans, more specific to key pests. *California Agriculture* 59, 29–34.

Hall, R.R. (1991) Pesticide application technology and integrated pest management (IPM). In: Pimentel, D. (ed.) *CRC Handbook of Pest Management in Agriculture*, 2nd edn., Vol. II., CRC Press, Boca Raton, Florida, pp. 135–167.

Hemingway, J. and Rauson, H.R. (2000) Insecticide resistance in insect vectors of human disease. *Annual Review of Entomology* 45, 371–391.

Higley, L.G. and Peterson, R.K.D. (1996) Environmental risk and pest management. Available at: http://ipmworld.umn.edu

Hilstrom, H. (2002) Government Orders: Pest Control Products Act, Canada. *Parliament House of Commons Debates* 137, 10107–10109.

Holm, R.E. and Baron, J.J. (2002) Evolution of the crop protection industry. In: Wheeler, W.B. (ed.) *Pesticides in Agriculture and the Environment*. Marcel Dekker, New York, pp. 295–326.

Holt, J.S., Powles, S.B. and Holtrum, J.A.M. (1993) Mechanisms and agronomic aspects of herbicide resistance. *Annual Review of Plant Physiology and Molecular Biology* 44, 203–229.

Hooten, T.M. and Levy, S.B. (2001) Antimicrobial resistance: a plan of action for community practice. *American Family Physician* 63, 1–14.

Horn, D.J. (1988) *Ecological Approach to Pest Management*. Guilford Press, New York.

Hoy, M.A. (1998) Myths, models and mitigations of resistance to pesticides. *Philosophical Transactions of Biological Sciences* 353, 1787–1795.

Huan, N.H. and Ang, D.T. (2000) Vietnam promotes solutions to pesticide risks. *AgroChemical Report* 11, 21–25.

Huang, J., Hu, R., Fan, C., Pray, C.E. and Rozelle, S. (2003) *Bt Cotton Benefits, Costs and Impacts in China*. IDS Working Paper 202, Institute of Development Studies (IDS), University of Sussex, Brighton, UK.

Ismael, Y., Bennett, R. and Morse, S. (2002) Benefits from Bt cotton use by small holder farmers in South Africa. *AgBioForum* 5, 1–5.

James, C. (2002) *Global Review of Commercialized Transgenic Crops 2001*. ISAAA Briefs 26, International Services in Advanced Agri-biotech Applications, Ithaca, New York.

James, C. (2004). *Global Status of Commercialized Biotech/GM Crops 2004*. ISAAA Briefs 32, International Services in Advanced Agri-biotech Applications, Ithaca, New York.

James, J.R., Tweedy, B.G. and Newby, L.C. (1993) Efforts by industry to improve the environmental safety of pesticides. *Annual Review of Phytopathology* 31, 423–439.

Kadir, A.A.S.A. and Barlow, H.S. (1993) *Pest Management and the Environment in 2000*. CAB International, Wallingford, UK.

Karlsson, S.I. (2004) Agricultural pesticides in developing countries: a multilevel governmental challenge. *Environment* 46, 22–41.

Kates, R.W. (2001) Queries on the human use of the earth. *Annual Review of Energy and Environment* 26, 1–26.

Knight, A.L. and Norton, G.W. (1989) Economics of agricultural pesticide resistance in arthropods. *Annual Review of Entomology* 34, 293–313.

Knight, S.C., Anthony, V.M., Brady, A.M., Greenland, A.J., Heaney, S.P., Murray, D.C., Powell, K.A., Schulz, M.A., Spinks, C.A., Worthington, P.A. and Youle, D. (1997) Rationale and perspectives on the development of fungicides. *Annual Review of Phytopathology* 35, 349–372.

Knutson, R.D., Taylor, C.R., Penson, J.B. and Smith, E.G. (1990) *Economic Impact for Reduced Chemical Uses*. Knutson and Associates, College Station, Texas.

Kogan, M. (1998) Integrated pest management: historical perspectives and contemporary developments. *Annual Review of Entomology* 43, 243–270.

Kogan, M. and Jepson, P. (2005) *Perspectives in Ecological Theory and Integrated Pest Management*. Cambridge University Press, Cambridge.

Kolcum, J., Dost, F., Moses, M. and Poje, G. (2003) Outline for chronic health effects of pesticides. Cornell Environment Risk Analysis Program, Ithaca, New York. Available at: http://pmep.cce.cornell.edu/issues/chronhealth.html

Koul, O., Dhaliwlal, G.S. and Cuperus, G.S. (2004) *Integrated Pest Management: Potential, Constraints and Challenges*. CAB International, Wallingford, UK.

Kovach, J., Petzoldt, C., Degni, J. and Tette, J. (1992) *A Method to Measure the Environmental Impact of Pesticides*. New York Food and Life Science Bulletin No. 139, New York State Agricultural Experimental Station, Geneva, New York.

Krattiger, A.F. (1997) *Insect Resistance in Crops: A Case Study of Bacillus thuringiensis (Bt) and its Transfer to Developing Countries*. ISAAA Briefs, No. 2, International Services for the Acquisition of Agri-Biotech Applications, Ithaca, New York.

Larson, L.L. (1998) Novel organic and natural product insect management tools. Available at: http://ipmworld.umn.edu

Lehman, H. (1993) New directions for pesticide use. In: Pimentel, D. and Lehman, H. (eds) *The Pesticide Question: Environment, Economics and Ethic*. Chapman & Hall, New York, pp. 3–9.

Levitan, L. (1997) *An Overview of Pesticide Risk Impact Systems (A.K.A. 'Pesticide Risk Indicators')*. Organization of Economic Cooperation and Development (OECD), Copenhagen, Denmark.

Levitan, L., Merwin, I. and Kovach, J. (1995) Assessing the relative environmental impacts of agricultural pesticides: the quest for a holistic method. *Agriculture Ecosystems and Environment* 55, 153–168.

Llewellyn, R.S., Lindner, R.K., Pannell, D.J. and Powles, S.B. (2001) Herbicide resistance and the decision to conserve the herbicide resource: review and framework. *Agribusiness Review* 29, 1–11.

Lupien, J.R. (2000) The Codex Alimentarious Commission: international science-based standards, guidelines, and recommendations. *AgBioForum* 3, 192–196.

Madsen, K.H. and Sandoe, P. (2005) Ethical reflections on herbicide-resistant crops. *Pest Management Science* 61, 318–325.

Marco, G.J., Hollingworth, R.M. and Durham, W. (1987) *Silent Spring Revisited*. American Chemical Society, Washington, DC.

Maredia, K.M. (2003) *Integrated Pest Management in the Global Arena*. CAB International, Wallingford, UK.

Maredia, K.M., Erbisch, F.H. and Sampaio, M.J. (2000) Technology transfer offices for developing countries. *Biotechnology and Development Monitor* 43, 15–18.

Mazur, B.J. and Falco, S.C. (1989) The development of herbicide resistant crops. *Annual Review of Plant Physiology and Molecular Biology* 40, 441–470.

McKinion, J.M., Baker, D.N., Whisler, F.D. and Lambert, J.R. (1989) Application of Gossym/Comax system to cotton crop management. *Agriculture Systems* 31, 55–65.

Mengech, A.N., Saxena, K.N. and Gopalan, H.N.B. (1995) *Integrated Pest Management in the Tropics: Current Status and Future Prospects.* John Wiley & Sons, New York.

Merrill, R.A. (1997) Food safety regulation: reforming the Delaney Clause. *Annual Review of Public Health* 18, 313–340.

Metcalf, R.L. and Luckmann, W.H. (1994) *Introduction to Insect Pest Management.* 3rd edn. John Wiley & Sons, New York.

Miller, A.S. and Holzman, L.R. (1994) Products liability and associated perceptions of risk: implications for technological innovation related to energy efficiency and environmental quality. *Annual Review of Energy and Environment* 19, 347–364.

Monaco, T.J., Ashton, F.M. and Weller, S.C. (2002) *Weed Science: Principles and Practice.* 4th edn. John Wiley & Sons, New York.

Mota-Sanchez, D., Bills, R.S. and Whalon, M.E. (2002) Arthropod resistance to pesticides: status and overview. In: Wheeler, W.B. (ed.) *Pesticides in Agriculture and the Environment.* Marcel Dekker, New York, pp. 241–272.

NAS (1980) *Regulating Pesticides.* National Academy Press, Washington, DC.

NAS (1987) *Regulating Pesticides in Food: The Delaney Paradox.* National Academy Press, Washington, DC.

NAS (2000) *The Future Role of Pesticides in US Agriculture.* National Academy Press, Washington, DC.

NAS (2004) *Pesticides in the Diets of Infants and Children.* National Academy Press, Washington, DC.

Nash, J. and Ehrenfeld, J. (1997) Codes of environmental management practice: assessing their potential as a tool for change. *Annual Review of Energy and Environment* 22, 487–535.

Naylor, R.L. (1996) Energy and resource constraints on intensive agricultural production. *Annual Review of Energy and Environment* 21, 99–123.

OECD (2001) *Survey of Best Practices in the Regulation of Pesticides in Twelve OECD Countries.* Organization for Economic Cooperation and Development, Series on Pesticides, No. 11 Paris, France.

Osteen, C.D. and Padgitt, M. (2002) Economic issues of agricultural pesticide use and policy in the United States. In: Wheeler, W.B. (ed.) *Pesticides in Agriculture and the Environment.* Marcel Dekker, New York, pp. 59–95.

Owen, M.D.K. and Zelaya, I.A. (2005) Herbicide-resistant crops and weed resistance to herbicides. *Pest Management Science* 61, 301–311.

PAN UK (1996) Pesticide use in China – A health and safety concern. *Pesticide news*, No. 34, December 1996.

Pedigo, L.P. (1996) *Entomology and Pest Management.* 2nd edn. Prentice-Hall, New York.

Penrose, L.J., Thwaite, W.G. and Bower, C.C. (1994) Rating index as a basis for decision making on pesticide use reduction and for accreditation of fruit produced under integrated pest management. *Crop Protection* 13, 146–152.

Perkins, J.H. and Holochuck, N.C. (1993) Pesticide: historical changes demand ethical choices. In: Pimentel, D. and Lehman, H. (eds) *The Pesticide Question: Environment, Economics and Ethics.* Chapman & Hall, New York, pp. 390–414.

Pilson, D. and Prendeville, H.R. (2004) Ecological effects of transgenic crops and the escape of transgenes into wild populations. *Annual Review of Ecology, Evolution and System* 35, 149–174.

Pimentel, D. (1991) *CRC Handbook of Pest Management in Agriculture.* Vols I, II and III, CRC Press, Boca Raton, Florida.

Pimentel, D. (1997) *Techniques for Reducing Pesticide Use: Economic and Environmental Benefits.* John Wiley & Sons, USA.

Pimentel, D. and Lehman, H. (1993) *The Pesticide Question: Environment, Economics and Ethic.* Chapman & Hall, New York.

Pimentel, D. and Levitan, L. (1991) Pesticides: amounts applied and amounts reaching pests. In: Pimentel, D. (ed.) *CRC Handbook of Pest Management in Agriculture*, Vol. I. CRC Press, Boca Raton, Florida, pp. 741–750.

Pimentel, D., Acquay, H., Biltonen, M., Rice, P., Silva, M., Nelson, J., Lipner, V., Giordano, S., Horowitz, A. and D'Amore, M. (1992) Environmental and economic costs of pesticide use. *Bioscience* 42, 750–760.

Prather, T.S., DiTomasco, J.M. and Holt, J.S. (2000) History, mechanisms, and strategies for prevention and management of herbicide resistant weeds. *Proceedings of the California Weed Science Society* 52, 155–163.

Ragsdale, N.N. (1999) The role of pesticides in agricultural crop protection. *Annals of New York Academy of Sciences* 894, 199–205.

Ragsdale, N.N. (2000) The impact of the food quality protection act on the future of plant disease management. *Annual Review of Phytopathology* 38, 577–596.

Ragsdale, N.N. and Kuhr, R.J. (1987) *Pesticides: Minimizing the Risks*. American Chemical Society, Washington, DC.

Raheja, A.K. (1995) Practice of IPM in South and Southeast Africa. In: Mengech, A.N., Saxena, K.N. and Gopalan, H.N.B. (eds) *Integrated Pest Management in the Tropics: Current Status and Future Prospects*. John Wiley & Sons, New York, pp. 69–120.

Reed, N.R. (2002) Risk assessment. In: Wheeler, W.B. (ed.) *Pesticides in Agriculture and the Environment*. Marcel Dekker, New York, pp. 97–125.

Reese, N. and Watson, D. (2000) *International Standards for Food Safety. International Food Safety Standards References*. Culinary and Hospitality Industry Publication Service, Aspen Publications, New York City, USA.

Reus, J., Leendertse, P., Bockstaller, C., Fomsgaard, I., Butshe, V., Lewis, K., Nilsson, C., Pursemier, L., Trevisan, M., van der Werf, H., Alfarroba, F., Blumel, S., Isart, J., McGrath, D. and Seppala, T. (2002) Comparison and evaluation of eight pesticide environmental risk indicators developed in Europe and recommendations for future use. *Agriculture Ecosystem and Environment* 90, 177–187.

Reynolds, J.D. (1997) International pesticide trade: is there any hope for the effective regulation of controlled substances? *Journal of Land Use and Environment Law* 13, 1–34.

Roberts, D., Karunarathna, A., Buckley, N., Manuweera, G., Rezvi Sheriff, M. and Eddleston, M. (2003) Influence of pesticide regulation on acute poisoning deaths in Sri Lanka. *Bulletin of the World Health Organization* 81, 789–798.

Ross, M.A. and Lembi, C.A. (1999) *Applied Weed Science*, 2nd edn. Prentice-Hall, New Jersey.

Roush, R.T. and Tabashnik, B.E. (1990) *Pesticide Resistance in Arthropods*. Chapman & Hall, New York.

Ruberson, J.R. (1998) *Handbook of Pest Management*. Marcel Dekker, New York.

Sagar, A.D. (2000) Capacity development for the environment: a view for the South, a view for the North. *Annual Review of Energy and Environment* 25, 377–439.

Salter, L. (1988) *Mandated Science: Science and Scientists in Making of Standards*. Kluwer Academic Publishers, Boston, Massachusetts.

Shaw, D.R., Rankins, A., Ruscoe, J.T. Jr and Byrd, J.D. Jr (1998) Field validation of weed control recommendations from HERB and SWC recommendations models. *Weed Technology* 12, 78–87.

Shelton, A.M., Zhao, J.Z. and Roush, R.T. (2002) Economics, ecological, food safety, and social consequences of the deployment of Bt transgenic plants. *Annual Review of Entomology* 45, 845–881.

Stark, J.D. and Banks, J.E. (2003) Population-level effects of pesticides and other toxicants on arthropods. *Annual Review of Entomology* 48, 505–519.

Stern, U.M., Smith, R.F., van den Bosch, R. and Hagen, K.S. (1959) The integration of chemical and biological control of the spotted alfalfa aphid; the integrated control concept. *Hilgardia* 29, 81–101.

Strange, R.N. (1993) *Plant Disease Control*. Chapman & Hall, London.

Sutton, T.B. (1996) Changing options for the control of deciduous fruit tree diseases. *Annual Review of Phytopathology* 34, 527–547.

Szmedra, P.I. (1991) Pesticide use in agriculture. In: Pimentel, D. (ed.) *CRC Handbook of Pest Management in Agriculture*, Vol. I. CRC Press, Boca Raton, Florida, pp. 649–677.

Tabashnik, B.E. (1994) Evolution of resistance to *Bacillus thuringiensis*. *Annual Review Entomology* 39, 47–79.

Tait, J. (2000) *Technology Transfer in Crop Protection: The Policy Context*. British Crop Protection Council Conference, Pests and Diseases, Brighton, UK.

Tait, J. (2001) Pesticide registration, product innovation, and public health. *Journal of Environment Monitoring*, 2, 64N–69N.

Tilman, D., Fargione, J., Wolff, B., D'Antonio, C., Dodson, A., Howarth, R., Schindler, D., Schlesinger, W.H., Simberloff, D. and Swackhamer, D. (2001) Forecasting agriculturally driven global environmental change. *Science* 292, 281–284.

Toenniessen, G. (2003) Opportunities for and challenges to plant biotechnology adoption in developing countries. In: Eagleshan, A., Ristow, S. and Hardy, R.W.F. (eds) *Biotechnology: Science and Society at a Crossroad*. National Agricultural Biotechnology Council, Report 15. Ithaca, New York, pp. 245–261.

Tomizawa, M. and Casida, J.E. (2003) Selective toxicity of neonicotinoids attributable to specificity of insect and mammalian nicotinic receptors. *Annual Review of Entomology* 48, 339–364.

Tren, R. and Bate, R. (2004) South Africa's war against malaria: lessons for the developing world. Cato Policy Analysis, No. 513, Cato Institutes, Washington, DC.

Van den Bosch, R. (1978) *The Pesticide Conspiracy.* University of California Press, Berkeley, California.

Van der Werf, H.M.G. and Zimmer, C.H. (1998) An indicator of pesticide environmental impact based on a fuzzy expert system. *Chemosphere* 36, 2225–2249.

Van Dyk, J.K. (2000) Impact of the Internet on extension entomology. *Annual Review of Entomology* 54, 795–802.

Waibel, H., Zadoks, J.C. and Fleisher, G. (2003) *Impact Assessment of Genetically Modified Plants: What Can We Learn from the Economics of Pesticides? Battling Resistance to Antibiotics and Pesticides.* Resources for the Future Press, Washington, DC.

Walker, J.T.S., Hodson, A.J., Batchelor, T.A., Manktelow, D.W. and Tomkins, A.R. (1997) A pesticide rating system for monitoring agrichemical inputs in New Zealand horticulture. In: O'Callaghan, M. (ed) *Proceedings of the 50th New Zealand Plant Protection Conference,* New Zealand Plant Protection Society, Inc., New Zealand, 529–534.

Wandiga, S.O. (2001) Use and distribution of organochlorine pesticides: the future in Africa. *Pure and Applied Chemistry* 73, 1147–1155.

Ware, G.W. and Whitacre, D.M. (2004a) *The Pesticide Book,* 6th edn. Meister Media Worldwide, Willoughby, Ohio.

Ware, G.W. and Whitacre, D.M. (2004b) *An Introduction to Insecticides,* 4th edn. Available at: http://ipmworld.umn.edu

Wargo, J. (1996) *Our Children's Toxic Legacy: How Science and Law Fail to Protect Us from Pesticides.* Yale University Press, New Haven, Connecticut.

Wenzel, R.P. and Edmond, M.B. (2000) Managing antibiotic resistance. *New England Journal of Medicine* 343, 1961–1963.

West, J.S., Bravo, C., Oberti, R., Lemaire, D., Moshou, D. and McCartney, H.A. (2003) The potential of optical canopy measurement for targeted control of field crop diseases. *Annual Review of Phytopathology* 41, 593–614.

Wheeler, W.B. (2002) *Pesticides in Agriculture and the Environment.* Marcel Dekker, New York.

Whitacre, D.M. and Ware, G.W. (2004) *An Introduction to Herbicides,* 2nd edn. Available at: http://ipmworld.umn.edu

Whitten, M.V. (1992) Pest management 2000: what we might learn from the 20th century. In: Kadir, A.A.S.A. and Barlow, H.S. (eds) *Pest Management and the Environment in 2000.* CAB International, Wallingford, UK, pp. 9–46.

WHO (1996) *Analysis of Government Responses to the Second Questionnaire on the State of Legislation of the Distribution and Use of Pesticides.* World Health Organization, Food and Agricultural Organization, United Nations, Rome.

WHO (1998) *Pesticide Residues in Food Evaluations:* Part II – *Toxicological and Environment.* World Health Organization, Geneva.

Whorton, J. (1974) *Before Silent Spring: Pesticides and Public Health in Pre-DDT America.* Princeton University Press, Princeton, New Jersey.

Wilkerson, G.G., Modena, S.A. and Coble, H.D. (1991) HERB: decision model for postemergence weed control in soybean. *Agronomy Journal* 83, 413–417.

Willson, H.R. (1996) *Pesticide Regulations.* Available at: http://ipmworld.umn.edu

Wolf, R.E. (2002) New technologies for the delivery of pesticides in agriculture. In: Wheeler, W.B. (ed.) *Pesticides in Agriculture and the Environment.* Marcel Dekker, New York, pp. 273–294.

Wolfenbarger, L.L. and Phifer, P.R. (2000) The ecological risks and benefits of genetically engineered plants. *Science* 290, 2088–2093.

Woteki, C.E. and Kineman, B.D. (2003) Challenges and approaches to reducing foodborne illness. *Annual Review of Nutrition* 23, 315–344.

Yanez, L., Ortiz, D. and Calderon, J. (2002) Chemical mixtures: overview of human health and chemical mixtures – problems facing developing countries. *Environmental Health Perspective, Supplement* 110, 901–909.

Yudelman, M., Ratta, A. and Nygaard, D. (1998) *Pest Management and Food Production: Looking to the Future.* 2020 Discussion Paper 25, International Food Policy Research Institute (IFPRI) Report, Vol. 20.

Zalom, F.G., Toscano, N.C. and Bryrne, F.J. (2005) Managing resistance is critical to future use of pyrethroids and neonicotinoids. *California Agriculture* 59, 11–15.

Zethner, O. (1995) Practice of integrated pest management in tropical and sub-tropical Africa: an overview of two decades (1970–1990). In: Mengech, A.N., Saxena, K.N. and Gopalan, H.N.B. (eds) *Integrated Pest Management in the Tropics: Current Status and Future Prospects.* John Wiley & Sons, New York, pp. 1–68.

Zimdahl, R.L. (1999) *Fundamentals of Weed Science,* 2nd edn. Academic Press, New York.

7 Sociology In Integrated Pest Management

Johann Baumgärtner[1,2], Achola O. Pala[1] and Pasquale Trematerra[3]

[1]International Centre of Insect Physiology and Ecology (ICIPE), Nairobi, Kenya;
[2]Center for Analysis of Sustainable Agricultural Systems (CASAS), Kensington,
USA; [3]Università degli Studi del Molise, Campobasso, Italy

Introduction

Agriculture, the study and practice of cultivating land for growing crops and rearing livestock (Anonymous, 1996), has been the subject of human activities since 10,000 years (Smith, 1995). In response to the needs of the society, agriculture in general and agricultural practices in particular have changed over the millennia and finally led to a gradual intensification of the production over a number of centuries (Dent, 1995 a, b). The increasing demand for food production during the second half of the 20th century has seen many developments in agricultural technology and practices that have greatly increased crop and livestock production (Anonymous, 1996). However, these advances had their drawbacks as exemplified by pest control technologies characterized by unilateral reliance on synthetic pesticides (Flint and van den Bosch, 1981). In response to these drawbacks and the increasing demand for agricultural goods, agriculture has been evolving into a more holistic discipline with ecosystem sustainability as a major objective (Altieri, 1991). In this process, relevant disciplines rely on an array of approaches and methodologies.

Pest is a label applied by man and has no ecological meaning (Luckmann and Metcalf, 1975). Pests are any of the various organisms, e.g. fungi, insects, rodents and plants, that harm crops or livestock or otherwise interfere with the well-being of humans (Anonymous, 1996). Hence, control programmes are developed and implemented against a wide array of organisms living in highly diverse environments. For the sake of simplicity, we focus on arthropod pests in agricultural production including crop and livestock systems.

In biological conservation programmes, the inclusion of institutions into ecosystem management is considered indispensable for successful project execution (Meffe et al., 2002; Brown, 2003). In design and implementation of integrated pest management (IPM) projects, Conway (1984) took into account a hierarchical arrangement of subjects of management (e.g. field, farm and regions) and decision-making institutions (e.g. farmers, extensionists and policymakers). According to him, there is a relationship among the type of intervention, the decision-makers and the subjects of management. For example, farmers undertake *tactical* measures to control a pest occurring at a particular moment in a particular field. Pest and crop management also relies on *strategic* measures, i.e. on application of general rules that farmers have developed within their cultural milieu or those that extensionists transfer to farmers. On national and international levels decisions are taken and implemented by policymakers and administrators. These institutions were considered by Kogan (1998), Kogan *et al.* (1999) and Baumgärtner *et al.* (2003a), who described the development and use of multilevel IPM systems.

Conway (1984) refers to decision makers when describing persons and institutions involved in IPM. Meffe and Caroll (1997) use the term 'inclusion' for those involved in making decisions and for what is being managed. Baumgärtner *et al.* (2003c) and Herren *et al.* (2004) refer to organized segments of the society and use the term 'institution' for characterizing individuals and organizations involved in IPM and ecosystem management. Waltner-Toews *et al.* (2003) emphasize interactions between humans and ecosystems and refer to 'people' and 'communities' firmly embedded as integral elements in ecosocial systems. Their approach inspired us to define, first, a social system acting on an ecological system, and, second, a social system interacting with an ecological system to form a unified whole.

Among others, Baumgärtner *et al.* (2003c), Gilioli *et al.* (2003, 2005) and Herren *et al.* (2004) also stress the importance of considering ecosystem complexity in ecosystem management. They rely on Jørgensen (2002) who considers, among others, complexity, multiple causalities and uncertainties as important attributes of ecosystems. To address such qualities, a systems approach and adaptive ecological system management procedures are appropriate (Huffaker and Croft, 1976; Holling, 1978; Walters, 1986). Comiskey *et al.* (1999) regard adaptive management as a systematic, cyclical process for continually improving management policies and practices (tactics and strategies) based on lessons learnt from operational activities. Adaptive management enables managers to act in a flexible and adaptable manner in the face of uncertainty and lack of knowledge (Haney and Boyce, 1996). The emphasis given to adaptive management

requires reconsideration of institutions, their relationships and responsibilities in IPM (Baumgärtner *et al.*, 2003c).

The purpose of this chapter is to review the history of IPM and present some elements of system theory relevant for representing structure and functioning of both social and ecological systems, in order to create a basis for assessing the role of institutions involved in IPM. The approach allows the schematization of social systems and permits the extension of IPM into comprehensive agricultural production management schemes, and subsequently into ecosocial management schemes. In ecosocial systems, the social and ecological systems interact and both become components of a unified whole. This chapter is not limited to social systems, but considers the structure and functioning of ecological systems as well. There are two reasons for this; first, the use of common concepts for representing both socio-economic and ecological systems facilitates research and implementation (Gutierrez, 1996) and secondly, it addresses the concerns of anthropologists who consider current socio-economic approaches insufficient because they do not satisfactorily take into account the complexities of ecological systems (Abell and Stepp, 2003). Some of these concepts are illustrated with examples of pest management in developing and industrialized agricultural systems.

IPM at the Interface Between Ecological and Economic Systems

Since the distant past, yield increase of agricultural crops was sought by selecting improved varieties and further developing cultural methods. Synthetic fertilizers and pesticides became available only a few decades ago and are increasingly used since then. However, unilateral reliance on synthetic pesticides had drawbacks that stimulated the interest in old methods and research and development of alternatives in the areas of pest forecasting, biological control, autocidal pest control, host plant resistance, cultural and mechanical methods as well as chemical control (Flint and van den Bosch, 1981). The latter was complemented with pest control substances originating from plants and with behaviour-modifying chemicals (Gut *et al.*, 2004). During the past few years, genetically modified cultivars that increase pest mortalities were made available for pest control. Among many others, Bernal *et al.* (2004) review the opportunities and challenges, and Professor Gutierrez evaluates their use in an ecological context (Professor A.P. Gutierrez, University of California, Berkeley, 2004, personal communication).

The drawbacks of unilateral reliance on traditional synthetic pesticides also stimulated the development of IPM. IPM has been defined as a pest management system that, in the context of the associated environment and the population dynamics of the pest species, utilizes all suitable techniques and methods in as compatible a manner as possible and maintains the pest populations at levels below those causing economic injury (Flint and van den Bosch, 1981). It is noteworthy that the consideration of pest population dynamics and thresholds introduced concepts of population ecology and ecosystem science as well as economics into pest management (Luckmann and Metcalf, 1975; Huffaker and Croft, 1976; Getz and Gutierrez, 1982; Delucchi, 1987a; Leslie and Cuperus, 1993; Dent, 1995a; Gutierrez, 1996; Koul *et al.*, 2004). Specifically, the IPM concept was further developed along the following lines.

Often, IPM is thought to be similar to threshold-based control. Accordingly, a farmer wants to maximize the difference between the value of production and the costs of reducing pest competition (Headley, 1975). So he takes into account a pest threshold at which the ratio of incremental benefits and incremental costs are equal to one. However, Headley (1975) has already pointed out that interdependence of decisions at different time periods and uncertainties, putting the farmer at risk, is disregarded in this concept. Today, risk–benefit analyses are considered very important in design and implementation

of IPM programmes, particularly when evaluating the cultivation of transgenic crops (Koul *et al.*, 2004). Roux and Baumgärtner (1998) evaluated uncertainties in the effects of control resources on potato tuberworm, *Phthorimaea operculella* (Zeller), in Tunisian rustic potato shelters. They concluded that farmers relying on biological control of existing natural enemies face a risk and recommended adjustment of the dosage of virus applications.

In the simplest situation, the individual farmer is not willing to allocate any more resources to pest control than can be justified by expected damages to be prevented by the use of such resources (Headley, 1975). However, Headley (1975) has already pointed out that IPM is not restricted to private economics of pest management, but deals with socio-economics as well and can have an impact on society as a whole. He stresses that all costs and benefits of pest control have to be considered and properly added, if not, society will not receive the maximum output from IPM activities. Since Headley's (1975) pioneering work, pest management has continued to change in response to the needs of the society. Today, farming systems are expected to be ecologically, economically and socially acceptable (Delucchi, 2001; Koul *et al.*, 2004). To meet such objectives, the economic standpoint of the early IPM area is too narrow and socio-economic considerations become important in IPM. Koul *et al.* (2004) emphasize extension of IPM concepts from the production into the storage and food processing sectors.

Traditional pest control generally dealt with the control of pests in specific fields, but IPM emphasized the undertaking of control operation in relation to population dynamics of pest populations. As a result of innumerable studies on pest population dynamics, the system for study and management could not be restricted to individual fields and short-term control operations. Instead, IPM was extended to include arrangements of crops and unmanaged land in space and the sequence of crop arrangements through time. Although the ecosystem context was emphasized from the beginning of the IPM era (Luckmann and Metcalf, 1975), decision makers often focused on single species populations occurring at some plant growth stages in a particular field. The management of single species pest populations is still the subject of many IPM programmes, but the simultaneous consideration of pest assemblages and the development of crop-specific IPM schemes has become more common than previous practice. According to van Lenteren (1995), IPM schemes in greenhouses are successful because they simultaneously deal with multiple constraints to crop production.

At the beginning of the IPM era, biological control was considered a control element similar to other control elements such as mechanical or chemical control. Now, it is considered as a key component in IPM programmes (van Driesche and Bellows, 1996). The inclusion of multiple pest species and natural enemy assemblages resulted in an extension of IPM towards more complex subjects of management (Gilioli et al., 2003). Hence, the subject of study and management shifted from single species pest control to ecological system management, i.e. the purposeful handling of natural units composed of populations, communities and their environment (Viggiani, 1977; Delucchi, 1987b; Gold, 1999; Baumgärtner et al., 2003c; Gilioli et al., 2003). Sustainability is considered the primary objective of ecosystem management (Christensen et al., 1996).

Sustainable agricultural systems have been developed to maximize internal resources of the farm, minimize environmental impacts and increase profits by reducing external inputs (Linker et al., 1991). In order to meet sustainability criteria, IPM must adopt current practices that are compatible with reduced input methodologies and rely on systems approach to farm management (Linker et al., 1991). Cuperus et al. (2004) recognize the essential role of IPM in promoting sustainability of agricultural systems. Their basic premise is that employment of principles of IPM is essential to optimize the sustainability of agricultural systems, and they support the concept that improving the sustainability of

production systems and implementation of IPM must be linked. To assure sustainability of agricultural production systems, economic, environmental and social mandates have to be addressed and contributions of IPM are of critical importance (Cuperus et al., 2004).

According to Conway (1984), farmers are undertaking tactical measures, the agricultural service or health board may be engaged in strategic measures, while regional, national and international institutions may be responsible for policy measures. At the early stages of IPM, the development and implementation of tactical measures might have received most attention by IPM specialists. In our assessment, further development of IPM leads to a more balanced consideration of institutions. Leslie and Cuperus (1993) list comparative advantages of IPM over traditional methods in major US crops and identify opportunities for further development in institutional, federal policy, regulatory and educational rather than technical areas.

In conclusion, the IPM concept has evolved on both ecological and socio-economic grounds, and thereby proved to be robust to include new concepts and answer to criticism regarding focus and methodologies (Kogan, 1998; Kogan et al., 1999; Royer et al., 1999). Basic concepts and philosophy of IPM have largely been accepted for some time, but the implementation is complex and faces numerous potential problems (Koul et al., 2004). To some extent, this is due to the complexities inherent in subjects of management and the adjustment to new management objectives. Specifically, the concept has been extended into time, space, subject of management and institutional dimensions. Remarkably, this extension is also observed in conservation ecology (Meffe and Carroll, 1997; Meffe et al., 2002). Hence, Baumgärtner et al. (2003c), Gilioli et al. (2003) and Herren et al. (2004) observe a convergence of agricultural and ecological disciplines at the ecosystem management level. Figure 7.1 shows the extension of traditional resource management into time, space, subject of management and institutional dimensions as described by Meffe and Carroll (1997).

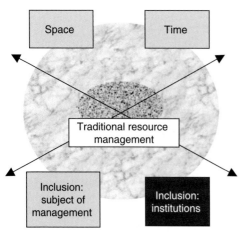

Fig. 7.1. The dimensional framework considered in conservation ecology. Accordingly, there is an extension of traditional resource management into space, time and inclusion dimensions. (Excerpted from Meffe and Carroll, 1997. Reproduced and modified with permission.)

Hierarchical organization and dynamics of ecological and socio-economic systems

Conceptually, IPM operates at the interface of the ecological and the socio-economic systems (Kogan *et al.*, 1999). Both systems are characterized by high complexity. The complexity of an ecosystem is not formed only by a high number of interacting components; it is far more complex (Jørgensen, 2002). Among the nine forms of complexity, there is a high degree of heterogeneity in space and time. An extremely high number of feed-backs and regulations make it possible for living organisms and populations to survive and reproduce in spite of changes in external conditions; the components and their related processes are organized hierarchically.

The extension into the aforementioned four dimensions (Fig. 7.1) requires consideration of scales, i.e. spatial or temporal extensions of an object or process (Turner and Gardner, 1991). Scaling theory suggests that hierarchical levels may occur at distinct breaks in the scale continuum (Wiens, 1989; Kemp *et al.*, 2001; Baumgärtner *et al.*,

2002). Hierarchies are typical classification methods (Hári and Müller, 2000). Since the hierarchical approach seems to be founded in the general pattern of man's structured thinking, it was tempting and very natural to use the conceptions in theoretical approaches (Hári and Müller, 2000). The concept of hierarchy theory was developed as a describing tool by general system theory in the 1960s (Hári and Müller, 2000). According to general system theory a system cannot be understood by looking at a single component – one must consider both the system and its elements. Consequently, one defines elements as well as the system and its boundaries, to develop computational expressions that apply to the system as a whole (Scheurer *et al.*, 2001).

The term 'hierarchy' originally referred to a vertical authority structure in human organization, meaning that one system contains another, smaller one, which contains an even smaller system and so on; however, the hierarchical order is not a straight line but branched like a tree (Hári and Müller, 2000). On each level, there are subsystems hierarchically and vertically connected with each other that comprise horizontal interactions between subunits at the same level. Each of the subsystems is at the same time part of a superior level and an organizational unit for the inferior level, which represents a kind of 'janus-faced' duality of single levels (Koestler, 1967; Hári and Müller, 2000). This means that all properties of the lower levels can also be found on the next higher one where their sum is the feature of the whole. However, new properties emerge on the higher level. Subsystems interact, and there are two kinds of interaction: a vertical one on the same level and a horizontal one between different levels of hierarchy. Interactions on the same level usually have minor significance for the system's behaviour and development, while relationships between different levels are more significant. Among other features of particular interest to this work is the response: superior levels have broad spatial extent and they act or react very slowly, while inferior levels have a small spatial extent and operate at faster rates (Hári and Müller, 2000). As a consequence, large and slow-acting levels provide constraints for small and fast-acting levels (Hári and Müller, 2000).

In socio-economics, hierarchical structures and their applications have been known for a long time in social communities and social sciences. Hári and Müller (2000) refer to Mesarovic *et al.* (1970), Pattee (1973) and Simon (1973), and observe that the concept has been widely utilized in the discussion of evolution and functioning of systems. IPM specialists follow Conway (1984), who included an organized social system (decision makers) in the hierarchical arrangement of pest control systems. According to Röling and Jiggins (1998), the institutions involved in IPM can be seen as potentially forming a soft system, i.e. a system that does not 'exist'. The construct of soft social systems, however, allows the exploration of possibilities for collaboration and enhancement of synergism and innovative performance.

In ecology, Allen and Starr (1982) developed the hierarchy theory. They extended the traditional hierarchy of organisms, populations and communities to a variety of ecological phenomena (Scheurer *et al.*, 2001). IPM specialists assert that the ecological systems under management have a hierarchical structure (Conway, 1984; Kogan *et al.*, 1999; Baumgärtner *et al.*, 2002, 2003a). However, the hierarchical organization of nature is considered a basic presumption in modern ecology (Naveh and Liebermann, 1984). Moreover, the classical notion of hierarchy involves discrete levels, which may be recognized by convenience, not truth (Allen and Starr, 1982), and is considered a useful construct in ecosystem research (Scheurer *et al.*, 2001). Allen and Starr (1982) emphasize that the discussion of levels is concerned with experience and restricted to what is known. Accordingly, levels are the subjects of epistemological discourse rather than ontological discourse referring to existence in terms of objective external reality. However, Hári and Müller (2000) analysed gradients and structures in forest ecosystems and concluded that spatial hierarchies exist and provide great influence on ecosystem performances.

Some integration levels relevant to integrated management of ecosystems are summarized in Fig. 7.2 (Herren *et al.*, 2004). The arrows originate at levels that have traditionally been the subject of integrated pest control. They point to levels that have received increased interest in the past decades.

Hierarchy theory refers to the structure of systems and has been applied to both social and ecological systems. The emphasis given to the structure of social and ecological systems does not imply undermining the importance of dynamical system properties. Here, we refer the reader to the relevant literature and exclusively focus on an attempt that has been made to formulate a common approach for representing the functioning of these systems. Gutierrez (1996) applied a supply/demand approach to population systems. Accordingly, all populations must face the problem of resource acquisition and allocation, referred to as functional and numerical responses in ecology. In ecological systems, the shapes of the resource acquisition functions are similar at all trophic levels, and allocation of resources has the same priorities. In economic models, the benefit or resource acquisition function has the same yield–effort relationship, and the cost of resource acquisition increases as the effort required to obtain them increases. The institution involved may set priorities for resource allocation. This approach facilitates coupling of ecological and socio-economic systems.

Kogan *et al.* (1999) place IPM at the interface of socio-economic and ecological systems, both of which are hierarchically structured into levels of organization. Traditionally, the two systems have been separated, and the socio-economic system is seen as the manager of the ecological system and needs a separate discussion.

IPM from an institutional standpoint

Dent (1995c) described the inclusion of socio-economics into pest management. Briefly, a farmer was narrowly seen as a recipient of the benefits derived from IPM

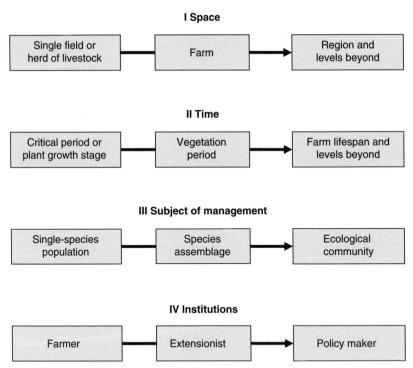

Fig. 7.2.   Some levels considered in integrated pest management (IPM). (Based on the work by Conway, 1984; Kogan, 1998; Kogan *et al.*, 1999; Baumgärtner *et al.*, 2002, 2003a; and Herren *et al.*, 2004.)

research. Now, farmers participate in IPM project preparation, research, development and implementation. Therefore, farmers' perceptions, needs, objectives and resource constraints are taken into account. Moreover, IPM is placed in the context of the overall farming system and the social and political forces acting on them. Socio-economic information on current pest control practices, costs and returns and other internal aspects of farmers' circumstances are indispensable for guiding project preparations and surveys that are undertaken (Dent, 1995c). Hence, the one-sided relationship between specialists and farmers has been changed into an interaction, and the survey becomes an opportunity for learning and exchanging ideas, skills and knowledge (Chambers *et al.*, 1989; Norton and Mumford, 1993; Dent, 1995c). The availability of pest control means should also be assessed, as lack of capital for purchasing control means and poor infrastructures restricting access to

control means can be impediments for IPM projects (Kenmore, 1997).

Farmers rely on fundamental information, historical information, real-time information as well as forecast information, and their perceptions influence their decisions (Dent, 1995c). Many of the aspects discussed so far are valid for industrial agricultural enterprises, but not necessarily applicable to subsistence farming in general and to African conditions in particular. According to Zethner (1995), African agriculture is characterized by a very large majority of small-scale farmers who cultivate small landholdings of less than 1 ha to a few hectares. Importantly, there are great differences with regard to landownership and tenure; in most countries farmers do not have title to the land they cultivate, but only the right to farm the land, which they and their ancestors have customarily cultivated. Although the majority of small-scale farmers control the land they cultivate, some are tenants or sharecroppers on land controlled by others. Tenants and sharecroppers are, however, far less abundant than in South Asia. Zethner (1995) points out that ownership and/or rights to farm are important once it comes to implementing IPM; only people who are sure of reaping the fruits of their efforts will be willing to invest the extra work and sometimes money necessary for implementing IPM. In our experience, African farmers are exposed to higher risks of pest damage than their colleagues in the developed world; in response, they may rely on diversification of crops, mixed cropping and low inputs. Moreover, they are poor and the cash they earn is not enough to invest in cultivation (Zethner, 1995).

The structure of institutions

At different levels in the hierarchical organization of the social systems there are horizontally integrated institutions. Here, we present a non-exhaustive list of institutions with respect responsibilities, and make some tentative remarks regarding their vertical integration at the end of this section.

FARMERS. Within a policy framework, farmers are assumed to follow strategic rules and undertake tactical measures justified on ecological, economic and social grounds (Conway, 1984; Delucchi, 2001). In the past, they have often been seen as users of technologies made available by agricultural research through the extension service; today, they are considered as participants in technology research and implementation (Röling, 1995). Their education is seen as a key policy instrument for governments seeking to improve the productivity of agriculture while protecting the environment (Feder et al., 2004). Some studies on economic viability of farmer training yield high rates of return while others observe poor performance. They recommend participatory approaches to alleviate inherent weaknesses in knowledge transfer systems (Feder et al., 2001, 2004).

Some of the problems may be overcome in an apprenticeship system that combines learning at public schools with practical work on farms operated by 'master' farmers who have been recognized by public institutions on the basis of their knowledge and ability to successfully manage a farm. Among the approaches and technologies on resource management for poverty alleviation and technologies, Ethiopian farmers and institutions showed great interest in the US 4-H Youth Development Program and Future Farmers of America (FFA) programmes (Aseffa et al., 2003). Therefore, a public education programme engages a diverse group of American young people in 'hands-on' learning experiences. The experimental youth development programme is a successful strategy to effectively engage youth and develop their academic, leadership and vocational capacities for practical community problem solving and overall sustainable development (Brenner and Desmond, 2003).

GROUP LEADERS. Groups of farmers may have a stake in production technologies including organic farming and IPM. Membership in such a group provides advantages in marketing farm products, access to information and opportunities for obtaining subsidies. For example, the Swiss Confederation and Swiss Regional Governments support environment-friendly farming including biodiversity conservation and integrated farming (Baumgärtner and Hartmann, 2000). In the Canton of the Grisons, groups of about 20 farmers have been formed. Each group elects a leader responsible for checking farming practices to conform to the rules using standardized questionnaires, working at the interface between the group and the cantonal supervisory authorities, and advising farmers in case of non-compliance.

An emerging new development in Western Kenya of local social organizations known as cereal banks is also proving to be a new organizational mechanism for farmer groups to control maize grain quality and storage pests, and to aggregate sufficient grain to enable them to access favourable market opportunities (Kelly et al., 2003; Rockefeller Foundation, 2003).

EXTENSIONISTS. Agricultural extension is an additional key policy instrument for governments seeking to improve the productivity of agriculture while protecting the environment (Feder *et al.*, 2004). However, the experience of extension systems over the past decades has been mixed: some studies estimate high rates of return to the investment in extension (Birkhaeuser *et al.*, 1996), while many observers document poor performance in the operation of extension (Feder *et al.*, 2004). That, in part, is the tendency of many public officers dealing with knowledge transmission in a 'top–down' manner, according to which extension transfers the findings of agricultural research to farmers. This is perceived as a less effective method when compared to more participatory approaches (Röling, 1995; Feder *et al.*, 2004).

The transfer of technologies from research to practice has been referred to as a linear model and qualified as having serious limitations (Röling, 1995). In fact, innovations are not seen as unchanging commodities that pass from one to another, but are reinvented as they are adopted (Rogers, 2003). Moreover, new ideas often spring from practice rather than research (Kline and Rosenberg, 1986).

ECOTRAINERS. Often, the results obtained at research centres are not efficiently passed to end-users. In collaboration with BioVision foundation, the Nairobi-based International Centre of Insect Physiology and Ecology (ICIPE) relies on ecotrainers to disseminate new pest control technologies (Dr. H.R. Herren, International Centre of Insect Physiology and Ecology, Nairobi, 2004, personal communication). Their roles are to: (i) propagate and inform potential end-users about new ICIPE technologies and information; (ii) identify interested farmers and groups for training; (iii) identify potential collaborators; (iv) carry out needs assessment for the identified farmer groups and develop a joint plan of work; (v) mobilize required resources to facilitate the organization, planning and implementation of training for the trainers; and (vi) organize for joint (all partners) monitoring, evaluation and follow-up activities with the farmer

groups. The facilitators stationed at ICIPE will also assist as mass communication specialists (BioVision, 2004).

The difference between extensionists and ecotrainers is that the latter seek ways to enable farmers to combine their own indigenous knowledge with scientific advances in agricultural sciences (BioVision, 2004). Ecotrainers build the technical and leadership capacity of farmers to make them more aware of the ecological conditions surrounding them so that they can build on ecological possibilities for enhanced results. In general, extensionists tend to be rule-oriented and at times inflexible because they are obliged to follow the demands of a hierarchical decision-making system that relies on technological/formulaic solutions to agricultural problems, sometimes regardless of the agroecological constraints faced by the farmers they seek to advice (BioVision, 2004). However, there is a great variability among extension services and extensionists and many of them meet the expectations of farmers. The degree to which scientists are willing to undertake large-scale experiments with farmers as partners will greatly boost the potential of ecotrainers.

FACILITATORS AND FARMER FIELD SCHOOLS. To overcome the problems of traditional education and extension, the farmer field schools (FFS) approach has been developed and implemented as an effective approach to extend science-based knowledge and practices to farmers. The approach is an intensive participatory training programme emphasizing IPM (Feder *et al.*, 2004), which relies on participatory methods to help farmers develop their analytical skills, critical thinking and creativity, and help them make better decisions (Kenmore, 1997). Rather than working with teachers and trainers, the farmers are guided by a facilitator who, according to Röling and van de Fliert (1994), represents a paradigm shift in extension work on the basis of non-formal education of farmers, which is itself a 'learner-centred' discovery process. It seeks to empower people to solve living problems actively by fostering participation, self-confidence, dialogue, joint decision making and self-determination. The 'discovery learning' by farmers on the basis of 'agroecosystem analysis', which uses their own field observation, is science-informed.

A facilitator, typically a government employee, but in some cases a non-governmental organization (NGO) or especially trained farmer, heads this village-level programme, focusing initially on problem-solving approaches in pest management, but also conveying knowledge pertaining to overall good crop management procedures and practices (Feder et al., 2004).

The FFS approach was introduced in Indonesia in 1989 for disseminating IPM technology among rice growers (van de Fliert, 1993). Selection of participants takes place at a meeting led by the IPM Field School facilitator with the members of the Farmers' Group from which participants will be drawn. The facilitator explains the approach to prospective participants, that they are expected to attend every week for the duration of the season. Prospective participants are given an opportunity to either agree the 'learning contract' or withdraw. The basic format of an IPM FFS consists of three activities: agroecosystem observation, analysis and presentation of results; a 'special topic'; and a 'group dynamics' activity. Agroecosystem analysis is the FFS core activity, and other activities are designed to support it. The agroecosystem analysis process sharpens farmers' skills in the areas of observation and decision making and helps develop their powers of critical thinking.

There are conflicting results of impact studies on the FFS approach (Feder et al., 2004). On one hand, studies carried out in South-east Asia claimed that pesticide applications decreased with more IPM knowledge and FFS training, while rice yields increased as much as 25%, and profits on Asian and African farms increased up to 40%; on the other hand, large-scale and long-term studies of farm-level outputs (yields and pesticide use) in Indonesia indicate that the programme did not have significant impacts on the performance of graduates and their neighbours. In Malawi, limited knowledge on yield-reducing factors

and pest biology were among the aspects responsible for the failure of an IPM project with an FFS component (Orr and Ritchi, 2004). According to Feder et al. (2004), there are opportunities for reorienting the FFS initiative to improve the likelihood of economic viability. Among them is to focus the training on highest priority topics, while simplifying the presentation of the information.

COMMUNITIES AND COMMUNITY COMMITTEES. Community IPM, oriented towards agricultural system management, has emerged from training programmes organized by Government agencies and NGOs in various parts of Asia. Community IPM has been established by graduates of FFS and relies on farmers who become producers rather than recipients of IPM (The Field Alliance, 2003).

Examples of community IPM activities where training is prominent are: FFS conducted by IPM farmers for other farmers, incorporation of IPM into the curricula of local schools and IPM as part of functional literacy programmes. Examples of activities where experimentation is prominent are: insect zoos and compensation studies managed by farmers as part of FFS organized by the government or NGOs, field studies that are organized and implemented by groups of FFS graduates and action research facilities involving a number of studies carried out by IPM farmers over a number of cropping seasons (The Field Alliance, 2003).

Local agricultural research committees (CIALs), composed of farmers, have been created to strengthen the capacity of farmers and rural communities as decision makers and innovators of agricultural solutions. Moreover, they empower rural communities to exert pressure on the formal research system and link local research with the formal system by providing access to new skills, information and research products that can be useful at the local level (Brown et al., 2000). This corresponds to the standard protocol in the US extension service for nearly 100 years (Rogers, 2003).

Community project committees, oriented towards ecosystem management

(Baumgärtner *et al.*, 2003b), operate on the basis of institutional diversity as a prerequisite for sustainability (Becker and Ostrom, 1995) and sustainability objectives (Goodland, 1995). In Ethiopia, an on-going ecosystem management project aiming at human health improvement and poverty alleviation relies on existing organizational structure of communities (Aseffa *et al.*, 2003; Herren *et al.*, 2004). To deal with particular constraints the community committee created subcommittees, e.g. a tsetse control subcommittee responsible for carrying out mass-trapping operations to control tsetse and reduce cattle disease prevalence. Importantly, the subcommittee members were not restricted to farmers, but were open to other groups of the society. Moreover, the responsibility of these subcommittees was not limited to IPM activities but included general resource management. The creation and activities of these subcommittees were guided after the design principles for sustainable agriculture identified by Becker and Ostrom (1995).

POLICYMAKERS. Adopting *Webster*'s (1976) definition to IPM, policy refers to a high-level overall plan embracing the general goals and acceptable procedures of executive bodies operating at the community, local, national and international levels. Policymaking is thus the high-level elaboration of municipal, governmental or international organizations' policy. A brief visit to the Internet readily provides examples of policymaking on these three levels. The Canadian municipality of Saanich has adopted a pesticide policy, and according to that all pest controls within the jurisdiction of the municipality must be conducted through an IPM approach (Saanich, 2003). The Ministry of Agriculture of the Government of the People's Republic of Bangladesh drafted a National Integrated Pest Management Policy according to which IPM will be the main policy for controlling pests and diseases (Bangladesh Department of Agricultural Extension, 2002). The International Food Policy Research Institute identifies and analyses policies for sustainably meeting the food

needs of the developing world (IFPRI, 2004). A global multilateral institution such as the United Nations supports agricultural policy development for all the three levels through its specialized agencies such as the Food and Agriculture Organizations (FAO) within a global multilateral context.

RESEARCHERS. At the beginning of the IPM era, researchers were focusing on pest control technology research, though emphasis was also given to the ecosystem (Luckmann and Metcalf, 1975). Further IPM development clearly showed the need to shift from technology research to ecological systems management, and IPM specialists advocated a systems approach (Huffaker and Croft, 1976; Getz and Gutierrrez, 1982; Kogan *et al.*, 1999). Traditionally, IPM research was primarily relying on laboratory work and field experimentation, while the complexities of ecological systems and the recent extension into space, time and inclusion dimensions (Fig. 7.1) stimulated the adoption of adaptive management approaches (Holling, 1978; Meffe *et al.*, 2002; Baumgärtner, *et al.*, 2003c).

At the beginning of the IPM era, researchers passed the results of their work through the extension system to end-users. This linear technology transfer has been replaced by more participative approaches.

Research and development of organic and sustainable agricultural systems rely on adequate infrastructures and team work at university institutes (e.g. Institute of Organic Farming, University of Agricultural Sciences, Vienna, Austria) or specialized centres (e.g. Research Institute of Organic Agriculture, Frick, Switzerland; Leopold Center for Sustainable Agriculture, Ames, Iowa, USA; Kerr Center for Sustainable Agriculture, Poteau, Oklahoma, USA; Kenya Institute for Organic Farming, Nairobi, Kenya).

OTHERS. An array of legislative executives and judicial bodies, relying on administrators, are involved in the design and implementation of IPM programmes and projects. A presentation of their respective responsibilities goes beyond the scope of this chapter.

The brief overview on the role and responsibilities of institutions involved in IPM indicates both horizontal and vertical linkages. Increasingly, an organized structure of institutions appears to emerge and deal with ecological systems in a more coordinated way than in the past. This may enhance innovations that may be seen as emergent properties of soft social systems (Röling and Jiggins, 1998). Horizontally, African and Asian farmers controlling the land may be joined by tenants or sharecroppers in IPM activities (Zethner, 1995). Vertically, farmers create groups to adjust farming operations to changing agricultural policies. Participatory approaches have replaced the linear transfer of agricultural research findings, through extension, to the end-users (Röling, 1995). The FFS approach illustrates a development, in that facilitators replace trainers to guide IPM programme activities (Feder *et al.*, 2004). The institutions involved no longer deal with agricultural systems, but develop multidisciplinary structures for managing the ecosystems of which some of them are part. In the latter case, the institutions included are national administrative and political structures that operate according to a national development agenda. In summary, IPM has been developed into space and time dimensions requiring the involvement of a broader range of institutions than at the beginning of the IPM era. The subject of management is increasingly seen as a complex system that can only be successfully managed by an organized social system. At each level of Conway's (1984) hierarchical arrangement of decision makers (Fig. 7.2), organized actors or institutions deal with a component of the ecological system. Vertically, the interactions among farmers, extensionists, researchers and policymakers lead to priority setting in agricultural research. The diversity of institutions is seen as a prerequisite for sustainability (Becker and Ostrom, 1995).

In IPM, little attempt appears to have been made so far to rigorously analyse structure and functioning of social systems. Presumably, the application of the hierarchy theory to the social system would pave the road to better structure decision making with respect to IPM. For example, the definition of levels with responsibilities should include lower-level elements and take into account constraints by upper levels. Furthermore, the flow of information across levels could be rendered more efficient when considering the needs and terms of references of individual levels. Therefore, socio-economists should be aware that the social system manages a complex ecological system. The tendency to simplify it is seen as a major hindrance in the design of adequate ecosystem management procedures (Abell and Stepp, 2003). The consideration of self-organization in social and economic systems (Fang and Sanglier, 1994) may furthermore allow improvement in quality of social systems in IPM. Sustainable agriculture, on the other hand, has been approached from the point of view of social science by Röling and Wagemakers (1998b). According to them, sustainability is an emergent property of a soft system and the outcome of collective decision making that arises from the interaction between natural resource users and managers. In Röling and Wagemakers (1998a), Pretty (1998) states that policies provide a crucial context that can affect the realization of regenerative agriculture both in positive and negative ways. Policymaking is often seen as a sufficient condition for bringing about societal change, especially fiscal policy and regulatory measures (Röling and Wagemakers, 1998b). Learning is seen as an important aspect in developing a sustainable agriculture and becomes efficient by facilitation and institutional support (Röling and Wagemakers, 1998b). Fisk *et al.* (1998) emphasize that achieving a sustainable agriculture requires integrated farming systems that involve many diverse individuals and institutions in rural communities.

The supply–demand approach to population systems permitted the analysis of ecological population dynamics and extended to the socio-economic system, thus providing the basis for efficiently analysing the functioning of an ecosocial system (Gutierrez, 1996).

The extension into the inclusion dimension (Fig. 7.1) led to both a wider array of institutions involved and to more complex subjects of management than considered at the beginning of the IPM era. For example, IPM activities cover the range from field-specific and single species pest control to the management of community and even region-specific ecosystems. In such systems, institutions do not act as pest control managers, but increasingly interact with ecological systems. The following section combines the social and the ecological systems into an ecosocial system and deals with complex interactions between the two components.

Ecosocial System Management

According to the Brundtland Report (World Commission on Environment and Development, 1987), sustainable development meets the current needs without compromising the ability of future generations to meet their own needs. Gutierrez (2003) observes that the present generation is spending the endowment of future generations and urges us to seek ecological sustainability in agriculture. According to Altieri (1991), sustainability is the capacity of an agroecosystem to maintain the production through the time facing ecological limitations and long-term socio-economic pressures. As mentioned earlier, sustainable agricultural systems have been developed to maximize the internal resources of the farm, minimize environmental impacts and increase profits by reducing external inputs; these production systems require fewer off-farm inputs and mimic the stability of natural systems (Linker *et al.*, 1991). We rely on Goodland (1995), who differentiates among environmental (maintenance of natural capital), economic (economic capital should be stable) and social (maintenance and replenishment of social capital) sustainability. Environmental sustainability, linked to economic sustainability, is the basis for social sustainability and is emphasized in our work. According to Goodland (1995), environmental sustainability is achieved by keeping the scale of the human economic system within the biophysical limits of the ecosystem. This means, among other

issues, holding waste emissions within the assimilation capacity of the system on the sink side, and keeping harvest traits of renewable resources within the regeneration rate on the source side. Goodland's (1995) concept provides a framework for evaluating ecological, economic and social qualifiers of modern agriculture (Delucchi, 2001) and addressing ecological, economic and social mandates assuring sustainability of agricultural systems (Cuperus *et al.*, 2004). The concept may also encompass the broad goals of sustainable agriculture and provide a basis for the great variety of strategies developed by farmers (SARE, 2004).

The purposeful management of ecosystems requires the specification of management objectives. As they may differ from case to case, the complexity of the ecosystems requires careful definition of the system for study and management. However, this is difficult, as objectives may vary and many definitions have been given of ecosystems and ecosystem management (Meffe *et al.*, 2002). But a review of relevant concepts goes beyond the scope of this paper. Jørgensen (2002) defines an ecosystem as a biotic and functional system or unit that is able to sustain life and includes all biological and non-biological variables. In that unit, spatial and temporal scales are not specified a priori, but are entirely based upon the objectives of the ecosystem study. The 'Ecological Society of Committee on the Scientific Basis for Ecosystem Management' regards ecosystem management as management driven by explicit goals, executed by policies, protocols and practices, and made adaptable by monitoring and research based on our best understanding of the ecological interactions and processes necessary to sustain ecosystem composition, structure and function (Christensen *et al.*, 1996). We follow their recommendation and consider sustainability as the primary objective, and levels of commodity and amenity provisions must be adjusted to meet that goal.

In the previous section, IPM has been placed at the interface between ecological and socio-economic systems. In this section we include both systems into a single ecosocial system as defined by Waltner-Toews *et al.* (2003) according to which humans are no longer external to the system but are considered integrative elements that interact with other systems elements.

Some principles

The inclusion of institutions and consequences for ecosystem management has recently been described by Waltner-Toews *et al.* (2003). According to them, environmental managers are no longer managing systems to which they are external, but they are part of ecosocial systems; they can no longer manage ecosystems per se, but they learn how to manage their interactions with the ecological system. Moreover, Waltner-Toews *et al.* (2003) state that this view incorporates notions of multiple, interacting, nested hierarchies, feedback loops across space and time and radical uncertainty with regard to prediction of system behaviour, and requires rethinking of ecosystem management. In summary, the following aspects are important for Waltner-Toews *et al.* (2003): environmental management has traditionally considered humans as external to 'pristine' ecosystems, but growing awareness of the complexity of ecosocial systems shows that this is unrealistic; when management decisions are made, people must be considered integral elements of ecological systems; by involving local communities, the new approach allows better understanding and more efficient decision making.

Costanza (2000) observed that long-term management of natural resources is poor because scales are not adequately addressed. Local ecosystems may suffer if superseded by national and international practices. The challenge is to match ecosystems and governance systems in ways that maximize the compatibility of these systems.

To describe additional principles we briefly refer to an on-going project in Ethiopia aiming at ecosocial system sustainability enhancement (Fig. 7.3), while Baumgärtner *et al.* (2003b) and Herren *et al.* (2004) have given a more detailed description. The project deals with a human population engaged in subsistence farming who live in absolute poverty and suffer from diseases including mosquito-transmitted malaria and malnutrition. Cattle productivity is low and drought power limited because of tsetse-transmitted diseases. Moreover, time constraints appear to limit possibilities for income-generating activities including bee keeping and vegetable production. Thus, the human population is not seen as managers of ecological systems, but as a system component that strongly interacts with the ecological system. Weaker interactions may occur between the ecological system and other components of the social system including Ethiopian National Institutions. Although the situation is different in industrialized societies, many of the principles may be of general validity.

We relied on Becker and Ostrom (1995), who identified diversity among institutions as an important attribute of long-term sustainable agricultural system. Accordingly, we created community committees for dealing with specific issues and built the project into Ethiopia's national development agenda (Aseffa *et al.*, 2003). The latter aspect has been found indispensable in biodiversity conservation (Brown, 2003). Next, we selected adaptive management procedures for placing the system on a trajectory towards environmental sustainability by emphasizing human health improvement (Baumgärtner *et al.*, 2003b; Gilioli *et al.*, 2003, 2005). This was done by seeking two levels of understanding. A heuristic procedure, i.e. a procedure of immediate utility, employed out of pragmatism but arbitrarily conceived (Allen and Starr, 1982) was applied first. Therefore, rapid improvement was sought by precision targeting interventions in time and space, as illustrated by tsetse control discussed in subsequent section. The accumulated experience combined with ecological knowledge allows study and management of ecosystems on a deeper level of understanding. Rather than controlling tsetse, we aim at understanding the epidemiology of tsetse-transmitted diseases. Human and cattle health is setting the stage for rural development where crop management is developed by seeking balanced land use (Herren *et al.*, 2004).

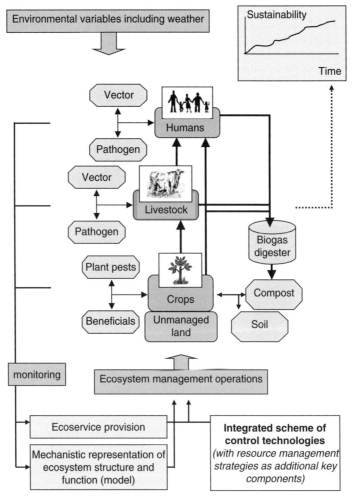

Fig. 7.3. Adaptive management concept for an ecosystem placed on a trajectory towards sustainability. (From Herren *et al.*, 2004.)

Ecosocial system sustainability may be enhanced at the level of ecosystem service provision, i.e. by increasing the array of conditions and processes through which ecosystems, and their biodiversity, confer benefits on humanity; these include the production of goods, life support functions and life-fulfilling conditions such as aesthetic beauty, cultural, intellectual and spiritual inspiration, existence values, scientific discovery, serenity and preservation of options (Daily, 1997; Daily and Dasgupta, 2001). However, there may be trade-offs and the relevant processes of energy, matter including

water, and biodiversity dynamics may be a more promising strategy (Gilioli *et al.*, 2003, 2005). In IPM, the subjects of management are pest assemblages that might negatively interfere with ecosystem service provision and constrain enhancement of sustainability.

From a social standpoint, two aspects appear to be important and undoubtedly require further research. First, a hierarchically structured human population strongly or weakly interacts with a complex ecological system. The human population operates within national poverty alleviation agenda, interacts with administrators and extensionists and seeks assistance by researchers who, through adaptive management, attempt to obtain better insight into socio-economic and ecological processes. Second, the aim of management is ecosocial system sustainability, divided into environmental, economic and social sustainability (Goodland, 1995). In our opinion, current knowledge provides little guidance on how to connect structured institutions to these interdependent sustainability components.

Integrated Pest Management Case Studies

We selected the following case studies because of our familiarity with the population systems under management and because they appear suitable for a presentation of the social system involved. Moreover, they represent both pest management in animal husbandry (tsetse) and crops (orchard tortricid pests) as addressed in a developing (Ethiopia) and an industrialized (Italy) country. The population management schemes are simple in that they combine few control components, but efforts are underway to integrate additional technologies. Both cases rely on precision farming, a management strategy that employs detailed, site-specific information to manage production inputs (Brenner et al., 1998; Gold, 1999; ARS, 2002). Precision farming practices, e.g. pest control technology application, can reduce costs while minimizing environmental and ecological impacts, but slow adoption (CSREES, 2004). Simple IPM schemes with slow adoption are considered particularly appropriate case studies for a brief evaluation of the social system.

For more complex IPM programmes, the reader is referred to Delucchi (1987a), Dent (1995a), Kogan et al. (1999), Leslie and Cuperus (1993), Baumgärtner et al. (2003a) and Koul et al. (2004). Importantly, the selection of the case studies does not reflect the level of achievement in IPM development, implementation or success in comparison to other agricultural pest control systems.

IPM of *Cydia funebrana* and *Cydia pomonella*

With the development of IPM, there has been an increase in the use of a variety of management tools to assist in the reduction of chemical control and in the development of more efficient monitoring strategies. These strategies depend on understanding specific insect population behaviours and need precision targeting of interventions to maximize their efficacy.

Ecological considerations

In this connection, geostatistical methods can be a powerful tool for the understanding of many spatially related phenomena. This analysis can provide crucial information for improving sustainable pest control techniques, particularly in the context of precision IPM (Fleischer et al., 1999; Sciarretta et al., 2001; Sciarretta and Trematerra, 2004; Trematerra et al., 2004).

The methodology used and the results obtained in spatial analysis of pheromone trap catches of *Cydia funebrana* (Treitschke), *C. pomonella* (Linnaeus), *Grapholita molesta* (Busck) and *Anarsia lineatella* (Zeller) have been reported previously (Sciarretta et al., 2001; Sciarretta and Trematerra, 2004; Trematerra et al., 2004). Here, we present the two former cases only. The research has been conducted from 1999 to 2003, in the Molise region of Central Italy.

The flight phenology of *C. funebrana* was monitored in a plum orchard and in the surrounding area covering a zone of about 250 ha. Cereal and sunflower fields surround the orchard; vegetables are grown between the orchard and a ravine with riparian vegetation including shrubs and blackthorn trees (Fig. 7.4). Sampling of *C. pomonella* was carried out in two heterogeneous agroecosystems (50 ha each), located 750–850 m above sea level, in hilly and mountainous landscapes. The ecosystem I is characterized by the presence

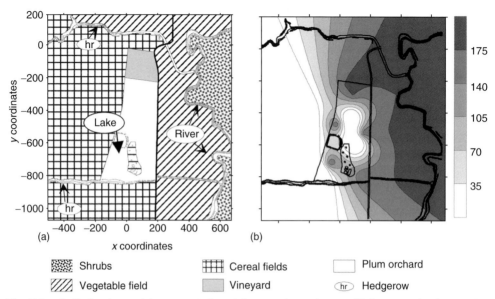

Fig. 7.4. *Cydia funebrana*: (a) representation of the experimental area with important landscape elements; (b) spatial distribution of male annual catches in 1999; *x* and *y* coordinates are expressed in metres. (From Sciarretta *et al.*, 2001.)

of two apple orchards, with different cultivars, about 350 m apart; the rest of the area comprises cereal and vegetable fields, uncultivated fields and woodlots (Fig. 7.5). In agroecosystem II, there are two adjacent apple orchards, surrounded by cereal and vegetable fields, where abandoned apple, pear, service and walnut plants grow.

Spatial analysis was carried out using Surfer Version 8 or 8.1 (Golden software, Golden, CO, USA) with x and y representing the coordinates and z the trap counts. The interpolation algorithm was kriging: exponential models were used for *C. pomonella*, and linear models with zero nugget in the other cases. The interpolation grid obtained is represented graphically by a contour map, which shows the configuration of the surface by means of isolines representing equal z-values; a base map showing the experimental area, with the same coordinate system, was placed on top of the contour map. For a full explanation of methods, refer Brenner *et al.* (1998).

High levels of *C. funebrana* trap catches inside the orchard occurred usually near the borders, while lowest catches were obtained in the interior part of the field (Fig. 7.4). The zones outside the orchard with highest trap catches are located in and around the ravine with shrubs, including blackthorn, follow irrigation channels and hedgerows and arrive at the contact between irrigation channels and the orchard; the lowest trap catches occurred on the opposite side of the orchard. Observed distribution of *C. funebrana* indicates high dispersal capability throughout the whole flight period. In particular, irrigation channels and hedgerows appear to serve as ecological corridors, where adults move from one zone to another. On the opposite side, arable fields act as barriers to the movements of the pest.

Trap catches of *C. pomonella* were clumped during almost all sampling weeks. Spatial characterizations obtained illustrate that the main trap capture foci were confined to the productive apple orchards for both ecosystems I and II (Fig. 7.5). High trap catch foci occurred also in the group of apple trees next to the orchards and in the agroecosystem II in the zone with pear trees, service trees and walnut trees. The highest pest density zones were isolated from each

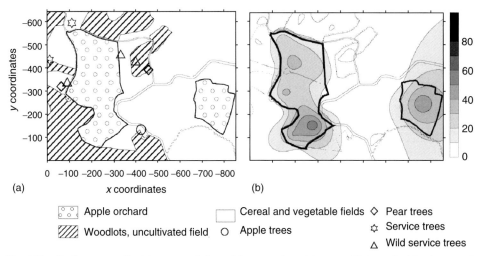

Fig. 7.5. *Cydia pomonella*: (a) representation of the experimental area with important landscape elements; (b) spatial distribution of male annual catches in ecosystem I in 2002; *x* and *y* coordinates are expressed in metres. (From Trematerra *et al.*, 2004.)

other, and only in few sporadic cases males were counted in traps located outside these zones. This distribution seems to indicate a low dispersal capability of male codling moths during the flight period.

The observed distribution patterns appeared very different in the two *Lepidoptera* pests, *C. funebrana* and *C. pomonella*, and seemed to be determined mainly by the location of the most important host plants and by the propensity of males to move in the environment, inside and outside the breeding and mating sites. Important landscape elements such as riparian vegetation and hedgerows can act as ecological barriers or corridors depending on the specific behaviour of a species.

On the basis of our results, adequate knowledge of both behavioural aspects of moths and landscape features must be achieved to improve efficacy of precision IPM tactics. This is particularly important in heterogeneous landscapes, where potential sources of external infestation can invalidate some pest management methods employed in orchards. Indication obtained from spatial maps should be considered an essential prerequisite for IPM precision targeting programmes and incorporated in the development of monitoring and control strategies in adaptive management.

Socio-economic considerations

In the Molise region, small-scale farmers who receive training at technical agricultural schools or learn fruit production by tradition in their families manage the fruit orchards. Additional training is provided by the extension service of the Agricultural Development Agency funded by European Union programmes. Professional organizations also provide training to fruit growers. Precision farming technologies, designed by the University of Molise, have to reach farmers, their organizations, schools and the extension service in order to be effective. Moreover, the technology has been developed in on-farm observations, and hence is the result of information exchange between farmers and researchers. We anticipate that further improvements are possible once the university receives additional feedback from these institutions.

Fruits are sold to farmers' associations or small commercial enterprises that currently serve the local and national market. Recently, interest in fruit quality has increased and growers, farmer associations and commercial enterprises show interest in integrated and organic farming production systems, because they expect higher fruit prices and use of more cost-efficient production systems than known so far. Farmers and their associations are also interested in training programmes. In regional urban centres, consumers buy quality products for various reasons including health and the opinion that they are contributing to biodiversity conservation on the genetic, species and ecosystem levels. These institutions are aware that only high-quality fruits can eventually be sold for a higher price than obtained so far on international markets. All these institutions, including technical schools, stimulate the development of university research.

The tortricid control scheme results in a reduction in synthetic chemical insecticide input because of better knowledge of both temporal and spatial dynamics. This may bring the IPM scheme closer to meeting the objective of establishing an environmentally, economically and socially acceptable fruit production system (Delucchi, 2001; Koul et al., 2004) with advantages to most of the institutions involved in IPM, including producers and the society. These benefits are not restricted to good provision but the inclusion of neighbouring managed and unmanaged land may also contribute to biodiversity conservation, and hence increase ecosystem service provision (Daily, 1997; Daily and Dasgupta, 2001; Daily and Ellison, 2002). The enhancement of ecosystem provision and the design and implementation of sustainable ecosystem are challenges for all institutions involved.

Adaptive management of tsetse populations in Ethiopia

Ecological considerations

Tsetse-transmitted diseases are a serious constraint to cattle health and economic development in about 100,000 km^2 of Africa (Kettle, 1990). In the Ethiopian community under consideration, tsetse-transmitted diseases are an impediment to rural development und their reduction is one of the conditions that set the stage for rural development (Getachew Tikubet et al., 2003). From an ecosocial standpoint, the pest status of tsetse results from negatively affecting the provision of ecosystem goods, including dairy products, and constraining the enhancement of ecosocial system sustainability. To reduce cattle disease prevalence, an adaptive tsetse control system based on odour-baited traps was developed (Baumgärtner et al., 2003c; Sciarretta et al., 2005). Therefore, a cost-efficient use of relatively expensive traps was seen as a prerequisite for successful tsetse control operations.

In the intervention area (Fig. 7.6), four monitoring traps per square kilometre were deployed. The geostatistical analysis of the data showed that similar information on tsetse distribution can be obtained by a considerable reduction of monitoring traps (Sciarretta et al., 2005). Moreover, the analysis identified the occurrence of hot spots, i.e. areas with higher trap catches than the surroundings. Monitoring trap information was processed in a biweekly rhythm and the resulting maps were used to guide control trap deployment (Fig. 7.6). Adaptive management resulted in efficient guide control operations, and also provided insight into the spatiotemporal dynamics of tsetse (Sciarretta et al., 2005).

This case is part of a strategy that aims at tsetse control in four steps. In step 1, awareness for tsetse is created, the community committee elects a tsetse management subcommittee and the research team collects baseline data. In step 2, the research team sets up the adaptive management scheme in collaboration with the tsetse subcommittee; a monitoring scheme is designed and control traps are deployed in response to tsetse occurrences in time and space. In step 3, the adaptive management technology is taken over by the community and carried out in collaboration with the research team; the insight obtained into tsetse occurrences is expected to simplify the adaptive management system. In step 4, the operations are expected to be carried out by the tsetse subcommittees while the research team provides technical backstopping only.

Mass-trapping and drug administration to infested cattle are the first components of an emerging IPM system aiming at animal health improvement. Managing herd

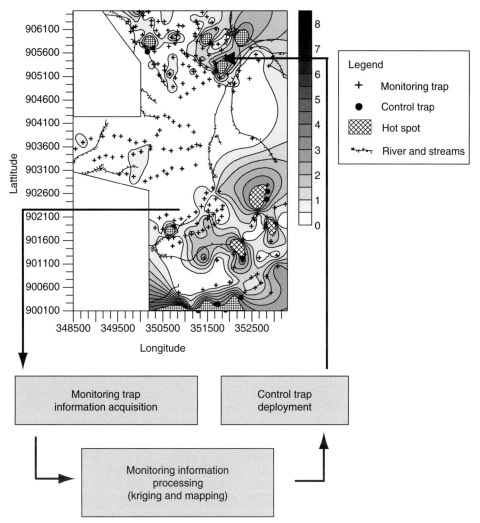

Fig. 7.6. Adaptive management of tsetse populations in Luke, Ethiopia., (From Baumgärtner *et al.*, 2003c; Sciarretta *et al.*, 2005. Courtesy of Dr A. Sciarretta, University of Molise, Italy.)

movements to avoid tsetse-infested areas and zero grazing for particularly productive livestock are other components under study.

Socio-economic considerations

The social system involved in tsetse management comprises several levels. Among the institutions there is a local tsetse control subcommittee with an important role in tsetse control operations, including trap deployment, trap servicing and mainte-nance. It reports to the elected community committee, and both institutions are linked to the ICIPE research team implementing and supervising project implementation. The farmers are advised by the Ethiopian Agricultural Research Organization (EARO) and by the local Ministry of Agriculture office that reports to the Regional and National Ministries of Agriculture. Moreover, these institutions are linked to the National Ethiopian Rehabilitation and

Development Fund (ESRDF) that is charged with community-driven poverty alleviation responsibilities and operates within a national poverty alleviation policy defined by the Ethiopian Government (Aseffa *et al.*, 2003). The involvement of these institutions has been found indispensable for successful project execution.

The control of tsetse results not only in increased animal productivity but also in drought power availability. In fact, animal health improvement to be complemented with human health improvement, including malaria control, appears to set the stage for sustainable rural development (Herren *et al.*, 2004). As a consequence, other resource management subcommittees have to be established and integrated into the institutional framework described for tsetse control. However, cautionary measures are required to assure that the intensification of agriculture does not interfere with ecosystem service provision including biodiversity conservation at genetic, species and ecosystem levels.

Conclusions

History of IPM and some elements of system theory that are used to represent the structure and functioning of social and ecological systems have been discussed. This approach allows the schematization of social systems including the composition and role of institutions engaged in IPM. In social systems, an increasing number of levels from farmers to policymakers are involved in IPM activities. The social systems are organized horizontally with institutions operating on the same level, and vertically with institutions assigned to different levels. The approach also permits extension of IPM into comprehensive agricultural production systems management, and subsequently into ecosocial management schemes. In traditional IPM schemes, the institutions are seen as managers of ecological systems. In ecosocial systems, however, the social systems interact with ecological systems and both become components of unified wholes. For two reasons this chapter is

not limited to social systems but considers the structure and functioning of ecological systems as well. First, the use of common concepts for representing both socio-economic and ecological systems facilitates research and implementation. Second, it addresses the concerns of anthropologists, who consider that current socio-economic approaches are insufficient because they do not satisfactorily take into account the complexities of ecological systems.

Examples of integrated crop and livestock pest management in industrialized and developing agricultural systems illustrate the role of institutions. In Italy, university research developed technologies that are made available to the extension service, farmer associations, technical agricultural schools and to small-scale fruit growers; the farmers benefit from precision target and more cost-efficient fruit tortricid pest management approaches, while reduced synthetic chemical input is an advantage to farmers and their associations, which compete with others on regional and national fruit markets. The society may benefit from enhanced ecosystem service provision. In Africa, livestock transmitting tsetse are highly mobile and can efficiently be controlled at the community level by precision targeting mass-trapping operations. To implement adaptive tsetse management technologies, Ethiopian communities have set up tsetse control teams that interact with researchers and operate within political and administrative organizations of communities. Their integration into regional and national institutions is a prerequisite to successful programme execution.

Acknowledgements

The Swiss Development Corporation (SDC) and BioVision Foundation provided financial and logistic support to the tsetse control project. Both the tsetse project and the orchard IPM work received support from the University of Molise, Italy. The collaboration of Dr A. Sciarretta, Dr Getachew Tikubet, Dr Melaku Girma, Dr Shifa Ballo and Lulseged Belayun is greatly appreciated. We are also grateful to the donors of ICIPE's core fund who supported the work of the first two authors. The comments of Professor Gerrit Cuperus, University of Oklahoma, USA, for drawing our attention to the relationship between IPM and sustainable agriculture are appreciated.

References

Abell, T. and Stepp, J.R. (2003) A new ecosystems ecology for anthropology. *Conservation Ecology* 7 [online].

Allen, T.F.H. and Starr, T.B. (1982) *Hierarchy. Perspectives for Ecological Complexity.* University of Chicago Press, Chicago.

Altieri, M.A. (1991) *Agroecologia. Prospettive scientifiche per una nuova agricoltura.* Franco Muzzio, Padova, Italy.

Anonymous (1996) *Concise Science Dictionary,* 3rd edn. Oxford University Press, Oxford.

ARS (2002) *National Programs Integrated Farming Systems. FY 2002.* Report of the Agricultural Research Service (ARS), US Department of Agriculture, Washington, DC. Available at: www.ars.usda.gov/research/programs/programs.htm.

Aseffa, A., Getachew, T. and Baumgärtner, J. (2003) *Resource Management for Poverty Reduction Approaches and Technologies.* Ethiopian Social Rehabilitation and Development Fund, Addis Ababa, Ethiopia.

Bangladesh Department of Agricultural Extension (2002) National Integrated Pest Management Policy, 9th draft, Dhaka, Bangladesh. Available at: www.daebd.org/Policy/National%20IPM%20Policy.pdf.

Baumgärtner, J. and Hartmann, J. (2000) The design and implementation of sustainable plant diversity conservation programs for alpine meadows and pastures. *Journal of Agricultural and Environmental Ethics* 14, 67–83.

Baumgärtner, J., Gilioli, G., Schneider, D. and Severini, M. (2002) The management of populations in hierarchically organized systems. *Notiziario Sulla Protezione delle Piante* 15, 247–263.

Baumgärtner, J., Schulthess, F. and Xia, Y.L. (2003a) Integrated arthropod pest management systems for human health improvement in Africa. *Insect Science and Its Application* 23, 85–98.

Baumgärtner, J., Getachew, T., Gilioli, G. and Bieri, M. (2003b) *Managing Ecosystems to improve human health and alleviate poverty.* In: Aseffa, A., Getachew, T. and Baumgärtner, J. (eds) *Resource Management for Poverty Reduction Approaches and Technologies.* Ethiopian Social Rehabilitation and Development Fund, Addis Ababa, Ethiopia, pp. 179–186.

Baumgärtner, J., Getachew, T., Melaku, G., Sciarretta, A., Shifa, B. and Trematerra, P. (2003c) Cases for adaptive ecological systems management. *Redia* LXXXVI, 165–172.

Becker, C.D. and Ostrom, E. (1995) Human ecology and resource sustainability: the importance of institutional diversity. *Annual Review of Ecology and Systematics* 26, 113–133.

Bernal, J.S., Prasifka, J., Sétamou, M. and Heinz, K.M. (2004) Transgenic insecticidal cultivars in IPM: challenges and opportunities. In: Koul, O., Dhaliwlal, G.S. and Cuperus, G. (eds) *Integrated Pest Management: Potential, Constraints and Challenges.* CAB International, Wallingford, UK, pp. 123–146.

BioVision (2004) Environmental advisors for Africa – Ecodissemination Unit. Available at: www.biovision.ch.

Birkhaeuser, D., Evanson, R. and Feder, G. (1996) The impact of agricultural extension: a review. *Economic Development and Cultural Change* 39, 607–650.

Brenner, J.S. and Desmond, D.J. (2003) 4-H and FFA: forging youth and adult partnerships for sustainable communities. In: Aseffa, A., Getachew, T. and Baumgärtner, J. (eds) *Resource Management for Poverty Reduction Approaches and Technologies.* Ethiopian Social Rehabilitation and Development Fund, Addis Ababa, Ethiopia, pp. 61–70.

Brenner, R.J., Focks, D.A., Arbogast, R.T., Weaver, D.K. and Shuman, D. (1998) Practical use of spatial analysis in precision targeting for integrated pest management. *American Entomologist* 44, 79–101.

Brown, K. (2003) Integrating conservation and development: a case of an institutional misfit. *Frontiers in Ecology and the Environment* 1, 479–487.

Brown, A.R., Thiele, G. and Fernández, M. (2000) Farmer Field Schools and Local Agricultural Research Committees: complementary platforms for integrated decision-making in sustainable agriculture. *Agricultural Research and Extension Network Paper 105.* DFID, London.

Chambers, R., Pacey, A. and Thrupp, L.A. (1989) *Farmer First – Farmer Innovation and Agricultural Research.* Intermediate Technology Publications, London.

Christensen, N.L., Bartuska, A.M., Brown, J.H., Carpenter, S., D'Antonio, C., Francis, R., Franklin, J.F., MacMahon, J.A., Noss, R.F., Parsons, D.J., Petersen, C.H., Turner, M.G. and Woodmansee, R.G. (1996) The report of the Ecological Society of America Committee on the Scientific Basis for Ecosystem Management. *Ecological Applications* 6, 665–691.

Comiskey, J.A., Dallmeier, F. and Alonso, A. (1999) Framework for assessment and monitoring of biodiversity. In: Levin, S. (ed.) *Encyclopedia of Biodiversity*, Vol. 3. Academic Press, New York, pp. 63–73.

Conway, G.R. (1984) Introduction. In: Conway, G.R. (ed.) *Pest and Pathogen Control. Strategic, Tactical, and Policy Models.* 13 International Series on Applied Systems Analysis. John Wiley & Sons, Chichester, UK, pp. 1–11.

Costanza, R. (2000) *Institutions, Ecosystems, and Sustainability.* Lewis Publishers, Boca Raton, Florida.

CSREES (2004) *Precision Farming.* Cooperative State Research, Education, and Extension Service, United States of Agriculture, Washington, DC. Available at: www.csrees.usda.gov.

Cuperus, G.W., Berberet, R.C. and Noyes, R.T. (2004) The essential role of IPM in promoting sustainability of agricultural production systems for future generations. In: Koul, O., Dhaliwlal, G.S. and Cuperus, G. (eds) *Integrated Pest Management: Potential, Constraints and Challenges.* CAB International, Wallingford, UK, pp. 265–280.

Daily, G.C. (1997) *Nature's Services Societal Dependence on Natural Ecosystems.* Island Press, Washington, DC.

Daily, G.C. and Dasgupta, S. (2001) Ecosystem services, concept of. In: Levin, S. (ed.) *Encyclopedia of Biodiversity*, Vol. 2. Academic Press, London, New York, pp. 353–361.

Daily, G.C. and Ellison, K. (2002) *The New Economy of Nature: The Quest to Make Conservation Profitable.* Island Press, Washington, DC.

Delucchi, V. (1987a) *Integrated Pest Management Protection Integrée Quo vadis?* Parasitis, Geneva.

Delucchi, V. (1987b) La protection integrée des cultures. In: Delucchi, V. (ed.) *Integrated Pest Management Protection Integrée Quo vadis?* Parasitis, Geneva, pp. 7–22.

Delucchi, V. (2001) La biodiversità e l'agricoltura ecocompatibile. *Atti dell'Accademia Italiana di Entomologia. Rendiconto, anno XLIX-2001*, 220–230.

Dent, D. (1995a) *Integrated Pest Management.* Chapman & Hall, London.

Dent, D. (1995b) Introduction. In: Dent, D. (ed.) *Integrated Pest Management.* Chapman & Hall, London, pp. 1–7.

Dent, D. (1995c) Programme planning and management. In: Dent, D. (ed.) *Integrated Pest Management.* Chapman & Hall, London, pp. 120–151.

Fang, F. and Sanglier, M. (1994) Complexity and self-organization in social and economic systems. *Proceedings of International Conference on Complexity and Self-organization in Social and Economic Systems*, Beijing, October 1994, Lecture Notes in Economics and Mathematical Systems 449. Springer-Verlag, Berlin.

Feder, G., Willett, A. and Zijp, W. (2001) Agricultural extension: generic challenges and the ingredients for solutions. In: Wolf, S. and Zilberman, D. (eds) *Knowledge Generation and Technical Change: Institutional Innovation in Agriculture.* Kluwer Academic Publishers, Boston, Massachusetts, pp. 311–316.

Feder, G., Murgai, R. and Quizon, J.B. (2004) Sending farmers back to school: the impact of farmer field school in Indonesia. *Review of Agricultural Economics* 26, 45–62.

Fisk, J.W., Hesterman, O.B. and Thorburn, T.L. (1998) Integrated farming systems: a sustainable agriculture learning community in the US. In: Röling, N.G. and Wagemakers, M.A.E. (eds) *Facilitating Sustainable Agriculture: Participatory Learning and Adaptive Management in Times of Environmental Uncertainty.* Cambridge University Press, Cambridge, pp. 217–231.

Fleischer, S.J., Blom, P.E. and Weisz, R. (1999) Sampling in precision IPM: when the objective is a map. *Phytopathology* 89, 1112–1118.

Flint, M.L. and van den Bosch, R. (1981) *Introduction to Integrated Pest Management.* Plenum Press, New York.

Getachew Tikubet, Shifa Ballo and Amare Birhanu (2003) Community-based tsetse control: a model project within a sustainable agriculture framework. In: Aseffa, A., Getachew, T. and Baumgärtner, J. (eds) *Resource Management for Poverty Reduction Approaches and Technologies.* Ethiopian Social Rehabilitation and Development Fund, Addis Ababa, Ethiopia, pp. 153–164.

Getz, W. and Gutierrez, A.P. (1982) A perspective on systems analysis in crop production and insect pest management. *Annual Review of Entomology* 27, 447–466.

Gilioli, G., Baumgärtner, J. and Vacante, V. (2003) Biological control as an ecosystem management tool for enhancing environmental sustainability. *Redia* LXXXVI, 173–185.

Gilioli, G., Baumgärtner, J. and Vacante, V. (2005). Biological control and ecosystem services. In: *Encyclopedia of Life Support Systems (EOLSS)*. EOLSS Publishers, UNESCO, Oxford. Available at: http://www.eolss.net.

Gold, M.V. (1999) *Sustainable Agriculture: Definitions and Terms*. Special Reference Briefs Series No. SRB99–02, US Department of Agriculture, Washington, DC.

Goodland, R. (1995) The concept of environmental sustainability. *Annual Review of Ecology and Systematics* 26, 1–24.

Gut, L.J., Stelinski, L.L., Thomson, D.R. and Miller, J.R. (2004) Behaviour modifying chemicals: prospects and constraints in IPM. In: Koul, O., Dhaliwlal, G.S. and Cuperus, G. (eds) *Integrated Pest Management: Potential, Constraints and Challenges*. CAB International, Wallingford, UK, pp. 73–122.

Gutierrez, A.P. (1996) *Applied Population Ecology. A Supply–demand Approach*. John Wiley & Sons, New York.

Gutierrez, A.P. (2003) Sustainable agriculture: the role of integrated pest management. In: Aseffa, A., Getachew, T. and Baumgärtner, J. (eds) *Resource Management for Poverty Reduction Approaches and Technologies*. Ethiopian Social Rehabilitation and Development Fund, Addis Ababa, Ethiopia, pp. 141–152.

Haney, A. and Boyce, M.S (1996) Introduction. In: Boyce, M.S. and Haney, A. (eds) *Ecosystem Management: Applications for Sustainable Forest and Wildlife Resources*. Yale University Press, New Haven, pp. 1–17.

Hári, S. and Müller, F. (2000). Ecosystems as hierarchical systems. In: Jørgensen, S.E. and Müller, F. (eds) *Handbook of Ecosystem Theories and Management*. Lewis Publishers, Boca Raton, Florida, pp. 265–280.

Headley, J.C. (1975) The Economics of Pest Management. In: Metcalf, R.L. and Luckmann, W. (eds) *Introduction to Insect Pest Management*. John Wiley & Sons, New York, pp. 75–99.

Herren, H.R., Baumgärtner, J. and Gilioli, G. (2004) From agricultural pest management to ecosystem sustainability enhancement: the way forward. In: Scherr, S.(ed.) International Ecoagriculture conference and Practitioners Fair, Nairobi, Kenya.

Holling, C.S. (1978) *Adaptive Environmental Assessment and Management*. International Institute for Applied Systems Analysis, Laxenburg, Austria. John Wiley & Sons, Chichester, UK.

Huffaker, C.B. and Croft, B.A. (1976) Integrated pest management in the US: progress and promise. *Environmental Health Perspectives* 14, 167–183.

IFPRI (2004) *Mission of the International Food Policy and Research Institute*, Washington, DC.

Jørgensen, S.E. (2002) *Integration of Ecosystem Theories: A Pattern*, 3rd edn. Kluwer, Dordrecht.

Kelly, V., Akinwumi, A. and Gordon, A. (2003) Expanding access to agricultural input in Africa: a review of recent market development experience. *Food Policy* 28, 379–404.

Kemp, W.M., Petersen, J.E. and Gardner, R.H. (2001) Scale-dependence and the problem of extrapolation. In: Gardner, R.H., Kemp, W.M., Kennedy, V.S. and Petersen, J.E. (eds) *Scaling Relations in Experimental Ecology*. Columbia University Press, New York, pp. 3–57.

Kenmore, P. (1997) *A Perspective on IPM*. Information Centre for Low-External-Input and Sustainable Agriculture Newsletter 13, 8–9.

Kettle, D.S. (1990) *Medical and Veterinary Entomology*. CAB International, Wallingford, UK.

Kline, S. and Rosenberg, N. (1986) An overview on innovations. In: Landau, R. and Rosenberg, N. (eds) *The Positive Sum Strategy, Harnessing Technology for Economic Growth*. National Academic Press, Washington, DC.

Koestler, A. (1967) *The Ghost in the Machine*. Hutchinson, London.

Kogan, M. (1998) Integrated pest management: historical perspectives and contemporary developments. *Annual Review of Entomology* 43, 243–270.

Kogan, M., Croft, B.A. and Sutherst, R.F. (1999) Applications of ecology for integrated pest management. In: Huffaker, C.B. and Gutierrez, A.P. (eds) *Ecological Entomology*. John Wiley & Sons, New York, pp. 681–728.

Koul, O., Dhaliwlal, G.S. and Cuperus, G. (2004) *Integrated Pest Management: Potential, Constraints and Challenges*. CAB International, Wallingford, UK.

Leslie, A.R. and Cuperus, G.W. (1993) *Successful Implementation of Integrated Pest Management for Agricultural Crops*. Lewis, Boca Raton, Florida.

Linker, M., Cuperus, G.W., Frisbee, R., Johnson, D., Sprenkle, R., Brittian, J. and Caron, R. (1991) *The Role of Integrated Pest Management in Sustainable Agriculture*. North Carolina State University Cooperation Extension Service, Raleigh, North Carolina.

Luckmann, W.H. and Metcalf, R.L. (1975) The pest-management concept. In: Luckmann, R.L. and Metcalf, R.L. (eds) *Introduction to Insect Pest Management*. John Wiley & Sons, New York, pp. 3–35.

Meffe, G.K. and Carroll, C.R. (1997) *Principles of Conservation Biology*. Sinauer, Sunderland, Massachusetts.

Meffe, G.K., Nielsen, L.A., Knight, R.S. and Schenborn, D.A. (2002) *Ecosystem Management. Adaptive Community-based Conservation*. Island Press, Washington, DC.

Mesarovic, M.D., Macko, D. and Takakara, Y. (1970) *Theory of Hierarchical, Multilevel Systems*. Academic Press, New York.

Naveh, Z. and Liebermann, A.S. (1984) *Landscape Ecology: Theory and Application*, 2nd edn. Springer-Verlag, Berlin.

Norton, G.A. and Mumford, J.D. (1993) *Decision Tools for Pest Management*. CAB International, Wallingford, UK.

Orr, A. and Ritchi, J.M. (2004) Learning from failure: smallholder farming systems and IPM in Malawi. *Agricultural Systems* 79, 31–54.

Pattee, H.H. (1973) *Hierarchy Theory. The Challenge of Complex Systems*. Braziller, New York.

Pretty, J.N. (1998) Supportive policies and practice for scaling up sustainable agriculture. In: Röling, N.G. and Wagemakers, M.A.E. (eds) *Facilitating Sustainable Agriculture. Participatory Learning and Adaptive Management in Times of Environmental Uncertainty*. Cambridge University Press, Cambridge, pp. 23–45.

Rockefeller Foundation (2003) *Annual Report 2003*. Rockefeller Foundation, Mahwah.

Rogers, E.M. (2003) *Diffusion of Innovations*, 5th edn. Free Press, New York.

Röling, N.G. (1995) What to think of extension? a comparison of three models of extension practice. In: Salamna, N. (ed.) *Article for the Francophone Issue of AERDD Bulletin*. ICRA, Montpellier Office.

Röling, N.G. and van de Fliert, E. (1994) Transforming extension into sustainable agriculture: the case of Integrated Pest Management in Rice in Indonesia. *Agriculture and Human Values* 11, 96–108.

Röling, N.G. and Jiggins, J. (1998) The ecological knowledge system. In: Röling, N.G. and Wagemakers, M.A.E. (eds) *Facilitating Sustainable Agriculture. Participatory Learning and Adaptive Management in Times of Environmental Uncertainty*. Cambridge University Press, Cambridge, pp. 283–311.

Röling, N.G. and Wagemakers, M.A.E. (1998a) *Facilitating Sustainable Agriculture. Participatory Learning and Adaptive Management in Times of Environmental Uncertainty*. Cambridge University Press, Cambridge.

Röling, N.G. and Wagemakers, M.A.E. (1998b) A new practice: facilitating sustainable agriculture. In: Röling, N.G. and Wagemakers, M.A.E. (eds) *Facilitating Sustainable Agriculture. Agriculture Participatory Learning and Adaptive Management in Times of Environmental Uncertainty*. Cambridge University Press, Cambridge, pp. 3–22.

Roux, O. and Baumgärtner, J. (1998) Evaluation of mortality factors and risk analysis for the design of an integrated pest management system. *Ecological Modelling* 109, 61–75.

Royer, T.A., Mulder, P.G. and Cuperus, G.W. (1999) Renaming (redefining) Integrated pest management: fumble, pass, or play. *American Entomologist* 45, 136–139.

Saanich (2003) *Integrated Pest Management*. Council Policy 03/166, Municipality of Saanich, Canada.

SARE (2004) *Exploring Sustainability in Agriculture*. Sustainable Agriculture Research and Education (SARE) Program of USDA's Cooperative State Research, Education, and Extension Service, Washington, DC. Available at: www.sare.org.

Scheurer, D.L., Schneider, D.C. and Sanford, L.P. (2001) Scaling issues in marine experimental ecosystems. In: Gardner, R.H., Kemp, W.M., Kennedy, V.S. and Petersen, J.E. (eds) *Scaling Relations in Experimental Ecology*, Columbia University Press, New York, pp. 331–360.

Sciarretta, A. and Trematerra, P. (2004) Spatial analysis of pheromone trap catches of *Cydia funebrana*, *Cydia pomonella*, *Cydia molesta* and *Anarsia lineatella*: contribution to IPM in fruit crops. *IOBC Bulletin* 28, 52–55.

Sciarretta, A., Trematerra, P. and Baumgärtner, J. (2001) Geostatistical analysis of *Cydia funebrana* (*Lepidoptera*: *Tortricidae*) pheromone trap catches at two spatial scales. *American Entomologist* 47, 174–184.

Sciarretta, A., Melaku, G., Lulseged, B., Getachew, T. and Baumgärtner, J. (2005) Development of an adaptive tsetse fly population management scheme for the Luke community, Ethiopia. *Journal of Medical Entomology* 42, 1006–1019.

Simon, H.A. (1973) The organization of complex systems. In: Pattee, H.H. (ed.) *Hierarchy Theory. The Challenge of Complex Systems*. Braziller, New York, pp. 1–27.

Smith, B.D. (1995) *The Emergence of Agriculture*. Scientific American Library, New York.

The Field Alliance (2003) Community Integrated Development, Concepts and Cases. Available at: www.communityipm.org.

Trematerra, P., Gentile, P. and Sciarretta, A. (2004) Spatial analysis of pheromone trap catches of codling moth *Cydia pomonella* L. (*Lepidoptera*: *Tortricidae*), in two heterogeneous agro-ecosystems, using geostatistical techniques. *Phytoparasitica* 32, 325–341.

Turner, M.G. and Gardner, R.H. (1991) Quantitative methods in landscape ecology: an introduction. In: Turner, M.G. and Gardner, R.H. (eds) *Quantitative Methods in Landscape Ecology*. Springer-Verlag, New York, pp. 4–14.

van de Fliert, E. (1993) *Integrated Pest Management: Farmer Field Schools Generate Sustainable Practices*. Wageningen Agricultural University papers 93–3, Wageningen.

Van Driesche, R.G. and Bellows, T.S. Jr (1996) *Biological Control*. Chapman & Hall, New York.

Van Lenteren, J.C. (1995) Integrated Management in protected crops. In: Dent, D. (ed.) *Integrated Pest Management*. Chapman & Hall, New York, pp. 311–343.

Viggiani, G. (1977) *Lotta Biologica ed Integrata*. Liguori Editoré, Napoli.

Walters, C. (1986) *Adaptive Management of Renewable Resources*, Macmillan, New York.

Waltner-Toews, D., Kay, J.K., Neudoerffer, C. and Gitau, T. (2003) Perspective changes everything: managing ecosystem from inside out. *Frontiers in Ecology and the Environment* 1, 23–30.

Webster's (1976) *Webster's New College Dictionary*. Merriam, Springfield, Massachusetts.

Wiens, J.A. (1989) Spatial scaling in ecology. *Functional Ecology* 3, 385–397.

World Commission on Environment and Development (1987) *Our Common Future*. Oxford University Press, Oxford.

Zethner, O. (1995) Practice of integrated pest management in tropical and sub-tropical Africa: an overview of two decades (1970–1990). In: Mengech, A.N., Saxena, K.N. and Gopalan, H.N.B. (eds) *Integrated Pest Management in the Tropics: Current Status and Future Prospects*. John Wiley & Sons, New York, pp. 1–67.

8 Economic Aspects of Ecologically Based Pest Management

George W. Norton

Department of Agricultural and Applied Economics, Virginia Tech,
Blacksburg, VA 24061, USA

Introduction

Why do farmers, homeowners or schools choose specific pest management practices or strategies? What are the impacts on society at large if ecologically based approaches are followed? How can public policies influence the nature and benefits of pest management strategies? The answers to these questions depend not only on pest densities and on the efficacy of specific strategies but also on economics. The focus of this chapter is on the economic dimension of why farmers (and others) make the decisions they do with respect to pest management, and the economic implications of those decisions.

Economic analysis can help farmers, government officials and others decide if specific pest management practices or programmes are worthwhile. It can indicate which practices or programmes are likely to give the greatest benefits and to whom, how large and certain those benefits are, and how public policies might influence them. Returns to ecologically based pest management can be financial, but they are also environmental or health-related, and attempts can be made to value those benefits as well.

Results of economic analyses of pest management practices or programmes can be both descriptive and prescriptive (Reichelderfer *et al.*, 1984). An analysis cannot say which is the best decision for a farmer, a homeowner or a government official, as that would depend on how each person subjectively weighs his or her multiple objectives, but it can indicate trade-offs and opportunity costs that may facilitate decisions.

Choice of Pest Management Practices

The most common economic analyses of pest management practices at the farm, household or school level focus on the choice of alternatives for pest control. Because cost or profitability is a factor in choosing a pest management practice, *enterprise budgets* or *partial budgets* are commonly used to assess relative costs and returns of a conventional practice vs. one or more ecologically based practices. An enterprise budget lists all estimated expenses associated with a particular enterprise to assess its cost or profitability. Partial budgets include only the changes in yield, prices and costs to assess whether benefits (due to increased revenues and reduced costs) exceed burdens (from reduced revenues or increased costs). Budgets for various pest management alternatives can be compared using data from replicated experiments or from producer surveys. The budgets assume typical conditions, no carry-over effects from one period to the next, predictable prices, costs and yield effects of pests, and that profits are all that matter to the decision maker (Swinton and Day, 2003).

When budgeting is used to compare pest management practices, statistical analyses are essential to ensure that the differences noted are important and significant at some predetermined level. For example, when experimental data are used to analyse the differences in costs and yields associated with alternative pest management practices, analysis of variance can be used in testing for significant differences. When samples are derived from populations of adopters and non-adopters of ecologically based pest management practices, regression analysis can be used to hold constant many of the non-pest management variables when testing for differences due to the practices. Masud et al. (1984) provide an example for delayed planting dates to control boll weevils while reducing pesticide use in the Texas rolling plains.

Once statistical significance is confirmed, results of budget analyses can be used to make recommendations to producers, homeowners or others about the most profitable or least expensive pest management practices. As noted several years ago by Stern et al. (1959), choice of pest management practice often involves choosing the right level and timing of a pest control action such as application of a pesticide or a biological control. Pest densities vary, plants compensate for damage and prices of inputs and outputs influence the revenues and costs associated with applying a specific pest management practice. The concept of an *economic or action threshold* was developed to help farmers identify when the benefits of applying a pest management practice (a pesticide) exceed the costs, so that it pays to apply it (Pedigo et al., 1986; Cousens, 1987). The concept has been applied not only to insects but also to weeds (Marra and Carlson, 1983).

Economic analysis of pest management risks

Although profitability is a key concern to producers and others when they consider the use of ecologically based pest management strategies, risk is also important. Risk may arise from biological, technical or economic factors, but pest infections or infestations often represent one of the largest risks to crop yields. Various types of economic analysis can be employed to aid farmers in decision making under risk. A pay-off matrix that lists projected net returns for various pest management practices and pest pressures (Norton and Mullen, 1994) can help. An example is provided in Table 8.1.

The decision to adopt a particular integrated pest management (IPM) practice will often have to be made before information is available on pest severity. Therefore, the decision will depend on the producer's ability and desire to absorb risk and on an assessment of the probabilities of light or severe pest attacks. If historical data were available to help in assessing the probabilities, expected monetary outcomes could be calculated for each pest management practice. This calculation would involve multiplying the pay-offs under each practice by the expected probabilities of light or severe pest attacks. For example, if the probability of a light pest severity (say of pecan scab) is 60% and heavy severity is 40%, and the expected pay-off (returns per acre or per block) for each practice for each pest pressure is the amount listed in Table 8.1, the expected value of the conventional spray alternative is $(200 \times 0.6) + (50 \times 0.4) = \$160$ and of a pheromone trap alternative is $(350 \times 0.6) + (-50 \times 0.4) = \$320$. If historical probabilities are unavailable, subjective probabilities based on experience of farmers or other experts could be used for the same purpose. Additional discussion of pay-off matrices is found in Reichelderfer et al. (1984).

In the example given above, the pheromone alternative has a higher return; but if a farmer is very risk-averse, he or she might still choose the spray approach. In addition, if a farmer's subjective probabilities of light

Table 8.1. Hypothetical monetary pay-off matrix for insect control (net returns per hectare). (From Norton *et al.*, 2001.)

	Pest management practices		
Pest severity	Conventional spray	Pheromone traps	Traps and spray
Light	$200	$350	$150
Heavy	$50	-$50	$100

or severe pest attacks are unduly skewed towards severe attacks, he or she might perform an inaccurate mental assessment of the gains from the various choices. In either case, the farmer might then apply the spray approach as insurance. Ecologically based approaches do not necessarily increase pest risk (Greene *et al.*, 1985; Napit *et al.*, 1988; Lamp *et al.*, 1991), although they sometimes can (Moffitt *et al.*, 1983; Szmedra *et al.*, 1990), or farmers might think they can. Risks arise not only because of actual variations in pest densities but also because of inaccurate scouting or poor model predictions of the effects of the densities.

One answer to the risk problem may be to provide producers with the opportunity to purchase crop insurance that will pay off if significant pest damage occurs (Feinerman *et al.*, 1992). Such insurance has begun to appear on the market for certain pest problems. For example, the Agricultural Conservation Innovation Center (ACIC) developed an insurance programme for IPM on corn rootworm in three states in the Midwest in 1999 and implemented it in conjunction with two insurance companies.

The economic attractiveness of alternative pest management practices in the presence of risk can also be assessed with a method called *stochastic dominance* (SD). With SD one can compare pairs of alternative pest management strategies for various sets of producers. These sets of producers are defined by their degrees of risk aversion. Examples of using SD in economic evaluation and comparison of IPM strategies with other strategies are found in Greene *et al.* (1985), Musser *et al.* (1981) and Moffitt *et al.* (1983).

Increasingly, pest-forecasting models are being used that predict the severity of

pest attack based on a series of biological and climatic factors. These models are especially well developed for certain diseases such as late blight on potatoes and tomatoes and early leafspot on groundnuts. For example, an early leafspot-forecasting model in Virginia has reduced fungicide use significantly on groundnuts (Mullen *et al.*, 1997). Perfect forecasting is of course unlikely, and therefore the value of any forecasting scheme depends on the probability of a correct forecast as well as on the probabilities of light and severe pest attacks. Producer surveys may be used to estimate how accurate the forecast must be before producers will adopt it. In some cases, pest alerts are publicized on crop websites such as the Oklahoma lucerne website (http://alfalfa.okstate.edu/index. htm).

Areawide IPM programmes such as those initiated by the US Department of Agriculture (USDA)–Agricultural Research Service (ARS) can also help reduce risk. For example, these programmes have reduced risk of codling moth in apples and pears in California, Oregon, Colorado and Washington; corn rootworm in the Midwest; pests in stored grains in Kansas and Oklahoma; fruit flies in Hawaii; and leafy spurge in the upper plains of the USA (http://www.ipmalmanac.com/articles/ areawide.asp, http://www.pswcrl.ars.usda. gov/AWPM2/index.htm and http://bru.gmprc. ksu.edu/proj/areawide/pubs/IPM_KS_ 1.3.pdf). A key motivation for an areawide programme is the fact that attacking a pest in one location alone is unlikely to suppress it effectively. Because an individual farmer or agribusiness manager realizes this possibility, he or she will be less likely to apply optimal IPM practices unless incentives are created through a coordinated effort. Therefore, in the USA these publicly supported areawide programmes have been used to at least initiate the coordination, thereby reducing pest risk for everyone (Flinn et al., 2003).

Economic analysis has been used to assess the trade-offs between income and health and environmental (HE) risks associated with pest management practices (Crissman et al., 1998). Measuring HE risks of pest management practices is difficult (and hence costly) because the physical and biological effects of pesticide use differ by the type of active ingredient (in terms of toxicity, mobility and persistence) as well as by the method, time and location applied. Location-specific models with data on soil type, irrigation system, slope and weather have been applied to provide information on fate of the chemicals applied, while non-location specific methods have been used to measure acute human health effects (dermatological, neurological, etc.) (Rola and Pingali, 1994; Pingali and Roger, 1995; Crissman et al., 1998).

Detailed studies of this type are expensive and may be best suited for policy prescription rather than for farmer decision making. However, these studies have led to conclusions that human health concerns may be more critical than broader environmental concerns with respect to pesticide use (Crissman et al., 1998), suggesting that application methods and the toxicity of specific chemicals should receive the greatest attention by farmers and others as they weigh income vs. HE risks.

Optimizing Pest Control Strategies

Applying an economic threshold in pest management involves identifying when incremental pest damage is equal to the cost of preventing it. However, many pest management practices have effects that carry over from one period of time to the next due to factors such as pesticide resistance, effects on beneficial predators and the simple fact that killing a pest today also reduces the offspring that may have caused damage in the future. For example, for a crop such as lucerne, pests on one cutting affect the next cutting and the yields throughout the stand life (Gutierrez et al., 1979). For other crops such as tree fruits and pecans, effects can carry over for years.

Dynamic thresholds can be developed that take into account these effects and the fact that future gains and losses must be discounted to bring them back to a present value (Taylor and Burt, 1984; Swinton and King, 1994). Various bioeconomic models, employing methods such as linear programming, non-linear programming, dynamic programming and dynamic simulation, have been used to provide information on optimal pest control strategies. For example, dynamic programming allows for examination of optimal pest control strategies when variables such as plant product, pest densities and pest susceptibility to pesticides are functions of time. An example of this type of analysis for corn rootworm and soybean cyst nematode control is found in Zacharias and Grube (1983). Martin et al. (1991) provide an example of using linear programming to analyse alternative tillage systems, crop rotations and herbicide use on east-central maize-belt farms, while Lazarus and Dixon (1984) utilize non-linear programming to assess potential gains from internalizing resistance through regional coordination. Swinton and King (1994) use a bioeconomic model that includes dynamic, whole-farm, stochastic simulation to assess the value of weed seedling and weed seed counts in maize and soybeans for weed control decisions. Their approach is suited to measuring returns to information on other types of pest populations as well.

Economics of resistance

The economics literature relating to pesticide resistance includes several dynamic optimizing models in which the optimal use of pest controls implies management of both the pest and its associated stock of susceptibility. Early dynamic models by

Shoemaker (1973a, b, c), Hueth and Regev (1974) and Taylor and Headley (1975) were primarily theoretical in nature, but empirical dynamic models became more commonplace with improvements in computational capacity (Regev *et al.*, 1983; Archibald, 1984; Plant *et al.*, 1985; Harper, 1986; Kazmierczak, 1991). Because the impact of pest management is often dynamic due to factors such as pest resistance to pesticides, the results of dynamic models may be more precise if sufficient complexity is incorporated. However, the inherent complexity of the biological processes, and hence the mathematical complexity of the models, continue to constrain the routine application of dynamic approaches for pest management decision making by producers. The models have increased our understanding of the economics of resistance and the potential for public policies to influence the adoption of practices that achieve both economic and ecological objectives. Further development and application of these models in an interdisciplinary environment are needed for policy analysis.

Evaluating Economic Benefits of Pest Management Programmes

The audience for economic analysis of pest management programmes includes research and extension administrators and other public officials who request information on the economic benefits of the programmes. Providing this information requires aggregating the individual-level economic analyses described earlier with estimates of the realized or projected level of adoption over time and with information on responsiveness of supply and demand to price changes. The models used in the analysis often involve calculation of *economic surplus* changes (described later). These changes are included in a *benefit–cost analysis* to account for discounting over time and to facilitate comparisons with other investments.

When widespread adoption of IPM occurs across large areas, changes in crop

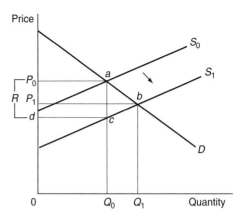

Fig. 8.1. Pest management benefits measured as changes in economic surplus.

prices, cropping patterns, producer profits and societal welfare can occur. These changes arise because costs differ and because supplies may increase, affecting prices for producers and consumers. The changes are illustrated in Fig. 8.1.

In this figure, S_0 represents the supply before adoption of an IPM strategy and D represents the demand. The initial price and quantity are P_0 and Q_0. Suppose that adoption of a new ecologically based pest management practice leads to savings of R in the cost of production, which is reflected in a shift down in the supply to S_1. This shift leads to an increase in production and consumption of Q_1 (by $\Delta Q = Q_1 - Q_0$) and the market price falls to P_1 (by $\Delta P = P_0 - P_1$). Consumers gain because they can consume more of the commodity at a lower price. They benefit from the lower price by an amount equal to their cost saving on the original quantity ($Q_0 \times \Delta P$) plus their net benefits from the increment to consumption. The area $P_0 abP_1$ represents their total benefit.

Although they may receive a lower price per unit, producers are better off too because their cost has fallen by R per unit, an amount greater than the fall in price. They gain the increase in profits on the original quantity ($Q_0 \times (R - \Delta P)$) plus the profits earned on the additional output, for a total producer gain of $P_1 bcd$. The distribution of benefits between producers and consumers depends on the size of the fall in price (ΔP) relative to the fall in costs (R) and on the nature of the supply shift. For example, if a commodity is traded and production in the area producing the commodity has little effect on price, most of the benefits would accrue to producers. If the supply curve shifts in more of a pivotal fashion as opposed to a parallel fashion as illustrated in Fig. 8.1, the benefits to producers would be reduced.

Examples of IPM evaluation using this economic surplus approach are found in Taylor and Lacewell (1977), Napit *et al.* (1988) and Eddleman *et al.* (1999). Formulas for calculating consumer and producer gains for a variety of market situations are found in Alston *et al.* (1995). It is possible to combine economic surplus models with the results from a geographic information system (GIS) to assess the spillover of IPM benefits across regions.

The most difficult aspect of this type of economic analysis is the calculation or prediction of the proportionate shift in supply that occurs with adoption of the pest management practices. Cost differences as well as adoption rates must be estimated. Information on cost and yield changes from producer surveys, field trials or other methods discussed earlier can be used. Methods for assessing adoption rates are described below. Once changes in economic surplus are calculated or projected over time, benefit–cost analysis can be used to calculate *net present values, internal rates of return or benefit–cost ratios.* The benefits are the changes in total economic surplus calculated for each year, and the costs are the public expenditures on the pest management programme. The primary purpose of the benefit–cost analysis is to take into account the fact that benefits and costs need to be discounted; the sooner they occur, the more they are worth.

Changes in economic surplus can also be embedded in mathematical programming models to further project interregional changes in production following the introduction of a widespread IPM programme, or to predict the impacts of IPM following policy changes that encourage or discourage IPM use. In instances where IPM adoption is believed to trigger impacts that touch other sectors in the economy, the impacts may be estimated or projected using models that incorporate cross-sector effects or even cross-country effects.

Economic impacts of ecologically based IPM programmes can also be assessed with *econometric* techniques that use survey data to assess on-farm effects (on pesticide use, yields, as well as on costs and profitability) of pest management practices. Fernandez-Cornejo *et al.* (2002) estimated the effects of herbicide-tolerant (transgenic) soybeans on herbicide use, yields and profits. The results of such econometric analyses can be incorporated in economic surplus models of the type described above to assess market level impacts as well.

Most agricultural biotechnology applications that involve transgenics have focused on pest problems, expecially weeds and insects, and also on diseases and viruses. The methods described earlier for evaluating direct economic benefits as well as the value of HE benefits of IPM are equally well suited for evaluating the impacts of biotechnologies. However, when conducting an economic assessment of biotechnologies, the relative importance of assessing the costs of meeting biosafety and other regulatory requirements is greater than for other IPM practices. In addition, the assessment models may need to be modified to capture the effects of imperfect competition resulting in part from ownership of intellectual property rights (Moshini and Lappan, 1997).

Assessing and Predicting Pest Management Adoption

A significant number of studies have been completed, especially in the USA, to assess why farmers adopt pest management practices, especially IPM (e.g. Napit *et al.*, 1988; Harper *et al.*, 1990; McNamara *et al.*, 1991; Vandeman *et al.*, 1994; Fernandez-Cornejo and Jans, 1998). The purpose of the studies is to evaluate what factors, economic and other, influence producer adoption, and how many producers have adopted or are likely to adopt the practices. The results of these studies are used to better target IPM practices to meet user needs and assist with economic assessment of IPM programmes.

Because IPM often implies a combination of pest management practices, adoption is often a manner of degree rather than with or without, and a commonly understood commodity- and location-specific definition of IPM is needed before its adoption can be measured. Because the desired impacts are usually in terms of improving the environment or income, practices considered part of an IPM programme should influence those factors.

Once practices are identified, grouping them to signify levels of IPM adoption (such as low, medium and high) or relating them to an adoption scale by assigning points to particular practices are common ways to measure the degree of adoption. Rajotte *et al.* (1987), Napit *et al.* (1988) and Vandeman *et al.* (1994) have developed crop-specific definitions of IPM levels based on sets of practices, while Hollingsworth *et al.* (1992) and Beddow (2000) have developed IPM point systems for various crops. Stakeholders can vary the points they assign depending on their weights on economic vs. environmental goals.

If adoption is based on levels representing groups of practices, rather than on a binary indicator or continuous scale, levels of adoption will show how well the groups of practices meet the goals of the IPM programme. For example, high adoption meets nearly all goals of the programme, while low adoption meets few. Such groupings are necessarily arbitrary.

After creating a measure of adoption, producer surveys, interviews of experts and analysis of secondary data from sources such as the USDA National Agricultural Statistics Service can be used to assess the extent of adoption. The method used will depend on the accuracy required, the availability of secondary data, resources available and whether the study is evaluating IPM practices already adopted or projecting future adoption. To predict long-term programme impacts, projections of future technology adoption are needed. Extension agents or other industry experts may provide reasonable estimates of IPM

adoption though reliability may vary. The diffusion of a new technology tends to follow a sigmoid or logistic curve shape over time and this knowledge can be used to help parameterize the adoption curve.

In studies that statistically assess the factors influencing adoption, a regression model (usually Probit) is used that relates adoption of IPM to a series of variables representing characteristics such as nature of the technology, physical environment, IPM user, economic environment and institutional environment. The dependent variable could be a binary adoption variable for a single level of IPM, a categorical variable for multiple levels of IPM or a continuous variable for some index of IPM practices adopted.

The explanatory variables would include one or more measures of exposure to the IPM programme. The significance of the IPM programme exposure variables would determine whether observed changes in IPM adoption are attributable to the programme or not. In general, past studies have found IPM adopters to be younger, more educated than average and having less farming experience. (Napit *et al.*, 1988; Leslie and Cuperus, 1993; Swinton and Day, 2003). However, results of adoption studies also tend to be site-specific, and the nature of the IPM technique or strategy itself can affect its adoption. The relative ease of application, the relative cost or other advantages compared to alternatives, compatibility with other practices and the complexity of the practice or strategy can all influence adoption decisions (Cuperus *et al.*, 1999).

Valuing Health and Environmental Benefits of Ecologically Based Pest Management

Economic analysis can be useful in assessing HE benefits of ecologically based pest management because (i) it can place a value on its contributions to one of its major goals and (ii) dollars provide a useful unit of measure when the contributions of pest management to multiple aspects of the HE need to be scored, weighted or combined. Because most HE benefits are not valued in the market, non-market assessment techniques are sometimes used to calculate HE costs and benefits associated with changes in pesticide use. Alternatively, costs of cleaning up polluted environments, costs of treating illnesses and costs of compensating for lost work time due to illnesses induced by HE problems can be used.

Changes in pesticide active ingredients that may be associated with changes in pest management practices are only very rough indicators of HE effects, and therefore models have been developed to formally link pesticide use to specific aspects of health and the environment. Location-specific models are mentioned above. Their use has been limited by high data and time (cost) requirements except in specialized applications. Non-location specific models that require information on the pesticides applied and the method of application have recently grown in use. These models produce indicators of risk by HE category as well as weighted total risk for the pest management application. Examples are the environmental impact quotient (EIQ) developed by Kovach et al. (1992), the pesticide index (PI) of Penrose et al. (1994) and the Benbrook (1997) index.

Each indexing scheme is a type of scoring model, and each inevitably involves subjective weightings of risks across environmental categories. In assessing the usefulness of specific indexes, it is important to recognize that they perform two tasks (Norton et al., 2001). One is to identify the risks of pesticides to the individual categories of health and the environment, such as groundwater, birds, beneficial insects, humans and so forth. The second is to aggregate across categories through a weighting scheme. Both of them are challenging tasks. The first is challenging because it is difficult to identify mutually exclusive categories, especially ones with data. The categories in most models contain a mixture of non-target organisms (e.g. humans, birds, aquatic organisms, beneficial insects and wildlife) and modes of exposure (e.g. groundwater and surface water). Aggregating across categories is challenging because of the need to weight categories. One means of addressing the weighting issue is to

weight categories with dollars as the unit of measure and willingness to pay (WTP) for risk reduction as the weights.

When a good is traded in the market, its value is easily determined. In the absence of market distortions, the market price provides an indicator of its value because it captures a great deal of information: consumer WTP, producer cost, scarcity and so on. Unfortunately, there is usually no market price for health effects and environmental damage. Therefore, alternative monetary valuation techniques have been developed, e.g. *contingent valuation* (CV), *hedonic pricing* and measuring costs of cleaning up pollution and treating illnesses. Each of these techniques is described briefly below.

Although it may seem odd to attempt to place a dollar value on health and the environment, there are advantages of doing so. Money is a familiar indicator of value and hence easily interpretable. It provides a common unit of measure when aggregating across categories or when comparing benefits of a programme to its costs. Both monetary and non-monetary indicators attempt to summarize a great deal of information using subjective weights. However, monetary techniques may be less subjective because people assign weights relative to other known values or market prices.

Approaches to monetary valuation of HE effects primarily focus on measuring cost or on WTP to avoid risks. The cost approaches attempt to directly measure the cost of environmental damage or illness. Pimentel (1978, 1980, 1991) used this approach to roughly assess environmental costs associated with pesticides. More careful studies have been completed on the cost of illness and productivity losses due to pesticides in the Philippines and Ecuador by Pingali et al. (1994), Antle and Pingali (1994) and Crissman et al. (1998). Data were collected on pesticide use, demographics and ailments that were possibly related to pesticide use for pesticide applicators and other relevant individuals. Medical doctors were used for health assessments. The costs of ailments were regressed on pesticide use, demographics and other variables so that the cost of illness related to pesticides could be estimated. Additional regression models were used to estimate the relationship between labour productivity and health problems related to pesticides.

An alternative cost approach is based on the cost of repairing environmental damage. For example, if pesticides render a supply of groundwater undrinkable, the repair cost would be the cost of treating the water to make it potable. Another cost-based approach is to measure the cost of avoiding exposure to environmental risk. For example, Abdalla et al. (1992) calculated the extra cost of purchasing drinking water in order to avoid drinking contaminated groundwater. Measures of expenditures avoided tend to underestimate total environmental impacts, because they focus only on avoiding human exposure.

WTP techniques are used to place values on environmental goods or amenities. The first of these, hedonic pricing, attempts to infer WTP for environmental amenities from the prices of other goods based on the characteristics of those goods. Beach and Carlson (1993) have used the approach to value safer pesticides.

A second, more common, approach for valuing pest management HE risk reduction is CV. This approach generally employs a survey to collect data on people's WTP to receive a benefit or their willingness to accept (WTA) compensation for a loss. A CV survey would typically ask respondents how much they would be willing to pay for a given improvement in an environmental asset, or how much they would be willing to accept in compensation for a given degradation of an environmental asset. In the context of pest management, respondents might be asked how much they would be willing to pay to reduce the risk of pesticides to various categories of environmental assets. The WTP data could later be linked to pesticide use data to arrive at a value for a change in pesticide use. Higley and Wintersteen (1992), Mullen et al. (1997), Swinton et al. (1999) and Cuyno et al. (2001) provide examples of using CV for such an assessment.

CV is subject to a number of potential biases and has therefore been controversial.

The more serious biases may arise due to the way the survey is designed or administered, or due to the hypothetical nature of the questions. The fear is that people will overstate WTP when they do not pay actual money. CV is one of the few procedures available for cost-effectively estimating environmental costs associated with pest management practices. While it is likely to remain controversial because of potential for bias, as a weighting technique it is likely to contain fewer biases than the other indexes mentioned earlier.

The study by Mullen et al. (1997) illustrates how to use CV to evaluate HE effects of an IPM programme. The study used secondary data to classify each relevant pesticide as high, moderate or low in risk for its effect on groundwater, surface water, acute and chronic human health, aquatic species, avian and mammalian species and non-target arthropods. The effect of the IPM programme on pesticide use was estimated using regression analysis. The value of pesticide reduction was estimated using a contingent valuation survey (CVS), the results of which were combined with the estimated reduction in pesticide use attributable to the programme to derive the value of the programme's HE benefits.

One technique that could be explored in the future for assessing WTP for reduced HE problems associated with reduced risk from pesticides is the experimental auction (EA) (Brookshire and Coursey, 1987). With an EA, people are given an initial endowment of money with which to bid on health or environmental improvements, and are then forced to pay their bids at the end of the experiment. The purpose of using EA for assessing WTP to reduce risks associated with pesticide use would be to reduce the hypothetical bias found in CV analysis.

Assessing Public Policies Related to Pest Management

Public policies can have a significant effect on development and adoption of ecologically based pest management practices. Public policies affect funding for pest management research and extension. They directly influence individual incentives through pesticide regulations, cost sharing of IPM practices, and credit or risk management programmes. They indirectly influence individual incentives through commodity price supports, macroeconomic policies, and IPM or organic certification programmes. Economic analysis is often used to help design or assess impacts of these policies.

Examples of economic analyses to assess economic impacts of IPM programmes are found in Napit *et al.* (1988), Mullen *et al.* (1997) and Beddow (2000). The latter study includes the value of environmental and health benefits. These studies use the economic surplus and CV methods described earlier. A large number of studies, mostly in the 'grey' literature of reports by and for government agencies such as the US Environmental Protection Agency (USEPA), provide estimates of the economic impacts of pesticide regulations. Results are often based on partial budgeting. A detailed study of the effects of pesticide regulations on pesticide resistance for a pest complex on apples is found in Kazmierczak (1991).

In general, pesticide policy in the USA has revolved around use restrictions, applied by USEPA under the Federal Insecticide, Fungicide and Rodenticide Act and the Food Quality Protection Act. These restrictions are a blunt tool for achieving environmental objectives, and economic analysis can help in designing policies such as cost sharing, credit and insurance programmes to encourage adoption of IPM and other ecologically based programmes.

An example of a cost-sharing programme is the Environmental Quality Incentives Programme (EQIP), which cost-shares adoption of IPM practices. Another example is the Integrated Crop Management (ICM) programme implemented in the 1980s in 45 states. An evaluation of that programme found little impact on pesticide use (Dicks *et al.*, 1993).

At times, pest management policies work counter to ecological objectives. For example, owners of federally subsidized housing in Virginia receive a subsidy per unit

for pest control, which has been primarily defined to mean pesticide applications for cockroach control. Recent economic analysis has found that IPM techniques are just as effective at roughly the same cost over time but with a slightly higher upfront cost (Miller and Meek, 2004). This cost-sharing programme can now be modified to include IPM techniques that substantially reduce environmental risk due to pesticides.

Pesticide policy in Europe (and increasingly in developing countries) also revolves around use restrictions. The Pesticide Action Network (PAN in Europe), an organization that coordinates non-governmental organizations (NGOs) interested in pesticide issues, has also lobbied governments to be more proactive in identifying alternatives to pesticides, recognizing that grower support is contingent on finding such alternatives (http://www.pan-europe.net/).

Indirect policies like federal price supports for crops such as wheat and cotton can reduce the threshold for pest control, and hence tend to encourage pesticide use (Reichelderfer and Hinkle, 1989). Even macroeconomic policies that influence items such as currency exchange rates can affect incentives to adopt pesticide-based as opposed to ecologically based pest management. For example, Tjornhom *et al.* (1998) found that policies in the Philippines that led to overvaluation of the exchange rate there were subsidizing pesticide use, as pesticides or their technical ingredients are all imported into the Philippines. A higher exchange rate makes imports (in this case, pesticides) cheaper.

Certification programmes such as IPM or organic certification can encourage ecologically based pest management by allowing sellers to charge a premium for products raised with those practices. Without certification with monitoring by a trusted source, such programmes are ripe for abuse and hence failure (Ward *et al.*, 1995). The public sector may be able to let the private sector take the lead in this type of programme, such as the case in the certification programme for potatoes in Wisconsin, with Wegmans' IPM label and with Oklahoma's lucerne IPM certification. Economic analysis can help in assessing the market for such programmes.

Conclusions

Economics plays a major role in the success of ecologically based pest management programmes. A wide variety of economic analyses of pest management practices and policies have been conducted since the first assessment of economic thresholds more than 40 years ago. Many of the analyses have involved projections of profitability, risk, HE effects, returns to research and implications for public policies affecting pest management decisions. Simple per acre budget analyses of IPM practices and analyses of factors influencing IPM adoption have been especially prevalent. Fewer analyses have addressed aggregate income and HE impacts, and early dynamic modelling of crop–pest–predator interactions have been slow to develop into routine analyses, though several academic papers concerning these issues have been published. Dynamic analyses are especially important for assessing pesticide resistance implications of public policies.

There continues to be an absence of basic economic analysis (simple partial budgeting) in some pest management research programmes. Well-designed biological research often fails to take the crucial step of establishing the profitability of recommended practices resulting from the research. The need for basic economic analysis stems both from the importance of recommending practices for which farmers will truly have an incentive to adopt, and from the need to design public policies or institutional arrangements to offset insufficient profitability incentives if an ecologically based practice helps in attaining HE objectives.

A growing number of studies have shown that ecologically based IPM programmes can reduce pesticide use while improving productivity (Antle and Pingali, 1994; Norton and Mullen, 1994; Fernandez-Cornejo, 1998). However, it has been almost 20 years since a relatively comprehensive economic impact evaluation of IPM programmes has been conducted in the USA. Since the report of the last study (Rajotte *et al.*, 1987) was published, numerous advances have been made in evaluation procedures, especially in incorporating HE effects. A major study in this area is long overdue given the resources devoted to IPM programmes. Since the last study, several large areawide programmes have been initiated and the economic effectiveness of these too should be assessed.

References

Abdalla, C.W., Roach, B.A. and Epp, D.J. (1992) Valuing environmental quality changes using averting expenditures: an application to groundwater contamination. *Land Economics* 68, 163–169.

Alston, J.M., Norton, G.W. and Pardey, P.G. (1995) *Science under Scarcity: Principles and Practice for Agricultural Research Evaluation and Priority Setting*. Cornell University Press, Ithaca, New York.

Antle, J.M. and Pingali, P.L. (1994) Pesticides, productivity and farmer health: a Philippine case study. *American Journal of Agricultural Economics* 76, 418–430.

Archibald, S.O. (1984) A dynamic analysis of production externalities: pesticide resistance in California cotton. PhD dissertation, University of California, Davis, California.

Beach, D. and Carlson, G.A. (1993) Hedonic analysis of herbicides: do user safety and water quality matter? *American Journal of Agricultural Economics* 75, 612–623.

Beddow, J. (2000) Protocols for assessment of integrated pest management programs. MS thesis, Virginia Tech, Blacksburg, Virginia.

Benbrook, C.M. (1997) *Pest Management at the Crossroads*. Consumers Union, Yonkers, New York.

Brookshire, D.S. and Coursey, D.L. (1987) Measuring the value of a public good: an empirical comparison of elicitation procedures. *The American Economic Review* 77, 554–566.

Cousens, R. (1987) Theory and reality of weed control thresholds. *Plant Protection Quarterly* 2, 13–20.

Crissman, C.C., Antle, J.M. and Capalbo, S.M. (1998) *Economic, Environmental and Health Tradeoffs in Agriculture: Pesticides and the Sustainability of Andean Potato Production.* Kluwer Academic Publishers, Dordrecht, The Netherlands.

Cuperus, G.W., Mulder, P.G. and Royer, T.A. (1999) Implementation of ecologically based IPM. In: Rechcigl, J. and Rechcigl, N. (eds) *Insect Pest Management.* Lewis Publishers, Boca Raton, Florida, pp. 171–206.

Cuyno, L.C.M., Norton, G.W. and Rola, A. (2001) Economic analysis of environmental benefits of integrated pest management: a Philippines case study. *Agricultural Economics* 25, 227–234.

Dicks, M.R., Norris, P.E., Cuperus, G.W., Jones, J. and Duan, J. (1993) Analysis of the 1990 integrated crop management practice. *Oklahoma Cooperative Extension Service* Circular E-925, Division of Agricultural Sciences and Natural Resources, Oklahoma State University, Stillwater, Oklahoma, 117.

Eddleman, B.R., Chang, C.C. and McCarl, B.A. (1999) Economic benefits from grain sorghum variety improvement in the United States. In: Wiseman, B.R. and Webster, J.A. (eds) *Economic, Environmental, and Social Benefits of Resistance in Field Crops.* Entomological Society of America, Lanham, Maryland, pp. 17–44.

Feinerman, E., Herriges, J.A. and Hottkamp, D. (1992) Crop insurance as a mechanism for reducing pesticide usage; a representative farm analysis. *Review of Agricultural Economics* 14, 169–186.

Fernandez-Cornejo, J. (1998) Environmental and economic consequences of technology adoption: IPM in viticulture. *Agricultural Economics* 18, 145–155.

Fernandez-Cornejo, J. and Jans, S. (1998) Issues in economics of pesticide use in agriculture: a review of empirical evidence. *Review of Agricultural Economics* 20, 462–488.

Fernandez-Cornejo, J., Klotz-Ingram, C. and Jans, S. (2002) Farm-level effects of adopting herbicide-tolerant soybeans in the U.S.A. *Journal of Agricultural and Applied Economics* 34, 149–163.

Flinn, P.W., Hagstrum, D.W., Reed, C. and Phillips, T.W. (2003) USDA–ARS stored-grain areawide integrated pest management program. *Pest Management Science* 59, 614–618.

Greene, C.R., Kramer, R.A., Norton, G.W., Rajotte, E.G. and McPherson, R.M. (1985) An economic analysis of soybean integrated pest management. *American Journal of Agricultural Economics* 67, 567–572.

Gutierrez, A.P., Regev, U. and Shalit, H. (1979) An economic optimization model of pesticide resistance: alfalfa and Egyptian alfalfa weevil. *Environmental Entomology* 8, 101–107.

Harper, C.R. (1986) Optimal regulation of agricultural pesticides: a case of chlordimeform in the Imperial Valley. PhD thesis, University of California, Berkeley, California.

Harper, J.K., Rister, M.E., Mjelde, J.W., Drees, B.M. and Way, M.O. (1990) Factors influencing the adoption of insect management technology. *American Journal of Agricultural Economics* 72, 997–1005.

Higley, L.G. and Wintersteen, W.K. (1992) A novel approach to environmental risk assessment of pesticides as a basis for incorporating environmental costs into economic injury levels. *American Entomologist* 38, 34–39.

Hollingsworth, C.S., Coli, W.M. and Hazzard, R.V. (1992) Massachusetts integrated pest management guidelines: crop specific definitions. *Fruit Notes* Fall, 12–16.

Hueth, D. and Regev, U. (1974) Optimal agricultural pest management with increasing pest resistance. *American Journal of Agricultural Economics* 56, 543–552.

Kazmierczak, R.F. Jr (1991) Pesticide regulatory actions and the development of pest resistance: a dynamic bioeconomic model. PhD dissertation, Virginia Tech, Blacksburg, Virginia.

Kovach, J., Petzoldt, C., Degni, J. and Tette, J. (1992) *A Method to Measure the Environmental Impact of Pesticides.* New York's Food and Life Sciences Bulletin, Number 139, Cornell University, New York State Agricultural Experiment Station, Geneva, New York.

Lamp, W.O., Nielson, G.R. and Dively, G.P. (1991) Insect pest–induced losses in alfalfa: pattern in Maryland and implications for management. *Journal of Economic Entomology* 84, 610–618.

Lazarus, W.F. and Dixon, B.F. (1984) Agricultural pests as common property: control of the corn rootworm. *American Journal of Agricultural Economics* 66, 456–465.

Leslie, A. and Cuperus, G. (1993) *Successful Implementation of Integrated Pest Management for Agricultural Crops.* Lewis Publishers, Boca Raton, Florida.

Marra, M.C. and Carlson, G.A. (1983) An economic threshold model for weeds in soybeans (*Glycine max*). *Weed Science* 31, 604–609.

Martin, M.A., Schreiber, M.M., Riepe, J.R. and Bahr, J.R. (1991) The economics of alternative tillage systems, crop rotations, and herbicide use on three representative east central corn belt farms. *Weed Science* 39, 299–397.

Masud, S.M., Lacewell, R.D., Boring, E.P. and Fuchs, T.W. (1984) *Economic Implications of a Delayed Uniform Planting Date for Cotton Production in the Texas Rolling Plains*, Bulletin 1489, Texas Agricultural Experiment Station, Texas.

McNamara, K.T., Wetzstein, M.E. and Douce, G.K. (1991) Factors affecting peanut producer adoption of integrated pest management. *Review of Agricultural Economics* 13, 129–139.

Miller, D.M. and Meek, F. (2004) Cost and efficacy comparison of integrated pest management strategies with monthly spray insecticide applications for German cockroach (*Dictyoptera: Blattellidae*) control in public housing. *Journal of Economic Entomology* 97, 559–569.

Moffitt, L.J., Tanagosh, L.K. and Baritelle, J.L. (1983) Incorporating risk in comparisons of alternative pest management methods. *Environmental Entomology* 12, 1003–1111.

Moshini, G. and Lappan, H. (1997) Intellectual property rights and the welfare effects of agricultural R&D. *American Journal of Agricultural Economics* 79, 1229–1242.

Mullen, J.D., Norton, G.W. and Reaves, D.W. (1997) Economic analysis of environmental benefits of integrated pest management. *Journal of Agriculture and Applied Economics* 29, 243–253.

Musser, W.N., Tew, B.V. and Epperson, J.E. (1981) An economic examination of an integrated pest management production system with a contrast between CV and stochastic dominance analysis. *Southern Journal of Agricultural Economics* 13, 119–124.

Napit, K.B., Norton, G.W., Kazmierczak, R.F. Jr and Rajotte, E.G. (1988) Economic impacts of extension integrated pest management programs in several states. *Journal of Economic Entomology* 81, 251–256.

Norton, G. and Mullen, J. (1994) *Economic Evaluation of Integrated Pest Management Programs: A Literature Review.* Virginia Cooperative Extension Publication 448–120, Virginia Tech, Blacksburg, Virginia.

Norton, G.W., Swinton, S.M, Riha, S., Beddow, J., Williams, M., Preylowski, A., Levitan, L. and Caswell, M. (2001) *Impact Assessment of Integrated Pest Management Programs.* Department of Agricultural and Applied Economics, Virginia Tech, Blacksburg, Virginia.

Pedigo, L.P., Hutchins, S.H. and Higley, L.G. (1986) Economic injury levels in theory and practice. *Annual Review of Entomology* 31, 341–368.

Penrose, L.J., Thwaite, W.G. and Bower, C.C. (1994) Rating index as a basis for decision making on pesticide use reduction for accreditation of fruit produced under integrated pest management. *Crop Protection* 13, 146–152.

Pimentel, D. (1978) Benefits and costs of pesticide use in U.S. food production. *Bioscience* 28, 772–790.

Pimentel, D. (1980) Environmental and social costs of pesticides: a preliminary assessment. *Oikos* 34, 126–140.

Pimentel, D. (1991) Environmental and economic impacts of reducing pesticide use. *Bioscience* 41, 403–409.

Pingali, P.L. and Roger, P.A. (1995) *Impact of Pesticides on Farmer Health in the Rice Environment.* Kluwer Academic Publishers, Norwell, Massachusetts.

Pingali, P.L., Marquez, C.B. and Palis, F.G. (1994) Pesticides and Philippine rice farmer health: a medical and economic analysis. *American Journal of Agricultural Economics* 76, 587–592.

Plant, R.E., Mangel, M. and Flynn, L.E. (1985) Multi-seasonal management of an agricultural pest. II. The economic optimization problem. *Journal of Environmental Economics Management* 12, 45–61.

Rajotte, E.G., Kazmierczak, R.F., Norton, G.W., Lambur, M.T. and Allen, W.E. (1987) *The National Evaluation of Extension's Integrated Pest Management (IPM) Programs.* Virginia Cooperative Extension Service Publication 491–010, Blacksburg, Virginia.

Regev, U., Shalit, H. and Gutierrez, A.P. (1983) On the optimal allocation of pesticides with increasing resistance: the cases of the alfalfa weevil. *Journal of Environmental Economic Management* 10, 86–100.

Reichelderfer, K.H. and Hinkle, M.K. (1989) The evolution of pesticide policy: environmental interests and agriculture. In: Kramer, C. (ed.) *Political Economy of U.S. Agricultural: Challenges for the 1990s.* Center for Food and Agricultural Policy, Resources for the Future, Washington, DC, pp. 147–177.

Reichelderfer, K.H., Carlson, G.A. and Norton, G.A. (1984) *Economic Guidelines for Crop Pest Control.* FAO Plant Production and Protection Paper No 58, FAO, Rome, 93.

Rola, A.C. and Pingali, P.L. (1994) *Pesticides, Rice Productivity, and Farmers' Health: An Economic Assessment.* International Rice Research Institute, Manila, The Philippines.

Shoemaker, C. (1973a) Optimization of agricultural pest management. I: Biological and mathematical background. *Mathematical Biosciences* 16, 143–163.

Shoemaker, C. (1973b) Optimization of agricultural pest management. II: Formulation of a control model. *Mathematical Biosciences* 17, 357–365.

Shoemaker, C. (1973c) Optimization of agricultural pest management. III: Results and extension of the model. *Mathematical Biosciences* 18, 1–22.

Stern, V.M., Smith, R.F., van den Bosch, R. and Hagen, K.S. (1959) The integrated control concept. *Hilgardia* 29, 81–101.

Swinton, S.M. and Day, E. (2003) Economics in the design, assessment, adoption, and policy analysis of IPM. In: Barker, K.R. (ed.) *Integrated Pest Management: Current and Future Strategies.* Council for Agricultural Science and Technology, June, R-140.

Swinton, S.M. and King, R.P. (1994) A bio-economic model for weed management in corn and soybeans. *Agricultural Systems* 44, 313–335.

Swinton, S.M., Owens, N.N. and van Ravenswaay, E.O. (1999) Health risk information to reduce water pollution. In: Casey, F., Schmitz, A., Swinton, S. and Zilberman, Z. (eds) *Flexible Incentives for the Adoption of Environmental Technologies in Agriculture.* Kluwer Academic Publishers, Boston, pp. 263–271.

Szmedra, P.I., Wetzstein, M.E. and McClendon, R. (1990) Economic threshold under risk: the case of soybean production. *Journal of Economic Entomology* 83, 641–646.

Taylor, C.R. and Burt, O.R. (1984) Near-optimal management strategies for controlling wild oats in spring wheat. *American Journal of Agricultural Economics* 66, 50–60.

Taylor, C.R. and Headley, J.C. (1975) Insecticide resistance and the evaluation of control strategies for an insect population. *Canadian Entomologist* 107, 237–242.

Taylor, C.R. and Lacewell, R.D. (1977) Boll-weevil control strategies: regional benefits and costs. *Southern Journal of Agricultural Economics* 9, 129–135.

Tjornhom, J.D., Norton, G.W and Gapud, V. (1998) Impacts of price and exchange rate policies on pesticide use in the Philippines. *Agricultural Economics* 18, 167–175.

Vandeman, A.J., Fernandez-Cornejo, J., Jans, S. and Lin, B. (1994) *Adoption of Integrated Pest Management in U.S. Agriculture.* USDA/ERS, Agriculture Information Bulletin No. 707, Washington, DC.

Ward, C.E., Huhnke, R.L. and Cuperus, G.W. (1995) *Alfalfa Hay Preference of Oklahoma and Texas Dairy Producers.* Oklahoma Cooperative Extension Service Circular E-936, Stillwater, Oklahoma.

Zacharias, T.P. and Grube, A.H. (1986) Integrated pest management strategies for approximately optimal control of corn rootworm and soybean cyst nematode. *American Journal of Agricultural Economics* 68, 704–715.

9 Economics of Host Plant Resistance in Integrated Pest Management Systems

Philip Kenkel

Department of Agricultural Economics, Oklahoma State University, Stillwater, OK 74075, USA

Introduction

Crop resistance is considered a fundamental component of integrated pest management (IPM). The use of a crop variety that is resistant or tolerant to pests has several inherent advantages. Protection is 'built into' the seed and it does not require monitoring or management. Plant resistance does not require complete elimination of the pest to be effective. Plant resistance is compatible with insecticides and with many IPM strategies. Plant resistance is not dependent on pest density. The effects of resistance are specific to the target pest. The effectiveness of the resistant cultivar can be fairly long-lasting. In the case of low-resource farmers, or low per acre value enterprises, plant resistance may be the most viable pest control strategy.

The importance of host plant resistance as a tool in IPM has been long documented (Snelling, 1941; Painter, 1951; Chesnokov, 1953; Russel, 1978; Lara, 1979; Panda, 1979; Smith, 1989; Dhaliwal and Dilawari, 1993; Smith *et al.*, 1994; Panda and Khush, 1995; Smith, 1999; Dhaliwal and Singh, 2004). Examples of how insect-resistant cultivars can be a highly cost-effective control mechanism abound. However, the overall economic return to the development of host plant resistance

remains elusive. In addition, many of the economic issues surrounding cultivar resistance are not well understood. A better understanding of these issues is important to encourage sufficient investment in host plant resistance.

Definitions of Host Plant Resistance

There are many varied definitions of host plant resistance. In a broad sense, a resistant plant has properties that make it less damaged by disease or insects relative to a non-resistant or 'susceptible' plant. Host plant resistance protects a crop by making it less suitable to pests or more tolerant of pests. From a practical point of view, resistance means that a cultivar will have a higher yield or higher quality of harvestable material relative to a non-resistant cultivar facing the same pest pressures and control measures. The expression of resistance may be mediated by environmental factors such as temperature, moisture or fertility. Therefore, the effectiveness of a variety's resistance may vary from year to year. Resistance may be the sole pest control strategy or it may represent a control tactic that, when combined with other tactics, will reduce pest losses to acceptable levels.

©CAB International 2007. *Ecologically Based Integrated Pest Management* (eds O. Koul and G.W. Cuperus)

Resistance Mechanisms

Host plant resistance involves several different mechanisms. Insect resistance mechanisms have been categorized as tolerance, non-preference and antibiosis. Tolerance is present if a variety survives or produces better than a standard variety under the same pest pressure. Tolerance includes a number of plant responses that allow the plant to withstand insect infestations and have a satisfactory yield in spite of injury levels that would damage the yields of non-resistant plants. Unlike non-preference and antibiosis, tolerance involves only the plant and not the insect pest. Plants may be tolerant because of general vigour, compensatory growth in infested plants, quick wound healing, changes in photosynthate partitioning or mechanical support in tissues and organs. Tolerance does not place any selective pressure on insect populations. Because of the lack of selection pressure, tolerance-based resistance may be long-lasting. However, tolerance may be strongly influenced by environmental extremes. Common examples of tolerance-based resistance include maize genotypes resistant to the western corn rootworm, *Diabrotica virgifera*.

Non-preference resistance may be based on any mechanism that channels the insect pest away from the resistant variety towards a non-resistant host. Non-preference may stem from the presence or absence of chemicals within the plant or from plant architecture (structural characteristics) that interfere with the behaviour of the insect pests. Structural-based non-preference does not cause selection pressure and may be long-lasting. However, the success of non-preference depends on the presence of non-resistant hosts. The resistance may break down if the resistant cultivars are planted to the extent that alternate preferred hosts are not available.

Antibiosis resistance mechanisms involve physical or chemical characteristics that inhibit the growth of a pest. Plants with tough stems or with chemicals that are difficult for a pest to digest are examples of antibiosis resistance. Insect-resistant (Bt) cotton varieties that produce an insecticidal protein from the naturally occurring soil bacterium *Bacillus thuringiensis* (Bt) are a successful example of antibiosis resistance.

Benefits of Insect-resistant Crop Varieties

The development and adoption of resistant crop cultivars provides numerous economical and ecological advantages. Crop yields can be improved while insect monitoring and management costs are reduced. Resistant crops typically require fewer insecticide treatments leading to reductions in insecticide-related costs and pesticide residues. The reduction in pesticide applications may also benefit non-target organisms including beneficial insect populations and reduce negative environmental impacts such as groundwater contamination.

Economic benefits

Although there are no comprehensive estimates of the economic benefits of insect-resistant crops, numerous crop-specific examples exist. Insect-resistant sorghum cultivars were estimated to increase US producer profits each year by several hundred million dollars (Eddleman *et al.*, 1999). Smith *et al.* (1999) estimated the total value of insect-resistant wheat cultivars at over $250 million per year while the value of a multiple insect-resistant rice cultivar was estimated at $1 billion per year (Khush and Brar, 1991).

Insect-resistant cotton varieties have been widely adopted that have provided numerous estimates of the economic advantage. Benefits to US farmers from Bt cotton were estimated at $20/acre equating to $103 million per year (Gianessi *et al.*, 2002). Net revenue increases from Bt cotton over non-Bt cotton in China ranged from $357/ha to $549/ha over a 3-year study period (Pray *et al.*, 2001). Benefits to smallholder farmers in the Makhathini region of South Africa growing Bt cotton were estimated at $25–51/ha (Ismael *et al.*, 2002).

Various authors have also attempted to estimate the returns from research on resistant cultivars. Eddleman *et al.* (1999) estimated the return on investment in US sorghum insect resistance research at 41.7%. Azzam *et al.* (1997) estimated a 9:1 return on investment for Hessian fly resistance research in Morocco.

Economic issues in value of host plant resistance

Although estimates of the value of crop resistance for particular crops or regions exist, there is a lack of comprehensive estimates on its economic value. This is unfortunate because a greater recognition of the value of host plant resistance could encourage increased investment in plant resistance research. Determining the value of plant resistance is a difficult task and requires cooperation between economists and scientists (Antle and Wagenet, 1995). Estimating the value of resistant plants or the potential return to research on plant resistance raises a number of economic issues. These include difficulties in measuring the benefits of resistant cultivars, limitations to adoption, estimating the time span of the benefits, incorporating the value of risk reduction and addressing non-economic measures such as ecological impacts or environmental advantages or disadvantages.

Estimating benefits

Measuring the value of plant resistance involves a comparison of the management costs and harvestable yield of a resistant plant relative to a non-resistant variety. Developing comprehensive measures of insect losses is difficult and expensive. In the USA, the coordinated programme of measuring small grain insect losses was discontinued due to budgetary cutbacks (Webster and Kenkel, 1999). In cases where resistant plants are widely adopted, valuing the economic performance of a non-resistant variety becomes more speculative. When resistant varieties are widely adopted, measuring resistance benefits requires estimating the insect damage that was prevented.

Other factors complicate estimates of resistance benefits. Pest damage levels and the value of resistance vary from year to year. For example, estimates of the damage from the European corn borer on the US maize crop demonstrated significant variations from $325 million in 1991 to $33.3 million in 1992 (Barry *et al.*, 1999). The adoption of resistant cultivars may also provide indirect benefits to non-adopters. Lunginbill (1969) documented how the adoption of sawfly-resistant wheat in Montano reduced sawfly populations to a level that susceptible cultivars could be grown without damage.

Comparisons between resistant and non-resistant cultivars are also complicated by yield and/or quality differences. For example, percentage of spring wheat planted in the USA to wheat stem sawfly-resistant varieties gradually decreased because of the yield disadvantages (Weiss and Morrill, 1992). There may also be trade-offs in plant resistance and a variety resistant to a particular pest may be more susceptible to another.

Rate of adoption

Another factor impacting on the returns to plant resistance research is the rate of adoption by farm decision makers. Farmers base their crop variety decisions on a number of factors including adaptability to the production region, yield and quality potential, management requirements and susceptibility to insects and diseases. In order to realize the benefits of plant resistance, producers must be aware of the differential performance. Even after producers are convinced of economic advantages, widespread adoption of a technology occurs only over a period of several years. For example, in the US state of Georgia, it took 10 years (1984–1994) for the adoption rate of Hessian fly-resistant wheat to reach 90% (Webster and Kenkel, 1999). Even the adoption of herbicide-resistant soybeans (often used as an example of rapid adoption), required 5 years to reach 51% adoption (Kalaitzandonakes, 1999).

Useful Life

The value of resistant plant varieties is realized over a period of time. Resistant benefits cannot be realized until the variety is adopted and continued as long as the resistance is effective and the variety is economically competitive. The useful life or 'durability' of a resistant variety can be limited either by the development of non-resistant cultivars with more attractive yield potential or by the loss of resistant effects due to selection pressures. The uncertainty in estimating the useful life of resistant properties adds to the difficulties in measuring the benefit of existing cultivars and projecting the potential return on resistance research.

Risk Impacts

The development and adoption of resistant plants also has another important, but widely ignored, dimension. Pest-resistant varieties influence the risks faced by both the farm decision maker and the broader society. As illustrated previously, pest damages vary widely from year to year. A producer adopting a resistant variety would presumably experience less year-to-year variation in yield losses and control costs. Plant resistance therefore provides an 'insurance value' that is omitted from most benefit estimates.

Issues with Transgenic Insect-resistant Plants

The development of transgenic insect-resistant plants has greatly broadened the discussion of the risks and benefits associated with insect-resistant plants. New biotechnologies have changed and expanded the ability to introduce insect-resistant traits into crops, expanding the capabilities to develop single or multiple pest-resistant plants. A discussion of the attitudes of consumers and the general public on the value and risks of transgenic technologies is beyond the scope of this chapter. It is obvious that scientists' perceptions of the risks and benefits of

biotechnologies are not always congruent with those of consumer and advocacy groups. The scientific community must explore new methods to create meaningful interchange with stakeholder groups over the issues surrounding transgenic insect-resistant crops.

Cook (1999) recognized the importance of this issue and discussed methods for developing scientifically based estimates of the risks of transgenic crops. The Entomological Society of America (ESA) has developed a position on the uses of genetic engineering in crop production. Key Points of the ESA position include:

- Genetically engineered crops that express insect pest resistance traits could facilitate a shift away from the reliance on broad-spectrum insecticides and towards biolo-gically based pest management.
- Numerous mechanisms exist by which transgenic plants may reduce insecticide use, including engineered resistance to insect-vectored viruses. Further improvements in molecular technologies will expand our capabilities to develop single or multiple pest-resistant plants.
- Transgenic plants that produce insec-ticidal substances should be subjected to careful testing to ensure safety and minimize environmental risks.
- Insect-resistant crops should be deployed according to scientifically based resistance-management plans to prevent the evolution of genetically adapted insect strains.
- The use of insect-resistant plants is not equally appropriate for all crops in all agricultural systems. Therefore, a case-by-case scientific analysis of risks and benefits should be conducted before commercial use.

Plant Resistance Management

Despite the success of resistant plants there is concern that widespread adoption of resistant varieties will lead to the selection and multiplication of resistant insects. Because of the widespread adoption of Bt crops the agricultural industry has been particularly interested in developing strategies to delay the onset of Bt resistance. Best management principles for Bt crops have been developed (EPA, 1999). These principles include a number of strategies such as the development of structured refuges to maintain susceptible insects in the population. The principles also focus on grower education, adoption and the importance of using plant resistance as a component of an IPM system. Resistance management is an important tool for extending the useful life and thus the economic benefits of plant resistance.

Future Issues

The main limitation to the deployment of resistant varieties has been the lack of varieties with both the useful agronomic characteristics and the resistance characteristics (CAST, 2003). Underlying causes for this deficit include underinvestment in host-resistant variety research, competing emphasis on yield and quality improvement in crop breeding programmes, difficulties in identifying sources of resistance characteristics and difficulties in transferring resistant traits to varieties with the most competitive commercial traits. Advances in biotechnology may provide the opportunity to address some of these limitations. Future investments in the development of insect-resistant crops will depend on communicating the full economic benefits of resistance to stakeholder groups and the general public. Measuring and communicating these benefits will continue to require cooperation between scientists and economists.

References

Antle, J.M. and Wagenet, R.J. (1995) Why scientists should talk to economists. *Agronomy Journal* 87, 1033–1040.

Azzam, A., Azzam, S., Lhaloui, S., Amri, A., El Bouhssini, M. and Moussaoui, M. (1997) Economic returns to research in hessian fly (*Diptera: Cecidomyidae*) resistant bread-wheat varieties in Morocco. *Journal of Economic Entomology* 90, 1–5.

Barry, B.D., Wisemand, B.R., Davis, F.M., Mihn, J.A. and Overman, J.L. (1999) Benefits of insect resistant maize. In: Wisemand, B.R. and Webster, J.A. (eds), *Economic, Environmental and Social Benefits of Resistance in Field Crops*. Thomas Say Publications, Entomological Society of America, Lanham, Maryland, pp. 59–85.

CAST (Council for Agricultural Science and Technology) (2003) *Integrated Pest Management: Current and Future Strategies*. Task Force Report, ISSN0194-4088, No. 140, pp. 29–34.

Chesnokov, P.G. (1953) *Methods of Investigating Plant Resistance to Pests*. National Science Foundation, Washington, DC.

Cook, R.J. (1999) Science-based risk assessment for the approval and use of plants in agricultural and other environments. In: Persley, G.J. and Lantin, M.M. (eds) *Agricultural Biotechnology and the Poor*. Proceedings of an International Conference, Consultative Group on International Agricultural Research (CGIAR), Washington, DC, pp. 123–130.

Dhaliwal, G.S. and Dilawari, V.K. (1993) *Advances in Host Plant Resistance to Insects*. Kalyani Publishers, New Delhi.

Dhaliwal, G.S. and Singh, R.P. (2004) *Host Plant Resistance to Insects: Concepts and Applications*. Panima Publishing Corporation, New Delhi.

Eddleman, B.R., Chang, C.C. and McCarl, B.A. (1999) Economic benefits from grain sorghum variety improvement in the United States. In: Wiseman, B.R. and Webster, J.R. (eds) *Economic, Environmental and Social Benefits of Resistance in Field Crops*. Thomas Say Publications, Entomological Society of America, Lanham, Maryland, pp. 17–24.

EPA (Environmental Protection Agency) (1999) *EPA and USDA Position Paper on Insect Resistance Management in Bt Crops*. Washington, DC.

Gianessi, L., Silvers, C., Sankula, S. and Carpenter, J. (2002) Plant biotechnology: current and potential impact for improving pest management. In: *US Agriculture: An Analysis of 40 Case Studies*, Washington, DC: National Center for Food and Agricultural Policy. Available at: http://www.ncfap.org/40CaseStudies/ NCFAB% 20Exec%20Sum.pdf

Ismael, Y., Bennett, R. and Morse, S. (2002, July). *Bt cotton, pesticides, labour and health: a case study of smallholder farmers in the Makhatini Flats, Republic of South Africa.* Paper presented at the 6th International ICABR Conference, Ravello, Italy.

Kalaitzandonakes, N. (1999). A farm level perspective on agrobiotechnology: how much value and for whom? *AgBioForum* 2, 61–64. Available at: http://www.agbioforum.org

Khush, G.S. and Brar, D.S. (1991) Genetic resistance to insects in crop plants. *Advances in Agronomy* 45, 223–274.

Lara, F.M. (1979) *Principios de Resistancia de Plantas a Insectos* (in Portugese), Livroceres Ltda. Piracicaba, SP, Brazil.

Lunginbill, P. Jr (1969) *Developing Resistant Plants – The Ideal Method of Controlling Insects.* US Department of Agriculture Products Research Report, 111.

Painter, R.H. (1951) *Insect Resistance in Crop Plants.* Macmillan, New York.

Panda, N. (1979) *Principles of Host Plant Resistance to Insect Pests.* Allanheld, Osmun & Co. and Universal Books, New York.

Panda, N. and Khush, G.S. (1995) *Host Plant Resistance to Insects.* CAB International, Wallingford, UK.

Pray, C., Ma, D., Huang, J. and Qiao, F. (2001) Impact of Bt cotton in China. *World Development* 29, 1–34.

Russel, G.E. (1978) *Plant Breeding for Pest and Disease Resistance.* Butterworths, London.

Smith, C.M. (1989) *Plant Resistance to Insects – A Fundamental Approach.* John Wiley & Sons, New York.

Smith, C.M. (1999) Plant resistance to insects. In: Rechcigl, J. and Rechcigi, N. (eds) *Biological and Biotechnological Control of Insects.* Lewis Publishers, Boca Raton, Florida, pp. 171–205.

Smith, C.M., Khan, Z.R. and Pathak, M.D. (1994) *Techniques for Evaluating Insect Resistance in Crop Plants.* Lewis Publishers, Boca Raton, Florida.

Smith, C.M., Quisenberry, S.S. and du Toit, F. (1999) The value of conserved wheat germplasm possessing arthropod resistance. In: Clement, S.L. and Quisenberry, S.S. (eds) *Global Plant Genetic Resources for Insect Resistant Crops.* CRC Press, Boca Raton, Florida, pp. 25–49.

Snelling, R.O. (1941) Resistance of plants to insect attack. *Botany Reviews* 7, 543–586.

Webster, J.A. and Kenkel, P. (1999) Benefits of managing small-grain pests with plant resistance. In: Wiseman, B.R. and Webster, J.R. (eds) *Economic, Environmental and Social Benefits of Resistance in Field Crops.* Thomas Say Publications, Entomological Society of America, Lanham, Maryland, pp. 87–114.

Weiss, M.J. and Morrill, W.L. (1992) Wheat stem sawfly (*Hymenoptera: Cephidae*) revisited. *American Entomologist* 38, 241–245.

10 Integrated Pest Management with the Sterile Insect Technique

Donald B. Thomas

United States Department of Agriculture, Agricultural Research Service, Kika de la Garza Subtropical Agricultural Research Center, 2413 E. Hwy 83, Weslaco, TX 78596, USA

Introduction

The sterile insect technique (SIT) is a form of biological control that exploits the mate-seeking behaviour of the pest insect to deliver a lethal genetic load, breaking the reproductive cycle, thus causing the pest population to decline. The sex drive of the pest insect is the force that drives the SIT programme. The technique involves mass rearing the pest under factory-like conditions, sterilizing the adults, then releasing the sterile insects into the environment or area targeted for protection. Once released into the wild, the sterilized insects find their wild counterparts and copulate with them; this mating is infertile and fails to produce progeny. As the proportion of sterile copulations increases, the reproductive rate declines. Downward pressure on the reproductive rate, if applied continuously, eventually results in elimination of the pest from the area to which the technique is applied (Knipling, 1955).

In theory, at least, SIT is supposed to work in this manner. In practice, there are situational contingencies that impinge on the actual result. For example, if there is substantial migration of fertile individuals into the area from non-target regions, sterility rates may be insufficient to achieve complete eradication. In some cases, competitive mating behaviour can put the sterile insects at a selective disadvantage even where they have a numerical advantage (Knipling, 1960). Alternatively, in cases where the pest population responds to density-dependent regulation, the effects of increased population density can drive out the native population (Monro, 1966) or induce an increase in predator pressure, causing a decrease in the native pest population even without effective sterile mating (Mangan, 1985).

The advantages of SIT are manifest, most particularly from the reduced dependence on pesticide applications. Issues associated with traditional pesticide use, such as drift, crop residues, environmental contamination, acquired resistance and accidental toxicity to applicators, are reduced or avoided altogether. Also, SIT is narrowly targeted to a specific pest, thus minimizing impacts on beneficial or other non-target species. But there are also disadvantages, mainly associated with the start-up costs of mass production, dispersal of insects and licensing and maintenance of a radiation facility. If the programme successfully controls an economically important pest over a wide area, the benefits are generally worth the cost. Of course, if the sterilized insects are still capable of damaging the commodity, spreading disease or otherwise creating a nuisance, such species might not be suitable candidates for a mass release programme.

In mass-rearing programmes gamma radiation is the most efficient means of sterilizing the insects prior to release even though competitiveness of the sterilized insects can be affected by irradiation. The ionizing rays that induce sterility can also cause cell death in the somatic tissues (Grosch, 1962). Because the technique requires that the rate of sterility be high (more than 99.99%), the irradiated insects will inevitably suffer some damage in the somatic tissues. Consequently, the ability to disperse, survive in the wild and be competent to find and mate with a conspecific may be impaired to some degree (Moreno *et al.*, 1991). Thus, it is critical to establish an optimum radiation dose that ensures both sterility and viability, and tolerance will vary from one species to another. In many species, a courtship precedes copulation and even a slightly weakened individual may be at a competitive disadvantage if the target insect, e.g. a lekking female, has a choice of mates. If the less vigorous male is consistently selected against, the SIT programme may not be successful or may require higher sterile/fertile ratios to achieve the desired results (Kaneshiro *et al.*, 1993).

A contingency that can be important involves the number of times that an insect copulates. In theory, SIT is most effective against insect species that normally allow a single insemination. In multiple mating species, a single fertile insemination may be sufficient to negate the effects of multiple sterile copulations (Bloem *et al.*, 1993). Alternatively, a single sterile insemination can largely negate multiple fertile mating if the sterile mating is the last. Because the propagating female is the target of SIT, it is clearly important to consider its mating behaviour as it relates to the reproductive physiology of the pest.

In most instances it is believed that the sterile males achieve the bulk of suppression because, typically, males will copulate more often than females. Again, this will vary from one species to the next. If the sterile released males have no greater propensity to mate than the female, the males would account for only 50% of sterile copulations, all else being equal. But, if male propensity to mate was ten times that of females, the males would account for 90% of sterile copulations. In the latter case it would clearly be an advantage to release only the sterile males. Sterile females may distract the sterile males from seeking the target of the SIT programme, the wild fertile females, thus reducing the efficiency of the programme (Whitten and Taylor, 1970).

Because it is not always practical to produce male-only adults for release, it is important to compensate by overflooding the target population, i.e. by inundating the target zone with steriles to achieve and maintain a high ratio of steriles to fertiles. As a rule of thumb, programme managers generally try to achieve a 100:1 ratio as a working number (Curtis, 1985). This can be achieved by producing sufficient numbers of steriles or by reducing the pest population to manageable numbers prior to the sterile releases using more traditional control methods. Hence, the effectiveness of an SIT programme can be enhanced by an integrated approach, and, actually, all successful SIT programmes are integrated pest management (IPM) programmes. We are speaking now of vertical integration, i.e. coordinating a number of different control elements against the target pest. Clearly, the SIT method is superior to chemical applications from the standpoint of horizontal integration, i.e. compatibility with other control elements within a particular cropping system against an array of pests. Some examples of successful SIT programmes are described herein with an emphasis on the vertical integrative methods used to optimize or enhance the SIT programme to achieve efficacious control.

Screwworm Eradication

The first application of the SIT concept, and to date the most successful, was the programme to eradicate the New World screwworm, *Cochliomyia hominivorax* (Coquerel), from North America. The screwworm is a blowfly (*Calliphoridae*) that is an obligatory parasite of warm-blooded animals. It is considered to be the most economically important veterinary pest in the western hemisphere (Steelman, 1976). Livestock, pets, wild mammals and even humans become infested when the female fly lays its eggs on, or adjacent to, a wound or sore on the body surface of its victim (Fig. 10.1). Maggots hatching from the eggs invade and enlarge the wound by feeding on the flesh of the living host (Laake *et al.*, 1936). The screwworm was selected as a target for SIT because it was relatively easy to rear, the adult population tended to be small in nature and there was evidence that the females mate only once (Bushland and Hopkins, 1951). All of these factors were favourable for achieving high sterile/fertile ratios. In a pilot programme, the screwworm was eradicated from the Caribbean island of Curacao using adults (males and females) reared and sterilized at a facility in Florida (Baumhover *et al.*, 1955). Encouraged by this experiment, a mass-rearing programme was established at Mission, Texas, and within a few years the screwworm was eradicated from the USA. Following the successful eradication, a fly-free barrier zone was established

and maintained by releasing the sterilized adults from an airplane along the border of Mexico. Yet, the barrier was eventually breached, the USA was reinvaded and only at great cost was the programme ramped up and the pest newly eradicated.

Several factors contributed to the failure of the barrier. Firstly, in the region south of the barrier zone, populations unaffected by control programmes naturally waxed and waned, tracking environmental conditions. SIT depends on a high sterile/fertile ratio, so when the populations became very high, they were able to penetrate and bypass the barrier zone. Before eradication, when the pest was prevalent in the cattle-producing areas of south-western USA, inspection and treatment of livestock was routine. Such routine inspections were curtailed when the insect was thought to be absent, and with less vigilance, the reinvasion spread rapidly before it was detected (Bushland, 1985). Secondly, under mass-rearing conditions there tends to be a selection for behaviours that favour survival and reproduction in crowded, artificial conditions, and these adaptations are not always conducive to survival and

Fig. 10.1.   An obligate parasite, the female screwworm, oviposits on the margin of lesions in the skin of living, warm-blooded animals.

competitiveness in the wild (Bush *et al.*, 1976). All of these factors contributed to the breakdown of the barrier.

To reinvigorate the programme, managers took steps to alleviate the mass-rearing selection problem, but also instituted protocols to enhance the effectiveness of SIT in the field. First, the programme was moved to southern Mexico to establish a new barrier zone at the Isthmus of Tehuantepec. By reducing the width of the barrier zone it was easier to maintain high overflooding ratios, and easier to maintain surveillance and detection on the eradicated side of the barrier. A larger rearing facility was built close to the barrier zone in the state of Chiapas (Fig. 10.2). New rearing procedures were implemented to minimize the problem of laboratory adaptation. Importantly, new genetic lines were introduced periodically into the breeding colony to slow selection for undesirable traits. The new genetic material was usually collected from the areas targeted for eradication. These are methodologies that are now routinely employed in the mass-rearing aspects of SIT (Krafsur *et al.*, 1987). But supplemental methods were developed in the delivery aspect of the technique that contributed substantially to the success of the programme.

Reproductive behaviour of the screwworm

Because the sexually active female is the target of the SIT programme, it is important to understand the mating behaviour of the pest. The age at first oviposition for mated females is 4–5 days (Hightower *et al.*, 1972). Thereafter, she lays eggs in clutches at 3-day intervals with an average clutch size of 200 eggs (Thomas and Mangan, 1989). The greatest recorded longevity for a female in the laboratory is 56 days (Laake *et al.*, 1936), though life expectancy in the wild is closer to 21 days (Thomas and Chen, 1990). Screwworm males aggregate at sites from which they strike out at passing females (Krafsur, 1978). The courtship is essentially a rapacious conquest of the female by the male. The male and female struggle, and if the male is vigorous enough to hold fast and subdue the female by tumbling to the ground, the copulation is accomplished (Alley and Hightower, 1966). Observations

Fig. 10.2.   Inside the sterile screwworm production facility in Chiapas, Mexico. The trays mounted on the conveyor contain puparia in sawdust.

indicate that a high mortality is associated with this behaviour, especially in tropical areas where predatory ants tend to dominate the fauna on the ground surface (Thomas, 1991). Baumhover (1965) and Spates and Hightower (1967) demonstrated that in mass production there is selection for aggressive male mating behaviour and that this results in higher female mortality. With field studies, Mangan (1985) found that oviposition rates drop sharply under SIT applications even though sterility rates plateau. Hence, it is suspected that release of a large, aggressive, sterile male population will cause high mortality rates among the native population even if insemination is not highly successful.

Livestock IPM

An IPM programme for livestock pests in Latin America has not been developed to any real extent. The pest complex on livestock includes a diverse array of arthropod parasites, which inflict sufficient damage to require control. The primary screwworm (C. hominivorax), secondary screwworm (C. macellaria Fabricius), ticks and another myiasis-causing fly, the torsalo, Dermatobia hominis (Linnaeus Jr), are the primary pests on cattle. Lice and other botflies (nasal and stomach bots) are problems in horses, sheep, goats and swine. Biting flies such as the horn fly and stable fly are not as damaging in tropical Latin America as they are in temperate zones, but mosquitoes and tabanids are abundant. Because ticks are vectors of disease in cattle, ranchers regularly dip or spray their livestock, and this treatment also kills screwworms if present. Although pesticide resistance has inevitably arisen in Mexican cattle ticks, it has not yet been reported in the screwworm. The torsalo is another hide maggot that infests livestock, and it was thought to be important because the resulting wound could become infested by screwworms. However, evidence suggests that this is not the case, probably because the screwworm prefers bloody wounds to suppurating wounds induced by the torsalo (Thomas, 1987). Ranchers treat for torsalo by expressing the maggot manually, or more frequently by ignoring them. Although the torsalo is not uncommon, the prevalent Indo-African breeds are generally tolerant of this, and other external parasites (Roncalli, 1984). The secondary screwworm is the more serious problem. It is so called because it oviposits only on wounds infested by the primary screwworm. But unlike the primary screwworm, which is a host-adapted parasite, the secondary screwworm maggots lack bacteriostatic compounds that prevent serious wound infections (Erdmann and Khalil, 1986). Untreated secondary screwworm infestations produce wounds that tend to fester, leading to septicemia. Sickly animals typically 'hole up' in the brush and will die if not sought out for treatment (Fig. 10.3). Economic losses result from deaths to livestock, weight reduction from illness and from the extra labour costs required to monitor and treat animals with infested wounds. In the absence of primary screwworms, the secondary screwworm is not a pest.

The screwworm has few natural enemies other than generalist predators, such as spiders (Welch, 1993) and the aforementioned ants. Apart from the in coitu adults, the larvae are briefly vulnerable to ants when they egress the wound to seek pupariation sites in the ground or under litter. The puparia are susceptible to a pteromalid wasp, Nasonia vitripennis (Walker), a parasitoid of calliphorid pupae, and at times a serious pest in the mass-rearing facility. But no measurable mortality due to this or other pupal parasitoids have been detected in the wild (Thomas, 1989). Opportunities for classical biological control of the screwworm have not shown much promise (Bay et al., 1976).

Screwworm adult suppression system

Screwworm adult suppression system (SWASS) is an 'attract and kill' technology designed to reduce the fertile adult population, making the SIT programme more effective. Although the attractant bait, swormlure, was also attractive to the sterile flies, it was designed to be selective for

Fig. 10.3. Livestock with wounds infested by screwworm must be sought out and treated.

females. Essentially, the bait was designed to mimic the odour of an infested wound, as it was discovered that infested wounds are much more attractive to gravid females than are freshly opened wounds. Males are not attracted to wounds. By mixing the attractant with a pesticide, dichlorvos, SWASS works theoretically by selectively attracting and killing gravid females. In practice, the material seems to mimic certain floral odours (Broce, 1980) and attracts both sexes. None the less, the method was deemed successful where it was first applied in south-western USA and northern Mexico. The attractant material was pelletized and distributed by aircraft over the targeted areas (Goodenough *et al.*, 1979). In size and appearance the pellets resemble the spoor of the American marsupial, *Didelphis virginana* Kerr. But as the eradication programme moved southward into the tropics it became difficult to demonstrate an impact (Brown and Mackley, 1983). Adult screwworm populations are much more prevalent in the rainy season. In the tropics where it rains almost every day in the rainy season, the effective life of an exposed pellet was greatly reduced due to the pellet being dissolved and washed away. It is thought that under conditions of dense vegetation, high humidity and generally low air movement, the material seems not to produce an effective plume of attractant volatiles (Mackley and Brown, 1985). Subsequent research has been directed at improving the 'attract and kill' technology to be effective in the humid tropics. Research and development of stations where the toxic bait could be rain-sheltered were successful (Coppedge *et al.*, 1981), but not practical for areas without road access. Another such study on an artificial wound has also made progress, but as a practical application it suffers from the same impediments.

Control of myiasis

SIT targets the adult population. At any given point in time, the majority of the population is immature in the parasitic stage. The immature stages were also targeted during the successful phase of the eradication effort. Each regional campaign included teams of inspectors whose duty was to assist livestock producers in the inspection and treatment of infested animals. Unlike south-western USA where livestock are generally free-ranging, in tropical regions it is common practice to bring cattle and sheep into a corral or pen at night. This practice reduces losses from predators. To take advantage of this cultural technique, a propaganda campaign directed at livestock owners encouraged the inspection and treatment of livestock and reporting of screwworm cases. Treatment kits were distributed free to livestock owners. Each treatment kit included a collection tube, a report form and an easily applicable dose of pesticide. The standard disinfestation treatment involved manual removal of the worms and salting the wound with an insecticide of low mammalian toxicity, the chemical of choice being Coumaphos.

The successful maintenance of a barrier also requires vigilance. Quarantine stations were established along all highways to restrict the movement of livestock between infested and eradicated areas. The stations provided another interface where potential hosts could be inspected and treated.

With coordinated elements such as quarantines, surveillance, inspection and treatment of infested livestock, genetic management of the factory strain, SWASS dispersal and massive releases of the sterile adults in place, the screwworm programme successfully eradicated the pest from the USA, Mexico and Central America.

In 1988 the screwworm invaded North Africa soon after shipments of live sheep were begun from Latin America to Libya. Fortunately an international effort quickly eradicated the outbreak before it could spread to sub-Saharan Africa. Although SIT is given the bulk of the credit for the success of this eradication, the fact remains that the procedures to suppress the larval incidence were also an integral part of the programme. In Libya the invasion was first detected in human cases (Showler, 1991). Before a sterile release programme could be implemented, involving trans-Atlantic shipment of the factory-reared and sterilized puparia, steps were taken to control the incidence of myiasis. Of the more than 100 cases of human infestation, almost all were invalids in hospitals. Improved hygiene and patient care ameliorated the incidence. Secondly, stray dogs that provided a reservoir of hosts in the cities were eliminated. Thirdly, Egyptian and Tunisian authorities at the border with Libya established quarantine stations. Pets were only subject to inspection, but all livestock passing the stations (mainly sheep) were doused with an insecticide before entry was allowed. With the known population confined to the coastal area, and with the advantage that there is no rainy season in Libya, the sterile insect releases quickly eradicated the African outbreak (Reichard, 1999).

Overall, screwworm has now been successfully eradicated from North and Central America and a barrier zone has been established in the Republic of Panama at the Darien Gap (Wyss, 2000). Plans now call for eradication from the Caribbean Islands (Grant et al., 2000). The significance of the screwworm eradication is that it serves as the exemplar of a successful SIT programme.

Fruit Fly Suppression

Largely because of the success of the screwworm eradication, SIT programmes have been established against invasive species of fruit flies (Tephritidae). The major pests in this family are host generalists that damage an array of fruit crops, and countries where fruit production is important to the economy will typically mount an eradication programme against these species when detected. In the past, eradication was most often accomplished by the intensive application of pesticides. Such applications increasingly face strong public opposition, especially in urban or residential settings. SIT provides an alternative to chemical sprays but, with a few exceptions, these programmes have not been as successful as the screwworm SIT programme or the insecticide spray campaigns. The oriental fruit fly, *Bactrocera dorsalis* (Hendel), was eradicated from the Mariana Islands (Steiner et al., 1970), and the melon fly, *Bactrocera cucurbitae* (Coquillet), was eradicated from Okinawa (Kuba et al., 1990) using SIT. However, an island setting has distinct advantages over the continental situation faced by most pest management specialists, and both of the aforementioned programmes included pesticide applications as a key component.

Arguably one of the most successful of the continental-based programmes has been the Mexican fruit fly, *Anastrepha ludens* (Loew), suppression effort jointly managed by the USA and Texas Departments of Agriculture. The mexfly is native to northern Mexico where its sylvatic breeding host is a wild rutaceous plant, *Casimiroa greggi* (S. Wats.) (Plummer *et al.*, 1941). When *Citrus* spp., also *Rutaceae*, were introduced into Mexico as an orchard crop, the mexfly moved easily into this new host. And as a host generalist, it found the mango (*Mangifera indica* L.) to be equally suitable for breeding sites (Fig. 10.4). Consequently, the mexfly is considered to be the most important insect pest of commercially grown fruit in Mexico (Gutierrez-Samperio *et al.*, 1990). Because the targeted area in Texas is located on the border, and proximal to uncontrolled native populations of the mexfly, permanent eradication is deemed unfeasible, at least over the short term. The objective of the Texas SIT programme is rather to suppress the population to a level below the 'maximum pest limit' for quarantine security (Mangan *et al.*, 1997). Under this concept, the prevalence of infestation is sufficiently low so as to constitute an insignificant risk, and export of fruit is permitted from the production area without costly fumigation or other restrictions (Nilakhe, 1991). Because of the success of the Texas SIT programme,

and because of the equity requirements of international trade agreements, the Texas protocols have served as a model for SIT programmes against fruit flies in other countries. However, in some situations the model may not be appropriate to that purpose, and therefore it is important to define and understand the parameters of the programme and the way it integrates with the cropping system in Texas.

Under current procedures, an average of 30 million sterile flies per week are released by aircraft over the Texas side of the Lower Rio Grande Valley, an area of about 2900 km^2, of which about 12,000 ha are covered by commercial *Citrus*. No significant amount of fruit other than *Citrus* is grown commercially in the area. Both groves and urban areas are targeted by the releases year round because dooryard fruit trees provide breeding sites and a haven for the pest. A grid of 2200 surveillance traps placed in fruit trees (two traps per square kilometre) is in constant operation with a ratio of about 40% of the traps in dooryards (Thomas *et al.*, 1999). The export of *Citrus* from the region is allowed as long as wild mexflies are not detected during, or immediately prior to, the harvest season. When the mexfly in any stage is detected, either as adults in the traps or larvae in the fruit, the quarantine restrictions are triggered. Although untreated shipments are allowed to continue to non-*Citrus*-growing states, once the quarantine is triggered, the fruit must be fumigated with methyl bromide before it can be shipped to, or through, other *Citrus*-growing areas.

In the Lower Rio Grande Valley of Texas the primary export commodity is grapefruit grown for the fresh fruit market. Prior to the implementation of the suppression programme, mexfly larvae (Holler *et al.*, 1984) frequently infested Texas *Citrus*. Since the implementation of the SIT programme in 1983, no commercially grown *Citrus* has been found to be infested with larvae. Although the programme is deemed successful because the product flow to the market is largely uninterrupted, and no infestations have been spread from Texas, the concepts on which the export protocols

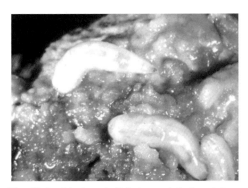

Fig. 10.4. Mexican fruit fly larvae infest and damage commercial fruit especially, *Citrus* and mango.

are based rely more on history than on science. Absence of larvae in the fruit, or flies in the surveillance traps, is taken as evidence that the area is essentially pest-free. It is true that wild flies are almost never detected during the harvest and shipping season in the autumn and winter, or in the months immediately preceding the harvest. But this is due in large part to the fact that the mexfly overwinters in the immature stages. Since the implementation of the SIT programme, virtually all wild adult captures have occurred in the springtime, usually after all, or most, of the commodity has been harvested and sent to market. At first these springtime detections were attributed to reinfestations from smuggled fruit (Nilakhe, 1991). But, as the life cycle and behaviour of the insect became better known, it was realized that the annual springtime detections were the result of an adult emergence of a small but persistent population present in the valley.

Life cycle of the mexfly

At the northern limits of its range the mexfly is bivoltine, with distinct breeding peaks in the spring and fall (Thomas, 2003). The spring peak is always much larger than the autumn peak. In the areas of northern Mexico where the mexfly is indigenous, the population overwinters in the immature stages. Early in the year the adults of this generation emerge and breed in fruit, producing the large peak of adults found in the mid- to late spring. Because of the low prevalence, only the spring peak was detected in Texas. In northern Mexico there is often a second peak in the early summer followed by very low population activity until the fall. Evidently, a combination of high temperatures, low humidity, a dearth of fruit for oviposition and a paucity of food sources such as nectar and honeydew naturally suppresses breeding at this time in the *Citrus* groves. When high temperatures abate with the arrival of autumn, adult activity increases. These autumnal adults, mostly survivors from the late spring and early summer generations, breed to

produce the overwintering generation (Fig. 10.5). Since the implementation of the SIT programme, no adults have ever been captured in the summertime in Texas. But this low ebb in adult activity also occurs in areas of Mexico where the fly is indigenous and there is no control programme. Hence, managers believe that a population survives the dearth season in havens, such as urban dooryards, where the requisites for adult survival, e.g. water, shade and nectar, are abundant and largely beyond the control of cultural practices used by commercial growers.

Operating under low prevalence, rather than an eradication mode, may not be the ideal situation preferred by agricultural officials. It is relevant to ask what is the difference between the tephritid fruit fly and the screwworm such that only suppression and not eradication is being achieved. In the case of the mexfly the males attract females to a lek (courtship arena) by a complex combination of posture displays, acoustical signalling and a weak contact pheromone (anastrephin) (Robacker *et al.*, 1991). Unlike the screwworm, where male sexual aggressiveness was – if only inadvertently – selected for in the factory strain, the female mexfly controls mating success, and no selective advantage accrues in the mass-rearing process that becomes manifest in the SIT programme (Horng and Plant, 1992). Perhaps the most important biological difference between the carnivorous screwworm and the frugivorous tephritids is the availability of breeding sites. Lindquist (1955) estimated that screwworm populations never exceeded more than a few hundred individuals per square kilometre in the USA. By contrast, more than 200 mexflies can emerge from a single grapefruit (Thomas, 1997). Removal of abandoned groves and fruit stripping from commercial orchards is costly and practised only by the most fastidious growers. Because of the enormous amount of fruit as potential breeding material, it has been difficult to integrate an effective larval control programme against fruit flies. Historically, the areawide fruit fly campaigns have not been vertical IPM programmes, but instead

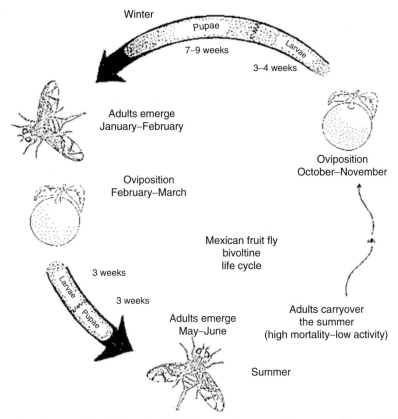

Fig. 10.5. The life cycle of the Mexican fruit fly in the northern part of its range, *Citrus*-producing areas of Mexico and the USA.

have been heavily reliant on adult control, either by chemicals or now by SIT. Unlike insecticides, whose efficacy is density-independent, SIT by itself is unlikely to achieve eradication, even with high-ratio, inundative releases of steriles.

Citrus IPM in Texas

Although the appeal of SIT is that it is an alternative to insecticidal control, the reality is that the mexfly is only one of a complex of pests in *Citrus* (Table 10.1), and growers still use synthetic chemicals to control other pests. The primary pests of *Citrus* in Texas and northern Mexico are mites, scales and the citrus black fly, *Aleurocanthus woglumi* Ashby (Dean *et al.*, 1983). Under current practices, a minimum

of two and usually three acaricidal sprays per year are applied to control the citrus rust mite, *Phyllocoptruta oleivora* (Ashmead), and spider mites, *Brevipalpus* spp. For resistance management, available chemicals are rotated. Producers have become reliant on chemicals because they give effective control of the mites and because acaricides generally have minimal effects on biological control agents (French and Villareal, 1990; French and Rakha, 1994). By contrast, broad-spectrum organophosphate compounds have caused increases in secondary pest populations by disrupting natural enemy populations (Dean *et al.*, 1977). Secondary pests on *Citrus* are predominantly sessile insects such as scales, mites, mealybugs, whiteflies and leaf miners. The citrus blackfly problem in Texas was brought under control by releasing a combination of

Table 10.1.   Pests of *Citrus* in Texas important to IPM programmes. (Adapted from Dean *et al.*, 1983.)

Pest	Management level	Pesticide use[a]
Mexican fruit fly	Quarantine pest	Moderate
Sapote fruit fly	Quarantine pest	Minor
Citrus rust mite	Primary pest	Intensive
Citrus black fly	Primary pest	Moderate
Chaff scale	Primary pest	Moderate
California red scale	Primary pest	Moderate
Florida red scale	Secondary pest	Moderate
Purple scale	Secondary pest	Moderate
Citrus mealybug	Secondary pest	Minor
Citrus whitefly	Secondary pest	Minor
False spider mite	Secondary pest	Minor
Citrus red mite	Secondary pest	Minor
Citrus leaf miner	Secondary pest	Minor
Cloudy winged whitefly	Sporadic pest	Minor
Brown soft scale	Sporadic pest	Minor

[a]Usage: intensive = most growers apply multiple treatments per season; moderate = most growers apply one treatment per season; minor = some growers apply occasional treatments.

two hymenopterous parasitoids, *Encarsia opulenta* Silvestri (*Platygasteridae*) and *Amitus hesperidum* Silvestri (*Aphelinidae*) (Summy and French, 1988; French *et al.*, 1990). Unfortunately, the latter species is unable to survive the harsh summers in south Texas; hence, continued control requires sustained releases of this insect. In the absence of natural controls, the growers revert to chemical treatments. Even in organic systems, growers apply insecticides and organic pesticides such as sulphur, oils and microbials to control citrus rust mite and citrus black fly.

Classical biological control

Any mortality factor impacting on the immature stages will enhance an SIT programme and Knipling (1992) has outlined the advantages of complementing SIT with the rearing and release of natural enemies. Accordingly, fruit fly programmes have made substantial efforts to introduce and augment biological control agents of fruit flies. Although initial efforts included predators, subsequent introductions exclusively involved hymenopterous parasitoids. In the western hemisphere, all of the

successful introductions of parasitic wasps were first established in Hawaii on the invasive species there, and later released on the American mainland against the native fruit flies (Wharton, 1989). A total of 17 exotic wasp species have been released in Latin America, of which three have become established: *Diachasmimorpha longicaudata* (Ashmead) (*Braconidae*), *Fopius arisanus* (Sonan) (*Braconidae*) and *Aceratoneuromyia indica* (Silvestri) (*Eulophidae*) (Ovruski *et al.*, 2000). A fourth species, *Diachasmimorpha tryoni* (Cameron), has been recovered and may also be established in Central America, but this has not been confirmed. In addition to these exotics, seven species of native parasitoids have been introduced against *Anastrepha* spp., but none of these have become established outside their country of origin. Unfortunately, even in the cases where establishment has been successful, there is scant evidence that the wasps are significantly controlling the pest fruit fly populations (Sivinski, 1996).

Somewhat better results have been attained by sustained, augmentative releases. Augmentative biocontrol involves mass rearing for inundative releases of the parasitoids in the area targeted for suppression. An augmentative release programme is a logical extension of SIT because the mass-reared sterile pests provide the supply of hosts for mass-rearing the parasitoids. Such a combined programme was first successfully applied in Hawaii against the medfly, *Ceratitis capitata* Wiedemann, using *Diachasmimorpha tryoni* (Wong *et al.*, 1992). On the American mainland one of the more successful trials was carried out in the Indian River region of Florida in 1993 against the caribfly, *Anastrepha suspensa* (Loew). *D. longicaudata* was reared and released at a rate of 20,000 wasps per square kilometre biweekly over a three-county area beginning in June 1993. The following spring, the number of caribflies captured in traps was found to be reduced by 73% (*n* = 30) compared to that in the spring prior to parasitoid releases (*n* = 111) (Burns *et al.*, 1996).

In southern Mexico an industrial scale–rearing facility is maintained at Metapa, Chiapas. The plant has the capacity to produce 50 million *D. longicaudata* wasps per week (Rull-Gabayet *et al.*, 1996). But field trials have given uneven results. Montoya and Liedo (2000) report on a project conducted in a mango orchard in which releases of 1000 wasps per hectare were made weekly for 35 weeks. Although infestation rates were high, up to 92% of fruit fly pupae recovered, the adult population detected in traps was not reduced and the economic damage from fruit fly larval infestations remained high.

Some researchers (Wharton, 1989; Sivinski, 1996) theorize that natural enemies may be more efficacious in natural and suburban dooryard situations, as opposed to the monocultures found in commercial groves. In biotically diverse habitats hosts tend to be available year round, as are nectar and other food sources; microclimatic extremes tend to be less stressful, and such settings are generally freer from pesticide applications. It may be that augmentative releases of natural enemies are most compatible with SIT-IPM as a complementary control method in suburban situations, which might otherwise be havens for the pest species.

'Attract and kill' technology

The operational concept of augmented SIT is that a higher sterile/fertile ratio can be achieved most efficiently by lowering the fertile population rather than rearing and releasing more steriles. Consequently, it is tempting to apply pesticides to knock down the population and then use the sterile releases to complete the eradication. Apart from the usual problems associated with pesticide use, the concern for the SIT programme is that the chemicals targeted at the pest adults will also kill the sterile insects. But with judicious management, adulticides can be integrated with fruit fly SIT.

The eradication of the melon fly from the island of Okinawa and the oriental fruit fly from the Marianas Island was achieved by integrating SIT with pesticidal control. In both cases the population was first suppressed by intense application of the male annihilation technique. Bait stations consisting of cotton wicks soaked with sex pheromone (cuelure or methyl-eugenol, respectively) and an insecticide (naled) were dropped by aircraft every 20 days, except in urban areas where the 'attract and kill' stations were placed manually. On Okinawa, trap monitoring indicated that the male population of melon flies had been suppressed by 95% before the sterile releases were initiated. Using a heavy release rate (60–87 million flies per week) over a relatively small area of coverage (143,000 ha), the SIT programme achieved eradication quickly. The suppression effort began in May 1986 and the last melon fly was detected in November 1989 (Kakinohana *et al.*, 1993).

Unlike the dacine fruit flies, the Mexican and Mediterranean fruit flies do not respond to volatile sex pheromones. Consequently, managers have experimented with methods to reduce the population with analogous 'attract and kill' stations, but baited with a feeding attractant. Moreno and Mangan (2000) have reported on elaborate trap-like stations that may have application in urban settings. The problem with food-based bait stations is the limited distance of attraction, and competition with natural food sources. For comparison, with the food-baited (yeast slurry) surveillance traps at the standard programme trapping density of two traps per square kilometre, about one trap per grove, only about one-tenth of 1% of the sterile released flies are trapped back, a proportion which is negligible in terms of affecting the population. Consequently, the stations would have to be placed in nearly every host tree to be effective, reducing its practical application in commercial groves.

Ultra-low volume, toxic bait sprays

An alternative to a bait station in every tree is to disperse the toxicant–bait mixture in ultra-low volumes as droplets, applied by orchard spray rigs, or dispersed over the groves by aircraft. The standard formulation for fruit fly eradication programmes is 175 ml technical malathion (91% a.i.) to 600 ml of Nulure, a liquid protein bait applied at 875 ml/ha (Lopez-Davila et al., 1969). The targeted dispersion rate is about 1 droplet per square decimetre of leaf surface. Such sprays effectively eradicated the medfly from California and Florida (Penrose, 1993). None the less, follow-up studies have documented impacts on beneficial insects, pollinators and predators, within the spray areas (Harris et al., 1980; Troetschler, 1983; Hoelmer and Dahlsten, 1993), and in some cases, outbreaks of secondary pests that are attributable to the adverse impact on their natural enemies (Ehler and Endicott, 1984).

Proceeding on the assumption that the theory is sound, researchers have tested alternative combinations of bait–toxicant formulations. Phloxine-B is a photoactive compound equal in insecticidal efficacy to Malathion but with a mammalian toxicity 100,000 times lower (Heitz and Downum, 1995); unlike Malathion, Phloxine-B has little contact toxicity. It has to be ingested to kill the target insect. In a sugar/protein formulation called Suredye, control was obtained in side-by-side field trials equivalent to that obtained by Malathion–Nulure against mexfly, medfly and the West Indian fruit fly, *Anastrepha obliqua*

(Macquart) (Moreno et al., 2001). Studies on the impact of Suredye on beneficials were inconclusive. Several factors confound the results. Ideally the bait material must be attractive, acceptable and phagostimulating to the target pest, but not to beneficials. Acidity, salinity and concentration are as important as the ingredients, both for ingestion by the insect and for activity by the toxin. A systematic search for bait compounds by Moreno and Mangan (2002) resulted in a formulation based on maize protein lysate, called SolBait, that is superior to previous fruit fly baits, yet repellent to honey bees and perhaps to other non-targets (Moreno et al., 2000).

On the toxicant front, a novel compound, Spinosad, was found to be active against the medfly as both a stomach and contact poison (Adan et al., 1996). An advantage of this compound is that it is a natural microbial product and, consequently, it is registered as an organic insecticide. Combined with SolBait, the formulation is now commercially available as GF-120 (DowAgrosciences), which can be applied at very low volumes, e.g. 3.6 kg/ha. Field trials in Florida demonstrated effectiveness (equal to that of Malathion–Nulure) against medfly and caribfly (Burns et al., 2001). Moreover, the sprays did not affect beehives placed in the treatment areas. Data on other non-targets were insufficient to obtain statistically significant results, but appeared to show no effect. Consequently, GF-120 applications have been incorporated into the SIT programme against mexfly in Texas. Applications are made only in groves or dooryards where feral mexflies have been detected in surveillance traps. Now in its third year, the number of wild flies captured in the springtime 2002–2004, has decreased from 715 to 305 to 69, respectively. Because in the 10 previous years numbers varied from a low of 21 to a high of 1348, the reduction could be a happy coincidence. But managers are planning on continued applications and tests are underway to ascertain non-target effects.

The Mexican fruit fly SIT programme in Texas is thus successful in the sense that it is the key component of a protocol that allows the flow of product to the market for most of the season. But quarantine restrictions are still triggered, curtailing exports at the end of the season, causing the importing states and countries to consider market alternatives. Hence, there is pressure from exporters and importers to expand the programme to make it an eradication rather than a suppression effort. For eradication to be successful the fruit fly SIT programme must be successful as an IPM programme. The elements of a successful IPM programme will have to include either intensive toxic bait sprays, a larvicidal programme such as fruit stripping from dooryards and groves, and/or collaboration with Mexico such that the fruit-growing areas in the northern states of Tamaulipas and Nuevo Leon are covered in an areawide effort implementing SIT.

Codling Moth IPM in Western North America

The codling moth, *Cydia pomonella* (Linnaeus), is a major pest of pome fruits throughout the world. It also attacks stone fruits and walnuts. The insect hibernates over the winter (Fig. 10.6), with adults appearing in the spring. Females lay eggs on or near developing fruit. A single moth can lay more than 100 eggs on as many as 60 separate fruits. The eggs hatch in 8–14 days. The hatching larvae bore into the fruit and, in the case of pomes, burrow deep into the pulp and feed on the seeds and core. Larval development is completed in 3–4 weeks and pupal development in 2–3 weeks, giving rise to a second generation of moths. This second generation is difficult to control, so suppression of the first generation is vital. In apple orchards of the western USA and Canada, uncontrolled populations can render an entire crop unmarketable. Prior to the advent of synthetic pesticides, damage levels of 10–20% were considered low (Johansen, 1971). When DDT became available, production of a crop free of wormy apples was not just possible, but commonplace. But as growers

Fig. 10.6. The immature stages of the codling moth form a hibernaculum for surviving the winter.

became more reliant on pesticides, the problems associated with pesticide use, e.g. resistance, environmental degradation, food contamination and secondary pest outbreaks, have worsened (Croft and Hull, 1992). Moreover, the codling moth is just one of an array of economically important pests that attack apples (Table 10.2). Consequently, apples are one of the most intensively managed crops and are among the crops with the highest amount of pesticide use in the USA. Croft (1982) estimated that 95% of the US apple crop receives a pesticide spray each year, with 2–5 applications being typical, and in the western USA it is not unusual to apply 10–15 treatments in a single season, more than half of which are directed at the codling moth. In an effort to reduce reliance on pesticides, the United States Department of Agriculture (USDA) funded an areawide IPM programme against the codling moth in the western USA.

Table 10.2. Pests of apples in western USA. Subject to IPM programmes. (Adapted from Kogan, 1994.)

Pest	Management level	Pesticide use
Codling moth	Key pest	Intensive
San Jose Scale	Key pest	Intensive
Apple rust mite	Secondary pest	Moderate
Pear Psylla	Secondary pest	Minor
Apple maggot	Secondary pest	Moderate
Plum Curculio	Secondary pest	Moderate
Wooly apple aphid	Secondary pest	Moderate
European red mite	Secondary pest	Minor
Two-spotted spider mite	Secondary pest	Minor
Apple aphid	Secondary pest	Minor
Rosey apple aphid	Secondary pest	Minor
Apple grain aphid	Secondary pest	Minor
Obliquebanded leafroller	Sporadic pest	Moderate
Tentiform leafminer	Sporadic pest	Moderate
Fruit tree leafroller	Sporadic pest	Moderate
Apple pandemis	Sporadic pest	Moderate
White apple leafhopper	Sporadic pest	Moderate

Mating disruption

Initially, the US IPM programme was designed to reduce pesticide applications by substituting a technology called 'mating disruption' (MD). For MD, sex pheromone (Isomate-C) dispensers are hung throughout the orchard. Male moths are attracted to the dispensers rather than to a female. Saturation of the orchard with sex pheromone confounds the signal of emitting females so that males cannot locate them; mating behaviour is disrupted and reproduction fails (Vickers and Rothschild, 1991). To achieve saturation, programme cooperators were asked to place pheromone dispensers at a rate of 1000/ha. At a typical tree density of 500 trees per hectare, this required two dispensers per tree, a labour-intensive system. And because the system works best against low population densities, the MD was combined with pesticide cover sprays (Calkins et al., 2000). Efficacy of the programme was reduced by reluctance of some growers to participate because of the greater costs (compared to pesticides) and labour involved. Participants often experienced failure attributable to influx of moths from neighbouring uncontrolled orchards. The MD technique works best when large area blocks are covered by the treatment. In 1999 in Washington state alone 65,000 acres were under MD protection, mostly large commercial growers. US government subsidies for this programme ended in 1999.

Biological control with natural enemies

The codling moth is native to central Asia, the aboriginal home of apples, *Malus* spp., and pears, *Pyrus* spp. In its native range natural enemies heavily parasitize the codling moth larvae, with rates exceeding 60% (Zlatanova and Tarabaev, 1985). In the western USA, however, the native parasitoids have switched to attacking this exotic pest in very low numbers, in rates less than 5%. Using classical biological control, this rate was increased by the introduction of a braconid egg parasite, *Ascogaster quadridentatus* Wesmael, to as high as 20% among the overwintering larvae. Experimentally, augmentative biological control was found to be efficacious and cost-effective in California by mass rearing and releasing the native egg parasite, *Trichogramma platneri* Nagaraja.

In these tests, fruit infestation by codling moth was reduced by 58–86% in treated orchards compared to non-treated control orchards. In the northwest another species, *Trichogramma minutum* Riley, is sold commercially and is an option available to individual growers.

SIT for codling moth

The development of an SIT programme against the codling moth has presented challenges that are unique in several respects. Lepidopterans require a heavier dosage of radiation than dipterans to obtain adult sterility. The radiation dosage for the codling moth SIT programme is 330 Gys (Bloem *et al.*, 1998). By comparison, the sterilizing dosage used in tephritid fruit fly programmes is around 80 Gys (Toledo, 1993). The higher required dosage is most likely due to the unusual cytology of lepidopterans, which have holokinetic chromosomes, an arrangement that makes them more tolerant of DNA strand breakage of the kind associated with ionizing radiation (Robinson, 1971). But higher tolerance by the chromosomes does not mean that other cell elements are more tolerant to ionizing radiation; therefore, the quality of the sterilized insects in terms of competitiveness and longevity is an issue of concern. To abate this problem some researchers have investigated the use of substerilizing dosages, around 100 Gys, for lepidopteran pests. For example, although a 100 Gy dosage does not sterilize the corn earworm, *Helicoverpa zea* (Boddie), the irradiated moths are highly competitive with the wild type for mates, and the F_1 progeny of the irradiated generation inherit sterility (Carpenter, 1991). Although the F_1 larval stage can still inflict some damage to the crop plant, they provide field hosts for biological control agents, and Carpenter (2000) suggests that the complementarity of the substerile releases and augmented biological control can be synergistic. Anisimov *et al.* (1989) advocated the use of F_1 sterility for the codling moth SIT programme. But the objective of the SIT

programme in Canada, initially at least, was eradication, and not just suppression.

Another challenge faced by the codling moth SIT project is the natural winter interruption in the population. As in most other insects, the number of generations per season is temperature-driven. Even with tropical insects there is a dearth season, usually the dry season, when the populations are at low ebb. Because SIT works most efficaciously when the populations are low, releases during the dearth season are important. In northern temperate areas where the hosts are deciduous fruit trees, the codling moth enters a winter diapause and the population is completely inactive, and thus sterile releases over most of the year are inconsequential. Mating populations of the moth are present in the wild for only about 20 weeks of the year (Bloem *et al.*, 2000). As a practical matter it means that the mass-rearing facility must continue to produce moths and expend resources during most of the year when releases are futile, or induce diapause in the moth, maintaining the colony in hibernation in parallel with the wild population. In either case, the annual upswing in the production of the sterile moths must be timed to match biofix, the first date of persistent springtime emergence. Therefore, methods were developed to adapt production to include maintenance of hibernacula. One advantage of diapause maintenance is that the cold-reared moths survive better in the wild compared to moths reared at the high temperatures used in standard production (Bloem *et al.*, 1998). In 1992 an SIT project was added to the codling moth IPM programme in the 8000 ha where apple and pear were growing in British Columbia, with the first releases made in 1994. Between 6 and 18 million sterile moths are released during the mating season in the Okanagan Valley (Bloem and Bloem, 2000). It is estimated that the sterile releases would be about double the cost of pesticide applications. But the benefits of the reduction of insecticidal sprays confer both environmental and economic advantages, and hold the promise of eliminating economic losses from the key pest in the region. The SIT programme is vertically integrated with complementary control techniques.

Cultural controls

Unharvested fruit is stripped from the trees to eliminate residual host material. Urban encroachment among the orchards also provides uncontrolled host material. With the cooperation of local nurseries, homeowners are encouraged to replace their dooryard pome fruit trees with non-host alternatives offered at reduced prices. Between 1997 and 2000 under this programme, nearly 74,000 pome fruit trees were removed from residences and abandoned groves. Also, used fruit bins are notorious as hibernation sites for the codling moth. Sanitation of these bins and their separation from the orchards after the harvest season are important mitigations.

Larval control

Corrugated bands are placed around the trunk near the base of the trees. Larvae enter the corrugations to pupate, which can then be removed and destroyed. The bands also provide a method for detection and surveillance of the larval population.

Mating disruption

As in the USA, this method is employed in the valleys where the sterile releases are not yet conducted. Of course, the two techniques MD and SIT are mutually exclusive (Barclay and Van den Driessche, 1989), and although they could be applied sequentially against different generations, such a strategy is unlikely to be efficacious because neither method is optimal against high-density populations.

Supplementary pesticide sprays

Some cover applications are still necessary, but overall there has been a 62% reduction in pesticide use in the Canadian programme since the initiation of SIT. However, there is a problem faced by managers of the Canadian

Table 10.3. Reduction in codling moth damage to fruit in the Okanagan Valley. (From Anon, 2001.)

Year	Orchards sampled	Percentage free of damage
1995	633	42.0
1996	624	79.0
1997	623	91.0
1998	1528	84.3
1999	1414	91.1
2000	1068	96.0

project, which is analogous to the one faced by managers of the Texas mexfly project, i.e. the constant threat of reinfestation from populations south of the border. An SIT programme was not undertaken in the USA because of the large acreage and high cost, around $200 per acre. An advantage favouring the Canadian programme is the geographical location. The apple-growing areas are in isolated valleys surrounded by high mountains. Because the moth does not migrate appreciable distances, the isolation of the populations creates an island-like situation that is more favourable than the continental situations faced by US producers.

When instituted in 1992, the goal of the Canadian SIT project was to eradicate the codling moth from Canada by the year 2000. Although that goal was not realized, the project was sufficiently successful (Table 10.3) to continue the programme, at least through 2005. The long-term goal of the programme is to further decrease reliance on pesticides by reducing and maintaining populations of the codling moth below economic threshold levels (Anon., 2001). There are now plans in place for construction of sterile codling moth–rearing facilities in the apple-growing regions of South Africa and Argentina.

Conclusions

It is a primary mission of the USDA to find alternatives to chemical controls against agricultural pests, especially those impacting food crops. Although the goal of complete elimination of pesticides is not realistic, at least for the near future, drastic reductions are certainly possible. Alternative technologies have shown promise but have not always received acceptance. The concept of genetically modified crops has not been welcomed by consumers and has been rejected under the official guidelines defining organic products. Distressingly, even classical biological control has come under strong criticism by environmentalists (Simberloff and Stiling, 1996; Lockwood, 1997) and is now strictly regulated by the federal government.

SIT is an alternative that is likely to become increasingly attractive, even at its higher cost, as production systems become more organically oriented and/or organized (as the proportion of corporate farming or government subsidies increases). The great success of the screwworm SIT programme is a mixed blessing because on the one hand it serves as a model of an efficacious technique. And on the other, it was atypical in that there were advantages that accrued from the unique biology of this particular pest that required a less integrated approach to be successful than would be necessary for most situations. Achieving a high sterile/ fertile ratio is necessary for success, but simply releasing more steriles is not always, or even usually, the best way to achieve that goal. Knipling (1992) explained this principle when he stated:

> Uniform suppressive pressure applied against the total population of the pest over a period of generations will achieve greater suppression than a higher level of suppression on most, but not all, of the population each generation.

In most circumstances this will mean application of a non-density-dependent or complementary suppression method to lower the pest population and enhance the SIT effect. Although many IPM programmes may succeed without including an SIT project, an SIT project that is not part of an IPM programme, is less likely to succeed.

Acknowledgements

I am most grateful to John Pruett, USDA–ARS, Kerrville, Texas, and Allan T. Showler, USDA–ARS, Weslaco, Texas, for their help in improving the manuscript. Mention of a proprietary product does not constitute an endorsement by the USDA.

References

Adan, A., Del Estal, P., Budia, F., Gonzalez M. and Vinuela, E. (1996) Laboratory evaluation of the novel naturally derived compound spinosad against *Ceratitis capitata*. *Pesticide Science* 48, 261–268.

Alley, D.A. and Hightower, B.G. (1966) Mating behavior of the screw-worm fly as affected by differences in strain and size. *Journal of Economic Entomology* 59, 1499–1502.

Anisimov, A.I., Lazurkina, N.V. and Shedov, A.N. (1989) Influence of radiation induced genetic damage on the suppressive effect of inherited sterility in the codling moth (Lepidoptera: Tortricidae). *Annals of the Entomological Society of America* 82, 769–777.

Anon. (2001) *Okanagan Kootenay Sterile Insect Release Program Strategic Plan 2001–2005*. British Columbia, Canada.

Barclay, H.J. and Van den Driessche, P. (1989) Pest-control models of combinations of sterile releases and trapping. *Insect Science and Its Application* 10, 107–116.

Baumhover, A.H. (1965) Sexual aggressiveness of male screwworm flies measured by effect on female mortality. *Journal of Economic Entomology* 58, 544–548.

Baumhover, A.H., Graham, O.H., Bitter, B.A., Hopkins, D.E., New W.D., Dudley, F.H. and Bushland, R.C. (1955) Screwworm control through release of sterilized flies. *Journal of Economic Entomology* 48, 462–466.

Bay, E.C., Berg, C.O., Chapman, H.C. and Legner, E.F. (1976) Biological control of medical and veterinary pests. In: Huffaker, C.B. and Messenger, P.S. (eds) *Theory and Practice of Biological Control*. Academic Press, New York, pp. 457–479.

Bloem, K.A. and Bloem, S. (2000) SIT for codling moth eradication in British Columbia, Canada. In: Tan, K.H. (ed.) *Area-wide Control of Fruit Flies and Other Insect Pests*. Penerbit University, Penang, Malaysia, pp. 207–214.

Bloem, S., Bloem, K.A. and Calkins, C.O. (2000) Incorporation of diapause into codling moth mass rearing: production advantages and insect quality issues. In: Tan, K.H. (ed.) *Area-wide Control of Fruit Flies and Other Insect Pests*. Penerbit University, Penang, Malaysia, pp. 329–335.

Bloem, S., Bloem, K.A. and Knight, A.L. (1998) Assessing the quality of mass-reared codling moths (Lepidoptera: Tortricidae) by using field release-recapture tests. *Journal of Economic Entomology* 91, 1122–1130.

Bloem, K., Bloem, S., Rizzo, N. and Chambers, D. (1993) Female medfly refractory period: effect of male reproductive response. In: Aluja, M. and Liedo, P. (eds) *Fruit Flies: Biology and Management*. Springer-Verlag, New York, pp. 189–192.

Broce, A.B. (1980) Sexual behavior of screwworm flies stimulated by swormlure-2. *Annals of the Entomological Society of America* 73, 386–389.

Brown, H.E. and Mackley, J.W. (1983) Changes in attractancy and chemical composition of the screwworm (Diptera: Calliphoridae) chemical attractant, Swormlure-2, under field conditions. *Journal of Economic Entomology* 76, 1273–1278.

Burns, R.E., Diaz, J.D. and Holler, T.C. (1996) Inundative release of the parasitoid *Diachasmamorpha longicaudata* for the control of the Caribbean fruit fly, *Anastrepha suspensa*. In: McPheron, B. and Steck, G. (eds) *Fruit Fly Pests: A World Assessment of their Biology and Management*. St Lucie Press, Delray Beach, Florida, pp. 377–381.

Burns, R.E., Harris, D.L., Moreno, D.S. and Eger, J.E. (2001) Efficacy of spinosad bait sprays to control Mediterranean and Caribbean fruit flies (Diptera:Tephritidae) in commercial citrus in Florida. *Florida Entomologist* 84, 672–678.

Bush, G.L., Neck, R.W. and Vitto, G.B. (1976) Screw-worm eradication: inadvertent selection for noncompetitive ecotype during mass-rearing. *Science* 212, 563–575.

Bushland, R.C. (1985) Eradication program in the southwestern U.S. In: Graham, O.H. (ed.) *Symposium on Eradication of the Screwworm from the United States and Mexico*. Miscellaneous Publications of the Entomological Society of America, No. 62, Maryland, pp. 12–14.

Bushland, R.C. and Hopkins, D.E. (1951) Experiments with screwworm flies sterilized by X-rays. *Journal of Economic Entomology* 44, 725–731.

Calkins, C.O., Knight, A.L, Richardson, G. and Bloem, K.A. (2000) Area-wide suppression of codling moth. In: Tan, K.H. (ed.) *Area-wide Control of Fruit Flies and other Insect Pests*. Penerbit University, Penang, Malaysia, pp. 215–219.

Carpenter, J.E. (1991) Effect of radiation dose on the incidence of visible chromosomal aberrations in *Helicoverpa zea* (Lepidoptera: Noctuidae). *Environmental Entomology* 20, 1457–1459.

Carpenter, J.E. (2000) Area-wide integration of Lepidopteran F1 sterility and augmentative biological control. In: Tan, K.H. (ed.) *Area-wide Control of Fruit Flies and other Insect Pests*. Penerbit University, Penang, Malaysia, pp. 193–200.

Coppedge, J.R., Brown, H.E, Snow, J.W. and Tannahill, F.H. (1981) Bait stations for suppression of screwworm populations. *Journal of Economic Entomology* 74, 168–172.

Croft, B.A. (1982) Arthropod resistance to insecticides: a key to pest control failures and success in North American apple orchards. *Entomologia Experimentalis et Applicata* 31, 88–110.

Croft, B.A. and Hull, L.A. (1992) Tortricid pests of pome and stone fruits, chemical controls and resistance to pesticides. In: van der Geest, L.P. and Evenjuis, H.H. (eds) *Tortricid Pests: Their Biology, Natural Enemies and Control*. Elsevier Science Publishers, Amsterdam, The Netherlands, pp. 473–486.

Curtis, C.F. (1985) Genetic control of insect pests: growth industry or lead balloon. *Biological Journal of the Linnean Society* 26, 359–374.

Dean, H.A., French, J.V. and Meyerdirk, D. (1983) *Development of Integrated Pest Management in Texas Citrus*. Texas A&M Agricultural Experiment Station, Bulletin 1434, Weslaco, Texas.

Dean, H.A., Hart, W.G. and Ingle, S.J. (1977) Pest management considerations of the effects of pesticides on Texas citrus pests and certain parasites. *Journal of the Rio Grande Valley Horticultural Society* 31, 37–44.

Ehler, L.E. and Endicott, P.C. (1984) Effect of malathion bait sprays on biological control of insect pests of olives, citrus and walnut. *Hilgardia* 52, 1–47.

Erdmann, G.R. and Khalil, S.K. (1986) Isolation and identification of two antibacterial agents produced by a strain of *Proteus mirabilis* isolated from larvae of the screwworm (*Cochliomyia hominivorax*) (Diptera: Calliphoridae). *Journal of Medical Entomology* 23, 208–211.

French, J.V. and. Rakha, M.A. (1994) False spider mite: damage and control on Texas citrus. *Subtropical Plant Science*, 46, 16–19.

French, J.V. and Villareal, J. (1990) Agri-Mek (Abamectin): a new miticide for control of citrus rust mite. *Journal of the Rio Grande Valley Horticultural Society*, 43, 9–15.

French, J.V., Meagher, R.L. and Esau, K.L. (1990) Release of two parasitoid species for biological control of citrus blackfly in south Texas. *Journal of the Rio Grande Valley Horticultural Society* 43, 23–27.

Goodenough, J.L., Coppedge, J.R., Broce, A.B., Petersen, H.D. and Higgins, A. (1979) Screwworm flies: a system for baiting and distributing screwworm adult suppression units. *Transactions of the American Society of Agricultural Engineers* 22, 1260–1263.

Grant, G.H., Snow, J.W. and Vargas-Teran, M. (2000) The new world screwworm as a pest in the Caribbean and plans for its eradication from Jamaica and the other infested Caribbean islands. In: Tan, K.H. (ed.) *Area-wide Control of Fruit Flies and Other Insect Pests*. Penerbit University, Penang, Malaysia, pp. 87–94.

Grosch, D.S. (1962) Entomological aspects of radiation as related to genetics and physiology. *Annual Review of Entomology* 7, 81–106.

Gutierrez-Samperio, J., Reyes, J. and Villaseñor, A. (1990) National Plan against fruit flies in Mexico. In: Aluja, M. and Liedo, P. (eds) *Fruit Flies: Biology and Management*. Springer-Verlag, New York, pp. 419–423.

Harris, E.J., Hafraoui, A. and Toulouti, B. (1980) Mortality of nontarget insects by poison bait applied to control the Mediterranean fruit fly, *Ceratitis capitata* (Diptera: Tephritidae), in Morocco. *Proceedings of the Hawaiian Entomological Society* 23, 227–229.

Heitz, J.R. and Downum, K.R. (1995) *Light Activated Pest Control*. American Chemical Society Symposium. Series 616, American Chemical Society, Washington, DC.

Hightower, B.G., O'Grady, J.J. and Garcia, J.J. (1972) Ovipositional behavior of wild-type and laboratory-adapted strains of screwworm flies. *Environmental Entomology* 1, 227–229.

Hoelmer, K.A. and Dahlsten, D.L. (1993) Effects of malathion bait spray on *Aleyrodes spiraeoides* (Homoptera: Aleyrodidae) and its parasitoids in northern California. *Environmental Entomology* 22, 49–56.

Holler, T.C., Davidson, J.L., Suarez, A. and Garcia, R. (1984) Release of sterile Mexican fruit flies for control of feral populations in the Rio Grande Valley of Texas and Mexico. *Journal of the Rio Grande Valley Horticultural Society* 37, 113–121.

Horng, S.B. and Plant, R.E. (1992) Impact of lek mating on the sterile insect technique: a modeling study. *Research on Population Ecology* 34, 57–76.

Johansen, C. (1971) Insect pests of tree fruits. In: Pfadt, R. (ed.) *Fundamentals of Applied Entomology*, 2nd edn. Macmillan, Toronto, Canada, pp. 407–438.

Kakinohana, H., Kuba, H., Yamagishi, M., Kohama, T., Kinjyo, K., Tanahara, A., Sokei, Y. and Kirihara, S. (1993) The eradication of the melon fly from the Okinawa Islands, Japan: II. Current control program. In: Aluja, M. and Liedo, P. (eds) *Fruit Flies: Biology and Management*. Springer-Verlag, New York, pp. 467–469.

Kaneshiro, K.Y., Kanegawa, K.M. and Whittier, T. (1993) Sexual selection in tephritid fruit flies and its implication in the sterile insect release method. In: Aluja, M. and Liedo, P. (eds) *Fruit Flies: Biology and Management*. Springer-Verlag, New York, pp. 177–179.

Knipling, E.F. (1955) Possibilities of insect control or eradication through the use of sexually sterile males. *Journal of Economic Entomology* 48, 467–469.

Knipling, E.F. (1960) Use of insects for their own destruction. *Journal of Economic Entomology* 53, 415–420.

Knipling, E.F. (1992) *Principles of Insect Parasitism Analyzed from New Perspectives*. Agriculture Handbook No. 693, USDA–ARS, Washington, DC.

Kogan, M. (1994) *Codling Moth IPM: An Areawide Program for Fruit Crops in the Western US*. Integrated Plant Protection Center, Oregon State University, Corvallis, Oregon.

Krafsur, E.S. (1978) Aggregations of male screwworm flies, *Cochliomyia hominivorax* (Coq.) in south Texas (Diptera: Calliphoridae*). Proceedings of the Entomological Society of Washington* 80, 164–170.

Krafsur, E.S., Whitten, C.J. and Novy, J.E. (1987) Screwworm eradication in North and Central America. *Parasitology Today* 3, 131–137.

Kuba, H., Kakinohana, H. and Kawasaki, K. (1990) Eradication of the melon fly from the Okinawa Islands in Japan: I. Estimation of population density and number of sterile flies required for eradication. In: Aluja, M. and Liedo, P. (eds) *Fruit Flies: Biology and Management*. Springer-Verlag, New York, pp. 335–338.

Laake, E.W., Cushing, E.C. and Parish, H.E. (1936) *Biology of the Primary Screw Worm Fly, Cochliomyia americana and a Comparison of Its Stages with C. macellaria.* USDA Technical Bulletin No. 500, USA.

Lindquist, A.W. (1955) The use of gamma radiation for control or eradication of the screwworm. *Journal of Economic Entomology* 48, 467–469.

Lockwood, J.A. (1997) Competing values and moral imperatives: an overview of ethical issues in biological control. *Agriculture and Human Values* 14, 205–210.

Lopez-Davila, F., Chambers, D.L, Sanchez-Rivelo, M. and Kamasaki, H. (1969) Control of the Mexican fruit fly by bait sprays concentrated at discrete locations. *Journal of Economic Entomology* 62, 1255–1257.

Mackley, J.W. and Brown, H.E. (1985) Screwworm research 1976–84: attractants, behavior, ecology, and development of survey and control technologies. In: Graham, O.H. (ed.) *Symposium on Eradication of the Screwworm from the United States and Mexico.* Miscellaneous Publications of the Entomological Society of America No. 62, Maryland, pp. 49–55.

Mangan, R.L. (1985) Reproductive response of native screwworm (*Cochliomyia hominivorax* (Coq.)) populations to sterile fly release. *Proceedings of the Entomological Society of Washington* 87, 717–739.

Mangan, R.L., Frampton, E.R., Thomas, D.B. and Moreno, D.S. (1997) Application of the maximum pest limit concept to quarantine security standards for the Mexican fruit fly (Diptera: Tephritidae). *Journal of Economic Entomology* 90, 1433–1440.

Monro, J. (1966) Population flushing with sexually sterile insects. *Science* 151, 1536–1538.

Montoya, P. and Liedo, P. (2000) Biological control of fruit flies (Diptera: Tephritidae) through augmentative releases: current status. In: Tan, K.H. (ed.) *Area-wide Control of Fruit Flies and Other Insect Pests.* Penerbit University, Penang, Malaysia, pp. 719–723.

Moreno, D.S. and Mangan, R.L. (2000) Novel insecticide strategies such as phototoxic dyes in adult fruit fly control and suppression programmes. In: Tan, K.H. (ed.) *Area-wide Control of Fruit Flies and Other Pest Insects.* Penerbit University, Penang, Malaysia, pp. 421–432.

Moreno, D.S. and Mangan, R.L. (2002) Bait matrix for novel toxicants for use in control of fruit flies (Diptera: Tephritidae). In: Hallman, G.J. and Schwalbe, C.P. (eds) *Invasive Arthropods in Agriculture.* Science Publishers, Enfield, New Hampshire, pp. 333–362.

Moreno, D.S., Celedonio, H., Mangan, R.L., Zavala, J.L. and Montoya, P. (2001) Field evaluation of a phototoxic dye, Phloxine B, against three species of fruit flies (Diptera: Tephritidae). *Journal of Economic Entomology* 94, 1419–1427.

Moreno, D.S., Mangan, R.L., Harris, D.L. and Burns, R.E. (2000) Novel chemical approaches for the control of fruit flies in citrus orchards. *Proceedings of the International Society for Citriculture* IX Congress, Orlando, Florida, 846–847.

Moreno, D.S., Sanchez, M., Robacker, D.C. and Worley, J. (1991) Mating competitiveness of irradiated Mexican fruit fly (Diptera: Tephritidae). *Journal of Economic Entomology* 84, 1227–1234.

Nilakhe, S.S. (1991) Mexican fruit fly protocol helps export Texas citrus. *Subtropical Plant Science* 44, 49–52.

Ovruski, S., Aluja, M., Sivinski, J. and Wharton, R. (2000) Hymenopteran parasitoids on fruit-infesting Tephritidae (Diptera) in Latin America and the southern United States: Diversity, distribution, taxonomic status and their use in fruit fly biological control. *Integrated Pest Management Reviews* 5, 81–107.

Penrose, R. (1993) The 1989/1990 Mediterranean fruit fly eradication program in California. In: Aluja, M. and Liedo, P. (eds) *Fruit Flies: Biology and Management.* Springer-Verlag, New York, pp. 401–406.

Plummer, C.C., McPhail, M. and Monk, J.W. (1941) *The Yellow Chapote, a Native Host of the Mexican Fruit Fly.* USDA Technical Bulletin No. 775. USDA, Washington, DC.

Reichard, R. (1999) Case studies of emergency management of screwworm. *Revue Scientifique et Technique de l'Office International des Epizooties* 18, 145–163.

Robacker, D.C., Mangan, R.L., Moreno, D.S. and Tarshis-Moreno, A. (1991) Mating behavior and male mating success in wild *Anastrepha ludens* (Diptera: Tephritidae) on a field-caged host tree. *Journal of Insect Behavior* 4, 471–487.

Robinson, R. (1971) *Lepidoptera Genetics.* Pergamon Press, Toronto, Canada.

Roncalli, R.A. (1984) The biology and control of *Dermatobia hominis*, the tropical warble fly of Latin America. *Preventive Veterinary Medicine* 2, 569–578.

Rull-Gabayet, J.A., Reyes-Flores, J. and Enkerlin-Hoeflich, W. (1996) The Mexican national fruit fly eradication campaign: largest fruit fly industrial complex in the world. In: McPheron, B. and Steck, G. (eds) *Fruit Fly Pests: A World Assessment of Their Biology and Management.* St Lucie Press, Delray Beach, Florida, pp. 561–563.

Showler, A.T. (1991) The new world screwworm fly in north Africa: implications and eradication plans. *Agriculture, Ecosystems and Environment* 36, 235–239.

Simberloff, D. and Stiling, P. (1996) How risky is biological control. *Ecology* 77, 1965–1974.

Sivinski, J.M. (1996) The past and potential of biological control of fruit flies. In: McPheron, B. and Steck, G. (eds) *Fruit Fly Pests: A World Assessment of Their Biology and Management.* St Lucie Press, Delray Beach, Florida, pp. 369–375.

Spates, G.E. and Hightower, B.G. (1967) Sexual aggressiveness of male screwworm flies affected by laboratory rearing. *Journal of Economic Entomology* 60, 752–755.

Steelman, C.D. (1976) Effects of external and internal arthropod parasites on domestic livestock production. *Annual Review of Entomology* 21, 155–178.

Steiner, L.F., Hart, W.G., Harris, E.J., Cunningham, R.T., Ohinata, K. and Kamahaki, D.C. (1970) Eradication of the oriental fruit fly from the Mariana Islands by the methods of male annhilation and sterile insect release. *Journal of Economic Entomology* 63, 131–135.

Summy, K.R. and French, J.V. (1988) Biological control of agricultural pests: concepts every producer should understand. *Journal of Rio Grande Valley Horticultural Society* 41, 119–133.

Thomas, D.B. (1987) Incidence of screwworm (Diptera: Calliphoridae) and torsalo (Diptera: Cuterebridae) myiasis on the Yucatan peninsula of Mexico. *Journal of Medical Entomology* 24, 498–502.

Thomas, D.B. (1989) Survival of the pupal stage of the screwworm, *Cochliomyia hominivorax* (Coquerel) (Diptera: Calliphoridae) in subtropical Mexico. *Journal of Entomological Science* 24, 321–328.

Thomas, D.B. (1991) Time-activity budget of adult screwworm behavior (Diptera: Calliphoridae). *Journal of Medical Entomology* 28, 372–377.

Thomas, D.B. (1997) Degree-day accumulations and seasonal duration of the pre-imaginal stages of the Mexican fruit fly (Diptera: Tephritidae). *Florida Entomologist* 80, 71–79.

Thomas, D.B. (2003) Reproductive phenology of the Mexican fruit fly, *Anastrepha ludens* (Loew) (Diptera: Tephritidae) in the Sierra Madre Oriental, northern Mexico. *Neotropical Entomology* 32, 385–397.

Thomas, D.B. and. Chen, A.C. (1990) Age distribution of adult female screwworms (Diptera: Calliphoridae) captured on sentinel animals in the coastal lowlands of Guatemala. *Journal of Economic Entomology* 83, 1422–1429.

Thomas, D.B. and Mangan, R.L. (1989) Oviposition and wound-visiting behavior of the screwworm fly, *Cochliomyia hominivorax* (Diptera: Calliphoridae). *Annals of the Entomological Society of America* 82, 526–534.

Thomas, D.B., Worley, J.N., Mangan, R.L., Vlasik, R.A. and Davidson, J.L. (1999) Mexican fruit fly population suppression with the sterile insect technique. *Subtropical Plant Science* 51, 61–71.

Toledo, J. (1993) Optimum dosage for irradiating *Anastrepha obliqua* pupae to obtain highly competitive sterile adults. In: Aluja, M. and Liedo, P. (eds) *Fruit Flies: Biology and Management.* Springer-Verlag, New York, pp. 297–300.

Troetschler, R.G. (1983) Effects on nontarget arthropods of malathion bait sprays used in California to eradicate the medfly, *Ceratitis capitata* (Wiedemann) (Diptera: Tephritidae*). Environmental Entomology* 12, 1816–1822.

Vickers, R.A. and Rothschild, G.H. (1991) Use of sex pheromones for control of codling moth. In: van der Geest, L.P. and Evenjuis, H.H. (eds) *Tortricid Pests: Their Biology, Natural Enemies and Control.* Elsevier Science Publishers, Amsterdam, The Netherlands, pp. 339–354.

Welch, J.B. (1993) Predation by spiders on ground-released screwworm flies, *Cochliomyia hominivorax* (Diptera: Calliphoridae) in a mountainous area of southern Mexico. *Journal of Arachnology* 21, 23–28.

Wharton, R.A. (1989) Classical biological control of fruit-infesting Tephritidae. In: Robinson, A.S. and Hooper, G. (eds) *Fruit Flies, Their Biology, Natural Enemies and Control.* Elsevier Science, Amsterdam, The Netherlands, pp. 303–313.

Whitten, M.J. and. Taylor, W.C. (1970) A role for sterile females in insect control. *Journal of Economic Entomology* 63, 269–272.

Wong, T.T., Ramadan, M.M., Herr, J.C. and McInnis, D.O. (1992) Suppression of a Mediterranean fruit fly (Diptera: Tephritidae) population with concurrent parasitoid and sterile fly release in Kula, Maui, Hawaii. *Journal of Economic Entomology* 85, 1671–1681.

Wyss, J.H. (2000) Screw-worm eradication in the Americas – Overview. In: Tan, K.H. (ed.) *Area-wide Control of Fruit Flies and Other Insect Pests.* Penerbit University, Penang, Malaysia, pp. 79–86.

Zlatanova, A.A. and Tarabaev, C.K. (1985) Entomophagous species and their importance for reducing the numbers of codling moth *Laspeyresia pomonella* L. (Lepidoptera: Tortridicae) in south-east Kazakhstan. *Entomologicheskoye Obozreniye* 3, 510–515.

11 Ecology of Predator–prey and Parasitoid–host Systems: Its Role in Integrated Pest Management

Geoff M. Gurr,[1] Peter W. Price,[2] Mauricio Urrutia,[3] Mark Wade,[3]
Steve D. Wratten[3] and Aaron T. Simmons[1]
[1]Pest Biology and Management Group, School of Rural Management, Charles Sturt
University, PO Box 833, Orange, New South Wales, 2800, Australia; [2]Department of
Biological Sciences, Northern Arizona University, Flagstaff, AZ 86011-5640, USA;
[3]National Centre for Advanced Bio-protection Technologies, PO Box 84, Lincoln
University, Canterbury, New Zealand

Introduction

The topic of this chapter is a huge field
and the subject of whole texts, e.g. Price
(1997) and Fisher *et al.* (1999). Accordingly,
the aim of this contribution is to provide
an overview of the area, illustrating major
themes with recent examples and identifying
areas of especially important progress.
We consider the ecology of predators and
parasitoids, the 'natural enemies' of pests,
in three sections:

- First, the key aspects of their biology
  will be reviewed. This covers areas such
  as biodiversity, taxonomic groups of
  importance, multitrophic interactions,
  life cycles and reproductive strategies,
  feeding and foraging behaviour.
- Second, we consider the major ways
  in which natural enemies are used in
  biological control programmes. This
  involves the range of targets of biological control and the major biological
  control approaches – classical, inundative and conservation – as well as
  the concept of integrated biocontrol.

Within these sections, the major taxa
and functional groups of biological
control agents are reviewed along with
comments about levels of success.

- Third, we cover the important practical
  area of the ecology of natural enemies
  in integrated pest management (IPM)
  systems. This involves an examination
  of their interactions with host plant
  resistance and with pesticides.

Biology of Predators and Parasitoids

Biodiversity

In almost any landscape on earth, the richness
of species and interactions of natural enemies
of herbivorous insects are staggering. In a
well-studied fauna, such as in the British
Isles, over 26% of species may be parasitoids
and another 4% predators, making up 30% of
the insect fauna (Price, 1980). Another 35%
is composed of herbivorous insects, and of
course these interact with a large number
of host plant species, which may represent

22% of any biota (Strong *et al.*, 1984). In the worldwide fauna, one family of parasitoids, the Ichneumonidae, is thought to have about 60,000 species (Townes, 1969). On a global scale, natural enemies and insect herbivores must number in several hundreds of thousands of species. In just one field in England, 25 species of aphids were recorded that were attacked by 18 species of primary parasitoids and several families of predators. The primary parasitoids were attacked by 28 species of secondary parasitoids (Godfray and Müller, 1998). The numbers of parasitoid species per host species may range from 0 to 22 in northern temperate regions (Hawkins, 1988) and even up to 40 parasitoid species further south (Tomov, 2002). Understandably, concern for maintaining and increasing biodiversity of natural enemies in natural and agricultural systems is increasing (Altieri *et al.*, 1993; Thies and Tscharntke, 1999; Thies *et al.*, 2003).

Taxonomic affiliations

Many families of insects include predatory species. Important examples are the Anthocoridae, Asilidae, Carabidae, Coccinellidae, Corixidae, Dytiscidae, Mantidae, Pentatomidae, Sphecidae, Staphylinidae and Syrphidae. Predatory mites are also important, as are other arachnids including harvestmen, scorpions and spiders.

The parasitoids include, in the dipterans, the large family Tachinidae, some sarcophagids and conopids, and a very large number of hymenopteran families. Most species among the wasp families are in the Ichneumonidae and Braconidae, and the chalcidoid families, Eulophidae, Encyrtidae, Pteromalidae and the Eurytomidae.

Impact of natural enemies on herbivore populations

Not surprisingly, with many species of natural enemies attacking most insect herbivores, the mortality inflicted can be high. Demographic studies have repeatedly shown that natural enemies inflict the largest proportion of mortality when compared with other factors such as competition, weather and plant effects, although 'bottom–up' effects probably have been underestimated (Wratten *et al.*, 1988; Edwards *et al.*, 1992; Cornell and Hawkins, 1995). Bottom–up effects are based on the nutrition, architecture, and physical and chemical defences of plants.

Means of 60% mortality of a cohort caused by natural enemies are observed commonly, and in long-term population dynamics studies, natural enemies were of major importance in 40% of the cases (Price, 2003). Also, when natural enemies are employed as biological control agents, mortality of hosts may reach 90–100% in numerous cases, and successful biological control is brought about even when only about 40% mortality is achieved in some studies (Hawkins, 1994). However, assessments of natural-enemy effects need to be made at the most appropriate stage of the herbivore or prey life cycle. Large mortality rates of herbivorous insects can arise from plants' physical and/or chemical defences, the latter being constitutive or induced (Wratten *et al.*, 1988; Edwards *et al.*, 1992). Care is needed if predator effects are quantified on later herbivore stages, after plant-based ('bottom–up') processes may have already led to significant mortality (Varley and Gradwell, 1970).

Multitrophic-level interactions

All terrestrial organisms, especially plants, their insect herbivores and the natural enemies of herbivores exist in a complex *milieu* of multitrophic interactions (Fig. 11.1). Taking a single host plant as an example, many interactions are mediated by chemical signals. Specialized herbivores are attracted to the host plant by plant stimuli, and similar plant compounds (see 16 in Fig. 11.1) may attract even parasitoids. Other herbivores are repelled by phytochemicals (see 3 in Fig. 11.1), and in turn herbivores may produce chemicals or other defensive repellents to deter their natural enemies (see 13 in Fig. 11.1).

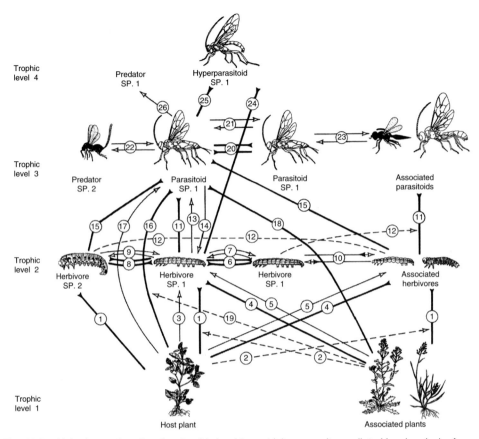

Fig. 11.1. Major interactions in a four-trophic-level terrestrial community mediated by chemicals. Arrows are placed against the responding organism, with thick lines and solid arrows indicating attraction to a stimulus (e.g. 1, 4, 11 and 24), and thin solid lines and open arrows showing repellency (e.g. 3, 13, 17 and 26). Thin dashed lines illustrate indirect effects such as interference with another response (e.g. 2, 12 and 19). (From Price, 1981. With permission of John Wiley & Sons. Copyright 1981.)

Phytochemicals may also be repellent to certain carnivores (see 17 in Fig. 11.1). In addition, plants in the community, growing with the host plant, may have associated effects on herbivores and carnivores, acting as either attractants (see 4 and 18 in Fig. 11.1) or repellents (see 5 in Fig. 11.1), or they may interfere with direct interactions among the host plant and herbivores (see 2 in Fig. 11.1) and carnivores (see 19 in Fig. 11.1). All interactions in Fig. 11.1 are explained, with specific examples (Price, 1981), and the broad sweep of interactions in multitrophic systems was introduced by Price *et al.* (1980), and covered more recently by Gange and Brown (1997) and Tscharntke and Hawkins (2002). It follows that food web linkages and organization provide one way of systematizing the rich variety of species and interactions in terrestrial vegetation. With the first trophic level represented by plants, the second by herbivores, the third by primary carnivores and the fourth by secondary carnivores, food webs have become a major focus in ecology. But even the interactions in Fig. 11.1 oversimplify reality, because underground processes are important for all trophic levels. Nutrients affect plant quality, and therefore herbivores, which in turn impact natural enemies. Mycorrhiza impact on plant quality, which may influence even parasitoids (Gange, 2002), and plant species may influence the bacterial gut community of insect herbivores (Broderick *et al.*, 2004). Also, work by Bardgett *et al.* (1999) has shown that herbivore damage can induce the production of plant root exudates that stimulate the growth of adjacent plants of the same species. These compounds may induce defences and attract natural enemies to neighbouring, uninfested plants. The ecological-community consequences of such interactions are likely to be profound. In general, it is safest to assume that, where there are food web connections, there are likely to be multitrophic-level interactions involved. These may well range from the roots of plants (see above) to their shoots and up to the higher trophic levels.

Functional groups

Each component in a food web plays a role in the community, or has a specific function. Of course, plants are autotrophic and fix the sun's energy. Herbivores are primary consumers, eating plants. Carnivores are secondary consumers and the insect component can be divided into predators and parasitoids. Predators kill their prey, either using their mandibles to consume an insect (e.g. mantids, lady beetles, ground beetles and wasps) or using a proboscis to suck fluids from their prey (e.g. assassin bugs, anthocorids and minute pirate bugs). Predators are usually generalized feeders exploiting many different species of prey.

Parasitoids lay eggs in or on host insects and the larvae feed parasitically on or in the living host, and eventually kill it. The adults are free-living, like typical predators, with females spending their time feeding, searching for hosts and ovipositing. Because of the larva's intimate relationship with a living host, species tend to be host-specific to varying degrees. For example, 53% of ichneumonid and 60% of braconid parasitoid species are recorded from a single host species in the British Isles (Price, 1980), although some species are recorded from over 80 host species.

The wide range of host specificity in parasitoids can be accounted for by the considerable differences in life history. Some parasitoid larvae live with the mobile host, which continues to live, move and defend itself. These are mostly internal (or endo-) parasitoids of larvae and adults, and are called koinobionts, meaning that host and parasitoid live together (Askew and Shaw, 1986). Such intimacy is likely to result in specialization of the parasitoid species. Other parasitoid larvae live in or on a host that does not develop further. These are called idiobionts, indicating an individual life style without an active host (Askew and Shaw, 1986). Such insects include parasitoids attacking and emerging from eggs. They also include species that paralyse the host, which does not develop further, and feed on it externally or internally. Idiobionts are likely to be able to attack several or many hosts because many are ectoparasitoids and are not exposed to the immune defences of their hosts.

Parasitoids have been categorized further, as have other insect groups, into guilds, depending on the host stage attacked, the mode of parasitism and the form of development (Mills, 1992, 1994). The number of such guilds may differ among host types such as moths (11 guilds) and weevils (12 guilds), but one example is the *egg parasitoid guild*, composed of idiobiont endoparasitoids that complete development within the host egg. Another guild is the *early parasitoid guild*, with koinobiont endoparasitoids that complete development within feeding host larvae. Also, towards the end of the host life cycle are the *cocoon parasitoid guild* and the *pupal parasitoid guild*, which are, respectively, idiobiont ectoparasitoids that attack hosts in cocoons and idiobiont endoparasitoids that exploit host pupae. This rich variety of guilds helps to account for the fact that so many species of parasitoids can coexist on or in a single host species. Another necessary categorization of parasitoid functions is whether they are *primary parasitoids*, which attack the host directly, or *secondary parasitoids*, which attack the primary parasitoids. As explained in the discussion of biodiversity, secondary parasitoid species may outnumber primary parasitoids, and especially idiobionts may act as facultative secondary parasitoids because of their more general ability to utilize a diverse array of hosts, attacking both herbivore hosts and primary parasitoid hosts. Therefore, parasitoids play important roles in both the third and fourth trophic levels (Fig. 11.1).

Life Cycles and Reproductive Strategies

Oviposition

The ways in which predators and parasitoids find oviposition sites and deposit eggs are fascinating and extraordinarily diverse. The full range of alternative strategies is too rich to cover here, so only generalizations will be discussed. In both groups, prey or hosts may be discovered by visual or chemical cues (Fig. 11.1), or eggs may be laid on foliage independent of host presence, if there is a probability that prey or hosts will move within the vicinity of eggs while the eggs are developing. Plants damaged by herbivores may provide volatile cues to searching carnivores (Vinson, 1975; Nordlund, 1981; Dicke *et al.*, 1990; Turlings *et al.*, 2002). Eggs of natural enemies may be laid singly or in groups, for some predators gain increased protection when grouped during early instars, and parasitoids may feed on hosts gregariously.

Sex determination

Most predators have the normal XY sex determination system, while the parasitic Hymenoptera share with the whole order a system called *arrhenotoky* or *haplodiploidy*, in which males are haploid and females are diploid (an 'X, XX' system). It is a form of parthenogenesis in which an unfertilized egg results in a haploid male. Thus, an ovipositing female can 'choose' to lay a male egg by withholding sperm and a female egg by releasing sperm as the egg passes down

the oviduct. Females may then allocate male progeny to small hosts and female progeny to larger hosts. Sex allocation theory is well advanced, with many complexities (Charnov, 1982; Waage, 1986) including the fact that adult female parasitoids that have had access to plant nectar fertilize a higher proportion of their eggs, leading to more females in the F_1 generation (Berndt and Wratten, 2005).

Ovigenesis and fecundity

Females may emerge with all the eggs that they will ever lay mature and ready for oviposition (*pro-ovigenesis*), or they may have no mature eggs upon emergence, with maturation proceeding gradually during their life (*syn-ovigenesis*) (Flanders, 1950; Jervis *et al.*, 2001, 2004). As is usual in nature, there is a continuum between these two extremes that can be captured by estimating the *ovigeny index*, which is the fraction of all eggs laid by a female that is mature when she first emerges (complete pro-ovigenesis = 1, complete syn-ovigenesis = 0) (Jervis *et al.*, 2001). As can be expected, females will be closer to being fully pro-ovigenic when they can find prey or hosts rapidly, or lay many eggs on foliage, awaiting ingestion by hosts and other means of host discovery or infection. Among parasitoids, these are more likely to be koinobiont species, which attack hosts early in the life cycle when a host cohort is relatively abundant on the host survival curve (Fig. 11.2). Parasitoid species that must search for hidden hosts, such as those existing as pupae, in cocoons or under bark, are likely to discover hosts slowly and at irregular intervals, and evolve with a syn-ovigenic reproductive strategy (see Jervis *et al.*, 2001, 2004 for other correlates).

Fecundity in predators and parasitoids varies enormously, but is more predictable in parasitoids than in predators. In parasitoids, the ovigeny index and overall fecundity tend to be inversely related to the stage of the host attacked, and therefore to prey/host abundance within a generation, particularly in species that emerge from host pupae or cocoons (Fig. 11.2). For example,

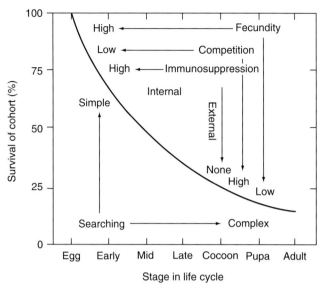

Fig. 11.2. A generalized survivorship curve of an insect herbivore from egg to the adult, showing a typical decline in survival as the cohort progresses through the generation. This pattern imposes an inevitable set of ecological conditions, noted in the figure, on parasitoid species attacking hosts at different stages. Immunosuppression refers to suppression of the host in immune response by many endoparasitoid koinobionts, whereas ectoparasitic idiobionts do not need to depend upon such suppression. (From Price, 1994. With permission of Oxford University Press.)

koinobionts that attack early host stages may live in the host for the rest of the life cycle, and are exposed to the same mortality factors as is the host. In the Ichneumonidae (Hymenoptera) alone, fecundity ranges from about 1000 eggs per female to about 10, and in the Tachinidae (Diptera), from 4000+ eggs to 40 (Price, 1975). At the high end of the range, eggs are laid on foliage and at the low end on adult insects or in a pupa, puparium or cocoon.

Feeding and Foraging

A traditional view of insect predators is that adults and immatures of predatory species feed on the same kinds of prey, although the size ranges of prey items are likely to differ considerably. Examples include those coccinellid species that concentrate on aphids, mantids that may take any kind of moving insect, anthocorids sucking eggs and pentatomids sucking haemolymph from caterpillars. In fact, many insect 'predators'

should be regarded more as omnivores because pollen, seeds, pods, leaves and nectar are consumed (Eubanks and Styrsky, 2005), and such plant foods may play a critical role in the ecology and use of natural enemies (Wackers, 2004).

Some predators and all parasitoids exhibit 'life history omnivory' (Polis and Strong, 1996; Tylianakis et al., 2004), in which some life stages, usually the adults, feed only on floral resources (e.g. hoverflies: Diptera: Syrphidae and many parasitoids) while their larvae are predatory or are endo- and ectoparasitoids. Coccinellid adults take pollen, and are predatory, but their larvae are predatory only. The use of these floral 'resource subsidies' (Tylianakis et al., 2004) in this way can markedly increase predator or parasitoid fitness (e.g. by increases in longevity or fecundity), and can make them more effective biological control agents (Barbosa, 1998; Landis et al., 2000; Gurr et al., 2004a,b).

In fact, adult parasitoids take mostly soluble food in the form of nectar or honeydew, damaged or rotten fruits and host haemolymph (Olson *et al.*, 2005). In some species, the females pierce a host with the ovipositor and then feed on the exuding haemolymph (Jervis and Kidd, 1986).

Foraging by predators and parasitoids is often complex, as searching individuals modify their behaviour as they search and according to experience. They may learn to associate the discovery of prey or hosts with particular odours, or they may increase the intensity of search when prey or hosts have been discovered. Prey or host patches may be searched for longer periods as rewards increase (e.g. Van Alphen and Vet, 1986; Casas, 2000). Video techniques can be useful in quantifying this, especially for nocturnal predators (Wratten, 1994; Merfield *et al.*, 2004).

In summary, predation and parasitism processes are complex, implicate at least four trophic levels and require a detailed knowledge of invertebrate community ecology. To maximize the likelihood of their incorporation into IPM systems, these components need to be understood, including the natural enemies' responses to full- or reduced-rate pesticide use (Poehling, 1989; van Emden, 1990). The concept of 'integrated biological control' (Gurr and Wratten, 1999) captures this idea. With 3 billion kg of pesticides currently used each year but with no evidence that this use has led to sustainable pest population reductions (Pimentel, 2004), the approaches outlined in this chapter, if practised more widely, should minimize the problems associated with frequent and/or prophylactic pesticide use. The use of predators and parasitoids in pest management is explored in the following section.

Methods of Biological Control

Biological control is an important component of an IPM programme. It involves utilizing entomophagous predators, parasitoids, parasites and pathogens that attack the various life stages of pests. The three key methods recognized in biological control are classical, augmentation and conservation (Van Driesche and Bellows, 1996).

Classical

Classical biological control involves the introduction of the natural enemies of an invasive pest species to a region where they were previously absent, to achieve establishment and long-term control of the pest. The underlying principle is that the population size of many species is limited by the presence of their natural enemies, and that in a new region the pest species can develop rapidly in the absence of these natural enemies. The aim of the introduction is to restore the balance created by their natural enemies (Van Driesche and Bellows, 1996; Hoddle, 2004), notwithstanding the role played by indigenous natural enemies and abiotic mortality factors (Michaud, 2002). A highly successful example of classical biological control was the importation of the vedalia beetle, *Rodolia cardinalis* (Mulsant), from Australia to California in the USA to control the cottony-cushion scale, *Icerya purchasi* Maskell, in *Citrus* in 1888 (Doutt, 1964). However, there are legitimate concerns relating to 'non-target' effects with classical biological control (e.g. Howarth, 2000; Barron *et al.*, 2003; Hoddle, 2004).

Augmentation

Augmentation biological control involves timely releases of mass-produced organisms to an area where existing populations are low, in order to suppress pest populations, which may be indigenous or introduced to the area. The two recognized release approaches are inoculative and inundative. An inoculative release is used with entomophages that reproduce sufficiently rapidly to check the growth of the pest population. Usually small numbers of entomophagous arthropods are introduced early in the crop cycle, which gives them time to establish and reproduce, and for their offspring to achieve effective pest suppression. An inundative release is used to obtain an immediate control, as the entomophages are unlikely to reproduce sufficiently to keep pace with the growth of the pest population. Large numbers of entomophagous arthropods are released on several occasions during the crop cycle, and pest suppression is primarily achieved by these arthropods rather than their offspring: e.g. *Trichogramma brassicae* Bezdenko releases in cabbage (Lundgren *et al.*, 2002). The limitations of augmentation biological control are restricted availability, high costs to rear, transport and release the arthropods (Van Driesche and Bellows, 1996), concern for the quality of these arthropods (Van Driesche and Bellows, 1996; van Lenteren *et al.*, 2003) and risks of non-target effects (Babendreier *et al.*, 2003; van Lenteren and Tommasini, 2003).

Conservation

Conservation biological control involves adopting practices that manipulate the environment to maintain or enhance the 'fitness' of entomophagous arthropods, in order to increase their effectiveness (Landis *et al.*, 2000). It assumes that the entomophagous arthropods are already present, which may be indigenous or introduced to the area, and that their effectiveness can be increased. Conservation biological control tactics include the use of narrow-spectrum insecticides and the provision of artificial or natural supplementary foods, such as sugar sprays (Ewert and Chiang, 1966; Evans and Swallow, 1993; McEwen and Kidd, 1995) and nectar from flowering plants (Berndt *et al.*, 2002; Fitzgerald and Solomon, 2004; Tylianakis *et al.*, 2004). With conservation biological control, it is possible to achieve wider suppressive action by an entire assemblage of entomophagous arthropods against a diverse pest complex, rather than a single target pest species. Unfortunately adoption rates and economic benefits of conservation biological control have seldom been quantified (Gurr *et al.*, 2000a).

Integrated biological control

There is scope to integrate conservation biological control tactics with classical and augmentative methods, potentially to improve the success of the release strategies, and this approach was dubbed 'integrated

biological control' (IBC) (Gurr and Wratten, 1999). For example, the South American encyrtid, *Copidosoma koehleri* Blanchard, has become widely established in temperate Australia as a classical biological control agent of the potato moth, *Phthorimaea operculella* (Zeller). Its densities and effectiveness are, however, low. Baggen and Gurr (1998) explored the scope for use of nectar-rich plants to be sown in the field margins of potato crops and went on to identify 'selective' food plants that whilst fed upon by the parasitoid were not used by the pest. Field experiments demonstrated the effectiveness of this IBC approach (Baggen *et al.*, 1999). A contrasting example shows the potential for conservation biological control to support natural enemies released in an augmentative fashion. The Australian endemic egg parasitoid, *T. carverae* Oatman and Pinto, is reared commercially for mass release in vineyards for control of the tortricid, *Epiphyas postvittana* (Walker). However, the parasitoid's longevity tends to be very short, limiting its impact on the target and necessitating accurate monitoring so that releases coincide with the peak in host egg densities. Recent work by Begum *et al.* (2004 a,b) has shown that the longevity and fecundity of this natural enemy can be improved by access to nectar-producing plants such as alyssum.

Targets of Biological Control

A large range of entomophagous and pest arthropods found in arable, fruit and glasshouse crops are the target of biological control programmes. Concise descriptions of the biology and ecology of these arthropods are provided by Smith *et al.* (1997), Hagen *et al.* (1999) and Helyer *et al.* (2003).

Classical

The BIOCAT database contains records of the introduction of entomophagous insects for the control of insect pests worldwide since 1880 (Greathead and Greathead, 1992). A review of the database up to March 2004 reveals that there have been 5634 recorded releases of 2119 species of entomophagous arthropods to control 597 pest species. However, 'substantial' (other control measures rarely needed; $n = 419$) or 'complete' (no other control measures required; $n = 212$) control was recorded in only 631 out of 5634 cases (11.2%). The primary target pests in classical biological control programmes have belonged to the orders Homoptera ($n = 2420$ releases), Lepidoptera ($n = 1772$), Coleoptera ($n = 596$), Diptera ($n = 533$), Hymenoptera ($n = 137$) and Heteroptera ($n = 96$); the Homoptera have been the most successful group controlled (429 releases or 17.7% were successful).

The primary agents in classical biological control programmes have belonged to the orders Coleoptera ($n = 1168$), Diptera ($n = 457$) and Hymenoptera ($n = 3887$); the success rates recorded were 9.2%, 8.1% and 12.3% for these groups, respectively (D. Greathead, CABI, Ascot, UK, personal communication, 2004; see also Greathead and Greathead, 1992).

Augmentation

More than 125 species of entomophagous arthropods are commercially used in augmentative biological control programmes, including 37 commonly used species, such as the spider mite predator, *Phytoseiulus persimilis* Athias-Henriot, the whitefly parasitoid, *Encarsia formosa* Gahan and the moth egg parasitoid, *Trichogramma* spp. (van Lenteren, 2003). Based on sales figures of the natural enemies used, the main pest species targeted for augmentative biological control in glasshouse crops are whiteflies (33%), thrips (22%), spider mites (16%), aphids (13%) and others (16%) (van Lenteren, 2003).

Conservation

The target species of entomophagous and pest arthropods involved in conservation biological control programmes, and the levels of adoption and success of these programmes, have seldom been investigated. Unlike classical biological control, where

there are databases present such as BIOCAT, and augmentative biological control, where annual sales figures for biological control agents exist, there is no equivalent readily accessible source of information on conservation biological control tactics (Gurr *et al.*, 2000a). Critical reviews of the success of various habitat manipulation strategies (a form of conservation biological control) based on hierarchical approaches showed that increased profitability has seldom been shown (Gurr *et al.*, 2000a; Wade *et al.*, 2004).

Types of Biological Control Agents

Entomophagous arthropods can be categorized in three ways: taxonomic affiliation, functional groups and life history type.

Taxonomic groups of biological control agents

Biological control agents are drawn from many of the taxa named in the Taxonomic Affiliations section. Taxa of particular importance in classical and augmentative approaches include Acarina (Anystidae, Erythraeidae, Phytoseiidae); Coleoptera (Coccinellidae); Hymenoptera (Aphelinidae, Braconidae, Encyrtidae, Eulophidae, Formicidae, Ichneumonidae); and Neuroptera (Chrysopidae) (Van Driesche and Bellows, 1996; Smith *et al.*, 1997; Helyer *et al.*, 2003). Other taxa such as Araneida (Arachnidae, Clubionidae, Oxyopidae, Salticidae, Theridiidae, Thomisidae); Coleoptera (Carabidae, Staphylinidae); Diptera (Cecidomyiidae, Syrphidae, Tachinidae); Hemiptera (Anthocoridae, Lygaeidae, Miridae, Nabidae, Pentatomidae, Reduviidae); and Neuroptera (Hemerobiidae) are important in conservation biological control.

Functional groups

A more informative classification system is based on the ecology of the organism, whereby organisms are categorized according to various functional groups or guilds. The different types of functional groups of predators and parasitoids have been previously discussed (i.e. specialist and generalist predators, and primary and secondary parasitoids). Although many predatory species are classified as 'generalists' because they feed on a taxonomically diverse range of prey, distinct preferences are likely to exist and their fitness may be separately linked to a lesser number of species (Eubanks and Denno, 2000; Michaud, 2000; Evans *et al.*, 2004). For example, the 'suitability' of Citrus aphids, *Aphis spiraecola* Patch and *Toxoptera citricida* (Kirkaldy), differs as prey for the coccinellid *Harmonia axyridis* Pallas. Although consumption rates of the two aphid species by larval or adult coccinellids did not differ, females fed on *A. spiraecola* for 30 days never produced eggs, while those fed on *T. citricida* laid 289 eggs over 17 days (Michaud, 2000).

Life history type

A third approach for categorizing entomophagous arthropods is concerned with the extent of ontogenic changes in diet (Polis and Strong, 1996; Coll and Guershon, 2002); the immature and adult life stages of species utilize different diets. For example, the larvae of hymenopteran parasitoids, hoverflies (Syrphidae) and some species of green lacewings (Chrysopidae) are parasitic or predacious, while the adults are herbivorous, feeding on nectar and/or pollen. In some coccinellid species such as *Coleomegilla maculata* DeGeer, the larval stage is strictly predacious (feeds on aphids alone), while the adult stage is omnivorous (feeds on both aphids and pollen) (Harmon *et al.*, 2000). In the Hemiptera, for example, both the immature and adult life stages may be omnivorous. This 'life history omnivory' information has led to the planting of flowering plants, such as buckwheat *Fagopyrum esculentum* Bentham (Polygonaceae), to provide potential food sources for the herbivorous or omnivorous life stages of these insects (e.g. Baggen *et al.*, 1999; Berndt *et al.*, 2002; Tylianakis *et al.*, 2004).

Biological Control and Integrated Pest Management

Although classical biological control and augmentation biological control can be so successful that no additional techniques are required for adequate suppression of pests (Gurr and Wratten, 2000), natural enemies are used most often in conjunction with host plant resistance or pesticides in IPM strategies. As the following sections explain, however, the use of predators and parasites may not always be fully compatible with resistant plant traits and chemical control. An understanding of the possible positive and negative effects is clearly critical for the optimal design of IPM.

Host plant resistance and biological control

Price (1997) describes several selection pressures within a herbivore population for the development of an ability to use a toxic food plant. First, competition for food is reduced because few other herbivores are able to use the same resource; second, the plant compounds that are responsible for the original toxicity may be used as location cues by the adapted herbivore; and third, the toxins may provide the herbivore with protection from its agonists, whether pathogens, parasitoids or predators. The last of these suggests that host plant resistance and biological control may not always be compatible. Notwithstanding the scope for antagonism between host plant resistance and biological control, a considerable body of literature suggests that these two, essentially ecological, pest management approaches can be compatible or even synergistic (Dicke, 1996). Evidence for the various effects is reviewed briefly below.

In a review of the evidence for positive and negative effects of host plant resistance on biological control, Groot and Dicke (2002) list 57 arthropod systems involving crops as diverse as cucumber, potatoes, tomato, grasses, brassicas, cotton, gerbera and apple. Of these, 10 systems showed no apparent effect of host plant resistance on biological control. In the remaining 47 systems host plant resistance operating through physical, chemical, a combination of both or unknown mechanisms had an effect on natural enemies (Table 11.1). Although simple numerical summaries of such evidence mask the severity of effects on natural enemies, it is clear that adverse consequences may arise for hymenopteran and dipteran parasitoids and for predators in taxa as diverse as Arachnida, Neuroptera, Dictyoptera and Hemiptera. The most intensively studied taxon, Hymenoptera, was subject to adverse effects far more frequently than beneficial effects. Glandular trichomes were an especially important feature responsible for adverse effects, which may be simultaneously physical (i.e. entrapment in exudates) and chemical (i.e. irritant, toxic and antixenosis effects of exudates). Physical traits such as entrapment in glandular and non-glandular trichomes were antagonistic almost as frequently as were chemical effects such as glandular trichomes, lack of extrafloral nectar and hydroxamic acids.

Synergistic effects

As observed earlier, plant trichomes may exert a negative effect on natural enemies if glandular (Fig. 11.3); however, non-glandular forms that lack the toxic heads may have a positive effect on some natural enemies through the provision of micro-refugia. It is considered that leaf surface structures alter mite abundance and influence predator–prey interactions (Walter, 1996; Walter and Proctor, 1999), the result of an evolutionary relationship that favours smaller mites with specific morphological characteristics, e.g. narrow body (Walter, 1996). Although it is yet to be confirmed that trichomes alone directly provide habitat for small beneficial mites, the predatory mites *Typhlodromus pyri* Scheuten and *Amblyseius aberrans* (Oudemans) are more abundant on hirsute than glabrous grape leaves (Duso, 1993). In addition, leaf domatia on grape leaves are also known to harbour fungivorous mites (English-Loeb *et al.*, 1999). Another important mechanism through which host plant resistance may favour biological control is the effect of plant digestibility reducers, which slow the developmental

Table 11.1. Summary of evidence for effects of host plant resistance traits on parasitoids and predators. (From Groot and Dicke, 2002.)

Host plant resistance form	Number of cases in the literature	
	Positive effect	Negative effect
Physical plant trait	Arachnida: 2	Hymenoptera: 8
		Diptera: 2
		Coleoptera: 2
		Neuroptera: 1
		Arachnida: 2
Chemical plant trait	Hymenoptera: 2	Hymenoptera: 11
	Diptera: 1	Diptera: 2
	Coleoptera: 1	Coleoptera: 2
		Hemiptera: 8
		Neuroptera: 1
		Dictyoptera: 1
		Arachnida: 1
Unknown plant trait	Hymenoptera: 2	Hymenoptera: 1
	Hemiptera: 1	Coleoptera: 1
		Hemiptera: 1

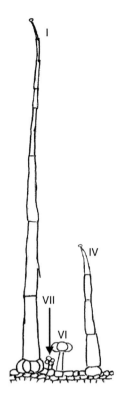

Fig. 11.3. The four major types of glandular tri-chomes on *Lycopersicon* spp. (Based on Luckwill, 1943.)

rate of herbivores (Feeny, 1975, 1976). A slower developmental rate can lead to a herbivore remaining in the larval stage for longer, prolonging the length of time that it is susceptible to attack by larval parasitoids. Greater susceptibility can increase the impact of this important guild of natural enemies (e.g. Haggstrom and Larsson, 1995), an effect that may be evident in the harvested seed of the resistant plants. For example, Schmale *et al.* (2003) studied the impact of a parasitoid, *Dinarmus basalis* Rondani, on bruchid pests of bean cultivars with varying levels of the anti-feedant protein arcelin. Bruchid development was slowed by 15% in the arcelin-rich beans.

An important plant defence mechanism that operates by the action of natural enemies is the production of host insect–induced synomones (HIS). Nordlund (1981) defined a synomone as 'a chemical substance produced or acquired by an organism that, when it contacts an individual of another species in the natural context, evokes in the receiver a behavioural or physiological response that is adaptively favorable to both the emitter and the receiver'. HIS are produced by plants as a response to attack by many herbivores and are herbivore-specific, allowing host-specific parasitoids to utilize them for the detection of their prey. A plant attacked by a herbivore systemically produces HIS in other, uninfested, plant parts (Du *et al.*, 1996) and, most remarkably, the response may be induced in uninfested conspecific plants. This effect has been shown to operate through the rhizosphere in the well-studied broad bean/pea aphid *Acyrthosiphon pisum* Harris/*Aphidius ervi* Haliday system (Guerrieri *et al.*, 2002). The compound *cis*-jasmone has been identified as a plant signal in such systems and offers scope as a new class of crop-protectant chemicals (Powell and Pickett, 2003).

Antagonistic effects

Biological control may be impaired directly by the toxins and other factors, e.g. surface structures, responsible for host plant resistance. Natural enemies may come into direct contact with the toxins as they forage over plant surfaces. The predatory mite, *P. persimilis*, taps its palps on the leaf surface whilst foraging for prey and subsequent cleaning of the palps is thought to be responsible for ingestion of plant toxins, leading to adverse effects on this natural enemy (Krips *et al.*, 1999). More intimate effects may result when natural enemies use plant products such as pollen or plant sap (Dicke, 1999). This has clear implications for conservation biological control in which workers seek to maximize the fitness and performance of biological control agents by methods such as provision of alternative foods. Tomatine levels in tomato, *Lycopersicon esculentum* Mill., affect the parasitoid *Hyposoter exiguae* (Viereck) more severely than the herbivore *Heliothis*

(*Helicoverpa*) *zea* (Boddie) (Campbell and Duffey, 1979).

Morphological and chemical characteristics of plants may adversely affect the third trophic level (Fig. 11.4). Several systems are known in which glandular trichomes adversely affect predators and parasitoids. Baggen and Gurr (1995) demonstrated lethal effects by glandular trichomes of solanaceous plants on the encyrtid parasitoid (*C. koehleri*) (Fig. 11.5), and in follow-up studies Gooderham *et al.* (1998) demonstrated sublethal effects such as reduced searching efficiency and realized parasitism of the host, *P. operculella*. Results from a study on *T. exiguum* Pinto, Platner and Oatman by Keller (1987) showing that walking speed is affected by leaf trichomes suggest that reductions in parasitism on hirsute leaf surfaces may be a result of slower walking speeds. Leaf trichomes of *Gerbera* had a similar effect on the walking speed of *P. persimillis* (Krips *et al.*, 1999) and lacewing, *Mallada signata* (Schneider), larvae may also be influenced by the density of trichomes on the leaf surface of wild *Lycopersicon* spp. (Simmons and Gurr, 2004). Cuticular waxes constitute a further type of plant character that can disrupt the activity of natural enemies. Work by Eigenbrode *et al.* (1995, 1996) established that the failure of another neuropteran, *Chrysoperla carnea* (Stephens), to control diamond back moth, *Plutella xylostella* (Linnaeus), on normal, waxy-leaved cabbage cultivars – as they were able to do on glossy-leaved cultivars – was the consequence of waxy debris impairing the mobility of the lacewing larvae, increasing grooming time and decreasing the time spent searching.

Some antagonistic effects can influence the behaviour of natural enemies indirectly. Apart from being smaller, herbivores feeding on resistant plants may contain higher levels of toxins (van Emden, 1990). This could result in a reduction in parasitoid emergence (Kashyap *et al.*, 1991) or make prey less palatable, though specialist carnivores are better able to cope with herbivore-sequestered plant toxins than are generalist predators (Price, 1991). These negative effects could offset any

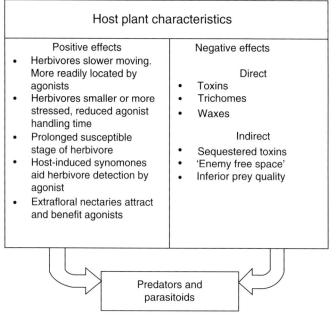

Fig. 11.4. Summary of mechanisms by which host plants may exert positive and negative influences on biological control agents.

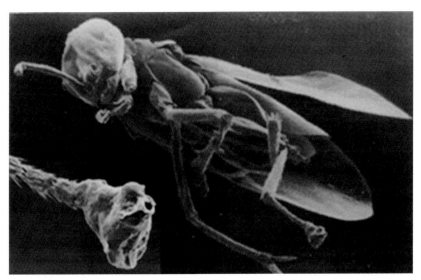

Fig. 11.5.   Glandular trichome exudate on tarsus of the parasitoid *Copidosoma koehleri*.

positive results associated with an extended period of vulnerability. In addition to the toxic effects of glandular trichome exudates mentioned above, trichome exudates may repel parasitoids. As Romeis *et al.* (1999) reported, two constituents, malic and oxalic acids, of pigeonpea glandular trichome exudates had a repellency effect on *T. chilonis* Ishii.

Genetically engineered crops

Recent years have seen the advent of several species of genetically engineered crops and insect resistance is a commonly introduced type of trait (Altieri *et al.*, 2004). The use of genetically modified insect-resistant crops is estimated to have reduced pesticide used by 22.3 million kg of formulated product (Phipps and Park, 2002) and, given the adverse impact of pesticides on non-target invertebrates, these reductions are likely to favour biological control. There is, however, evidence of negative implications (Table 11.2). Plants genetically modified to express *Bacillus thuringiensis* (Bt) toxins have been the subject of significant research to test for adverse effects on predators and parasitoids. For example, the emergence of healthy adult *Myiopharus doryphorae*

(Riley), a Tachinid fly, was reduced by Bt-treated hosts (Lopez and Ferro, 1995). *Spodoptera littoralis* (Boisduval) feeding on Bt maize were less preferred prey for *C. carnea* larvae than those feeding on non-Bt maize, though a similar effect was not reported for aphids (Hilbeck, 2001). In contrast, research has shown that the proportion of the parasitoids *Eriborus terebrans* Gravenhorst and *Macrocentrus grandii* Goidanich (Orr and Landis, 1997) and *Diadegma insulare* (Cresson) emerging from both Bt-treated and untreated hosts did not differ (Ulpuh and Kok, 1996).

The extent of any adverse effects of Bt crops on natural enemies may be offset by the use of non-modified plants in the crop as 'refugia'. Early work predicted the development of resistance to Bt cotton (Fitt *et al.*, 1994). Non-modified cotton is now planted to produce *Helicoverpa* adults that have not been exposed to Bt toxins and that will mate with the resistant survivors emerging from the nearby genetically modified crop. The use of such refuges may favour the conservation of natural enemies in farmscapes.

Plants have also been genetically modified to resist pests by the production of lectins. It has been reported that *Parallorhogas*

Table 11.2. Summary of possible beneficial and negative effects of insect-resistant genetically engineered crops on biological control. (From Altieri *et al.*, 2004.)

Possible beneficial effect	Possible negative effect
Reduced pesticide-induced mortality of natural enemies	Effects of gene products on natural enemies: directly via polyphagous pests feeding on plant material (e.g. pollen) or indirectly via their prey or hosts
Resistance management refugia may provide resources for natural enemies	Effects of gene products on soil fauna: food web consequences for natural enemies
Depressed development rate of pests provides more time for predation or parasitism to occur	
Morphological resistance traits, e.g. trichomes, may provide habitat for natural enemies	Pest resistance to Bt from widespread use of GE crops precludes use of Bt sprays and forces use of insecticides that suppress natural enemy activity

pyralophagus (Marsh) was less likely to parasitize hosts fed on modified sugarcane (Tomov *et al.*, 2003) and that lectin influenced the length of life cycles (Tomov and Bernal, 2003). Romeis *et al.* (2003) found that lectins reduced survival of *Cotesia glomerate* L., *T. brassicae* and *A. colemani* Viereck. Any effects may be genus-dependent, as Bell *et al.* (2001) reported no effect on the parasitoid *Eulophus pennicornis* (Nees). Although lectin had no toxic effects on the predator *Adalia bipunctata* L., it did influence the length of life cycles (Birch *et al.*, 1999; Down *et al.*, 2000). Some crops have been modified to produce inhibitors of enzymes that pests need in their guts. When the host's diet contained proteinase inhibitor the size and fecundity of the parasitoid *E. pennicornis* increased (Ashouri *et al.*, 2001), whereas cowpea trypsin inhibitor in the host diet reduced parasitism by *E. pennicornis* (Bell *et al.*, 2001). Although a relatively new area of plant

protection, genetic modification of plants has a strong foothold, and the widespread use of such crops demands that interactions with natural enemies be better understood (Altieri *et al.*, 2004).

Pesticides

Several authors (Mazzone and Viggiani, 1988; Babu and Ramanamurthy, 1999; Bostanian *et al.*, 2001; Colignon *et al.*, 2003) have recognized the lethal and sublethal effects of agrochemicals such as herbicides and fungicides on arthropod natural enemies. However, in the present section, the term pesticide will be used for any compound intended to control arthropod pests such as insects or mites, and this section will consider only that agrochemical group from now on. Implicit in the IPM concept is the maximum use of natural enemies, supplemented with insecticides when necessary (Croft and Brown, 1975). Unfortunately, parasitoid and predator populations can be severely affected by pesticides (Theiling and Croft, 1988). Considering that on a world scale, more than 3 billion kg of pesticides are applied annually to crops (Pimentel, 2004), and the fact that only 0.1% of the applied formulated product actually reaches the target pest, a large proportion of insecticide is available in the environment to affect non-target species (Metcalf, 1994). So, IPM reaches its greatest potential when pesticide effects on natural enemies are minimized.

Pesticide effects on natural enemies

Pesticide use can have direct and indirect effects on natural enemies. The former are produced after exposure by direct spraying or contact with toxicant residues (Banken and Stark, 1998), and can be short term (up to 24 h) or long term. Acute mortality is the most common end point measured in short-term effect studies, and this can be expressed as median lethal dose or concentration that kills 50% of the population treated (LD_{50} or LC_{50}, respectively) (Stark and Banks, 2003).

The lower the LD_{50} or LC_{50} value, the more lethal or toxic a pesticide is considered to be (Flint and Gouveia, 2001). These studies are usually made with uniform cohorts of insects exposed to toxicants in the laboratory, and do not consider effects on reproduction of survivors or survival in the F_1 generation (Robertson and Worner, 1990), which are density-dependent regulatory mechanisms that alter the toxicological impacts at the population level (Stark et al., 1998). For instance, the phytophagous mite, *Tetranychus urticae* (Koch), and its mite predator, *Iphiseius degenerans* (Berlese), exhibited almost identical acute susceptibility (LD_{50}) to the pesticide dicofol, but sublethal effects (i.e. lower fecundity) rendered the predator more susceptible than its prey when population growth was evaluated (Stark et al., 1998). Sublethal effects can be expressed by long-term changes in developmental rate (Rumpf et al., 1998), longevity (Ibrahim and Yee, 2000), fecundity (Stark et al., 1992a,b), sex ratio (Vinson, 1974) and behaviour (Wiles and Jepson, 1994). The consequences of these are generally detrimental, since natural enemies that survive toxicant exposure may still suffer one or more of the above sublethal effects (Stark et al., 1998), even after several generations (Rumpf et al., 1998). In contrast, many reports show that sublethal doses of pesticides are capable of increasing some aspects of the performance of natural enemies (Croft, 1990a), a phenomenon called hormoligosis (Luckey, 1968). This is brought about by an increase in fecundity (Holland et al., 1994) or body weight (Luckey, 1968).

Pesticides can also indirectly affect natural enemies by their consumption of contaminated prey and hosts (Ahmed, 1955), and a change in host or prey distribution after application, which may decrease exploitation of surviving pest populations (Waage, 1989). Also, the synchronization of natural enemy and pest can be altered, with gaps of up to one pest generation in the availability of the necessary pest stages for natural enemy reproduction. This especially affects natural enemies, which have a generation time that is much shorter than that of the pest (Waage, 1989). Pesticides can also, of course, severely reduce pest populations and hence induce

starvation in natural enemies, and this can lead to resurgence of primary pests or to secondary pest outbreaks. Pest populations may be reduced in this way to levels below the predator's oviposition threshold (Wratten, 1973). After spraying, pests can rapidly increase in numbers, because natural enemies die of starvation and the pest is set free from density-dependent factors (e.g. competition for plants) (Flint and Gouveia, 2001). Indirect effects can also alter third and fourth trophic-level interactions, since differences in behaviour between these trophic levels can influence the exposure and hence susceptibility to pesticides (Longley and Jepson, 1996). Finally, indirect effects acting through other species can alter the composition of the natural enemy species complex, caused by changes in the quantities and directions of energy and nutrient flow in the ecosystem (Croft, 1990b).

Minimizing natural enemy mortality

Bartlett (1964) defined pesticide selectivity as the capacity of a pesticide treatment to spare natural enemies while destroying the target pest. This view would now be considered rather simplistic. Pest populations do not need to be 'destroyed' and predator mortality from pesticides does not need to be zero to effect useful shifts in the prey/predator ratio, as recent modelling has illustrated in the context of conservation biological control (Kean et al., 2003). Selectivity can be achieved through physiological and ecological means. Physiological selectivity results from differences in uptake, detoxification and excretion of pesticides (Croft, 1990c). Ecological selectivity arises from differences in biology, behaviour and exposure of natural enemies with respect to pests and through pesticide dose, distribution, timing and operational aspects of spraying (Graham-Bryce, 1987).

Monitoring and thresholds

Probably the best practice to reduce the impact of pesticides on natural enemies is to minimize pesticide applications altogether. It may be possible to reduce pesticide applications through the use of monitoring techniques that are well founded on research (Johnson and Tabashnik, 1999; Fadamiro, 2004) to avoid calendar spraying and by setting more realistic economic thresholds (Poehling, 1989), which growers follow. However, while pesticides continue to be a relatively low variable cost, the likelihood of farmers adhering to thresholds remains low, given the financial risks sometimes associated with no such intervention (Mann *et al.*, 1991). Farmers often consider regular, prophylactic application of pesticide a 'cheap insurance'. However, convincing growers to abandon a 'pesticide treadmill' (Tait, 1980) approach is a complex operation, requiring a combination of education and persuasion. The works by Mann *et al.* (1991) and Roduner *et al.* (2003) are useful cases of study in this respect.

Selective pesticides

Among insecticide classes, a trend of increasing toxicity to natural enemies is present from the early 'inorganics' to the synthetic pyrethroids. More recent 'microbials' (e.g. *B. thuringiensis*) and insect growth regulators are less toxic to natural enemies and more selective against pests (Theiling and Croft, 1988). Unfortunately, in multiple-pest systems, selective compounds will control only the target pest, and may leave other pest species free of competition. Another way to exploit physiological differences is through the use of resistance of natural enemies to pesticides. The best-documented cases of resistance among natural enemies are those for the predatory phytoseiid mites. For example, *A. fallacis* (Garman) shows resistance or cross-resistance to at least 29 pesticides from several classes (Croft, 1990e).

Dose reduction

The dose suggested by the manufacturer is generally higher than the dose actually necessary to control the pest effectively (Poehling, 1989). Reduced-rate (below 'label' recommendations) applications of

pesticides can exert dynamic changes in pest/predator survival (van Emden, 1990). Figure 11.6 shows that as dose rates are reduced, predator (P) mortality declines sharply but that of the herbivorous pest (H) remains relatively high. This is because herbivores have evolved enzymes to detoxify plant poisons, giving a slow increase in mortality rate with increased pesticide dose. Predators, in contrast, show a steep response to increased dose. It follows that reduced-rate pesticide application can, counterintuitively, enhance relative natural enemy number by a higher survival rate (lower mortality) of predators than of pests at reduced doses (van Emden, 1990; Wiles and Jepson, 1995; Longley *et al.*, 1997). In fact, models have shown that the chances of a polyphagous predator population persisting in a sprayed field are improved if the rate of pesticide application is low (Sherratt and Jepson, 1993). Furthermore, in winter wheat in Germany, reduced-rate applications tended to be more profitable than full-rate applications, especially in years of low aphid numbers (Mann *et al.*, 1991). However, it may not be feasible to reduce pesticide dosage for pests for which economic thresholds are extremely low or for vectors of diseases. In addition, legal restrictions should also be taken into account before a reduced-dose approach is used. For example, in some states in the USA, such as

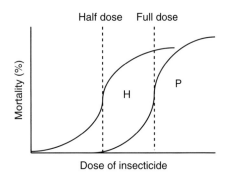

Fig. 11.6. Theoretical selectivity of pesticide application on a resistant variety. H = dose mortality curve of a herbivore; P = dose mortality curve of a predator or parasitoid. (Based on van Emden, 1990.)

Oklahoma, it is legally forbidden to spray less than the dose specified in the label (G.W. Cuperus, Oklahoma State University, USA, 2004, personal communication).

Spatial discrimination

Spatial discrimination may be achieved by spraying where the pest is present but the natural enemy is not. Applications can be made to a portion of the plant (e.g. canopy periphery or tree trunks) (Johnson and Tabashnik, 1999; Flint and Gouveia, 2001). Baits, soil drenches and seed treatments can greatly concentrate chemicals where they are needed (Graham-Bryce, 1987). Trap crops planted on the borders can attract specific pests, helping to conserve the natural enemies in the rest of the field (Rea *et al.*, 2002). Pheromones can be deployed with insecticides in order to 'lure and kill' specific pests (Jones, 1998; Khan and Pickett, 2004). In addition, systemic pesticides have a particular selective action, especially against piercing–sucking pests (e.g. mites and aphids) (Graham-Bryce, 1987) without harming predatory coccinellids and syrphids (Metcalf, 1994). Although the use of systemic pesticides may eliminate widespread residues, toxicity to natural enemies may still occur through food-chain uptake (Croft, 1990d), and through starvation of natural enemies through pest depletion (Johnson and Tabashnik, 1999).

When the natural enemies are present in the same habitat as the pest, it is possible to establish refugia from pesticide applications by the use of spot treatments, or of alternate strips in row crops or orchards (Ripper, 1956), avoiding specially created predator refuges (Thomas *et al.*, 1991), or spraying just the perimeter of the field (Sivasubramaniam *et al.*, 1999).

Temporal discrimination

Arthropod pest and natural enemy populations may be asynchronous due to behaviour, development and population dynamics, even though they reside in the same area. Timing of the co-occurrence of the species is particularly critical when highly specialized natural enemies and their hosts or preys are involved (Croft, 1990d). For example, some aphid parasitoid species are protected from pesticides while within the mummified cases of their hosts (Shean and Cranshaw, 1991; Longley *et al.*, 1997). Temporal discrimination by proper timing can also be achieved through multigenerational or seasonal cycles with parasitoids. For example, over several generations, parasitism rates of the leafminer, *Leucoptera meyricki* (Ghesquière), can be higher in sprayed than in unsprayed fields (Cuperus and Radcliffe, 1984; Croft, 1990d). However, for generalist predators or natural enemies, which have many overlapping generations, achieving ecological selectivity through generational asynchrony is less feasible (Poehling, 1989).

Non-persistent pesticides that are rapidly degradable can be used to achieve substantial selectivity if applications are properly timed. In this way, natural enemies can better survive pesticide applications in their less-susceptible life stage or in refugia (Metcalf, 1994). A single application of a persistent and broad-spectrum pesticide can affect population recovery in the short term (Duffield *et al.*, 1996); similarly, repeated applications of short-persistence chemicals, which do not allow recovery of exposed populations between seasons, can produce long-term detrimental effects on natural enemies (Cuperus and Radcliffe, 1984; Burn, 1989).

Operational measures

Pesticide formulation can promote ecological selectivity by limiting or directing the distribution of pesticide residues. The contact activity of pesticides can be reduced without decreasing the stomach-poisoning activity by coating or microencapsulating them (Graham-Bryce, 1987). Application technology (e.g. spray volume) has received less attention with respect to its impacts on natural enemies; however, it is important to keep this option in mind, since broadcast application is a very inefficient process, with usually only 10–20% of insecticide applied being deposited on plant surfaces (Metcalf, 1994).

Conclusion

Biological control can be 'elegant, sustainable, non-polluting and inexpensive' (Gurr *et al.*, 2000b) but its success rate has not been particularly high, especially for biological control of arthropods by arthropods (Gurr *et al.*, 2000a). Better understanding of the ecology of predators and parasitoids is vital for improving the performance of predators and parasitoids. As outlined in the foregoing sections, there are many complexities involved in achieving sufficient understanding. Already, however, the literature suggests that simple two and three trophic level concepts of the functioning of natural enemies in agricultural systems are being superseded. Life history omnivory and food web interactions are now underpinning practical approaches such as habitat manipulation to enhance natural enemy fitness.

Reflecting such advances in theoretical understanding, the practice of biological control increasingly involves a consideration of the agent's ecological requirements for resources other than the target host or prey species. Further, the IBC concept offers scope for conservation biological control methods to support classical and augmentative approaches. With increasing attention being devoted to the ecology of predators and parasitoids, such natural enemies offer great potential for the effective and sustainable management of pests in crops, forests and other managed landscapes. As pest invasions increase with 'globalization', and pesticide use continues to increase (Pimentel, 2004), greater expectations for natural enemies to reduce pest numbers will result (Loope and Howarth, 2003).

References

Ahmed, M.K. (1955) Comparative effect of systox and schradan on some predators of aphids in Egypt. *Journal of Economic Entomology* 48, 530–532.

Altieri, M.A., Cure, J.R. and Garcia, M.A. (1993) The role and enhancement of parasitic Hymenoptera biodiversity in agroecosystems. In: LaSalle, J. and Gauld, I.D. (eds) *Hymenoptera and Biodiversity.* CAB International, Wallingford, UK, pp. 257–275.

Altieri, M.A., Gurr, G.M. and Wratten, S.D. (2004) Genetic engineering and ecological engineering: a clash of paradigms or scope for synergy? In: Gurr, G.M., Wratten, S.D. and Altieri, M.A. (eds) *Ecological Engineering for Pest Management: Advances in Habitat Manipulation for Arthropods.* CSIRO Publishing, Collingwood, Victoria, Australia, pp. 13–31.

Ashouri, A., Michaud, D. and Cloutier, C. (2001) Recombinant and classically selected factors of potato plant resistance to the Colorado potato beetle, *Leptinotarsa decemlineata*, variously affect the potato aphid parasitoid *Aphidius nigripes. Biocontrol* 46, 401–418.

Askew, R.R. and Shaw, M.R. (1986) Parasitoid communities: their size, structure and development. In: Waage, J. and Greathead, D. (eds) *Insect Parasitoids.* Academic Press, London, pp. 225–264.

Babendreier, D., Kuske, S. and Bigler, F. (2003) Parasitism of non-target butterflies by *Trichogramma brassicae* Bezdenko (Hymenoptera: Trichogrammatidae) under field cage and field conditions. *Biological Control* 26, 139–145.

Babu, T.R. and Ramanamurthy, G. (1999) Residual toxicity of pesticides to the adults of *Cryptolaemus montrouzieri* Mulsant (Coccinellidae: Coleoptera). *International Pest Control* 41, 137–138.

Baggen, L.R. and Gurr, G.M. (1995) Lethal effects of foliar pubescence of solanaceous plants on the biological control agent *Copidosoma koehleri* Blanchard (Hymenoptera: Encyrtidae). *Plant Protection Quarterly* 10, 116–118.

Baggen, L.R. and Gurr, G.M. (1998) The influence of food on *Copidosoma koehleri*, and the use of flowering plants as a habitat management tool to enhance biological control of potato moth, *Phthorimaea operculella. Biological Control* 11, 9–17.

Baggen, L.R., Gurr, G.M. and Meats, A. (1999) Flowers in tri-trophic systems: mechanisms allowing selective exploitation by insect natural enemies for conservation biological control. *Entomologia Experimentalis et Applicata* 91, 155–161.

Banken, J.A. and Stark, J.D. (1998) Multiple routes of pesticide exposure and the risk of pesticide to biological controls: a study of neem and the seven-spotted lady beetle (Coleoptera: Coccinellidae). *Journal of Economic Entomology* 91, 1–6.

Barbosa, P. (1998) *Conservation Biological Control.* Academic Press, San Diego, California.

Bardgett, R.D., Denton, C.S. and Cook, R. (1999) Below-ground herbivory promotes soil nutrient transfer and root growth in grassland. *Ecology Letters* 2, 357–360.

Barron, M.C., Barlow, N.D. and Wratten, S.D. (2003) Non-target parasitism of the endemic New Zealand red admiral butterfly (*Bassaris gonerilla*) by the introduced biological control agent *Pteromalus puparum*. *Biological Control* 27, 329–335.

Bartlett, B.R. (1964) Integration of chemical and biological control. In: DeBach, P. (ed.) *Biological Control of Insect Pests and Weeds.* Chapman & Hall, London, pp. 489–514.

Begum, M., Gurr, G.M., Wratten, S.D. and Nicol, H.I. (2004a) Flower color affects tri-trophic-level biocontrol interactions. *Biological Control* 30, 584–590.

Begum, M., Gurr, G.M., Wratten, S.D., Hedberg, P.R. and Nicol, H.I. (2004b) The effect of floral nectar on the efficacy of the grapevine leafroller parasitoid, *Trichogramma carvarae*. *International Journal of Ecology and Environmental Sciences* 30, 3–12.

Bell, H.A., Fitches, E.C., Down, R.E., Ford, L., Marris, G.C., Edwards, J.P., Gatehouse, J.A. and Gatehouse, A.M.R. (2001) Effect of dietary cowpea trypsin inhibitor (CpTI) on the growth and development of the tomato moth *Lacanobia oleracea* (Lepidoptera : Noctuidae) and on the success of the gregarious ectoparasitoid *Eulophus pennicornis* (Hymenoptera: Eulophidae). *Pest Management Science* 57, 57–65.

Berndt, L.A. and Wratten, S.D. (2005) Effects of alyssum flowers on the longevity, fecundity and sex ratio of the leafroller parasitoid *Dolichogenidea tasmanica*. *Biological Control* 32, 65–69.

Berndt, L.A., Wratten, S.D. and Hassan, P.G. (2002) Effects of buckwheat flowers on leafroller (Lepidoptera: Tortricidae) parasitoids in a New Zealand vineyard. *Agricultural and Forest Entomology* 4, 39–45.

Birch, A.N.E., Geoghegan, I.E., Majerus, M.E.N., McNicol, J.W., Hackett, C.A., Gatehouse, A.M.R. and Gatehouse, J.A. (1999) Tri-trophic interactions involving pest aphids, predatory 2-spot ladybirds and transgenic potatoes expressing snowdrop lectin for aphid resistance. *Molecular Breeding* 5, 75–83.

Bostanian, N.J., Larocque, N., Chouinard, G. and Coderre, D. (2001) Baseline toxicity of several pesticides to *Hyaliodes vitripennis* (Say) (Hemiptera: Miridae). *Pest Management Science* 57, 1007–1010.

Broderick, N.A., Raffa, K.F., Goodman, R.M. and Handelsman, J. (2004) Census of the bacterial community of the gypsy moth larval midgut by using culturing and culture-independent methods. *Applied and Environmental Microbiology* 70, 293–300.

Burn, A.J. (1989) Long-term effects of pesticides on natural enemies of cereal crop pests. In: Jepson, P.C. (ed.) *Pesticides and Non-target Invertebrates.* Intercept, Wimborne, UK, pp. 177–194.

Campbell, B.C. and Duffey, S.S (1979) Tomatine and parasitic wasps: potential incompatibility of plant antibiosis with biological control. *Science* 205, 700–702.

Casas, J. (2000) Host location and selection in the field. In: Hochberg, M.E. and Ives, A.R. (eds) *Parasitoid Population Biology.* Princeton University Press, Princeton, New Jersey, pp. 17–26.

Charnov, E.L. (1982) *The Theory of Sex Allocation.* Princeton University Press, Princeton, New Jersey.

Colignon, P., Haubruge, E., Gaspar, C. and Francis, F. (2003) Effects of reducing recommended doses of insecticides and fungicides on the non-target insect *Episyrphus balteatus* (Diptera: Syrphidae). *Phytoprotection* 84, 141–148.

Coll, M. and Guershon, M. (2002) Omnivory in terrestrial arthropods: Mixing plant and prey diets. *Annual Review of Entomology* 47, 267–297 + online supplemental table.

Cornell, H.V. and Hawkins, B.A. (1995) Survival patterns and mortality sources of herbivorous insects: some demographic trends. *American Naturalist* 145, 563–593.

Croft, B.A. (1990a) Sub-lethal effects. In: Croft, B.A. (ed.) *Arthropod Biological Control Agents and Pesticides.* John Wiley & Sons, New York, pp. 157–184.

Croft, B.A. (1990b) Ecological influences. In: Croft, B.A. (ed.) *Arthropod Biological Control Agents and Pesticides.* John Wiley & Sons, New York, pp. 185–218.

Croft, B.A. (1990c) Physiological selectivity. In: Croft, B.A. (ed.) *Arthropod Biological Control Agents and Pesticides.* John Wiley & Sons, New York, pp. 221–246.

Croft, B.A. (1990d) Ecological selectivity. In: Croft, B.A. (ed.) *Arthropod Biological Control Agents and Pesticides.* John Wiley & Sons, New York, pp. 247–268.

Croft, B.A. (1990e) Pesticide resistance: documentation. In: Croft, B.A. (ed.) *Arthropod Biological Control Agents and Pesticides.* John Wiley & Sons, New York, pp. 357–382.

Croft, B.A. and Brown, W.A. (1975) Responses of arthropod natural enemies to insecticides. *Annual Review of Entomology* 20, 285–330.

Cuperus, G.W. and Radcliffe, E.B. (1984) Effect of trichlorfon sprays and alfalfa (*Medicago sativa* L.) cultivars on pea aphid, *Acyrthosiphon pisum* (Harris). *Crop Protection* 3, 199–208.

Dicke, M. (1996) Plant characteristics influence biological control agents: implications for breeding for host plant resistance. *Bulletin Oilb/Srop* 19, 72–80

Dicke, M. (1999). Direct and indirect effects of plants on performance of beneficial organisms. In: Ruberson, J.R. (ed.) *Handbook of Pest Management*. Marcel Dekker, New York, pp. 105–153.

Dicke, M., van Lenteren, J.C., Minks, A.K. and Schoonhoven, L.M. (1990) Proceedings of the international symposium: Semiochemicals and pest control – prospects for new applications. *Journal of Chemical Ecology* 16, 3017–3212.

Doutt, R.L. (1964) The historical development of biological control. In: DeBach, P. (ed.) *Biological Control of Insect Pests and Weeds*. Chapman & Hall, London, pp. 21–42.

Down, R.E, Ford, L., Woodhouse, S.D., Raemaekers, R.J.M., Leitch, B., Gatehouse, J.A. and Gatehouse, A.M.R. (2000) Snowdrop lectin (GNA) has no acute toxic effects on a beneficial insect predator, the 2-spot ladybird (*Adalia bipunctata* L.). *Journal of Insect Physiology* 46, 379–391.

Du, Y.J., Poppy, G.M. and Powell, W. (1996) Relative importance of semiochemicals from first and second trophic levels in host foraging behavior of *Aphidius ervi*. *Journal of Chemical Ecology* 22, 1591–1605.

Duffield, S.J., Jepson, P.C., Wratten, S.D. and Sotherton, N.W. (1996) Spatial changes in invertebrate predation rate in winter wheat following treatment with dimethoate. *Entomologia Experimentalis et Applicata* 78, 9–17.

Duso, C. (1993) Factors affecting the potential of phytoseiid mites (Acari: Phytoseiidae) as biocontrol agents in North-Italian vineyards. *Experimental and Applied Acarology* 17, 241–258.

Edwards, P.J., Wratten, S.D. and Parker, E.A. (1992) The ecological significance of rapid wound-induced changes in plants: insect grazing and plant competition. *Oecologia* 91, 266–272.

Eigenbrode, S.D., Moody, S. and Castagnola, T. (1995) Predators mediate host plant resistance to a phytophagous pest in cabbage with glossy leaf wax. *Entomologia Experimentalis et Applicata* 77, 335–342.

Eigenbrode, S.D., Castagnola, T., Roux, M.B. and Steljes, L. (1996) Mobility of three generalist predators is greater on cabbage with glossy leaf wax than on cabbage with a wax bloom. *Entomologia Experimentalis et Applicata* 81, 335–343.

English-Loeb, G., Norton, A.P., Gadoury, D.M., Seem, R.C. and Wilcox, W.F. (1999) Control of powdery mildew in wild and cultivated grapes by a tydeid mite. *Biological Control* 14, 97–103.

Eubanks, M.D. and Denno, R.F. (2000) Health food versus fast food: the effects of prey quality and mobility on prey selection by a generalist predator and indirect interactions among prey species. *Ecological Entomology* 25, 140–146.

Eubanks, M.D. and Styrsky, J.D. (2005) Effects of plant feeding on the performance of omnivorous "predators". In: Wäckers, F., van Rijn, P. and Bruin, J. (eds) *Plant-provided Food for Carnivorous Insects: A Protective Mutualism and Its Applications*. Cambridge University Press, Cambridge, pp. 148–177.

Evans, E.W. and Swallow, J.G. (1993) Numerical responses of natural enemies to artificial honeydew in Utah alfalfa. *Environmental Entomology* 22, 1392–1401.

Evans, E.W, Richards, D.R. and Kalaskar, A. (2004) Using food for different purposes: female responses to prey in the predator *Coccinella septempunctata* L. (Coleoptera: Coccinellidae). *Ecological Entomology* 29, 27–34.

Ewert, M.A. and Chiang, H.C. (1966) Dispersal of three species of coccinellids in corn fields. *Canadian Entomologist* 98, 999–1003.

Fadamiro, H.Y. (2004) Monitoring the seasonal flight activity of *Cydia pomonella* and *Argyrotaenia velutinana* (Lepidoptera : Tortricidae) in apple orchards by using pheromone-baited traps. *Environmental Entomology* 33, 1711–1717.

Feeny, P.P. (1975) Biochemical coevolution between plants and their insect herbivores. In: Gilbert, L.E. and Raven, P.H. (eds) *Coevolution of Animals and Plants*. University of Texas Press, Austin, Texas, pp. 3–19.

Feeny, P.P. (1976) Plant apparency and chemical defense. In: Wallace, J.W. and Mansell, R.L. (eds) *Biochemical Interaction Between Plants and Insects*. Plenum Press, New York, pp. 1–40.

Fisher T., Bellows, T., Caltagirone, L., Dahlsten, D., Huffaker, C. and Gordh, G. (eds) (1999) *Handbook of Biological Control*. Elsevier, Amsterdam, The Netherlands.

Fitt, G.P., Mares, C.L. and Llewellyn, D.J. (1994) Field evaluation and potential ecological impact of transgenic cottons (*Gossypium hirsutum*) in Australia. *Biocontrol Science and Technology* 4, 535–548.

Fitzgerald, J.D. and Solomon, M.G. (2004) Can flowering plants enhance numbers of beneficial arthropods in UK apple and pear orchards? *Biocontrol Science and Technology* 14, 291–300.

Flanders, S.E. (1950) Regulation of ovulation and egg disposal in the parasitic Hymenoptera. *Canadian Entomologist* 82, 134–140.

Flint, M.L. and Gouveia, P. (2001) Management methods for IPM programs. In: Flint, M.L. and Gouveia, P. (eds) *IPM in Practice: Principles and Methods of Integrated Pest Management*. University of California, Oakland, California, pp. 91–148.

Gange, A.C. (2002) Arbuscular mycorrhizal fungi affect phytophagous insect specialism. *Ecology Letters* 5, 11–15.

Gange, A.C. and Brown, V.K. (1997) *Multitrophic Interactions in Terrestrial Systems*. Blackwell Science, Oxford.

Godfray, H.C.J. and Müller, C.B. (1998) Host–parasitoid dynamics. In: Dempster, J.P. and McLean, I.F.G. (eds) *Insect Populations in Theory and in Practice*. Kluwer Academic Publishers, Dordrecht, The Netherlands, pp. 135–168.

Gooderham, J., Bailey, P.C.E., Gurr, G.M. and Baggen, L.R. (1998) Sub-lethal effects of foliar pubescence on the egg parasitoid *Copidosoma koehleri* and influence on parasitism of potato moth. *Phthorimaea operculella. Entomologia Experimentalis et Applicata* 87, 115–118.

Graham-Bryce, I.J. (1987) Chemical methods. In: Burn, A.J., Coaker, T.H. and Jepson, P.C. (eds) *Integrated Pest Management*. Academic Press, London, pp. 113–160.

Greathead, D.J. and Greathead, A.H. (1992) Biological control of insect pests by insect parasitoids and predators: the BIOCAT database. *Biocontrol News and Information* 13, 61N–68N.

Groot, A.T. and Dicke, M. (2002) Insect-resistant transgenic plants in a multi-trophic context. *Plant Journal* 31, 387–406.

Guerrieri, E., Poppy, G.M., Powell, W., Rao, R. and Pennacchio, F. (2002) Plant-to-plant communication mediating in-flight orientation of *Aphidius ervi. Journal of Chemical Ecology* 28, 1703–1715.

Gurr, G.M. and Wratten, S.D. (1999) 'Integrated biological control': a proposal for enhancing success in biological control. *International Journal of Pest Management* 45, 81–84.

Gurr, G.M. and Wratten, S.D. (2000) *Biological Control: Measures of Success*. Kluwer Academic Publishers, Dordrecht, The Netherlands.

Gurr, G.M., Barlow, N.D., Memmott, J., Wratten, S.D. and Greathead, D.J. (2000a). A history of methodological, theoretical and empirical approaches to biological control. In: Gurr, G.M. and Wratten S.D. (eds) *Biological Control: Measures of Success*. Kluwer Academic Publishers, Dordrecht, The Netherlands, pp. 3–37.

Gurr, G.M., Wratten, S.D. and Barbosa, P. (2000b) Success in conservation biological control of arthropods. In: Gurr, G.M. and Wratten, S.D. (eds) *Biological Control: Measures of Success*. Kluwer Academic Publishers, Dordrecht, The Netherlands, pp. 105–132.

Gurr, G.M., Scarratt, S.L., Wratten, S.D., Berndt, L. and Irwin, N. (2004a) Ecological engineering, habitat manipulation and pest management. In: Gurr, G.M., Wratten, S.D. and Altieri, M.A. (eds) *Ecological Engineering for Pest Management: Advances in Habitat Manipulation for Arthropods*. CSIRO Publishing, Collingwood, Victoria, Australia, pp. 1–12.

Gurr, G.M., Wratten, S.D. and Altieri, M.A. (2004b) Ecological engineering for enhanced pest management: towards a rigorous science. In: Gurr, G.M., Wratten, S.D. and Altieri, M.A. (eds) *Ecological Engineering for Pest Management: Advances in Habitat Manipulation for Arthropods*. CSIRO Publishing, Collingwood, Victoria, Australia, pp. 219–225.

Hagen, K.S., Mills, N.J., Gordh, G. and McMurtry, J.A. (1999) Terrestrial arthropod predators of insect and mite pests. In: Bellows, T.S. and Fisher, T.W. (eds) *Handbook of Biological Control: Principles and Applications of Biological Control*. Academic Press, San Diego, California, pp. 383–503.

Haggstrom, H. and Larsson, S. (1995) Slow larval growth on a suboptimal willow results in high predation mortality in the leaf beetle, *Galerucella lineola. Oecologia* 104, 308–315.

Harmon, J.P., Ives A.R., Losey, J.E., Olson, A.C. and Rauwald, K.S. (2000) *Colemegilla maculata* (Coleoptera : Coccinellidae) predation on pea aphids promoted by proximity to dandelions. *Oecologia* 125, 543–548.

Hawkins, B.A. (1988) Species diversity in the third and fourth trophic levels: patterns and mechanisms. *Journal of Animal Ecology* 57, 137–162.

Hawkins, B.A. (1994) *Pattern and Process in Host–parasitoid Interactions*. Cambridge University Press, Cambridge.

Helyer, N., Brown, K. and Cattlin, N.D. (2003) *A Colour Handbook of Biological Control in Plant Protection*. Manson Publishing, London.

Hilbeck, A. (2001) Implications of transgenic, insecticidal plants for insect and plant biodiversity. *Perspectives in Plant Ecology Evolution and Systematics* 4, 43–61.

Hoddle, M.S. (2004) Restoring balance: using exotic species to control invasive exotic species. *Conservation Biology* 18, 38–49.

Holland, J.M., Chapman, R.B. and Penman, D.R. (1994) Effects of fluvalinate on two-spotted spider mite dispersal, fecundity and feeding. *Entomologia Experimentalis et Applicata* 71, 145–153.

Howarth, F.G. (2000) Non-target effects of biological control agents. In: Gurr, G.M. and Wratten, S.D. (eds) *Biological Control: Measures of Success*. Kluwer Academic Publishers, Dordrecht, The Netherlands, pp. 369–403.

Ibrahim, Y.B., and Yee, T.S. (2000) Influence of sub-lethal exposure to abamectin on the biological performance of *Neoseiulus longispinosus* (Acari: Phytoseiidae). *Journal of Economic Entomology* 93, 1085–1089.

Jervis, M.A. and Kidd, N.A.C. (1986) Host-feeding strategies in hymenopteran parasitoids. *Biological Reviews* 61, 395–434.

Jervis, M.A., Heimpel, G.E., Ferns, P.N., Harvey, J.A. and Kidd, N.A.C. (2001) Life-history strategies in parasitoid wasps: a comparative analysis of 'ovigeny'. *Journal of Animal Ecology* 70, 442–458.

Jervis, M.A., Lee, J.C. and Heimpel, G.E. (2004) Use of behavioural and life-history studies to understand the effects of habitat manipulation. In Gurr, G.M., Wratten, S.D. and Altieri, M.A. (eds) *Ecological Engineering for Pest Management: Advances in Habitat Manipulation for Arthropods*. CSIRO Publishing, Collingwood, Victoria, Australia, pp. 65–100.

Johnson, M.W. and Tabashnik, B.E. (1999) Enhanced biological control through pesticide selectivity. In: Bellows, T.S. and Fisher, T.W. (eds) *Handbook of Biological Control*. Academic Press, San Diego, California, pp. 297–317.

Jones, O. (1998) Lure and kill. In: Howse, P., Stevens, I. and Jones, O. (eds) *Insect Pheromones and Their Use in Pest Management*. Chapman & Hall, London, pp. 300–313.

Kashyap, R.K., Kennedy, G.G. and Farrar, R.R. Jr (1991) Mortality and inhibition of *Helicoverpa zea* egg parasitism rates by *Trichogramma* in relation to trichome/methyl ketone-mediated insect resistance of *Lycopersicon hirsutum* f. *glabratum*, Accession PI 134417. *Journal of Chemical Ecology* 17, 2381–2395.

Kean, J., Wratten, S.D., Tylianakis, J. and Barlow, N. (2003) The population consequences of natural enemy enhancement, and implications for conservation biological control. *Ecology Letters* 6, 604–612.

Keller, M.A. (1987) Influence of leaf surfaces on movements by the hymenopterous parasitoid *Trichogramma exiguum*. *Entomologia Experimentalis et Applicata* 43, 55–59.

Khan, Z.R. and Pickett, J.A. (2004) The 'push–pull' strategy for stemborer management: a case study in exploiting biodiversity and chemical ecology. In Gurr, G., Wratten, S.D. and Altieri, M. (eds) *Ecological Engineering for Pest Management: Advances in Habitat Manipulation for Arthropods*. CSIRO Publishing, Collingwood, Victoria, Australia, pp. 155–164.

Krips, O.E., Kleijn, P.W., Willems, P.E.L., Gols, G.J.Z. and Dicke, M. (1999) Leaf hairs influence searching efficiency and predation rate of the predatory mite *Phytoseiulus persimilis* (Acari: Phytoseiidae). *Experimental and Applied Acarology* 23, 119–131.

Landis, D.A., Wratten, S.D. and Gurr, G.M. (2000) Habitat management to conserve natural enemies of arthropod pests in agriculture. *Annual Review of Entomology* 45, 175–202.

Loope, L.L. and Howarth, F.G. (2003) Globalization and pest invasion: where will we be in five years? *Proceedings of the 1st International Symposium on Biological Control of Arthropods*. USDA Forest Service Health Technology Enterprise Team, *FHTET-03-05*, pp. 34–39.

Longley, M. and Jepson, P.C. (1996) The influence of insecticide residues on primary parasitoid and hyperparasitoid foraging behaviour in the laboratory. *Entomologia Experimentalis et Applicata* 81, 259–269.

Longley, M., Jepson, P.C., Izquierdo, J. and Sotherton, N. (1997) Temporal and spatial changes in aphid and parasitoid populations following applications of deltamethrin in winter wheat. *Entomologia Experimentalis et Applicata* 83, 41–52.

Lopez, R. and Ferro, D.N. (1995) Larviposition response of *Myiopharus doryphorae* (Diptera: Tachirridae) to Colorado potato beetle (Coleoptera: Chrysomelidae) larvae treated with lethal and sublethal doses of *Bacillus-Thuringiensis* Berliner subsp. *tenebrionis*. *Journal of Economic Entomology* 88, 870–874.

Luckey, T.D. (1968) Insecticide hormoligosis. *Journal of Economic Entomology* 61, 7–12.

Luckwill, L.C. (1943) *The Genus Lycopersicon; An Historical, Biological and Taxonomic Survey of the Wild and Cultivated Tomatoes,* Aberdeen University Press, Aberdeen, UK.

Lundgren, J.G., Heimpel, G.E. and Bomgren, S.A. (2002) Comparison of *Trichogramma brassicae* (Hymenoptera: Trichogrammatidae) augmentation with organic and synthetic pesticides for control of cruciferous Lepidoptera. *Environmental Entomology* 31, 1321–1339.

Mann, B.P., Wratten, S.D., Poehling, M. and Borgemeister, C. (1991) The economics of reduced-rate insecticide applications to control aphids in winter wheat. *Annual of Applied Biology* 119, 451–464.

Mazzone, P. and Viggiani, G. (1988) The side-effects of pesticides on beneficial arthropods. *Bollettino del Laboratorio di Entomologia Agraria 'Filippo-Silvestri'* 45, 59–66.

McEwen, P.K. and Kidd, N.A.C. (1995) The effects of different components of an artificial food on green lacewing (*Chrysoperla carnea*) fecundity and longevity. *Entomologia Experimentalis et Applicata* 77, 343–346.

Merfield, C.N., Wratten, S.D. and Navntoft, S. (2004) Video analysis of predation by polyphagous invertebrate predators in the laboratory and field. *Biological Control* 29, 5–13.

Metcalf, R.L. (1994) Insecticides in pest management. In: Metcalf, R.L. and Luckmann, W.H. (eds) *Introduction to Insect Pest Management.* John Wiley & Sons, New York, pp. 245–314.

Michaud, J.P. (2000) Development and reproduction of ladybeetles (Coleoptera: Coccinellidae) on the citrus aphids *Aphis spiraecola* Patch and *Toxoptera citricida* (Kirkaldy) (Homoptera: Aphididae). *Biological Control* 18, 287–297.

Michaud, J.P. (2002) Classical biological control: a critical review of recent programs against citrus pests in Florida. *Annals of the Entomological Society of America* 94, 531–540.

Mills, N.J. (1992) Parasitoid guilds, life styles, and host ranges in the parasitoid complexes of tortricoid hosts (Lepidoptera: Tortricidae). *Environmental Entomology* 21, 320–329.

Mills, N.J. (1994) Parasitoid guilds: a comparative analysis of the parasitoid communities of tortricids and weevils. In: Hawkins, B.A. and Sheehan, W. (eds) *Parasitoid Community Ecology.* Oxford University Press, Oxford, pp. 30–46.

Nordlund, D.A. (1981) Semiochemicals: a review of the terminology. In: Nordlund, D.A., Jones, R.L. and Lewis, W.J. (eds) *Semiochemicals: Their Role in Pest Control.* John Wiley & Sons, New York, pp. 13–28.

Olson, D.M., Takasu, K. and Lewis, W.J. (2005) Food needs of adult parasitoids: behavioural adaptations and consequences. In: Wäckers, F., van Rijn, P. and Bruin, J. (eds) *Plant-provided Food for Carnivorous Insects: A Protective Mutualism and Its Applications.* Cambridge University Press, Cambridge, pp. 137–147.

Orr, D.B. and Landis, D.A. (1997) Oviposition of European corn borer (Lepidoptera: Pyralidae) and impact of natural enemy populations in transgenic versus isogenic corn. *Journal of Economic Entomology* 90, 905–909.

Phipps, R.H. and Park, J.R. (2002) Environmental benefits of genetically modified crops: global and European perspectives on their ability to reduce pesticide use. *Journal of Animal and Feed Sciences* 11, 1–18.

Pimentel, D. (2004) Foreword. In: Gurr, G., Wratten, S.D. and Altieri, M. (eds) *Ecological Engineering for Pest Management: Advances in Habitat Manipulation for Arthropods.* CSIRO Publishing, Collingwood, Victoria, Australia, p. vi.

Poehling, H.M. (1989) Selective application strategies for insecticides in agricultural crops. In: Jepson, P.C. (ed.) *Pesticides and Non-target Invertebrates.* Intercept, Wimborne, UK, pp. 151–176.

Polis, G.A. and Strong, D.R. (1996) Food web complexity and community dynamics. *American Naturalist* 147, 813–846.

Powell, W. and Pickett, J.A. (2003) Manipulation of parasitoids for aphid pest management: progress and prospects. *Pest Management Science* 59, 149–155.

Price, P.W. (1975) Reproductive strategies of parasitoids. In: Price, P.W. (ed.) *Evolutionary Strategies of Parasitic Insects and Mites.* Plenum Press, New York, pp. 87–111.

Price, P.W. (1980) *Evolutionary Biology of Parasites.* Princeton University Press, Princeton, New Jersey.

Price, P.W. (1981) Semiochemcials in evolutionary time. In: Nordlund, D.A., Jones, R.L. and Lewis, W.J. (eds) *Semiochemicals: Their Role in Pest Control.* John Wiley & Sons, New York, pp. 251–279.

Price, P.W. (1991) Evolutionary theory of host and parasitoid interactions. *Biological Control* 1, 83–93.

Price, P.W. (1997) *Insect Ecology.* John Wiley & Sons, New York.

Price, P.W. (2003) *Macroevolutionary Theory on Macroecological Patterns.* Cambridge University Press, Cambridge.

Price, P.W., Bouton, C.E., Gross, P., McPheron, B.A., Thompson, J.N. and Weiss, A.E. (1980) Interactions among three trophic levels: influence of plants on interactions between insect herbivores and natural enemies. *Annual Review of Ecology and Systematics* 11, 41–65.

Rea, J.H., Wratten, S.D., Sedcole, R., Cameron, P.J., Davis, S.I. and Chapman, R.B. (2002) Trap cropping to manage green vegetable bug *Nezara viridula* (L.) (Heteroptera: Pentatomidae) in sweetcorn in New Zealand. *Agricultural and Forest Entomology* 4, 101–107.

Ripper, W.E. (1956) Effect of pesticides on balance of arthropod populations. *Annual Review of Entomology* 1, 403–438.

Robertson, J.L. and Worner, S.P. (1990) Population toxicology: suggestions for laboratory bioassays to predict pesticide efficacy. *Journal of Economic Entomology* 83, 8–12.

Roduner, M., Cuperus, G., Mulder, P. and Stritzke, J. (2003) Successful biological control of the musk thistle in Oklahoma using the musk thistle head weevil and the rosette weevil. *American Entomologist* 49, 112–120.

Romeis, J., Babendreier, D. and Wackers, F.L. (2003) Consumption of snowdrop lectin (*Galanthus nivalis agglutinin*) causes direct effects on adult parasitic wasps. *Oecologia,* 134, 528–536.

Romeis, J., Shanower, T.G. and Peter, A.J. (1999) Trichomes on pigeonpea (*Cajanus cajan* (L.) Mil. sp.) and two wild *Cajanus* spp. *Crop Science* 39, 564–569.

Rumpf, S., Frampton, C. and Dietrich, D.R. (1998) Effects of conventional insecticides and insect growth regulators on fecundity and other life-table parameters of *Micromus tasmaniae* (Neuroptera: Hemerobiidae). *Journal of Economic Entomology* 91, 34–40.

Schmale, I., Wackers, F.L., Cardona, C. and Dorn, S. (2003) Combining parasitoids and plant resistance for the control of the bruchid *Acanthoscelides obtectus* in stored beans. *Journal of Stored Products Research* 39, 401–411.

Shean, B. and Cranshaw, W.S. (1991) Differential susceptibilities of green peach aphid (Homoptera: Aphididae) and two endoparasitoids (Hymenoptera: Encyrtidae and Braconidae) to pesticides. *Journal of Economic Entomology* 84, 844–850.

Sherratt, T.N. and Jepson, P.C. (1993) A metapopulation approach to modelling the long-term impact of pesticides on invertebrates. *Journal of Applied Ecology* 30, 696–705.

Simmons, A.T. and Gurr, G.M. (2004) Trichome-based host plant resistance of *Lycopersicon* species and the biocontrol agent *Mallada signata*: are they compatible? *Entomologia Experimentalis et Applicata* 113, 95–101.

Sivasubramaniam, W., Wratten, S.D. and Frampton, C.M. (1999) Modifying the location and application rate of insecticides for carrot rust fly (*Psila rosae* F.) control in New Zealand. *International Journal of Pest Management* 45, 161–166.

Smith, D., Beattie, G.A.C. and Broadley, R.H. (1997) *Citrus Pests and Their Natural Enemies: Integrated Pest Management in Australia*. Department of Primary Industries, Brisbane, Australia.

Stark, J.D. and Banks, J.E. (2003) Population-level effects of pesticides and other toxicants on arthropods. *Annual Review of Entomology* 48, 505–519.

Stark, J.D., Banken, J.A. and Walthall, W.K. (1998) The importance of the population perspective for the evaluation of side-effects of pesticides on beneficial species. In: Haskell, P.T. and McEwen, P. (eds) *Ecotoxicology: Pesticides and Beneficial Organisms*. Kluwer Academic Publishers, Dordrecht, The Netherlands, pp. 348–362.

Stark, J.D., Vargas, R.I., Messing, R.H. and Purcell, M. (1992a) Effects of cryomazine and diazinon on three economically important Hawaiian tephritid fruit flies (Diptera: Tephritidae) and their endoparasitoids (Hymenoptera: Braconidae). *Journal of Economic Entomology* 85, 1687–1694.

Stark, J.D., Wong, T.T., Vargas, R.I. and Thalman, R.K. (1992b) Survival, longevity, and reproduction of tephritid fruit fly parasitoids (Hymenoptera: Braconidae) reared from fruit flies exposed to azadirachtin. *Journal of Economic Entomology* 85, 1125–1129.

Strong, D.R., Lawton, J.H. and Southwood, R. (1984) *Insects on Plants: Community Patterns and Mechanisms*. Harvard University Press, Cambridge, Massachusetts.

Tait, J. (1980) Treadmills and catastrophes in pest control. *New Scientist* 86, 254–256.

Theiling, K.M. and Croft, B.A. (1988) Pesticide side-effects on arthropod natural enemies: a database summary. *Agriculture, Ecosystems and Environment* 21, 191–218.

Thies, C. and Tscharntke, T. (1999) Landscape structure and biological control in agroecosystems. *Science* 285, 893–895.

Thies, C., Steffan, D.I., Tscharntke, T., Bjorkman, C. and Hamback, P. (2003) Context dependence in plant–herbivore interactions. *Oikos* 101, 18–25.

Thomas, M.B., Wratten, S.D. and Sotherton, N.W. (1991) Creation of 'island' habitats on farmland to manipulate populations of beneficial insects. *Journal of Applied Ecology* 28, 906–917.

Tomov, R. (2002) Species composition of parasitoids (Hymenoptera) on apple feeding *Phyllonorycter* (Lepidoptera: Gracillariidae) in Bulgaria. In: Melika, G. and Thuróczy, C. (eds) *Parasitic Wasps: Evolution, Systematics, Biodiversity and Biological Control.* Agroinform, Budapest, pp. 437–442.

Tomov, B.W. and Bernal, J.S. (2003) Effects of GNA transgenic sugarcane on life history parameters of *Parallorhogas pyralophagus* (Marsh) (Hymenoptera: Braconidae), a parasitoid of Mexican rice borer. *Journal of Economic Entomology* 96, 570–576.

Tomov, B.W., Bernal, J.S. and Vinson, S.B. (2003) Impacts of transgenic sugarcane expressing GNA lectin on parasitism of Mexican rice borer by *Parallorhogas pyralophagus* (Marsh) (Hymenoptera: Braconidae). *Environmental Entomology* 32, 866–872.

Townes, H. (1969) The genera of Ichneumonidae, Part I. *Memoirs of the American Entomological Institute* 11, 1–300.

Tscharntke, T. and Hawkins, B.A. (2002) *Multitrophic Level Interactions.* Cambridge University Press, Cambridge.

Turlings, T.C.J., Gouinguené, S., Degen, T. and Fritzshe-Hoballah, M.E. (2002) The chemical ecology of plant–caterpillar-parasitoid interactions. In: Tscharntke, T. and Hawkins, B.A. (eds) *Multitrophic Level Interactions.* Cambridge University Press, Cambridge, pp. 148–173.

Tylianakis, J.M., Didham, R.K. and Wratten, S.D. (2004) Improved fitness of aphid parasitoids receiving resource subsidies. *Ecology* 85, 658–666.

Ulpuh, S. and Kok, L.T. (1996) Interrelationship of *Bacillus thuringiensis* Berliner to the diamondback moth (Lepidoptera: Noctuidae) and its primary parasitoid, *Diadegma insulare* (Hymenoptera: Ichneumonidae). *Journal of Entomological Science* 31, 371–377.

Van Alphen, J.J.M. and Vet, L.E.M. (1986) An evolutionary approach to host finding and selection. In: Waage, J. and Greathead, D. (eds) *Insect Parasitoids.* Academic Press, London, pp. 23–61.

van Driesche, R.G. and Bellows, T.S. (1996) *Biological Control.* Chapman & Hall, New York.

van Emden, H.F. (1990) The interaction of host plant resistance to insects with other control measures. *Proceedings of the Brighton Crop Protection Conference, Pests and Diseases,* 939–948.

van Lenteren, J.C. (2003) Commercial availability of biological control agents. In: van Lenteren, J.C. (ed.) *Quality Control and Production of Biological Control Agents: Theory and Testing Procedures.* CAB International, Wallingford, UK, pp. 167–179.

van Lenteren, J.C. and Tommasini, M.G. (2003) Regulation of import and release of mass-produced natural enemies: a risk-assessment approach. In: van Lenteren, J.C. (ed.) *Quality Control and Production of Biological Control Agents: Theory and Testing Procedures.* CAB International, Wallingford, UK, pp. 181–204.

van Lenteren, J.C., Hale, A., Klapwijk, J.N., van Schelt, J. and Steinberg, S. (2003) Guidelines for quality control of commercially produced natural enemies. In: van Lenteren, J.C. (ed.) *Quality Control and Production of Biological Control Agents: Theory and Testing Procedures.* CAB International, Wallingford, UK, pp. 265–303.

Varley, G.C. and Gradwell, G.R. (1970) Recent advances in insect population dynamics. *Annual Review of Entomology* 15, 1–24.

Vinson, S.B. (1974) Effect of an insect growth regulator on two parasitoids developing from treated tobacco budworm larvae. *Journal of Economic Entomology* 67, 335–336.

Vinson, S.B. (1975) Biochemical coevolution between parasitoids and their hosts. In: Price, P.W. (ed.) *Evolutionary Strategies of Parasitic Insects and Mites.* Plenum Press, New York, pp. 14–48.

Waage, J.K. (1986) Family planning in parasitoids: adaptive patterns of progeny and sex allocation. In: Waage, J. and Greathead, D. (eds) *Insect Parasitoids.* Academic Press, London, pp. 63–95.

Waage, J. (1989) The population ecology of pest–pesticide–natural enemy interactions. In: Jepson, P.C. (ed.) *Pesticides and Non-target Invertebrates.* Intercept, Wimborne, UK, pp. 81–94.

Wackers, F.L. (2004) Assessing the suitability of flowering herbs as parasitoid food sources: flower attractiveness and nectar accessibility. *Biological Control* 29, 307–314.

Wade, M.R., Zalucki, M.P. and Wratten, S.D. (2004) Use of artificial food supplements in conservation biological control. *Proceedings of the 4th California Conference on Biological Control.* University of California, Berkeley, San Francisco, pp. 145–149.

Walter, D.E. (1996) Living on leaves: mites, tomenta, and leaf domatia. *Annual Review of Entomology* 41, 101–114.

Walter, D.E. and Proctor, H.C. (1999) *Mites; Ecology, Evolution and Behaviour.* University of New South Wales Press Limited, Sydney.

Wiles, J.A. and Jepson, P.C. (1994) Sub-lethal effects of deltamethrin residues on the within-crop behaviour and distribution of *Coccinella septempunctata. Entomologia Experimentalis et Applicata,* 72, 33–45.

Wiles, J.A. and Jepson, P.C. (1995) Dosage reduction to improve the selectivity of deltamethrin between aphids and coccinellids in cereals. *Entomologia Experimentalis et Applicata,* 76, 83–96.

Wratten, S.D. (1973) The effectiveness of the coccinellid beetle, *Adalia bipunctata* (L.) as a predator of the lime aphid, *Eucallipterus tiliae* L. *Journal of Animal Ecology,* 42, 785–802.

Wratten, S.D. (1994) *Video Techniques in Animal Ecology and Behaviour.* Chapman & Hall, New York.

Wratten, S.D., Edwards, P.J. and Winder, L. (1988) Insect herbivory in relation to dynamic changes in host plant quality. *Biological Journal of the Linnean Society,* 35, 339–350.

12 Ecological Considerations for the Use of Entomopathogens in Integrated Pest Management

Leslie C. Lewis

USDA–ARS Corn Insects and Crop Genetics Research Unit, Iowa State University, Genetics Laboratory, Ames, IA 50011, USA

Introduction

Entomopathogens, insect pathogens, are microbes representing numerous taxa in microbiology that cause a disease in insects. They have been researched for more than 100 years in an effort to define their mode of action, ecological range, efficacy as a crop or commodity protectant and potential as a component of integrated pest management (IPM). Pathogens were considered only as an alternative to a chemical insecticide, and were therefore used merely as substitutes for chemical insecticides. Early work to test the efficacy of entomopathogens was conducted by placing a candidate pathogen on a plant and returning days, weeks or months later to evaluate the amount of damage to the plant caused by the targeted insects. Testing was conducted on rates, formulations (liquid or granule) and application (ground or aerial). The uniqueness of the pathogens was either not recognized or ignored. One significant difference between pathogens and chemical insecticides was that most initial work with a pathogen would be conducted with product produced and formulated by the investigator instead of provided by a commercial firm. In the 1970s, commercial products formulated with *Bacillus thuringiensis* became available along with a

few other bacterium-based products. Some insect viruses also became commercially available at this time. Once the uniqueness of pathogens was recognized, evaluations became focused not only on assessing the number of dead insects and plant damage, but also on quantifying the presence of the pathogen remaining in the environment and fitness of the surviving insects.

Entomopathogens are living organisms that become a part of the ecosystem in which they are placed and not merely a toxicant to kill a targeted pest insect. They have the potential to influence the environment beyond their initial point of impact. Because they are biotic, all insect pathogens at a given point in time and space were indigenous. They, however, may have been introduced into an environment different from that in which they were originally isolated. These pathogens have a natural history of influencing and being influenced by both biotic and abiotic entities in the environment. The dynamics of entomopathogens in the environment will be discussed later in the chapter. IPM practitioners and researchers are challenged to 'know the indigenous pathogens', and design IPM programmes to harness their assets. Often, insect pathogens are thought of as synthetic insecticides by growers and agriculturalists, and nothing could be further from the truth. As will be reiterated throughout this chapter, entomopathogens are living organisms that have a documented history of causing disease in insects. They can be successfully used as crop protectants – entities to kill a pest insect on a given plant. They cannot by synthesized and many can be produced only *in vivo*. However, some bacteria and fungi can be produced by fermentation with varying degrees of success. A listing of examples of commercially available microbial insecticides (pathogens) is presented in Appendix. In this chapter specifics on application methods, timing, dosages and spectra of activity will not be covered; I refer readers to the volumes edited by Lacey (1997) and Lacey and Kaya (2000) for information on these issues.

Classical Description of Mode of Action of Pathogens

Protozoan/protozoan-like

There are several examples of this group of insect pathogens. They range from the well-known Microsporida with representatives such as *Nosema apis*, pathogen of the honey bee, *N. locustae*, pathogen of grasshoppers, and *N. pyrausta*, pathogen of the European corn borer (ECB), to *Sporozoa*, rather benign organisms which are disease-causing in laboratory populations of pests such as *Diabrotica* spp.

Microsporida have a relatively complex vegetative life cycle resulting in a spore that functions as the resting stage and the stage that initiates infection. *N. pyrausta*, an example of Microsporida, is an obligatory intracellular pathogen of the ECB. Infected adult females pass the spore to the egg. As the egg embryonates it becomes infected, so at eclosion the neonatal larva has an infection primarily throughout the midgut, malpighian tubules, tracheal matrix and fat body cells. As the larva feeds and develops, the spores are multiplying within its cells causing them to rupture, disseminating spores into the haemolymph where they infect other susceptible cells. Ruptured mid-gut cells slough their contents into the gut lumen. When the larva defecates, viable spores are deposited in the feeding arena. Any non-infected larva feeding in the same arena consumes frass along with the feeding substrate and is infected with *N. pyrausta*. This type of transmission from insect to insect within a generation is horizontal transmission (Lewis, 1978).

As the infected larvae develop to pupae and adults, spores continue to multiply and infect developing tissues. The spores infect epithelial layers and stroma cells of larval ovarian tissues and trophocytes and oocytes of the adult (Sajap and Lewis, 1988). Infection in these tissues results in transovarial transmission, transmission from adult to offspring or vertical transmission.

While the previous description of a life cycle represents many Microsporida, there are multiple variations. Two examples are *Amblyospora connecticus*, a microsporidium of the mosquito *Aedes cantator*, which has an obligatory cycle in a copepod (Andreadis, 1990), and *Duboscquia dengihilli*, a microsporidium from *Anopheles hilli*, which also has horizontal transmission between mosquitoes and a copepod, *Apocyclops dengizicus* (Sweeney *et al.*, 1993). Other Microsporida have modified life cycles depending on the temperature at which the host is reared, e.g. *Vairimorpha necatrix* (Pilley, 1976). Further, some Microsporida like *Vairimorpha* are generally not vertically transmitted because infections are primarily in the fat body instead of in the reproductive tissue (Lewis *et al.*, 1982).

Bacteria

In insect pathology, the most recognized bacteria are in the genus *Bacillus*. The most recognized species is *thuringiensis* with several recognized subspecies. Most of these bacteria are indigenous to ecologically diverse environments and are more readily associated with an insect in a mass-rearing system in an insectary than in the wild. For descriptive purposes, the life cycle of *B. thuringiensis* involves a vegetative stage, sporulation and formation of a protein crystal. To become affected, an insect must consume the crystalline protein, which is digested by proteases in the gut into active proteins that bind to sites on the midgut of the insect. Once bound, the protein causes disruption of the basement membrane allowing leakage of gut contents into the haemocoel, which results in septicaemia and death of the insect. Rarely does *B. thuringiensis* cycle through insects in nature, but rather in an insectary environment such as that in which Berliner (1915) discovered this bacterium in the Mediterranean flour moth. Currently, *B. thuringiensis* is commercially manufactured for several commodities and this bacterium produces the protein used in many transgenic crops (Gordon-Kamm *et al.*, 1990). Other bacteria within the genus *Bacillus* that are significant entomopathogens are *B. sphaericus*, (Ramoska *et al.*, 1977), a crystalline toxin producer that kills mosquitoes, and *B. popillae*, which causes 'milky disease' in larvae of the Japanese beetle *Popilla japonica* (Dutky, 1963). This latter bacterium does not produce a crystalline protein but the spore penetrates the gut wall.

Viruses

There are many viruses infecting insects. In general, they are either DNA- or RNA-infective units surrounded by a protein occlusion body (OB). The protein body protects the infective agent, the virion, from degradation by the environment. When a susceptible insect larva eats an OB, the OB protein is degraded by gut proteases, freeing virions, which penetrate the midgut cells of the insect. The virions invade the cell nucleus, and multiply causing cell rupture. Virions are then passed into the haemocoel, invade and form OBs in additional susceptible tissues including the tracheal matrix, fat body and hypodermis. An infection in the hypodermis causes rupture of the intersegmental membranes and oozing of the larval contents to the exterior of the insect. In nature once this fracture begins, the larva 'melts' and contents containing OB and free virions coat plant surfaces, infect additional

larvae feeding in the arena and complete the viral cycle. Agip MNPV from the black cutworm and AfMNPV from the celery looper are notable representatives of the nucleopolyhedroviruses (Hostetter and Putler, 1991; Boughton *et al.*, 2001). The nucleopolyhedrosis viruses, the cytoplasmic viruses, which infect the cytoplasm, and the granulosis viruses are all occluded viruses. Others such as iridescent viruses, picornaviruses and parvoviruses are non-occluded and less known in suppressing insect populations.

Fungi

Fungi are ubiquitous indigenous organisms and are predominantly soil-borne. Entomopathogenic fungi can be significant mortality factors or they can occasionally cause incidental, academically curious infections of many species of insects, both pests and non-pests including beneficial insects. A conidium, or 'resting stage', infects an insect by landing on the cuticle, germinating, secreting chitinases and invading the haemocoel. Once in the haemocoel, mycelial growth proliferates and the fungus consumes the contents of the insect. Once the nutrients are depleted, the mycelia grow back out through the cuticle, develop conidiophores, which in turn produce conidia covering the cadaver. These conidia can continue the infection cycle by coming in to contact with additional insects in the vicinity, or possibly be moved to greater distances by wind and rain and form additional reservoirs of inoculum. *Beauveria bassiana* is common in insects in row crops, *Nomuraea rileyi* in soybean-feeding *Lepidoptera* and *Zoopthora phytonomi* in the lucerne weevil.

Ecosystem Interactions with Pathogens

Entomopathogens being living organisms are a dynamic part of the ecosystem in which they occur, regardless of whether they were indigenous, introduced or augmented. In this section trophic interactions of entomo-pathogens will be discussed.

Protozoan/protozoan-like

Microsporida are unique among pathogens. They are obligatory intracellular parasites, chronic in their effect on a host insect and analogous to the common cold in humans. An infection renders an individual less fit and vulnerable to environmental stresses, biotic and abiotic.

N. pyrausta significantly compromises an ECB larva's capacity to feed on maize plants. ECB larvae are a primary pest of maize, feeding on unfurling leaves (whorl) and also behind the leaf-sheath collar in later-stage maize, and subsequently boring into the stalk. Some plants have native resistance to such feeding, whereas others are intermediate to very susceptible. An insect infected with N. pyrausta feeding on a susceptible plant damages the plant equivalent to a 'healthy' insect feeding on a resistant plant. There is little to no feeding by infected larvae on resistant plants (Lewis and Lynch, 1976; Lynch and Lewis, 1976). Although data are not readily available, most likely fitness to establish on host plants by other phytophagous insects is readily compromised by Microsporida.

Several biotic factors may influence the action of pathogens. In a study that simulated cool, wet conditions during oviposition on ECB fecundity, it was determined that cold temperatures greatly increased the impact of N. pyrausta. ECB adults infected with N. pyrausta held at 16°C for 7 days prior to oviposition produced 50% fewer eggs than those not infected and held at 16°C for 7 days. In the most extreme case, infected adults produced 97% fewer eggs when held at 16°C for 7 days than non-infected adults held at 27°C. This illustrates the effect of the environment (temperature), further compromising the fecundity of an infirmed insect. These conditions most likely occur in nature and with other microsporidian-infected insects.

Infection of 'non-target' species is minimal as Microsporida are generally host-spe-cific. There are, however, trophic interactions involving hosts, predators and parasitoids. Empirical data quantifying such interactions are predominantly laboratory-generated. In laboratory experiments ECB and black cutworm eggs are suitable prey for *Chrysoperla carnea* (Obrycki *et al.*, 1989). When *C. carnea* larvae feed on ECB eggs infected with N. pyrausta, there is no detrimental effect on the lacewing; however, the N. pyrausta spores pass through the *C. carnea* gut remaining infective to neonatal ECB (Sajap and Lewis, 1989). This is one scenario where an inoculum of *Nosema* can be moved away from the vicinity of the phytophagous insect depending on the searching behaviour of the predator.

Parasitoid introduction to manage ECB has had limited success. Some do, however, exist and provide some reduction of ECB populations (Lewis, 1982). The critical point is that they can coexist with N. pyrausta. *Macrocentrus cingulum* (= *M. grandii*), a polyembrionic parasitoid, can develop within a host infected with N. pyrausta with varying impacts on the parasitoid ranging from little to no effect (Cossentine and Lewis, 1987), to reducing the *M. cingulum* population (Andreadis, 1980, 1982) to being transmitted during the act of parasitism by an infected parasitoid to an ECB in the filial generation (Siegel *et al.*, 1986). N. pyrausta coexists with an egg parasitoid *Trichogramma nubilale*. *T. nubilale* will parasitize N. pyrausta–infected eggs, which embryonate and produce viable adults, which in turn parasitize the filial generation (Sajap and Lewis, 1988). There are no data supporting transmission of N. pyrausta by an infected egg parasitoid.

Bacteria

In general *B. thuringiensis* is not dynamic unless associated with insects in a mass-rearing environment; it kills the target insect and the cycle ends. An exception is *B. thuringiensis* subsp. *dendrolimus* applied to control the Siberian silkworm in coniferous forests in Russia (Talalaev, 1958, 1959; Talalayeva, 1966a). Insects that received a sublethal dosage developed into mature larvae and/or pupae before dying. The dead insects decomposed the following spring and the bacterial spores were washed to lower tree branches. These remaining spores were consumed by young larvae of the subsequent generation and the cycle continued. A naturally occurring epizootic has been reported in larvae of *Selenephera lunigera*, also a forest lepidopteran (Talalayeva, 1966b).

Viruses

Nucleopolyhedroviruses are maintained and facilitated in an ecosystem by several biotic factors. Maintenance of the gypsy moth virus in the forest ecosystem is an excellent example. It is documented that neonatal larvae obtain infection from OBs left on tree bark from previous generations (Woods *et al.*, 1989), from egg mass surfaces and from infected larvae (Woods *et al.*, 1990). Abiotic factors including wind, rain and surface water provide mechanical movement and are instrumental in dispensing the virus. This is well documented in a review by Hostetter and Bell (1985).

Fungi

The dynamics of fungi are frequently cryptic, i.e. an epizootic will occur, and in some instances be unnoticed, whereas others are obvious. The green clover worm is an occasional economic pest of soybean (Pedigo *et al.*, 1983). In outbreak years the fungus *N. rileyi* is a key mortality factor. Studies show that such epizootics occurred too late in the cropping season to prevent economic damage to full-grown soybeans. *N. rileyi* overwinters in southern climes; thus epizootics are initiated earlier in the growing season. Under controlled experimentation, the fungus does overwinter in the northern soybean-growing areas; thus the potential is for earlier epizootics (Thorvilson *et al.*, 1985).

B. bassiana is widely distributed in nature, infecting many insects. Field ento-mologists and IPM practitioners likely have seen many insects that have died from an infection of this fungus. Documentation of interactions with the ecosystem is mainly with dependency on temperature and moisture for maximum efficiency (Carruthers *et al.*, 1985). *B. bassiana* is present in the soil and in a maize system the tillage practice has little influence (Bing and Lewis, 1993), although ground cover does influence the quantity of conidia dispersed by rainfall onto the phylloplane of vegetative-stage plants (Bruck and Lewis, 2002a).

Entomophaga maimaiga, a fungal pathogen of the gypsy moth, is very dependent on moisture for expression (Elkinton *et al.*, 1991). Others have reported that *E. maimaiga* has narrow abiotic parameters, contact time between conidia and host, temperature and moisture for maximum expression (Hajek *et al.*, 1993). Most likely these requirements were paramount in initial limited success in establishing this fungus in the gypsy moth forest ecosystem.

Copious species of insects other than pests inhabit any ecosystem. Little is known about dynamics of insect or fungi occurrences unless an insect with mycosis is collected. Experimentally, members of the fungus beetle group *Nitidulidae* will consume *B. bassiana* and excrete viable conidia in their faecal pellets (Bruck and Lewis, 2002b). Ostensibly, there are many undocumented occurrences in nature in which a pathogen is fortuitously vectored.

Pathogens and IPM

Performance of entomopathogens is very dependent on the species of an organism. All pathogens are living organisms and are as mentioned in this chapter vulnerable to environmental conditions of temperature, moisture and radiation. In general, longevity of activity compared with that of a chemical insecticide is shorter, and efficacy ranges from less efficacious to equal and sometimes greater.

Data on efficacy are published in peer-reviewed journals and retrievable when searched as a certain pathogen or sometimes by trade name. Comparisons with chemical insecticides are not included in all publications. Numerous research efforts are underway to develop new approaches, new methods of insect management not merely tests to determine percentage of insect kill relative to a chemical insecticide.

Timing of application is critical; however, in many instances it is not that different from conventional means of control. Timing, as in any crop protection scenario, is dependent on occurrence of pest, age of pest and whether or not the pest will contact the pathogen. Most groups of pathogens must be consumed to be effective, the fungi being an exception.

Pathogens as crop protectants

A multitude of insect pathogens is indigenous to many ecosystems including row crop and other production and recreational systems. As one examines the concept of IPM, the definition is usually expressed as using available components in concert to suppress a pest before economic levels are reached. The use of a pathogen as a component of IPM is dependent on acceptability. Assessing the attitudes of producers on the use of pathogens (microbial insecticides) is a difficult task; however, a survey conducted in 1991 gives some insight. Producers' responses were highly variable, ranging from no knowledge of pathogens to being enthusiastic about their use. A general finding of the survey is that the average producer would use pathogens if they were no more expensive than chemical insecticides, and would provide equal control (Pingel, 1991). The main interest in using pathogens was to reduce the environmental contamination. There is a renewed interest in organic produce, not only for fresh produce but also for meat and dairy products. Ostensibly these producers will use microbial insecticides in production of produce and production of animal feed.

For several years many pathologists and entomologists have spent numerous hours and dollars isolating, identifying, modifying, formulating, applying and evaluating pathogens to be components for use in this type of approach. There are several success stories using this approach. Examples of selected success stories follow.

Microsporida as crop protectants

Microsporida have been investigated as possible components for plant protection, with majority of the work in small plots with non-sophisticated formulation and application. Using backpack sprayers as application equipment, researchers demonstrated efficacy of N. pyrausta in reducing ECB damage in maize (Lewis and Lynch, 1978). Other experiments in small plots using Microsporida are N. varivestis and N. epilachnae to suppress the Mexican bean beetle (Brooks, 1986), N. locustae against the mormon cricket (Henry, 1982), V. necatrix against the black cutworm (Cossentine and Lewis, 1986; Grundler et al., 1987) and Nosema spp. against a stored-product beetle (Khan and Selman, 1989).

In experiments covering several hectares, N. locustae was mass-produced from infected grasshoppers, formulated in a bait and applied to rangeland to reduce the number of grasshoppers (Henry et al., 1973). In some trials up to 60% reduction of grasshoppers was obtained. Also, N. locustae has been evaluated for sustainability, applied and evaluated yearly for three consecutive years (Bomar et al., 1993). Initially significant reductions were obtained; however, after 3 years no differences in density of grasshoppers were recorded between treatment and control plots. Microsporida have potential as pest suppressants. They are generally host-specific but can have negative effects to other biotics in the trophic system, i.e. parasitoids. They are sustainable but commercialization is limited due to lack of efficient in vitro production technology.

Bacteria as crop protectants

For many years, B. thuringiensis was the bacterium evaluated for, and utilized as, a crop protectant. Study reports range from very primitive procedures of harvesting

B. thuringiensis from an agar plate, mixing with talc and applying to maize (Metalnikov and Chorine, 1929) to commercial formulations of granules that were applied to maize using conventional application equipment (Lynch *et al.*, 1977a, b). Formulation technology developed such that improved efficacy was obtained by addition of feeding stimulants and UV protectants (McGuire *et al.*, 1990, 1994). These developments in formulation and application technology provided a product that was equal to recommended insecticides in control of ECB on field maize (McGuire *et al.*, 1994).

Similar developments occurred in forest entomology, i.e. selection of isolates, formulation and application. Morris (1982) documents these accomplishments. More recent advancements include defining susceptibility of instars of spruce budworm to *B. thuringiensis* (Masse *et al.*, 2000), to research defining timing and potency of application against the spruce budworm (Bauce *et al.*, 2004).

B. thuringiensis has also been a dependable plant protection product in the vegetable industry. It is a choice of producers because of efficacy and safety to personnel and non-targets. As with other commodities, research has focused on improved formulations, timing and application technology (Behle *et al.*, 1997; Tamez-Guerra *et al.*, 2000).

B. thuringiensis has an excellent success record of controlling insects. Of the pathogens discussed herein, *B. thuringiensis* has been used on the most crops and insects. It is of course similar to a chemical insecticide in that a toxin is formulated and applied to the plant, the difference being that the toxin is produced by a bacterium instead of chemical synthesis.

B. thuringiensis has attributes of narrow host range, allowing non-targets such as insects and mite predators to remain in the system to contribute to pest suppression. It is, however, most similar to a synthetic insecticide of any pathogen group. It is not, however, indigenous or sustainable. An application lasts for a few days to a week with no potential of remaining in the ecosystem unless it is an enclosed environment.

Conventional wisdom was that insects would not develop resistance to insect pathogens; thus, regardless of the pressure put on an insect population by use of a microbial insecticide, the product would be available indefinitely. In 1985, McGaughey (1985a, b) reported resistance to a commercial formulation of *B. thuringiensis* in the Indianmeal moth feeding in grain bins. Resistance to *B. thuringiensis* has also been reported in the almond moth (McGaughey and Beeman, 1988). The first report – at the time of this writing the only confirmed report of resistance in the open field – was in the diamondback moth (Tabashnik *et al.*, 1990). Laboratory selection of resistance to *B. thuringiensis* has been reported in additional species of *Lepidoptera*, and in species of *Coleoptera* and *Diptera* (Tabashnik, 1994).

As mentioned earlier, gene(s) from subspecies of *B. thuringiensis* are the basis for the insecticidal properties of several species of transgenic plants. Several entomologists are conducting research to determine the genetic possibility of insects to develop resistance to these plants.

Viruses as crop protectants

Insect viruses have been extensively used in insect management, although more often in a forest system than with conventional row crops. As with many microbials, the economic issue is critical, specifically cost of product compared with a recommended chemical insecticide. For environmental reasons and because *Helicoverpa zea* was developing resistance to the recommended insecticide, the virus isolated from the earworm was commercialized to control *H. zea* on cotton. The *H. zea* virus came to market at the time synthetic pyrethrins were being introduced; because of high costs the virus was never widely used.

The virus of the gypsy moth has been used to manage this pest insect in a forest system (Webb *et al.*, 1999). There are numerous examples of viruses being used to control insect pests, primarily in small-scale experiments with highly variable results, e.g. beet armyworm (Kolodny-Hirsch *et al.*,

1997), ECB (Lewis and Johnson, 1982) and black cutworm (Johnson and Lewis, 1982).

Novel approaches have been used with viruses to increase efficacy. The use of optical brighteners was introduced by Shapiro (1992) to protect the gypsy moth virus from radiation, and has been used by others to enhance activity of viruses of fall armyworm (Hamm and Shapiro, 1992), western spruce budworm (Li and Otvos, 1999) and black cutworm (Boughton *et al.*, 2001). Several insect viruses are commercially produced either *in vivo* or *in vitro*. Viruses have a narrow host range with a few exceptions. In a cropping season a virus can cycle several times, and is thus sustainable.

Fungi as crop protectants

Fungi have a poor track record as crop protection agents. Their mode of action is such that they kill a pest after damage to a commodity occurs. Thus their use in rescue treatments is very limited. However, recent successes using *B. bassiana* to suppress ECB on maize are known. Commercially produced and formulated *B. bassiana* applied as a granule significantly reduced plant damage (Lewis *et al.*, 2002). Efficacy of *B. bassiana* and most likely other entomopathogenic fungi is dependent on environmental temperature and humidity (Carruthers *et al.*, 1985), age of insect (Feng *et al.*, 1985) and isolate of fungus.

Fungi are an option for protecting commodities from insects, although with limited use and success. Most fungi are sustainable to some extent in that cadavers from an infection serve as a substrate and provide conidia that are available to infect subsequent insect populations. Some fungi are produced by *in vitro* methods and are commercially available, e.g. *B. bassiana*.

Alternative View of Pathogens in IPM

The above description of dynamics of pathogens in ecosystems documents their suppressant activities. As a practitioner or researcher plans an IPM programme for a pest insect, it behoves them to become as aware as possible of pathogens that infect the target pest and whether or not these pathogens exist within the ecosystem where they plan to manage the pest. A practitioner must first be aware of whether the pathogen is indigenous, what governs its cycles, when to expect an epizootic, how to conserve the pathogen and possible ways of facilitating contact between the pathogen and the target insect.

The dynamics of the pathogen must be learned in order to design an IPM programme that utilizes the indigenous pathogen. In general, the questions 'What do I know about the pathogen?' 'Is it compatible with my proposed approach?' 'Will a certain production practice conserve or maybe increase the pathogen?' need to be researched. If the data are available, practitioners should team with an investigator and define researchable problems and proceed.

The following are scenarios or descriptions of existing ecosystems comprising both documented and hypothetical situations in which a pest insect is, or could be, controlled by a pathogen. These scenarios are useful to stimulate one to develop a new paradigm: instead of conducting research to modify a pathogen to more closely resemble a chemical insecticide, conduct research within the following regimes to utilize indigenous pathogens to manage indigenous pests. Bottom line – how does one increase the probability of contact between pathogen and pest at the optimum time to maximize insect suppression?

Protozoan/protozoan-like

Microsporida, as noted earlier, are obligate intracellular parasites, and are thus dependent on their host for propagation and survival. There are, however, opportunities for trophic interactions in many ecosystems. The maize system will be used as an example.

Once an ECB oviposits an egg infected with *N. pyrausta*, that egg becomes a reservoir of infection not only for the developing embryo but also for any predator that uses the egg as prey. A lacewing larva can consume the egg with *N. pyrausta*, digest and defecate leaving viable spores, which if consumed may infect other ECB larvae feeding in the vicinity with no harm to the lacewing larva.

This only breaks the surface of what most likely occurs with interactions with Microsporida. As mentioned heretofore, cold, wet weather greatly magnifies the effects of an infection with *N. pyrausta* in ECB. With this in mind consider that in nature there are tremendous fluctuations in temperature and moisture that influence insects already infirmed by an infection of a microsporidium. Although not supported by empirical data it is witnessed annually that newly emerged ECB adults are 'grounded' because of cool spring conditions. As they await ambient temperatures sufficient for flight and success in mating and finding oviposition substrates, these insects are constantly plagued by increasing titers of Microsporida that consume energy reserves, infect ovarian tissues and accessory organelles and decrease the ovipositional potential. Once they do oviposit, the magnitude of oviposition and fitness of eggs is greatly reduced.

Horizontal transmission of a microsporidium is key to the survival of Microsporida, and to suppressing insect populations. In the maize system, ECB larvae routinely migrate from plant to plant within the row. In so doing, when a larva infected with *N. pyrausta* moves to an adjacent plant, it will deposit faeces contaminated with viable spores, which in turn are available to infect other larvae. Also, non-infected larvae do move to adjacent plants inhabited by infected larvae and obtain an infection of *N. pyrausta* (Lewis, 1978).

Horizontal transmission by definition is among insects in a generation. There are times when transmission is among generations. Faecal material from the larva of a first-generation ECB infected with *N. pyrausta* remains in a maize plant and is an inoculum to larvae of the second generation when eggs are oviposited on this plant (Lewis and Cossentine, 1986).

These interactions are important but one must also be cognizant of compatibility with other biological control components. Predators and parasitoids, which are members of any ecosystem, are subject to exposure to Microsporida by consuming microsporidian-infected prey and parasitizing infected hosts, respectively. Predators generally do not obtain an infection from infected prey, but they can mechanically move inoculum when defecating following consumption of infected prey. Parasitoids readily parasitize infected hosts and they may move Microsporida to additional hosts on a contaminated ovipositor, most likely not a large inoculum source, but it can nevertheless occur. As research continues in this fascinating arena of trophic interactions, additional phenomena are documented, e.g. *V. necatrix* spores accumulating in the midgut of the parasitoid *Meteorus gyrator* developing in an infected host, *Lacanobia oleracea* and subsequently *V. necatrix* spores are excreted in the parasitoid meconium upon pupation. These spores are then available to infect another lepidopteran host (Down *et al.*, 2005).

Bacteria

The dynamics of entomopathogenic bacteria within an ecosystem, especially row crop agriculture, are limited. Once the product is placed on a plant it is inactive in a matter of days. Of course many subspecies of *B. thuringiensis* have been isolated from the environment primarily from the soil; however, definitive data on environmental interactions are not known. The bacteria as indigenous organisms to be exploited in an IPM plan are of little significance.

Viruses

Insect viruses for the most part are specific to a given insect or at the most an additional insect in the same genus or family. This concept of very narrow host range changed when a nucleopolyhedrosis virus was isolated from the lucarne looper, *Autographa californica* (Vail and Jay, 1973). Cross-infectivity will occur between susceptible species, i.e. *A. californica* to the cabbage looper, *Trichoplusia ni*, if insects are feeding on the same substrate.

Viremic insects rupture and OBs contaminate plant parts and soil. In some cases, the inoculum remains in the environment until another insect feeds on the same plant and becomes infected, or until a new crop is seeded. In the latter, a seedling could be contaminated as it emerges or by splashing during a rain (Jaques, 1985).

Insect viruses are also moved within, and out of, an ecosystem by vertebrate predators. English sparrows are known to feed on cabbage looper larvae infected with the cabbage looper virus. Faeces from these sparrows contained virus infective to third instar cabbage loopers (Hostetter and Biever, 1970). Several species of birds too are documented predators of ECB: robin, crow, blackbird, grackle, chickadee and ring-necked pheasant (Baker *et al.*, 1949). Although not reported, these birds most likely mechanically move pathogens of ECB and of other insects they consume.

Fungi

Some fungi have a narrow or even species-specific host range, whereas some have a broad host range. *B. bassiana* is one of these. It is a common mortality factor of *Diabrotica* adults, ECB larvae and sap beetle adults (nitidulids) in the maize ecosystem. Also, this fungus routinely infects other insects, e.g. coccinellids, and forms an endophytic relationship with maize, killing ECB larvae within the stalk and overwintering in maize residue. Thus, the plant serves as a substrate, increasing inoculum of *B. bassiana*. This inoculum kills a large percentage of ECB postharvest and reduces other maize insect pest densities. In most fields in most years, *B. bassiana* is a major mortality agent of maize pests throughout the Midwestern maize belt.

Both the western maize rootworm, *Diabrotica virgifera virgifera*, and the northern maize rootworm, *D. barberi*, are primarily pests of maize, and both species have similar life cycles (Chiang, 1973). Adult female beetles oviposit in the soil in the fall. The eggs overwinter in diapause, hatch in the spring and larvae feed on the roots of maize plants. The most severe damage caused by rootworms is usually by larval feeding on maize roots. Larvae pupate in the soil, and adults emerge in late summer, mate and oviposit in the soil. Adult feeding on maize silks can be a severe problem in seed maize production, but is less a concern in field maize. Until 1994, adult beetles were thought to oviposit primarily in maize fields, and the rotation of maize and soybeans was an effective strategy for reducing or eliminating the possibility of rootworm damage to maize in rotation with soybeans. In 1995, many documented cases of rootworm oviposition in soybean fields and subsequent damage to first-year maize occurred in eastern Illinois (Gray and Levine, 1996). Therefore, maize–soybean rotations may be a much less successful control strategy for rootworms in the future. Without rotation, most Midwestern IPM programmes would collapse and rootworm 'management' would be reduced to a calendar–insecticide application approach, with dramatic and predictable environmental damage.

B. bassiana has been applied experimentally to the soil as a microbial insecticide for control of rootworm larvae (J.V. Maddox and S.T. Jaronski, 1985, unpublished results) and released inoculatively in an attempt to initiate an epizootic in adult rootworm populations (Day, 1986). Neither of these attempts resulted in acceptable rootworm control. Nevertheless, *B. bassiana* epizootics occur regularly in adult rootworm populations, and it is not uncommon to see over 50% of the adult beetles in a maize field infected with *B. bassiana* (J.V. Maddox and L.C. Lewis, 1985, unpublished observations). Observations such as these are anecdotal; we do not know where the adult beetles acquired the *B. bassiana* infections or the ecological significance of the infections.

The ECB, the primary above-ground pest of maize, has two generations in many parts of the US maize belt. The adults of the first generation emerge in late May to early June and oviposit on whorl-stage maize and weeds. Young larvae feed externally and late fourth and early fifth instars bore into stalks of weeds and maize where they pupate. The second-generation adult emerges in late July to early August. Again, eggs are oviposited primarily on maize and weeds, and the larvae feed externally until they enter the maize stalk. The ECB overwinters in a state of diapause as a mature fifth instar.

There are many studies of the association between ECB and *B. bassiana*. Eggs and neonatal larvae are susceptible to fungal infections. Later instars are routinely infected with *B. bassiana* late in the growing season, dying from a mycosis either in late summer and in early autumn or prior to pupation the following spring (Lewis, 1995).

An endophytic relationship between *B. bassiana* and maize defines a mechanism in which the fungus growing within the maize plant penetrates and kills ECB inside the stalk. This relationship increases the *B. bassiana* inoculum in crop residue and subsequently the mortality suffered by overwintering ECB larvae (Lewis, 1995). *B. bassiana* also occurs naturally in the soil of different tillage systems (plow, chisel, no-tillage) (Bing and Lewis, 1993).

The picnic beetle, *Glischrochilus quadrisignatus*, is common to the maize ecosystem. Eggs are oviposited in the soil in May to June, and the neonatal larvae develop on decomposing plant material (Luckman, 1963). Larvae pupate in the soil. Adults emerge in July to August and feed on maize silks, rotting fruits and occupy tunnels made by ECB (McCoy and Brindley, 1961; Luckman, 1963; Pree, 1968; Carlson and Chiang, 1973; Windels *et al.*, 1976). The beetle is associated with *B. bassiana*. In a recent study in Wisconsin, up to 37% of the beetles were infected with *B. bassiana* (Schell, 1994). Adults contaminated with *B. bassiana* conidia can transmit the fungus to maize plants allowing an endophytic relationship to develop. The role of *G. quadrisignatus* in the infection of ECB and *Diabrotica* sp. with *B. bassiana* is not well understood.

Several *Coccinellidae* are routinely found in the maize ecosystem. Adults oviposit on vegetation within and outside the maize field. Once the larvae eclose, they feed on maize pollen, small arthropods and insect eggs. Pupation occurs primarily on leaf surfaces and the adults feed on prey similar to that consumed by the larvae. Coccinellids are known to be susceptible to *B. bassiana* (Goettel *et al.*, 1990) but different isolates (pathotypes) vary in their ability to infect coccinellids (Magalhaes *et al.*, 1988; Pingel and Lewis, 1996).

Several species of insects that routinely inhabit the maize ecosystem have close and enigmatic relationships with *B. bassiana*. Nitidulids, rootworms and coccinellids may be adversely affected but may also be instrumental in mechanically transferring the fungus within the ecosystem. ECB also directly infected by the fungus may act as a mechanical vector within the maize stalk but is less likely to transfer fungal conidia to other fields. In the future we need to further define the roles of these insects in the ecology of *B. bassiana*, the role of *B. bassiana* as a natural mortality factor of pest insects, and the effect of changing cropping systems and ovipositional behavioural changes in *Diabrotica* spp. on these interactions.

Conclusion

There are documented methods to utilize indigenous pathogens to manage a pest. Some require slight changes in a production practice, which in turn conserves the pathogen. Populations of the lucerne weevil are subject to epizootics of *Zoophthora phytonomi* (Giles and Obrycki, 1997). By leaving uncut strips of lucerne following the first cutting of the crop, an environment is provided in which mycotic weevils die and reservoirs of inoculum are established in the field. These reservoirs produce fungal conidia, which are dispersed by abiotic factors and infect weevils in the regrown lucerne.

As noted earlier, some fungi are soil inhabitants. In experimentation to evaluate recombinant *Metarhizium anisopliae* in soil it was discovered that this fungus was rhizosphere-competent (Hu and St Leger, 2002). This concept has been employed by Bruck (2005) to develop protocols to use

M. anisopliae in potting media to suppress the black vine weevil, a significant pest in the nursery industry.

The concept that an entomopathogenic fungus can colonize the rhizosphere was the impetus for research to quantify colonization of the maize rhizosphere with the ultimate goal of suppressing the larval Corn rootworm. Additional studies are underway to evaluate the possibility of rhizosphere colonization leading to the formation of an endophyte in the above-ground portion of the plant to suppress ECB.

The original work on *B. bassiana* as an endophyte was published in the early 1990s (Lewis and Bing, 1990; Bing and Lewis, 1991). There are reports of endophytism by *B. bassiana* in potato (Jones, 1994), tomato (Leckie, 2002), a relative of cocoa (Evans *et al.*, 2003), and it has recently been isolated from coffee plants in Columbia (Posada and Vega, 2006).

As a final point, keep in mind that there is chaos in almost any ecosystem. For example, a phytophagous insect colonizing a plant is vulnerable not only to any microbes in the environment but also predators, parasitoids and other pathogens. The biotic organisms are also vulnerable to each other. In addition, abiotic forces such as temperature, moisture, radiation, agronomic practices and the unknown influence all possible interactions among biotic organisms. Rather than developing pathogens to be applied components of IPM through molecular and other technologies, one should understand the ecology of the indigenous pathogens and exploit their attributes to manage a pest insect.

Acknowledgements

I thank Miriam Lopez, USDA–ARS, for helpful commentary on ideas, concepts and content; Maggie Lewis, Northwestern University, Department of Learning and Organizational Change, for editorial comments; and Edwin C. Berry, Iowa State University, for technical comments.

References

Andreadis, T.G. (1980) *Nosema pyrausta* infection in *Macrocentrus grandii*, a braconid parasite of the European corn borer, *Ostrinia nubilalis*. *Journal of Invertebrate Pathology* 35, 229–233.

Andreadis, T.G. (1982) Impact of *Nosema pyrausta* on field populations of *Macrocentrus grandii*, an introduced parasite of the European corn borer, *Ostrinia nubilalis*. *Journal of Invertebrate Pathology* 39, 298–302.

Andreadis, T.G. (1990) Epizootiology of *Amblyspora connecticus* (Microsporida) in filial populations of the saltmarsh mosquito, *Aedes cantator*, and the cyclopoid copepod, *Acanthocyclopas vernalis*. *Journal of Protozoology* 37, 174–182.

Baker, W.A., Bradley, W.G. and Clark, C.A. (1949) Biological control of the European corn borer in the United States. *US Department of Agriculture Technical Bulletin* 983, USDA, Washington, DC, 185.

Bauce, E., Carisey, N., Dupont, A. and Van Frankenhuyzen, K. (2004) *Bacillus thuringiensis* subsp. *kurstaki* aerial spray prescriptions for balsam fir stand protection against spruce budworm (Lepidoptera: Tortricidae). *Journal of Economic Entomology* 97, 1624–1634.

Behle, R.W., McGuire, M.R. and Shasha, B.S. (1997) Effects of sunlight and simulated rain on residual activity of *Bacillus thuringiensis* formulations. *Journal of Economic Entomology* 90, 1560–1566.

Berliner, E. (1915) Über die Schlaffsucht der Mehlmottenraupe (*Ephestia kuhniella*, Zell.) und ihren Erreger *Bacillus thuringiensis*, n. sp. *Zeitschrift Angewandte Entomologie* 2, 29–56.

Bing, L.A. and Lewis, L.C. (1991) Suppression of *Ostrinia nubilalis* (Hübner) (Lepidoptera: Pyralidae) by endophytic *Beauveria bassiana* (Balsamo) Vuillemin. *Environmental Entomology* 20, 1207–1211.

Bing, L.A. and Lewis, L.C. (1993) Occurrence of the entomopathogen *Beauveria bassiana* (Balsamo) Vuillemin in different tillage regimes and in *Zea mays* L. and virulence towards *Ostrinia nubilalis* (Hübner). *Agriculture, Ecosystem and Environment* 45, 147–156.

Bomar, C.R., Lockwood, J.A., Pomerinke, M.A. and French, J.D. (1993) Multiyear evaluation of the effects of *Nosema locustae* (Microsporida: Nosematidae) on rangeland grasshopper (Orthoptera: Acrididae) population density and natural biological controls. *Environmental Entomology* 22, 489–497.

Boughton, A.J., Lewis, L.C. and Bonning, B.C. (2001) Potential of *Agrotis ipsilon* nucleopolyhedrovirus for suppression of the black cutworm (Lepidoptera: Noctuidae) and effect of an optical brightener on virus efficacy. *Journal of Economic Entomology* 94, 1045–1052.

Brooks, W.M. (1986) Comparative effects of *Nosema epilachnae* and *Nosema varivestis* on the Mexican bean beetle, *Epilachna varivestis*. *Journal of Invertebrate Pathology* 48, 344–354.

Bruck, D.J. (2005) Ecology of *Metarhizium anisopliae* in soilless potting media and the rhizosphere: implications for pest management. *Biological Control* 32, 155–163.

Bruck, D.J. and Lewis, L.C. (2002a) Rainfall and crop residue effects on soil dispersion and *Beauveria bassiana* spread to corn. *Applied Soil Ecology* 20, 183–190.

Bruck, D.J. and Lewis, L.C. (2002b) *Carpophilus freemani* (Coleoptera: Nitidulidae) as a vector of *Beauveria bassiana*. *Journal of Invertebrate Pathology* 80, 188–190.

Carlson, R.E. and Chiang, R.C. (1973) Reduction of an *Ostrinia nubilalis* population by predatory insects attracted by sucrose sprays. *Entomophaga* 18, 205–211.

Carruthers, R.I., Feng, Z., Robson, D.S. and Roberts, D.W. (1985) *In vivo* temperature-dependent development of *Beauveria bassiana* (Deuteromycotina: Hyphomycetes) mycosis of the European corn borer, *Ostrinia nubilalis* (Lepidoptera: Pyralidae). *Journal of Invertebrate Pathology* 46, 305–311.

Chiang, R.C. (1973) Bionomics of the northern and western corn rootworms. *Annual Review of Entomology* 18, 47–72.

Cossentine, J.E. and Lewis, L.C. (1986) Impact of *Vairimorpha necatrix* and *Vairimorpha* sp. (Microspora: Microsporida) on *Bonnetia comta* (Diptera: Tachinidae) within *Agrotis ipsilon* (Lepidoptera: Noctuidae) hosts. *Journal of Invertebrate Pathology* 47, 303–309.

Cossentine, J.E. and Lewis, L.C. (1987) Development of *Macrocentrus grandii,* Goidanich, within microsporidian-infected *Ostrinia nubilalis* (Hübner) host larvae. *Canadian Journal of Zoology* 65, 2532–2535.

Day, E.R. (1986) Studies on the susceptibility of maize rootworms to infection by *Beauveria bassiana*. MS thesis, Department of Entomology, University of Illinois, Urbana-Champaign, Illinois.

Down, R.E., Smethurst, F., Bell, H.A. and Edwards, J.P. (2005) Interactions between the solitary endoparasitoid, *Meteorus gyrator* (Hymenoptera: Braconidae) and its host, *Lacanobia oleracea* (Lepidoptera: Noctuidae), infected with the entomopathogenic microsporidium, *Vairimorpha necatrix* (Microspora: Microsporidia). *Bulletin Entomological Research* 95, 133–144.

Dutky, S.R. (1963) The milky diseases. In: Steinhaus, E.A. (ed.) *Insect Pathology*. Academic Press, New York, pp. 75–115.

Elkinton, J.S., Hajek, A.E., Boettner, G.H. and Simons, E.E. (1991) Distribution and apparent spread of *Entomophaga maimaiga* (Zygomycetes: Entomophthorales) in gypsy moth (Lepidoptera: Lymantriidae) populations in North America. *Environmental Entomology* 20, 1601–1605.

Evans, H.C., Holmes, K.A. and Thomas, S.E. (2003) Endophytes and mycoparasites associated with an indigenous forest tree, *Theobroma gileri*, in Ecuador and a preliminary assessment of their potential as biocontrol agents of cocoa diseases. *Mycology Progress* 2, 149–160.

Feng, Z., Carruthers, R.I., Roberts, D.W. and Robson, D.S. (1985) Age-specific dose–mortality effects of *Beauveria bassiana* (Deuteromycotina: Hyphomycetes) on the European corn borer, *Ostrinia nubilalis* (Lepidoptera: Pyralidae). *Journal of Invertebrate Pathology* 46, 259–264.

Giles, K.L. and Obrycki, J.J. (1997) Reduced insecticide rates and strip-harvesting effects on lucerne weevil (Coleoptera: Curculionidae) larval populations and prevalence of *Zoophthora phytonomi* (Entomophthorales: Entomophthoraceae). *Journal of Economic Entomology* 90, 933–944.

Goettel, M.S., Poprawski, T.I., Vandenberg, I.D., Li, Z. and Roberts, D.W. (1990) Safety to nontarget invertebrates of fungal biocontrol agents. In: Laird, M., Lacey, L.A. and Davidsojn, E.W. (eds) *Safety of Microbial Insecticides*. CRC Press, Boca Raton, Florida, pp. 210–231.

Gordon-Kamm, W.J., Spencer, T.M., Mangano, M.L., Adams, T.R., Daines, R.J., Start, W.G., O'Brien, J.V., Chambers, S.A., Adams, W.R., Willets, N.G., Rice, T.B., Mackey, C.J., Krueger, R.W., Kausch, A.P. and Lemaux, P.G. (1990) Transformation of maize cells and regeneration of fertile transgenic plants. *The Plant Cell* 2, 603–618.

Gray, M. and Levine, E. (1996) *First-year Maize Rootworm Injury: East-central Illinois Research Progress to Date and Recommendations for 1996*. Integrated Crop Management Conference Proceedings, Ames, Iowa.

Grundler, J.A., Hostetter, D.L. and Keaster, A.J. (1987) Laboratory evaluation of *Vairimorpha necatrix* (Microspora: Microsporidia) as a control agent for the black cutworm (Lepidoptera: Noctuidae). *Environmental Entomology* 16, 1228–1230.

Hajek, A.E., Larkin, T.S., Carruthers, R.I. and Soper, R.S. (1993) Modeling the dynamics of *Entomophaga maimaiga* (Zygomycetes: Entomophthorales) epizootics in gypsy moth (Lepidoptera: Lymantriidae) populations. *Environmental Entomology* 22, 1172–1187.

Hamm, J.J. and Shapiro, M. (1992) Infectivity of fall armyworm (Lepidoptera: Noctuidae) nuclear polyhedrosis virus enhanced by a fluorescent brightener. *Journal of Economic Entomology* 85, 2149–2152.

Henry, J.E. (1982) Experimental control of the mormon cricket, *Anabrus simplex*, by *Nosema locustae* (Microspora: Microsporida), a protozoan parasite of grasshoppers (Orthoptera: Acrididae). *Entomophaga* 27, 197–201.

Henry, J.E., Tiahrt, K. and Oma, E.A. (1973) Importance of timing, spore concentrations, and levels of spore carrier in applications of *Nosema locustae* (Microsporida: Nosematidae) for control of grasshoppers. *Journal of Invertebrate Pathology* 21, 263–272.

Hostetter, D.L. and Bell, M.R. (1985) Natural dispersal of baculoviruses in the environment. In: Maramorosch, K. and Sherman, K.E. (eds) *Viral Insecticides for Biological Control*. Adademic Press, New York, pp. 249–284.

Hostetter, D.L. and Biever, K.D. (1970) The recovery of virulent nuclear-polyhedrosis virus of the cabbage looper, *Trichoplusia ni*, from the feces of birds. *Journal of Invertebrate Pathology* 15, 173–176.

Hostetter, E.L. and Putler, B. (1991) A new broad spectrum nuclear polyhydrosis virus isolated from a celery looper, *Anagrapha falcifera* (Lepidoptera: Noctuidae). *Environmental Entomology* 20, 1480–1488.

Hu, G. and St Leger, R. (2002) Field studies using a recombinant mycoinsecticide (*Metarhizium anisopliae*) reveal that it is rhizosphere competent. *Applied and Environmental Microbiology* 68, 6383–6387.

Jaques, R.P. (1985) Stability of insect viruses in the environment. In: Maramorosch, K. and Sherman, K.E. (eds), *Viral Insecticides for Biological Control*. Adademic Press, New York, pp. 285–360.

Johnson, T.B. and Lewis, L.C. (1982) Evaluation of *Rachiplusia ou* and *Autographa californica* nuclear polyhedrosis viruses in suppressing black cutworm damage to seedling maize in greenhouse and field. *Journal of Economic Entomology* 75, 401–404.

Jones, K.D. (1994) Aspects of the biology and biological control of the European corn borer in North Carolina. PhD thesis, North Carolina State University, Raleigh, North Carolina.

Khan, A.R. and Selman, B.J. (1989) *Nosema* spp. (Microspora: Microsporida: Nosematidae) of stored-product Coleoptera and their potential as microbial control agents. In: Russell G.E. (ed.) *Management and Control of Invertebrate Crop Pests*. Andover, Hampshire, UK, pp. 133–163.

Kolodny-Hirsch, D.M., Sitchawat, T., Jansiri, T., Chenrchaivachirakul, A. and Ketunuti, U. (1997) Field evaluation of a commercial formulation of the *Spodoptera exigua* (Lepidoptera: Noctuidae) nuclear polyhedrosis virus for control of beet armyworm on vegetable crops in Thailand. *Biocontrol Science and Technology* 7, 475–488.

Lacey, L.A. (1997) *Manual of Techniques in Insect Pathology*. Academic Press, San Diego, California.

Lacey, L.A. and Kaya, H.K. (2000) *Field Manual of Techniques in Invertebrate Pathology*. Kluwer Academic Publishers, Dordrecht, The Netherlands.

Leckie, B.M. (2002) Effects of *Beauveria bassiana* mycelia and metabolites incorporated into synthetic diet and fed to larval *Helicoverpa zea*, and detection of endophytic *Beauveria bassiana* in tomato plants using PCR and ITS. MS thesis, The University of Tennessee, Knoxville, Tennessee.

Lewis, L.C. (1978) Migration of larvae of *Ostrinia nubilalis* (Lepidoptera: Pyralidae) infected with *Nosema pyrausta* (Microsporida: Nosematidae) and subsequent dissemination of this microsporidium. *Canadian Entomologist* 110, 897–900.

Lewis, L.C. (1982) Present status of introduced parasitoids of the European corn borer. *Ostrinia nubilalis* (Hübner), in Iowa. *Iowa State Journal of Research* 56, 429–436.

Lewis, L.C. (1995) Role of insect diseases in managing the European corn borer. *7th Annual Integrated Crop Management Conference*, November 29–30, 1995. Iowa State Universtiy.

Lewis, L.C. and Bing, L.A. (1990) *Bacillus thuringiensis* Berliner and *Beauveria bassiana* (Balsamo) Vuillemin for European corn borer control: potential for immediate and season-long suppression. *Canadian Entomologist* 123, 387–393.

Lewis, L.C. and Cossentine, J.E. (1986) Season long intraplant epizootics of entomopathogens, *Beauveria bassiana* and *Nosema pyrausta*, in a corn agroecosystem. *Entomophaga* 31, 363–369.

Lewis, L.C. and Johnson, T.B. (1982) Efficacy of two nuclear polyhedrosis viruses against *Ostrinia nubilalis* in the laboratory and field. *Entomophaga* 27, 33–38.

Lewis, L.C. and Lynch, R.E. (1976) Influence on the European corn borer of *Nosema pyrausta* and resistance in maize to leaf feeding. *Environmental Entomology* 5, 139–142.

Lewis, L.C. and Lynch, R.E. (1978) Foliar application of *Nosema pyrausta* for suppression of populations of European corn borer. *Entomophaga* 23, 83–88.

Lewis, L.C., Lublinkhof, J., Berry, E.C. and Gunnarson, R.D. (1982) Response of *Ostrinia nubilalis* (Lepidoptera: Pyralidae) infected with *Nosema pyrausta* to insecticides. *Entomophaga* 27, 211–218.

Lewis, L.C., Bruck, D.J. and Gunnarson, R.D. (2002) On-farm evaluation of *Beauveria bassiana* for control of *Ostrinia nubilalis* in Iowa, USA. *BioControl* 47, 167–176.

Li, S.Y. and Otvos, L.S. (1999) Optical brighteners enhance activity of a nuclear polyhedrosis virus against western spruce budworm (Lepidoptera: Tortricidae). *Journal of Economic Entomology* 92, 335–339.

Luckman, W.H. (1963) Observations on the biology and control of *Glischrochilus quadrisignatus*. *Journal of Economic Entomology* 56, 681–686.

Lynch, R.E. and Lewis, L.C. (1976) Influence on the European corn borer of *Nosema pyrausta* and resistance in maize to sheath-collar feeding. *Environmental Entomology* 5, 143–146.

Lynch, R.E., Lewis, L.C., Berry, E.C. and Robinson, J.F. (1977a) European corn borer *Ostrinia nuilalis*: granular formulations of *Bacillus thuringiensis* for biological control. *Journal of Economic Entomology* 70, 389–391.

Lynch, R.E., Lewis, L.C., Berry, E.C. and Robinson, J.F. (1977b) European corn borer control with *Bacillus thuringiensis* standardized as corn borer international units. *Journal of Invertebrate Pathology* 30, 169–174.

Magalhaes, B.P., Lord, J.C., Wraight, S.P., Daoust, R.A. and Roberts, D.W. (1988) Pathogenicity of *Beauveria bassiana* and *Zoophthora radicans* to the cocconellid predators *Coleomegilla maculata* and *Eriopis connexa*. *Journal of Invertebrate Pathology* 52, 471–473.

Masse, A., Van-Frankenhuyzen, K. and Dedes, J. (2000) Susceptibility and vulnerability of third-instar larvae of the spruce budworm (Lepidoptera: Tortricidae) to *Bacillus thuringiensis* subsp. *kurstaki*. *Canadian Entomologist* 132, 573–580.

McCoy, C.E. and Brindley, T.A. (1961) Biology of the four-spotted fungus beetle, *Glischrochilus quadrisignatus*, and its effect on European corn borer populations. *Journal of Economic Entomology* 54, 713–717.

McGaughey, W.H. (1985a) Insect resistance to the biological insecticide *Bacillus thuringiensis*. *Science* 229, 193–195.

McGaughey, W.H. (1985b) Evaluation of *Bacillus thuringiensis* for controlling Indian meal moths (Lepidoptera: Pyralidae) in farm grain bins and elevator silos. *Journal of Economic Entomology* 78, 1089–1094.

McGaughey, W.H. and Beeman, R.W. (1988) Resistance to *Bacillus thuringiensis* in colonies of Indian meal moth and almond moth (Lepidoptera: Pyralidae). *Journal of Economic Entomology* 81, 28–33.

McGuire, M.R., Shasha, B.S., Lewis, L.C., Bartelt, R.J. and Kinney, K. (1990) Field evaluation of granular starch formulations of *Bacillus thuringiensis* against *Ostrinia nubilalis* (Lepidoptera: Pyralidae). *Journal of Economic Entomology* 83, 2207–221.

McGuire, M.R., Shasha, B.S., Lewis, L.C. and Nelson, T.C. (1994) Residual activity of granular starch-encapsulated *Bacillus thuringiensis*. *Journal of Economic Entomology* 87, 631–637.

Metalnikov, S. and Chorine, V. (1929) Experiments on the use of bacteria to destroy the corn borer. *International Maize Borer Investigation Science Reports* 2, 54–59.

Morris, O.N. (1982) Bacteria as pesticides: forest applications. In: E. Kurstak (ed.) *Microbial and Viral Pesticides*. Marcel Dekker, New York, pp. 239–287.

Obrycki, J.J., Hamid, M.N., Sajap, A.S. and Lewis, L.C. (1989) Suitability of maize insect pests for development and survival of *Chrysoptera carnea* and *Chrysopa oculata* (Neuroptera: Chrysopidae). *Environmental Entomology* 18, 1126–1130.

Pedigo, L.P., Bechinski, E.J. and Higgins, R.A. (1983) Partial life tables of the green cloverworm (Lepidoptera: Noctuidae) in soybean and a hypothesis of population dynamics in Iowa. *Environmental Entomology* 12, 186–195.

Pilley, B.M. (1976) A new genus *Vairimorpha* (Protozoa: Microsporida), for *Nosema necatrix*. *Journal of Invertebrate Pathology* 28, 177–183.

Pingel, R.L. (1991) The perceptions of Iowa farmers toward integrated pest management and *Beauveria bassiana* (Balsamo) Vuillemin in the maize ecosystem – its effect on *Coleomegilla maculata* DeGeer. MS thesis, Iowa State University, Ames, Iowa.

Pingel, R.L. and Lewis, L.C. (1996) The fungus *Beauveria bassiana* (Balsamo) Vuillemin in a maize ecosystem: its effect on the insect predator *Coleomegilla maculata* De Geer. *Biological Control* 6, 137–141.

Posada, F. and Vega, F.E. (2006) Inoculation and colonization of coffee seedlings (*Coffea arabica* L.) with the fungal entomopathogen *Beauveria bassiana* (Ascomycetes: Hypocreales). *Mycoscience* (In press).

Pree, D.J. (1968) Control of *Glischrochilus quadrisignatus* (Say) (Coleoptera: Nitidulidae), a pest of fruit and vegetables in southwestern Ontario. *Proceedings Entomological Society of Ontario* 99, 60–64.

Ramoska, W.A., Singer, S. and Levy, R. (1977) Bioassay of three strains of *Bacillus sphaericus* on field-collected mosquito larvae. *Journal of Invertebrate Pathology* 30, 151–154.

Sajap, A.S. and Lewis, L.C. (1988) Effects of the microsporidium *Nosema pyrausta* (Microsporida: Nosematidae) on the egg parasitoid, *Trichogramma nubilale* (Hymenoptera: Trichogrammatidae). *Journal of Invertebrate Pathology* 52, 294–300.

Sajap, A.S. and Lewis, L.C. (1989) Impact of *Nosema pyrausta* (Microsporida: Nosematidae) on a predator, *Chrysoperla carnea* (Neuroptera: Chrysopidae). *Environmental Entomology* 18, 172–176.

Schell, K.K. (1994) Studies on biological and cultural control of the European corn borer, *Ostrinia nubilalis* (Lepidoptera: Pyralidae). MS thesis, Library, University of Wisconsin, Madison.

Shapiro, M. (1992) Use of optical brighteners as radiation protectants for gypsy moth (Lepidoptera: Lymantriidae) nuclear polyhedrosis virus. *Journal of Economic Entomology* 85, 1682–1686.

Siegel, J.P., Maddox, J.V. and Ruesink, W.G. (1986) Impact of *Nosema pyrausta* on a Braconid, *Macrocentrus grandii*, in central Illinois. *Journal of Invertebrate Pathology* 47, 271–276.

Sweeney, A.W., Doggett, S.L. and Piper, R.G. (1993) Life cycle of a new species of Duboscqia (Microsporida: Thelohaniidae) infecting the mosquito *Anopheles hilli* and an intermediate copepod host, *Apocyclops dengizicus*. *Journal of Invertebrate Pathology* 62, 137–146.

Tabashnik, B.E. (1994) Evolution of resistance to *Bacillus thuringiensis*. *Annual Review of Entomology* 39, 47–79.

Tabashnik, B.E., Cushing, N.L., Finson, N. and Johnson, M.W. (1990) Field development of resistance to *Bacillus thuringiensis* in diamondback moth (Lepidoptera: Plutellidae). *Journal of Economic Entomology* 83, 1671–1676.

Talalaev, E.V. (1958) Induction of epizootic septicaemia in the caterpillar of Siberian silkworm moth, *Dendrolimus sibericus* Tschtv. (Lepidoptera: Lasiocampidae). *Entomologicheskoe Obozrenie* 37, 641.

Talalaev, E.V. (1959) A bacteriological method of controlling *Dendrolimus sibericus*. In: (Trans.) *1st International Conference on Insect Pathology and Biological Control*, Parha, 1958, 51–57.

Talalayeva, E. (1966a) On the Siberian *Bombyx* septicaemic epizootic artificially provoked by Talalayen *Bacillus dendrolimus*. In: *9th International Congress of Microbiology*, Moscow (abstract) 324.

Talalayeva, E. (1966b) Artificial and natural reservoirs of *Bacillus dendrolimus* Talalyeva in eastern Siberia. In: *9th International Congress of Microbiology*, Moscow (abstract), 317.

Tamez-Guerra, P., McGuire, M.R., Behle, R.W., Shasha, B.S. and Galn Wong, L.J. (2000) Assesment of microencapsulated formulations for improved residual activity of *Bacillus thuringiensis*. *Journal of Economic Entomology* 93, 219–225.

Thorvilson, H.G., Lewis, L.C. and Pedigo, L.P. (1985) Overwintering potential of *Nomuraea rileyi* (Fungi: Deuteromycotina) from *Plathypena scabra* (Lepidoptera: Noctuidae) cadavers in central Iowa. *Journal of Kansas Entomology* 58, 662–667.

Vail, P.V. and Jay, D.L. (1973) Pathology of a nuclear polyhedrosis virus of the alfalfa looper in alternate hosts. *Journal of Invertebrate Pathology* 21, 198–204.

Webb, R.E., Thorpe, K.W., Podgwaite, J.D., Reardon, R.C., White, G.B. and Talley, S.E. (1999) Field evaluation of an improved formulation of Gypchek (a nuclear polyhedrosis virus product) against the gypsy moth (Lepidoptera: Lymantriidae). *Journal of Entomological Science* 34, 72–83.

Windels, C.E., Windels, M.B. and Kommedahl, T. (1976) Association of *Fusarium* species with picnic beetles on corn ears. *Phytopathology* 66, 328–331.

Woods, S.A., Elkinton, J.S. and Podgwaite, J.D. (1989) Acquisition of nuclear polyhedrosis virus from tree stems by newly emerged gypsy (Lepidoptera: Lymantriidae) larvae. *Environmental Entomology* 18, 298–301.

Woods, S.A., Elkinton, J.S. and Shapiro, M. (1990) Factors affecting the distribution of a nuclear polyderosis virus among gypsy moth (Lepidoptera: Lymantriidae) egg masses and larvae. *Environmental Entomology* 19, 1330–1337.

Appendix

Database of commercial insect pathogen products

Name	Trade name	Spectrum of control	Country of registration	Availability
Bacteria				
Bacillus popilliae	Doom™, Japademic™	Larvae of Japanese beetles, chafers, some May and June beetles	USA	(i) Fairfax Biological Laboratory Inc., Electronic Road, PO Box 300, Clinton Corners, NY 12514, USA (ii) Arizona Biological Control, Inc. (1-800-827-2847)
B. sphaericus (strain 2362)	VectoLex CG (*B. sphaericus* Serotype H5a5b)	Specific kinds of mosquitoes (especially Culex), active against the larvae of *Culex*, *Psorophora* and *Anopheles*; less effective against *Aedes*	USA	Valent BioSciences Corporation, 870 Technology Way, Suite 100, Libertyville, IL 60048, USA Phone: 800-323-9597
B. thuringiensis subsp. *tenebrionis*	Norodor™, M-Trak™	Colorado potato beetle and some other leaf beetles	USA	Mycogen Corporation, 5501 Oberlin Drive, San Diego, CA 92121, USA
B. thuringiensis delta endo-toxins (Cry1Ac and Cry1C) encapsulated in killed *Pseudomonas fluorescens*	MVP™, MVPII™	Larvae of many species of moths on agricultural crops turf, forestry, ornamentals, landscape trees and nursery crops	USA Australia, Canada	Mycogen Corporation, 5501 Oberlin Drive, San Diego, CA 92121, USA
B. thuringiensis subsp. *israelensis*	Bactimos, Mosquito attack, Skeetal, Teknar, Vectobac	Mosquito larvae		
B. thuringiensis subsp. *tenebrionis*	Trident	Larvae of Colorado potato beetle	Pending in USA	
B. thuringiensis subsp. *san diego*	M-One	Larvae of Colorado potato beetle, elm leaf beetle adults	USA	Mycogen Corporation, 5501 Oberlin Drive, San Diego, CA 92121, USA

continued

Appendix   *Continued*

Name	Trade name	Spectrum of control	Country of registration	Availability
B. thuringiensis subsp. *aizawai*	Agree™, Design™, XinTari™, Matttch™	*Lepidoptera* in vegetables and maize	USA	Mycogen Corporation, 5501 Oberlin Drive, San Diego, CA 92121, USA
B. thuringiensis subsp. *kurstaki*	Dipel™, Larvo-BT™, Thuricide™, JavelinWG™, Vault™, Raptor™, Bactec, Bernan™, BMP 123™, Condor™, Cutlass™, Foil BFC™, Raven™, Forwabit™, MVP, MVP II, Biobit HP™, Biobit 16K, P™, Biobit 320B FC™, Foray™, Foray 48B™, Foray 68B™, Furtura™, Bactosis K™, Agrobac™, Troy-BT™, Biocot™	Most *Lepidoptera* larvae with high gut pH	USA	
B. thuringiensis subsp. *israelensis*	VectoBac™, Gnatrol™, Technar™, Bactimos™, Aquabac™, BMP 144™, Aquabac, Primary Powder™, Bactis™, Teknar™	Larvae of mosquitoes, black flies and midges	USA Europe, Japan, Australia	(i) Mycogen Corporation, 5501 Oberlin Drive, San Diego, CA 92121, USA (ii) Sandoz, Inc., Crop Protection, 480 Camino Del Rio South, San Diego, CA 92108, USA (iii) Biochem Products, PO Box 264, Montchanin, DE 19710, USA (iv) Reuter Laboratories, 14540 John Marshall Highway, Gainesville, VA 22065, USA
Serratia entomophila	Invade™	Grass grub (*Costelytra zealandica* (White)) (*Coleoptera*: *Scarabaeidae*)	Pending in New Zealand	AgResearch NZ Pastoral Agriculture Research Institute Ltd, New Zealand

Appendix   *Continued*

Name	Trade name	Spectrum of control	Country of registration	Availability
Microsporida				
Nosema locustae	Locucide, NOLO Bait	Grasshoppers; Mormon cricket		A-1 Unique Insect Control 5504 Sperry Drive, Citrus Heights, CA 95621, USA Phone: (916) 961–7945 Fax: (916) 967–7082
Viruses				
Anagrapha falcifera Nucleopoly-hedrosis virus	None at present	Celery looper	USA	N A
Autographa californica Nucleopoly-hedrosis virus	Gusano biological pesticide	Alfalfa looper	USA	Thermo Trilogy, 9145 Guilford Road, Columbia, MD 21046, USA Phone: 1-301-604-7340 Fax: 1-301-604-7015
A. californica Nucleopoly-hedrosis virus	VFN80™	Alfalfa looper (*A. californica*)	USA Central America	(i) Thermo Trilogy, 9145 Guilford Road, Columbia, MD 21046, USA Phone: 1-301-604-7340 Fax: 1-301-604-7015
Helicoverpa zea Nucleopoly-hedrosis virus	Gemstar LC, Biotrol, Elcar	*Helicoverpa zea* and *Heliothis virescens*	USA	Thermo Trilogy, 9145 Guilford Road, Columbia, MD 21046, USA Phone: 1-301-604-7340 Fax: 1-301-604-7015
Hekuciverpa zea Nucleopoly-hedrosis virus	Gemstar LC™	Lepidoptera	USA	(i) Thermo Trilogy, 9145 Guilford Road, Columbia, MD 21046, USA Phone: 1-301-604-7340 Fax: 1-301-604-7015
Lymantria dispar Nucleopoly-hedrosis virus	Gypchek	Gypsy moth	USA	SDA Forest Service, 51 Mill Pond Road, Hamden, CT 06514, USA Phone: (203) 230–4325
Mamestra brassicae Nucleopoly-hedrosis virus	Mamestrin™	*Trichoplusia, Heliothis, Diparopsis, Phthorimaea operculella* (potato tubermoth)	Europe	Natural Plant Protection, Pau-Pyrenees, Avenue Leon Blum, 6400 PAU, France, Phone: +33 559 609292 Fax: + 33 559 609219
Mamestra configurata Nucleopoly-hedrosis virus	Virosoft	Aphids	Pending in USA Canada	BioTEPP Inc. of Charlesbourg, Quebec, Canada

Continued

Appendix   *Continued*

Name	Trade name	Spectrum of control	Country of registration	Availability
Orgyia psuedot-sugata Nucleopoly hedrosis virus	TM Biocontrol	Douglas fir tussock moth	USA	For more infrormation: USDA Forest Service, 180 Canfield Street, Morgantown, WV 26505, USA Phone: (304) 285–1584
Spodoptera exigua Nucleopoly-hedrosis virus	Otienem-S™, Spod-X™	Beet armyworm (*S. exigua*)		Thermo Trilogy, 9145 Guilford Road, Columbia, MD 21046, USA Phone: 1-301-604-7340 Fax: 1-301-604-7015

Fungi

Beauveria bassiana	Naturalis-L™, Naturalis-O™, (ornamentals), Naturalis-T™, (turf), Ostrinil™, Mycotrol™, Botanigard™	Adults and larvae of many kinds of insects; eggs of lepidopteran pests such as moths. Mole cricket, chiggers, white grubs, fire ants, flea beetles, boll weevils, whiteflies, plant bugs, grasshoppers, thrips, aphids, mites and many others	USA Europe	Mycotech, 117 South Parkmont, PO Box 4109, Butte, MT 59702, USA Phone: (406) 782–2386 Fax: (406) 782–9912
Metarhizium anisopliae var. *acridium*	Green muscle™	Specific to species of short-horned grasshoppers (*Acridoidea: Acrididae* and *Pyrgomorphidae*)		LUBILOSA (LUtte BIologique contre les LOcustes et Sauteriaux: biological control of locusts and grasshoppers)
Metharizium anisopliae	Bay Bio 1020™, BioBlast™	Soil-inhabiting beetle, termites		EcoScience Corporation, 10 Alvin Ct., East Brunswick, NJ 08816, USA Phone: (732) 432–8200
Paecilomyces fumosoroseus	PFR97™	Whiteflies, aphids and thrips		Thermo Trilogy, 9145 Guilford Road, Columbia, MD 21046, USA Phone: 1-301-604-7340 Fax: 1-301-604-7015

Modified from: http://ippc.orst.edu/biocontrol/biopesticides/

13 Role of Biotechnological Advances in Shaping the Future of Integrated Pest Management

A.M. Shelton[1] and R.R. Bellinder[2]

[1]Department of Entomology, Cornell University/New York State Agricultural Experiment Station, Geneva, NY 14456, USA; [2]Department of Horticulture, Cornell University, Ithaca, NY 14850, USA

Introduction

Beginnings of IPM

The term and practice of integrated pest management (IPM) has a long history (Kogan, 1998). Prior to the advent of effective pesticides, pest management relied primarily on an understanding of pest and host ecology, and tactics that could be used for pest management were largely cultural and biological controls. With the introduction of effective pesticides in the 1940s, there was a shift in pest management tactics to emphasize pesticides. However, two decades after their introduction, problems of pest resistance, pest resurgence and long-term environmental and human health effects occurred. The public outcry over the environmental and human issues described by Carson (1962) in *Silent Spring* provided a major impetus for the development of pest management programmes that focused on reducing the use of more harmful pesticides through more integrated approaches.

Kogan (1998) cites the work by Michelbacher and Bacon (1952) on codling moth as being one of the first to use the term 'integrated control' when they stated the need to look at the 'entire entomological picture in developing a treatment for a particular pest'. Stern *et al.* (1959) first mentioned a more formalized definition of integrated control of pests in a 1959 article published in *Hilgardia*. Since then there have been at least 64 recorded definitions of integrated control (Kogan, 1998). IPM has evolved from earlier concepts of pest control and is now commonly referred to as a philosophy of pest control that utilizes a variety of compatible tactics to maintain pest population levels below those causing economic injury. The term IPM has been adopted by the disciplines of entomology, plant pathology and weed science, and has become part of the political and administrative structure at the local, state, national and international level. The complexity of biological systems can cause fits to those who must make management decisions in the field, but it is the recognition of the complexity of pest management that is perhaps the most important and humbling aspect of IPM.

More information has become available on many of the tactics in the IPM toolbox. This includes a better understanding of the relationship between pest populations and crop losses, more accurate prediction of the effect of climate on pest populations and increased availability of biological and chemical pesticides. Biotechnology offers the possibility of developing new tools for pest management to explore on a more fundamental level how particular agricultural systems operate, such as following changes in the genetic structure of a pest population. Additionally, biotechnology can have, and has had, a profound influence on IPM practices through the development of transgenic plants for pest management. While some may argue that use of the latter continues an emphasis on single-tactic approaches, we see such plants having the potential to be used in more biologically based and more effective IPM programmes than previously possible.

In this chapter we briefly discuss some of the ways in which biotechnology has been used to advance our knowledge about pest biology in a more rapid and efficient manner, and to produce genetically engineered plants that can be used in IPM systems.

Concepts and practices of biotechnology

Like IPM, biotechnology is a term that has evolved over time. In a broad sense it means using living organisms or their products for some end. A more refined definition focuses on the use of techniques of modern biology to study living organisms (or parts of organisms) and make or modify products, improve plants or animals, or develop microorganisms for specific uses. Genetic engineering is one form of biotechnology and may involve copying a gene from one living organism (plant, animal or microorganism) and adding it to another organism, silencing specific genes or creating extra copies of genes. Products of genetic engineering are now widely used in agriculture for insect control (e.g. maize and cotton plants expressing insecticidal proteins from the bacterium *Bacillus thuringiensis* (Bt)), weed management (maize, cotton, soybean and canola plants that are resistant to specific herbicides) and disease management (papaya and squash, which are resistant to viruses). In 2003, plants genetically engineered for pest management were grown on 68 million ha worldwide (James, 2005).

Besides specific genetically engineered crops, the tools of biotechnology allow researchers in plant breeding, agronomy, entomology, plant pathology and weed science to approach traditional pest management problems with new strategies. Detection and diagnosis of pests and their damage is facilitated by using immunodetection and DNA-hybridization. These techniques allow growers and crop consultants to detect pathogens in plants prior to symptom expression, identify weed biotypes and determine insect resistance to particular insecticides. Techniques such as DNA microarrays have the power to monitor the whole genome of an organism on a single chip and allow scientists to understand the interactions of thousands of its genes simultaneously. We believe such powerful techniques can help advance IPM.

Tools of biotechnology have also allowed scientists to develop the area of genomics in which genes and their functions are identified. From an understanding of the genome it is therefore theoretically possible to design new molecules that can target a specific site, rather than rely on massive screening methods that provide a more 'shotgun' approach to finding new pesticides. In addition to finding new target sites, understanding an organism's genome can lead to the ability to alter it for the purposes of pest management, such as introducing a lethal gene into an insect or a section of a plant virus gene into a plant to confer immunity to the virus. While this chapter does not describe all the tools or techniques of biotechnology that can be used in IPM, we hope to highlight some of the applications that can be used to manage insects, pathogens and weeds within the conceptual framework of IPM.

Insects

Host plant resistance

Whenever possible, plants that are resistant to one or more key pests in a system should be used in IPM. Although breeding plants resistant to insects has provided a number of successes (Smith, 1989), most notably in rice and other field crops, advancement has been hampered because resistance to insects is most often a quantitatively inherited trait that makes it difficult for breeders to work with. One of the most significant advances of the last decade in the development of plants that are resistant to insect pests arises from the use of one form of biotechnology – the use of molecular markers to identify and track genes of interest for breeders (Yencho et al., 2000). In addition to helping identify genes associated with resistance traits, molecular markers can facilitate understanding the genetic basis for resistance, and be used to map biochemical and/or physical mechanisms. Direct measures of insect resistance can be simultaneously mapped (Yencho et al., 2000). Molecular markers have been used to map genes for insect resistance in most major crop species and have greatly facilitated breeding plants that are resistant to insect pests.

However, the most dramatic use of biotechnology in plant resistance to insects has been achieved with cloned insect resistance genes being engineered into plants. Genes from cowpea that code for a trypsin inhibitor, which interferes with insect digestion, were moved into tobacco, and imparted resistance to Heliothis virescens (Fabricius) (Hilder and Gatehouse, 1990). It has also been suggested that plants could be engineered to express semiochemicals to inhibit host finding or host feeding (Pickett, 1998). The most profound development to date for insect-resistant plants has been achieved by incorporating insecticidal genes based on those in the bacterium Bacillus thuriugiensis (Bt). The insecticidal activity of Bt has been known for decades and was first commercialized in France in the late 1930s with the product Sporeine. Bt is an aerobic, motile, gram-positive, endospore-forming Bacillus whose insecticidal activity comes from endotoxins in crystals formed during sporulation. Ingested Bt toxins kill susceptible insects by binding to, and disrupting, their midgut membranes (Schnepf et al., 1998), although the specific molecules involved in the binding process are not completely understood. The crystals of different strains of most Bts contain varying combinations of insecticidal crystal proteins (ICPs), and different ICPs are toxic to different groups of insects. Although there have been dozens of Bt foliar products marketed, the total sales of Bt products is less than 2% of the total value of all insecticides (Shelton et al., 2002). Despite their safety for humans and many beneficial insects, the relatively high cost, narrow pest range and lack of persistence of Bt products are limitations to wide use as foliar sprays. Furthermore, foliar sprays of Bt are generally not very effective at controlling tissue-boring Lepidoptera such as the European corn borer, Ostrinia nubilalis (Hübner), tobacco budworm, H. virescens, cotton bollworm, Helicoverpa zea (Boddie) and pink bollworm, Pectinophora gossypiella (Saunders).

The future of Bt now appears inexorably linked to the development of transgenic plants expressing insecticidal proteins from Bt, and these plants are revolutionizing IPM programmes for insects (Shelton et al., 2002). Bt was first introduced into tobacco plants in 1987 (Vaeck et al., 1987). However, plants that used much more effective synthetic genes, modelled on those from Bt but designed to be more compatible with plant expression (Perlak et al., 1990; Perlak et al., 1991; Koziel et al., 1993; Carozzi and Koziel, 1997), were introduced a few years later. Bt, which had limited use as a foliar insecticide, has become a major insecticide because genes that produce Bt toxins have been engineered into major crops, primarily maize and cotton. Of the $8.1 billion spent annually on all insecticides worldwide, it has been estimated that nearly $2.7 billion could be substituted with Bt biotechnology products (Krattiger, 1997). Thus, Bt plants can have a profound impact on insect management. Lepidopteran larvae are the primary targets in more than 99% of the acreage of Bt crops grown. The primary ICPs used in Bt plants for control of Lepidoptera are Cry1Ab and Cry1Ac, both very similar in structure and activity. Recently, Cry1F has been introduced in maize (Dow's Herculex I™), and in October 2004, Dow received full registration by the US Environmental

Protection Agency (EPA) for WideStrike™, a cotton product that expresses both Cry1F and Cry1Ac. This registration followed the registration of Monsanto's Bollgard II™, which expresses Cry1Ac and Cry2Ab2. A Monsanto-affiliated company, NatureMark, marketed Cry3A potatoes in the USA, Romania and Canada under variations of the trade name NewLeaf™ as a control for Colorado potato beetle (CPB). However, in 2001 the company stopped marketing Bt potatoes. Corn with a cry3Bb gene or a binary toxin genetic system for control of the corn rootworm (*Diabrotica*) complex has been developed. More than 30 types of Cry proteins have now been described and over 100 genes have been sequenced that can allow for more diverse opportunities for engineering Bt plants (AAM, 2002). More than 16 million ha of Bt crops were grown worldwide in 2005 (James, 2005), making Bt plants one of the most rapidly adopted agricultural technologies ever. Insecticidal plants using Bt technologies are also being developed for apples, crucifers, tobacco and other crops, but currently the only commercialized Bt crops are cotton and maize. While the use of Bt crops has profoundly changed maize and cotton systems, the use of Bt plants has raised issues about the potential for insect resistance and potential effects on non-target organisms.

As with other insecticides, the risk of evolution of resistance to Bt toxins produced by Bt plants depends on the genetic basis of resistance, the initial frequency of resistance alleles in the population, the competitiveness of resistant individuals in the field and a resistance management strategy (Shelton et al., 2002). There are at least five possible ways to slow the development of resistance by transgenic plants. Of these, four utilize plants with constitutive expression of Bt toxins: (i) express toxin genes only moderately strongly, so that not all susceptible individuals are killed; (ii) provide refuges for susceptible insects while expressing the genes as high as possible but within acceptable limits to avoid deleterious effects on yield, health or the environment; (iii) deploy different toxins individually in different varieties; and

(iv) deploy plants expressing a mixture of toxins. Among these options, the refuge–high dose (ii) and pyramiding (iv) strategies seem most promising (Roush, 1997, 1998; Gould, 1998). The US EPA mandates a resistance management strategy for the use of Bt crops, based on plants expressing a high dose of the toxin and the use of a refuge to conserve susceptible alleles within the population. Recent greenhouse studies have shown that plants containing two dissimilar Bt toxin genes (stacked) have the potential to delay resistance more effectively than single-toxin plants used sequentially or in mosaics (Zhao et al., 2003). As noted above, dual Bt gene plants have been introduced in cotton. A fifth general option, much less studied, is to control the expression of the Bt genes in each plant so that they are expressed only when or where needed. This could be done through tissue-specific, temporal-specific or inducible gene promoters. In principle, specific promoters could be used to express genes only in the most important tissues such as fruit or other reproductive tissues ('tissue- or structure-specific' expression), or at critical growth periods ('temporal-specific' expression) (Gould, 1998). Alternatively, the genes could be turned on, perhaps by spraying an environmentally benign chemical (Bates et al., 2004). These tactics are not necessarily exclusive: a temporal-specific promoter may be effectively structure-specific if it is turned on only when needed (e.g. late in the season and affecting only the top of a plant). The use of inducible Bt plants for insect resistance management has been recently explored by Cao et al. (2001) and Bates et al. (2004) using the diamondback moth system.

Monitoring insect populations for the evolution of resistance to Bt proteins expressed in plants is essential for a sustainable IPM programme that uses Bt plants. In order to study Bt resistance evolution, it is necessary to have field-derived insect populations that have evolved resistance. However, after ten years of extensive use of Bt crops in the field, there have been no reported cases of field failure or increases in resistance allele frequency due to Bt crops

(Tabashnik *et al.*, 2003). This conclusion is based on surveys done largely without the use of biotechnology-derived monitoring methods. Lack of field-derived resistance has made the identification of Bt-resistant genes and understanding the biochemical mechanism of such resistance difficult (Ferré and Van Rie, 2002).

Considerably more public attention has been paid to the issue of the potential effect of Bt plants on non-target organisms and their potential effect on IPM. Regulatory agencies require information on the environmental risk of Bt plants but considerable debate continues on what and how to measure (Conner *et al.*, 2003). Perhaps the maximum work has been done using Bt maize (Orr and Landis, 1997; Pilcher *et al.*, 1997; Lozzia, 1999; Manachini, 2000; Wold-Burkness *et al.*, 2001; Bourget *et al.*, 2002; Dutton *et al.*, 2003; Musser and Shelton, 2003), but discussion continues on which species should be used as indicators. There is also concern about how species may be affected by Bt plants (or alternative strategies) as the potential exposure can occur in several ways (Schuler *et al.*, 1999; Groot and Dicke, 2002). A considerable number of publications have been generated on the effects of Bt plants on non-target organisms (See Wolfenbargar, 2003) e.g. Sears *et al.*, 2001; Naranjo Ellsworth, 2002; and Wolfenbarger, 2003 and show the overall safety of Bt plants. In 2005 a series of 13 field studies were published that demonstrated the safety of Bt plants to nontarget organisms (Naranjo *et al.*, 2005). One of the most thorough of these studies was a six-year field study in cotton (Naranjo, 2005). In this study, no chronic long-turn effects of Bt cotton were observed over multiple generations of nontarget taxa. Additionally, natural enemy function, measured as rates of predation and parasitism on two key pests of cotton, was unaffected by Bt cotton.

Dutton *et al.* (2003) compiled a list of the most frequently cited predators and parasitoids on maize in central and Western Europe and focused on the seven most commonly occurring natural enemies. From these data they selected four predators – *Coleomegilla maculata* (DeGeer), *Orius insidiosus* (Say), *O. majusculus*, *Chrysop-erla carnea* (Stephens) – most likely to be exposed to Bt toxins and identified studies that assessed their sensitivity to Cry1Ab presented in various ways (Dutton *et al.*, 2003). For all but *O. majusculus*, the predators were fed pollen or silk from Bt plants. While no direct effects were seen on the first three species, the experimental design did not eliminate secondary effects such as could occur if predators fed on larvae that had fed on Bt plants, a situation likely to occur in the field. Nor did it compare the lack of effects in these three species to results using a conventional technology for insect control (e.g. pyrethroid). With *C. carnea*, the results seemed to vary according to the methods used. However, none of the methods really allowed a critical and realistic evaluation that separated out the effect of the prey (European corn borer in this case) from the Cry1Ab toxin. A follow-up study in which *C. carnea* were directly fed Cry1Ab toxin revealed that the larvae were not sensitive to the toxin (Romeis *et al.*, 2004). Only with a resistant prey species could this be sorted out. Schuler *et al.* (2004) had access to Cry1Ac-resistant diamondback moth larvae and Cry1Ac-expressing oilseed rape plants. In a laboratory assay, they tested whether the diamondback moth parasitoid, *Cotesia plutellae*, would be killed by feeding on Cry1Ac-resistant larvae and whether the behaviour of the adult wasps would be affected. Their results indicated no effect on either parameter. Both sets of studies (Romeis *et al.*, 2004; Schuler *et al.*, 2004) did not compare the effects using a pyrethroid insecticide, a common treatment, to obtain the information needed for a risk–benefit comparison. In a comparative study examining the effect of Bt and commonly used foliar insecticide sprays on pest populations, predator abundance and predation levels, Bt maize provided the best control of the targeted pest species with the least amount of disruption to the natural enemy complex (Musser and Shelton, 2003). Collectively, the studies published to date indicate that Bt plants can contribute to integrated pest management systems with a strong biological control component (Romeis *et al.*, 2006).

Novel insecticides

Biotechnology is aiding in the discovery and utilization of novel insecticides, both synthetic and biological that can be used in IPM systems. According to Cassida and Quistad (1998), genetic engineering has provided the means to facilitate insecticide production (by improving yields of biologicals) and discovery (by producing target enzymes or receptors for screens). Sequencing of cDNA libraries and genomes of pest organisms is regarded as the key to gene discovery and should enable the identification of novel targets for new insecticides (Heckel, 2003). Besides the discovery of new target sites for insecticides, biotechnology also has the capacity to improve insect pathogens or to help them serve as vectors for insecticidal traits. Baculoviruses have been the dominant entomopathogenic virus group used in insect pest management and have been considered safe and effective bioinsecticides against Lepidoptera. However, they have had limited use because of their narrow host range, slow killing speed and technical and economic difficulties for *in vitro* commercial production (Moscardi, 1999). To overcome some of these limitations, viruses have been genetically engineered to express genes coding for foreign proteins such as insect-specific toxins (arthropod venoms or Bt toxins), insect hormones (diuretic, eclosion and prothoracic hormones), juvenile hormone esterase, as well as for deletion of ecdysteroid UDP-glycosyltransferase (Moscardi, 1999). However, with the introduction of these foreign genes there is concern about their persistence in the environment. Genetically engineering the virus to also lack its protective occlusion body reduces the time that recombinant viruses remain active and thus decreases environmental risk (Wood *et al.*, 1994). Genetically engineered viruses for insect control were the subjects of considerable work in the 1990s. However, shifting priorities by companies, institutes and universities, past players in developing viruses for insect control, appear to have halted much of the development work.

Worldwide there are more than 700 species of entomopathogenic fungi in approximately 100 genera. However, only 10 species have been, or are currently being, developed for insect control (Hajek and St Leger, 1994). The majority of control programmes, which rely on fungi as the principle management tool, use fungi essentially as a frequently applied biological insecticide (Shelton and Roush, 2000). Use of biotechnology for fungal pathogens of insects appears to focus on identifying more effective strains and on understanding the molecular basis of pathogenicity, a complex process involving many steps. For example, the cuticle serves as a multilayered barrier to fungi. The failure of a fungus to penetrate this barrier could be caused by the presence of inhibitory compounds in the cuticle, a lack of factors needed for recognition on the cuticle or a lack of proper nutrients in the cuticle. Once the fungus has entered the insect's body, the host may be killed by some combination of mechanical damage caused by the fungus, depletion of the host's nutrients or toxins produced by the fungus. It appears that the tools of biotechnology may be most useful in the near future, more as aids in understanding the infection process of fungi and the potential usefulness of fungal metabolites rather than genetic engineering and release of fungal strains.

Entomopathogenic nematodes are lethal obligatory parasites of insects that carry specific pathogenic bacteria, either *Photorhabdus* in the case of the nematode family *Heterorhabditae* or *Xenorhabdus* in the nematode family *Steinernematidae* (Liu *et al.*, 2000). An increased understanding of the biology of the nematodes and their associated bacteria can be greatly aided by the tools of biotechnology. For example, understanding the differences in *Heterorhabditis* isolates, as determined by restriction fragment length polymorphism (RFLP) or *ND4* gene sequence analyses, contributed significantly to elucidating the population genetic structure and identifying isolates and species that were more effective against pests such as the strawberry root weevil and CPB (Liu *et al.*, 2000). Besides using the nematode directly, another approach has been to use the 'toxin A' isolated from *Photorhab-*

dus luminscens. Arabidopsis thaliana was recently transformed to express the 'toxin A' gene at extremely high levels, and transformed plants had good insecticidal activity against at least one lepidopteran pest, and moderate activity against a coleopteran pest (Liu *et al.*, 2003).

Introducing lethal or beneficial traits into insects

The spectacular success of the sterile insect technique (SIT) for screwworm control (see Chapter 10 for details) in cattle validated the concept of genetic control of insect species. SIT involves mass rearing of the pest species, sterilizing it and then releasing it in the field in very high proportions to the wild population to mate, thereby leading to a reduction in fertility and population suppression (Dent, 1991). The most commonly used SIT method involves dominant lethal mutations that are introduced by irradiation or chemosterilization, both techniques that can cause fitness costs and decrease the probability of mating. Biotechnology can play a role in helping refine rearing techniques for SIT programmes. For example, in SIT cases where only males will be released, lethal genes can be introduced into females so that they die when reared under certain conditions. This results in reduced rearing expenses, both for materials and labour. Another way biotechnology can help is to offer an alternative to irradiation and chemical sterilization, which cause random chromosome breakages, often leading to insects being less fit once they are released in the field (Miller, 2002). For example, the *Notch* gene family has provided a proof of concept using biotechnology in an SIT programme. A vinegar fly strain carrying two extra copies of a *Notch* gene caused complete collapse of a wild strain of vinegar flies in only three generations when introduced in equal numbers (Fryxell and Miller, 1995).

Efforts are underway to apply biotechnology for control of insect-borne diseases that will provide a novel approach for IPM in medical entomology. There has been an emergence and resurgence of vector-borne medically important diseases worldwide, and an estimated annual 1.5–2.7 million deaths caused by malaria (Gratz, 1999). In late 2002, the genome sequence of *Anopheles gambiae* Giles, the major mosquito vector of the malaria parasite, and the parasite itself (*Plasmodium falciparum*) were published. Knowledge of the genome of these two organisms, together with the human genome, can provide critical genetic information relevant to all stages of the malaria transmission cycle, and offer unprecedented opportunities to the scientific and public health communities engaged in the fight against malaria (Aultman *et al.*, 2002). Recent technology has resulted in the creation of transgenic mosquitoes using insect-derived transposable element–based vectors as part of disease prevention programmes (O'Brochta, 2002). One of the outcomes of this genetic engineering will be the creation of mosquitoes that block the development of pathogenic parasites such as the one that causes malaria. Similar efforts are underway with the parasitic protozoan, *Trypanosoma cruzi*, the causal agent of Chagas disease that is transmitted by insects in the family *Reduviidae* (Beard *et al.*, 2002). The long-term goal of this work is to introduce transgenic genotypes successfully into wild populations to achieve public health benefits. While the potential benefits of genetically engineering insects or the pathogens they transmit could be substantial, so are the challenges of conducting experiments on human subjects and developing a suitable regulatory framework for the implementation of large-scale programmes for IPM of medically important insects.

In addition to combating insect-borne diseases, biotechnology may also be used to introduce desirable traits into insect populations. For example, increased parasitism of a pest population may be achieved by infecting parasitic wasps with the bacterium *Wolbachia*. Infected parasites will produce only females, the parasitizing sex, and in some insects this bacterium will also favour the production of infected offspring (Miller, 2002). Furthermore, *Wolbachia* may serve as a vector to introduce other traits, such as increased tolerance to environmental conditions, into a population. The effectiveness of these enhanced natural enemies needs to be determined in the field, but limited studies have been conducted. The first experimental release of a transgenic natural enemy, the predatory mite *Metaseiulus occidentalis* (Nesbitt), carried a molecular marker (lacZ construct with a *Drosophila heatshock 70* promoter). Although this release took place in 1996 without any environmental harm (Hoy, 2002), additional releases of transgenic organisms have been few because of concern about horizontal gene transfer, changes in host range or specificity, and the lack of guidelines on methods of containment. As noted by Hoy (2002), the deployment of transgenic arthropod natural enemies, although having great promise, will remain a challenge because of the need for comprehensive risk assessments, very detailed knowledge of population genetics, as well as coordinated efforts of geneticists, regulatory agencies and pest management specialists.

Monitoring

A more immediate use of biotechnology is for the identification of insect species, the dynamics of predation and parasitism in agricultural settings, insect damage to host plants and insect genotypes in the field. Techniques developed through biotechnology can help gain insights into important components in an IPM programme. Enzyme electrophoresis, RFLP, random amplified polymorphic DNA (RAPD), polymerase chain reaction (PCR) and DNA sequencing have proved invaluable for species identification and separation of biotypes, especially where distinctive morphological characteristics are lacking (Menken and Raijmann, 1996). These procedures have been particularly useful in confirming pest species and biological control agents, especially small parasitic Hymenoptera that can be essential components of a biological control programme. Such techniques can also be used to measure population structure in time and space and thus add to our knowledge of the populations of pest and beneficial arthropods.

Important contributions to our understanding of predator activity and ecological relationships have been made through techniques such as enzyme-linked immunosorbent assay (ELISA), monoclonal antibody production and nucleic acid probes. As an example of the use of these biotechnology applications, Hagler and Naranjo (1996) reported on an extensive study conducted in cotton in which they used immunoassays to evaluate gut contents of biological control agents. Pest-specific monoclonal antibodies were used to identify key predators of the sweet potato whitefly and the pink bollworm and monitor the efficacy of an augmentative predator released against these two pests. The gut contents of more than 10,000 individual predators were examined for whitefly and pink bollworm egg antigens using ELISA, and from these data they were able to determine the major predator species of each pest. An additional test was used to evaluate the performance of commercially reared *O. insidiosus* (Say) compared with native *O. tristicolor* (White). Such extensive datasets could not have been collected through observations or other methods outside of biotechnology.

Techniques such as immunodetection and DNA-hybridization have also been used to identify insect eggs in cotton (Greenstone, 1995) and cryptic damage to other crops. An example of using these techniques to study the impact of a pest in conifers is described by Bates (2002), who used polyclonal antibodies to detect residual salivary proteins of the western conifer seed bug in seeds of seven conifer species. As aborted seeds are indistinguishable from seeds damaged by the western conifer seed bug, it was only after the development of these molecular techniques that a true assessment of the importance of this pest could be made, and this led to identifying it as a key pest in the lodgepole pine ecosystem. Recently, our laboratory has used PCR analysis to help confirm the first infestation of swede midge in the USA. As larvae cannot be identified through morphological characteristics and it is extremely difficult to rear them from host material collected in the field, molecular methods may be the best method in future survey work for this important pest of crucifers.

An important use of biotechnology in pest management is, and will continue to be, detecting resistance to insecticides. Traditionally, resistance to insecticides has been evaluated using bioassays of whole insects, but biochemical and DNA methods are now in use or being developed (Daly and Trowell, 1996). These methods include enzyme electrophoresis, enzyme assays and immunoassays that allow more rapid detection of the evolution of resistance and provide more insight into gene flow and fitness costs. It is appropriate that the tools of biotechnology used to detect insecticide resistance are also being developed to detect resistance to crops produced through biotechnology. A number of monitoring techniques for resistance detection have been proposed and include: grower reports of unexpected damage; systematic field surveys of Bt maize; discriminating concentration assays; the F_2 screen, sentinel plots of Bt maize and isolation of resistance genes from laboratory-selected colonies for development of molecular diagnostic techniques. The most common method, and the one that has been used the longest, is the discriminating concentration assay in which one looks for increases in percentage of survival at a toxin concentration derived from baseline susceptibility of previously unexposed populations. However, this method is not very sensitive at low allele frequencies and may lead to a significant underestimation. Identification of specific genes and/or gene products that confer resistance would provide the means necessary for development of sensitive biochemical or molecular techniques to identify the genotype of field-collected individuals, but this has so far proved elusive. Mutations affecting a Bt Cry1A-binding midgut cadherin protein (Vadlamudi et al., 1995; Nagamatsu et al., 1999) have been shown to be tightly linked with laboratory-selected mode 1 resistance in two major pests, H. virescens (Gahan et al., 2001) and P. gossypiella (Morin et al.,

2003). Thus, the cadherin gene has been considered the prime target for DNA-based screening for resistance. However, additional studies with other insect species have indicated that the cadherin gene may not be involved (see Bates et al., 2004). Still the search for molecular markers associated with resistance will likely provide the most reliable method of following resistance allele frequency to Bt plants, as well as conventional insecticides.

Weeds

IPM for weed control is a collection of strategies that starts at the level of the cropping system and includes tillage, crop rotations, weed species prevalence and densities, as well as crop emergence patterns and competitiveness. Typically, crop-specific IPM programmes are based on combinations of strategies, e.g. mechanical and chemical controls. Frequent scouting identifies weed species that are present and their size, and this allows as-needed use of optimum rates of post-emergence herbicides. It is expected that IPM programmes will differ widely with crops; thus, introducing herbicide resistance into crops will have significantly different impacts on IPM programmes for each cropping system.

Scientists largely agree that the potential benefits that will accrue with the adoption of genetically modified, herbicide-resistant (GMHR) crops include: a shift to more environment-friendly production systems (reduced or zero tillage that decrease soil erosion), water conservation, lower fuel consumption, reduction in greenhouse gas emissions, decreased pesticide run-off, and as-needed use of herbicides (Carpenter et al., 2002). If adoption occurs in less developed countries, improved farm economies, reduced drudgery and improved life quality for farm families are additional expected benefits. In countries where the use of GMHR crops is allowed, adoption is likely to occur only in crops where growers expect to realize significant economic benefits and increased simplicity and flexibility in their weed management programmes. This has been evident in the USA, where 80% of the soybean acreage is now planted with GMHR varieties compared with maize where only 15% of the acreage is in GMHR varieties (Economic Research Service, 2004; Sankula and Gianessi, 2004).

While there are numerous benefits that may result from GMHR adoption, two of the most commonly expressed negative outcomes are the development of crops as weeds and the transfer of the resistance trait to weedy relatives, which would then become 'super weeds'. Long before the advent of GMHR crops, the potential for the development of herbicide resistance in weeds was first identified in the late 1940s when differences in weed selectivity to 2,4-D were reported (Albrecht, 1947). Ryan (1970) made one of the earliest reports of atrazine resistance, and Heap (2004) reported that there are now 291 documented herbicide-resistant (HR) weed biotypes. Since the late 1990s, rigid ryegrass (*Lolium rigidum* Gaud.) (Powles *et al.*, 1998; Feng *et al.*, 1999), goosegrass (*Eleusine indica* (L.) Gaertn) (Lee and Ngim, 2000), field bindweed (*Convolvulus arvensis* L.) (DeGennaro and Weller, 1984) and marestail (*Conyza canadensis* L.) (VanGessel, 2001) have been identified as resistant to glyphosate. Thus, while resistance development and its management have been of more intense and public concern with insects and diseases, growers and scientists have been dealing with resistant weeds for many years.

Initially, weed scientists resisted the concept of herbicide resistance. Insect and disease resistance was understood, largely because of the rapidity of population turnover with multiple generations every growing season, something that rarely occurs with agricultural weeds. The concept of resistance development through 'selection pressure', i.e. the gradual increase of resistant biotypes in fields subjected to long-term use of the same herbicide, came to be accepted. It is now understood that herbicide resistance may also develop by rapid detoxification processes, alteration of enzyme binding sites and decreased uptake or translocation.

Development of new herbicides with highly specific modes of action in the 1980s

and 1990s led to more rapid resistance development in multiple weed species. In answer to this, the Herbicide Resistance Action Committee (HRAC, 2004) was formed in 1989 to develop and encourage the use of 'resistance management strategies'. The group's members now include herbicide manufacturers, weed scientists and regulatory representatives, and there are branches in several countries. Their mission has been to educate growers in the strategies needed to avoid herbicide resistance development in weeds. Commonly employed strategies include crop rotation, alternating herbicide chemistries and cultivation. Thus, managing weeds that develop resistance to herbicides is not a new problem, is already being practised and would require no additional control measures if resistance was to arise through gene flow from an HR crop.

Similarly, HR crops derived through traditional breeding methods have been grown for many years. Some of these include imidazolinone-resistant maize, rice, wheat and sunflower, and sethoxydim-resistant maize. In the 10 years since their introduction, no evidence of gene transfer from these crops to weeds has been reported. In terms of agronomic weed management practices, GMHR crops do not pose additional or more complex problems than those of HR crops produced through conventional breeding. An example of this is using highly effective graminicides to control volunteer Roundup-Ready maize plants following Roundup-Ready soybean crop. For virtually all GMHR crops there are alternative methods available for controlling volunteers.

Clearly, hybrid formation and increased 'weediness' will not occur with adoption of HR or GMHR crops in areas where there are no natural wild relatives. The resistant crops may survive in fields and would be managed the way any volunteer crop plant is dealt with. Such crop plants would not survive in a natural, undisturbed environment. However, caution should be exercised when introducing any resistant crop into its 'center of origin' without also introducing barriers to gene flow (e.g. sterility and maternal inheritance), as gene flow between closely related species, followed by introgression in the population, could with time contaminate the original gene pool.

It is important to note that the use of GMHR and HR crops will facilitate adoption of, and improve, current IPM strategies related to weed management. Fundamentally, having resistant crops will move current herbicide programmes away from routine, prophylactic, soil-applied residual products towards the use of as-needed total post-emergence applications to crop foliage. Many of the new herbicides being developed have more environmentally benign profiles than the older herbicides. Plant tissues, translocated to sites of action that impact only plants and not mammals or invertebrates, and have extremely short breakdown periods, largely absorb these new products. The use of non-selective herbicides (glyphosate, glufosinate) in resistant crops will additionally enable timing of applications to match what is identified as the 'critical weed-free period' for individual crops. This period is the point in time when a failure to remove weeds from the crop will result in significant yield reductions. This critical period varies significantly for different crops, and is a function of the timing of seed development. In the case of maize, this occurs by the time 7–8 leaves have emerged, i.e. early in the growing season. In soybeans, the critical point occurs over a longer period, beginning with flowering, but extending for a period into pod development, much later than in maize. The implication of this is that while weeds in soybeans might be managed with a single glyphosate application, maize would require two applications to prevent substantial yield losses. Knowing that one has such tools will encourage and expand scouting activities in many crops, leading to accurate identification of existing weed species, and facilitating the correct choice of appropriate herbicides and rates needed for effective control, resulting ultimately in reduced herbicide use.

The use of GMHR crops will enhance IPM development and significantly improve agricultural production systems in water-poor areas of the world. A case in point is the rice/wheat production system common to the Indo-Gangetic Plain in South Asia. In this region soils are puddled (compacted) and flooded annually for the production of paddy or transplanted rice. The fields are flooded or saturated throughout the 4-month rice season. Following rice harvest, the fields are cultivated 6–8 times and the cloddy soils are crushed to create an aerated soil for wheat. This system has led to serious deterioration in soil quality, decreasing water tables and salinization. Between 1997 and 2003, many farmers in the region had adopted zero tillage for wheat production; however, the fields were still being puddled for the rice crop. Weeds are the most significant problem in the production of non-puddled rice and are the primary deterrent to adoption of direct-seeded or zero till-age rice. An HR or GMHR rice crop would enable moving the entire system to one that would restore soil quality, conserve water, reduce fuel use and overall improve the sustainability of agriculture in the region. This would increase the production of essential grain crops necessary as population numbers are expected to increase until 2050.

Another area where use of HR and GMHR crops will significantly benefit growers in less developed countries is in crop diversification. This is an essential component of sustainable agriculture as it enables crop rotation, which in turn prevents the long-term deleterious effects of monocultures (e.g. the rice/wheat system). However, traditional markets have long been closed to new crops and this has limited growers' options. Presently there is a need to develop a market for maize to sustain increased poultry production in less developed countries. As incomes have increased in these regions, poultry is becoming the meat of choice. Feed currently being grown is of poor and inconsistent quality. HR maize (imidazolinone, glufosinate, glyphosate) would enhance the development of this market, which is virtually essential as the number of farm workers available for hand weeding these crops is rapidly decreasing.

Agriculture in developed countries will also benefit from the use of HR or GMHR crops through enhanced weed management. The USA currently uses the labour of hundreds of thousands of immigrants to manage weeds manually in many labour-intensive vegetable and fruit crops. The condition under which much of this work is done is onerous and difficult. In 2001, labour rights activists in California suggested that hand weeding should be made illegal, to protect workers. Allowing the development of herbicide resistance in poorly competitive crops like onions, lettuce, strawberries and carrots would remove the drudgery associated with producing these crops. This would enable greater crop rotation possibilities for producers of these crops.

Biotechnology, evidenced here as GMHR crops, has the potential to become a valuable tool for the management of weeds. It will have numerous benefits to large and small growers in both developed and less developed countries. However, it is one tool in a box that has many tools, all of which need to be used with care. HR crops, like other weed management strategies, must be integrated with other sound agronomic practices to enhance agricultural sustainability worldwide.

Plant Pathogens

With the increasing globalization of agriculture and changing climatic conditions, there has been an emergence or re-emergence of several important agricultural diseases. These include late blight of potato and tomato caused by *Phytophthora infestans*, citrus canker caused by the bacterium *Xanthomonas axonopodis* pv. *citri*, black sigatoka of banana caused by the fungus *Mycosphaerella fijiensis*, and a host of plant viruses, especially tospoviruses and geminiviruses that are often transmitted by insects (Scholthof, 2003). Many of the affected crops are important sources of calories, vitamins and minor elements necessary for human health throughout the world. The need for disease management within the framework of IPM has taken on new importance for plant pathologists worldwide and the tools of biotechnology have become key elements.

The traditional components of maintaining plant health, as identified by Cook

(2000), are important components of IPM and can be categorized as: the use of disease-free seeds and other planting materials; sanitized and optimally fertilized soil; proper irrigation practices; and the use of practices and products to protect the crop against hazards imposed through the air. Biotechnology can help in all these stages of plant protection by practices as diverse as using diagnostic methods to certify the cleanliness of seeds and planting materials, detection of pesticide-resistant strains of pathogens and the creation of plants resistant to pathogens. There are many similarities between the uses of biotechnology for plant pathogens and for insect pests.

Molecular analyses of plant pathogens

Accurate identification of plant pathogens is a fundamental need in disease management, but is challenging because traditional methods have relied on disease expression in the host and/or culturing the organism extracted from the plant or the soil. Molecular methods that allow the extraction of DNA and/or RNA from plant tissue or soil may provide far more rapid and accurate methods of diagnosing the presence of a pathogen. As noted by Martin *et al.* (2000), advances in virus detection have been especially useful for certification programmes for plants such as grapevines, where the labour-intensive practice of grafting host tissue on to indicator species has been the standard. With the introduction of ELISA in the late 1970s, this technique became a standard method for detecting the major diseases of grapevines caused by viruses. Later PCR and Western blots were used and then reverse transcriptase (RT) and immunocapture (IC) RT-PCR tests were developed and used to detect the complex of eight closteroviruses associated with grapevine leafroll disease. The results of using these laboratory tests, especially RT-PCR and ELISA, have been dramatic. What once took 2 years for biological indexing of grapevine leafroll or rugose wood virus complexes can now be done in 2 days or less, using molecular methods for virus detection.

Molecular diagnostic tools have also been used for fungal detection and identification of new isolates (Gold *et al.*, 2001), e.g. PCR-based detection, immunological detection and microscopy techniques such as fluorescent *in situ* hybridization (FISH). These techniques have also contributed to an understanding of fungal pathogenesis needed to develop new tools for IPM. Perhaps the most dramatic use of molecular diagnostic tools has been work conducted with late blight disease of potato and tomato. This pest has become particularly devastating due to the development of aggressive strains of *P. infestans*, which are resistant to the systemic fungicide metalaxyl (Scholthof, 2003). The establishment of resistant strains in many parts of the world points out the need to 'genetically fingerprint' pathogens so that appropriate control strategies can be implemented for specific strains (Ristanio and Gumpertz, 2000).

Accurate detection of plant pathogens is becoming more important in the increasingly globalized food chain where plant quarantine regulations play a greater role. To enforce regulations, detection methods and accurate sampling strategies must be devised, but they must also be efficient and workable within the context of the global food system. Molecular diagnostic methods can provide in many cases a more accurate detection method for bacteria, fungi, nematodes, viruses, phytoplasma and viroids than visual examination of plant material. Cost and accuracy of detection are critical elements, especially in large-scale diagnostic surveys, and molecular-based methods can be a great aid. However, they are sometimes limited because laboratories may lack equipment or trained personnel, or the sensitivity of the tests may not be sufficient. In many cases, ELISA methods have proven to be superior to some of the more sophisticated molecular methods because of their simplicity, robustness and minimal risk of cross-contamination or false positives (Martin *et al.*, 2000).

Besides detection of specific organisms, techniques developed through biotechnology are playing a key role in detecting fungicide resistance. The first case of resistance to a foliar fungicide was reported in 1968 for dodine used in the post-infection control of apple scab, *Venturia inaequalis*. Since then, all major fungal pathogens managed with pathogen-specific fungicides have developed resistance to at least one of the pathogen-specific fungicides (Koeller, 2004). Resistance is usually temporally and spatially specific, so knowledge is needed about the susceptibility in a particular field or region at a particular time. A common technique for assessing resistance is to conduct bioassays using field-collected populations of the fungus but this is labour-intensive and time-consuming. Advancements have been made using molecular analysis that allows rapid documentation of resistance. The strobilurin fungicides are a case in point.

This class of fungicides has broad-spectrum activity against many pathogens on key crops worldwide. Additionally, they are not persistent in the environment and very suitable for IPM programmes (Ishii *et al.*, 2001). However, resistance to strobilurin fungicides has occurred in several pathogens. Using PCR and DNA sequence analyses, it was suggested that a single point mutation (GGT to GCT) in the mitochondrial cytochrome *b* gene, resulting in substitution of glycine by alanine at position 143, was a major cause for rapid development of high strobilurin resistance in apple scab (Zheng *et al.*, 2000) and cucumber powdery mildew and downy mildew (Ishii *et al.*, 2001). This same point mutation was also found for strobilurin-resistant populations of Alternaria blight, caused by *Alternaria* spp. in the *alternaria*, *tenuissima* and *arborescens* species-groups in pistachio in California (Ma and Michailides, 2004). Using allele-specific PCR assays, coupled with rapid DNA extraction, it is now possible to detect strobilurin resistance in a few hours, rather than the 7 days needed for the conventional spore germination method. Fraaije *et al.* (2002) were able to take this one step farther by developing a quantitative allele-specific real-time PCR measurement for strobilurin-resistant isolates of the causal agent of wheat powdery mildew. Such techniques can be used to monitor resistance when resistance genes are at low frequency and thus provide a proactive resistance management programme for this important class of fungicides.

Besides the use of molecular tools to detect plant pathogens, disease progression or resistance to fungicides, biotechnology approaches have also been valuable for clarifying the taxonomy of pathogens. For example, the taxonomy of closteroviruses is only now being sorted out with the help of molecular genetics (Karasev, 2000). These viruses are transmitted by aphids, whiteflies and mealybugs in a semi-persistent fashion and cause serious damage to several crops of economic importance. Because of changing agricultural practices and the spread of new insect vectors, this family of viruses has become increasingly important in IPM. Recent studies have demonstrated an astonishing genetic diversity in closteroviruses that continues to change as the viruses coevolve with their insect vectors (Karasev, 2000). Understanding the genetic structure of pathogen populations is fundamentally important for breeding durable resistance to many plant pathogens (McDonald and Linde, 2002).

Host plant resistance

As with breeding for resistance to insects mentioned above, marker-assisted breeding has proven to be of tremendous value for breeding plants that are resistant to many of the world's major diseases, revolutionizing plant breeding. At least 19 single dominant genes for resistance to viruses, nematodes and fungi have been mapped for potato using DNA markers, and thus enhanced breeding efforts in potatoes (Gebhardt and Valkonen, 2001). New techniques in molecular biology are also allowing plant pathologists to develop a better understanding of the disease infection cycle, which holds promise for genetic manipulation of plants and disease causing organisms. For example, the *Pto* gene, which confers resistance to *Pseudomonas syringae* pv *tomato*, the causative agent of tomato bacterial speck, was the first plant gene cloned that participates in gene interaction with a pathogen (Pedley and Martin, 2003). Results from this model system have allowed scientists to develop a better understanding of how some plants have evolved sophisticated recognition systems to detect proteins produced by specific races of pathogens (Martin *et al.*, 2003). A better understanding of the interaction could lead to the development of novel control strategies. Genetic engineering for disease-resistant plants has also been accomplished and commercialized, but to a lesser extent than for insect or weed management.

For decades, insect-transmitted viruses have been a recalcitrant problem at the interface of plant pathology and entomology. Insecticides have been the major control strategy but have had variable success and have elicited environmental concerns (Perring *et al.*, 1999). Parasite-derived resistance (PDR) (Sanford and Johnston, 1985), a strategy in which portions of a virus genome are inserted into a plant to protect it from later infection, has been commercially implemented in two crops, papaya and squash. Although virus-resistant papaya and squash have never been planted on more than 0.1 million ha in a given year, compared with a total of 67.7 million ha of total biotech crops (James, 2005), there are unique aspects and considerable potential in this technology. All commercial genetically engineered virus-resistant plants to date have been for control of the papaya ringspot virus (PRSV). Papaya is grown worldwide and is an important plant for small backyards and larger plantations. PRSV is the most widespread and damaging virus affecting papaya. Beginning in 1986, the Gonsalves laboratory began to utilize the PDR concept by cloning the coat protein gene of PRSV HA 5-1, a mild mutant of the virus (Gonsalves, 1998). What is unique about the work on papaya is that it was neither developed nor commercialized by a large corporation; rather the task of creating, deregulating and commercializing transgenic papaya was done largely through university scientists and the Hawaiian papaya growers. Transgenic papaya is now widely grown in Hawaii and is credited with saving the industry. Although papaya and squash are the only crops genetically engineered for disease control that are currently being produced commercially, significant efforts are underway to produce a wide range of disease-resistant plants, including many fruit, vegetable and field crops (Khachatourians *et al.*, 2002). As with plants genetically engineered for insect protection and weed management, there are technical, ecological, social and trade issues involved with the development and deployment of genetically engineered disease-resistant plants (De Boer, 2003). However, such plants offer tremendous opportunities in the management of the world's major diseases affecting crops in the field and in storage.

Conclusions

IPM is a philosophy of pest management that relies on information about pest biology, economics of the system and appropriate management tactics that are environmentally sound and economically feasible. Because IPM is a philosophy, it is difficult to say whether one particular tactic used in a particular cropping system is an IPM tactic. One must look at the whole system and assess how that tactic affects the system. In reality, however, those who practice IPM would readily admit that they cannot know all the interactions, so they mainly focus on better understanding how the pest they wish to manage interacts with its environment, and use that knowledge rather than prophylactic pesticide sprays as a starting point. It is hoped that the philosophy of IPM will help implement more ecologically based tactics and thus avoid the scenario depicted in *Silent Spring*. However, to do this requires knowledge of the system, and biotechnology can help shape the future of IPM as a knowledge-based system. The tools developed through biotechnology can help growers; consultants and scientists obtain in a rapid and detailed fashion the needed information about pest status in the field, or understand the molecular basis of the infection process. At least as important, however, is its ability to help create new strategies that

can be used in a variety of pest management programmes, including novel techniques to control insect-transmitted pathogens of medical and agricultural importance. But for now, the only commercialized genetically engineered products of biotechnology used in agriculture have been those developed for weed, insect and disease management. Since their introduction in 1996 until 2005, a cumulative total of over 475 million ha of transgenic crops were grown in 21 countries (James, 2005). In 2005, the estimated total of transgenic plants grown for pest management was 90 million ha, an 11% increase over 2002 (James, 2005).

The development of transgenic plants has not been without controversy about its risks and benefits. There seems to have been little, if any, controversy about using the tools of biotechnology to assess the genetic characteristics of a population (e.g. frequency of resistance alleles), to facilitate breeding through marker-assisted breeding technologies, or to understand the molecular basis of a disease cycle. However, there is far more controversy about using it to create transgenic insects for the control of human diseases (e.g. malaria) or to create plants that resist insects or diseases, or using it in weed management. Plants engineered for pest management have generated the most controversy. The controversy has largely revolved around safety for humans and the environment, and social issues such as who develops, owns, markets and regulates the plants (Shelton, 2004). A detailed discussion of the technical, environmental and social risks and benefits of biotechnology is beyond the scope of this chapter, and for further discussion the reader is referred to a recent article by Shelton (2004). However, if one focuses on the role of genetically engineered plants for IPM, it is clear to us that substantial benefits will occur, although each case must be evaluated individually. In the case of Bt plants for insect control, the use of Bt cotton has been successful in reducing the number of sprays and allowing natural enemies to flourish. In China, Bt cotton plants have provided a 60–80% decrease in the use of foliar insecticides (Xia *et al.*, 1999), and in Australia and the USA, similar reductions have occurred (Shelton, 2004). In addition to environmental benefits such as reduced harm to natural enemies, a recent report (Hossain *et al.*, 2004) provided the first evidence of a direct link between the adoption of a genetically engineered crop (Bt cotton) and improvements in human health achieved through the reduced use of more harmful insecticides.

Although genetically engineered crops have been adopted by both small-scale and large-scale growers in several countries, their rapid adoption has been hindered by social concerns. For example, because of restrictions on importing transgenic crops into Japan – the main market for Hawaiian papaya prior to widespread infection by PRSV – transgenic papaya is not shipped to Japan where prices are highest, but rather to the mainland states. Ironically, Japanese visitors, a major tourist source, likely consume genetically engineered papaya while in Hawaiian restaurants. Efforts are underway between Hawaiian papaya producers and regulators in Japan to allow the importation of genetically engineered papaya, but this process has taken longer than anticipated. The acceptance of Bt potatoes, first commercialized in 1996, was hindered by concerns of processors about a potential backlash from consumers concerned about the environment or human health (Ken-nedy, 2003). As a substitute for Bt potatoes for the control of CPB, growers began to use a new class of insecticides (neonicotinoids). One could argue that the use of Bt plants would have been more in line with the philosophy of IPM.

There is a revolution occurring in biology and agriculture, and the tools of biotechnology are playing an important role (Shelton *et al.*, 2002). The expectations for biotechnology to affect IPM have been great and some successes have already been achieved. Some uses of biotechnology have elicited little if any controversy (e.g. monitoring for resistance to a pesticide), while others such as genetically engineered plants for pest control will require additional educational efforts, a longer time frame and a regulatory framework before they reach the same level of acceptance. Over time we are convinced that the broad range of tools of biotechnology will play an even more important role in the development of more economically and environment-friendly IPM programmes.

Acknowledgements

The authors gratefully acknowledge C. Smart for reviewing the section on plant pathogens and H. Collins for assistance in editing.

References

Albrecht, H.R. (1947) Strain differences in tolerance to 2,4-D in creeping-bent grasses. *Journal of the American Society of Agronomy* 39, 163–165.

AAM (American Academy of Microbiology) (2002) Hundred years of *Bacillus thuringiensis*: a critical evaluation. Available at: http://www.asmusa.org/

Aultman, K.S., Gottlieb, M., Giovanni, M.Y. and Fauci, A.S. (2002) *Anopheles gambiae* genome: Completing the malaria triad. *Science* 298, 13.

Bates, S.L. (2002) Detection, impact and management of the western conifer seed bug, *Leptoglossus occidentalis*, in lodgepole pine seed orchards. PhD thesis, Simon Fraser University, Burnaby, British Columbia, Canada.

Bates, S.L., Zhao, J.Z., Roush, R.T. and Shelton, A.M. (2004) Insect resistance management in GM crops: past, present and future. *Nature Biotechnology* 23, 57–62.

Beard, C.B., Cordon-Rosales, C. and Durvasula, R. (2002) Bacterial symbionts of the Triatominae and their potential use in control of Chagas disease transmission. *Annual Review of Entomology* 47, 123–141.

Bourget, D., Chaufaux, J., Micoud, A., Delos, M., Naibo, B., Bombarde, F., Marque, G., Eychenne, N. and Pagliari, C. (2002) *Ostrinia nubilalis* parasitism and the field abundance of non-target insects in transgenic *Bacillus thuringiensis* corn (*Zea mays*). *Environmental Biosafety Research* 1, 49–60.

Cao, J., Shelton, A.M. and Earle, E.D. (2001) Gene expression and insect resistance in transgenic broccoli containing a *Bacillus thuringiensis* cry1Ab gene with the chemically inducible PR-1a promoter. *Molecular Breeding* 8, 207–216.

Carozzi, N. and Koziel, M. (1997) *Advances in Insect Control: The Role of Transgenic Plants.* Taylor Francis, London.

Carpenter, J., Felsot, A., Goode, T., Haming, M., Onstad, D. and Sankula, S. (2002) *Comparative Environmental Impacts of Biotechnology-derived and Traditional Soybean, Maize and Cotton Crops.* Council for Agricultural Science and Technology, Ames, Iowa.

Carson, R. (1962) *Silent Spring.* Fawcett World Library, New York.

Cassida, J.E. and Quistad, G.B. (1998) Golden age of insecticide research: past, present or future. *Annual Review of Entomology* 43, 1–16.

Conner, A.J., Glare, T.R. and Nap, J.P. (2003) The release of genetically modified crops into the environment. II. Overview of ecological risk assessment. *The Plant Journal* 33, 19–46.

Cook, R.J. (2000) Advances in plant health management in the twentieth century. *Annual Review of Phytopathology* 38, 95–116.

Daly, J.C. and Trowell, S. (1996) Biochemical approaches to the study of ecological genetics: the role of selection and gene flow in the evolution of insecticide resistance. In: Symondson, W.O.C. and Liddell, J.E. (eds) *The Ecology of Agricultural Pests.* Chapman and Hall, London, pp. 73–92.

De Boer, S.H. (2003) Perspective on genetic engineering of agricultural crops for resistance to disease. *Canadian Journal of Plant Pathology* 25, 10–20.

DeGennaro, R.P. and Weller, S.C. (1984) Differential sensitivity of field bindweed (*Convolvulus arvensis*) biotypes to glyphosate. *Weed Science* 32, 472–476.

Dent, D. (1991) *Insect Pest Management.* CAB International, Wallingford, UK.

Dutton, A., Romeis, J. and Bigler, F. (2003) Assessing the risks of insect resistant transgenic plants on entomophagous arthropods: Bt-maize expressing Cry1Ab as a case study. *BioControl* 48, 611–636.

Economic Research Service (2004) Adoption of genetically engineered crops in the United States. Available at: http://www.ers.usda.gov/data/biotechcrops/

Feng, P.C.C., Pratley, J.E. and Bohn, J.A. (1999) Resistance to glyphosate in *Lolium rigidum.* II. Uptake, translocation, and metabolism. *Weed Science* 47, 412–415.

Ferré, J. and Van Rie, J. (2002) Biochemistry and genetics of insect resistance to *Bacillus thuringiensis. Annual Review of Entomology* 47, 501–533.

Fraaije, B.A., Butters, J.A., Coelho, J.M., Jones, D.R. and Hollomon, D.W. (2002) Following the dynamics of strobilurin resistance in *Blumeria graminis* f.sp. *tritici* using quantitative allele-specific real-time PCR measurements with the fluorescent dye SYBR Green I. *Plant Pathology* 51, 45–54.

Fryxell, K.J. and Miller, T.A. (1995) Autocidal biological control: a general strategy for insect control based on genetic transformation with a highly conserved gene. *Journal of Economic Entomology* 88, 1221–1232.

Gahan, L.J., Gould, F. and Heckel, D.G. (2001) Identification of a gene associated with Bt resistance in *Heliothis virescens. Science* 293, 857–860.

Gebhardt, C. and Valkonen, J. (2001) Organization of genes controlling disease resistance in potatoes. *Annual Review of Phytopathology* 39, 79–102.

Gold, S.E., Garcia-Pedrajas, M.D. and Martinez-Espinoza, A.D. (2001) New (and used) approaches to the study of fungal pathogenicity. *Annual Review of Phytopathology* 39, 337–365.

Gonsalves, D. (1998) Control of papaya ringspot virus in papaya: a case study. *Annual Review of Phytopathology* 36, 415–437.

Gould, J. (1998) Sustainability of transgenic insecticidal cultivars – integrating pest genetics and ecology. *Annual Review of Entomology* 43, 701–726.

Gratz, N.G. (1999) Emerging and resurging vector-borne diseases. *Annual Review of Entomology* 44, 51–75.

Greenstone, M.H. (1995) Bollworm or budworm? Squashblot immunoassay distinguishes eggs of *Helicoverpa zea* and *Heliothis virescens. Journal of Economic Entomology* 88, 112–119.

Groot, A.T. and Dicke, M. (2002) Insect resistant transgenic plants in a multi-trophic context. *The Plant Journal* 31, 387–406.

Hagler, J.R. and Naranjo, S.E. (1996) Using gut content immunoassays to evaluate predaceous biological control agents: a case study. In: Symondson, W.O.C. and Liddell, J.E. (eds) *The Ecology of Agricultural Pests.* Chapman and Hall, London, pp. 383–400.

Hajek, A.E. and St Leger, R.J. (1994) Interactions between fungal pathogens and insect hosts. *Annual Review of Entomology* 39, 293–322.

Heap, I.M. (2004) International survey of herbicide-resistant weeds. Available at: http://www.weedscience. com

Heckel, D.G. (2003) Genomics in pure and applied entomology. *Annual Review of Entomology* 48, 235–260.

HRAC (Herbicide Resistance Action Committee) (2004) Available at: http://www.plantprotection.org/HRAC/

Hilder, V.A. and Gatehouse, A.M.R. (1990) Transforming plants as a means of crop protection against insects. *Outlook on Agriculture* 19, 179–183.

Hossain, F., Pray, C.E., Lu, Y., Huang, J., Fan, C. and Hu, R. (2004) Genetically modified cotton and farmers' health in China. *International Journal of Occupational and Environmental Health* 10, 296–303.

Hoy, M.A. (2002) Genetic improvement of biological control agents. In: Pimentel, D. (ed.) *Encyclopedia of Pest Management.* Marcel Dekker, New York, pp. 329–332.

Ishii, H., Fraaije, B.A., Sugiyama, T., Noguchi, K., Nishimura, K., Takeda, T., Amano, T. and Hollomon, D.W. (2001) Occurrence and molecular characterization of strobilurin resistance in cucumber powdery mildew and downy mildew. *Phytopathology* 91, 1166–1171.

James, C. (2005) Preview: global status of commercialized transgenic crops: 2003. *ISAAA Briefs* No. 34, ISAAA, Ithaca, New York.

Karasev, A.V. (2000) Genetic diversity and evolution of closteroviruses. *Annual Review of Phytopathology* 38, 293–324.

Kennedy, G.G. (2003) The rise and fall of Bt potatoes: why and where next? Available at: http://www. nysaes.cornell.edu/ent/faculty/shelton/BiotechVeg/ppt.html

Khachatourians, G.G., McHughen, A., Scorza, R., Nip, W.-K. and Hui, Y.H. (2002) *Transgenic Plants and Crops.* Marcel Dekker, New York.

Koeller, W. (2004) Pathogens. In: *Management of Pest Resistance: Strategies Using Crop Management, Biotechnology, and Pesticides.* Council for Agricultural Science and Technology Special Publication No. 24, Ames, Iowa, pp. 17–19.

Kogan, M. (1998) Integrated pest management: historical perspectives and contemporary developments. *Annual Review of Entomology* 43, 243–270.

Koziel, M.G., Beland, G.L., Bowman, C., Carozzi, N., Crenshaw, R., Crossland, L., Dawson, J., Desai, N., Hill, M., Kadwell, S., Launis, K., Lewis, K., Maddox, D., McPherson, K., Meghji, M.R., Merlin, E., Rhodes, R., Warren, G.W., Wright, M. and Evola, S.V. (1993) Field performance of elite transgenic maize plants expressing an insecticidal protein gene derived from *Bacillus thurengiensis. Bio/Technology* 11, 195–200.

Krattiger, A.F. (1997) Insect resistance in crops: a case study of *Bacillus thuringiensis (*Bt*)* and its transfer to developing countries. *ISAAA Briefs* No. 2. ISAAA, Ithaca, New York, p. 42.

Lee, L.J. and Ngim, J. (2000) A first report of glyphosate-resistant goosegrass (*Eleusine indica* (L) Gaertn) in Malaysia. *Pest Management Science* 56, 336–339.

Liu, J., Poinar, G.O. and Berry, R.E. (2000) Control of insect pests with entomopathogenic nematodes: the impact of molecular biology and phylogenetic reconstruction. *Annual Review of Entomology* 45, 287–306.

Liu, D., Burton, S., Glancy, T., Li, Z., Hampton, R., Meade, T. and Merlo, D.J. (2003) Insect resistance conferred by 283-kDa *Photorhabdus luminescens* protein TcdA in *Arabidopsis thaliana. Nature Biotechnology* 21, 1222–1228.

Lozzia, G.C. (1999) Biodiversity and structure of ground beetle assemblages in Bt corn and its effects on nontarget insects. *Bollettino di Zoologia Agraria e di Bachicoltura* 31, 37–58.

Ma, Z. and Michailides, T.J. (2004) An allele-specific PCR assay for detecting azoxystrobin-resistant *Alternaria* isolates from pistachio in California. *Journal of Phytopathology* 152, 118–121.

Manachini, B. (2000) Ground beetle assemblages and plant dwelling non-target arthropods in isogenic and transgenic corn crops. *Bollettino di Zoologia Agraria e di Bachicoltura* 32, 181–198.

Martin, R.R., James, D. and Levesque, C.A. (2000) Impacts of molecular diagnostic technologies on plant disease management. *Annual Review of Phytopathology* 38, 207–239.

Martin, G.B., Bodganove, A.J. and Sessa, G. (2003) Understanding the functions of plant disease resistance proteins. *Annual Review of Plant Biology* 54, 23–61.

McDonald, B.A. and Linde, C. (2002) Pathogen population genetics, evolutionary biology and durable resistance. *Annual Review of Phytopathology* 40, 349–379.

Menken, S.B.J. and Raijmann, L.E.L. (1996) Biochemical systematic: principles and perspectives for pest management. In: Symondson, W.O.C. and Liddell, J.E. (eds) *The Ecology of Agricultural Pests.* Chapman and Hall, London, pp. 7–30.

Michelbacher, A.E. and Bacon, O.G. (1952) Walnut insect and spider mite control in northern California. *Journal of Economic Entomology* 45, 1020–1027.

Miller, T.A. (2002) Lethal genes for insect control. In: Pimentel, D. (ed.) *Encyclopedia of Pest Management.* Marcel Dekker, New York, pp. 448–450.

Morin, S., Biggs, R.W., Sisterson, M.S., Shriver, L., Ellers-Kirk, C., Higginson, D., Holley, D., Gahan, L.J., Heckel, D.G., Carriere, Y., Dennehy, T.J., Brown, J.K. and Tabashnik, B.E. (2003) Three cadherin alleles associated with resistance to *Bacillus thuringiensis* in pink bollworm. *Proceedings of the National Academy of Sciences USA* 100, 5004–5009.

Moscardi, F. (1999) Assessment of the application of baculoviruses for control of Lepidoptera. *Annual Review of Entomology* 44, 257–289.

Musser, F.R. and Shelton, A.M. (2003) Bt sweet corn and selective insecticides: their impacts on sweet corn pests and predators. *Journal of Economic Entomology* 96, 71–80.

Nagamatsu, Y., Koike, T., Sasaki, K., Yoshimoto, A. and Furukawa, Y. (1999) The cadherin-like protein is essential to specificity determination and cytotoxic action of the *Bacillus thuringiensis* insecticidal Cry1Aa toxin. *FEBS Letters* 460, 385–390.

Naranjo, S.E. (2005) Long-term assessment of the effects of transgenic Bt cotton on the abundance of nontarget arthropod natural enemies. *Environmental Entomology* 34, 1193–1210.

Naranjo, S.E. and Ellsworth, P.C. (2002) Looking for functional non-target differences between transgenic and conventional cottons: implications for biological control. In: *2002 Arizona Cotton Report Series P-130.* Available at: http://ag.arizona.edu/pubs/crops/az1283/contents.html

Naranjo, S.E., Head, G. and Dively, G.P. (2005) Field studies assessing arthropod nontarget effects in Bt transgenic crops: Introduction. *Environmental Entomology* 34, 1178–1180.

O'Brochta, D. (2002) Transgenic mosquitoes: the state of the art. In: Takken, W. and Scott, T. (eds) *Proceedings of the Frontis Workshop on Ecological Challenges Concerning the Use of Genetically Modified Mosquitoes for Disease Control.* Wageningen, The Netherlands. Available at: http://library.wur.nl/frontis/malaria/index.html

Orr, D.B. and Landis, D.A. (1997) Oviposition of European corn borer and impact of natural enemy populations in transgenic versus isogenic corn. *Journal of Economic Entomology* 90, 905–909.

Pedley, K.F. and Martin, G.B. (2003) Molecular basis of *Pto*-mediated resistance to bacterial speck disease in tomato. *Annual Review of Phytopathology* 41, 215–243.

Perlak, F.J., Deaton, R.W., Armstrong, T.A., Fuchs, R.L. and Sims, S.R. (1990) Insect resistant cotton plants. *Bio/Technology* 8, 939–943.

Perlak, F.J., Fuchs, R.L., Dean, D.A., McPherson, S.L. and Fischhoff, D.A. (1991) Modification of coding sequence enhances plant expression of insect control protein genes. *Proceedings of the National Academy of Sciences USA* 88, 3325–3328.

Perring, T.M., Gruenhagen, N.M. and Farrar, C.A. (1999) Management of plant viral diseases through chemical control of insect vectors. *Annual Review of Entomology* 44, 457–481.

Pickett, J.A. (1998) The future of semiochemicals in pest control. In: *Aspects of Applied Biology No. 17, Environmental Aspects of Applied Biology.* The Association of Applied Biologists, Wellsbourne, UK, pp. 397–406.

Pilcher, C.D., Obrycki, J.J., Rice, M.E. and Lewis, L.C. (1997) Preimaginal development, survival and field abundance of insect predators on transgenic *Bacillus thuringiensis* corn. *Environmental Entomology* 26, 446–454.

Powles, S.B., Lorraine-Colwill, D.F., Dellow, J.J. and Preston, C. (1998) Evolved resistance to glyphosate in rigid ryegrass (*Lolium rigidum*) in Australia. *Weed Science* 46, 604–607.

Ristanio, J.B. and Gumpertz, M.L. (2000) New frontiers in the study of dispersal and spatial analyses of the epidemics caused by species in the genus *Phytophthora*. *Annual Review of Phytopathology* 38, 541–576.

Romeis, J., Dutton, A. and Bigler, F. (2004) *Bacillus thuringiensis* toxin (Cry1Ab) has no direct effect on larvae of the green lacewing *Chrysoperla carnea*. *Journal of Insect Physiology* 50, 175–183.

Romeis, J., Meissle, M. and Bigter, F. (2006) Transgenic crops expressing *Bacillus Thuringiensis* toxins and biological control. *Nature Biotechnology* 24, 63–71.

Roush, R.T. (1997) Managing resistance to transgenic crops. In: Carozzi, N. and Koziel, M. (eds) *Advances in Insect Control: The Role of Transgenic Plants.* Taylor and Francis, London, pp. 271–294.

Roush, R.T. (1998) Two-toxin strategies for management of insect resistant transgenic crops: can pyramiding succeed where pesticide mixtures have not? *Philosophical Transactions of the Royal Society of London B* 353, 1777–1786.

Ryan, G.F. (1970) Resistance of common groundsel to simazine and atrazine. *Weed Science* 18, 614–616.

Sanford, J.C. and Johnston, S.A. (1985) The concept of parasite-derived resistance-deriving resistance genes from the parasite's own genome. *Journal of Theoretical Biology* 113, 395–405.

Sankula, S. and Gianessi, L. (2004) Impact of biotechnology-derived herbicide-resistant crops on the NE United States. *Proceedings of the Northeast Weed Science Society* 58, 11.

Schnepf, E., Crickmore, N., van Rie, J., Lereclus, D., Baum, J., Feitelson, J., Zeigler, D.R. and Dean, D.H. (1998) *Bacillus thuringiensis* and its pesticidal crystal proteins. *Microbiology and Molecular Biology Review* 62, 775–806.

Scholthof, K.G. (2003) One foot in the furrow: linkages between agriculture, plant pathology, and public health. *Annual Review of Public Health* 24, 153–174.

Schuler, T.H., Potting, R.P., Denholm, I. and Poppy, G.M. (1999) Parasitoid behaviour and Bt plants. *Nature* 400, 825.

Schuler, T.H., Denholm, I., Clark, S., Stewart, C.N. and Poppy, G. (2004) Effects of Bt plants on the development and survival of the parasitoid *Cotesia plutellae* in susceptible and Bt-resistant larvae of the diamondback moth *Plutella xylostella*. *Journal of Insect Physiology* 50, 435–443.

Sears, M.K., Hellmich, R.L., Stanley-Horn, D.E., Oberhauser, K.S., Pleasants, J.M., Mattila, H.R., Siegfried, B.D. and Diverly, G.P. (2001) Impact of Bt corn pollen on monarch butterfly populations: a risk assessment. *Proceedings of the National Academy of Sciences USA* 98, 11937–11942.

Shelton, A.M. (2004) Risks and benefits of agricultural biotechnology. In: Ahmed, F. (ed.) *Testing of Genetically Modified Organisms in Foods*. Haworth Press, Binghamton, New York, pp. 1–53.

Shelton, A.M. and Roush, R.T. (2000) Resistance to insect pathogens and strategies to manage resistance. In: Lacey, L.A. and Kaya, H.K. (eds) *Field Manual of Techniques in Invertebrate Pathology*. Kluwer Academic Press, Dordrecht, The Netherlands, pp. 829–846.

Shelton, A.M., Zhao, J.Z. and Roush, R.T. (2002) Economic, ecological, food safety and social consequences of the deployment of Bt transgenic plants. *Annual Review of Entomology* 47, 845–881.

Smith, C.M. (1989) *Plant Resistance to Insects*. John Wiley & Sons, New York.

Stern, V.M., Smith, R.F., van den Bosch, R. and Hagen, K. (1959) The integrated control concept. *Hilgardia* 29, 81–101.

Tabashnik, B.E., Carriere, Y., Dennehy, T.J., Morin, S., Sisterson, M.S., Roush, R.T., Shelton, A.M. and Zhao, J.Z. (2003) Insect resistance to transgenic Bt crops: lessons from the laboratory and field. *Journal of Economic Entomology* 96, 1031–1038.

Vadlamudi, R.K., Weber, E., Ji, I., Ji, T.H. and Bulla, L.A. (1995) Cloning and expression of a receptor for an insecticidal toxin of *Bacillus thuringiensis*. *Journal of Biological Chemistry* 270, 5490–5494.

Vaeck, M., Reynaerts, A., Hofte, H., Jansens, S. and Beuckeleer, M.D. (1987) Transgenic plants protected from insect attack. *Nature* 328, 33–37.

VanGessel, M.J. (2001) Glyphosate-resistant horseweed from Delaware. *Weed Science* 49, 703–705.

Wold-Burkness, S.E., Hutchison, W. and Venetee, R. (2001) In-field monitoring of beneficial insect populations in transgenic corn expressing a *Bacillus thuringiensis* toxin. *Journal of Entomological Science* 36, 177–187.

Wolfenbarger, L. (2003) Annotated bibliography on environmental and ecological impacts from transgenic plants. II. Unintended effects. Available at: http://www.isb.vt.edu/eeito_bibs/plant_Unintended_Effects.cfm

Wood, H.A., Hughes, P. and Shelton, A. (1994) Field studies of the co-occlusion strategy with a genetically altered isolate of the *Autographa californica* nuclear polyhedrosis virus. *Environmental Entomology* 23, 211–219.

Xia, J.Y., Cui, J.J., Ma, L.H., Dong, S.X. and Cui, X.F. (1999) The role of transgenic Bt cotton in integrated pest management. *Acta Gossypii Sinica* 11, 57–64.

Yencho, G.C., Cohen, M.B. and Byrne, P.F. (2000) Applications of tagging and mapping insect resistance loci in plants. *Annual Review of Entomology* 45, 393–422.

Zhao, J.Z., Cao, J., Li, Y., Collins, H.L., Roush, R.T., Earle, E.D. and Shelton, A.M. (2003) Plants expressing two *Bacillus thuringiensis* toxins delay insect resistance compared to single toxins used sequentially or in a mosaic. *Nature Biotechnology* 21, 1493–1497.

Zheng, D., Olaya, G. and Koeller, W. (2000) Characterization of laboratory mutants of *Venturia inaequalis* resistant to the strobilurin-related fungicide kresoxim-methyl. *Current Genetics* 38, 148–155.

14 Grower Perspectives on Areawide Wheat Integrated Pest Management in the Southern US Great Plains

Sean P. Keenan,[1] Kristopher L. Giles,[1] Norman C. Elliott,[2] Tom A. Royer,[1] David R. Porter,[2] Paul A. Burgener[3] and David A. Christian[3]

[1]Department of Entomology and Plant Pathology, Oklahoma State University, Stillwater, OK 74078, USA; [2]USDA–ARS, Plant Science and Water Conservation Laboratory, Stillwater, OK 74075, USA; [3]The University of Nebraska Panhandle Research & Extension Center, Scottsbluff, NE 69361, USA

Introduction

Areawide wheat integrated pest management (IPM) is a program of the United States Department of Agriculture–Agricultural Research Service (USDA–ARS). To implement the program, the USDA–ARS established cooperative agreements with research and extension professionals from five universities: Colorado State University, Kansas State University, Oklahoma State University, Texas A&M University and the University of Nebraska. The 4-year implementation phase of this program began in the fall of 2002. The main theme of the program is to collaborate with growers of dryland winter wheat to evaluate and demonstrate non-chemical pest management techniques, with particular emphasis on the management of the Russian wheat aphid *Diuraphis noxia* (Mordvilko) and the greenbug *Schizaphis graminum* (Rondani).

Chemical control of insect pests in dryland winter wheat is undesirable for several reasons. One of these is that very little can be spent on managing pests to maintain acceptable profit margins in winter wheat

(Dhuyvetter *et al.*, 1996; Holtzer *et al.*, 1996). Another is that widespread outbreaks of aphid pests are infrequent. Two consequences of these factors are that growers do not typically budget for insecticide applications or carefully monitor their fields for aphid outbreaks. When outbreaks do occur, insecticide applications are more likely to be made after substantial crop damage has occurred.

For the Russian wheat aphid and greenbug, host plant resistance, cultural practices and biological control by natural enemies have the greatest potential for preventive pest management (Elliott *et al.*, 1998a; Peairs, 1998; Souza, 1998). A central theme from the literature is that preventive tactics are most likely to succeed when implemented in a diversified cropping system; in this case, where winter wheat is grown in rotation with spring or summer crops. These tactics, if incorporated into the management system as routine practices, can prevent economic infestations of aphid pests from occurring. Successful preventive pest management maintains pests below economic thresholds most of the time. Analogous to human medicine, a comprehensive and cost-effective pest management system for wheat will include both low-cost preventive tactics, which are built into the cropping system, and curative tactics (e.g. insecticide applications), which are used only when preventive tactics fail.

In this chapter we review aspects of the areawide wheat program after its first year of implementation, paying particular attention to results obtained from focus groups with cooperating wheat growers between January and March 2003. We observe that some growers are finding success with cropping systems that may prevent economically significant aphid infestations. By increasing awareness of grower success stories with these cropping systems, the areawide program can contribute to improved pest management and agricultural profitability on a large geographical scale. The program also represents an opportunity to coordinate education and technology transfer efforts regarding benefits of resistant wheat varieties, pest and natural enemy identification, and simplified methods of field scouting.

Program Development

Background

Winter wheat is well suited to the growing conditions of the central and southern US Great Plains. It is resilient to stress as a dryland crop and its water needs fit well with the tendency for most rainfall to occur in the fall and spring. Another advantage of winter wheat in southern portions of the Great Plains is its 'dual purpose' as fall/winter cattle forage and subsequent grain production in the following spring from the same planting.

Wheat production in the region traditionally involves winter wheat followed by a year of fallow in semi-arid locales and continuous winter wheat (planted every fall) in areas where precipitation levels are adequate to support it (Lyon and Baltensperger, 1995; Ali *et al.*, 2000; Royer and Krenzer, 2000). From an ecological stand-

point, wheat monoculture habitats increase reliance on insecticide-based management of aphids by decreasing the presence and effectiveness of the natural factors that regulate pest insects (Andow, 1983, 1991; Way, 1988; Elliott *et al.*, 1998b; French and Elliott, 1999; Brewer *et al.*, 2001; French *et al.*, 2001; Elliott *et al.*, 2002). The Russian wheat aphid and greenbug thrive in the simplified wheat–fallow system, and pest problems in general have increased in this system. The widespread use of more intensive crop rotations could help alleviate this problem (Holtzer *et al.*, 1996). Dry, hot summer weather presents a challenge to growers interested in summer field crops like maize, soybeans, grain sorghum and cotton, though with proper crop management these and other summer crops can be grown successfully in appropriate locales.

Fundamental aspects of 'integrated wheat management' or 'best management practices' for wheat have been promoted for a significant period of time (Heyne, 1987; Cuperus *et al.*, 1993; Watson, 1994). Even though effective IPM tools have been developed, they have not been integrated into a comprehensive program and tested. There are also a number of ways that adoption of IPM differs from research-based outcomes and recommendations. Agricultural innovations are often developed with the implicit assumption that end users – the farmers – will adopt these technologies out of necessity to remain competitive and to achieve acceptable profits from production. In reality, there are many circumstantial influences on the adoption decision. Growers vary in the extent to which they can assume risk involved with adoption of a new practice and in their ability to continue the practice if not successful in the short term.

One way for IPM education to be effective in this context is to involve the end user from the beginning of the effort to develop and disseminate program information. Towards this end, an objective of areawide pest management is to foster partnerships with government agencies, farmers and the private sector. This is a principal goal of USDA–ARS areawide IPM programs:

Social, political, and economic factors must come together with science before an areawide program can succeed. In addition, scientific challenges include defining the appropriate geographic area, selecting the control approaches to test and combine, and addressing the different life cycles of the target pest as well as secondary pests.

(Faust, 2001, p. 2)

In the areawide wheat program, we use several methods to monitor pests in dryland cropping systems, determine grower needs, transfer technology and assess program success. The maximum impact of a program based on these technologies will be achieved when implemented over a broad geographical area. Farms in wheat-growing areas of the Great Plains are typically several thousand acres in size. From the standpoint of logistics and cost, an individual farm is the most practical spatial unit for implementing this preventive pest management strategy.

Due to these factors, growers are the key to implementation of areawide wheat IPM.

The areawide wheat program focuses on regions of the central and southern US Great Plains where the Russian wheat aphid and greenbug are significant pests. The Russian wheat aphid is the predominant aphid pest of winter wheat in the northern and westernmost areas depicted in Fig. 14.1, stretching from Wyoming and Nebraska down to the Texas Panhandle. The greenbug is a significant pest in portions of Kansas, Oklahoma, Texas and Colorado. The ranges of these pests overlap in south-eastern Colorado and portions of the Oklahoma and Texas Panhandle regions.

The areawide project team established field demonstration sites in three zones, based on the two pest ranges and their overlapping range. We also recruited growers as program participants in clusters of counties within these zones. Zone 1 includes counties in south-eastern Wyoming, north-eastern Colorado and the Nebraska Panhandle.

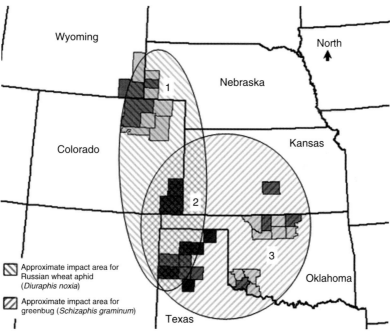

Fig. 14.1. Three zones of implementation for areawide wheat IPM, including approximate ranges of the Russian wheat aphid and greenbug and clusters of counties with participating growers (darker shaded counties contain field demonstration sites).

Demonstration field sites are located in Laramie County, Wyoming, Banner County, Nebraska and Weld County, in northern Colorado. Some of the typical rotation crops grown with winter wheat in these counties are sunflower, maize and Proso millet. Zone 2 includes counties within, or proximate to, the overlapping ranges of the Russian wheat aphid and greenbug. Demonstration field sites are located in Prowers and Baca counties in south-eastern Colorado, and in Ochiltree, Hutchinson, Moore, Deaf Smith and Swisher counties in the Texas Panhandle. Grain sorghum (or 'milo') is often grown in rotation with dryland winter wheat in this zone, along with other crops. Zone 3 includes wheat-growing counties in eastern and southern regions where the greenbug is a significant pest. Demonstration field sites are located in Reno County, Kansas, and in Alfalfa, Kay and Jackson counties in Oklahoma. Dryland crop rotations in this zone vary by location, but may include grain sorghum, maize, soybeans or cotton among other crops. Throughout the project region, researchers and growers alike are experimenting with other crops that may be grown in rotation with dryland winter wheat.

The three zones also differ significantly in terms of average rainfall, length of growing season and average farm size. In general, average annual rainfall increases significantly west to east, and length of growing season increases north to south. As a rule, farmers in northern and western areas tend to grow crops on larger acreages at lower per-acre yields compared with farmers in southern and eastern areas. In both zone 2 and zone 3, winter wheat is frequently used for grazing in addition to harvesting for grain. Growers participating in the areawide wheat program are generally representative of these patterns, as illustrated below.

Demonstration methods and grower recruitment

Demonstration is widely used in Cooperative Extension work for developing and disseminating information about agricultural innovations, and it is a significant component of the areawide wheat program. Research and extension personnel working on the program established a total of 23 demonstration field sites with cooperating wheat growers distributed around the project region. Roughly half of these demonstration fields are 'wheat-only' systems, farmed either as continuous wheat or as a wheat–fallow rotation. The other systems have one or more spring or summer crops grown in rotation with winter wheat. The project team developed an intensive field-monitoring protocol to monitor pest and natural enemy abundance in wheat and in rotational crops. The work began in the fall of 2002 with the planting of the winter wheat crop for harvest in the following spring. Researchers will continue to monitor wheat and rotation crops on these fields for a 4-year period, completing with the 2006 harvest.

Economic and sociological evaluations of farming systems are being conducted through annual cost-of-production interviews with growers and through focus group discussions. A series of 20 focus groups was the medium through which we initiated grower involvement in the program and obtained baseline information about growers' decisions regarding cropping systems and pest management. Following focus group discussions, we have maintained education and technology transfer efforts through a quarterly newsletter and an Internet website. One goal in these efforts is to incorporate the end user in the process of program implementation.

In addition to 23 growers who farm demonstration fields, we recruited 123 growers as participants in focus groups and cost-of-production surveys. They were from counties with demonstration field sites and adjacent counties wherever feasible. As we were interested in recruiting growers who had similar farm sizes and characteristics to growers of demonstration field sites, we primarily relied on county Cooperative Extension personnel to identify possible participants. We were particular in identifying equal numbers of growers who have and have not had success (or interest) in growing alternative crops in a rotation with dryland winter wheat.

In focus group discussions, we explored grower decisions regarding cropping systems and pest management. We used the same question route for all focus groups. In contrast to a questionnaire or structured personal interview, a focus group question route provides a general direction for discussion. One or two initial questions are presented in a 'round robin' fashion, whereby the moderator asks the group to 'go around the table' to get acquainted and help everyone feel comfortable to speak. Once the participants appear at ease, the moderator poses subsequent questions to the group as a whole, allowing anyone to initiate responses and others to provide follow-up responses or clarifying questions. The moderator interjects to probe for details, solicit responses from reserved individuals, or move the discussion onto the next topic (Morgan, 1997; Morgan and Krueger, 1998; Greenbaum, 2000; Krueger and Casey, 2000).

Our focus group question route, displayed in Fig. 14.2, was designed to broadly explore growers' decisions regarding winter wheat crop production, selection of alternative crops for rotation with winter wheat (or other alternative enterprises, such as livestock production), and finally growers' experiences with weeds, insects and plant diseases in winter wheat. For synthesis and comparison, focus group discussions were audio-recorded and transcribed. These texts were indexed by focus group number, location and speech segment (excluding any personally identifying information). As focus group questions are loosely structured and participants may respond to one another, useful information regarding topics covered may occur at any point in the discussion, and not just in response to a specific question posed by the moderator. Consequently, the typed transcript is invaluable in reassembling the discourse later to evaluate information obtained and to compare focus group sessions.

Table 14.1 summarizes the number of participants we recruited by project zone and centralized locations where we held focus groups. A total of 138 growers participated in 1 of 20 focus group discussions between January and March 2003. Following focus group discussions, we personally interviewed all participants and 8 additional growers who were not able to attend one of the focus groups, for a total of 146 growers involved with the project in the first year of implementation (2002–2003). Consistent with the larger number of demonstration sites in Colorado and Oklahoma, we had more participating growers in those states – a total of 42 in Oklahoma locations and 37 in Colorado locations (with Colorado split between northern and southern areas). Growers in Nebraska and Wyoming combined for a total of 28 participants in that part of the program area. We had the least number of growers in Kansas because we had only two pairs of demonstration sites in that state, located in Reno County.

In interviews, we obtained baseline data on farm acreage, crop yields, livestock and other operation characteristics. In the areawide wheat program, we will be conducting annual interviews to obtain this information for 4 successive years of crop production (crop years 2003–2005). Table 14.2 summarizes average acreages farmed and head of livestock per farm operator, based on our 2002 interview. As our project was designed to work with commercial operators of dryland wheat farms in the US Great Plains, these operators tended to have a large number of dryland acres (an average of 2367 dryland acres per farm) and comparatively few irrigated acres (an average of 319 irrigated acres per farm). However, the project includes some growers who have a large number of irrigated acres in addition to dryland, and many growers who have no irrigated acres, depending mostly on the prevalence of irrigation in particular locales. As we noted earlier, farms tend to be somewhat larger in terms of acres farmed in the more arid areas of the Great Plains (including zones 1 and 2 of our project) and comparatively smaller in less arid areas (including zone 3 of our project). This tendency is evident for the averages by zone

1. Briefly tell us about yourself:
   - Who you are
   - The place you consider home
   - How long you have been farming
   - Crops that you currently grow, including cattle if you run them

   You do not need to tell us how many acres you produce or head of cattle you stock. Instead, just give us a sense of what you produce.

2. Let us go around the room one more time. Tell us about:
   - Crops you have grown in the past but no longer grow
   - Any new crops you are thinking about growing (or new cropping practices)
   - Anything else you would like to add

3. If a grower is thinking about a new crop (or new cropping practice) here, what are the greatest challenges or limitations in being able to do that?

4. How does your wheat look this year? (recently planted crop in your area)
   - Follow-up: How do you make decisions?
   - What were some decisions you made in planting your current wheat crop?
   - Are these the decisions you typically make?
   - If anything different, whom did you talk to about it? (what information did you consult?)

5. Now I am going to ask about weeds, plant diseases and insect problems for wheat in this area. (Create a list on your index card as we mention some.)
   - What are some problem weeds for wheat fields in this area?
   - What are some wheat diseases you find here?
   - What are some insects you find in wheat fields here?

6. We have mentioned several types of pests in wheat, including insects, weeds and plant diseases. With all of these in mind, what have been your biggest pest concerns over the past year or two?
   - Follow-up: How have you dealt with these?
   - Whom did you ask for advice? (What source of information did you consult, if any?)

7. Thinking back over a longer time period (the last 10 years), what have been the biggest pest problems for wheat production in this area?

8. What do you like most about your farm operation (wheat/cattle/crop rotations)? (Use index cards to list 2–3 things you like most.)

9. What do you like least?

10. We have discussed many questions about your agricultural production today. Is there anything you would like to add?

Fig. 14.2. Focus group question route.

reported in Table 14.2. It is also important to note that rangeland and livestock are significant aspects of the overall farm enterprises of many growers involved in the areawide wheat program (here, too, growers varied greatly – some were heavily invested in livestock enterprises while others had none). The influence of livestock enterprises on farm decision making for some growers was evident in focus groups, as we discuss later in the chapter.

In the next section we review current research regarding components of the areawide wheat program as well as general wheat management practices appropriate to the project area. For each practice, we explore grower comments from our focus groups. Here our goal is to simultaneously explore both the

Table 14.1. Number of demonstration field sites and participating growers by project zones and focus group locations, 2003.

Project zones	Focus groups numbers and locations (conducted January–March, 2003)	Demonstration field sites	Participating growers
1	01 and 02 – Scottsbluff, Nebraska	2	14
	03 and 04 – Pine Bluffs, Wyoming	2	14
	05 and 06 – Brush, Colorado	2	18
2	07 and 08 – Lamar, Colorado	4	19
	09 and 10 – Etter and Perryton, Texas	3	12
	11 and 12 – Umbarger and Claude, Texas	2	14
3	13 and 14 – South Hutchinson, Kansas	2	13
	15 and 16 – Cherokee, Oklahoma	2	17
	17 and 18 – Blackwell, Oklahoma	2	12
	19 and 20 – Altus, Oklahoma	2	13
Total		23	146

scientific and the practical rationales for the adoption or non-adoption of IPM practices in dryland winter wheat. In doing so, we also aim to identify important areas for future IPM education and outreach.

Elements of Areawide Wheat IPM

Environmental factors such as yearly rainfall and soil type vary broadly across the project area. These factors determine the feasibility of various cultural and fertility management approaches. However, all systems involving winter wheat share certain agronomic needs and production characteristics that may increase or decrease potential damage to wheat from aphid infestations. The areawide wheat program focuses on demonstrating the effectiveness of IPM practices in dryland winter wheat systems. The most important elements of this demonstration are crop rotation, host plant resistance, biological control and field scouting. Other wheat practices that are important for a complete IPM program will be evaluated as well. These include fertility management, planting date and wheat grazing considerations, control of weeds and volunteer wheat, and conservation tillage. In this section, we review current research regarding pest management benefits of each

Table 14.2. Averages of acres and livestock per farm operator among project growers (figures from 2002 cost-of-production interviews).

	Project areas			
	Zone 1	Zone 2	Zone 3	All areas
Averages per farm operator of				
Dryland acres farmed	2916.0	2491.2	1804.9	2367.4
Irrigated acres	163.0	766.7	75.4	319.1
Uncultivated acres in Conservation Reserve Programme (CRP)	508.0	560.3	76.7	363.0
Range-pasture acres	1215.5	1484.5	805.5	1146.3
Total head of livestock[a]	416.2	248.4	412.6	362.3
Number of farm operators	46	45	55	146

[a]Total livestock included all categories of on-farm livestock such as cattle, hogs, sheep and horses. The average for zone 1 includes some feedlot operations that significantly increased the average figure for this zone.

of these practices. We also evaluate grower perspectives on each of these practices from our baseline focus groups and interviews.

Crop rotation

Diversification of cropping systems within a field or farm can have several desirable consequences for farmers, one of which is increasing the abundance and effectiveness of natural enemies in biological control of the Russian wheat aphid and greenbug (Elliott *et al.*, 1998b; Ahern and Brewer, 2002; Elliott *et al.*, 2002). These systems can also increase water use efficiency and stabilize or increase farm profits compared with wheat–fallow and wheat-only cropping (Dhuyvetter *et al.*, 1996; Peterson *et al.*, 1996). Furthermore, they facilitate more effective and economical weed control.

A typical dryland diversified cropping system in the Great Plains is a 3-year rotation of winter wheat with an additional crop, such as sorghum, millet, maize or sunflower, typically followed by a fallow period for one growing season prior to planting winter wheat again (Lyon and Baltensperger, 1995; Krall and Schuman, 1996; Peterson *et al.*, 1996). Variations on this theme may involve 'double-cropping' the summer crop immediately after wheat harvest for a shorter rotation cycle, or extending production of summer crops for 2 or more successive years for a longer rotation cycle. Long-term studies confirm that these intensive rotations increase net returns and reduce financial risk compared with the traditional wheat–fallow rotation. As an example, annualized grain production from 1987 to 1993 in dryland wheat–maize–fallow and wheat–maize–millet–fallow was 72% higher than for wheat–fallow, with a 25–40% increase in net annual income (Dhuyvetter *et al.*, 1996).

In subhumid areas in the Great Plains, a fallow growing season is not required between crops used in rotation, and the choice of crops for use in rotation is typically greater than in semi-arid areas. These intensive crop rotations are often grown

without tillage, which reduces soil erosion (Peterson *et al.*, 1996). In more humid areas less study has been conducted on the economics of diversified wheat production systems, and it has not been demonstrated whether greater net profits occur by diversifying the production system. The major goal of the areawide program is to work with growers in evaluating enterprise and whole-farm profitability during the period of program implementation.

In focus groups, growers widely acknowledged benefits of crop rotation, though they varied in the extent to which they have found crop rotation successful or appropriate to their own farming opportunities and constraints. Using crop rotation may help growers to observe the development of pest problems, and consequently they may monitor their fields for insects more than growers who do not rotate crops with winter wheat. As an example, an Oklahoma grower made the following comment regarding chinch bugs:

> FG15-Oklahoma [#60]: Anytime I plant milo along a wheat field I know that I've really got to watch it. I mean just checking about every day when that wheat starts getting ripe because they are going to start coming out of that wheat into your milo. It's just an ongoing learning – like all of you know. The more you are out in your field, watching it, you know what's going on and a lot of times you can head off problems.

One of the most articulate statements about insect management benefits of crop rotation was from another grower in the same focus group:

> FG15-Oklahoma [#42]: In terms of insect control, I think it is a very simple theory of breaking the food chain a little bit for the insects. When you go out of winter wheat for three years on a place, you've changed that food supply for, say, greenbugs.

Later, the same grower continued the thought:

> FG15-Oklahoma [#98]: I've found that you need to stay out of your winter grass production, three years is best, minimum

of two years, to really break cycles of insects and competitive weeds.

Others generally agreed whenever someone in a focus group made statements like these, citing insect management benefits of crop rotation; however, it is important to note that such statements were relatively rare. Growers were much more likely to cite weed management as a benefit of crop rotation, as the following examples show:

FG15-Oklahoma [#96]: When I first started looking at rotation out of winter wheat into some kind of summer crop, it was mainly because of cheat grass infestation, as well as lots of winter weed problems – henbit, mustard, and just go on and on with the list.

FG05-Colorado [#21]: Back in the late '80's I could see the storm clouds on the horizon with the cheat grass and goat grass starting to show up in the wheat and just decided we needed the benefits of a rotation to break the cycles.

FG13-Kansas [#168]: We were experiencing some problems with the weeds and by lengthening our rotation and stacking our crops we have a better handle on some of these weeds.

Some focus group participants were adamant that other growers needed to adopt crop rotation to manage weed problems:

FG15-Oklahoma [#358]: I rotate to keep the fields cleaned up, like the rye problem that we've got (in our area). People don't rotate crops and it's getting out of control.

FG01-Nebraska [#213]: One of my neighbors took a 92 cent-a-bushel dockage fee because of jointed goat grass last year. That's almost a third of the price of his wheat crop – a third of it went to the landlord and a third to the elevator [in dockage]. People have to get serious about this! Crop rotation – we can sit here and discuss whether you want to grow sunflowers, corn, millet, whatever I don't care – but crop rotation is part of the answer to getting these things cleaned up.

One grower from a Kansas focus group expanded on the idea to mention broader environmental and public perception benefits of weed control involving crop rotation:

FG13-Kansas [#417]: Crop rotation is good for soil and environment. We hear so much from the public about how farmers are ruining the land and water, and I think one point we can talk about is crop rotation. I think long-term we'll use less herbicide because weeds can be dealt with easier when there's more crop rotation.

Another benefit of crop rotation noted by growers was labour and equipment efficiency. In particular, growers who have adopted crop rotation as part of a 'no-till' farming operation seemed most likely to cite labour and equipment savings:

FG17-Oklahoma [#72]: Labor is what pushed me into no-till and crop rotation, I guess. Five years ago I got tired of fighting the labor problem (lack of available hired hands) and I turned lose about half of my rented farm ground. I'm down to nine hundred and sixty acres now – what I can take care of myself. I had to go to rotation to break up my harvesting and planting efforts to three times a year instead of all of it at one time.

FG15-Oklahoma [#354]: I guess my first motivation in entering no-till crop rotation was machinery cost per acre, as I'm going to put it. I had a line of machinery that needed updating, and based on my previous twenty-five years of farming wheat, the money wasn't there to do it. So, in a no-till farming operation, I felt I could do more with much less equipment.

FG15-Oklahoma [#358]: Crop rotation splits the workload up. I do a lot of custom work – I custom cut and I custom plant for people – and it allows me to have more time to do that if I'm not farming all of my ground at the same time.

Again, statements similar to these tended to be generally accepted in focus groups, but some individuals were quick to explain why they had not found crop rotation beneficial or appropriate to their own situation: some had experienced failures with summer crops; some were older farmers who were not interested in investing in new crops; some were oriented towards livestock production and primarily used wheat as pasture rather than as a grain crop.

Overall, we found a diverse range of views regarding crop rotation, and growers were more likely to cite weed management or cost savings benefits of crop rotation over insect management benefits. An important objective of the areawide program is to increase grower awareness of insect management benefits of crop rotation. As crop rotation is also closely related to other wheat IPM practices, we examine additional details regarding growers' experiences with crop rotation at the end of the chapter.

Host plant resistance

Growers have several options when choosing a cultivar with disease resistance; however, only a few aphid-resistant cultivars are available throughout the Great Plains (Quick *et al.*, 1996; Kindler *et al.*, 2002; Peairs, 2003). 'TAM-110' was bred for resistance to greenbug biotypes C and E and is adapted to growing conditions in the semi-arid southern Great Plains. It is widely utilized in Texas and the Oklahoma Panhandle. Resistant cultivars for the Russian wheat aphid incorporate the Dn4 resistance gene, which was bred into 'Halt', 'Yumar', 'Prairie Red' and 'Prowers 99'. The resistance mechanism in these cultivars is primarily tolerance with some antibiosis; all cultivars support aphid infestations, but at reduced levels and suffer less injury compared to susceptible cultivars.

In focus groups, it was primarily growers in Colorado who commented about aphid resistance properties of wheat varieties. Some Colorado growers felt that resistant varieties have helped prevent Russian wheat aphid damage, as these examples indicate:

FG05-Colorado [#281]: The Russian wheat aphid hasn't developed to be near the problem we thought it would be ten years ago and we've gone to resistant wheat [during that time].

FG06-Colorado [#260]: We have a piece of ground where we had Russian wheat aphids every year until we planted aphid-resistant wheat. I quit looking for aphids after that. I don't even think about planting any other kind of wheat.

FG08-Colorado [#267]: I think our biggest pest problem was Russian wheat aphids until we got the resistant varieties.

Evident in the middle comment above, some growers may have become overly reliant on resistance as a method of preventing damage to the exclusion of considering other IPM methods. Shortly after we conducted focus groups in 2003 a new strain of Russian wheat aphid capable of damaging previously resistant wheat (labelled biotype B) was discovered in Colorado (Peairs *et al.*, 2003). This population is virulent to cultivars with the Dn4 resistance gene; however, the biological and economic attributes of this new biotype are currently unknown. Researchers suggest that it is important to manage aphids using a combination of management tactics to reduce the incidence of new biotypes and to avoid crop damage when they do occur.

From focus groups, it is also important to note that Colorado growers varied widely in their opinions and apparent knowledge about resistant wheat varieties. The range of viewpoints expressed in Colorado focus groups included the following:

FG05-Colorado [#69]: I think in Colorado the real aphid hot spots are pretty well converted to resistant wheat varieties, but where I am, there are very few acres in resistant wheat planted, simply because we've never had a huge aphid problem.

FG07-Colorado [#80]: I've stayed with the Russian wheat aphid resistant varieties because we got hammered awfully hard. And I've had pretty good yields with them, too, so that's all I've planted.

FG07-Colorado [#33]: I've raised a lot of the different Russian aphid resistant varieties, but when you're talking about mites – which have been a big problem for us – it doesn't solve the problem. So I've kind of moved away from some of them, because I can get a better yield from some of these other varieties. If I have to spray the mites, I get the aphids as well.

FG07-Colorado [#368]: I think the TAM varieties and the Jagger wheat and similar wheat that makes real fast [establishes a stand early], are able to outgrow the bugs. I think that's why those varieties have proven to be good yielding varieties.

FG06-Colorado [#247]: The Russian wheat aphid is kind of a strange insect – we'll go for three or four years without it and then all of a sudden we've got a full-blown case. We raise seed wheat and it seems like always the year after you have Russian wheat aphid, you sure as heck can sell the resistant varieties a lot better – that's just the way human nature works. (group laughter)

FG05-Colorado [#39]: I think we need to be very cognizant of what we're using to control infestations and bugs, be it pesticides or be it varieties that may or may not be determined to be genetically modified organisms – and whether they are or are not GMO really isn't the question, it's what people's perceptions might be – because perception is reality. Bug infestations are also a reality, but there's a balance there somewhere ... we need to be very cognizant of how things are perceived throughout the world with our customers.

In our Texas focus groups, very few growers commented about resistant wheat when we asked about their experiences in coping with agricultural pests. In one of four focus groups we conducted in Texas, two growers commented about the susceptibility of varieties to greenbug damage. These growers referred to TAM varieties generally, not specifically mentioning 'TAM-110':

FG12-Texas [#250]: I had one variety of wheat, jagger, which I had to spray last year. I didn't have to spray the TAM wheat.

FG12-Texas [#290]: The TAM has greenbug resistance, but jagger doesn't. Jagger has leaf rust resistance, but the TAM is susceptible. So we don't have the perfect wheat variety yet.

Growers in other areas included in the areawide wheat program did not comment about insect resistance in wheat. Instead, growers in these areas seemed to be primarily familiar with disease resistance attributes of various wheat varieties. The inconsistency and the absence of opinions about host plant resistance highlight some important educational needs. Evident in the above comments is that some growers may be unaware of resistance properties of wheat; others may misunderstand the basis of host plant resistance. Some may be using resistant varieties sporadically, perhaps only after experiencing a significant loss, while others may be relying on plant resistance to the exclusion of other IPM methods. Some growers may also be concerned about wheat breeding and genetic technology as well as public perceptions in global markets. Also, the selection of cultivars with resistance to Russian wheat aphid and greenbug is currently limited. The widespread adoption of resistant cultivars may be dependent on the availability of locally adapted, high-yielding varieties that combine insect and disease resistance traits. New cultivars are under development in breeding programs that will fill the needs of growers, and they should be made aware of these new products as they become available.

The areawide wheat program will help address these educational needs in several ways. First, cooperating researchers in Texas, Colorado and Nebraska are providing resistant wheat varieties to growers for planting in demonstration fields. By monitoring aphid and natural enemy populations in these fields over a 4-year period, the project team will be able to evaluate resistance in both simplified and diverse crop rotation settings. Second, through periodic updates sent to participants and through extension programs, the project team will disseminate information about host plant resistance in wheat.

Biological control

Natural enemies play an important role in regulating greenbug populations in wheat in the Great Plains, sometimes eliminating the need for insecticides (Kring et al., 1985;

Rice and Wilde, 1988; Giles *et al.*, 2003). Evidence is accumulating that diversifying dryland–cropping systems by replacing wheat–fallow and continuous-wheat systems by intensive rotations increases natural enemy populations and consequently increases the effectiveness of biological control (Brewer and Elliott, 2004). For example, relay intercropping involves the use of crops with overlapping growing seasons to maintain resources needed by natural enemies in the cropping system. In a winter wheat, sorghum and cotton relay intercropping study (approximately 4 m wide strips) in the Texas Great Plains, densities of coccinellids, lygaeids, chrysopids, anthocorids and nabids were greater, and cotton aphid, *Aphis gossypii* Glover, densities were lower in the relay intercropped system than in a monocrop (Parajulee *et al.*, 1997). If larger, more stable natural enemy populations are maintained within agricultural landscapes by cropping system diversification, natural enemies can move among crops to respond to locally abundant prey populations in a particular crop (Elliott *et al.*, 1998b, 2002).

In focus groups, few growers discussed natural enemies. When prompted, at least some growers mentioned that adult lady beetles and parasitic wasps are known beneficial insects. A few felt that beneficial insects were important in their area or on their farm:

> FG12-Texas [#257]: It seems to me like in [local] areas where farmers practice more irrigation or [practice less crop rotation], they use a lot more chemicals than we do here. And my theory is that we're keeping a beneficial population that's helping us with biological control.

> FG15-Oklahoma [#60.01]: I watch the plants and you'll have some insect pressure. But, if you get to watching it closely, first you will see the problem insect and then a couple of days later you start seeing a big increase in the lacewings, the ladybugs, and a lot of your parasites. If you give it time, it usually works.

However, even growers who were aware of natural enemies did not seem to be systematically monitoring pest or beneficial popu-

lations. Some growers were uncertain about the effectiveness of beneficial insects based on past experience:

> FG02-Nebraska [#114]: We did try [releasing] ladybugs one year. Couldn't tell if they made a difference or not – the aphids were growing faster than the ladybugs were growing is basically what it looked like to me. We had a good population of ladybugs, I guess it might have made a difference, but we couldn't see it, a true difference out there because of that. And then we didn't want to spray because we didn't want to kill the ladybugs we had spent all this money on, so where do you go from there?

In actuality, the need for curative insecticide applications depends on aphid intensity, size and health of the wheat plant, environmental and/or agronomic conditions (irrigated wheat can withstand aphid infestations better than dryland wheat) and the knowledge about the effectiveness of natural enemies. Aphid populations are often prevented from increasing to economic levels or are quickly reduced by predators and parasites, including lady beetles, parasitic wasps, spiders, damsel bugs, lacewing larvae and syrphid fly larvae (Patrick and Boring, 1998; Jones, 2001; Giles *et al.*, 2003). Cool temperatures may limit the activity of these beneficial insects; however, reliable natural enemy thresholds are available that allow growers to predict population suppression from ladybeetles or parasitoid wasps and avoid unnecessary insecticide applications. In fact, parasitism and aphid suppression can be predicted from binomial counts of mummified aphids on each tiller. Most growers are probably unaware of these methods.

In focus groups, some growers acknowledged their desire to learn more about insect populations in their crops. This Oklahoma grower also speculated that crop rotation probably helped prevent economically significant pest problems:

> FG15-Oklahoma [#10]: I am probably a little bit embarrassed of my lack of knowledge of insects and entomology. I do think I probably experienced more

problems with greenbugs and spider mite infestations when I farmed continuous wheat. With the rotation plan I'm using now, I do have some insects out there, but I have not applied chemicals for insects now for about five years. Some of that may be just coincidence and some of that I attribute to crop rotation.

As we argued earlier, practicing crop rotation likely creates a situation in which growers are more likely to observe insect–crop interactions. In fact, all four of the quotes above regarding beneficial insect observations came from growers who were practicing crop rotation.

Field scouting

Growers could also benefit from new information and education regarding simplified methods of field scouting. Historically, two aphid-sampling approaches have been recommended for wheat. These sampling plans rely on counting the number of aphids in a foot row of wheat, or calculating the average number of aphids per tiller and estimating potential for damage (Royer *et al.*, 1998). Both techniques have been useful, in that workable treatment thresholds are available; however, each approach requires a significant time commitment when aphid populations are increasing and the treatment thresholds are inflexible and do not account for differences in yield potential or cost of control.

When we inquired about field scouting or decision thresholds in focus groups, some growers acknowledged that there were methods of counting aphid densities to determine thresholds. Some stated that these methods were difficult to use, but most simply acknowledged that they only look for aphids when 'we hear planes flying' in the area, or when they heard about significant crop damage 'at the coffee shop'. Some growers responded wryly:

FG14-Kansas [#266]: As long as you drive by at fifty miles an hour, you don't have to worry about what's out there. (group laughter)

FG06-Colorado [# 298]: We used to get a couple of beers and go out and look at wheat fields for aphids – we didn't find as many aphids after the second beer. (group laughter)

Behind comments like these, growers were probably frustrated about making decisions for treating fields for insects. The second grower quoted above continued with a serious response regarding his past efforts in scouting fields for Russian wheat aphids:

FG06-Colorado [#298.01]: Sometimes you wouldn't appear to have much of an infestation, yet you had tremendous damage and then other times you'd have tremendous infestation and not the damage. I guess it depended on the condition of the crop more than a lot of things – drier years, obviously they hurt you worse.

Another grower in the same focus group expressed a similar experience:

FG06-Colorado [#288]: I used to spend a lot of waking hours driving around and walking the fields because it didn't take a lot of aphids where it could do some economic damage. You might go a month in field without building up to an economic damage threshold and then I've seen fields that went from nothing to disaster in a week, just depended. I sprayed fields and got excellent kill, then went back in ten days and you couldn't tell that you had sprayed because the conditions were right and they were back. That's not something that I miss, spending all my time wandering around a wheat field looking for Russian wheat aphids.

Also evident in these comments, and in earlier observations regarding natural enemies, is the impression that many growers probably do not scout fields as a routine practice. One rotational grower in an Oklahoma focus group acknowledged that he probably should:

FG15-Oklahoma [#266]: I get busy with other things and I put off (field scouting). You know, I'm always going to do that tomorrow. Anymore, I'm farming almost year round with the rotation. I do all of my own spray work now, I put on all of my fertilizers myself. There are not too many weeks out of the year that I'm not doing something with my farming operation other than out scouting my fields like I need to.

The current sampling approaches in winter wheat utilize binomial (presence–absence) counts and appropriate thresholds based on crop development stage to make decisions on whether aphids require suppression to profitably save yields (Hein *et al.*, 1995; Giles *et al.*, 2000a, b; Royer *et al.*, 2002a, b; Elliott *et al.*, 2003). Based on validated research the sampling schemes and thresholds vary for each aphid species. Presence–absence sampling allows for an estimation of aphid density (number per stem or tiller) from the frequency of infested tillers. Thresholds (number per tiller) are based on documented relationships between aphid density and grain yields at different growth stages (Kindler *et al.*, 2002; Elliott *et al.*, 2003).

In the areawide program, we are disseminating information about these simplified field-scouting methods and about aphid and natural enemy identification. By conducting on-farm demonstrations and encouraging crop rotation, we also hope to increase growers' awareness of insect–crop interactions. In doing so, it will be important to convey the relative advantage and ease of the newer scouting method over conventional decision-making practices (Cuperus and Berberet, 1994).

Other elements of wheat IPM

Some of the most widely applicable wheat IPM practices relate to fertility management, planting date and grazing considerations, control of weeds and volunteer wheat, and conservation tillage. On-farm demonstrations in the areawide program evaluate the impact of these agronomic factors on pest and natural enemy densities. Growers also expressed viewpoints about these factors in our focus group discussions.

Fertility management

In focus groups we found that wheat growers generally understood the importance of efficient incorporation of nutrients to promote optimal plant growth and development, though they differed in approach and extent to which they feel they successfully manage crop fertility on their farm. Some expressed the belief that healthy plants are able to resist injury from aphids.

Efficient incorporation of added nutrients does promote optimal plant growth, and healthier plants are more likely to resist injury from aphids (Royer and Krenzer, 2000). However, excessive nitrogen will result in succulent plant growth that may allow for rapid early build-up of cereal aphids and other pests. Of greatest concern, however, is when wheat plants become stressed by receiving insufficient nutrients. These stressed plants are more susceptible to injury by aphids and the diseases they transmit.

Planting date and wheat grazing considerations

In the Great Plains, winter wheat is grown as: (i) forage for livestock; (ii) forage for livestock plus grain production (dual-purpose); or (iii) exclusively for grain production (Royer and Krenzer, 2000). The need for increased forage requires early planting during late summer or early fall (August–September), whereas optimal grain production systems require that wheat planting occur during late fall (October–November). Use of wheat for grazing is important among growers who are participating in the areawide wheat program, primarily among those in Texas and Oklahoma, where the majority of growers participating in the program typically graze at least some of their wheat.

One common observation holds true regardless of location: early-planted winter wheat is more likely to be colonized by cereal aphids that can quickly build up to damaging levels (Hein *et al.*, 1996; Royer and Krenzer, 2000). Management of aphids can be achieved by delaying planting as much possible; however, early planting must occur if significant forage is required for cattle (Royer and Krenzer, 2000; Ismail *et al.*, 2003).

Early infestations are most severe when aphids are able to build up in nearby crops and/or grassy weeds prior to migrating into wheat. These alternative host plants include sorghum, oats, Johnsongrass, downy brome, jointed goatgrass, wild rye and several other cool-season grasses (Hein *et al.*, 1996; Patrick and Boring, 1998; Peairs, 2003). Aphids that colonize wheat in the early fall may also transmit barley yellow dwarf virus (BYDV), which can significantly reduce grain yields (Royer and Krenzer, 2000; Ismail *et al.*, 2003).

Growers and researchers have observed significant reductions of aphids on grazed wheat. In dual-purpose systems in Oklahoma and Texas, aphids and their effects on grain yields are nearly eliminated for the growing season when cattle consume and trample wheat (Ismail *et al.*, 2003). This approach is limited to managing low to moderate non-economic aphid numbers at grazing initiation during the fall, and preventing subsequent build-up during winter and spring growth. However, when aphids colonize seedling wheat and increase in numbers above economic thresholds, damage to forage and grain yield cannot be prevented by grazing cattle.

In focus groups, some growers pointed out that forage was important for their operation and this necessitated earlier planting date, in spite of the pest management risks. Others also acknowledged that they were familiar with research regarding reduced aphid pressure in grazed wheat, or that they had observed this on their own fields.

Many growers discussed benefits of raising cattle with wheat pasture and forage crops vs. producing wheat for grain. They also discussed risks and other decision factors in evaluating the value of wheat for forage vs. wheat for grain, as illustrated in this sequence of discussion from a Colorado focus group:

FG07-Colorado [#198]: We're pretty heavy in cattle. I think that we have made more money in cattle than we have farming.

FG07-Colorado [#204]: [Y]ou also float out your danger of hailstorms. Your cattle will make you some money. So if you get a hail you've already got something out of it.

FG07-Colorado [#205]: [S]ame thing with grazing volunteer wheat. If you've got your own cattle out there and you're grazing that volunteer wheat, it's probably going to hurt your yield the next time around – but you might be better off doing that. You might take more dollars off it and not have the risk involved.

The perspective of these growers seemed to be that profitability of cattle and wheat used for forage was more predictable than wheat for grain. When pressure from drought, weeds and/or insects becomes too much of a problem in the wheat crop, growers may consider graze-out and possible use of fallow, cover crop, forage grasses or grain sorghum on a temporary basis before going back to wheat:

FG18-Oklahoma [#167]: Something I try on some of my winter ground where the jointed goatgrass has been bad is I'll lay that out a year and plant hybrid Sudan grass. That helps seemingly because you wait until spring climate and that way you work the ground ... try to kill all of that out and then plant your feed [hay] and put it up as early as you can and then have it ready for fall for wheat again. So you just missed one year that way.

FG07-Colorado [#175]: I've always had pretty good luck with feed [hay crops] when I couldn't raise milo – the milo would develop a good plant, but no head – but I have had pretty good luck with feed crops.

The Oklahoma grower quoted above points out the importance of gaining some predictable profit or benefit from the farm while trying to address a weed problem. Since grazing is important to so many growers involved in the areawide program, we will evaluate aphid and natural enemy densities in grazed wheat and in wheat–hay crop rotations in addition to rotations with summer cash crops. We will also return to the issue of grazing and crop rotation at the end of the chapter when we examine grower considerations in adopting crop rotations with winter wheat.

Control of weeds and volunteer wheat

Jointed goatgrass, downy brome, volunteer rye and volunteer wheat are serious weeds in winter wheat production in the Great Plains (Lyon and Baltensperger, 1995; Royer and Krenzer, 2000; Croissant *et al.*, 2002). Jointed goatgrass is very difficult to control because its seasonal life cycle and biochemistry are very similar to that of wheat (Colorado Department of Agriculture, 1990, 1991; Royer and Krenzer, 2000; Croissant *et al.*, 2002). Emerging volunteer wheat can serve as an in-field green-bridge host for several pest arthropods (including aphids and mites), diseases transmitted by these arthropods (BYDV and wheat streak mosaic virus) and other diseases such as leaf rust (Royer and Krenzer, 2000; Peairs, 2003).

Use of a second crop in a 3-year rotation permits cheaper, less chemical-intensive control of weeds in wheat (Westra and D'Amato, 1995). Rotation from wheat to a broadleaf crop facilitates economical control of jointed goatgrass and other grassy weeds because the jointed goatgrass can be controlled when the field is fallow or during the growing season of the broadleaf crop using herbicides that target grassy weeds. It is generally important to rotate herbicide modes of action to prevent future cases of herbicide resistance, which is accomplished more easily and economically when crops are rotated.

Effective control of volunteer wheat prior to planting winter wheat is also an important management activity. Generally, volunteer wheat needs to be destroyed at least 2 weeks before crop emergence. Appropriately delayed planting along with suppression of volunteer wheat can significantly reduce the opportunity for aphids and mites to colonize production fields. Volunteer wheat management is also more easily achieved in diverse agroecosystems; when crops are rotated there are more years between wheat plantings (Croissant *et al.*, 2002).

As we noted in the previous section, using wheat for forage is one way that some growers have adjusted to severe weed problems in wheat. Other growers in focus groups were very dedicated to working through weed management problems with crop rotation, as with these examples:

FG02-Nebraska [#163]: We started rotating crops to get rid of jointed goatgrass and downy brome. Over time it has, but it's been a learning experience. Initially we were trying to do a wheat–millet–fallow rotation, [but] it looked like we needed a longer break before coming back to wheat. So then we did a wheat–corn–millet–fallow rotation. We did that for a couple of years and we couldn't figure out why we were fallowing, so we took the fallow out and added another year of corn in there. And what we have now is pretty much a clean farm for goatgrass. We still have some downy brome problems, and we developed other problems in those rotations – like broadleaf weeds – so we're working on those now.

FG08-Colorado [#66]: I guess one of the reasons I went away from the wheat–fallow was trying to economically spray the bindweed – like after a wheat harvest and getting a lot of money in it [during a fallow period]. I found myself skipping one of the applications of the herbicide and then not gaining anything on bindweed control, so it was pretty obvious we were going to have to do something different. Well, introducing another crop like milo, or feed hay, or something else where you could justify some expense has really helped in the cost aspect of it.

Conservation tillage

The choice of reduced tillage or no-till farming methods is closely related to crop rotation decisions. Growers who follow a wheat–fallow rotation will typically till the soil before planting and may use tillage during fallow periods to keep weeds under control. Various forms of reduced tillage or no-till farming in winter wheat are usually combined with a planned crop rotation to manage crop residue and soil moisture throughout the production cycle (O'Brien *et al.*, 1998; Croissant *et al.*, 2002).

Minimum till or no-till can have beneficial consequences for pest management. Reduced tillage increases stored soil moisture, which can lead to healthier plants that are able to resist injury from aphid pests. However, increased surface residue with reduced tillage alters the habitats of both pest insects and natural enemies in complex ways (Burton and Burd, 1994). Research focusing on cereal aphids suggests that populations of some species are favoured while others are suppressed with reduced tillage (Burton and Krenzer, 1985; Andersen, 2003; Hesler and Berg, 2003). Minimum till or no-till can also increase the incidence of diseases like cephalosporium stripe, take-all and tan spot (Watkins and Boosalis, 1994; Royer and Krenzer, 2000), but crop rotation reduces the incidence of these diseases because wheat is absent for one or more years, allowing disease titers in fields to decline between wheat plantings (Croissant et al., 2002). A rotation plan that keeps a field out of wheat for up to 3 years helps to avoid many disease problems that can arise in a reduced tillage situation.

A 1994 survey of wheat farms by the USDA found that adoption of conservation tillage practices in the central and southern Great Plains was low when compared with wheat farms in the nation as a whole (Ali et al., 2000). Among the nationwide wheat farms 18% reported using mulch tillage and 10% used no-till. Corresponding percentages for farms in the central and southern plains were 12% mulch tillage and less than 5% no-till. We observed that a higher proportion of growers participating in the areawide program used no till or some form of 'mulch tillage' or 'conservation tillage' (about 27% of program participants). This was because we selected growers who have had some success with crop rotation, and many of these growers were rotating crops in a no-till system.

Many of the no-till growers participating in focus groups were quick to point out benefits of no-till:

FG09-Texas [#19]: Since we have started doing no-till we have not taken a zero and a lot of guys around have flat burned up with conventional till. From 1990 to now we have been either bone dry or swamped wet – nothing around normal. So the key to making those acres do something besides blow away or eat more money is to save that water from when it is wet and get through the dry – keep cover on it and keep the weeds down.

FG17-Oklahoma [#352]: When you think of the two or three weeks you spend dragging dirt in a conventional tillage system, the no-till is a big plus. I would worry about my other ground when it would rain and after I planted it. In the no-till, I knew it wasn't washing out and I knew it was going to come up.

FG15-Oklahoma [#354]: I now have moisture conservation for summer crops, and my long-term goal is to build that organic matter up – get that humus level back up to that four or five percent like it was in the old prairie days before the mold board plow started eroding it. Hopefully when I reach the organic matter problem, get that addressed, a lot of other problems will be erased.

FG03-Wyoming [#21]: We started doing some no-till probably 4 or 5 years ago. We use a wheat–corn–summer fallow rotation on some ground and some we have a wheat–millet–sunflower–summer fallow – depending on the ground and the weed problem, I mean rye, goatgrass, or whatever the situation – but we saw some unique changes in the ground when we started no-till.

FG13-Kansas [#405]: Since I started no-tilling, it put the fun back into farming. I enjoy the changes that I'm seeing in the soil on a yearly basis, the improvements that come along with the crop diversity that we have, makes things interesting.

Some growers were finding successive years of drought to be very challenging for maintaining their no-till system. Some were talking about converting back to a minimum tillage system and/or a simpler crop rotation system. As discussed in the next section, not all growers see benefits, opportunities and constraints of no-till or crop rotation in the same way.

Grower Perspectives on Crop Rotation

In focus groups, we learned that growers considered a number of factors when evaluating systematic crop rotations with winter wheat. Some barriers to crop rotation evident in focus groups included: past experience with crop rotation and/or personal preference for growing 'wheat only'; landlord preference for wheat; changes in federal farm program support; experience with extended periods of drought; and the age of the primary farm operator and other farm decision makers. One alternative that growers described in some focus groups was opportunistic crop rotation. This involves, as examples, only planting summer crops when the grower believes there is adequate soil moisture, or planting a summer crop when confronted with a severe weed problem on a particular wheat field. A second alternative was to focus the use of wheat for grazing and hay rather than wheat for grain. With this second alternative, growers may use summer hay crops in a rotation with winter wheat on limited acreage.

Past experience and personal preference

With respect to experience and personal preference as a barrier, some growers in Colorado focus groups debated the relative merits of crop rotation vs. wheat–fallow. A southern Colorado grower who had adopted no-till and crop rotation began by reflecting on his perceptions of change in local dryland farming practices:

> FG08-Colorado [#31]: Where I farm it's very sandy ground. It was traditionally farmed to grass-type row crops – broom-corn originally, and then milo. And it moved from a monoculture basically in milo to where we kind of went the other way to wheat monoculture. We started introducing wheat because we were starting to get problems with the monoculture of milo. And also the tillage systems changed and we needed to get more residue on the ground so we went to wheat instead of milo. Then, later, we saw that we were getting some problems with grassy weeds and things when we went to no-till so we introduced sunflowers into a no-till rotation.

However, this grower also noted that soil moisture was a challenge in his region, and growers who use crop rotation often do so opportunistically. Concluding his thought about the evolution of present-day farming practices, he said:

> FG08-Colorado [#31.01]: If we have moisture we need to use it, one way or the other. We don't wait in a fallow period. We just really hate fallow. Sometimes we're forced into fallow, like last year, but we don't like it. So we try to choose a crop that will complement whatever moisture that we have available to us.

A grower from a northern Colorado focus group summarized this philosophy by saying:

> FG05-Colorado [#26]: About all I can say about our rotation is that until I sink a soil probe in the spring I have absolutely no idea what we are going to do where. Because it seems like the best-laid plans get changed by Mother Nature somewhere along the way.

Other growers in southern Colorado argued that the risks of growing continuous wheat in a conventional tillage system were preferable to growing summer crops opportunistically:

> FG08-Colorado [#43]: For me, decisions are a lot easier to make on dryland wheat. You don't have to be right on the game all the time, that's part of it. We put about 40 pounds of nitrogen on summer fallow ground and then you put out 30 pounds of seed – you don't have a big input in it. And wheat is a 'tough' crop [resilient to stress]. You don't have to have the rain quite as much or as timely as you do for some of the other stuff.

Later, this grower continued his thought in response to additional comments from other growers who were advocating opportunistic crop rotation:

> FG08-Colorado [#64]: [T]he risk factor for me is a lot different. If you're going to use everything that's there every time [meaning soil moisture], you better be

right on your game or you'll get behind. On a summer fallow period for us, you just do things kind of right, you know, just within a time period and it all works out – Mother Nature takes care of you. And that's kind of the philosophy I use – just the opposite of what others are talking about.

Growers in another south-eastern Colorado focus group echoed this dryland wheat philosophy, reflecting on how their grandfathers had thought about it. In part, these comments were in response to growers' frustration with an extended period of drought in recent years:

FG07-Colorado [#41]: Thinking back on the last couple of years, one of the things grandpa always said was: 'I really don't care if I raise any wheat, I just want to raise some stubble'. (laughter) The last couple of years I've kept that thought in the back of my mind. (laughter) You know, just cover the ground.

FG07-Colorado [#167]: We're pretty much locked in with wheat. My grandpa always said: 'If it's going to be a wheat year you can grow wheat in a parking lot. If it's not, it doesn't matter what you do, you can't grow wheat!' There are a few things we do different [from wheat–fallow], but it seems like the last few years we've been a salvage operation.... We've tried the sunflowers and it's just so hard on our ground – seems like we paid for it the next two years. We're always looking for something new, but when you're in Rome you better do as the Romans it seems like.

Federal farm programs and drought

Growers gave contrasting accounts of how changes in federal support payments have influenced their decisions regarding crop rotation and no-till farming. Some growers in a Colorado focus group expressed frustration that recent years of drought and regulations for federal crop insurance and yield payments were limiting their farming choices:

FG07-Colorado [#167]: [Y]ou know the crop insurance game you gotta' play, that's determined a lot of what we can plant.

FG07-Colorado [#315]: The last couple of years have been one of those deals where you go see what the insurance will cover, what the payments are at the FSA office [referring to the USDA Farm Services Agency].

FG07-Colorado [#319]: [T]o me that's kind of sad. Instead of going out and doing what we do best, which is farming, we're relying off of what the government's going to pay us or what the insurance will cover.

Growers in a Texas focus group expressed similar concerns, but debated the extent to which changes in federal crop support and recent years of drought have influenced their use of crop rotation:

FG11-Texas [#17]: We used to be in a pretty good rotation with wheat – a three-year rotation with wheat–milo–fallow but we started going to cotton, dryland and irrigated. So I really don't know what my rotation is right now. We haven't been able to grow any cover crops. We used to grow no-till but the last three to four years we haven't been able to grow much cover due to the drought so my rotation is totally messed up.

FG11-Texas [#30.01]: I'm kind of like you, we were in a good rotation and then this drought came along and has really messed things up. And since the new Farm Bill came out, everyone is trying to maximize profits. We have been in this drought, so you see your opportunity and you plant fence row to fence row just trying to make some money. It has messed up the rotation.

FG11-Texas [#35]: Well we've stayed pretty much with our three-year rotation even through these drought years. I don't think it would have made any difference what we did; we weren't going to raise anything anyway. So I don't think we gained much and probably hurt ourselves a little bit with this new farm program – if we had had any base acres to update. I am not too excited about this farm program unless we can start raising something. 'Freedom to Farm' [the previous farm program] had been really good to me in dry years. We had had some pretty decent yields on most of our farming.

Another grower in this focus group tried to summarize the group's observations by describing how his farming efforts have changed in recent years:

> FG11-Texas [#36.01]: The older I get, the less I know. When I first started farming I had a lot of interest in new crops, but the older I get the less I am interested in. Used to be a three-year rotation on dryland, wheat–milo–fallow. Plant all no-till, grow milo back in the wheat stubble. In this part of the world that is (an) ideal (rotation) … if you get the rain. The trouble is sufficient stubble in the wheat crop. We had twelve inches of rain last year and five inches of that came in September and October. You can't grow anything that way. If you don't have any stubble, no-till won't work, period.

> So, we have just been … the cash flow … when your ground is bare, you've got to put something on it, milo or whatever. … So, right now we are trying to learn to grow cotton dryland. Had an exceptional year for one year and then [not so good] the last two years. [But] economics is what is driving that because, dryland cotton, if you make even one bale of cotton to the acre you are looking at three hundred dollars compared to a good milo crop at one hundred. And that is possible for dryland cotton in this country.

As these growers have struggled with summer cash crop rotations, many of them noted that the focus of their wheat operation was on grazing. Several growers from the Kansas focus group cited earlier felt that using wheat for grazing was preferable to wheat for grain in their situation. These growers typically utilized hay crops for rotation on some acres:

> FG14-Kansas [#66]: I want to do some spring crops, but I tried the corn and the soybeans and tried milo probably the most. I haven't had the best luck with it. It's just wheat country right here. … So for a rotation I plant some Sudan grasses and stuff, so I can rotate a little bit of ground, but just for what I can utilize myself (for hay), I don't do it for the cash crops.

> FG14-Kansas [#69.01]: I'm glad to hear you say that, 'cause that's been my feeling, too. You hear about using milo in a rotation, but milo is an 'iffy' deal. It depends a lot, too. If you go north of us the soil is different and it works better there. Where we are, we try and rotate the Sudan with something, because at least you can harvest everything that's out there.

This last grower concluded by reflecting on generational changes in the perceived value of wheat for grazing vs. wheat for grain:

> FG14-Kansas [#69.02]: It's just my observation that people in my dad's generation were farmers that had cattle, but I think over the last twenty years our focus has changed to the fact that we have cattle and we farm. Not that we've changed numbers or acres, but I was still in high school when the price of wheat was decent. Ever since then you can't make anything on wheat, or at least that's the way that I look at it. So, wheat is okay, but what really makes it work is if you can graze it. So that's our emphasis.

Landlord preference for wheat

Some growers cited landlord preference for wheat as a reason for continuing to grow 'wheat only', or not wanting to take chances on spring/summer crops on rented farm ground:

> FG08-Colorado [#77]: I think in my operation, it makes a lot of difference whether I own the land or rent it. You know, like on the dryland wheat. On rented land I like to have just wheat–fallow. The stuff I own we do a little different, I'm more into taking risks on it.

A group of Kansas growers also noted concerns about landlord attitudes and preferences as a reason for growing conventionally tilled wheat for grain only. These growers were partially concerned about use of wheat for grazing, no-till practices and limited alternatives for rotation crops:

> FG14-Kansas [#78]: Well, the landlord wants some grain in the elevator. I

compensate them for the graze, but they like grain in the elevator.

FG14-Kansas [#80]: I think a lot of the older guys – retired guys that I farm for – they like to play a little bit in the market. So it's about the same in our area.... Some guys have gone to cash rent. You can spend quite a bit for cash rent and what's a fair value for that? I mean every piece of ground seems to be different around our area ... but you know how the coffee shop talk is: 'I heard so-and-so is getting up to twenty-five an acre, how come you're not paying more?' You know, so you've got that kind of competition and other guys maybe want to pay a little more for it.

FG14-Kansas [#89]: And there's been a big push towards no-till, and I can see advantages of the no-till, moisture saving and all that, but I'm still wondering about these other things that we have to consider. And again, I don't see too many options on rotating. And, what happens if you graze it and try no-till and end up with soil compaction and so on? I'm just pretty skeptical.

FG14-Kansas [#90]: [L]ots of my neighbors with no-till won't allow cattle on their property.

FG14-Kansas [#91]: Exactly. That's it exactly. So, options about doing things different, I'll sure look at them, but I haven't seen anything yet.

Life-stage considerations

Perhaps the most important consideration in evaluating growers' interest in adopting crop rotations with winter wheat is that many wheat farmers of the Great Plains are near or beyond retirement age. Some of the older operators in our focus groups had experience with crop rotation, but over the years have settled into wheat-only production. Some were continuing to farm wheat primarily as an income supplement and for lifestyle considerations. In the two comments below from older focus group participants, the first grower notes that slow time during winter months allows him to be involved with church missionary work. The second grower had also mentioned the importance of down-time during winter

months and did not want to make new farm investments:

FG16-Oklahoma [#41]: I raise wheat! They talk about how it separates your workload by growing all those alternative crops, but I think it just kind of keeps you working all year. (laughter from the group) And I like to go on mission trips in the wintertime so I'm wheat only.

FG10-Texas [#20]: I've been farming all my life ... and I love what I'm doing or I wouldn't be doing it. I have my own equipment. I operate strictly on a cash basis. If I can't afford it, I don't buy it. I don't like to run cattle because you get too many problems when they take them off the field, so I plant late and strictly for wheat production. That's about it. If I have a big problem, I fallow and if I don't, I just keep planting wheat.

Life stage is relevant in a number of ways when we consider not only the primary operator but other farm decision makers as well – spouses, parents, siblings, children and landlords for rented farm ground, as we have noted above. However, life stage does not exert a simple or consistent influence on farm decisions – some growers have successfully negotiated relationships with landlords, and many of the most 'progressive' growers associated with the areawide program were among the older growers. One grower in an Oklahoma focus group nicely summarized how multigenerational influences were couched within other challenges to successful wheat farming in his experience:

FG17-Oklahoma [#338]: You ask about change – what limits change? Well the number one thing is rainfall, as others have said. And equipment, if you are thinking about a different crop or no-till, when I look at a thirty or forty thousand dollar no-till drill – whatever they cost – I think, 'Boy, that's going to slow me down'.

Then you've got the landlords, which others have talked about. People who are resistant to change. You try to talk a ninety-four-year-old man into going into a no-till operation and he's not going to be impressed with it! But that is what I have been doing – even though I'm fifty-seven, I'm a youngster in my community. Everybody that I rent from is in their eighties or above. They don't see things quite the same. They are a little resistant to change.

Well, I suppose I'm a little resistant to change. My kid thinks I am. He has been talking about purchasing a GPS system to put on our sprayer – I listen to him and I think: 'I sure hope he knows what he's doing!' (group laughter)

Achieving success with crop rotations

Some of the growers who had been using crop rotation successfully have overcome limitations described above. Some have learned through experience and trial and error, some have used crop consultants and some are younger operators who implemented diversified farming when they became independent farm operators.

Some growers have taken the risk of experimenting with crop rotation where they have been able to capitalize on a higher-valued, seed wheat crop:

> FG17-Oklahoma [#177]: I've been selling a lot of seed wheat and that's kind of what has kept me farming, I feel. I'm getting a little bit extra for my grain and that's kind of why I went to rotation – to keep my fields clean and I really like what that is doing on those acres.

> FG18-Oklahoma [#68]: We do some rotation – we put in Austrian winter peas in the fall, just to enrich our soils with nitrogen. I know a lot of people use them for hay, but my grandfather always said we need to put those tops back in the ground – that's where all of your nutrients and things are. Being in the seed business and with new varieties coming out, I try to put those varieties on the soil behind Austrian winter peas to have clean ground.

So experimenting with a crop rotation on limited acreage is one way that growers have mitigated the risk of growing a new crop. Growers who have experimented with crop rotations for an extended period of time have come to see long-term vs. short-term benefits. Some growers who have become strongly committed to crop rotation made comments like:

> FG15-Oklahoma [#60]: People, if they find something that works they want to keep going back to it instead of rotating out – like growers who are producing a lot of cotton-behind-cotton right now – eventually they are going to learn that if they go to milo or wheat after their cotton it's going to be a big benefit to them. They're not seeing the whole picture right now.

A second grower in the same focus group responded at a later point by saying:

> FG15-Oklahoma [#98]: (Name) really hit the nail on the head on your rotation plans. The first crop I rotated as a summer crop was milo, and I had a super good milo crop. So what did I do? I just kept planting milo year after year on this place until I had a Johnson grass infestation, and now I've been five years struggling to control that problem!

A Wyoming grower argued that it was necessary to stick with the rotation plan when confronted with crop failures or temporary setbacks:

> FG03-Wyoming [#149]: You have to stick with your crop! We don't give up on wheat when we have a bad year. That is what you have to do with your rotation.

Others have overcome perceived risks of adopting crop rotation through assistance from a crop consultant or crop advisor:

> FG13-Kansas [#123]: One thing that kept me from trying rotations or new crops was confidence that I had enough information to make good choices. [Now] I hire a crop consultant – of course I end up spending quite a few dollars to have that assistance – but it gives me the confidence to try something new. And now our fieldsman has been servicing us for about ten years and is knowledgeable. That helps a great deal to make up for what I don't know. There are a lot of things I don't know.

Finally, some younger growers have adopted crop rotations as they began farming independently:

> FG04-Wyoming [#39]: I started farming with my dad. We used to always do wheat

and summer fallow, but it got to where there just wasn't much money in it. So I heard about no-till and I thought it might help me to utilize the water better. We also had some problems with a little bit of rye and joint grass coming in and I thought that giving it the rotation would help that type of stuff, too.

FG19-Oklahoma [#19]: When I was in high school I was getting sick and tired of growing cotton – all we did was spray the stuff around the clock! We used to have a huge operating note all year and we'd finally be able to pay it down in January for about two days and then start all over. I like when we brought in peanuts, the way it has allowed us to diversify and rotate. And since I keep more wheat, and I keep milo now, I can keep my overall debt running lower because I have income coming in all year long. I like the structure of my soil with crop rotation. I've been able to make yields that I never dreamed I could make despite a lot of the bad weather problems we've had. So that is what I like about diversified crop production.

Conclusions

The areawide program will be beneficial to all participating wheat growers by improving their knowledge of insect pests, natural enemies and benefits of improved field-scouting methods. The areawide program will also demonstrate insect management benefits of crop rotations.

Through focus groups, we observed that many growers acknowledged pest management benefits of crop rotation, but were mainly attuned to weed management benefits. So increasing grower awareness of insect management advantages is an important educational goal for the program. But we also observed that adoption of crop rota-

tion involves a broad range of decisions. From the grower's perspective, crop rotation involves a reallocation of assets to various possible farm enterprises. The decision will be influenced by grower's assessments of financial risk and related factors like federal farm program support, weather trends, markets for various commodities and leasing arrangements for rented farmland.

It is uncertain how the various factors influencing crop rotation adoption will play out in the future. None the less, some growers associated with the areawide wheat project have seen ongoing benefits of crop rotation in spite of many sources of volatility in winter wheat production. A significant outcome of the areawide wheat IPM program will be to help growers see how their peers have coped with successful implementation of diversified winter wheat production in the central US Great Plains.

Acknowledgements

The authors acknowledge contributions of many individuals to the program, Areawide Pest Management for Russian Wheat Aphid and Greenbug (USDA–ARS, Project Number: 500-44-012-00). The authors wish to thank USDA–ARS national program leader, Robert Faust. We also acknowledge contributions to the design and implementation of the areawide wheat program from many individuals, including: Gary Hein and Drew Lyon, University of Nebraska, Lincoln; Thomas Holtzer, Frank Peairs and Gary Peterson, Colorado State University; Phillip Sloderbeck and Gerald Wilde, Kansas State University; Gerald Michels, Texas A&M University; Gerrit Cuperus and Thomas Peeper, Oklahoma State University; and Christine Johnson, Director of the Bureau for Social Research, Oklahoma State University.

References

Ahern, R.G. and Brewer, M.J. (2002) Effect of different wheat production systems on the presence of two parasitoids (Hymenoptera: Aphelinidae; Braconidae) of the Russian wheat aphid in the North America Great Plains. *Agriculture Ecosystem and Environment* 92, 201–210.

Ali, M.B., Brooks, N.L. and McElroy, R.G. (2000) *Characteristics of U.S. Wheat Farming: A Snapshot*. Statistical Bulletin No. 968. Resource Economics Division, Economic Research Service, US Department of Agriculture, Washington, DC.

Andersen, A. (2003) Long-term experiments with reduced tillage in spring cereals. II. Effects on pests and beneficial insects. *Crop Protection* 22, 147–152.

Andow, D.A. (1983) The extent of monoculture and its effects on insect pest populations with particular reference to wheat and cotton. *Agriculture Ecosystem and Environment* 9, 25–35.

Andow, D.A. (1991) Vegetational diversity and arthropod population response. *Annual Review of Entomology* 36, 561–586.

Brewer, M.J. and Elliott, N.C. (2004) Biological control of cereal aphids in North America and mediating effects of host plant and habitat manipulations. *Annual Review of Entomology* 49, 219–242.

Brewer, M.J., Nelson, D.J., Ahern, R.G., Donahue, J.D. and Prokrym, D.R. (2001) Recovery and range expansion of parasitoids (Hymenoptera: Aphelinidae and Braconidae) released for biological control of *Diuraphis noxia* (Homoptera: Aphididae) in Wyoming. *Environmental Entomology* 30, 578–588.

Burton, R.L. and Burd, J.D. (1994) Effects of surface residues on insect dynamics. In: Unger, P.W. (ed.) *Managing Agricultural Residues*. Lewis Publishers, Boca Raton, Florida, pp. 245–260.

Burton, R.L. and Krenzer, E.G. Jr (1985) Reduction of greenbug (Homoptera: Aphididae) populations by surface residues in wheat tillage studies. *Journal of Economic Entomology* 78, 390–394.

Colorado Department of Agriculture (1990) *Jointed Goatgrass Task Force Report*, February.

Colorado Department of Agriculture (1991) *Annual Grass Weed Contamination Survey*, August.

Croissant, R.L., Peterson, G.A. and Westfall, D.G. (2002) *Dryland Cropping Systems*. Fact Sheet No. 0.516. Colorado State University Cooperative Extension. Available at: http://www.ext.colostate.edu/pubs/crops/00516.html

Cuperus, G.W. and Berberet, R.C. (1994) Training specialists in sampling procedures. In: Pedigo, L.P. and Buntin, G.D. (eds) *Handbook of Sampling Methods of Arthropods in Agriculture*. CRC Press, Boca Raton, Florida, pp. 669–681.

Cuperus, G.W., Johnson, G.V. and Morrison, W.P. (1993) Integrated wheat management. In: Leslie, A.R. and Cuperus, G.W. (eds) *Successful Implementation of Integrated Pest Management for Agricultural Crops*. Lewis Publishers, Boca Raton, Florida, pp. 33–55.

Dhuyvetter, K.C., Thompson C.R., Norwood, C.A. and Halvorson A.D. (1996) Economics of dryland cropping systems in the Great Plains: a review. *Journal of Production Agriculture* 9, 216–222.

Elliott, N.C., Hein, G.L., Carter, M.C., Burd, J.D., Holtzer, T.J., Armstrong, J.S. and Waits, D.A. (1998a) Russian wheat aphid (Homoptera: Aphididae) ecology and modeling in Great Plains agricultural landscapes. In: Quisenberry, S.S. and Peairs, F.B. (eds) *Response Model for an Introduced Pest, the Russian Wheat Aphid*. Thomas Say Publications in Entomology, Entomological Society of America, Lanham, Maryland, pp. 31–64.

Elliott, N.C., Kieckhefer, R.W., Lee, J.H. and French, B.W. (1998b) Influence of within-field and landscape factors on aphid predator populations in wheat. *Landscape Ecology* 14, 239–252.

Elliott, N.C., Kieckhefer, R.W., Michels, G.J. Jr and Giles, K.L. (2002) Predator abundance in alfalfa fields in relation to aphids, within-field vegetation, and landscape matrix. *Environmental Entomology* 31, 253–260.

Elliott, N.C, Constein, K.L., Royer, T.R., Giles, K.L., Kindler, S.D. and Waits, D.A. (2003) *Greenbug Pest Management Decision Support System*, Agricultural Research Service, US Department of Agriculture, Washington, DC. Available at: http://www.entoplp.okstate.edu/gbweb

Faust, R. (2001) Forum-invasive species and areawide pest management: what we have learned. Available at: http://www.ars.usda.gov/is/AR/archive/nov01/form1101.htm

French, B.W. and Elliott, N.C. (1999) Temporal and spatial distribution of ground beetle (Coleoptera: Carabidae) assemblages in grasslands and adjacent wheat fields. *Pedobiologia* 43, 73–84.

French, B.W., Elliott, N.C., Kindler, S.D. and Arnold, D.C. (2001) Seasonal occurrence of aphids and natural enemies in wheat and associated crops. *Southwest Entomologist* 26, 49–61.

Giles, K.L., Royer, T.A., Elliott, N.C. and Kindler, S.D. (2000a) Development and validation of a binomial sequential sampling plan for the greenbug (Homoptera: Aphididae) infesting winter wheat in the Southern plains. *Journal of Economic Entomology* 93, 1522–1530.

Giles, K.L., Royer, T.A., Elliott, N.C. and Kindler, S.D. (2000b) Binomial sequential sampling of *Rhopalosiphum padi* in winter wheat in the southern plains. *Southwest Entomologist* 25, 191–199.

Giles, K.L., Jones, D.B., Royer, T.A., Elliott, N.C. and Kindler, S.D. (2003) Development of a sampling plan in winter wheat that estimates cereal aphid parasitism levels and predicts population suppression. *Journal of Economic Entomology* 96, 975–982.

Greenbaum, T.L. (2000) *Moderating Focus Groups*. Sage Publications, Thousand Oaks, California.

Hein, G.L., Elliott, N.C., Michels, G.J. Jr and Kieckhefer R.W. (1995) A general method for estimating cereal aphid populations in small grain fields based on frequency of occurrence. *Canadian Entomologist* 127, 59–63.

Hein, G.L., Kalisch, J.A. and Thomas, J. (1996) Cereal Aphids. Publication G96-1284-A. University of Nebraska Cooperative Extension Service NebGuide. Available at: http://ianrpubs.unl.edu/insects/g1284.htm

Hesler, L.S. and Berg, R.K. (2003) Tillage impacts cereal-aphid (Homoptera: Aphididae) infestations in spring small grains. *Journal of Economic Entomology* 96, 1792–1797.

Heyne, E.G. (1987) *Wheat and Wheat Improvement*. American Society of Agronomy, Madison, Wisconsin.

Holtzer, T.O., Anderson, R.L., McMullan, M.P. and Peairs, F.B. (1996) Integrated pest management of insect pests, weeds, and plant diseases in dryland cropping systems of the Great Plains. *Journal of Production Agriculture* 9, 200–208.

Ismail, E.A., Giles, K.L., Coburn, L., Royer, T.A., Hunger, R.M., Verchot, J., Horn, G.W., Krenzer, E.G., Peeper, T.F., Payton, M.E., Michels, G.J. Jr and Owings, D.A. (2003) Effects of aphids, barley yellow dwarf, and grassy weeds in grazed winter wheat. *Southwest Entomologist* 28, 121–130.

Jones, D.J. (2001) Natural enemy thresholds for greenbug, *Schizaphis graminum* (Rondani), on Winter wheat. MS thesis, Oklahoma State University, OK.

Kindler, S.D., Elliott, N.C., Giles, K.L., Royer, T.A., Fuentes-Granados, R. and Tao, F. (2002) Effect of greenbugs on winter wheat yield. *Journal of Economic Entomology* 95, 89–95.

Krall, J.M. and Schuman, G.E. (1996) Integrated dryland crop and livestock production systems on the Great Plains: extent and outlook. *Journal of Production Agriculture* 9, 187–191.

Kring, T.J., Gilstrap, F.E. and Michels, G.J. Jr (1985) Role of indigenous coccinellids in regulating greenbugs (Homoptera: Aphididae) on Texas grain sorghum. *Journal of Economic Entomology* 78, 269–273.

Krueger, R.A. and Casey, M.A. (2000) *Focus Groups*, 3rd edn. Sage Publications, Thousand Oaks, California.

Lyon, D.J. and Baltensperger, D.D. (1995) Cropping systems control winter annual grass weeds in winter wheat. *Journal of Production Agriculture* 8, 535–539.

Morgan, D.L. (1997) *Focus Groups as Qualitative Research*. Sage Publications, Thousand Oaks, California.

Morgan, D.L. and Krueger, R.A. (1998) *The Focus Group Kit*. Sage Publications, Thousand Oaks, California.

O'Brien, D., Sartwelle, J. and Thompson, C.R. (1998) *Managing Intensive Nonirrigated Cropping Systems in Western Kansas*. Publication MF2317. Kansas State University Research and Extension, Kansas.

Parajulee, M.N., Montandon, R. and Slosser, J.E. (1997) Relay intercropping to enhance abundance of insect predators of cotton aphid (*Aphis gossypii* Blover) in Texas cotton. *International Journal of Pest Management* 43, 227–232.

Patrick, C.D. and Boring, E.P. III (1998) *Managing Insect and Mite Pests of Texas Small Rains*. Texas Agricultural Extension Service Publication B-1251, College Station, Texas.

Peairs, F.B. (1998) Cultural control tactics for management of the Russian wheat aphid (Homoptera: Aphididae). In: Quisenberry, S.S. and Peairs, F.B. (eds) *Response Model for an Introduced Pest, the Russian Wheat Aphid*. Thomas Say Publications in Entomology, Entomological Society of America, Lanham, Maryland, pp. 288–296.

Peairs, F.B. (2003) Aphids in Small Grains. Fact Sheet no. 5.568. Colorado State University Cooperative Extension. Available at: http://www.ext.colostate.edu/pubs/insect/05568.html

Peairs, F.B., Haley, S. and Johnson, J. (2003) Russian wheat aphid infestations in Prairie Red. Wheat Breeding and Genetics Program, Soil and Crop Sciences Department, Colorado State University. Available at: http://wheat.colostate.edu/links.html

Peterson, G.A., Schlegel, A.J., Tanaka, D.L. and Jones, O.R. (1996) Precipitation use efficiency as affected by cropping and tillage systems. *Journal of Production Agriculture* 9, 180–186.

Quick, J.S., Ellis, G.E., Norman, R.M., Stromberger, J.A., Shanahan, J.F., Peairs, F.B., Rudolph, J.B. and Lorenz, K. (1996) Registration of 'Halt' wheat. *Crop Science* 36, 210.

Rice, M.E. and Wilde, G.E. (1988) Experimental evaluation of predators and parasitoids in suppressing greenbugs (Homoptera: Aphididae) in sorghum and wheat. *Environmental Entomology* 17, 836–841.

Royer, T.A. and Krenzer, E.G. (2000) *Wheat Management in Oklahoma: A Handbook for Oklahoma's Wheat Industry*. Publication E-831. Oklahoma Cooperative Extension Service, Oklahoma.

Royer, T.A., Giles, K.L. and Elliott, N.C. (1998) *Small Grain Aphids in Oklahoma*. Publication F-7183. Oklahoma Cooperative Extension Service, Oklahoma.

Royer, T.A., Giles, K.L. and Elliott, N.C. (2002a) *Glance 'n' Go Sampling for Greenbugs in Winter Wheat*, Spring edition, Extension Facts, L-306. Oklahoma Cooperative Extension Service, Oklahoma.

Royer, T.A., Giles, K.L. and Elliott, N.C. (2002b) *Glance 'n' Go' Sampling for Greenbugs in Winter Wheat*, Fall edition, Extension Facts, L-307. Oklahoma Cooperative Extension Service, Oklahoma.

Souza, E.J. (1998) Host plant resistance to the Russian wheat aphid (Homoptera: Aphididae) in wheat and barley. In: Quisenberry, S.S. and Peairs, F.B. (eds) *Response Model for an Introduced Pest, the Russian Wheat Aphid*. Thomas Say Publications in Entomology, Entomological Society of America, Lanham, Maryland, pp. 122–147.

Watkins, J.E. and Boosalis, M.G. (1994) Plant disease incidence as influenced by conservation tillage systems. In: Unger, P.W. (ed.) *Managing Agricultural Residues*. Lewis Publishers, Boca Raton, Florida, pp. 261–283.

Watson, S. (1994) *Best Management Practices for Wheat: A Guide to Profitable and Environmentally Sound Production*. National Association of Wheat Growers Foundation, Washington, DC.

Way, M.J. (1988) Entomology of wheat. In: Harris, M.K. and Rogers, C.E. (eds) *The Entomology of Indigenous and Naturalized Systems in Agriculture*. Westview Press, Boulder, Colorado.

Westra, P. and D'Amato, T. (1995) Issues related to kochia management in western agriculture. *Proceedings Western Society Weed Science* 48, 39–40.

15 Integrated Pest Management of Rice: Ecological Concepts

Gary C. Jahn,[1] James A. Litsinger,[2] Yolanda Chen[1] and Alberto T. Barrion[1]

[1]International Rice Research Institute (IRRI), DAPO Box 7777, Metro Manila, Philippines; [2]1365 Jacobs Place, Dixon, CA 95620, USA

Introduction

Rice is one of the oldest domesticated crops and due to its importance as a food crop humans took it with them when settling in new areas further removed from flooded river plains extending to the uplands. For thousands of years, rice, with the exception of some dryland systems, was grown in monoculture. This traditional, low-yielding, rice production system was more sustainable than any other crop in human history (Bray, 1986; von Uexkuell and Beaton, 1992; Reichardt *et al.*, 1998). Rice has morphed into many forms as a result of its domestication by humans, initially along the large rivers of monsoon Asia, where it was selected to be tall to tolerate seasonal flooding but also could survive periods of drought, and then further up the watershed to fertile valleys and finally into mountainous areas. The major rice ecosystems are *wetland* (also known as 'paddy' or 'lowland'), *dryland* (or 'upland') and *deepwater*. Wetland rice is grown under flooded field conditions, and can be divided into irrigated, rainfed and recession rice. Recession rice is grown when rice seedlings are transplanted into receding water such as occurs when a lake dries up. Earth embankments, called bunds, to retain standing water surround wetland rice

fields. Dryland rice is grown without standing water. Deepwater rice is flooded deeper than 50 cm for 1 month or longer during the growing season. It does not include recession rice (Catling, 1992).

Insect pests have adapted to each of the major rice ecosystems as well as to new rice varieties, cultural practices, fertilizers and pesticides. Among the wide array of rice pests, some transferred from wild rice, which are monophagous to *Oryza* spp. such as yellow stem borer (YSB), *Scirpophaga incertulas* (Walker), green leafhopper, *Nephotettix virescens* (Distant), brown planthopper (BPH), *Nilaparvata lugens* (Stål) and Asian gall midge, *Orseolia oryzae* (Wood-Mason). To this day, the distribution of gall midge coincides with the distribution of wild rice varieties. For example, neither the gall midge nor its wild rice hosts are found in the Philippines. Other pests such as *Chilo* and *Sesamia* stem borers and many species of leaffolders and butterflies [(skippers, *Parnara guttata* (Bremer and Grey) and *Pelopidas mathias* (Fabricius), and greenhorned caterpillar, *Melanitis leda ismene* (Cramer)] transferred from grasses. Various species of armyworms, *Mythimna separata* (Walker), and cutworms, *Spodoptera litura* (Fabricius), attack the pre-flooded crop. As the fields flood, aquatic species enter the ecosystem, such as black bugs, *Scotinophara* spp., and rice caseworm,

Nymphula depunctalis (Guenée). The lar-vae of caseworms have gills, thus standing water is required for survival (Litsinger *et al.*, 1994a). Thrips, *Stenchaetothrips biformis* (Bagnall), and mealybugs, *Brevennia rehi* (Lindinger), thrive in rice during droughts. Thrips numbers are normally held in check by heavy rainfall (Mochida *et al.*, 1987). Mealybugs tap the rich flows of soluble nitro-gen (N) that are available in rice plants under drought stress (Jahn, 2004). Rice bugs, *Lepto-corisa* spp., are one of the few pests that feed directly on developing seeds, resulting in reduced yields, poor grain quality and lower seed germination rates (Jahn *et al.*, 2004).

Over time rice became adapted to the more deeply flooded areas near rivers in the deepwater environment. Here rice crops are first seeded in dry soil before the monsoon seasonal flood. The insect pests at this stage are the same as in rainfed wet-land environments, which typically adjoin deepwater rice areas (Catling, 1992). Rice elongates with the rising floodwater up to several centimeters per day. There are four rice pests that have adapted to these gruell-ing conditions: (i) YSB, (ii) ufra nematode *Aphelenchoides besseyi* Christie, (iii) rice hispa *Dicladispa armigera* (Olivier) and (iv) rats. Bandicoot rats *Bandicota indica* (Bechstein) and *B. bengalensis* (Gray and Hardwicke) are dominant in the Indian subcontinent with *Rattus* spp. in South-east Asia. Once fields are flooded popu-lations of leafhoppers, leaf feeders and planthoppers decline markedly. Floodwa-ters bring riverine invertebrate pests such as crabs and chironomids into rice fields. This system also has a close association with fish culture providing additional nat-ural enemies of insect pests. The photope-riod-sensitive cultivars flower during short days marking the end of the flood season so that the crop is harvested in dry condi-tions once again. YSB is highly adapted to deepwater rice. Its eggs can withstand 2 days of submergence. The larval stage can be passed inside submerged rice stems and before pupation the larva cuts an exit hole for the moth to escape, which is sealed watertight in silk. When the moth emerges it floats to the surface.

International Rice Research Institute (IRRI) developed rapid maturing, photope-riod-insensitive, semi-dwarf rice varieties in the mid-1960s. In most irrigated systems and some rainfed areas, these new high-yielding varieties (HYVs) enabled a change from the single-season rice culture to two crops per year. This dramatic shift in agricul-ture was the basis of the Green Revolution in Asia. The more stable irrigated systems reduced losses from drought or flooding, and the shorter stiffer stems were less likely to lodge (i.e. topple over) when N was added. The more nutrient rich crop fuelled greater pest populations (stem borers, planthoppers, leafhoppers and leaffolders) particularly under conditions of indiscriminate insecti-cide usage. New pests emerged such as the whorl maggots and lepidopterous defolia-tors. In more temperate climates leaf beetles, *Oulema oryzae* (Kuwayama), occurred and the striped stem borer (SSB), *Chilo sup-pressalis* (Walker), survived over winter in dormancy. In Asia, there are hundreds of species of flora and fauna living in a typi-cal rice field each season (Way and Heong, 1995; Schoenly *et al.*, 1996). When taken to Africa new pests such as the stalk-eyed fly, *Diopsis* spp. transferred from wild rice and grasses to domestic rice, but local species of stem borers, leafhoppers, planthoppers, gall midge and seed bugs filled similar niches as their Asian cohorts. The stalk-eyed fly's pre-ferred habitat appears to be marshes and is thus easily adapted to irrigated rice. Fewer new pest species emerged when HYVs were introduced to Latin America, but the stan-dard complement of stem borers, planthop-pers, leafhoppers, water weevils and seed bugs colonized rice fields.

As humans moved from the alluvial river valleys to mountainous regions rice was carried with them. In some extreme cases the wetland systems were replicated as rice ter-races in the Himalayas, Indonesia or the Phil-ippines. River water was diverted to carve out the terraces and irrigate the crop. New pests appeared under these conditions. At times annelid worms attacked young seedlings and were carried by eroded riverbanks because of deforestation (Barrion and Litsinger, 1997). Rats and mole crickets bore into the terrace walls and became pests by creating leaks causing the water to drain out.

As pressure for agricultural land increased, mountainous forests were burned and dibbled rice was planted in its place. In slash-and-burn agriculture rice is planted the first year in the nutrient rich ash. These small plantings attract a menagerie of vertebrate pests that dwell in the forest and, depending on the location, can include rats, birds, wild pigs, squirrels, monkeys, elephants and even rhinoceroses. With further logging, forests become smaller, causing vertebrate pests to become concentrated and any nearby plantings become overwhelmed. As Sumatran farmers revealed (Fujisaka et al., 1991), such high populations become difficult to control. Damage is most severe usually near harvest time. The main control method is the use of dogs and noise makers hung on strings across fields, with family members taking shifts in making noise over 24 h as pigs attack by night and monkeys by day. The larger animals do more damage milling around than actually eating the crop. At night pigs, rhinos and elephants bed in the fields and trample plants underfoot.

Over time, agriculture typically causes grasslands to replace forest. Slash-and-burn agriculture is then slowly replaced with dryland rice fields, which are ploughed. Seed is dibbled or broadcast in plough furrows. This more sedentary culture brings an array of seed and soil pests virtually unknown in flooded rice (Litsinger et al., 1987a). Among the multitude of vertebrate pests only rats and birds tend to remain but in less dense numbers. Pests such as ants remove sown seed to nearby nests. A host of other grassland pests transfer to rice to attack the roots such as white grubs, black beetles, termites, field crickets, mole crickets, root aphids, mealybugs and root bugs. Seedling maggot (Atherigona oryzae Malloch) attacks young seedlings tunnelling into developing tillers causing deadhearts. Grasshoppers, thrips and seed bugs also transfer from grasses and cultivated crops such as sugarcane, sorghum and maize. A wide array of wetland pests seasonally disperse into the uplands from the lowlands including leafhoppers, planthoppers, stem borers, leaffolders and

rice bugs. As more of the grasslands turn into plough agriculture, the severity of the grassland pests declines and the system stabilizes. Frequent tillage for annual crops, which are planted before or after rice, suppresses the soil pests.

With impending shortages of rice predicted in the 1970s, irrigation systems were expanded to take advantage of HYVs. In Asia, a nation's food requirements are largely dependent on these rice bowls. Along with improved yields came frequent pest epidemics. Entire crops may be lost to rodents, BPH or combinations of pests. To recover from such losses farmers often borrow money to purchase seeds for replanting. The interest on village credit commonly ranges from 30% to 100% per year. If the same farmer requires any additional inputs (e.g. labour, fertilizer or pesticides), he or she must borrow additional money. If the inputs are not made, due to the high cost, the farmer will produce less rice and have less income to pay off debts. Pest outbreaks, therefore, can throw farmers into a cycle of poverty from which it is difficult to escape (Jahn et al., 1996, 1999).

Fields receiving supplemental water or full irrigation tend to have more stable yields than rainfed wetlands. In irrigated rice systems, pests affect sustainability in a different way. Where yields are relatively high and predictable, farmers have enough disposable income to purchase pesticides. Often the pesticides are used routinely as a form of insurance, rather than a graduated response. In other words, farmers may protect their crops by applying a pesticide at the first signs of pest damage, regardless of the incidence or severity of the injury. In lesser developed countries, insecticides are generally used on rice in an inefficient, ineffective and unsafe manner (Heong and Escalada, 1997; Jahn et al., 1997a, b).

These differences in attitude towards pests among farmers in the rainfed and irrigated systems are reflected in their pesticide use patterns. Pesticide use is generally higher in irrigated rice than in rainfed conditions. Pest management is perhaps the weakest link in developing a sustainable agriculture system. This is particularly evident when insecticides are involved. Prolonged insecticide use often leads to well-known problems such as destruction of natural enemies, secondary pest outbreaks, pesticide resistance and ultimately entry into the 'pesticide treadmill', described by Litsinger (1989) and van den Bosch (1980), where continually greater amounts of insecticide are required to achieve the same level of control.

What is a Rice Pest?

In this chapter rice pest is defined as any organism or microbe with the potential to reduce rice yields, grain quality, seed viability or rice farm profits. Chronic pests are those that normally occur at low levels in rice fields. Managing chronic pests is generally a matter of managing the crop in such a way that pest damage is minimized. Acute pests are those that occasionally reach such high populations that they significantly reduce yields or damage crops beyond the normal situation, usually destroying a quarter of the crop or more. Depending on conditions, some chronic pests can reach outbreak proportions and become acute pests. Rice bugs, *Leptocorisa oratorius* (Fabricius), are examples of chronic pests that become acute pests under certain conditions. Other acute pests, such as the BPH, do not reduce yields unless present at outbreak levels, at which times they can destroy entire crops with "hopperburn". In dryland rice, acute pests include white grubs and mole crickets. In wetland rice, acute pests can include gall midge, thrips, BPH, rice black bug and armyworms. Acute deepwater rice pests include YSB and stem nematode *Ditylenchus angustus*. For acute pests that cause medium to high yield loss, but only infrequently, pest-resistant rice varieties are an ideal addition to the farming system.

Adaptation of pests to their changing environment

Changes in the environment may favour one pest group but not another (Loevinsohn, 1994). The most dramatic aspect of the rice environment is whether the crop is dryland or wetland. Flooding a rice field eliminates soil pests whereas aquatic species thrive in ponded conditions. Droughts cause outbreaks of armyworms, locusts, mealybugs and thrips. Prolonged El Niño droughts in the mid-1980s in Indonesia and the southern Philippines (Glanz, 1996) led to a shift in stem borer species from the non-aestivating YSB to the aestivating white stem borer (WSB), *S. innotata* (Walker), as seen in annual kerosene light trap collections (Fig. 15.1). Shifts in dominance have occurred several times in the last century but not in all areas. WSB was the most abundant rice stem borer in Java (van der Goot, 1925) as well as central and southern Philippines (Cendaña and Calora, 1967) from the earliest records in the last century. The first shift from WSB to YSB occurred in Java in the early 1970s and WSB was confined to Indramayu (W. Java) and Gresik (E. Java) by 1978 (Hattori and Siwi, 1986). On other islands, WSB remained abundant only in Sulawesi.

WSB and YSB adults are morphologically distinct, but the WSB and YSB larvae are indistinguishable. There are colouration differences in the larvae; some appear whiter while others are yellowish. But these colour changes have nothing to do with distinguishing the two species. The adult female YSB moth has a black spot on each front wing, while WSB has none. Cendaña and Calora (1967) commented on the distribution of the three most prevalent stem borers in the Philippines finding YSB and SSB in all parts of the Philippines with WSB only in the Visayas and Mindanao. In a survey in the early 1960s, YSB was most dominant in Central Luzon (>95% of species) and SSB most prevalent in Laguna province (>85%). The changing distributions of stem borers can be explained by climate, construction of irrigation systems, introduction of new rice varieties and insect size. YSB is the most highly adapted to flooding and, like WSB, is narrow-bodied. Both feed only on *Oryza* spp. In terms of climatic preferences SSB can adapt well to temperate climate. YSB is prevalent in monsoonal tropical Asia

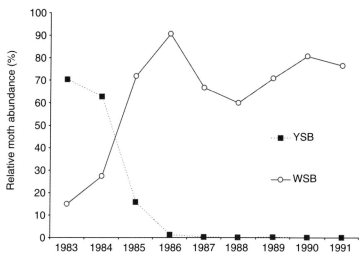

Fig. 15.1. Annual catches in kerosene light traps in Koronadal, Philippines, showing the shift in rice stem borer species from yellow stem borer (YSB) to white stem borer (WSB) probably in response to the El Niño drought of 1982–1983.

while WSB is prevalent along the Intertropical Convergence Zone. Records of WSB in Luzon are suspect and may be confused with *S. nivella* (Fabricius), a pest of sugarcane. In the last 30 years there are no confirmed records of WSB in Luzon. Monsoon rain leads to flooding and Central Luzon is often under water for weeks after strong typhoons. In some cases irrigation systems exacerbate flooding when canals prevent natural drainage. Expansion of irrigation systems and the popularity of double-cropped irrigated rice also favoured YSB. YSB's distribution mirrors areas that flood. It does not survive in very cold winter climates such as northern Japan and China. Areas outside of the monsoonal tropics favour WSB, especially those that are affected by the El Niño drought. WSB is highly adapted to drought and can survive in aestivation for 10–12 months (van der Goot, 1925). WSB evolved in single-cropped rainfed rice with a long dry season as the larvae aestivated at the base of stubble. Early monsoon rain brought large emergences of adult moths after diapause was broken. El Niño droughts are of long duration favouring WSB. YSB can also aestivate during the dry season but normally does not if rice fields are found. YSB populations are favoured by double cropping. The larvae of both WSB

and YSB are slender and long and only one is typically found per tiller, whereas SSB larvae are larger and may amass in over 20 in one tiller (Kanno, 1962). Thus tillers with larger diameters favour SSB. Tiller dissections show that it is relatively more abundant in older and thicker tillers (Table 15.1). It is more prevalent in thicker *japonica* rice than *indica* rice (Barrion *et al.*, 1987). Further evidence of this is its resurgence with the introduction of a new plant type in the 1990s that was larger-stemmed. SSB also has a broader host range and infests a wide variety of grassy weeds.

Effects of neighbouring crops on pests

Rice grown near maize is prone to *Chilo* stem borer infestations, and rice grown near sugarcane is prone to pink stem borer, *Sesamia inferens* (Walker), infestations. Grasslands near rice fields can serve as a source of armyworms, grasshoppers, white grubs and seedling maggots. Asian gall midge survives in wild rice in the off season. For polyphagous species local floral composition determines relative preference among possible alternate plant hosts (Litsinger *et al.*, 1993). Many of the same ant species inhabit both rice fields

Table 15.1. Stem borer species abundance determined from dissections in three rice cultures and three island groups in the Philippines.

| Site | | | | | Crop age[a] | Crops (no.) | Tiller dissections — Relative abundance (%) | | | | | | Sample unit | n |
Island	Province	Municipality	Years	Culture			Scirpoph-aga[b]	Chilo sup-pressalis	Sesamia inferens	C. auricilius	C. poly-chrysus	Maliarpha[a]		
Mindanao	Misamis Oriental	Claveria	1988–1989	Dryland	Post harvest	2	24.3	67.7	7.7	0.1	0.1	0.1	1-m row	33 m rows
Luzon	Laguna	Tanauan	1977–1979	Dryland	21–77 DE	3	81.2	1.2	12.8	4.8	0	0	Damaged tillers	1020 tillers
Luzon	Laguna	Los Baños	1986	Dryland	Ripening	1	42.4	45.8	9.1	1.2	0	1.4	150 tillers/ variety	5250 tillers
Visayas	Iloilo	Oton	1976	Rainfed wetland	35–70 DT	3	60.1	24.1	8.3	7.5	0	0	50 hills	150 hills
Luzon	Laguna	Calauan	1988–1990	Irrigated wetland	4–13 WT	2	97.4	2.3	0.3	0	0	0	20 hills	160 hills
			1986	Irrigated wetland	Ripening	2	64.3	28.2	7.5	0	0	0	20 half hills	480 half hills
Luzon	Nueva Ecija	Guimba	1985–1989	Irrigated wetland	4–11 WT	10	100	0	0	0	0	0	20 half hills	6000 half hills
Luzon	Nueva Ecija	Zaragoza	1985–1989	Irrigated wetland	4–11 WT	10	100	0	0	0	0	0	20 half hills	6000 half hills
Mindanao	South Cotabato	Koronadal	1983–1991	Irrigated wetland	Vegetative	11	92.5	6.1	1.3	0	0	0	20 half hills	1320 half hills
					Reproductive	11	63.2	35.4	1.3	0	0	0	20 half hills	1320 half hills
					Ripening	11	55.0	41.1	3.8	0	0	0	20 half hills	1320 half hills
					Koronadal mean	11	70.8	27.6	2.1	0	0	0	20 half hills	

[a]DE = days after crop emergence; DT = days after transplanting; WT = weeks after transplanting.
[b]Mixture of S. innotata and S. incertulas in Mindanao and Visayas sites, the larvae of each species could not be determined.
[c]Undetermined species.

and pineapple fields, though most of these species are beneficial in rice and pineapple (Jahn and Beardsley, 1996; Way *et al.*, 1998; Jahn *et al.*, 2003).

Pest rankings

The stable traditional rice ecosystem with coevolved pests that developed over millennia was dramatically altered in the mid-1960s with the advent of HYVs, year-round irrigation, and heavy pesticide and fertilizer usage becoming an open laboratory providing evolutionary lessons on the adaptive capabilities of rice pests. Few of the changes were predicted and even today there is not a consensus of opinion on the cause and effect of many of these relationships. Host plant records have at times been solely based on the presence of the pest on a rice crop. Thus, the planthopper *Nisia atrovenosa* was thought to be a pest of rice but actually does not survive on rice. Its host is *Cyperus rotundus*, which grows on rice bunds and when people walk on the bunds, the planthopper jumps away landing on rice plants (dela Cruz and Litsinger, 1986). This habit led it to be claimed a rice pest. Another insect, the caseworm *Parapoynx fluctuosalis* (Zeller) feeds on the aquatic weed *Hydrilla* and was thought to be a pest because its numbers increased at rice planting times (Litsinger and Chantaraprapha, 1995). It lives in irrigation canals and flourishes when irrigation water is delivered, giving the impression that it feeds on rice. When forced to feed on rice in greenhouse trials, it succumbed. Two 'biotypes' of BPH were found living side by side, one on rice and the other on the weed *Leersia hexandra*, and were infertile when crossed (Saxena *et al.*, 1983).

Loevinsohn (1984) compared insect pest abundance in three rice environments by setting up a transect of kerosene light traps between an irrigated wetland area running into a dryland area with a rainfed wetland area in between. The most common pests encountered were leafhoppers, planthoppers, leaffolders and stem borers. Densities dramatically increased as one travelled from dryland to irrigated wetland areas. Regression analysis revealed that aspects of the environment most influencing population increase were (in declining importance): (i) greater number of rice crops per year (short dry fallow periods); (ii) greater area planted to rice; and (iii) greater asynchrony of planting. Fertilizer and insecticide usage were insignificant in the relationship that examined community-wide changes.

Another kerosene light trap data-set of discrete locations showed lowest insect numbers in the slash-and-burn and eroded acid soil drylands (Table 15.2). In the Siniloan, Laguna slash-and-burn site in the Sierra Madre Mountains in eastern Luzon with a soil pH of 4.0, a crop could only be grown if organic matter were applied in dibbled seed holes along with low amounts of nitrogen/phosphorus/potassium (NPK). Only about 2% of the hillside was planted with rice, causing a great concentration of pests. A second dryland site established in Claveria, North Mindanao, was converted from rainforest to grassland over the last 5–6 decades. Less than 10% of the area was planted with rice in any given year. The acid soils were highly eroded on sloping lands but the land was prepared by animal-driven plough. Most of the area that was formerly grassland pasture lay fallow, partly settled due to civil strife. These sites are the most isolated. Still insects such as BPH, green leafhoppers and *Cyrtorhinus* predator were the most abundant in the sites, demonstrating their migratory powers. Data from two intensively cropped sites completed the dryland environment: both were on rather fertile volcanic soils and non-sloping terrain. Tanauan, Batangas, is mainly a vegetable and sugarcane area near the Manila markets. Tupi in South Cotabato is near Mt Matutum volcano, and generates its own rainfall pattern, so a few farmers even attempt to double-crop dryland rice. These favourable dryland sites matched the rainfed wetland sites in terms of insect abundance. The Solana rainfed wetland site in the Cagayan Valley in North Luzon was prone to the deepest floods and longest droughts of the three rainfed wetland sites.

Table 15.2. Annual mean kerosene light trap collections from sites in three rice ecosystems, Philippines.

Species	Dryland						Rainfed wetland				Irrigated wetland			
	Acid soils/sloping hills			Favourable soils/ploughed			Flood-prone	Favourable			Favourable			
	Siniloan, Laguna, 1984–1985	Claveria, M. Oriental, 1984–1985	Avg	Tanauan, Batangas, 1980–1981	Tupi, S. Cotabato, 1989–1991	Avg	Solana, Cagayan, 1980–1983	Manaoag, Pangasinan, 1979–1981	Oton, Iloilo, 1979–1981	Avg	Calauan, Laguna, 1985–1986	Zaragoza, N. Ecija, 1981–1982	Koronadal, S. Cotabato, 1983–1991	Avg
Green leafhoppers *Nephotettix* spp.	421	75	248	860	1,421	1,141	863	870	2,500	1,411	1,860	9,380	4,731	5,324
Zigzag leafhopper *Recilia dorsalis*	0	190	95	710	1,958	1,334	313	535	1,600	816	690		4,255	2,473
Brown planthopper *Nilaparvata lugens*	185	260	223	130	3,183	1,657	330	410	2,450	1,063	740	15,655	13,764	10,053
Whitebacked planthopper *Sogatella furcifera*	0	134	67	1,700	556	1,128	1,393	260	1,290	981	6,300		10,267	8,284
Mirid predatory bug *Cyrtorhinus lividipenis*	425		425	600	1,788	1,194	260	1,870	265	798	620		6,526	3,573
Leaffolder *Cnapha locrocis medinalis*	105	39	72	70	63	67	82	135	155	124	60	980		520
Yellow and white stem borers *Scirpophaga* spp.	41	49	45	50	190	120	223	280	515	339	540	440	3,670	1,550
Striped stem borer *Chilo suppressalis*	0	111	56	0	61	31	17		45	11	160		615	388

Dark-headed stem borer *C. auricilia*	0	52	26	0		0	12	20	0
Pink stem borer *Sesamia inferens*	0	12	6	0	6	48	45	46	51	43	47
No. sites	2	4	2	2	3	3	3	2	4	4
Cropping intensity										
Rice crops (no./year)	1.0	1.0	1.0	1.0	1.0	1.1	1.5	2.0	2.0	2.4
Rice area (%)	2	9	20	12	60.0	85.0	85	90.0	80.0	70.0

The traditional cultivar was Wagwag while the modern variety was IR36. The other two sites were in more favourable locations, one in Manaoag, Pangasinan, in North Luzon and the other in Iloilo in the Visayas, with both IR28 and IR36 rice varieties. The Iloilo and Tupi sites were near large irrigated rice areas. The three irrigated sites had the highest insect collections due to the double-cropping, as the other sites were single crops. Calauan is near to Los Baños in Laguna along Laguna de Bay Lake where 90% of the 1 km wide strip of land along the lake is irrigated by small rivers off volcanic slopes draining into the lake. Zaragoza is at the tail end of the largest irrigation system in the Philippines, in Central Luzon. Koronadal, the site with the highest insect numbers, had areas that cultivated five crops in 2 years due to abundant rains and artesian irrigation.

Irrigation and HYVs accelerated rice intensification both spatially and temporally. The pests that benefited most were monophagous to *Oryza* spp. (Loevinsohn, 1994). At the other extreme, dryland pests were mainly polyphagous and survived the non-rice period by: (i) feeding on other plants, (ii) entering aestivation, or (iii) dispersing to other locations (Litsinger *et al.*, 1987a). With year-round irrigation, four rice crops can be grown in a year in a system called the rice garden. Essentially the field is divided into plots, the number being equal to the number of weeks to maturity from transplanting. Each week a new plot is transplanted and another one harvested, thus all stages are present 1 week apart. On a small scale the system worked well and encouraged natural enemy build-up (Pantua and Litsinger, 1981). But large-scale attempts to carry this out in the Philippines failed due to extreme pest abundance, principally BPH (and associated viral diseases) and rats.

Yield Loss

Understanding yield loss is a key to rice integrated pest management (IPM) but it has been difficult to correlate insect pest densities to loss (Litsinger *et al.*, 1987b; Jahn, 1992). At times high losses were associated with low pest densities based on a data-set of yield losses and pest densities employing the insecticide check method (Litsinger, 1991) in the three main rice environments in the Philippines: irrigated areas, rainfed wetlands and drylands.

Yield loss in dryland rice

Within the Philippines drylands there are examples of sites fitting the three progressive stages of rice ecosystem development from the pioneering slash-and-burn to grassland, and finally to intensified permanent agriculture. In the Siniloan slash-and-burn site undergoing deforestation, grain harvest could only be measured in one of the three study years as rats destroyed the crops in the other years. Of the 1 year when harvest was possible, the highest losses from insect pest damage were recorded in all 12 test sites, averaging 58% from the collective damage of ants, seedling maggot, stem borers, thrips and rice bug (Table 15.3). The yield potential was less than 0.5 t/ha. Seedling maggot, the most abundant insect pest, averaged 37% deadhearts. A higher yield potential of 3 t/ha was evident in Claveria, a highly eroded dryland site, due to the high tillering UP-Ri 5 variety and the only sporadic occurrence of rats as most of the forests were logged. The dominant pests transferred from the grasslands that replaced the forests – ants, termites, seedling maggot, white grubs, root aphids, root mealybugs, flea beetle and seven species of stem borers. Yield loss averaged 25% over 7 years, with maximum loss in the vegetative stage from seedling maggot and soil pests. Some of the lowest yield losses occurred in the two favourable dryland sites of Tanauan, Batangas, and Tupi, South Cotabato, ranging from 2% to 12%. Least loss occurred in Tanauan on the low-yielding traditional cultivar Dagge, prized for its eating quality. The main rice pests were stem

Table 15.3. Yield loss determination using the insecticide check method in farmers' fields by rice environment, Philippines.

	Dryland					Rainfed wetland						Irrigated wetland – modern							
	Acid soils, sloping hills		Favourable soils/ploughed			Flood-prone		Favourable			Flood-prone	Favourable							
	Slash-and-burn	Ploughed																First crop	Second crop
Variety	Traditional	Modern	Traditional	Modern	Modern	Traditional	Modern	Traditional	Modern	Modern	Modern								
Town	Siniloan	Claveria	Tanauan		Tupi	Solana		Manaoag		Oton	Gadu	Zaragoza		Guimba		Calauan		Koronadal	
Province	Laguna	Misamis Oriental	Batangas		S. Cotabato	Cagayan		Pangasinan		Iloilo	Cagayan	Nueva Ecija				Laguna		S. Cotabato	
Island	Luzon	Mindanao	Luzon		Mindanao	Luzon		Luzon		Visayas	Luzon	Luzon						Mindanao	
Year	1984–1985	1985–1991	1976–1980	1978–1980	1989–1991	1978–1980		1979–1980	1976–1980	1976–1979	1982–1983	1979–1991		1984–1991		1982–1991		1983–1991	
Seasons											WS	WS	DS	WS	DS	WS	DS		
Yield (t/ha)																			
Protected	0.48	3.0	2.90	4.01	3.58	1.84	1.99	2.58	4.72	4.52	2.38	5.09	6.23	4.39	4.80	4.61	4.79	5.16	4.85
Check	0.20	2.3	2.85	3.61	3.16	1.73	1.65	1.94	3.61	3.84	1.83	4.42	5.5	3.67	4.03	4.27	4.38	4.55	4.1
Yield loss (t/ha)	0.28	0.7	0.05	0.4	0.42	0.11	0.34	0.64	1.11	0.68	0.55	0.70	0.63	0.72	0.77	0.3	0.39	0.6	0.75
Yield loss (%)	58	25	2	10	12	6	17	25	24	15	23	13	10	21	18	6	8	11	15
Vegetative	13	13	1	10	1	3	7	11	17	7	14	5	4	9	5	4	2	4	7

Continued

Table 15.3. *Continued*

Variable	Dryland — Acid soils, sloping hills, Slash-and-burn, Traditional	Dryland — Acid soils, sloping hills, Ploughed, Modern	Dryland — Favourable soils/ploughed, Traditional	Dryland — Favourable soils/ploughed, Modern	Dryland — Favourable soils/ploughed, Modern	Rainfed wetland — Flood-prone, Traditional	Rainfed wetland — Flood-prone, Modern	Rainfed wetland — Favourable, Traditional	Rainfed wetland — Favourable, Modern	Rainfed wetland — Favourable, Modern	Rainfed wetland — Flood-prone, Modern	Irrigated wetland – modern — Favourable							
Town	Siniloan	Claveria	Tanauan	Tanauan	Tupi	Solana	Solana	Manaoag	Manaoag	Oton	Gadu	Zaragoza	Zaragoza	Guimba	Guimba	Calauan	Calauan	Koronadal	Koronadal
Province	Laguna	Misamis Oriental	Batangas	Batangas	S. Cotabato	Cagayan	Cagayan	Pangasinan	Pangasinan	Iloilo	Cagayan	Nueva Ecija	Nueva Ecija	Nueva Ecija	Nueva Ecija	Laguna	Laguna	S. Cotabato	S. Cotabato
Island	Luzon	Mindanao	Luzon	Luzon	Mindanao	Luzon	Luzon	Luzon	Luzon	Visayas	Luzon	Luzon	Luzon	Luzon	Luzon	Luzon	Luzon	Mindanao	Mindanao
Year	1984–1985	1985–1991	1976–1980	1978–1980	1989–1991	1978–1980	1978–1980	1979–1980	1976–1980	1976–1979	1982–1983	1979–1991	1979–1991	1984–1991	1984–1991	1982–1991	1982–1991	1983–1991	1983–1991
Seasons											WS	WS	DS	WS	DS	WS	DS	First crop	Second crop
Reproductive	8	8	0.5	8	11	2	7	5	7	4	4	5	4	7	8	1	4	4	5
Ripening	8	8	0.5	5	0	1	3	3	6	4	5	3	2	5	5	1	2	3	3
(no. crops) Input usage	2	7	5	3	3	3	3	2	5	4	2	11	12	7	6	9	8	7	8
Fertilizer (kg N/ha)	0	0	60		70	0			20	35	40	60	90	110	80	70	80	35	45

Species	Dryland					Rainfed wetland										Irrigated wetland				
Province / Town	S. Cotabato Tupi	Misamis Oriental Claveria	Batangas Cale	Laguna Siniloan		Pangasinan Carmen			Cagayan Bangag				Iloilo Buray			Laguna Santa Maria	Nueva Ecija Caba-natuan	Zara-goza		S. Cotabato
Year	1989–1991	1984–1985	1980–1981	1984–1985	Avg	1979–1980	1980–1981	Avg	1980–1981	1981–1982	1982–1983	Avg	1979–1980	1980–1981	Avg				Avg	
Nephotettix virescens	975	75	650	421	530	1,240	500	870	1,800	330	210	780	3,400	1,600	2,500	160	360	18,400	9,380	2,765
N. nigropictus	446		210						140	90	20	83				1,700				1,966
Nephotettix spp.	1,421	75	860	421	694															4,731
Recilia dorsalis	1,958	190	710	0	715	180	890	535	630	140	170	313	2,400	800	1,600	690				4,255
N. lugens	3,183	260	130	185	940	300	520	410	750	180	60	330	3,300	1,600	2,450	740	310	31,000	15,655	13,764
Sogatella Furcifera	556	134	1,700	0	598	310	210	260	3,300	610	270	1,393	980	1,600	1,290	6,300				10,267
Cyrtorhinus lividipennis	1,788		600	425	938	540	3,200	1,870	290	160	330	260	120	410	265	620				6,526
Scirpophaga spp.	190	49	50	41	83	410	150	280	180	220	270	223	1,000	30	515	540	610	270	440	3,670
C. medinalis	63	39	70	105	69	70	200	135	140	100	6	82	250	60	155	60	60	1,900	980	
C. suppressalis	61	111	0	0	43				30	20	0	17	6	3	4.5	160				615
Chilo auricilius		52	0	0	17				20	3	60	28	20	3	11.5	0				
Sesamia inferens	11	12	0	0	6				80	60	3	48	10	80	45	51				43

borers, leaffolders and white grubs. In Tupi the soils and rainfall are so favourable that wetland varieties are grown, but the variety in the trials was UP-Ri5. The main pests in Tupi were stem borers and white grubs. The yield potential is higher in both these sites where farmers apply 60–70 kg N/ha, highly unusual for dryland rice in Asia. Insecticide usage is almost non-existent in dryland rice sites in the Philippines although most farmers would have tested them in the past (Litsinger *et al.*, 1987a).

Yield loss in rainfed wetland rice

In the Philippines, losses ranging from 6% to 25% with both traditional and modern cultivars were recorded from three rainfed wetland sites (Litsinger, 1991). Yield potential in both sites was over 4.5 t/ha with the modern high tillering rice varieties. In Iloilo in favourable years it was possible to double-crop using modern rice varieties. Losses were mainly from YSB and WSB, whorl maggots, caseworms and leaffolders. Whorl maggots, caseworms and leaffolders only damage rice leaves. Before rice reaches the reproductive stage, i.e. before panicle initiation, leaf damage in otherwise healthy plants rarely reduces yields. Leaf damage experiments indicate that removal of all leaves from rice seedlings and 50% of leaves from tillering-stage rice does not significantly reduce rice yields (Khiev *et al.*, 2000). However, if the plant is also suffering from other stresses such as additional pest pressure, insufficient nutrients or insufficient water, lower levels of vegetative-stage leaf damage could reduce yields. During the reproductive stage of rice the flag leaf emerges. The flag leaf is the leaf at the top of the rice plant, just below the panicle. It is the last leaf to emerge. Removal of the flag leaf can reduce yields by 25% (CIAP, 2000; CARDI, 2001).

In Cambodia, 104 farmers' fields of rainfed wetland rice were each divided into six plots and studied for variety, fertilizer and pest interactions (CIAP, 1998, 1999). Fields were spread over 13 provinces. In fields of early duration varieties major constraints related to variation in yield included deadhearts (stem borer), brown spot, rodents, rice hispa, false smut, whorl maggot and leaf blight. Gall midge, stem borers and cutworms had the greatest impact on yields of medium duration variety. Armyworms, stem borers, rodents and sheath rot were the major causes of crop loss in late duration varieties. Farmers' traditional Cambodian rice varieties were more susceptible to pest damage than pure line strains of traditional Cambodian rice varieties. This may be due to the off-types in farmers' varieties, which lead to highly asynchronous development. HYVs were less competitive with weeds than pure line or farmers' traditional varieties. A Cambodian study of stem borer damage in rainfed rice found that crop loss per hill ranged from 0% to 33% at two whiteheads per hill (CIAP, 1995). Typically, 4–5% of tillers suffer gall midge damage in Cambodian farmers' fields of traditional varieties in rainfed wetland rice (CIAP, 1995).

Yield loss in irrigated rice

Irrigated sites have the highest yield potential in the most stable environmental conditions. Irrigated fields in Zaragoza, Philippines, have yields that peak at over 6 t/ha in the dry season (Litsinger, 1991). Losses in five irrigated sites ranged from 6% to 23%. Lowest losses were in Calauan, Laguna, where farmers grew longer maturing C1 variety. Generally two rice crops were grown but in Koronadal, South Cotabato, farmers in some areas grew five crops in 2 years with IR60 as the most favoured variety under more favourable irrigation, some from artesian wells flowing year-round except in El Niño years. The site suffered from BPH as well as green leafhopper and tungro. Main rice pests were whorl maggots, defoliators, caseworms, stem borers, leaffolders and occasionally rice bug and whitebacked planthopper.

Pest Management

Biological control

Realization of the importance of natural enemies in rice

Until the 1980s parasitoids were thought to be the only manageable biological control agents against rice insect pests. The successes with introduction and augmentation of parasitoids against stem borers in tropical sugarcane in many parts of the world prompted a similar effort in tropical rice. Greater results in rice accrued against introduced rice pests rather than indigenous ones (Ooi and Shepard, 1994). Early work at IRRI likewise focused on the introduction of exotic parasitoids against stem borers. One parasitoid was established against the pink stem borer (*S. inferens*) in the Philippines but no beneficial effect was observed (IRRI, 1971). In the Philippines the spread of the rice black bug to the island of Palawan in the 1980s led to the recent efforts to control a rice pest through introduction of natural enemies. Two egg parasitoids indigenous to Luzon were introduced; however, none of the species was established (Ooi and Shepard, 1994).

The sudden epidemics of planthoppers and leafhoppers (and diseases they vectored) after modern rice varieties were introduced were blamed on genetic susceptibility of the varieties rather than insecticide-induced resurgence. Control tactics against these epidemic pests therefore relied on genetic resistance and developing more effective chemical control. Earlier work in Japan had concluded that natural enemies (parasitoids) were not a major mortality factor for this pest group (Ooi and Shepard, 1994). Low-interest credit programmes designed to spread the adoption of modern rice varieties included recommendations of 3–4 insecticide applications per season (Fajardo *et al.*, 2000). The use of insecticides was thought to be needed just as fertilizer was to obtain yield potential. This high-input usage was later found to cause resurgence, demonstrating the importance of natural enemies.

Excessive insecticide usage killed off parasitoids and predators, which caused the damaging insect epidemics (Gallagher *et al.*, 1994). The high populations accelerated the evolution of new biotypes that survived on newly released resistant varieties. Instead of an insecticide treadmill, there was a genetic treadmill with the need to replace new resistant varieties every 2–3 years. The system was unsustainable. The concept of insecticide-induced resurgence was difficult for many to accept despite its widespread occurrence on such crops as cotton, due to a large degree of the failure to understand the importance of natural enemies. A turn around in understanding stemmed from the novel biological control studies at IRRI using exclusion and prey enrichment to quantify the role of predators in the natural control of BPH (Kenmore *et al.*, 1984). Further work using these techniques expanded our knowledge on the effect of predators (e.g. Canapi *et al.*, 1988; van den Berg *et al.*, 1988; Rubia *et al.*, 1990).

The genetic resistance programme at IRRI in the 1970s–1980s focused initially on stem borers, with some modest success. Greater success occurred with planthoppers and leafhoppers, but if insecticides were not considered essential for modern rice varieties, perhaps the breeding programme would only have focused on stem borers. Leafhoppers and planthoppers were minor pests before the introduction of HYVs and have once more returned to minor status with the decline of insecticide usage. Outbreaks can occur without the influence of insecticides but the conditions have to include a highly susceptible variety and high fertilizer usage or a weather event. A case documented in the Cagayan Valley produced a localized BPH outbreak after flooding removed natural enemies on a susceptible variety with no fertilizer usage (Litsinger *et al.*, 1986).

Constraints to biological control research in national programmes

Given the importance of natural enemies, more research should be forthcoming in this area. Constraints that limit more biological control research in tropical Asia are understaffed taxonomic services and local libraries with limited entomological holdings. The success of the biological control programmes at IRRI began with setting up a taxonomic unit in 1977 under the care of A.T. Barrion and an entomological literature collection in 1976 under the supervision of M.A. Austria, each supported by a modest budget. Biological control studies flourished over the next 2–3 decades. Rice arthropod collections are present in most countries but the specimens are protectively held waiting for local taxonomists to identify them. There are far too many arthropod groups to expect local taxonomists to handle. In IRRI's case the specimens were sent to taxonomists worldwide who were first identified through the literature as having expertise and then contacted to determine their interest. Upon agreement specimens were sent in a manner prescribed by the specialist who could retain all specimens, as duplicates were kept at IRRI. No payment was made for identification services. The role of the IRRI taxonomist was to collect, catalogue and send specimens. Contact with worldwide specialists and access to the world literature kept IRRI staff updated on nomenclatural changes, which continue to occur.

Accurate nomenclature is the key to the world literature. Early work on rice leaffolder pheromones was thwarted by the lack of taxonomic knowledge of the number of species involved. The leaffolder was thought to be one species but in the Philippines it is a complex of three species, with *Cnaphalocrocis medinalis* (Guenée) 46%, *Marasmia patnalis* Bradley 34% and *M. exigua* 20% based on larval collections throughout the Philippines (Barrion *et al.*, 1991). The community ecological studies in the next section could not have been carried out without a sound taxonomic foundation. Collaboration with the Boyce Thompson Institute for Plant Research at Cornell University, New York, and IRRI supported a large research effort to catalogue and use entomopathogens for control of rice insect pests (Rombach *et al.*, 1994). This focus eventually led to the isolation of indigenous *Bacillus thuringiensis* (Bt)

strains from the Philippines (Theunis *et al.*, 1998). Attempts to augment entomopathogens other than Bt in rice unfortunately did not live up to expectations.

Food webs illustrate the stability of arthropod communities

Spiders are the most diverse in terms of species in arthropod food webs that have been compiled over the last three decades in rice ecosystems (Schoenly *et al.*, 1996). A study in the Philippines used spiders as a surrogate for natural enemies in general to compare densities across rice ecosystems. Greatest numbers occurred in the most nutrient-rich irrigated wetlands, followed by rainfed wetlands, and the least in the low-energy dryland systems (Barrion and Litsinger, 1984). Deepwater rice does not occur in the Philippines but this ecosystem is most similar to rainfed wetlands (Catling, 1992). Despite the differences in densities, spider species compositions were remarkably similar. Of the ten most prevalent irrigated rice spiders, nine also occurred in the drylands.

Perturbations in rice agroecosystems produce effects at the community level that are best understood at that level (Cohen *et al.*, 1994). The number of distinct food chains within a food web correlates with the stability of the system. IPM of biological control agents must be considered at the community level due to the high potential number of interactions. A case in point was the work by Cohen *et al.* (1994) on the effect of insecticides on the food web. A number of studies have elucidated the complexity of the natural biodiversity in rice ecosystems with the general hypothesis that the more complex the food web, the greater the level of natural biological control (Settle *et al.*, 1996). The following will summarize the components of food webs for key pest groups. The food webs under discussion are cumulative rather than time-specific (Schoenly *et al.*, 1996) and are all from the Philippines.

The whorl maggot food web comprises 114 taxa including 71 predators, 13 parasitoids and 62 secondary natural enemies (IRRI, 1987). Dolichopodids (*Capsicnemus*,

Medetera) and other generalist predators prey on eggs. Egg predation went from 5% during the first week after transplanting to a peak of about 20% after 5 weeks in one study in Laguna province (van den Berg *et al.*, 1988). In contrast, more than double the egg predation rate occurred on the green hairy caterpillar *Rivula atimeta* Swinhoe in the same study, indicating some natural defence probably by the heavily mineralized whorl maggot egg. A single-egg parasitoid (*Ootetrastichus* spp.) occurs but is only rarely recovered. This may be a niche for classical biological control. Larval parasitoids include the braconid (*Opius*) as the most abundant but also include the eulophids (*Tetrastichus* spp.) and pteromalids (*Trichomalopsis* spp.). Parasitization rates are generally low but fluctuate from 14% to 71% in high host densities (Ferino, 1968). The dominant adult predators include an ephyrid fly (*Ochthera*) and generalist spiders (17 hunting species, 12 orb-weavers and 3 space web species). Damselflies (*Ischnura*, *Pseudagrion*, *Agriocnemis*) also prey on adults. Rice whorl maggot attains high levels of damage because it colonizes very early, before natural enemies accumulate. *Ochthera sauteri* is a highly voracious predator that captures whorl maggot flies in mid-air killing up to 20 adults per day (Barrion and Litsinger, 1987). But it is most effective when the field is drained. When flies were introduced into 1 m³ screen cages, whorl maggot egg densities were found to

be lowest in drained plots or when Azolla fern covered the water surface (Table 15.4). High egg densities and damage, and thus high adult whorl maggot survival, occurred in flooded plots or plots covered with aluminium foil sheeting. The conclusion was that whorl maggot could detect the presence of *O. sauteri* adults from their reflection in the water surface or aluminium foil surface. This may be the reason why whorl maggot became recognized as a pest only in the 1960s with the introduction of HYVs and irrigation as water levels in rainfed rice fluctuate often.

The semi-aquatic rice caseworm food web has 99 taxa with 78 predators, only 4 parasitoids, but 39 secondary natural enemies (IRRI, 1990). There is no egg parasitoid as the eggs are laid on the undersides of floating leaves, but aquatic snails passively scrape them off while moving underwater (Litsinger *et al.*, 1994b). There are two parasitic braconid wasps (*Dacnusa*), a nematode and a nucleopolyhedrosis virus, both attacking larvae but incidence is low. Aquatic beetle larvae (hydrophilids, dytiscids) prey voraciously on larvae, and the larger beetles attack larvae within their protective cases. Older larvae moving across the water surface are most vulnerable. Hunting spiders prey on larvae that crawl on to plants. Adult predation is usually high from spiders, dragonflies and damselflies, and birds. As a result of the activity of natural enemies caseworm is only a vegetative stage pest.

Table 15.4. Effect of water management on the efficiency of the whorl maggot (WM) adult predator *Ochthera sauteri* (OS).

Treatment	Whorl maggot eggs (no./25 hills)			Damaged leaves (%)		
	Flooded	Drained	Difference	Flooded	Drained	Difference
WM adults	301 ± 29 a	296 ± 35 a	ns	96 ± 7 a	91 ± 7 a	ns
WM adults + OS	261 ± 32 a	28 ± 8 b	<0.0001	91 ± 4 a	30 ± 13 c	<0.0 001
WM adults + Al foil	249 ± 38 ab	229 ± 91 a	ns	93 ± 2 a	86 ± 7 a	ns
WM adults + Azolla	144 ± 48 b	35 ± 25 b	<0.0001	79 ± 12 b	58 ± 19 b	<0.0001

In a column, means ± SEM followed by a common letter are not significantly different ($p \leq 0.05$) by LSD test.

The worldwide listing of rice stem borer natural enemies can be found in Khan *et al.* (1991). Stem borer food webs are similar between species; in the Philippines that of YSB comprises 118 species including 57 predators, 38 parasitoids and 23 secondary natural enemies (IRRI, 1984). The food web of *Chilo* stem borers includes 111 species with 57 as predators, 33 parasitoids, 18 hyperparasitoids and 28 secondary predators (IRRI, 1989). Dominant predators were spiders and beetles. Principal egg parasitoids are scelionids (*Telenomus*), eulophids (*Tetrastichus*) and trichogrammatids (*Trichogramma*). *Telenomus* wasps locate the female stem borer by its sex pheromone. The phoretic parasitoid actually attaches to the moth and accompanies it to the site where the eggs will be laid, parasitizing them during oviposition. *Tetrastichus* plays a dual role as both parasitoid and predator after emerging from the stem borer egg as a mid-sized larva. It is the largest of the parasitoid species and cannot mature on just one egg. After eggs are laid, a hair mat scraped from anal tufts by the female covers the egg masses protecting them against all but the larger or more specialized predators. *Tetrastichus* wasps have elongated ovipositors and can penetrate the mat. Egg parasitization rates can be very high (Ooi and Shepard, 1994).

Dominant larval parasitoids are braconids (*Bracon*, *Tropobracon*) and ichneumonids (*Temelucha*, *Xanthopimpla*). The latter is a pupal parasitoid while the others attack the larvae. Normally incidence is low due to the protected habitat of the larvae inside the tillers (Ooi and Shepard, 1994). Egg predators are coccinellids and orthopterans (*Metioche*, *Anaxipha* and *Conocephalus*). The latter is so large that it consumes both the eggs and hair mat. *Metioche* and *Anaxipha* cannot penetrate the hair mat, and thus are more important on unprotected *Chilo* and *Sesamia* eggs. A wide range of predators feed on neonate larvae before they enter the plant. These include coccinelid beetles (*Micraspis*, *Harmonia*) and the carabid (*Ophionea*). Neonate stem borer larvae often disperse by silk threads and many fall into the water where they are soon detected by surface swimming veliids (*Microvelia*) and mesoveliids (*Mesovelia*). Two pathogens (*Beauveria* and *Cordyceps*) and two mermithid nematodes (*Agamermis* and *Hexamermis*) occur minimally on larvae. Mainly spiders feed upon moths either while resting on the plant or in webs during flight. Dragonflies and birds are also effective during daytime when moths are startled into taking flight. Otherwise the moths are active only at night to prevent such predation.

A study of the WSB in Mindanao, Philippines, illustrates the high levels of egg parasitization. As many as 50 egg masses were collected weekly in each of two sites representing synchronous and asynchronous planting schedules in Koronadal, South Cotabato and held for parasitoid emergence. With artesian irrigation the asynchronous site had year-round rice cropping. Egg parasitization rates were generally over 50% (Fig. 15.2). Times of markedly lower parasitization rates appeared to be associated with larger egg masses. Eggs are more likely to escape parasitization and predation when masses are larger. Larger masses have multiple rows of eggs and those at the bottom and centre are more protected. Of the four parasitoid species only *Tetrastichus* increased mortality rates on larger egg masses due to its dual role as a predator. The other species had decreased rates, because with short ovipositors and the thick hair mat covering, they could only reach the eggs on the edges of the mass. Due to marking pheromones, interspecific competition between egg parasitoids is less than intraspecific competition. The intensity of the competition is seen from the examples of high multiple parasitism of egg masses (Table 15.5). Egg mass parasitization reached 82% while egg parasitization reached 65% during the 9-month study. This shows that the action of egg parasitoids alone cannot quell the WSB and that contributions from other biological control agents will be needed. This role was shown to be amply filled by spiders, coccinellids and orthopterans whose predation rates steadily increased during the crop against WSB on the same crop (Fig. 15.3). Despite outbreaks reported in other regions, the WSB

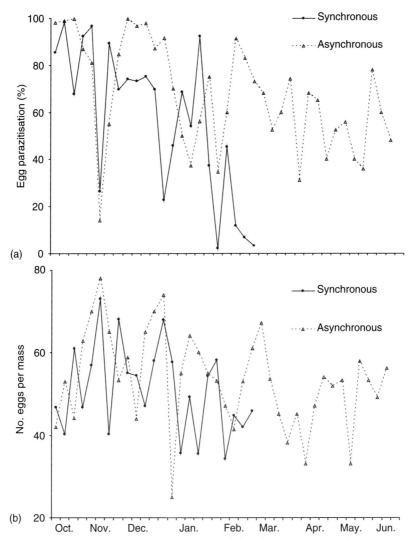

Fig. 15.2. Effect of egg parasitoids on white stem borer (WSB) in synchronous and asynchronous planting areas on parasitization of (a) eggs and on (b) egg mass size in the second rice crop (1990–1991) in Koronadal, Philippines.

in Koronadal has not become an alarming pest. This may be due to the interspersal of asynchronously cropped rice areas around the more synchronously planted crops fed by the Marbel River Irrigation System.

Leaffolders are a group that is heavily attacked by natural enemies as evidenced from the worldwide natural enemy listing (Khan *et al.*, 1988). Predators alone have been shown to reduce leaffolder numbers by 50–70% in field studies (Ooi and Shepard, 1994). Stud-

ies on parasitoids affected 40% larval mortality in a season-long study that produced 15 hymenopterans (Ooi and Shepard, 1994). It is to be noted that many of the prey that predators consumed were parasitized. Their impact is inferred by finding little leaf damage a month after seeing literally clouds of leaffolder moths taking flight while walking through the fields. The food web from the Philippines comprises 163 taxa with 77 predators, 51 parasitoids and 24 secondary natural

Table 15.5. Breakdown of white stem borer (WSB) egg and egg mass parasitization levels by single and multiple parasitoid species, second crop rice, 1990–1991, Koronadal, Philippines.

	Egg mass parasitization (%)	Egg parasitization (%)
Single species/egg mass		
Tetrasticus schoenobii	17.3	16.6 ± 1.4 a
T. japonicum	7.0	3.2 ± 0.6 c
T. rowani	6.1	3.1 ± 0.4 c
T. dignus	1.7	1.4 ± 0.4 e
Two species/egg mass		
T. row–T. dig	14.5	11.6 ± 1.1 b
T. sch–T. row	5.4	5.1 ± 0.6 c
T. sch–T. jap	5.2	4.5 ± 0.8 c
T. row–T. jap	3.5	2.4 ± 0.7 d
T. sch–T. dig	1.4	1.2 ± 0.3 e
T. dig–T. jap	0.8	0.6 ± 0.5 e
Three species/egg mass		
T. row–T. dig–T. Jap	7.1	5.0 ± 1.4 c
T. sch–T. row–T. dig	5.6	4.9 ± 1.2 c
T. sch–T. row–T. jap	2.5	2.1 ± 0.9 de
T. sch–T. jap–T. Dig	0.8	0.6 ± 0.3 e
Four species/egg mass	3.0	2.7 ± 1.4 cd
Average	82.2	65.0
n	2016	
p		<0.0001
F		26.35
d.f.		1651

In a column, means ± SEM followed by a common letter are significantly different (*p* < 0.05) by LSD test.

enemies. Predators exert a great impact to restrict leaffolders (Barrion *et al.*, 1991). Eggs are laid unprotected on leaves and are ravaged by crickets (*Metioche*), coccinellids (*Micraspis, Synharmonia*) and mirids (*Cyrtorhinus*). Spiders and carabids (*Ophionea*) prey on the larvae even within folded leaves. Eggs are parasitized by encrytids (*Copidosomopsis*) and trichogrammatids (*Trichogramma*); larvae are parasitized by bethylids (*Goniozus*), braconids (*Cotesia, Cardiochiles, Macrocentrus*) and ichneumonids (*Trichomma, Temelucha*), and pupae are attacked by ichneumonids (*Xanthopimpla*) and eulophids (*Tetrastichus*). Epizootics of pathogens on larval populations are common under favourable weather conditions with the fungus *Zoophthora* producing flattened cadavers. Other pathogens include *Beauveria* and *Nomurea*.

The two dominant butterflies, the green-horned caterpillar *M. leda ismene* and rice skipper *P. mathias*, rarely reach economic densities but due to their large size are highly recognized by farmers who apply insecticide when they see only a few in a field. A list of natural enemies of both species worldwide and in the Philippines is given by Litsinger *et al.* (1997). The worldwide list includes more parasitoids, which reflects the emphasis in the early years on parasitoids over predators. A more wide-ranging collection in the Philippines recorded 83 predators, 30 parasitoids and 4 pathogens. Among the predators, 52% attacked larvae, 28% eggs, 19% adults and 1% pupae. Among parasitoids 66% were larval or larval–pupal, with 21% on eggs and 13% pupal. Parasitoids alone ranged from 10% to 80% parasitization using the prey enrichment method.

The common rice leafhoppers in the genera of *Nephotettix, Recilia* and *Cofana* have a wide array of shared natural enemies. The food web of the rice white leafhopper, *Cofana spectra* (Distant), lists 107 taxa, with predators outnumbering (76) the parasit-

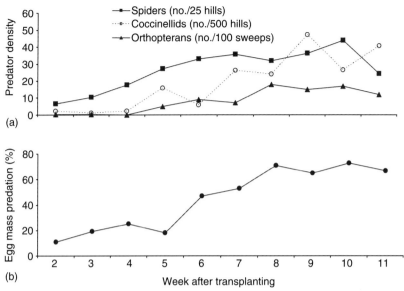

Fig. 15.3. Time series of (a) densities of selected predator guilds sampled by D-Vac suction machine averaged over eight fields each from 16 crops, 1983–1991, and (b) egg mass predation rates determined by the sentinel egg mass method in the second rice crop (1990–1991), against the white stemborer, in Koronadal, Philippines.

oids (9) and secondary natural enemies (30) (IRRI, 1983). Spiders represented 43 species. There was only one primary pathogen. The food web of green leafhoppers (*Nephotettix* spp.) includes 205 taxa with 129 predators, 35 parasitoids, 6 pathogens and 63 secondary natural enemies (IRRI, 1986). Leafhoppers are under heavy pressure from natural enemies (Ooi and Shepard, 1994) beginning with trichogrammatids (*Paracentrobia*, *Oligosita*) and mymarids (*Gonatocerus*, *Anagrus*) egg parasitoids. Nymphs and adults are parasitized by pipunculid flies (*Pipunculus*, *Tomasvaryella*), which are key mortality factors. Nymphs and adults are also parasitized by dryinids (*Haplogonatropus*) and strepsipterans (*Halictophagus*). Predators may be the most important mortality factor beginning with the mirid *Cyrtorhinus lividipennis* Reuter that preys on eggs as well as young nymphs. Spiders represent the most important mortality factor, especially wolf spider *Pardosa* as well as *Tetragnatha* and *Oxyopes*. Other predators include aquatic veliid bugs (*Microvelia*), damselflies (*Agriocnemis*) and dragonflies (*Crocothe-*

mis). Epizootics of fungal diseases of *Entomophthora* and *Beauveria* also occur under favourable environmental conditions.

Planthoppers share a similar natural enemy complex. Of the BPH food web only 11 of the 76 taxa are parasitoids and 11 secondary natural enemies with the balance as predators dominated by spiders (50 species) (IRRI, 1980). A food web of the whitebacked planthopper, *Sogatella furcifera* (Horváth), lists 199 species with 139 predators (63 spiders) and 33 parasitoids (IRRI, 1988). Predators play the most important regulatory role beginning with *C. lividipennis*, which prefers planthoppers to leafhoppers. The wolf spider (*Pardosa pseudoannulata*) may be the key natural enemy, which was highly correlated to planthopper abundance by Kenmore *et al.* (1984). A wolf spider can consume 7–45 hoppers per day (Ooi and Shepard, 1994). Other important spiders include the small micryphantid (*Atypena*), araneids (*Neoscona*, *Araneus*) and clubionids (*Clubiona*). Aquatic veliids prey on fallen hoppers, as it is common for hoppers to fall off tillers on to the paddy water surface. Their impact on the water surface is detected by the veliids, which race to their prey. Up to 67% of hoppers naturally fall from plants, usually as a result of strong winds. Coccinellids (*Synharmonia*), staphylinids (*Paederus*) and carabids (*Ophionea*) prey on nymphs and adults. Eggs are parasitized by mymarids (*Anagrus*), trichogrammatids (*Paracentrobia*, *Oligosita*) and eulophids (*Tetrastichus*). Major parasitoids of nymphs and adults are dryinids (*Pseudogonatopus*, *Haplogonatopus*, *Echthrodelphax*), strepsiptera (*Elenchus*) and dryinids (*Pseudogonatopus*). Common pathogens are *Beauveria*, *Erynia*, *Metarhizium* and *Hirsutella*. Parasitoids generally cause about 10% mortality in the population during low numbers on resistant rice varieties but percentages rise as the planthopper and leafhopper numbers rise (Ooi and Shepard, 1994).

Rice bug *L. oratorius* (Fabricius) has specific and effective scelionid (*Gryon*) and polyphagus encyrtid (*Ooencyrtis*) egg parasitoids with general parasitism rates of 10–45%, more so in larger egg masses and higher host densities (Sands, 1977). There are no recorded adult or nymphal parasitoids. As determined from precipitin tests, tettigoniid, gryllid and coccinellid generalist egg predators are more important (Rothschild, 1970). The bad odour emitted by rice bug nymphs and an adult is undoubtedly a defensive mechanism against bird predation but nymphs and adults fall prey to a wide range of generalist arthropod predators. Rothschild (1970) determined that 45% nymphal mortality occurred before adulthood from the activity of Orthoptera and web building araneid and tetragnathid spiders. The fungus *Beauveria* is recorded in epizootics of nymphs and adults.

Other pest groups have few recorded natural enemies either because of little research or their habitat. Black bugs (*Scotinophara*) are under attack by scelionid egg parasitoids (*Telenomis*) and some generalist predators such as frogs and carabid beetles. *Metarhizium* fungus can be seen issuing from cadavers. No nymphal or adult parasitoids have been recorded. The Asian gall midge is held in check by several specialist larval parasitoids (*Neanastatus*, *Propicroscystus*, *Playtgaster*) and a phytoseiid mite (*Amblyseius*) attacking the eggs. Rice hispa has few recorded natural enemies. Seedling maggot, *Atherigona oryzae* Malloch, has several trichogrammid and eulophid egg (*Trichogramma*, *Tetrastichus*) parasitoids and a braconid larval (*Opius*) parasitoid along with a number of spider predators. Natural enemies exert a minimal effect on these early colonizers in dryland rice. Tending ants protects root aphids and root mealybugs but have some coccinellid predators (*Coccinella*, *Menochilus*, *Harmonia*). Soil pests such as white grubs, mole crickets, termites, ants and root weevils have few natural enemies. The white *Leucopholis irrorata* is attacked by a host-specific scoliid, *Campsomeris*, but parasitization rates are low.

Cultural control

Cultural, mechanical and physical pest control can take place in a single field or through community-wide cooperation. Litsinger (1994) has catalogued these methods for rice pest control. Mechanical and physical methods involve motion and force such as handpicking, trapping or crushing insects by hand, tools or the use of light traps or burning. Grasshoppers and crickets are collected in community hunts by hand and with nets in some rice-growing villages of Cambodia. This method appears very effective. The grasshoppers and crickets are fried for eating, fed to farm animals or used in compost. Discussions with farmers in Thailand and Laos indicate that trapping provides adequate control of crabs, but Cambodian farmers report that trapping does not. Cultural methods involve crop husbandry practices that have a dual purpose of crop production and insect pest suppression. The following discussion will focus on the role of these methods in IPM programmes for three rice ecosystems: drylands, rainfed wetlands and irrigated wetlands.

Cultural control on acid soil drylands

IPM can only be seen in the context of integrated crop management for both slash-and-burn as well as plough-based agriculture.

Farmland cleared from former tropical rain-forests is typically acidic (pH 4–5.5) and often on highly eroded sloping land. In order to overcome the severe constraints that these environments present, a series of measures must be carried out as a package of practices before IPM can be considered. In slash-and-burn agriculture, seed holes are first dibbled by pointed sticks. Organic matter should be incorporated along with the seed to amelio-rate the soil and raise the pH to allow greater uptake of nutrients. Small amounts of inor-ganic N and P should also be incorporated. Erosion control is not necessary except on the steepest slopes, as the stumps and logs of the former forest generally slow down water movement. For permanent, plough-based dryland agriculture on sloping land where the forest has been totally cleared, the first order of business is to stabilize the soils through contour hedgerow technol-ogy to terrace the hillsides (MacLean et al., 2003). There are various ways to carry this out but the simplest is to use an A-frame as a tool to outline the contours of buffer strips of unploughed soil.

The hedgerows demarcate the terraces whose width needs to be narrower on steeper slopes. Multipurpose trees or peren-nial crops can be planted in the buffer strips and the rainfall will carve out the terraces over the succeeding years. Organic matter can be applied from the leaves of legumi-nous multipurpose trees in plough furrows within the terrace benches, again to mitigate the soil problems. Inorganic fertilizer in modest amounts should also be applied in the seed furrows. The terraces will greatly reduce erosion to allow fertile soil to accu-mulate, which will also have the benefit of storing rainfall. If the soil is highly eroded, the multipurpose trees should also be given organic matter and fertilizer the first few years while the agroforestry system is devel-oping. Eventually the roots of the trees will form a net under the terrace benches and capture leached nutrients as is done by the original tropical rainforests where nutrients are recycled through the leaves.

In one study in the Philippines (Mac-Lean et al., 2003), the result of increasing the organic matter and adding inorganic matter dramatically increased populations of seedling maggot, white grubs and stem borers, and up to sixfold in the case of white grubs from 1 to 6 per square metre. However, yield more than compensated the infestations without insecticide, especially on the most eroded soils where yield rose from 0.8 to 1.6 t/ha the first year on high-tillering rice UP-Ri5. It is important to point out that a low-tillering rice cultivar would not have resulted in such a dramatic effect. The capacity of HYVs to tolerate pests through tillering has been overshadowed by the accomplishments in genetic pest resis-tance (Khush, 1989), but both act comple-mentarily. If a tiller is severed by stem borers or other pests, a neighbouring one can take its place. HYVs also produce many more spikelets per area than traditional types, and thus have a larger physiological sink. When a whitehead occurs, the photosyn-thate can transfer to unfilled spikelets of an adjacent tiller. Trials showed optimal crop management enhances compensation with crop fertility at the top of the list (Litsinger, 1993; Rubia et al., 1996).

Selection of high-tillering rice cultivars and achieving optimal fertility represent the first line of defence against insect pests. The next step is to avoid the most damag-ing infestations by planting early. Dryland environments are punctuated by long dry seasons and both the pests and the farmer spring into action upon the resumption of rainfall. Aestivation in seedling maggots and white grubs is broken by early heavy rains (Litsinger et al., 2002). Farmers also begin land preparation at the same time. Early planting of rice can be defined either in absolute or relative terms. It can be abso-lute in regard to calendar dates, or relative in regard to: (i) season-initiating rainfall or (ii) one's neighbours (Litsinger et al., 2003). Planting in regard to calendar dates is less important than in relation to rain-fall. Planting earlier than one's neighbours only has merit if the interval is longer than the generation time of the pest. Pests, like the seedling maggot, first increase through successive generations on grasses or maize.

Overseeding is another cultural con-trol method that has merit against seed pests and seedling maggot. In a study on seedling maggot, the economically optimal seeding rate was 90 kg/ha without insecticide, which was equal to 50 kg/ha with insecticide seed treatment (Litsinger *et al.*, 2003). Highest yield occurred at 120 kg/ha, but above this interplant competition decreased yield.

Cultural control in favourable drylands

The same measures can be followed on favourable dryland areas without eroded soils or pH problems. This rice cultural type has fewer insect pest problems than in other dryland systems. A good crop stand will compete better with weeds, and thus high seeding rates are recommended. Farmers often till the soil during the dry season for weed control, particularly nutsedge. The constant tillage operations before planting and during off-barring for weed control during early crop growth take a natural toll on soil pests such as cutworms and white grubs. Early planting before several generations of seedling maggots build up on weeds is crucial. To avoid white grubs it is important to establish the crop before the third instar larva develops (Litsinger *et al.*, 1983). Due to drought most dryland varieties are short-maturing, limiting the compensation period. But very early cultivars should be avoided as they have the least time to compensate.

Cultural control in rainfed wetlands

Rainfed wetland rice crops are inherently unstable due to the unpredictable nature of rainfall. For example, in Cambodia, most rice crops are in the more than 20 cm/year rainfall range – enough water for high yields if it fell in a predictable pattern. Since it rains in an inconsistent manner, Cambodian farmers prefer photoperiod-sensitive rice varieties because the growth period is not fixed and can be adjusted to the availability of water. Although these varieties have a low yield potential, they can be transplanted up to 5 months after sowing the seedbed, i.e. transplanting can be delayed if there is insufficient water (Nesbitt and Chan, 1997). The HYVs are photoperiod-insensitive and must be transplanted 20–35 days after sowing. In addition, HYVs are much shorter and therefore less likely to survive floods. As a result, in Cambodia today, HYVs are generally grown near a source of water (e.g. a pond, a receding river or an irrigation canal), while the traditional varieties are grown in the rainfed ecosystem – about 86% of the total rice-growing area in Cambodia (Javier, 1997). Drought stress is associated with outbreaks of mealybugs, BPH, thrips and armyworms in rainfed rice. Rainfed rice therefore suffers from more pest problems and crop stresses than irrigated rice; thus, yield losses are proportionally larger.

Cultural control in irrigated wetlands

It is important to select high-tillering cultivars with genetic pest resistance against local strains of diseases and epidemic insect pests as a first line of defence. High tillering is a key feature that enables the greatest compensation, which, when coupled with moderate maturity (110–130 days), will allow ample time for the crop to tolerate most stresses. The beneficial effect of longer maturity can be seen in Litsinger (1993; Table 3.2) or in a follow-up trial conducted in the same location, comparing IR58 (90-day variety) and IR70 (110-day variety) (Table 15.6). The beneficial role of N is also evident. The very early maturing IR58 had less time to compensate, and at every level of N tested, the non-insecticide-treated plots yielded significantly less than the treated ones. This was not the case for the longer-maturing IR70 at the 60 and 90 kg N/ha levels. The main pests were whorl maggots, defoliators and YSB. We conclude that a combination of longer maturity and increased fertility overcomes much insect pest damage.

Planting early in the season can also avoid pest build-up in many instances. A trial in four locations in the Philippines regressed insect pest abundance with planting date and showed that, for four pest groups, early planting had the lowest pest abundance in six site × pest combinations where regressions were significant (Table 15.7). Sites with the highest pest densities showed more beneficial effects from early planting.

Table 15.6. Interaction between varieties differing in crop maturity period with nitrogen (N) rates and insecticide protection, Zaragoza, Nueva Ecija, Philippines, 1991 dry season.

| | Yield (t/ha) | | | | | |
| | IR58 (90 days) | | | IR70 (110 days) | | |
N (kg/ha)	Insecticide-protected	Untreated	Difference p	Insecticide-protected	Untreated	Difference p
0	4.0 c	3.1 c	<0.0001	4.6 b	4.2 b	0.004
30	4.4 bc	3.5 bc	<0.0001	4.7 b	4.4 b	0.02
60	4.6 ab	3.8 b	<0.0001	5.4 a	5.3 a	ns
90	5.1 a	4.2 a	<0.0001	5.7 a	5.5 a	ns

Average of four replications; means in a column followed by a common letter are not significantly different ($p \leq 0.05$) by LSD test.

In irrigated rice, water management can be an important factor in controlling pests. Flooding seedbeds or fields during armyworm and thrips outbreaks is the safest and most reliable method of control. Caseworm and golden snails *Pomacea canaliculata* (Lamarck) are managed by draining the field for a few days (Litsinger and Estaño, 1993). Plant density also affects pests. Whorl maggot is a pest of transplanted rice as in direct seeded rice the high densities result in insignificant damage. A field study in the Philippines showed that

Table 15.7. Regression correlations between planting date and insect pest damage in four sites in the Philippines.[a]

Crops/site	Pest damage	Linear regression
High pest densities		
Zaragoza	Whorl maggot (damaged leaves)	$y = 18.1 + 4.6x$, $r = 0.376$, $p = $ <0.0001, d.f. = 106
	Defoliators (damaged leaves)	$y = 0.3 + 2.3x$, $r = 0.219$, $p = 0.03$, d.f. = 95
	Leaffolders (damaged leaves)	ns, d.f. = 112
	Stem borers (deadhearts, whiteheads)	ns, d.f. = 112
Koronadal	Whorl maggot (damaged leaves)	$y = 15.2 + 0.09x$, $r = 0.541$, $p = $ <0.0001, d.f. = 62
	Defoliators (damaged leaves)	$y = -3.9 + 4.3x$, $r = 0.424$, $p = 0.0005$, d.f. = 63
	Leaffolders (damaged leaves)	$y = 20.2 + 6.8x$, $r = 0.419$, $p = 0.0006$, d.f. = 63
	Stem borers (deadhearts, whiteheads)	ns, d.f. = 67
Low pest densities		
Guimba	Whorl maggot (damaged leaves)	ns, d.f. = 67
	Defoliators (damaged leaves)	ns, d.f. = 67
	Leaffolders (damaged leaves)	ns, d.f. = 68
	Stem borers (deadhearts, whiteheads)	ns, d.f. = 67
Calauan	Whorl maggot (damaged leaves)	ns, d.f. = 65
	Defoliators (damaged leaves)	ns, d.f. = 33
	Leaffolders (damaged leaves)	ns, d.f. = 42
	Stem borers (deadhearts, whiteheads)	$y = -8.0 + 3.7x$, $r = 0.371$, $p = 0.02$, d.f. = 41

[a]Pest damage is the dependent variable (y) measured weekly; planting date is the dependent variable (x) based on the number of elapsed days after the first planted field; level of significance ($p \leq 0.05$).

increasing plant density in an unevenly sown direct seeded crop resulted in significantly reduced oviposition and damage (Viajante and Heinrichs, 1986b).

Some ecologists advocate that continuously planted rice could be sustainable if cultured without insecticides, as natural enemies would be at an even better advantage than insect pests (Way and Heong, 1995). In the double rice system, Settle *et al.* (1996) found greater frequency of BPH outbreaks in areas of Java where there was a longer dry fallow between crops due to the delay of natural enemy build-up at the beginning of the first crop season. They advocated for encouraging natural enemies by organizing irrigation systems to keep dry fallow periods short and encourage asynchrony to produce heterogeneous landscapes. The results have shown that the key to ecosystem stability is to decrease insecticide usage. Synchronous planting is recommended for control of more pests than the BPH as it is an important method for disease and rodent control. It would be too risky to depend solely on natural enemies, no matter how abundant, and IPM managers should incorporate more control methods for greater stability.

As was the case in acid dryland soils, greatest benefits in irrigated rice occur from adoption of several control methods. It is important to select longer-maturing varieties of 120 days to maturity (particularly in the wet season): (i) employ genetic pest resistance (diseases and epidemic insect pests); (ii) use adequate seeding rates; (iii) plant early to minimize pest build-up; (iv) ensure adequate ponding; (v) do not allow weeds to build up in the seedbed and field; and (vi) use optimal fertility management (Savary *et al.*, 2000b; Litsinger *et al.*, 2005).

Chemical control

Types of chemical control

Chemical control is another means of increasing pest mortality. Chemical control includes pheromones, botanicals, repellents, attractants, antifeedants and insecti-cides. Pheromones are chemical substances secreted by an animal that elicits a specific behavioural or physiological response in another animal of the same species. A number of insect pheromones have been isolated, synthesized and commercially produced. In cotton, sex pheromones are applied on a broad scale to prevent males from finding females, thus disrupting mating. In rice, pheromones have not been used in this fashion, but as lures in traps to monitor insect pests. For example, in Bangladesh, pheromone traps are successfully used by farmers to monitor populations of YSB.

Farmers in Cambodia control crabs with botanicals and organophosphates. The organophosphate applications are effective but dangerous, and certainly not sustainable over a long period as they kill aquatic natural enemies and some types of fish. In northern Cambodia, some farmers use indigenous botanically derived pesticides like chopped *Euphorbia lactea* Haw for crab control in newly transplanted fields. Experiments indicate that there are dosages of *E. latea* that kill crabs, but not fish (CIAP, 1998). While the botanical toxin may have some advantages over an organophosphate (e.g. cost, biodegradability, compost contribution), it is still a non-selective poison that potentially kills natural enemies and must be handled with extreme caution (the sap of the plant can cause blindness in humans).

Action thresholds for insecticide decision making

A 13-year study in the Philippines on 68 crops tested action thresholds against the four prominent chronic insect pest groups: whorl maggots, defoliators (*Rivula, Naranga*), leaffolders and stem borers. Various characters and insecticide responses were tested over four sites in an iterative approach (Litsinger *et al.*, 2005). Pest populations were monitored weekly and yield loss was calculated in each crop for the three main crop growth stages following the insecticide check method. The best-performing characters and insecticide responses are given in Table 15.8. Criteria were estab-

Table 15.8. Best-performing action thresholds against chronic pests, Philippines.

Pest	Crop stage	Pest intensity	Character	Correct decisions (%)	Insecticide response	Control of insect damage (%)
Whorl maggot	Vegetative	Low	1–2 eggs/hill	75–90	Triazophos or azinphos ethyl foliar spray	25–30
		High	15–30% damaged leaves in earlier planted field	66–71	Isofenphos seedling root soak	45
Defoliators	Vegetative	Low	10% damaged leaves	87	Carbaryl foliar spray	
		High	10% damaged leaves in earlier planted field	87	Carbaryl foliar spray	30
Leaffolders	Vegetative	Low/high	10–15% damaged leaves or 1 larva/ hill	99	BPMC foliar spray	
	Reproductive	Low/high		98	BPMC foliar spray	50
	Ripening	Low/high	10–15% damaged leaves	93		
			10–15% damaged leaves		BPMC foliar spray	
Stem borer	Vegetative	Low/high	3–5% deadhearts	99	Chlorpyrifos (21%) + BPMC (11%) foliar spray	
	Reproductive	Low/high	15–25% deadhearts	99	Chlorpyrifos (21%) + BPMC (11%) foliar spray	35
	Ripening	Low/high	8–15 deadhearts	96	Chlorpyrifos (21%) + BPMC (11%) foliar spray	

lished to evaluate the performance of each threshold character on being able to predict when benchmark levels of both minimal pest abundance and yield loss would occur. Each character was based on four possible outcomes. Two outcomes were rated 'correct'. The first was a positive decision to respond when the pest density and yield loss benchmark conditions occurred. For whorl maggot the two benchmark conditions were 15% damaged leaves during the crop and 0.25 t/ha yield loss in the vegetative stage. The second outcome was a decision not to respond and was judged 'correct' if at least one of the conditions did not occur. There were also two 'incorrect' decisions: (i) a decision to respond was made but either of the two benchmark standards did not eventually occur; and (ii) a decision not to respond when they did. Threshold characters for defoliators, leaffolders and stem borers were all correct in more than 87–99% of occasions, while those for whorl maggot were less, particularly for high-density sites (66–71%). High-density sites were distinguished by needing to have characters that allowed earlier insecticide application for best results. Whorl maggot proved to be the most difficult to control with insecticides although the level of control was at most 50% for any of the pests. Thus, action threshold characters proved very accurate in detecting a situation leading to economic loss, but the insecticide response was poor. Insecticide trials that led to the determination of the recommended chemicals achieved more than 80% control (Litsinger *et al.*, 1980). Under farmers' field conditions each insecticide was applied only once and the foliar sprays at a dosage more affordable to farmers (0.4 kg a.i./ha) vs. the chemical company's recommended 0.75 kg. a.i. Farmers, however, applied insecticide at an average of 0.21 kg a.i./ha and consequently

achieved a slightly higher economic return over action thresholds (Table 15.9). The prophylactic regime of three insecticide applications was the least profitable. The data show that farmers accept low rates of return from insecticide usage, and risk aversion rather than profit appears to be the motivation for farmers to use insecticide.

Plant resistance in rice IPM systems

The development of resistant varieties in rice at IRRI began in 1962, when Dr M.D. Pathak established the IRRI insect genetic resistance-screening programme. Before this programme was established, it was thought that rice did not possess sufficient levels of insect resistance (Jennings *et al.*, 1979). By 1973, the first BPH-resistant variety, IR26, was released in Asia. At this point, more than 50,000 accessions had been screened for resistance against a number of key pests at IRRI. The basic methods for screening for resistance were adopted throughout Asia, to Japan (Choi *et al.*, 1979), Korea (Choi *et al.*, 1979), Taiwan (Cheng and Chang, 1979), India (Kalode and Krishna, 1979), Thailand (Pongpraset and Weerapat, 1979), Sri Lanka

Table 15.9. Economic analysis of the threshold treatment compared to other practices.

Site	Pest density	Marginal return from insecticide ($/ha)[a]					
		Prophylactic[b]	Threshold	Farmers' practice	*p*	*F*	d.f.
Zaragoza	High	−11.3 ± 9.8 B b	−1.7 ± 3.2 B ab	3.5 ± 5.7 B a	0.007	4.23	52
Koronadal	High	−33.7 ± 29.6 C b	−20.9 ± 24.9 C ab	−9.1 ± 5.5 B a	0.005	3.29	40
Guimba	Low	−34.4 ± 41.4 C b	−7.8 ± 9.6 B a	0.5 ± 0.4 B a	<0.0001	2.99	38
Calauan	Low	24.1 ± 20.5 A b	48.1 ± 27.1 A a	28.1 ± 23.4 A b	<0.0001	5.12	42
Average		−13.8	4.4	5.8			
p		<0.0001	<0.0001	0.003			
F		4.36	5.99	3.58			
d.f.		66	54	48			

In a column, means ± SEM followed by a common uppercase letter are not significantly different ($p \leq 0.05$) by LSD test.
In a row, means ± SEM followed by a common lower case letter are not significantly different ($p \leq 0.05$) by LSD test.
[a]Cost based on prices in 1986 for insecticide, unmilled rice $0.128/kg farm gate, interest on materials 60% per season, labour 8 h to spray 1 ha, labour $0.10/h, interest for labour 33% per season, pest monitoring 60 h per season for thresholds and 4 h per season for farmers' monitoring.
[b]0.5 kg a.i/carbofuran G/ha soil incorporated before transplanting plus applications of 0.4 kg a.i. chlorpyrifos/ha at 35 and 45 days after transplanting.

(Fernando *et al.*, 1977), Indonesia (Harahap, 1979) and the Solomon Islands (Stapley *et al.*, 1979). It is now accepted that rice has a rich diversity of germplasm that can be used for breeding pest-resistant cultivars. Screening for genetic variability in resistance has been done for most rice pests (Heinrichs *et al.*, 1985). However, efforts to develop resistant cultivars and a general understanding of the mechanisms of resistance have been mostly limited to the major pests in rice (planthoppers, leafhoppers, gall midge and stem borers) (Heinrichs *et al.*, 1985). Despite extensive efforts to identify resistant varieties, there is still limited knowledge about the mechanisms of conventional resistance and how resistant varieties perform in the field.

The field of host plant resistance has traditionally focused on interactions between insect pests and their host plants (Painter, 1951). If a plant is avoided by insects, the type of resistance is called antixenosis (Kogan and Ortman, 1978). If the plant adversely affects the survival, development or reproduction of an insect, resistance is characterized as antibiosis. Tolerance is the ability of a plant to withstand severe damage, and maintain good yields.

Antixenosis, antibiosis and tolerance are considered part of the intrinsic or direct defence in plants against insect herbivores, while natural enemies can be considered part of a plant's extrinsic defence. There is a large body of literature showing that chemical, physical and semiochemical plant traits directly influence the interactions between herbivores and their natural enemies (Price *et al.*, 1980; Price, 1986; Cortesero *et al.*, 2000). For example, defensive chemicals or lower nutritional value may slow larval development and lengthen their 'window of vulnerability' to natural enemies (Haggstrom and Larsson, 1995; Benrey and Denno, 1997). Physical resistance traits can also affect immatures by affecting their spatial distribution and the foraging behaviour of natural enemies (Chen and Welter, 2003). Finally, herbivore and natural enemy behaviour and host location can also be mediated by semiochemicals (Vet and Dicke, 1992). Understanding the relationship between host plant resistance and natural enemy activity in the field could be a major opportunity for developing ecologically based IPM strategies.

Biology of rice resistance to key pests

Both physical and chemical characteristics are important in the resistance to stem borers, and resistance appears to be under polygenic control (Chaudhary *et al.*, 1984). Most of the major stem borer species have the same general life history (Pathak, 1968; Chaudhary *et al.*, 1984). A volatile chemical, *p*-methylacetophenone, has been shown to attract *C. suppressalis* larvae and adults (Munakata *et al.*, 1959; Kawano *et al.*, 1968). Resistant and susceptible varieties appear to also differ in chemical resistance factors (Ishii *et al.*, 1962). For instance, benzoic acid, salicylic acid and fatty acids isolated from resistant plants appeared to slow stem borer development (Munakata *et al.*, 1959). Lower free amino acid content was found in resistant varieties (Das, 1976; Mishra *et al.*, 1990).

Physical plant characteristics also play a role in rice host plant resistance. In a path analysis, Shahjahan and Hossain (2003) found that plant height, number of tillers, leaf hairiness and length were the major morphological characteristics correlated with *S. incertulas* incidence in the field. Tight leaf sheaths that partially cover the internodes and ridged culms have been correlated with lower stem borer infestation (Patanakamjorn and Pathak, 1967). Stem diameter is inversely correlated with stem borer survival. Wild rice varieties appear to be more resistant to stem borers due to narrow stems (Padhi and Sen, 2002). On the other hand, stem borers may damage only stems of a certain thickness. For example, *S. incertulas* larvae do not severely damage the large stems of deepwater rice, because they feed on the pith and do not damage plant tissues involved with nutrient flow (Taylor, 1988). Rice plants are able to compensate for damage by early borers (Rubia *et al.*, 1996), but larval feeding during the booting significantly reduces yield and causes whiteheads (Chaudhary *et al.*, 1984; K.L. Heong,

Philippines, unpublished material). In the development of the rice crop there are two periods where the plant is more susceptible: during the elongation of the tillers and of the panicle (Bandong and Litsinger, 2005). During the other crop growth stages either hardening agents or tight wrapping of the leaf sheath infers resistance to penetration of neonate larvae. Plant silica content also appears to provide resistance against stem borers, but this appears only to be effective during the vegetative stage. Silica reduces larval survival, because it wears down larval mandibles (Djamin and Pathak, 1967; Subbarao and Perraju, 1976). Therefore, it is important to identify variation in stem borer resistance during plant stages when damage has the highest impact upon yield.

Resistance to planthoppers is thought to relate to the biochemical and nutritional compositions of the phloem. Low concentrations of asparagine may deter extended feeding (Sogawa and Pathak, 1970). BPH is unable to feed for extended periods of time on resistant plants, suggesting that feeding on the phloem is blocked (Kimmins, 1989), the plants lack essential nutrients or some chemical is inhibiting feeding. Silicic acid (Yoshihara *et al.*, 1979), oxalic acid (Yoshihara *et al.*, 1980) and β-sitosterol (Kaneda, 1982; Shigematsu *et al.*, 1982) have been proposed to be feeding inhibitors. Because silicic acid is found outside of the phloem, Yoshihara *et al.* (1979) suggested that silicic acid might function to localize BPH feeding. However, silicic acid has been found in both resistant and susceptible rice varieties (Saxena, 1986). Resistant varieties contain higher concentrations of three types of flavenoid glycosides, which were shown to inhibit feeding (Grayer *et al.*, 1994). As a minor source of resistance, cuticular waxes have been identified as repellents to BPH (Cook *et al.*, 1987; Woodhead and Padgham, 1988). Recently it has been discovered that some *japonica* rice varieties have ovicidal activity against BPH (Seino *et al.*, 1996; Yamasaki *et al.*, 1999). Egg hatch on resistant plants may be reduced by 80%; the ovicidal effect appears to be higher on the larger tillers and on the main stem (Suzuki *et al.*, 1996). Seino *et al.* (1996) showed that

benzyl benzoate was the active ingredient in the watery lesions.

The green leafhopper (*N. virescens*) is an important pest of rice because it transmits the tungro virus. Female survival on resistant varieties is generally correlated with tungro infection, and seedling preference by *N. virescens* for certain varieties also determines susceptibility (Heinrichs and Rapusas, 1983). Transmission of the rice tungro virus occurs only if leafhoppers feed on the phloem (Heinrichs and Rapusas, 1984). However, when feeding on resistant plants, leafhoppers excrete ten times more honeydew. Honeydew of leafhoppers feeding on resistant varieties lacks sugar (Kawabe, 1985). On resistant varieties, *N. virescens* pierce the sieve elements but do not feed on the phloem on resistant plants, suggesting that resistant plants contain a feeding deterrent.

For the Asian gall midge, the nature of resistance appears to be mostly chemical. Female midges do not exhibit an ovipositional preference for either resistant or susceptible varieties (Kalode *et al.*, 1983). Gall length is positively correlated with plant height (Joshi and Venugopal, 1985), suggesting that galls develop better on more vigorous shoots. Peraiah and Roy (1979) found that resistant varieties have more free amino acids, phenols and soluble sugars. On resistant varieties, inducible defences against *O. oryzae* may show hypersensitive reactions such as tissue necrosis at the apical meristem, which leads to death and mortality for larvae that feed subsequently (Bentur and Kalode, 1996). Amudhan *et al.* (1999) found that *O. oryzae* galls induced higher levels of phenolics in resistant varieties. Tolerance to the African rice gall midge (*O. oryzivora*) involves the ability to produce many tillers (Omoloye *et al.*, 2002).

Host plant resistance in rice can work synergistically with biological control (Cortesero *et al.*, 2000). BPHs are apparently more restless on resistant varieties and are more easily detected by lycosid spiders (Kartoharjono and Heinrichs, 1984). Additionally, Zhang *et al.* (2004) found that *N. lugens* fed on the upper parts of resistant plants rather than their usual location in the basal portion of the stem, suggesting that they were repelled by something in the lower part of the plant. However, by feeding higher on the plant, planthoppers may be more exposed to predators. Both lycosid spiders (Kartoharjono and Heinrichs, 1984) and mirid bugs, *C. lividipennis* (Myint *et al.*, 1986; Senguttuvan and Gopalan, 1990), consume prey at higher rates on pest-resistant rice cultivars than on susceptible ones. Quantitative and qualitative differences in volatile production may also function to attract natural enemies. Some pest-resistant cultivars appear to be more attractive than susceptible cultivars to certain predators (Sogawa, 1982; Rapusas *et al.*, 1996). For instance, *Anagrus nilaparvatae*, an egg parasitoid of BPH, is significantly attracted to rice plants that are damaged by BPH (Lou and Cheng, 1996). In olfactory tests, *C. lividipennis* can distinguish BPH-infested and uninfested rice plants, and between nymphs and gravid females (Lou and Cheng, 2003).

Performance of resistant varieties in the field

While the influence of natural enemies in controlling rice pests has been clearly demonstrated (Kenmore *et al.*, 1984; Gallagher *et al.*, 1994; Way and Heong, 1995), it is still less clear how plant resistance contributes to IPM, and it has been suggested that host plant resistance is not needed. Gallagher *et al.* (1994) recommended that even susceptible varieties could be planted on a large scale if pesticides are not overused. Cuong *et al.* (1997) studied the effect of varietal resistance against BPH on pest abundance, natural enemy abundance and yield in Vietnam. They found hopperburn in insecticide-treated susceptible varieties, but not in insecticide-free areas. The susceptible variety had lower yields than the resistant varieties in only one out of the four seasons. Also, the ratio of natural enemies to BPH was generally higher on resistant varieties. Therefore, plant resistance may play a role in IPM systems when pest populations are particularly high. In temperate zones, migrating pests may arrive in rice paddies before resident natural enemy populations have built up. Under these condi-

tions, delaying pest development by the use of host plant resistance will help natural enemy populations to build up and control pest populations. Therefore, under some circumstances, varietal resistance appears to contribute to pest control by directly affecting insect pests and enhancing natural enemy activity (i.e. tritrophic effects).

Sustainability of conventional resistance

Once varieties are deployed in the field, it is important that resistance is sustainable over time. Insect pest populations may adapt to resistant varieties, thereby reducing their efficacy. A particular illustration of this scenario is shown in the case of the BPH. Outbreaks of BPH became a major threat to rice production in the early 1970s (Dyck and Thomas, 1979) due to the overuse of pesticides (Kenmore *et al.*, 1984; Gallagher *et al.*, 1994). Host plant varietal resistance was considered a major strategy for controlling BPH. Several genes, such as *Bph1*, *bph2* and *bph3*, were identified and bred into the germplasm by IRRI and other national research institutes in Asia (Khush, 1979).

IR26 with the *Bph1* gene showed good results in the field initially, but planthopper outbreaks resumed within a few years of the release (Gallagher *et al.*, 1994). The pattern was repeated later with other resistant varieties. The new populations of BPH that developed on the resistant germplasm were so damaging in the field that they were labelled biotypes 1 and 2 (Pathak and Heinrichs, 1982). Whether these biotypes constitute genetically distinct and reproductively isolated populations became a heated source of debate in the literature (Den Hollander and Pathak, 1981; Claridge and Den Hollander, 1983; Saxena and Barrion, 1985; Shufran and Whalon, 1995). The virulence of BPH was highly plastic, and populations could quickly be selected to feed on resistant varieties (Claridge and Den Hollander, 1982). Given the high variability within each 'biotype' and how rapidly one biotype could adapt to another resistant variety, the concept of biotypes in BPH appeared to be more of selected populations rather than host races on their way to speciation

(Claridge and Den Hollander, 1983; Shufran and Whalon, 1995).

It is important to understand how resistance can be durable. In the field, BPH has been slower to adapt to IR64, which has the *bph1* major gene, but also contains minor genes for resistance (Alam and Cohen, 1998b). Also, the presence of natural enemies may slow the rate of pest adaptation to resistant varieties. Another cultural method for slowing insect adaptation to resistance is the use of varietal mixtures. In an experimental selection study, Nemoto and Yokoo (1994) found that mixtures of varieties with different resistance genes could delay BPH adaptation, but the effect was only temporary.

Transgenic HPR

The moderate levels of resistance within the rice germplasm to rice stem borers have motivated researchers to search for transgenic sources of resistance (Khan *et al.*, 1991). Public and private efforts have been underway since 1998 to evaluate Bt rice as a source of control for the major stem borers, YSB and *C. suppressalis*. It is important to determine whether transgenic crops can be incorporated into an ecologically based IPM programme. There is the potential that the introduction of Bt crops may reduce pesticide use by enabling the farmers to spray less. However, data on actual measurable pesticide reduction have not yet been collected (Benbrook *et al.*, 1996; Wolfenbarger and Phifer, 2000).

Despite the potential benefits for pest control, it is also important to consider the sustainability of transgenic host plant resistance. For Bt rice, one major threat to its sustained use is the development of stem borer resistance. A survey of Philippine populations of YSB found that alleles with a high level of resistance to the *Bt* Cry1Ab toxin was less than 3.6×10^{-3} (Bentur *et al.*, 2000). Studies on adult movement show that the majority of YSB and *C. suppressalis* moths disperse from the field after eclosion (Cuong and Cohen, 2003). However, first instars of both stem borer species move between multiple plants, indicating that seed mixtures of Bt and non-Bt plants could reduce the amount of toxin ingested by the larvae and hasten the development of resistance to Bt (Cohen *et al.*, 2000b). Feeding on the plant toxin with Bt rice appears to stimulate the larvae to disperse (Dirie *et al.*, 2000). Because the level of resistance in field populations is very low, Cohen *et al.* (2000a) recommended the 'high-dose/refuge' strategy in order to manage the development of resistance in the field. This strategy uses varieties with high levels of Bt toxin combined with spatially separated areas of non-Bt varieties as refuges. They also recommended the use of two Bt toxins, because alleles conferring resistance to both toxins in the stem borer populations would be exceedingly rare.

Many critical questions need to be answered in order to assess the environmental risks of transgenic rice and the future of IPM. Whether there is an impact on arthropod food webs depends upon the particular genes that are incorporated. Based upon the available evidence so far, non-lepidopteran pests appear not to be severely affected by Bt rice (Chen *et al.*, 2006). A study of the effects of Bt rice on BPH and a mirid predator showed that both were unaffected by Bt rice (Bernal *et al.*, 2002). Dispersal studies show that planthoppers, leafhoppers and their parasitoids mostly disperse towards non-transgenic rice (Mao *et al.*, 2003). The taxa likely to be more directly affected are lepidopterans and their natural enemies. *Cotesia chilonis*, a larval parasitoid of *C. suppressalis*, parasitized larvae on Bt rice less, had a shorter development time until pupation and reduced male wasp longevity (Jiang *et al.*, 2004). However, it is also difficult to predict how arthropod biodiversity in the rice fields may respond to Bt rice. Both the transgenic form of Bt and the spray formulation may affect hymenopteran parasitoids, but more direct studies are necessary to determine whether the differences are large enough to compromise biological control (Cheng *et al.*, 2003; Schoenly *et al.*, 2003).

The consequence of transgenic resistance is that it may cause genetic pollution of wild resources. Significant gene flow between cultivated and wild rice already occurs (Chen *et al.*, 2004). Wild rice continues to be a major source of useful plant traits. Also, non-target lepidopterans that feed on wild rice may be affected by transgenes if significant gene flow occurs (Letourneau *et al.*, 2003). Currently, there is little known about lepidopteran species that feed on wild *Oryza* spp. The database used by Letourneau *et al.* (2003) suggested that over 95% of lepidopteran species that feed on *Oryza* spp. feed only on *Oryza sativa*. The under-representation of lepidopteran species on other *Oryza* spp. in the database may reflect on the lack of studies examining arthropod communities in wild rice or the difficulty in differentiating *Oryza* spp. in the field. There are very limited data regarding the level of susceptibility among lepidopteran families to Bt toxin expressed in the plant, but foliar Bt sprays appear to have a highly variable effect even within families (Letourneau *et al.*, 2003).

Use of molecular tools in rice host plant resistance

Current advances in rice genetics are creating new opportunities for understanding the genetic basis for resistance. Because the rice genome has been sequenced, rice is currently considered a model system for studying functional genomics. Many special resources are available for studying host plant resistance in rice that are unavailable in other crop species such as: more than 250,000 Expressed Sequence Tags (ESTS); detailed sequencing and annotation for a large part of the genome (http:www.tigr.org); quantitative trait locus (QTL) maps of resistance traits (http://www.gramene.org/qtl/index.html); and extensive bioinformatics resources (http://www.tigr.org/tdb/e2k1/osa1). These new molecular tools allow more specific characterization of rice plant defences, and more precise determination as to where useful traits are physically located in the genome. For example, these tools can be used to identify the genes involved with resistance, under what conditions these genes are expressed and how the genes are linked. With these new genetic tools, it may be possible to design more sophisticated combinations of genes related to resistance.

Most of the efforts in developing insect resistance in rice have targeted planthoppers (BPH and *S. furcifera*) and stem borers. Many authors have proposed that polygenic resistance to BPH is more durable than single major genes (Heinrichs *et al.*, 1985; Bosque-Perez and Buddenhagen, 1992; Alam and Cohen, 1998b). However, in the past it has been difficult to detect and select for polygenic resistance. The use of molecular markers in QTL analysis has allowed researchers to map and characterize linkage of resistance traits. Two QTLs were related to BPH non-preference and one QTL was related to tolerance (Alam and Cohen, 1998a). Recently, there have been other efforts to determine and identify QTLs for resistance (Xu *et al.*, 2002; Yang *et al.*, 2002; Ren *et al.*, 2004).

A recent study of gene expression using 108 cDNA clones of BPH on susceptible and resistant plants showed that about 14 genes had significantly different levels of expression (Zhang *et al.*, 2004). Most of the genes were regulated in the same direction: expression levels in the susceptible plants were much higher. Proteomic analysis and subtractive hybridization have revealed protein and gene expression that appears to be induced by BPH (Chen *et al.*, 2002; Yuan *et al.*, 2004). An elegant study by Wang *et al.* (2004) describes the activity of *OsBi1*, a rice gene that is induced by the attack of BPH in resistant varieties. They designed an anti-sense RNA probe to perform *in situ* hybridization within sections of the plant tissue. They found that the anti-sense *OsBi1* probe could detect that signals increased at 36 h after BPH feeding, and staining revealed that signals were clustered around the bundle sheath and parenchyma. Although mechanical wounding and pricking damage did not induce *OsBi1* expression, they found that ethylene and water stress also induced *OsBi1* expression.

Although the new genetic technology can stimulate exciting ways of studying host plant resistance, it is still important to understand how host plant resistance works in the field and functions in an ecological context. A good understanding of insect–plant interactions is necessary to accurately describe resistance characteristics and the function of resistance genes. However, understanding gene function alone will not determine whether insect pests will be successfully controlled in the field. It will be important to determine the factors that influence gene expression and also whether gene expression can be linked to successful population control in the field.

Lycosid spiders (Kartoharjono and Heinrichs, 1984) and mirid bugs, *C. lividipennis* (Senguttuvan and Gopalan, 1990), consume prey at higher rates on pest-resistant rice cultivars than on susceptible ones. Some pest-resistant cultivars appear to be more attractive than susceptible cultivars to certain predators (Sogawa, 1982; Rapusas *et al.*, 1996). Other interactions among control methods are discussed in Litsinger (1993).

Why pest management can become unsustainable?

Combinations of control methods are required to manage some pests. Farmer participatory research in Cambodia indicates that no form of rat control by itself is completely effective (Cox and Mak, 1999). Various combinations of sanitation, cultural practices, hunting, digging of burrows, trapping, a trap-barrier system and baiting are the best way to manage rice field rats. The specific techniques recommended depend on the rice ecosystem, the risk of yield loss to rats, the economic status of the farmers and the preferences of villagers. The weather must also be considered an important factor in choosing rat management options. Rats tend to flee from flooded areas, so that infestations in higher rice fields are fairly predictable during floods. Control measures in small areas are rarely practicable because the rats migrate from untreated areas back into treated areas. Specific recommendations (Singleton and Petch, 1994; Singleton, 1997; Leung, 1998; Jahn *et al.*, 1999) for managing rats include:

- organizing two or three village rat hunts per season to keep rat populations low;
- destroying all rat nests in rice bunds;
- removing weeds on bunds and surrounding areas;
- transplanting large areas at the same time;
- surrounding the field with a plastic fence with traps, where cost-effective;
- placing poison baits in coconut shells or aluminium cans so the baits do not get wet; and
- trapping to capture some rodent species that attack stored grain and paddy rice.

Developing sustainable IPM systems

Rice pest management practices may not be sustainable for a number of reasons:

- Over time the control tactic is no longer effective due to selection against pests that are susceptible to the tactic. Examples include pests that develop resistance to pesticides (Tabashnik, 1992), pests that overcome resistant crop varieties (Cohen *et al.*, 1998) and pests that avoid traps. Susceptibility to a control measure is essentially a resource that must be managed (Cox and Forrester, 1992; Gould, 1998). The situation is that described by Hardin (1968) in his famous essay 'The Tragedy of the Commons': while individual growers have the incentive to capitalize on susceptibility as much as possible for short-term gains, this resource will then be squandered to the detriment of all growers.
- The control tactic leads to disruptions in the ecosystem (e.g. destruction of natural enemies) that result in further outbreaks of the target pest or outbreaks of new pests.
- The cost of the practice is too expensive to maintain indefinitely. This may be due to changes in the cost of labour, the price of the agricultural commodity or the cost of the control measure itself.
- The practice degrades the quality of human health, the environment or agronomic resources over time (Reichardt *et al.*, 1998).
- New pest problems arise due to pest introductions, introductions of natural

enemies that attack existing biological control agents and thereby increase pest populations (Rosenheim, 1998), pest immigration or a change in the environment including changes to the way that rice is cultivated. For example, new rice cultivars, designed for higher yields, may inadvertently impede the effectiveness of natural enemies (Bottrell *et al.*, 1998).

- As rice cultivation intensifies, the types and the abundance of pests change. Therefore, the previous management tactics may not adequately control pest populations.

If pesticides are part of the IPM system, a pesticide resistance management strategy may be necessary so that the target pest's susceptibility to the pesticide will not decline over time (Cox and Forrester, 1992). Organic pest control, the practice of using no synthetic pesticides on a crop, is sometimes claimed to be a sustainable system when used as part of an overall organic farming programme. However, organic pest control tends to rely heavily on botanical pesticides and other so-called natural pesticides. Over time, pests develop resistance to these natural chemicals, just as they develop resistance to synthetic pesticides. While generally biodegradable, many botanical pesticides (e.g. pyrethrum) are broad-spectrum insecticides, which carry the same risks as synthetic chemicals of disrupting the ecosystem by killing natural enemies.

Even if pesticides are not part of an IPM programme, there will still be issues of sustainability. Pest-resistant cultivars become susceptible over time. Newly introduced pests may require new natural enemies to manage them. Manual monitoring and control programmes become too costly as the cost of labour increases and crop production becomes more efficient. Cultural pest prevention techniques (e.g. relating to water or fertilizer management) are also not necessarily sustainable. For example, some species of rice pests (e.g. aphids, gall midge, planthoppers, stem borers) increase populations, and cause more damage, as N is applied to the crop (Uthamasamy *et al.*,

1983; Jahn *et al.*, 2001, 2005). Keeping N applications below the critical N concentration (Sheehy *et al.*, 1998) permits the plant to grow slower and develop harder tissue, making it less susceptible to stem borers and other pests (Nadarajan and Pillai, 1985; Litsinger, 1994). However, the rate of demand for N by the panicle exceeds the rate at which the above-ground portions of the plant acquire N (Sheehy *et al.*, 1998), so that keeping N applications below the critical levels will lower the yield potential of the crop. As markets and human population pressures change, a farmer may not be able to afford growing a crop with low yield potential.

Generally, control measures are disruptive to the ecosystem; so preventing the pest problem from arising in the first place is preferable to control and promotes sustainability. Depending on the type of pest, it is generally necessary to monitor pest levels, pest damage or some other indicator to determine whether it is time to apply a control measure.

Understanding the mechanisms that influence yield loss in rice is a key to determining IPM strategies. It has been difficult to relate insect pest densities to yield loss (Litsinger *et al.*, 1987b) probably for two reasons. First, rice is usually subjected to stresses from more than one pest, and to other non-insect pest stresses at the same time. The combination of stresses may act synergistically (Litsinger, 1991). The second reason is the great ability of high-tillering rice varieties to compensate from insect pest damage if well managed.

Synergistic losses have been recorded from 2 to 3 insect pests occurring at the same time, each at subeconomic damage levels (Table 15.10). Savary *et al.* (1994) undertook a large survey of farmers' fields in the Philippines and using correspondence analysis showed that low yields were related to certain combinations of crop production practices, seasons and kinds of stresses. From this information as well as that from Fajardo *et al.* (2000) and Litsinger (1993), a list of the most important crop stresses can be summarized: (i) early-maturing varieties (less time for compensation, namely IR58, the earliest-maturing rice tested suffered highest

Table 15.10. Yield as affected by single-pest infestations or in combinations showing synergistic yield loss from combinations.[a] (From IRRI, 1983, p. 204.)

Pest	Yield (g/m²)
Caseworm (CW)	515 a
Whorl maggot (WM)	514 a
Stem borer (SB)	457 b
CW + WM	426 bc
CW + SB	413 c
WM + SB	419 c
CW + WM + SB	402 c

[a]Artificial infestation of potted rice plants in the greenhouse in 2.25 m² cages. Infestation levels held the same with each pest despite the combination.

yield losses); (ii) low seeding rates (farmers double or triple recommendations); (iii) late planting (early planting escapes pest build-up with the season and in the wet season avoids the most cloudy weather); (iv) inadequate ponding (too high or too low water levels reduce rice growth); (v) weeds in the seedbed and field (transplanted weeds from the seedbed are worst as they cannot be hand-weeded); (vi) stem borers (particularly in conjunction with weediness); and (vii) N fertilizer applied at the wrong time or in wrong amounts (either too much or too little). We hypothesize that transplanting shock (slow recovery of transplanted seedlings because of root injury during pulling) due to poor seedbed conditions would be an overlooked eighth stress. On the other

hand, sheath blight was associated with the highest yields by Savary *et al.* (1994), and diseases in general were not powerful determinants probably due to the adequate levels of genetic resistance in HYVs in the Philippines.

There are reports of exceptional examples of HYVs' ability to completely compensate for abnormally high insect infestations: (i) crops suffering 50% and 82% rice whorl maggot–damaged leaves (Viajante and Heinrichs, 1986a; Shepard *et al.*, 1990); (ii) 67% damaged leaves from leaffolder (Miyashita, 1985); (iii) more than 50% defoliation of seedlings from caseworm (Bandong, 1981); (iv) 30% stem borer deadhearts and 10% whiteheads (Rubia *et al.*, 1996); and 3 whiteheads per hill (Litsinger, 1993). Yield loss studies in two sites showed that compensation can be highly seasonal. Compensation was measured as the degree to which yield loss was reduced with increasing yield levels in the Philippine yield loss trials. Compensation would occur if the rate of yield loss did not rise proportionally with increasing yield (i.e. slope of the linear regression equation was insignificant). Yield in two sites was examined for this effect. In Zaragoza, a high–pest density site due to its location` at the end of a large irrigation system, dry season yields are 20% higher than wet season yields. Compensation was registered in the dry season with an insignificant slope; however, it was not in the lower-yielding wet season (Fig. 15.4a and b). In Calauan, on the other hand,

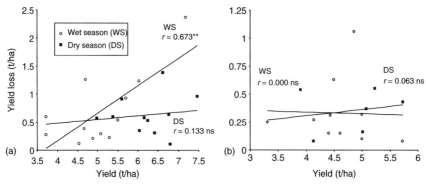

Fig. 15.4. Relationship between yield and insect pest–caused yield loss in four sites showing compensatory capacity in crops with significant slopes in the regression equation: (a) Zaragoza, and (b) Calauan, Philippines. **= p<0.01, ns=p>0.05.

there was not a great difference between the yields in both seasons and the slopes were insignificant in both crops. Calauan farmers plant longer-maturing rice varieties C1 and Malagkit in both seasons, allowing a longer period for compensation.

Farmers have utilized cultural control methods to enhance crop compensation, probably not knowing the exact reason, by trial and error where higher yields were noted. A comparison of farmers' crop husbandry practices with national recommendations illustrates the point as farmers generally developed their own practices (Table 15.11). Farmers who were direct seeding or transplanting used 2–3 times the recommended seeding rates. In Zaragoza farmers sowed seedbeds with 141 kg instead of 50 kg seed/ha as recommended. Similarly for direct seeding the recommended rate is 100 kg but the farmers averaged 168 kg/ha. Even larger differences were evident from nearby Guimba also in Nueva Ecija (Litsinger, 1993; Table 3.1), where farmers used

149 kg seed/ha for transplanted rice and 209 kg/ha for direct-seeded rice. Farmers transplanted older seedlings (30–31 days) than the 21 days recommended. They transplanted 10–14 seedlings per hill instead of 3–4 as recommended. Finally in Guimba farmers applied 84 kg N/ha in the wet season and 124 kg N/ha in the dry season when the recommended dosages were 60 kg and 90 kg N/ha, respectively. Zaragoza farmers, however, followed the recommendation closely.

Some farmers may have been overexuberant in using seeds and N but were probably on the right track (Litsinger, 1993). The researchers' methodologies that led to the recommendations undertook reductionist trials looking at only one factor at a time while keeping the others constant. The result was that the trials were managed to be stress-free, leading to recommendations that were too conservative. Fertility trials, for example, were conducted in a pest-free environment maintained by excessive insecticide protection. Water management

Table 15.11. Comparison of recommended and farmers' crop husbandry practices for irrigated rice, Zaragoza, Nueva Ecija, Philippines.

Practice	Recommended by researchers	Farmers' practice
Seeding rate (kg/ha)		
Transplanted	50	141
Direct seeded	100	168
Seedbed		
Fertilizer (kg N/ha)	None	93
Age of seedlings at transplanting (days after sowing)	21	30
Insecticide applications (no.)	None	1.4
Main crop		
Seedlings per hill (no.)	3–4	14.1
Timing of first fertilizer application	Basal (before transplanting)	13 days after transplanting
Nitrogen application (kg/ha)		
Wet season	60	55
Dry season	90	92
Fertilizer application frequency (no.)		
Wet season	2	1.5
Dry season	2–3	1.9
Insect pest assessment	Quantitative	Qualitative
Insecticide application frequency (no.)	Based on action thresholds (1.9)	2.4
Dosage per application (kg a.i./ha)	0.40	0.21

Average of 40 farmers interviewed.

was carried out under controlled conditions to eliminate flooding and drought. Weeds were totally eliminated. These are conditions that few farmers can achieve. But crop management and not the adoption of modern rice varieties per se is the key to high yields. This can be seen in the variation in yields in a farm community, which can range from 1 to 9 t/ha. An illuminative study by Pingali *et al.* (1990) in Luzon showed that yields of the top one-third of farmers matched those of research stations. The top one-third of farmers achieved yields greater than 5.5 t/ha in the dry season and 4.5 t/ha in the wet. Looking at the practices of those top yielding farmers provides clues to management to minimize the effects of pests. Under favourable environmental conditions increasing levels of crop management leads to higher yields and greater compensation from insect injury.

In light of the above discussion on compensation, one may conclude that the insecticide check trials (Table 15.3) overestimate loss caused by insect pests. The insecticide check method compares yields from insecticide-treated and untreated plots, and the yield difference between the full protection treatment and untreated measure loss. Releasing the insect pest stress in the full protection treatment, however, enables the crop to compensate against other stresses. At the same time, the untreated plots may not compensate for the combination of insect pests and other stresses. That combination would produce synergistic losses in the same way as in Table 15.10 for multiple insects or from the clusters of stresses measured in Savary *et al.* (1994). If the crop were only under insect pest stresses, the trials would have to measure losses from insect pest damage only. But this would be a rare situation as the rice crop is typically under a number of stresses at any one time.

An example with three farmers is given in Table 15.12 to elucidate the main points. The first farm suffers damage from whorl maggot and no other stresses and the crop can compensate, so no yield loss is registered as the difference between the full insecticide protection treatment and the untreated plot. In the second farm the crop

also suffers from transplanting shock, as the seedbed was dry when the seedlings were pulled and rootlets were torn off. The full protection treatment compensated from the transplanting shock, as this was the only stress. However, the untreated plot suffered from the combination of whorl maggot damage and transplanting shock and registered 6% yield loss. In the third farm where both whorl maggot and transplanting shock stresses occurred, the farmer applied 90 kg N instead of 60 kg as in the first two farms. The full protection plot compensates again from the transplanting shock but because the N rate is higher, yield is also higher. However, in the untreated plot with the dual stresses the increased N was not enough to fully compensate, and 11% loss was registered. When the full protection plot compensated relatively more than the untreated plot, a higher yield loss was registered from the same whorl maggot infestation level.

Highest losses occurred in fields under the greatest number of stresses, each at a height intensity of stress combined with the lowest level of crop management plus favourable weather which together afforded the least degree of compensation. The degree to which each crop compensated was a function of the variety used, the number and intensity of the stresses present, the management practices and the weather conditions during the crop. In the yield loss trials we did not manage farmers' agronomy except to choose the most popular rice variety at the time; thus farmers applied their own fertilizer regime, weed control, etc. In the Philippines this level of management can be highly variable. As farmers were randomly chosen in the yield loss trials, the results are representative of the variation in crop management and weather conditions that existed at the time. We conclude that losses measured by the insecticide check method accurately reflect what happens when insect pests are not controlled. The results, however, are highly conditional.

The highly dynamic and plastic relationship between combinations of pests and stresses on the one hand and management practices and weather affecting crop compensation on the other argue against the

Table 15.12. Hypothetical examples from three farmers in the same location explaining the complexities in measuring yield loss from insect pest damage based on synergistic losses from multiple stresses and crop compensation from different levels of management.

	Farmer 1[a]				Farmer 2[b]				Farmer 3[c]			
	Full-protection plot[d]	Untreated	Yield loss t/ha	%	Full-protection plot[d]	Untreated	Yield loss t/ha	%	Full-protection plot[d]	Untreated	Yield loss t/ha	%
Whorl maggot damage (%)	0	30			0	30			0	30		
Transplanting shock	No	No			Yes	Yes			Yes	Yes		
Total nitrogen (kg/ha)	60	60			60	60			90	90		
Yield (t/ha)	5.0	5.0	0.0	0.0	5.0	4.7	0.3	6.0	5.4	4.8	0.6	11.1

[a]Farmer 1 has only whorl maggot damage.
[b]Farmer 2 has both whorl maggot and transplanting shock stresses.
[c]Farmer 3 has both whorl maggot and transplanting shock stresses but higher level of nitrogen.
[d]Total of nine insecticide applications over the whole crop duration.

future development of economic thresholds based on damage functions relating yield loss to increasing insect pest abundance.

What are some of the crop management practices that provide for greater crop resilience to pest damage? Many are practices where farmers differed with researchers (Table 15.11). Follow-up research tested some of the farmers' practices and found that increasing the seeding rate from 3 seedlings per hill to 12 augmented the crop's tolerance for stem borer damage when infestation levels were between 1 and 3 whiteheads per hill (Litsinger, 1993; Figure 3.1). Ovipositing females do not penetrate into denser plantings. The role of N in augmenting tolerance was shown in Table 15.6 as well as by Litsinger (1993) but rate above 90 kg was not tested. The use of leaf colour charts to gauge when to apply N will assist farmers in achieving optimal N efficiency, as overapplying N can lead to lower yields. The use of N in seedbeds by farmers is carried out not so much with high yields in mind but to make it easier to pull seedlings from seedbeds. In the IRRI research station seedbeds are established in a location that has particularly soft soil with the consistency of porridge, whereas in farmers' fields seedbed soils are hard. Farmers said if they pulled 3-week-old seedlings, the tillers would rip in half, thus they allow the seedlings to grow older. Use of N in seedbeds encourages shallow rooting, making pulling easier. It is possible that transplanting shock, which is bound to be greater from more damaged root systems, may explain data from field trials where whorl maggot, despite very high populations on the IRRI experiment station, rarely causes yield loss, whereas outside of the station loss is common. The reason could be that transplanting shock exacerbates whorl maggot damage.

In general, what is good for the plant is good for the pest, and this is particularly true of N applications (Litsinger, 1994). This is why most pest populations will increase along with rice yields. Although there are exceptions among rice pests, N applications tend to promote greater survival, increased tolerance of stress, higher fecun-

dity, increased feeding rates and higher populations. N applications also make rice more attractive to many phytophagous insects (Maischner, 1995; Mattson, 1980). Rice diseases, e.g. sheath blight and blast, also respond favourably to increased N applications (Savary *et al.*, 1995, 2000a, b).

Rice production since the mid-1960s has gradually shifted towards semi-dwarf, N-responsive varieties. This change is believed by some to have caused increased pest problems (Kiritani, 1979; Litsinger, 1989; Mew, 1992). While N applications contribute to increased pest abundance, they should also aid plants in recovering from pest damage. There is no simple formula to describe how plant recovery from herbivory is related to the availability of soil nutrients. The 'continuum of responses' model (Maschinski and Whitham, 1989), for example, predicts that plants are best able to recover from pest damage when well fertilized, while the 'growth rate' model (Hilbert *et al.*, 1981) predicts that damaged plants will exhibit superior recovery from herbivory when grown under stress, i.e. low levels of nutrients. Meta-analysis suggests that basal meristem monocots, such as rice and other grasses, exhibit better recovery from pest damage when the plants are grown under high resource levels, while dicots show better recovery when grown under low resource levels (Hawkes and Sullivan, 2001). The degree of compensation and recovery from damage also depends on the stage and cultivar of rice (Khiev *et al.*, 2000).

How fertilizer applications affect the severity of pest damage depends not only on the amount of fertilizer but also on the composition and timing of the applications. Studies in India, China, Indonesia, the Philippines and Vietnam have found lower pest incidence in fields with site-specific nutrient management compared with the farmers' fertilizer practices. Most research has focused on the effects of specific nutrients on specific pests. N applications apparently decrease thrips populations (Ghose *et al.*, 1960), but most insect pests become more abundant on rice when N is applied (e.g. Oya and Suzuki,

1971; Chelliah and Subramanian, 1972; Litsinger, 1994). Several species of stem borer larvae exhibit significant weight gains when N is applied to the host plant. Heavier stem borer larvae presumably cause more damage to the host plant than lighter larvae (Ishii and Hirano, 1958; Ghosh, 1962; Soejitno, 1979; Rubia, 1994). More stem borer (*C. suppressalis*) eggs are found in fields with high N rates (Hirano, 1964). Rice stem borer (*C. suppressalis*) larval survival is enhanced by N applications (Alinia *et al.*, 2000).

How do fertilizer applications affect pest damage? N augments rice plant growth and results in softer plant tissues, which presumably allows for easier penetration of the rice plant by insects and pathogens (Oya and Suzuki, 1971; Nadarajan and Pillai, 1985). High N rates generally attract ovipositing insects and increase insect fecundity. P improves root development and tolerance for root pests (e.g. the root weevil, *Echinocnemus oryzae* Marshall) (Tirumala Rao, 1952). K applications may suppress pests by lowering plant sugar and amino acid levels, promoting thicker cell walls and increasing silicon uptake (Baskaran, 1985). Minor plant nutrients, such as silicon and zinc, can also contribute to pest suppression. In the proper quantities, silicon can increase the resistance of rice plants to blast, brown spot, bacterial blight, planthoppers and stem borers (Pathak *et al.*, 1971; Kim and Heinrichs, 1982; Prakash, 1999; Chang *et al.*, 2001). Damage by *Elasmopalpus*, a stem borer of dryland rice, is reportedly minimized by zinc applications (Reddy, 1967).

Understanding pest–nutrient interactions on rice have led to some simple IPM recommendations: e.g. to split N applications, plough straw into the soil to increase silicon uptake, apply K during planthopper outbreaks and apply N to promote recovery following stem borer and defoliator damage (Litsinger, 1994). Basal applications of fertilizer have been found to reduce golden apple snail populations (Cruz *et al.*, 2001), which are serious rice pests throughout Asia (Halwart, 1994; Jahn *et al.*, 1998). Still, more research is needed. The scientific literature contains numerous contradictory implications for the manipulation of soil nutrients to manage rice pests. For instance, N applications cause at least some rice cultivars to release oryzanone, which makes them more attractive to stem borers (Seko and Kato, 1950). On the other hand, the addition of N stimulates tiller production, which helps rice plants compensate for early stem borer damage (Rubia *et al.*, 1996). Planthopper incidence is increased by N applications (Uthamasamy *et al.*, 1983; Cook and Denno, 1994; Preap *et al.*, 2001; Lu *et al.*, 2004), but plants with high N content have an improved ability to compensate for damage caused by planthopper feeding (Rubia-Sanchez *et al.*, 1999). These apparent contradictions result from several factors: most of the studies are conducted on one or a few cultivars, soil properties are rarely considered and most of the research is done under artificial conditions where natural enemies and other contributing variables cannot affect the results. Currently, nutrient and pest management recommendations are inadequate for predicting edaphic effects on multiple pests of different rice cultivars.

References

Alam, S.N. and Cohen, M.B. (1998a) Detection and analysis of QTLs for resistance to the brown planthopper, *Nilaparvata lugens*, in a doubled-haploid rice population. *Theoretical and Applied Genetics* 97, 1370–1379.

Alam, S.N. and Cohen, M.B. (1998b) Durability of brown planthopper, *Nilaparvata lugens*, resistance in rice variety IR64 in greenhouse selection studies. *Entomologia Experimentalis et Applicata* 89, 71–78.

Alinia, F., Ghareyazie, B., Rubia, L., Bennett, J. and Cohen, M.B. (2000) Effect of plant age, larval age, and fertilizer treatment on resistance of a *cryIAb*-transformed aromatic rice to lepidopterous stem borers and foliage feeders. *Journal of Economic Entomology* 93, 484–493.

Amudhan, S., Rao, U.P. and Bentur, J.S. (1999) Total phenol profile in some rice varieties in relation to infestation by Asian rice gall midge *Orseolia oryzae* (Wood-Mason). *Current Science* 76, 1577–1580.

Bandong, J.P. (1981) Rice caseworm: mass rearing technique and its utilization. MSc thesis, The University of the Philippines, Los Baños, Laguna, Philippines.

Bandong, J.P. and Litsinger, J.A. (2005) Rice crop stage susceptibility to the rice yellow stemborer *Scirpophaga incertulas* (Walker) (Lepidoptera: Pyralidae). *International Journal of Pest Management* 51, 37–43.

Barrion, A.T. and Litsinger, J.A. (1984) The spider fauna of Philippine rice agroecosystems. II. Wetland. *Philippine Entomologists* 6, 11–37.

Barrion, A.T. and Litsinger, J.A. (1987) *Ochthera sauteri* Cresson (Diptera: Ephydridae), predator of rice whorl maggot (RWM) flies. *International Rice Research Newsletter* 12, 18.

Barrion, A.T. and Litsinger, J.A. (1997) *Dichogaster* nr. *curgensis* Michaelsen (Annelida: Octochaetidae): an earthworm pest of terraced rice in the Philippine Cordilleras. *Crop Protection* 16, 89–93.

Barrion, A.T., Litsinger, J.A., Sison, E. and Arraudeau, M. (1987) Stem borers in dryland and wetland rice. *International Rice Research Newsletter* 12, 16.

Barrion, A.T., Litsinger, J.A., Medina, E.B., Aguda, R.M., Bandong, J.P., Pantua P.C. Jr, Viajante, V.D., dela Cruz, C.G., Vega, C.R., Soriano J.S. Jr, Camañag, E.E., Saxena, R.C., Tryon, E.H. and Shepard, B.M. (1991) The rice *Cnaphalocrocis* and *Marasmia* (Lepidotera: Pyralidae) leaffolder complex in the Philippines: taxonomy, bionomics and control. *Philippine Entomologist* 8, 987–1074.

Baskaran, P. (1985) Potash for crop resistance to insect pests. *Journal of Potassium Research* 1, 81–94.

Benbrook, C.M., Groth, E., Halloran, J.M., Hansen, M.K. and Marquardt, S. (1996) *Pest Management at the Crossroads.* Consumers Union, Yonkers, New York.

Benrey, B. and Denno, R.F. (1997) The slow-growth–high-mortality hypothesis: a test using the cabbage butterfly. *Ecology* 78, 987–999.

Bentur, J.S. and Kalode, M.B. (1996) Hypersensitive reaction and induced resistance in rice against the Asian rice gall midge *Orseolia oryzae*. *Entomologia Experimentalis et Applicata* 78, 77–81.

Bentur, J.S., Andow, D.A., Cohen, M.B., Romena, A.M. and Gould, F. (2000) Frequency of alleles conferring resistance to a *Bacillus thuringiensis* toxin in a Philippine population of *Scirpophaga incertulas* (Lepidoptera: Pyralidae). *Journal of Economic Entomology* 93, 1515–1521.

Bernal, C.C., Aguda, R.M. and Cohen, M.B. (2002) Effect of rice lines transformed with *Bacillus thuringiensis* toxin genes on the brown planthopper and its predator *Cyrtorhinus lividipennis*. *Entomologia Experimentalis et Applicata* 102, 21–28.

Bosque-Perez, N.A. and Buddenhagen, I.W. (1992) The development of host-plant resistance to insect pests: outlook for the tropics. In: Menken, S.B.J., Visser, J.H. and Harrewijn, P. (eds) *Proceedings of the 8th International Symposium on Insect–plant Relationships.* Kluwer Academic Publishers, Dordrecht, The Netherlands, pp. 235–249.

Bottrell, D.G., Barbosa, P. and Gould, F. (1998) Manipulating natural enemies by plant variety selection and modification: a realistic strategy? *Annual Review of Entomology* 43, 347–367.

Bray, F. (1986) *The Rice Economies: Technology and Development in Asian Societies.* Blackwell, London.

Canapi, B.L., Rubia, E.G., Litsinger, J.A., Shepard, B.M. and Rueda, L.M. (1988) Predation by sword-tailed cricket *Anaxipha longipennis* (Servile) Gryllidae on eggs of three lepidopterous pests of rice. *International Rice Research Newsletter* 13, 40–41.

CARDI (2001) *Annual Research Report 2000.* Cambodian Agricultural Research and Development Institute, Phnom Penh, Cambodia.

Catling, D. (1992) *Rice in Deep Water.* Macmillan Press, London.

Cendaña, S.M. and Calora, F.B. (1967) Insect pests of rice in the Philippines. In: *Major Insect Pests of the Rice Plant.* Johns Hopkins Press, Baltimore, Maryland, pp. 591–616.

Chang, S.J., Li, C.C. and Tzeng, D.S. (2001) Effects of nitrogen, calcium and silicon nutrition on bacterial blight resistance in rice (*Oryza sativa* L.). In: Peng, S. and Hardy, B. (eds) *Rice Research for Food Security and Poverty Alleviation. Proceedings of the International Rice Research Conference, 31 March–3 April 2000, Los Baños, Philippines.* International Rice Research Institute, Los Baños, Laguna, Philippines, pp. 441–448.

Chaudhary, R.C., Khush, G.S. and Heinrichs, E.A. (1984) Varietal resistance to rice stem borers in Asia. *Insect Science and Its Application* 5, 447–463.

Chelliah, S. and Subramanian, A. (1972) Influence of nitrogen fertilization on the infestation by the gall midge, *Pachydiplosis oryzae* (Wood-Mason) Mani, in certain rice varieties. *Indian Journal of Entomology* 34, 255–256.

Chen, Y.H. and Welter, S.C. (2003) Confused by domestication: non-congruent behavioral responses of the sunflower moth, *Homoeosoma electellum* (Lepidoptera: Pyralidae) and its parasitoid, *Dolichogenidea homoeosomae* (Hymenoptera: Braconidae), towards wild and domesticated sunflowers. *Biological Control* 28, 180–190.

Chen, R.Z., Weng, Q.M., Huang, Z., Zhu, L.L. and He, G.C. (2002) Analysis of resistance-related proteins in rice against brown planthopper by two-dimensional electrophoresis. *Acta Botanica Sinica* 44, 427–432.

Chen, L.J., Lee, D.S., Song, Z.P., Suh, H.S. and Lu, B.R. (2004) Gene flow from cultivated rice (*Oryza sativa*) to its weedy and wild relatives. *Annals of Botany* 93, 67–73.

Chen, M., Ye, G.Y., Liu, Z.C., Yao, H.W., Chen, X.X., Shen, S.Z., Hu, C., and Datta, S.K. (2006). Field assessment of the effects of transgenic rice expressing a fused gene of caylAb and crylAc from *Bacillus thuringiensis* Berliner on nontarget planthopper and leafhopper populations. *Environmental Entomology* 35, 127–134.

Cheng, C.H. and Chang, W.L. (1979) Studies on varietal resistance to the brown planthopper in Taiwan. In: *Brown Planthopper: Threat to Rice Production in Asia*. International Rice Research Institute, Los Baños, Laguna, Philippines, pp. 251–271.

Cheng, L.Z., Ye, G.Y., Cui, H. and Datta, S.K. (2003) Impact of transgenic indica rice with a fused gene of cry1Ab/cry1Ac on the rice paddy arthropod community. *Acta Entomologica Sinica* 46, 454–465.

Choi, S.Y., Heu, M.H. and Lee, J.O. (1979) Status of varietal resistance to the brown planthopper in Korea. In: *Brown Planthopper: Threat to Rice Production in Asia*. International Rice Research Institute, Los Baños, Laguna, Philippines, pp. 219–232.

CIAP (1995) *Annual Research Report 1994*. Cambodia–IRRI–Australia Project, International Rice Research Institute, Phnom Penh, Cambodia.

CIAP (1998) *Annual Research Report 1997*. Cambodia–IRRI–Australia Project, International Rice Research Institute, Phnom Penh, Cambodia.

CIAP (1999) *Annual Research Report 1998*. Cambodia–IRRI–Australia Project, International Rice Research Institute, Phnom Penh, Cambodia.

CIAP (2000) *Annual Research Report 1999*. Cambodia–IRRI–Australia Project, International Rice Research Institute, Phnom Penh, Cambodia.

Claridge, M.F. and Den Hollander, J. (1982) Virulence to rice cultivars and selection for virulence in populations of the brown planthopper *Nilaparvata lugens*. *Entomologia Experimentalis et Applicata* 32, 222–226.

Claridge, M.F. and Den Hollander, J. (1983) The biotype concept and its application to insect pests of agriculture. *Crop Protection* 2, 85–95.

Cohen, M.B., Savary, S., Huang, N., Azzam, O. and Datta, S.K. (1998) Importance of rice pests and challenges to their management. In: Dowling, N.G., Greenfield, S.M. and Fischer, K.S. (eds) *Sustainability of Rice in the Global Food System*. Pacific Basin Study Center, Davis, California and International Rice Research Institute, Manila, Philippines, pp. 145–164.

Cohen, M.B., Gould, F. and Bentur, J.S. (2000a) Bt rice: practical steps to sustainable use. *International Rice Research Newsletter* 25, 4–10.

Cohen, M.B., Romena, A.M. and Gould, F. (2000b) Dispersal by larvae of the stem borers *Scirpophaga incertulas* (Lepidoptera: Pyralidae) and *Chilo suppressalis* (Lepidoptera: Crambidae) in plots of transplanted rice. *Environmental Entomology* 29, 958–971.

Cohen, J.E., Schoenly, K., Heong, K.L., Justo, H., Arida, G., Barrion, A.T. and Litsinger, J.A. (1994) A food web approach to evaluating the effect of insecticide spraying on insect pest population dynamics in a Philippine rice irrigated ecosystem. *Journal of Applied Ecology* 31, 747–763.

Cook, A.G. and Denno, R.F. (1994) Planthopper/plant interactions: feeding behavior, plant nutrition, plant defense, and host plant specialization. In: Denno, R.F. and Perfect, T.J. (eds) *Planthoppers, Their Ecology and Management*. Chapman and Hall, London, pp. 114–139.

Cook, A.G., Woodhead, S., Magalit, V. and Heinrichs, E.A. (1987) Variation in feeding behavior of *Nilaparvata lugens* on resistant and susceptible rice varieties. *Entomologia Experimentalis et Applicata* 43, 227–235.

Cortesero, A.M., Stapel, J.O. and Lewis, W.J. (2000) Understanding and manipulating plant attributes to enhance biological control. *Biological Control* 17, 35–49.

Cox, P.G. and Forrester, N.W. (1992) Economics of insecticide resistance management in *Heliothis armigera* (Lepidoptera: Noctuidae) in Australia. *Journal of Economic Entomology* 85, 1539–1549.

Cox, P. and Mak, S. (1999) Participatory farmer research for the management of rats in Cambodia. *Cambodian Journal of Agriculture* 2, 23–28.

Cruz, M.S. de la, Joshi, R.C. and Martin, A.R. (2001) The effect of fertilizer on damage by the golden apple snail. *International Rice Research Notes* 26, 20–21.

Cuong, N.L. and Cohen, M.B. (2003) Mating and dispersal behaviour of *Scirpophaga incertulas* and *Chilo suppressalis* (Lepidoptera: Pyralidae) in relation to resistance management for rice transformed with *Bacillus thuringiensis* toxin genes. *International Journal of Pest Management* 49, 275–279.

Cuong, N.L., Ben, P.T., Phuong, L.T., Chau, L.M. and Cohen, M.B. (1997) Effect of host plant resistance and insecticide on brown planthopper *Nilaparvata lugens* (Stål) and predator population development in the Mekong Delta, Vietnam. *Crop Protection* 16, 707–715.

Das, Y.T. (1976) Some factors of resistance to *Chilo suppressalis* in rice varieties. *Entomologia Experimentalis et Applicata* 20, 131–134.

Dela Cruz, C.G. and Litsinger, J.A. (1986) Host plant range of the planthopper *Nisia atrovenosa*. *International Rice Research Newsletter* 11, 26–27.

Den Hollander, J. and Pathak, P.K. (1981) The genetics of the "biotypes" of the rice brown planthopper, *Nilaparvata lugens*. *Entomologia Experimentalis et Applicata* 29, 76–86.

Dirie, A.M., Cohen, M.B. and Gould, F. (2000) Larval dispersal and survival of *Scirpophaga incertulas* (Lepidoptera: Pyralidae) and *Chilo suppressalis* (Lepidoptera: Crambidae) on cry1Ab-transformed and non-transgenic rice. *Environmental Entomology* 29, 972–978.

Djamin, A. and Pathak, M.D. (1967) Role of silica in resistance to Asiatic rice borer *Chilo suppressalis* (Walker) in rice varieties. *Journal of Economic Entomology* 60, 347–351.

Dyck, V.A. and Thomas, B. (1979) The brown planthopper problem. In: *Brown Planthopper: Threat to Asian Rice Production*. International Rice Research Institute, Los Baños, Laguna, Philippines, pp. 3–17.

Fajardo, F.F., Canapi, B.L., Roldan, G.V., Escandor, R.P., Moody, K., Litsinger, J.A. and Mew, T.W. (2000) Understanding small-scale rice farmers' pest perceptions and management practices as a foundation for adaptive research and extension: a case study in the Philippines. *Philippine Journal of Crop Science* 25, 55–67.

Ferino, M. (1968) The biology and control of the rice leaf-whorl maggot, *Hydrellia philippina* Ferino (Ephydridae: Diptera). *Philippine Agriculturist* 52, 332–383.

Fernando, H., Elikewela, Y., De Alwis, H.M., Senadheera, D. and Kudagamage, C. (1977) Varietal resistance to the brown planthopper *Nilaparvata lugens* (Stål) in Sri Lanka. In *Brown Planthopper: Threat to Rice Production in Asia*. International Rice Research Institute, Los Baños, Philippines.

Fujisaka, S., Kirk, G., Litsinger, J.A., Moody, K., Nasrul Hosen, Adly Yusef, Firdos Nurdin, Tasmin Naim, Farida Artati, Abdul Aziz, Winardi Kahatib and Yustisia (1991) Wild pigs, poor soils, and upland rice: a diagnostic survey of Sitiung Sumatra, Indonesia. *IRRI Research Paper Series No. 155*.

Gallagher, K.D., Kenmore, P.E. and Sogawa, K. (1994) Judicial use of insecticides deter planthopper outbreaks and extend the life of resistant varieties in Southeast Asian rice. In: Denno, R.F. and Perfect, T.J. (eds) *Planthoppers: Their Ecology and Management*. Chapman and Hall, New York, pp. 599–614.

Ghose, R.L.M., Ghatge, M.B. and Subrahmanyan, V. (1960). Pests of rice. In: *Rice in India*, 2nd edn. Indian Council of Agricultural Research, New Delhi, pp. 248–257.

Ghosh, B.N. (1962) A note on the incidence of stem borer *Schoenobius incertulas* (Walker) on boro paddy under nitrogen fertilizers. *Current Science* 31, 472–473.

Glanz, M.H. (1996) *Currents of Change: El Niño's Impact on Climate and Society*. Cambridge University Press, Cambridge, UK.

Gould, F. (1998) Sustainability of transgenic insecticidal cultivars: integrating pest genetics and ecology. *Annual Review of Entomology* 43, 701–726.

Grayer, R.J., Harborne, J.B., Kimmins, F.M., Stevenson, P.C. and Wijayagunasekera, H.N.P. (1994) Phenolics in rice phloem sap as sucking deterrents to the brown planthopper, *Nilaparvata lugens*. *Acta Horticulturae*, 691–694.

Haggstrom, H. and Larsson, S. (1995) Slow larval growth on a suboptimal willow results in high predation mortality in the leaf beetle *Galerucella lineola*. *Oecologia* 104, 308–315.

Halwart, M. (1994) The golden apple snail *Pomacea canaliculata* in Asian rice farming systems: present impact and future threat. *International Journal of Pest Management* 40, 199–206.

Harahap, Z. (1979) Breeding for resistance to brown planthopper and grassy stunt virus in Indonesia. In: *Brown Planthopper: Threat to Rice Production in Asia*. International Rice Research Institute, Los Baños, Laguna, Philippines, pp. 201–208.

Hardin, G. (1968) The tragedy of the commons. *Science* 162, 1243–1244.

Hattori, I. and Siwi, S.S. (1986) Rice stemborers in Indonesia. *Japan Agricultural Research Quarterly* 20, 25–30.

Hawkes, C.V. and Sullivan, J.J. (2001) The impact of herbivory on plants in different resource conditions: a meta-analysis. *Ecology* 82, 2045–2058.

Heinrichs, E.A. and Rapusas, H. (1983) Correlation of resistance to the green leafhopper, *Nephotettix virescens* (Homoptera: Cicadellidae) with tungro virus infection in rice varieties having different genes for resistance. *Environmental Entomology* 12, 201–205.

Heinrichs, E.A. and Rapusas, H.R. (1984) Feeding activity of the green leafhopper (GLH) and tungro (RTV) infection. *International Rice Research Newsletter* 9, 15.

Heinrichs, E.A., Medrano, F. and Rapusas, H.R. (1985) *Genetic Evaluation for Insect Resistance in Rice.* International Rice Research Institute, Los Baños, Laguna, Philippines.

Heong, K.L. and Escalada, M.M. (1997) A comparative analysis of pest management practices of rice farmers in Asia. In: Heong, K.L. and Escalada, M.M. (eds) *Pest Management Practices of Rice Farmers in Asia.* International Rice Research Institute, Manila, Philippines, pp. 227–242.

Hilbert, D.W., Swift, D.M., Detling, J.K. and Dyer, M.I. (1981) Relative growth rates and the grazing optimization hypothesis. *Oecologia* 51, 14–18.

Hirano, C. (1964) Nutritional relationship between larvae of *Chilo suppressalis* (Walker) and the rice plant with special reference to role of nitrogen in the nutrition of larvae. *Bulletin National Institute of Agricultural Science Japan, Series C* 17, 103–181.

IRRI (International Rice Research Institute) (1971) *Annual Report for 1970.* Biological control of stemborers. International Rice Research Institute, Los Baños, Laguna, Philippines, pp. 247–248.

IRRI (1980) *Annual Report for 1979.* Food web of the brown planthopper. International Rice Research Institute, Los Baños, Laguna, Philippines, pp. 224–225.

IRRI (1983) *Annual Report for 1982.* Food web of the rice white leafhopper. International Rice Research Institute, Los Baños, Laguna, Philippines, pp. 184–185.

IRRI (1984) *Annual Report for 1983.* Food web of the yellow stem borer. International Rice Research Institute, Los Baños, Laguna, Philippines, pp. 182–183.

IRRI (1986) *Annual Report for 1985.* Green leafhopper food web. International Rice Research Institute, Los Baños, Laguna, Philippines, pp. 156–157.

IRRI (1987) *Annual Report for 1986.* Rice whorl maggot food web. International Rice Research Institute, Los Baños, Laguna, Philippines, pp. 212–213.

IRRI (1988) *Annual Report for 1987.* Whitebacked planthopper food web. International Rice Research Institute, Los Baños, Laguna, Philippines, pp. 238–239.

IRRI (1989) *Annual Report for 1988.* Food web of chilo stem borers. International Rice Research Institute, Los Baños, Laguna, Philippines, pp. 246–247.

IRRI (1990) *Annual Report for 1989.* Rice caseworm food web and natural enemies. International Rice Research Institute, Los Baños, Laguna, Philippines, pp. 179–181.

Ishii, S. and Hirano, C. (1958) Effect of fertilizers on the growth of the larva of the rice stem borer, *Chilo suppressalis* (Walker). I. Growth response of the larvae to the rice plant cultured in different nitrogen level soils. *Japan Journal of Applied Entomology and Zoology* 2, 198–202.

Ishii, S., Hirano, C., Iwata, Y., Nakawasa, M. and Miyagawa, H. (1962) Isolation of benzoic and salicylic acids from the rice plants as growth-inhibiting factor for the rice stem borer, *Chilo suppressalis* (Walker) and some rice plant fungus pathogens. *Japan Journal of Applied Entomology and Zoology* 6, 281–288.

Jahn, G.C. (1992) Effect of neem oil, monocrotophos, and carbosulfan on green leafhoppers, *Nephotettix virescens* (Distant) (Homoptera: Cicadellidae) and rice yields in Thailand. *Proceedings of the Hawaiian Entomological Society* 31, 125–131.

Jahn, G.C. (2004) Effect of soil nutrients on the growth, survival and fecundity of insect pests of rice: an overview and a theory of pest outbreaks with consideration of research approaches. *International Organization for Biological Control Bulletin* 27, 115–122.

Jahn, G.C. and Beardsley, J.W. (1996) Effects of *Pheidole megacephala* (Hymenoptera: Formicidae) on the survival and dispersal of *Dysmicoccus neobrevipes* (Homoptera: Pseudococcidae). *Journal of Economic Entomology* 89, 1124–1129.

Jahn, G.C., Pheng, S., Khiev, B. and Pol, C. (1996) Baseline Survey Report No. 6: farmer's pest management and rice production practices in Cambodian lowland rice. Cambodia–IRRI–Australia Project, Phnom Penh, Cambodia.

Jahn, G.C., Khiev, B., Pheng, S. and Pol, C. (1997a) Pest management in rice. In: Nesbitt, H.J. (ed.) *Rice Production in Cambodia.* International Rice Research Institute, Manila, Philippines, pp. 83–91.

Jahn, G.C., Pheng, S., Khiev, B. and Pol, C. (1997b) Pest management practices of lowland rice farmers in Cambodia. In: Heong, K.L. (ed.) *Pest Management Practices of Rice Farmers in Asia.* International Rice Research Institute, Manila, Philippines, pp. 35–51.

Jahn, G.C., Pheng, S., Khiev, B. and Pol, C. (1998) Pest potential of the golden apple snail in Cambodia. *Cambodian Journal of Agriculture* 1, 34–35.

Jahn, G.C., Cox, P., Solieng, M., Chhorn, N. and Tuy, S. (1999) Rat management in Cambodia. In: Singleton, G., Hinds, L., Leirs, H. and Zhang, Z. (eds) *Ecologically-based Rodent Management.* ACIAR, Canberra.

Jahn, G.C., Sanchez, E.R. and Cox, P.G. (2001) The quest for connections: developing a research agenda for integrated pest and nutrient management. *IRRI Discussion Paper No. 42.* International Rice Research Institute, Los Baños, Laguna, Philippines.

Jahn, G.C., Beardsley, J.W. and González-Hernández, H. (2003) A review of the association of ants with mealybug wilt disease of pineapple. *Proceedings of the Hawaiian Entomological Society* 36, 9–28.

Jahn, G.C, Domingo, I., Almazan, L.P. and Pacia, J. (2004) Effect of rice bugs (Alydidae: *Leptocorisa oratorius* (Fabricius)) on rice yield, grain quality, and seed viability. *Journal of Economic Entomology* 97, 1923–1927.

Jahn, G.C., Almazan, L.P., and Pacia, J.B. (2005). Effect of nitrogen fertilizer on the intrinsic rate of increase of *Hysteroneura setariae* (Thomas) (Homoptera: Aphididae) on rice (*Oryza Sativa* L.) *Environmental Entomology* 34, 938–943.

Javier, E.L. (1997) Rice ecosystems and varieties. In: Nesbitt, H.J. (ed.) *Rice Production in Cambodia.* International Rice Research Institute, Manila, Philippines, pp. 39–81.

Jennings, P.R., Coffman, W.R. and Kauffman, H.E. (1979) *Rice Improvement.* International Rice Research Institute, Los Baños, Laguna, Philippines.

Jiang, Y.H., Fu, Q.A., Cheng, J.A., Ye, G.Y., Bai, Y.Y. and Zhang, Z.T. (2004) Effects of transgenic *Bt* rice on the biological characteristics of *Apanteles chilonis* (Munakata) (Hymenoptera: Braconidae). *Acta Entomologica Sinica* 47, 124–129.

Joshi, R.C. and Venugopal, M.S. (1985) Gall length and its relation to plant height in rice varieties susceptible to gall midge. *Oryza* 22, 132–133.

Kalode, M.B. and Krishna, T.S. (1979) Varietal resistance to brown planthopper in India. In: *Brown Planthopper: Threat to Rice Production in Asia.* International Rice Research Institute, Los Baños, Laguna, Philippines, pp. 187–199.

Kalode, M.B., Sain, M., Bentur, J.S., Pophaly, D.J. and Sreeramulu, M. (1983) New donors and mechanism of resistance to gall-midge in rice grown in the greenhouse. *Indian Journal of Agriculture Science* 53, 483–485.

Kaneda, C. (1982) Breeding approaches for resistance to BPH in Japan. In: *International Rice Research Conference.* International Rice Research Institute, Los Baños, Laguna, Philippines.

Kanno, M. (1962) On the distribution pattern of rice stem borer in the paddy field. *Japanese Journal of Applied Entomology and Zoology* 6, 85–89.

Kartoharjono, A. and Heinrichs, E.A. (1984) Populations of the brown planthopper, *Nilaparvata lugens* (Stål) (Homoptera: Delphacidae), and its predators on rice varieties with different levels of resistance. *Environmental Entomology* 13, 359–365.

Kawabe, S. (1985) Mechanism of varietal resistance to the rice green leafhopper (*Nephotettix cincticeps* Uhler). *Japan Agricultural Research Quarterly* 19, 115–124.

Kawano, T., Saito, T. and Munakata, K. (1968) Study on attractant of the rice stem borer, *Chilo suppressalis* (Walker). (In Japanese with English summary.) *Botyu-Kagaku* 33, 122–130.

Kenmore, P.E., Cariño, F.O., Perez, C.A., Dyck, V.A. and Gutierrez, A.P. (1984) Population regulation of the rice brown planthopper (*Nilaparvata lugens* Stål) within rice fields in the Philippines. *Journal of Plant Protection in the Tropics* 1, 19–37.

Khan, Z.R., Barrion, A.T., Litsinger, J.A., Castillo, N.P. and Joshi, J.C. (1988) Bibliography of rice leaffolders (Lepidoptera: Pyralidae). *Insect Science and Its Application* 9, 25–74.

Khan, Z.R., Litsinger, J.A., Barrion, A.T., Villanueva, F., Fernandez, N. and Taylo, L. (1991) World Bioliograpy of Rice Stemborers. IRRI Philippines and the International Center of Insect Physiology and Ecology, Keaya, 415.

Khiev, B., Jahn, G.C., Pol, C. and Chhorn, N. (2000) Effects of simulated pest damage on rice yields. *International Rice Research Newsletter* 25, 27–28.

Khush, G.S. (1979) Genetics of and breeding for resistance to the brown planthopper. In: *Brown Planthopper: Threat to Rice Production in Asia.* International Rice Research Institute, Los Baños, Laguna, Philippines, pp. 321–332.

Khush, G.S. (1989) Multiple disease and insect resistance for increased yield stability in rice. In: *Progress in Irrigated Rice Research,* International Rice Research Institute, Los Baños, Philippines, pp. 79–92.

Kim, H.S. and Heinrichs, E.A. (1982) Effects of silica on whitebacked planthopper. *International Rice Research Newsletter* 7, 17.

Kimmins, F.M. (1989) Electrical penetration graphs from *Nilaparvata lugens* on resistant and susceptible rice varieties. *Entomologia Experimentalis et Applicata* 50, 69–79.

Kiritani, K. (1979) Pest management in rice. *Annual Review of Entomology* 24, 279–312.

Kogan, M. and Ortman, E.F. (1978) Antixenosis – a new term proposed to define Painter's "nonpreference" modality of resistance. *Bulletin of the Entomological Society of America* 24, 175–176.

Letourneau, D.K., Robinson, G.S. and Hagen, J.A. (2003) Bt crops: predicting effects of escaped transgenes on the fitness of wild plants and their herbivores. *Environmental Biosafety Research* 2, 219–246.

Leung, L.K.P. (1998) *A Review of the Management of Rodent Pests in Cambodian Lowland Rice Fields.* A consultancy report for the Cambodia–IRRI–Australia Project. CSIRO Wildlife and Ecology, Canberra, Australia.

Litsinger, J.A. (1989) Second generation insect problems on high yielding rices. *Tropical Pest Management* 35, 235–242.

Litsinger, J.A. (1991) Crop loss assessment in rice. In: Heinrichs, E.A. and Miller, T.A. (eds) *Rice Insects: Management Strategies.* Springer-Verlag, New York, pp. 1–65.

Litsinger, J.A. (1993) A farming systems approach to insect pest management for upland and lowland rice farmers in tropical Asia. In: Altieri, M.A. (ed.) *Crop Protection Strategies for Subsistence Farmers. Westview Studies in Insect Biology.* Westview Press, Boulder, Colorado, pp. 45–101.

Litsinger, J.A. (1994) Cultural, mechanical, and physical control of rice insects. In: Heinrichs, E.A. (ed.) *Biology and Management of Rice Insects.* Wiley Eastern, New Delhi, pp. 549–584.

Litsinger, J.A. and Chantaraprapha, N. (1995) Developmental biology and host range of *Parapoynx fluctuosalis* and *P. diminutalis* ricefield caseworms. *Insect Science and Its Application* 16, 1–11.

Litsinger, J.A. and Estaño, D.B. (1993) Management of the golden apple snail *Pomacea canaliculata* (Lamarck) in rice. *Crop Protection* 12, 363–370.

Litsinger, J.A., Apostol, R.F. and Obusan, M.B. (1983) White grub, *Leucopholis irrorata* (Coleoptera: Scarabaeidae): pest status, population dynamics, and chemical control in a rice-maize cropping pattern in the Philippines. *Journal of Economic Entomology* 76, 1133–1138.

Litsinger, J.A., Alviola, A.L. III and Canapi, B.L. (1986) Effects of flooding on insect pests and spiders in a rainfed rice environment. *International Rice Research Newsletter* 11, 25–26.

Litsinger, J.A., Barrion, A.T. and D. Soekarna, (1987a) Upland rice insect pests: their ecology, importance, and control. *IRRI Research Paper Series No. 123.*

Litsinger, J.A., Canapi, B.L., Bandong, J.P., dela Cruz, C.G., Apostol, R.F., Pantua, P.C., Lumaban, M.D., Alviola, A.L. III, Raymundo, F., Libetario, E.M., Loevinsohn, M.E. and Joshi, R.C. (1987b) Rice crop loss from insect pests in wetland and dryland environments of Asia with emphasis on the Philippines. *Insect Science and Its Application* 8, 677–692.

Litsinger, J.A., Chantaraprapha, N. and Bandong, J.P. (1993) Alternate weed hosts of the rice caseworm *Nymphula depunctalis* (Guenée) (Lepidoptera: Pyralidae). *Journal of Plant Protection in the Tropics* 10, 63–75.

Litsinger, J.A., Bandong, J.P. and Chantaraprapha, N. (1994a) Mass rearing, larval behaviour, and effects of plant age on the rice caseworm *Nymphula depunctalis* (Guenée) (Lepidoptera: Pyralidae). *Crop Protection* 13, 494–502.

Litsinger, J.A., Chantaraprapha, N., Bandong, J.P. and Barrion, A.T. (1994b) Natural enemies of the rice caseworm *Nymphula depunctalis* (Guenée) (Lepidoptera: Pyralidae). *Insect Science and Its Application* 15, 261–268.

Litsinger, J.A., Barrion, A.T., Bumroongsri, V., Morrill, W.L. and Sarnthoy, O. (1997) Natural enemies of the rice greenhorned caterpillar *Melanitis leda ismene* Cramer (Lepidoptera: Satyridae) and rice skipper *Pelopidas mathias* Fabricius (Lepidoptera: Hesperiidae) in the Philippines. *Philippine Entomologist* 11, 151–181.

Litsinger, J.A., Libetario, E.M. and Barrion, A.T. (2002) Population dynamics of white grubs in the upland rice and maize environment of Northern Mindanao, Philippines. *International Journal of Pest Management* 48, 239–260.

Litsinger, J.A., Libetario, E.M. and Barrion, A.T. (2003) Early planting and overseeding in the cultural control of rice seedling maggot *Atherigona oryzae* Malloch in the Philippines. *International Journal of Pest Management* 49, 57–69.

Litsinger, J.A., Bandong, J.P., Canapi, B.L., dela Cruz, C.G., Pantua, P.C., Alviola, A.L. III and Batay-an, E. (2005) Evaluation of action thresholds against chronic insect pests of rice in the Philippines. I. Less frequently occurring pests and overall assessment. *International Journal of Pest Management* 51, 45–61.

Litsinger, J.A., Heinrichs, E.A., Valencia, S.L. and Feuer, R. (1980) Biological efficacy, cost, and mammalian toxicity of insecticides recommended for rice in the Philippines. *International Rice Research Newsletter* 5, 16.

Loevinsohn, M.E. (1984) The ecology and control of rice pests in relation to the intensity and synchrony of cultivation. PhD thesis, University of London, London.

Loevinsohn, M.E. (1994) Rice pests and agricultural environments. In: Heinrichs, E.A. (ed.) *Biology and Management of Rice Insects.* Wiley Eastern, New Delhi, pp. 487–513.

Lou, Y.G. and Cheng, J.A. (1996) Behavioural responses of *Anagrus nilaparvatae* Pang *et* Wang to the volatiles of rice varieties. *Entomology Journal of East China* 5, 60–64.

Lou, Y.G. and Cheng, J.A. (2003) Role of rice volatiles in the foraging behaviour of the predator *Cyrtorhinus lividipennis* for the rice brown planthopper *Nilaparvata lugens. BioControl* 48, 73–86.

Lu, Z.X., Heong, K.L., Yu, X.P. and Hu, C. (2004) Effects of plant nitrogen on ecological fitness of the brown planthopper *Nilaparvata lugens* Stål in rice. *Journal of Asia-Pacific Entomology* 7, 97–104.

MacLean, R.H., Litsinger, J.A., Moody, K., Watson, A. and Libetario, E.M. (2003) Impact of *Gliricidia sepium* and *Cassia spectabilis* hedgerows on weeds and insect pests of upland rice. *Agriculture Ecosystems and Environment* 94, 275–288.

Mao, C., GongYin, Y., Cui, H., Tu, J. and Datta, S.K. (2003) Effect of transgenic *Bt* rice on dispersal of planthoppers and leafhoppers as well as their egg parasitic wasps. *Journal of Zhejiang University (Agriculture and Life Sciences)* 29, 29–33.

Maischner, H. (1995) *Mineral Nutrition of Higher Plants*, 2nd edn. Academic Press, New York.

Maschinski, J. and Whitham, T.G. (1989) The continuum of plant responses to herbivory: the influence of plant association, nutrient availability, and timing. *American Naturalist* 134, 1–19.

Mattson, W.J. Jr (1980) Herbivory in relation to plant nitrogen content. *Annual Review of Ecology and Systematics* 11, 119–161.

Mew, T.W. (1992) Management of rice diseases: a future perspective. In: Aziz, A., Kadir, S.A. and Barlow, H.S. (eds) *Pest Management and the Environment in 2000.* CAB International, Wallingford, UK, pp. 54–66.

Mishra, B.K., Sontakke, B.K. and Mohapatra, H. (1990) Antibiosis mechanisms of resistance in rice varieties to yellow stem borer *Scirpophaga incertulas* (Walker). *Indian Journal of Plant Protection* 18, 81–83.

Miyashita, T. (1985) Estimation on the economic injury level in the rice leafroller *Cnaphalocrocis medinalis* Guenee (Lepidoptera: Pyralidae). I. Relation between yield loss and injury of rice leaves at heading or in the grain filling period. *Japanese Journal of Applied Entomology and Zoology* 29, 73–76.

Mochida, O.M., Joshi, R.C. and Litsinger, J.A. (1987) Climatic factors affecting the occurrence of insect pests. In: *Weather and Rice.* International Rice Research Institute, Manila, Philippines, pp. 149–164.

Munakata, K., Saito, S., Ogawa, T. and Ishii, S. (1959) Oryzanone, an attractant to the rice stem borer. *Bulletin of Agricultural Chemical Society of Japan* 23, 64–65.

Myint, M.M., Rapusas, H.R. and Heinrichs, E.A. (1986) Integration of varietal resistance and predation for the management of *Nephotettix virescens* (Homoptera: Cicadellidae) populations on rice. *Crop Protection* 5, 259–265.

Nadarajan, L. and Pillai, J.S. (1985) The role of crop nutrients in pest incidence. In: *Role of Potassium in Crop Resistance to Rice Pests.* Research Review Series 3, Potash Research Institute of India, Gurgaon, Haryana, India, pp. 35–41.

Nemoto, H. and Yokoo, M. (1994) Experimental selection of a brown planthopper population on mixtures of resistant rice lines. *Breeding Science* 44, 133–136.

Nesbitt, H.J. and Chan, P. (1997) Rice-based farming systems. In: Nesbitt, H.J. (ed.) *Rice Production in Cambodia.* International Rice Research Institute, Manila, Philippines, pp. 31–37.

Omoloye, A.A., Odebiyi, J.A., Williams, C.T. and Singh, B.N. (2002) Tolerance indicators and responses of rice cultivars to infestation by the African rice gall midge, *Orseolia oryzivora. Journal of Agricultural Science* 139, 335–340.

Ooi, P.A.C. and Shepard, B.M. (1994) Predators and parasitoids of rice insect pests. In: Heinrichs, E.A. (ed.) *Biology and Management of Rice Insects.* Wiley Eastern, New Delhi, pp. 584–612.

Oya, S. and Suzuki, T. (1971) Studies on the population increase of green rice leafhopper *Nephotettix cincticeps* Uhler. II. The larval growth and oviposition on rice plants under controlled sunshine and nitrogenous fertilizer. *Proceedings of Association of Plant Protection, Hokuriku* 19, 45–49.

Padhi, G. and Sen, P. (2002) Evaluation of wild rice species against yellow stem borer, *Scirpophaga incertulas* (Walker). *Journal of Applied Zoological Researches* 13, 147–148.

Painter, R.H. (1951) *Insect Resistance in Crop Plants.* University of Kansas Press, Lawrence, Kansas.

Pantua, P.C. and Litsinger, J.A. (1981) Arthropod abundance in continuous vs biannual rice cropping systems. *International Rice Research Newsletter* 6, 20–21.

Patanakamjorn, S. and Pathak, M.D. (1967) Varietal resistance of rice to Asiatic rice borer *Chilo suppressalis* (Lepidoptera: Crambidae) and its association with various plant characters. *Annals of the Entomological Society of America* 60, 287–292.

Pathak, M.D. (1968) Ecology of common insect pests of rice. *Annual Review of Entomology* 13, 257–294.

Pathak, P.K. and Heinrichs, E.A. (1982) Selection of biotype population-2 and population-3 of *Nilaparvata lugens* (Homoptera: Delphacidae) by exposure to resistant rice varieties. *Environmental Entomology* 11, 85–90.

Pathak, M.D., Andres, F., Galacgac, N. and Raros, R. (1971) Resistance of rice varieties to striped stem borers. *International Rice Research Institute, Technical Bulletin*, 11.

Peraiah, A. and Roy, J.K. (1979) Studies on biochemical nature of gall midge resistance in rice. *Oryza* 16, 149–150.

Pingali, P.L., Moya, P.F. and Velasco, L.E. (1990) The post-green revolution blues in Asian rice production. *IRRI Social Science Division Papers 90*, 1–33.

Pongpraset, S. and Weerapat, P. (1979) Varietal resistance to the brown planthopper in Thailand. In: *Brown Planthopper: Threat to Rice Production in Asia*. International Rice Research Institute, Los Baños, Laguna, Philippines, pp. 273–283.

Prakash, N.B. (1999) Recycling plant silicon in rice farming. *Far Eastern Agriculture* (Sept./Oct. 1999), 25–26.

Preap, V., Zalucki, M.P., Nesbitt, H.J. and Jahn, G.C. (2001) Effect of fertilizer, pesticide treatment, and plant variety on realized fecundity and survival rates of *Nilaparvata lugens* (Stål); generating outbreaks in Cambodia. *Journal of Asia-Pacific Entomology* 4, 75–84.

Price, P.W. (1986) Ecological aspects of host plant resistance and biological control: interactions among three trophic levels. In: Boethel, D.J. and Eikenbary, R. (eds) *Interactions of Plant Resistance and Parasitoids and Predators of Insects*. Ellis Horwood, Chichester, UK, pp. 11–30.

Price, P.W., Bouton, C.E., Gross, P., McPheron, B.A., Thompson, J.N. and Weis, A.E. (1980) Interactions among three trophic levels: influence of plants on interactions between insect herbivores and natural enemies. *Annual Review of Ecology and Systematics* 11, 41–65.

Rapusas, H.R., Bottrell, D.G. and Coll, M. (1996) Intraspecific variation in chemical attraction of rice to insect predators. *Biological Control* 6, 394–400.

Reddy, D.B. (1967) The rice gall midge *Pachydiplosis oryzae* (Wood-Mason). In: *The Major Insect Pests of the Rice Plant*. Johns Hopkins Press, Baltimore, Maryland, pp. 457–491.

Reichardt, W., Dobberman, A. and Georg, T. (1998) Intensification of rice production systems: opportunities and limits. In: Dowling, N.G., Greenfield, S.M. and Fischer, K.S. (eds) *Sustainability of Rice in the Global Food System*. Pacific Basin Study Center, Davis, California and International Rice Research Institute, Manila, pp. 127–144.

Ren, X., Weng, Q.M., Zhu, L.L. and He, G.C. (2004) Dynamic mapping of quantitative trait loci for brown planthopper resistance in rice. *Cereal Research Communications* 32, 31–38.

Rombach, M.C., Roberts, D.W. and Aguda, R.M. (1994) Pathogens of rice pests. In: Heinrichs, E.A. (ed.) *Biology and Management of Rice Insects*. Wiley Eastern, New Delhi, pp. 611–655.

Rosenheim, J.A. (1998) Higher-order predators and the regulation of insect herbivore populations. *Annual Review of Entomology* 43, 421–447.

Rothschild, G.H.L. (1970) Observations on the ecology of the rice-ear bug *Leptocorisa oratorius* (F.) (Hemiptera: Alydidae) in Sarawak (Malaysian Borneo). *Journal of Applied Ecology* 7, 147–167.

Rubia, E.G. (1994) The pest status and management of white stem borer *Scirpophaga innotata* (Walker) (Pyralidae: Lepidoptera) in West Java, Indonesia. PhD thesis, University of Queensland, Brisbane, Australia.

Rubia, E.G., Ferrer, E.R. and Shepard, B.M. (1990) Biology and predatory behavior of *Conocephalus longipennis* (de Haan) (Orthoptera: Tettigoniidae). *Journal of Plant Protection in the Tropics* 7, 47–54.

Rubia, E.G., Heong, K.L., Zalucki, M., Gonzales, B. and Norton, G.A. (1996) Mechanisms of compensation of rice plants to yellow stem borer, *Scirpophaga incertulas* (Walker), injury. *Crop Protection* 15, 335–340.

Rubia-Sanchez, E., Suzuki, Y., Miyamoto, K. and Watanabe, T. (1999) The potential for compensation of the effects of the brown planthopper *Nilaparvata lugens* Stål (Homoptera: Delphacidae) feeding on rice. *Crop Protection* 18, 39–45.

Sands, D.P.A. (1977) *The Biology and Ecology of Leptocorisa (Hemiptera, Alydidae) in Papua New Guinea*. Research Bulletin No. 18, Department of Primary Industry, Port Moresby, Papua New Guinea.

Savary, S., Elazequi, F.A., Moody, K., Litsinger, J.A. and Teng, P.S. (1994) Characterization of rice cropping practices and multiple pest systems in the Philippines. *Agricultural Systems* 46, 385–408.

Savary, S.F., Castillia, N.P., Elazegui, F.A., McLaren, C.G., Ynalvez, M.A. and Teng, P.S. (1995) Direct and indirect effects of nitrogen supply and disease source structure on rice sheath blight spread. *Phytopathology* 85, 959–965.

Savary, S., Willocquet, L., Elazegui, F.A., Teng, P.S., Du, P.V., Tang, D.Z.Q., Huang, S., Lin, X., Singh, H.M. and Srivastava, R.K. (2000a) Rice pest constraints in tropical Asia: characterization of injury profiles in relation to production situations. *Plant Disease* 84, 341–356.

Savary, S., Willocquet, L., Elazegui, F.A., Castillia, N.P. and Teng, P.S. (2000b) Rice pest constraints in tropical Asia: quantification of yield losses due to rice pests in a range of production situations. *Plant Disease* 84, 357–369.

Saxena, R.C. (1986) Biochemical-bases of insect resistance in rice varieties. *ACS Symposium Series* 296, 142–159.

Saxena, R.C. and Barrion, A.A. (1985) Biotypes of the brown planthopper, *Nilaparvata lugens* (Stål) and strategies in deployment of host plant resistance. *Insect Science and Its Application* 6, 271–289.

Saxena, R.C., Velasco, M.V. and Barrion, A.A. (1983) Interspecfic hybridization between *Nilaparvata lugens* (Stål) and *Nilaparvata bakeri* (Muir) collected from *Leersia hexandra Swartz*. *International Rice Research Newsletter* 8, 13–14.

Schoenly, K.G., Cohen, J.E., Heong, K.L., Litsinger, J.A., Aquino, G.B., Barrion, A.T. and Arida, G. (1996) Food web dynamics of irrigated rice fields at five elevations in Luzon, Philippines. *Bulletin of Entomological Research* 86, 451–466.

Schoenly, K.G., Cohen, M.B., Barrion, A.T., WenJun, Z., Gaolach, B. and Viajante, V.D. (2003) Effects of *Bacillus thuringiensis* on non-target herbivore and natural enemy assemblages in tropical irrigated rice. *Environmental Biosafety Research*, 2, 181–206.

Seino, Y., Suzuki, Y. and Sogawa, K. (1996) An ovicidal substance produced by rice plants in response to oviposition by the whitebacked planthopper, *Sogatella furcifera* (Horvath) (Homoptera: Delphacidae). *Applied Entomology and Zoology*, 31, 467–473.

Seko, H. and Kato, I. (1950) Studies on the resistance of rice plant against the attack of rice stem-borer (*Chilo simplex*). 1. Interrelation between the plant character and the frequency of egg-laying in the first generation of rice-stem borer. *Proceedings of the Crop Science Society of Japan* 19, 201–203.

Senguttuvan, T. and Gopalan, M. (1990) Predator efficiency of mirid bugs (*Cyrtorhinus lividipennis*) on eggs and nymphs of brown planthoppers (*Nilaparvata lugens*) on resistant and susceptible varieties of rice (*Oryza sativa*). *Indian Journal of Agricultural Science* 60, 285–287.

Settle, W.H., Hartiahyo Ariawan, Endah Tri Astuti, Widyastama Cahyana, Arief Lukman Hakim, Dadan Hindayana, Alifah Sri Lestari, Pajarningsih and Sartanto (1996) Managing tropical rice pests through conservation of generalist natural enemies and alternative prey. *Ecology* 77, 1975–1988.

Shahjahan, M. and Hossain, M. (2003) Identification of the morphological characters influencing the infestation rate of yellow stem borer. *Pakistan Journal of Scientific and Industrial Research* 46, 33–42.

Sheehy, J.E., Dionora, M.J.A., Mitchell, P.L., Peng, S., Cassman, K.G., Lemaire, G. and Williams, R.L. (1998) Critical nitrogen concentrations: implications for high-yielding rice (*Oryza sativa* L.) cultivars in the tropics. *Field Crops Research* 59, 31–41.

Shepard, B.M., Justo H.D. Jr, Rubia, E.G. and Estaño, D.B. (1990) Response of the rice plant to damage by the rice whorl maggot *Hydrellia philippina* Ferino (Diptera: Ephydridae). *Journal of Plant Protection in the Tropics* 7, 173–177.

Shigematsu, Y., Murofushi, N., Ito, K., Kaneda, C., Kawabe, S. and Takahashi, N. (1982) Sterols and asparagine in the rice plant, endogenous factors related to resistance against the brown planthopper (*Nilaparvata lugens*). *Agricultural and Biological Chemistry* 46, 2877–2879.

Shufran, K.A. and Whalon, M.E. (1995) Genetic analysis of brown planthopper biotypes using random amplified polymorphic dna-polymerase chain reaction (RAPD-PCR). *Insect Science and Its Application* 16, 27–33.

Singleton, G.R. (1997) Integrated management of rodents: a Southeast Asian and Australian perspective. *Belgium Journal of Zoology* 127 (Supplement 1), 157–169.

Singleton, G.R. and Petch, D.A. (1994) *A Review of the Biology and Management of Rodent Pests in Southeast Asia*. ACIAR, Canberra.

Soejitno, J. (1979) Some Notes on the Larval Behavior of *Tryporyza incertulas* (Walker) (Lepidoptera: Pyralidae). Kongres Entomologi I, Jakarta, Indonesia, 9–11 January 1979.

Sogawa, K. (1982) The rice brown planthopper: feeding physiology and host-plant interactions. *Annual Review of Entomology* 27, 49–73.

Sogawa, K. and Pathak, M.D. (1970) Mechanisms of brown planthopper resistance in Mudgo variety of rice (Hemiptera: Delphacidae). *Applied Entomology and Zoology* 5, 145–158.

Stapley, J.H., Way, Y.Y. and Golden, W.G. (1979) Varietal resistance to the brown planthopper *Nilaparvata lugens* (Stål) in the Solomon Islands. In: *Brown Planthopper: Threat to Rice Production in Asia*. International Rice Research Institute, Los Baños, Laguna, Philippines, pp. 233–239.

Subbarao, D.V. and Perraju, A. (1976) Resistance in some rice strains to first-instar larvae of *Tryporyza incertulas* (Walker) in relation to plant nutrients and anatomical structure of the plants. *International Rice Research Newsletter* 1, 14–15.

Suzuki, Y., Sogawa, K. and Seino, Y. (1996) Ovicidal reaction of rice plants against the whitebacked planthopper, *Sogatella furcifera* Horvath (Homoptera: Delphacidae). *Applied Entomology and Zoology* 31, 111–118.

Tabashnik, B.E. (1992) Resistance risk assessment: realized heritability of resistance to *Bacillus thuringiensis* in diamondback moth (Lepidoptera: Noctuidae), and Colorado potato beetle (Coleoptera: Chrysomelidae). *Journal of Economic Entomology* 85, 1551–1559.

Taylor, B. (1988) The impact of yellow stem-borer, *Scirpophaga incertulas* (Walker) (Lepidoptera: Pyralidae), on deep-water rice, with special reference to Bangladesh. *Bulletin of Entomological Research* 78, 209–225.

Theunis, W., Aguda, R.M., Cruz, W.T., Decock, C., Peferoen, M., Lambert, B., Bottrell, D.G., Gould, F.L., Litsinger, J.A. and Cohen, M.B. (1998) *Bacillus thuringiensis* isolates from the Philippines: habitat distribution, alpha-endotoxin diversity, and toxicity to rice stem borers (Lepidoptera: Pyralidae). *Bulletin of Entomological Research* 88, 335–342.

Tirumala Rao, V. (1952) The paddy root weevil (*Echinocnemus oryzae* Mshll.): a pest of paddy in the Deltaic Tracts of the Northern Circars. *Indian Journal of Entomology* 14, 31–37.

Uthamasamy, S., Velu, V.M., Gopalan, M. and Ramanathan, K.M. (1983) Incidence of brown planthopper *Nilaparvata lugens* Stål on IR50 at graded levels of fertilization at Aduthurai. *International Rice Research Newsletter* 8, 13.

van den Berg, H., Shepard, B.M., Litsinger, J.A. and Pantua, P.C. (1988) Impact of predators and parasitoids on the eggs of *Rivula atimeta, Naranga aenescens* (Lepidoptera: Noctuidae) and *Hydrellia philippina* (Diptera: Ephydridae) in rice. *Journal of Plant Protection in the Tropics* 5, 103–108.

van den Bosch, R. (1980) *The Pesticide Conspiracy*. Anchor Books, Garden City, New York.

van der Goot, P. (1925) Levenswijze en bestrijding van de witte rijstboorder op Java (Life history and control of the white rice-borer in Java). *Mededelingen Institute Plziekten Buitenzorg* No. 66, 1–308 (Dutch with English summary).

Vet, L.E.M. and Dicke, M. (1992) Ecology of infochemical use by natural enemies in a tritrophic context. *Annual Review of Entomology* 37, 141–172.

Viajante, V. and Heinrichs, E.A. (1986a) Rice growth and yield as affected by the whorl maggot *Hydrellia philippina* Ferino (Diptera: Ephydridae). *Crop Protection* 5, 176–181.

Viajante, V. and Heinrichs, E.A. (1986b) Influence of plant density on oviposition by whorl maggot. *International Rice Research Newsletter* 10, 19.

von Uexkuell, H.R. and Beaton, J.D. (1992) A review of fertility management of rice soils. In: Kimble, J.M. (ed.) *Characterization, Classification and Utilization of Wet Soils*. Proceedings of the 8th International Soil Correlation Meeting (VIII. ISCOM). Soil Conservation Service. United States Department of Agriculture, Lincoln, Nebraska, pp. 288–300.

Wang, X.L., Ren, X., Zhu, L.L. and He, G.C. (2004) OsBi1, a rice gene, encodes a novel protein with a CBS-like domain and its expression is induced in responses to herbivore feeding. *Plant Science* 166, 1581–1588.

Way, M.J. and Heong, K.L. (1995) The role of biodiversity in the dynamics and management of insect pests of tropical irrigated rice – a review. *Bulletin of Entomological Research* 84, 567–587.

Way, M.J., Islam, Z., Heong, K.L. and Joshi, R.C. (1998) Ants in tropical irrigated rice: distribution and abundance especially of *Solenopsis geminata* (Hymenoptera: Formicidae). *Bulletin of Entomological Research* 88, 467–476.

Wolfenbarger, L.L. and Phifer, P.R. (2000) The ecological risks and benefits of genetically engineered plants. *Science* 290, 2088.

Woodhead, S. and Padgham, D.E. (1988) The effect of plant surface characteristics on resistance of rice to the brown planthopper, *Nilaparvata lugens*. *Entomologia Experimentalis et Applicata* 47, 15–22.

Xu, X.F., Mei, H.W., Luo, L.J., Cheng, X.N. and Li, Z.K. (2002) RFLP-facilitated investigation of the quantitative resistance of rice to brown planthopper (*Nilaparvata lugens*). *Theoretical and Applied Genetics* 104, 248–253.

Yamasaki, M., Tsunematsu, H., Yoshimura, A., Iwata, N. and Yasui, H. (1999) Quantitative trait locus mapping of ovicidal response in rice (*Oryza sativa* L.) against whitebacked planthopper (*Sogatella furcifera* Horvath). *Crop Science* 39, 1178–1183.

Yang, H.Y., Ren, X.A., Weng, Q.M., Zhu, L.L. and He, G.G. (2002) Molecular mapping and genetic analysis of a rice brown planthopper (*Nilaparvata lugens* Stål) resistance gene. *Hereditas* 136, 39–43.

Yoshihara, T., Sogawa, K., Pathak, M.D., Juliano, B.O. and Sakamura, S. (1979) Soluble silicic acid as a sucking inhibitory substance in rice against the brown planthoppper (Delphacidae: Homoptera). *Entomologia Explicata et Applicata* 26, 314–322.

Yoshihara, T., Sogawa, K., Pathak, M.D., Juliano, B.O. and Sakamura, S. (1980) Oxalic acid as a sucking inhibitor of the brown planthopper in rice (Delphacidae: Homoptera). *Entomologia Explicata et Applicata* 27, 149–155.

Yuan, H.Y., Chen, X.P., Zhu, L. and He, G.C. (2004) Isolation and characterization of a novel rice gene encoding a putative insect-inducible protein homologous to wheat Wir1. *Journal of Plant Physiology* 161, 79–85.

Zhang, F., Zhu, L. and He, G.C. (2004) Differential gene expression in response to brown planthopper feeding in rice. *Journal of Plant Physiology* 161, 53–62.

16 Ecologically Based Integrated Pest Management in Cotton

Dale W. Spurgeon

United States Department of Agriculture, Agricultural Research Service,
Areawide Pest Management Research Unit, 2771 F and B Road, College Station,
TX 77845, USA

Introduction

Volumes have been written about integrated pest management (IPM) in cotton. Beyond the earliest reports that established the various conceptual frameworks of IPM, most works on the topic fall into two categories: (i) historical accounts of events, case histories or developments; or (ii) elaborations on specific ecological, economic or applied aspects of IPM. The following is neither. Rather, this chapter attempts to reconcile the philosophy surrounding ecologically based IPM with the more practical aspects of cotton pest management. The context for this task is set by briefly addressing the diversity of concepts regarded as IPM. Next is an examination of some real-world constraints that place limits on efforts to incorporate ecological principles into the practice of cotton production. Finally, case studies of selected management tactics are used to illustrate some characteristics of successful implementations of ecological principles in cotton IPM. Although the concepts discussed are well established, they are presented from a vantage point that reflects the author's personal experience. However, most, if not all, of the concepts are relevant to production systems or regions beyond those that are discussed.

IPM, by definition, denotes complexity. It is this complexity that makes coherent treatment of the ecological or behavioural aspects of IPM in cotton a difficult task. In many circles, IPM has become synonymous with management of *arthropod* pests. That perception was not the intent of the original progenitors of IPM. In keeping with the author's area of expertise, the emphasis on arthropod pests is echoed in this chapter. However, the constraints, approaches and principles espoused herein have application to classes of pests beyond the insects and their allies.

IPM: The Contrast Between Philosophy and Reality

There is little purpose in presenting an overly optimistic account of the prospects for development and implementation of ecologically or behaviourally oriented pest management tactics in cotton. This is not because such goals are not worthwhile or possible, but because a critical assessment is more useful. In IPM, as in other endeavours, the first and most important step in problem solving is accurate problem identification and delineation. By illustrating how little has been accomplished, how temporary some of those accomplishments have been and some of the constraints limiting our progress, future efforts will perhaps be better framed in reality. Our best opportunity for advancement comes when the problem is identified and the approach is chosen with real-world constraints in mind. Given the incredible diversity of management approaches represented as IPM, a logical first step is to review its definition.

IPM, sustainable agriculture and other nebulous terms

The definitions of IPM

Any assessment of the status of IPM in cotton is dependent on the definition of IPM. That confusion and disagreement exists regarding the status and even the desirable form of IPM is evidenced by the diversity of definitions. For example, Kogan and Bajwa (1999) documented 67 definitions proposed over a 40-year period. If IPM is viewed as the optimal integration of multiple tactics aimed at maintaining or increasing the profitability of production while avoiding undesirable impacts on non-target organisms and the environment, it has not been practised to any appreciable extent. This viewpoint was previously expressed regarding assessments of the degree of IPM adoption by United States agriculture (Ehler and Bottrell, 2000; Prokopy et al., 2000). As supporting evidence, Ehler and Bottrell (2000) cite the virtual absence of tactical integration, continued reliance on conventional pesticides, level of inadequacy of our ecological understanding and operational shortcomings of economic thresholds (ETs). Hutchins (1995) adopted a more pragmatic expectation for IPM when he noted that the failure to more completely develop and implement IPM was, to some degree, a matter of perception resulting from a failure to distinguish between theory and the real world.

Definitions of IPM often arise from personal views regarding what types of tactics are, or are not, characteristic of 'true' IPM. Many of these definitions are quite focused,

specifically stating goals such as minimizing reliance on pesticides and maximizing the contribution of biological (Thomas, 1999) or non-chemical techniques (Schaller, 1993), or qualifying IPM with descriptors such as 'biointensive' (Benbrook, 2000) or 'sustainable' (Hilbeck, 2001). While these goals have a place in the overall context of IPM, as definitions they tend to be restrictive. Thus, one of the major difficulties in developing a more universally accepted definition of IPM springs from the lack of consensus regarding 'appropriate' goals.

Sustainable agriculture and its role in IPM

Sustainable agriculture is closely related to IPM, both conceptually and from the standpoint of implementation (Kogan, 1998). Similar to IPM, definitions of sustainable agriculture have encompassed a variety of connotations, including organic farming, low-input farming and alternative agriculture (Geng et al., 1990). Thus, neither the term 'sustainable agriculture' nor the practices that constitute it lend themselves to precise definition (Schaller, 1993). As in the case of IPM, the more specialized concepts of sustainable agriculture tend to focus on alternatives to modern conventional agriculture, using a high level of ecological sophistication to eliminate adverse effects on the environment and society in general. Most schemes proposed to attain these goals seem too simplistic for successful application to large-scale, highly productive, highly mechanized agricultural systems. In addition, proponents tend to ignore or downplay that these schemes are often incompatible with the needs and goals of modern, economically viable production systems.

In its more general form, sustainable agriculture may be described as any system that provides for the profitable production of crops while minimizing adverse environmental impacts and utilizing natural resources to ensure agricultural production indefinitely. Conceptually, such a description is similar to the original concepts of IPM, retaining a focus on the entire agricultural system rather than on pests per se. Never the less, IPM and sustainable agriculture share more similarities than differences, and efforts to define either with precision serve mainly to place limits on the corresponding concepts and approaches.

Conceptual variants or alternatives to IPM

Other variants on the theme of integrated management include ecologically based pest management proposed by the National Research Council's Board of Agriculture (Kogan, 1998) and integrated crop management (Meerman *et al.*, 1996). Conceptually, these two terms are indistinguishable from existing ideas of IPM and sustainable agriculture. One paradigm that has managed to maintain an identity distinct from IPM is areawide pest management. The distinctions between IPM and areawide management have generally been exaggerated by the personalities involved. A commonly referenced distinction is the different scale of control tactic organization typical of IPM (field-by-field) and areawide management (community- or region-wide). This distinction is particularly evident when organized eradication programmes are considered representative of areawide management. However, eradication programmes are perhaps not the most fitting examples of areawide management. For example, Myers *et al.* (1998) explicitly distinguished areawide management from eradication, suggesting that the former was more sustainable than the latter. Further, Reynolds *et al.* (1975) provide numerous examples of IPM practices that are decidedly areawide in nature. Kogan (1998) suggests that the areawide approach provides opportunities for 'ecosystem level IPM'. Therefore, in the development of ecologically based IPM, it seems more useful to include areawide management as a potential IPM strategy than to exclude it semantically.

Regardless of the definition of IPM, the need for a holistic or systems approach is generally recognized. What is usually lacking is a specific, achievable, real-world plan for developing this approach. However, truly holistic approaches are complex and system-specific. The paradox is that definitions of IPM restricted to such approaches serve to impede progress. This hindrance arises

partly from our inadequate understanding of system ecological details and partly from the lack of conceptual bridges between existing production systems and those that are more ecologically sophisticated. Major accomplishments in the development and implementation of true IPM will most likely arise from a series of smaller, less obvious accomplishments. Therefore, IPM should be inclusive, both conceptually and tactically, in order to provide the flexibility to adapt to the biotic, sociological and economic constraints of modern crop production systems. Using this approach, IPM is viewed less as a real entity and more as a conceptual ideal we may approach as our ecological knowledge increases and real-world constraints are overcome.

Historical goals and motivations of cotton IPM

Pesticide resistance and management by crisis

Historically, cotton has been a pesticide-intensive crop. Pest control initially afforded by modern synthetic pesticides was described as 'spectacular', and the effectiveness of these pesticides diminished interest in alternatives (Smith, 1978). However, the rapid development of insecticide resistance illustrated the shortcomings of this single-tactic management system. Development of resistance was countered with increased application rates and shortened treatment intervals, followed by substitution with new pesticides. Responses to these actions included increased pesticide tolerance and pest resurgence, release of secondary pests and corresponding increases in production costs (Luckmann and Metcalf, 1975). The severe economic and ecological consequences of pesticide resistance, particularly well illustrated in cotton, provided a powerful incentive to develop alternative management philosophies such as IPM. In practice, the development and adoption of IPM has followed a pattern of 'management by crisis', in which alternatives to synthetic insecticides were most heavily favoured during intervals in which affordable and effective insecticides were unavailable (Luttrell, 1994). Although the promotion of treatment thresholds combined with regular monitoring (scouting) has lessened the frequency and scope of catastrophic insecticide failures, cotton remains a relatively high pesticide use crop (Pimentel, 1978; Pimentel *et al.*, 1993). In fact, insecticides remain the primary control tactic in most cotton production systems (Luttrell *et al.*, 1994). Consequently, numerous pests including the tobacco budworm, *Heliothis virescens* (Fabricius), tarnished plant bug, *Lygus lineolaris* (Palisot de Beauvois), cotton aphid, *Aphis gossypii* Glover, and silverleaf whitefly, *Bemisia argentifolii* Bellows and Perring, have developed resistance to one or more classes of insecticide (Elzen and Hardee, 2003). Continued reliance on insecticides necessitates continued reactions to resistance when it arises. Episodes of resistance have largely been met by registration of new toxicants and emphasis on better management of pesticides, but the new materials tend to be more expensive than those they replace. Thus, insecticide resistance and associated problems with secondary pests continue to increase the cost of pest control in cotton (Luttrell, 1994).

Considerable emphasis has been placed on the management of resistance in an effort to extend the usefulness of available chemical controls (Luttrell, 1994; Elzen and Hardee, 2003). Development of resistance is not unique to chemical insecticides (Pedigo and Higley, 1992). In particular, resistance management has become a focus of efforts to ensure sustainability of the newly available transgenic cottons expressing the insecticidal proteins of *Bacillus thuringiensis* Berliner (Bt) (Tabashnik, 1997; Hilbeck, 2001).

In some ways, research priorities related to management of resistance to pesticides or transgenic plants inhibit the development and implementation of more ecologically or behaviourally based management tactics because they enhance the sustainability of pesticide-based management systems. However, this quest for sustainability is justified in that effective, economical and easily implemented biologically based replace-ments for most uses of synthetic pesticides in cotton are largely unavailable.

Pesticide reduction and environmental protection

In addition to the obvious production difficulties caused by resistance to pesticides, deleterious environmental effects of heavy and widespread pesticide usage have prompted support for IPM from policymakers and the public. Despite the well-documented damage to wildlife and non-target organisms (Pimentel, 1978), most public concern stems from the potential health hazards posed by chemical pesticides (Schaller, 1993). Zilberman *et al.* (1991), citing the Environmental Protection Agency (EPA) Science Advisory Board, reported that worker occupational hazards and contamination of drinking water were among the major human health risks from pesticides in the USA, while risks from pesticide exposure in food were primarily problems of perception. Whether specific concerns are real or perceived, Geng *et al.* (1990) suggest that concern for the environmental effects of agricultural chemicals have turned public opinion against agriculture. Ragsdale (2000) expressed similar sentiments, stating that agriculture has not made the public sufficiently aware of the benefits of pesticides or of the impacts of their loss.

Environmental risks posed by pesticides are most often addressed through regulatory policies including bans, use restrictions, pesticide fees and requirements for protective clothing and application standards (Zilberman *et al.*, 1991). While these measures may have reduced the most acute environmental impacts of pesticides, regulation cannot address all environmental hazards (Higley and Wintersteen, 1992). Efforts of the agricultural community to reduce the environmental impacts of pesticides have mainly focused on minimizing effects on non-target organisms, especially natural enemies of pests. One approach to this goal has been to seek pesticides with more selective activity, or to improve application timing in order to reduce effects on natural enemies (Landis *et al.*, 2000). This approach has its own unique limitations. For example, selectivity is relative. Although selectivity based on reduced dosages may not result in rapid decreases in populations of beneficial or other non-target arthropods, behavioural influences, and effects on population dynamics caused by sublethal effects, are poorly known. Where the basis of selectivity is toxicological and the material is more toxic to the target pest than to natural enemies, adverse impacts may still occur if the pest population constitutes an important food resource for the natural enemies.

An alternate strategy for reducing dependence on pesticide use focuses on the development of 'biopesticides' or so-called 'biorational' pesticides. These materials may include living (usually pathogenic) organisms or non-living 'natural products' obtained through metabolic processes or extraction. Few of these materials are yet legitimate competitors of conventional pesticides because of one or more problems including inconsistent field results, limited shelf life, unfavourable costs or excessive selectivity (Copping and Menn, 2000). Although selected natural products may eventually find important roles in cotton IPM, this approach should be pursued with the knowledge that their 'natural' origin provides no assurance that they will be safer or less environmentally disruptive than current synthetic insecticides (Geng *et al.*, 1990; Lewis *et al.*, 1997; Copping and Menn, 2000). One likely exception is the relatively new Bt cottons (Perlak *et al.*, 2001). However, use of this technology is not without concern for potential impacts (Hilbeck, 2001).

Although no one would seriously question that reducing pesticide use in cotton and other crops should be a high-priority goal, there is currently no clear path to this end. In reference to the goal of substantial reduction or elimination of conventional pesticides in agricultural production, Gianessi (1993) points out that available alternatives to conventional pesticides are limited, and the prospects for development of substitutes for all uses of pesticides are not realistic.

Sustainability of farming profitability

Because of the political and philosophical popularity of potential environmental benefits of IPM, it is often overlooked that the prime motivation for development and implementation of IPM is sustainability of farming profitability. Except in the cases of legislative or regulatory action (usually quarantines, pesticide bans or use restrictions), the need for, and extent of, adoption of IPM practices is determined by the end user (cotton producer). This is why undesirable effects of pesticides with immediate financial implications, like development of pesticide resistance, have elicited more producer interest in IPM than have more 'global' effects such as environmental impacts. The economic conditions surrounding modern agriculture are harsh, with production costs continuing to climb amid flat or decreasing commodity prices (Pedigo, 1995; Prokopy *et al.*, 2000). These conditions have provided incentive for increased production, and corresponding increases in the use of agrichemicals (Grumbly, 1978; Pedigo, 1995). Regulatory effects intended to stabilize cotton production, such as price supports and acreage controls, have also encouraged maximum production, increased pesticide use and increased production costs (Pimentel, 1978). Although farm policy, crop insurance, subsidy and set-aside programmes frequently change, they generally continue to favour higher inputs.

The financial consequences of management cause farmers to be risk-averse (Gurr *et al.*, 2003). This aversion has at least two important consequences. First, pest management tactics that are perceived as successful are aimed at maintaining pest population levels well below damaging levels (Pedigo and Higley, 1992), and producer confidence is important in the selection of pest control tactics (Hutchins and Gehring, 1993). Because conventional pesticides are generally effective, reliable and familiar, their use instils a high level of confidence. Second, producers hesitate to make substantial investments in integration and implementation of management alternatives to pesticides

(Prokopy *et al.*, 2000). Often, such investments are not compatible with short-term continuance of the farming operation.

Some authors have argued that economic evaluations of production systems based on direct inputs and outputs should be modified to reflect environmental benefits of pesticide alternatives to producers and society (Geng *et al.*, 1990; Schaller, 1993). However, there is neither a mechanism nor a strong non-regulatory incentive for the producer to absorb the costs of environmental protection. In theory, a degree of producer altruism could be tolerable provided decreases in production resulting from reduced pesticide inputs are offset by increased commodity prices. However, an increase in crop value would, in effect, increase the corresponding economic losses from a given pest population level, which would then provide an incentive to increase production levels through increased chemical inputs.

Appropriate goals for ecologically based cotton IPM

Progress towards integrating ecologically based tactics into cotton IPM has been modest in part because the diversity of definitions of IPM has resulted in divergent goals. Much of the effort spent towards solving the problems of pest management in cotton has reflected competition, rather than cooperation, among philosophical approaches and individual researchers. How one approaches IPM, and perhaps whether the stated goals are widely embraced, depends on a frank assessment of the motivation for developing ecologically or behaviourally based IPM systems, and whether those ultimate goals exclude conventional pesticides by definition. The critical question may be: 'Are we advocating environmentally based IPM in an effort to eliminate all pesticides and their associated risks, however large or small they may be? Or will an ecological, holistic approach to pest management be truly necessary for long-term sustainability and profitability of cotton production?' The substantial reduction or elimination of conventional pesticides is clearly a goal

of many contemporary approaches to cotton IPM. While these goals are admirable, they may not currently be practical. If the goals of IPM are not practical in the context of the current situation or specific production system, it should not be surprising that they will not be achieved. Paramount to any other goals of IPM is production of the cotton crop per se; management of pests is a secondary goal subject to forces driving the overall economy of production. An implicit quality of appropriate goals is that they are reasonably attainable. 'Appropriate' does not imply the best, most efficient or most environment-friendly in any specific context; defined practically, its meaning should change as new knowledge and technology become available. Therefore, what constitutes appropriate goals for development and implementation of ecologically based cotton IPM is context-dependent and subject to current biotic, conceptual and system constraints.

Constraints to Development of Ecologically Based Cotton IPM

Biotic constraints

Dominant pests

Although the similarity of pest complexes across cotton production systems worldwide is striking, the most severe pest problems tend to occur in areas infested by specialized pests such as the boll weevil, *Anthonomus grandis* Boheman, and pink bollworm, *Pectinophora gossypiella* (Saunders) (Luttrell *et al.*, 1994). Not all cotton production regions are characterized by the presence of key pests that consistently requires application of controls. Also, the dominance of key pests has been reduced in certain production regions by adoption of the recently available transgenic Bt cottons or progress of boll weevil eradication programmes. Still, this reliance on pesticides for control of dominant pests remains a characteristic of many production systems. A consequence of this reliance is the increased difficulty with which minor or occasional pests may be studied in the field. Such pests, which may otherwise be amenable to cultural or biological controls, are difficult to address in systems that are constantly disrupted by pesticide treatments for key pests.

Real and perceived inefficiency of natural enemies

A key goal of most approaches to cotton IPM has been the substitution of biological controls for chemical pesticides. Success of classical biological control in modern cotton production systems has been limited (Luttrell, 1994; Lewis et al., 1997). Reasons for the lackluster performance of this tactic include an inadequate understanding of the roles of biocontrol organisms within the cotton agroecosystem (Reynolds et al., 1975; Lewis et al., 1997) and the highly disturbed nature of the production system (Landis et al., 2000). Pedigo (1995) suggested that the problem was more basic, as the density dependence of many natural enemy–pest interactions makes such agents incompatible with other tactics that suppress pest (host or prey) populations. In addition, if population levels of the target pest(s) tend to be high, as with many key pests, biological controls need to rapidly and effectively suppress pest populations in order to be competitive with other tactics. If such levels of efficacy could be obtained, the immediate consequence for specialist natural enemies might be local extinction from the exhaustion of prey resources. Generalist natural enemies may be easier to integrate into cotton production systems because they may subsist on alternate prey resources or by plant feeding. Several authors suggest that conservation of natural enemies may offer greater potential for integration into modern production systems than either importation or augmentation (Luttrell, 1994; Lewis et al., 1997; Landis et al., 2000). However, economic viability of natural enemy conservation, whether of specialists or generalists, requires that pest suppression be sufficient to avoid substantial crop losses.

Several authors have suggested methods to improve the effectiveness of natural enemies in agricultural systems, but few of these methods appear readily adaptable to modern cotton production. For example, Verkerk et al. (1998) suggest that the development of synomones that attract natural enemies to crop plants may be useful in enhancing biological control. However, any tactic that attracts predators and parasitoids may have deleterious effects if hosts are not present in sufficient numbers to support or maintain natural enemy populations. The usefulness of such a tactic may depend on the relationship between the host population level required to maintain populations of natural enemies and the level that can be economically tolerated. Also, the use of attracted natural enemies as a 'rescue' or remedial tactic would require population estimates and predictive abilities that are better than those currently available, assurance of an adequate pool of natural enemies amenable to manipulation and precise timing of the attractant.

A method often promoted as a means of increasing effectiveness of natural enemies is vegetational diversity (e.g. Altieri, 1999). Increased availability of food sources in diverse systems may favour success of many natural enemies, but tenure time in patches of vegetation may depend on the availability of suitable prey populations (Sheehan, 1986). Although vegetational diversity is generally considered a means of increasing stability of the agroecosystem, increased diversity does not benefit all species of natural enemies (Verkerk et al., 1998; Landis et al., 2000). The selective enhancement of natural enemies requires an adequate understanding of their behaviours and identification of the relevant elements of diversity (Landis et al., 2000). Also, diversity is relative; in many production system landscapes cotton occurs alongside numerous other cultivated crops, pastures and weedy areas. The vegetational diversity already present in these systems may buffer effects of pesticides on natural enemies in cotton without further enhancement. For example, in South Texas, Sparks et al. (1997) attempted to disrupt natural enemies in cotton by making multiple insecticide applications to field-sized plots. Although populations of natural enemies were suppressed immediately after treatment, in most cases their numbers recovered quickly, apparently because of immigration from surrounding vegetation. Apart from effects on natural enemies, vegetational diversity can exacerbate pest problems when pests of cotton and other crops benefit from the presence of alternate cultivated and weed hosts (Kennedy and Storer, 2000).

Failure to suppress cotton pests by manipulating populations of natural enemies is not evidence that existing populations of predators and parasitoids are of little benefit. We probably derive more benefit from natural enemies, especially generalist predators, than is realized because the contributions of these populations have received comparatively little study in most systems. Still, infestations of some pests can increase rapidly, especially those immigrating to cotton from other plants or production regions. Under such conditions, natural enemies, whether conserved or augmented, may be ineffective because their population and activity levels cannot respond in time to prevent economic damage. Therefore, tactics to enhance the activities of natural enemies may be most effective when used in conjunction with other tactics that lower pest population levels or dampen their fluctuations or when the production system does not feature one or more dominant key pests. Until the dynamics of natural enemy–pest interactions within specific production environments are better understood it does not appear financially prudent for cotton producers to take substantial risks in hopes of increasing the impacts of natural enemies.

Emphasis on pest population reduction

Perhaps because of our experience with chemical pesticides, which generally produce high levels of mortality rapidly, we have historically emphasized killing pests rather than understanding and manipulating them. Consequently, our primary accomplishment in cotton IPM has been transition from preventative pesticide application schedules to more therapeutic treatments in response to the presence of potentially damaging pest populations. This emphasis on control belies the ecological potential of IPM. Pedigo and Higley (1992) argue for moving away from a focus on inducing pest mortality, because of the potential for any mortality-based tactic to select for resistance. They also suggest that consideration of the proportion of a pest population killed by a management tactic, as a component of the economic injury level (EIL), reflects a narrow focus compared with consideration of the proportion of pest injury avoided by use of a tactic. The implication of a focus on inducing mortality is less consideration for preventative tactics that reduce the need for rescue or therapy. Pedigo (1995) suggests that an IPM strategy composed of both preventative and therapeutic tactics would provide an increased opportunity to integrate tactics into the overall management programme.

Continued reliance on insecticides

Management of arthropod pests in cotton is still dominated by insecticides (Luttrell et al., 1994). This dependence persists despite the known problems with pesticides and decades of ecological rhetoric. Although the goal of substituting non-chemical tactics for conventional pesticides is politically popular, effective alternatives are not widely available and the difficulty associated with developing such alternatives is seldom fully acknowledged (Gianessi, 1993). Incentive for adoption of alternatives is also lacking so long as the scouting-pesticide approach remains effective (Pedigo, 1995). Even if simple and effective non-chemical alternatives to pesticides were developed and adopted, the need for conventional pesticides would remain to address pest populations that escape or overcome the effects of alternative controls.

There appears to be little likelihood of wholesale movement to alternative control tactics in the near future. Synthetic pesticides are widely recognized as being relatively inexpensive and effective, and requiring a low level of management for generally acceptable results (Quisenberry and Schotzko, 1994; Atkinson and McKinlay, 1997; Ehler and Bottrell, 2000). Also, the incremental benefits of their use far exceed their costs (Zilberman *et al.*, 1991). More than most other controls, they can be used as a rescue tactic when the most opportune time for application of a control tactic is missed because of weather or poor monitoring. This characteristic sets conventional pesticides apart from other control or management tactics. From a practical perspective, these qualities have set a very high standard for comparison.

Conceptual constraints

Economic injury levels and economic thresholds

The EIL and ET are core operational concepts of IPM whose adoption has reduced pesticide use in cotton and other crops (Pedigo and Higley, 1992). Because of the complexity of technically accurate EILs and ETs, most current treatment thresholds in cotton are based mainly on historical and regional experience rather than research (Luttrell *et al.*, 1994). That is not to imply that these thresholds have been wholly unsatisfactory. Current thresholds can be viewed as the product of an informal expert system developed through extensive field experience. Depending on the capabilities of the user, such a system can produce highly appropriate recommendations that account for more complexity than would be possible with a reasonably useable formal model. This intuitive model can draw information from an informal database limited only by the experience level of the user, and can accommodate data in the form of visual inputs that would be difficult to qualify, let alone quantify, in a formal rule-based system. In practice, such systems have performed reasonably well, and have facilitated establishment of relatively robust treatment decision rules. This approach is, however, potentially deficient in the areas of technology permanence and information transfer, as the durability of the information is predicated on the continuing availability of qualified and experienced practitioners.

Several authors have identified development of more formal, research-based treatment thresholds as a critical need. Quisenberry and Schotzko (1994) indicated that the development and implementation of sound pest management strategies rely on the accuracy of EILs and ETs. Hutchins (1995) suggested that pest management decisions could be improved by developing dynamic thresholds that account for changing economic or biological variables, in preference to traditional, static thresholds. Headley (1975) argued that the problem with static ETs is that they ignore the interdependence of decisions at different times, and that the optimal decision method requires that multiple time periods be considered. Also, Higley and Wintersteen (1992) pointed out the importance of recognizing environmental risks posed by pesticides and proposed incorporating environmental costs into EILs. Despite arguments for more refined EILs and ETs, there are equally strong practical arguments against precise thresholds. This position can be illustrated by considering the parameters used to estimate the EIL. Of these, cost of control appears the most straightforward to estimate. However, the cost of control is more than the accumulated costs of materials and application or implementation. It also includes costs associated with the risk that a control tactic may produce additional problems that result in losses or require further controls. Our ability to quantify this risk is rudimentary at best. A second parameter, per unit value of the crop, is similarly difficult to define. The cotton producer has little control over the value of the goods produced. Thus, crop value typically changes both within and among production seasons and frequently cannot be estimated with precision at the time controls are applied. Both the amount of injury produced per pest individual and the relationship between levels of crop injury and economic losses are products of the interactions between complex climatological and biological factors. While the results of these relationships can be estimated empirically, the mechanisms themselves are poorly understood. The usefulness of predictions at a given decision point are further limited by our inability to accurately predict local weather patterns weeks or months in advance. Finally, the level of control, or alternatively, preservation of yield or quality potential, resulting from the control tactic is usually presumed to be high and fixed. In practice, recommending effective insecticides identified through screening efforts minimizes variation in the level of control. However, the level of control that is actually achieved is also dependent on the application method and conditions, crop phenology and timeliness of application. Once thresholds are established, their continued use usually assumes that substantial changes in efficacy have not occurred. Even when reasonable estimates of the parameters used to calculate the EIL have been obtained, they cannot adequately reflect the possible ranges of these values, or changes in the values caused by changing markets, cultural practices, varieties, weather patterns or selection pressure on the target pest. Information to devise truly dynamic EILs is not currently available. Were this information available, adequately sophisticated dynamic EILs would likely be too expensive (because of the cost of model input collection) or too complicated for use by most producers or IPM practitioners. Incorporation of multiple time frames into decisions as suggested by Headley (1975) would add an additional layer of complexity to the problem. Furthermore, inclusion of environmental costs, as suggested by Higley and Wintersteen (1992), is not practical because those costs cannot be objectively estimated and there is no mechanism for their recovery by producers.

The inadequacies of current EILs and ETs are unlikely to be effectively addressed in the near future, especially for minor, occasional or early-season pests. In the case of occasional pests, their infrequent and unpredictable occurrence complicates efforts to study relationships between pest populations and crop yield. Although some early-season pests occur with considerable regularity, efforts to quantify their influences are typically hampered by the confounding effects of weather, management and other pests that occur in the interven-

ing period before harvest. In addition, these pests frequently impact on crop earliness or uniformity rather than yield per se, and useful estimates of their impact must include measures of crop quality and increased risk from delayed harvest.

EILs and ETs that more accurately reflect system dynamics and account for additional factors such as the effectiveness of natural enemies, varietal susceptibility or tolerance, and the accuracy of population estimates are clearly conceptually desirable. However, they are not practical as yet. What is practical is to establish approximate EILs or ETs, based on demonstrated plant growth, physiological or yield responses, that can be refined or adjusted in use. These approximations can serve as rough guidelines to be modified based on experience and individual circumstances, much as the current nominal thresholds are used. These estimates represent progress over nominal thresholds if they are 'accurate' from the standpoint of representing levels of a pest population that will normally produce substantial economic damage exceeding reasonable cost of a remedial treatment. In essence, this approach advocates reasonable and necessary incremental improvements to the nominal thresholds that are currently available, until the mechanisms influencing crop losses from pests are both better understood and more predictable.

Limitations of sampling methodology

The difficulty associated with accurate estimation of arthropod populations in cotton is often overlooked. From a purely mathematical standpoint, much of this difficulty arises from the nature of the sampled populations. These populations are typically diverse (composed of multiple species), variably distributed throughout a complex and heterogeneous physical environment and constantly changing. That available sampling methods impose limitations has long been recognized (Byerly et al., 1978; Phillips et al., 1980; Ellington and Southward, 1999). These limitations impact on cotton IPM in two distinct areas: research to develop or evaluate population management tactics, and population monitoring for making control decisions.

Research to develop or evaluate management tactics typically requires meaningful measurements of populations of pests and/or beneficial arthropods. These measurements must be both sufficiently accurate (indicative of the true population densities, intensities or trends) and precise (lending an acceptable degree of confidence in the estimates) to permit the researcher to unambiguously distinguish the effects of the studied factors. Mathematical precision is usually improved by increasing the number of samples. Formulas are available for calculating necessary sample sizes for specified levels of precision (e.g. Karandinos, 1976). However, sample sizes corresponding to fixed levels of precision vary with the population levels of interest. This means that a fixed sample size that provides adequately precise population estimates at a low population level is unnecessarily large (and thus expensive) when population levels are much higher. In addition, it is frequently impractical to collect enough samples to precisely estimate very low population levels. In most cases, sample sizes are dictated by the availability of resources (either time or money), which are typically limited. Thus, the researcher usually collects as many samples as time and fiscal resources allow. Therefore, the investigator has no idea, or only a very limited one, of the level of sampling precision achieved in most studies.

Sampling methods intended to directly estimate population densities ('absolute' methods) tend to be laborious and expensive. Numerous 'relative' sampling methods that do not directly estimate populations, such as the sweep net, drop cloth or D-vac, have been devised to reduce the cost of sampling. Although a few of these methods have been compared with more 'absolute' methods to determine whether they accurately indicate population trends (Garcia et al., 1982; Fleischer et al., 1985; Snodgrass, 1993), any single method is typically inadequate for at least some of the species or stages sampled. Thus, the sampling problem is compounded in studies in which more than one pest, or

pests combined with complexes of beneficial arthropods, must be sampled. In these situations sample sizes may be limited because resources must be partitioned among multiple sampling methods.

Restricting sampling to segments of the production cycle, or plant phenological stages, during which the pest of interest is problematic sometimes reduces monitoring costs. In some cases, specialized sampling methods have been developed for specific pests such as thrips or aphids on seedling cotton (Burris et al., 1990; Micinski et al., 1995), or different sampling methods are recommended at different times during the production season (e.g. Hardee et al., 1994). In many cases, direct population estimation is not reasonably possible (short- and long-range migrants, overwintering pests, populations inhabiting widely dispersed wild hosts), and estimates must be based on natural or artificial markers, trap captures or even qualitative observations. These estimates require assumptions that are made with considerable risk.

Limitations of sampling methodology are particularly relevant to efforts for refining ETs, which are based on anticipated decreases in crop yield or quality in response to pest activity. Cotton yields are inherently variable, and poor precision of yield estimates coupled with either poor accuracy or precision of estimated pest populations limits the accuracy of crop response estimates. Regarding our limited sampling abilities, a saving grace is that given the normal variations in both cotton yield and pest population levels in space (within and among field) and time (especially among years), we are usually interested in relatively large population changes or crop responses. Sampling is also problematic in treatment decision making because producers and crop consultants are more sensitive to sampling costs than are researchers. Cullen et al. (2000) reported that university-recommended sampling schemes were often too expensive for commercial uses, and that the expectation for precise pest population estimates in minimal sampling times was unfeasible. Although the sampling problem studied by Cullen et al. (2000) was not in cotton, their conclusions are as applicable to decision making in cotton as to any other crop.

In most cotton production regions of the USA general sampling recommendations are available through Extension Service guides or circulars. In a few cases, standardized sampling schemes have been developed to provide input to management or predictive models such as rbWHIMS (Williams *et al.*, 1995) or SIRATAC (Dillon and Fitt, 1995). In the author's experience, at least minimally acceptable guidelines are followed in formal crop scouting programmes and by a few commercial consultants. However, most consultants and producers use their own sampling procedures and plans, or modify recommended plans by reducing sample sizes or the portions of fields that are examined. The use of non-standard plans or methods adds an additional level of uncertainty to decision making because the results of these methods can be difficult to relate to established treatment thresholds.

Binomial (presence–absence) sampling plans (Binns and Nyrop, 1992) have received considerable attention for treatment decision making, and a number of such plans have been developed and evaluated in cotton (Wilson *et al.*, 1981; Rothrock and Sterling, 1982a, b; Naranjo *et al.*, 1996). Their major advantage is that sampling time and cost are substantially reduced when pest populations are well below or above the treatment threshold. However, their use assumes a consistent relationship between population density and the proportion of sampling units (e.g. leaves, plants) infested, and the sampling plan must change when the treatment threshold is changed. Furthermore, there is little evidence that most producers and consultants understand or use binomial sampling plans. One of the least desirable but widely used methods of simplifying sampling relies on estimates of the percentages of flower buds (squares) or fruit (bolls) infested or damaged. Estimates obtained by these methods are often meaningless unless they are adjusted to account for changing populations of the sampled units. For example, an estimate of 10% damaged squares during mid-season when square populations are very high has a very different meaning

than similar levels of square damage early or late in the production cycle.

An important consequence of our limited ability to support treatment decisions with precise population estimates is the propensity for consultants and producers to offset this lack of precision with what Atkinson and McKinlay (1997) termed 'insurance applications'. In this case, the cost of a potentially unnecessary chemical pesticide application is perceived as being less than either the unknown risk of crop damage or the cost of adequate sampling. Another implication of the limitations posed by current sampling methodology is the need to develop treatment thresholds in conjunction with sampling schemes that are acceptable to the user.

A two-tiered approach to cotton IPM

Thomas (1999) suggested that the orientation of IPM towards quick fixes, or the single solution 'magic bullet' approach, has impeded the development of true IPM. This seems a popular view widely held by academicians. However, this view in itself may serve as an impediment by neglecting the unpopular fact that the current ecological knowledge base does not permit the construction of adequate 'biologically based' pest management systems. Furthermore, if such information were available, producers' aversion to risk would make timely adoption unlikely. While the totally ecologically based management system may be an admirable and even an appropriate long-term goal, the most relevant issue regards the best path to that goal.

IPM in cotton is a two-tiered proposition. One tier is represented by the theoretical optimum of a biointensive, ecologically based management system, while the other tier encompasses the realities of current production systems and the modifications needed to move management methods towards the desired goals. One limitation to the development of ecologically or behaviourally based cotton IPM has been the tendency to place inordinate emphasis on complex and sweeping changes to the production system in order to eliminate the use of pesticides. Luckmann and Metcalf (1975) implied this by suggesting that pest management programmes need not be complete to begin implementation. Other authors also noted the limitations imposed by current agricultural production systems. Pedigo (1995) advocated a preventative management approach, but noted that therapeutic tactics are also necessary. In a similar vein, Huffaker *et al.* (1978) and Hutchins (1995) indicated that in concept, IPM does not exclude the use of therapeutic tactics such as insecticides. The salient point is that IPM represents an ultimate goal, but progress towards that goal must accommodate current management systems. Incremental advances that are compatible with current practices and constraints hold more promise for durable adoption than do those that require drastic change. Therefore, in order to accurately judge which components of the management system are most amenable to change we need to understand not only the ecological parameters that control the system but also their context in the production system.

Constraints of system complexity

Scientists simplify and generalize in order to conceptualize and understand complex systems. They also tend to assess system complexity in reference to external standards, such as natural ecosystems, and on that basis often characterize current, monoculture production systems as 'simple'. The underlying complexity of these systems, which involve innumerable components, many of which cannot be easily identified or characterized let alone quantified, is commonly underestimated. A consequence of this oversimplification is that efforts to develop improved pest management systems often neglect the overall cropping context. Accommodation of the nature and goals of a specific production system is essential not only for development of new management methods but also to increase the probability of their adoption.

Cotton as a framework for IPM

Theoretical approaches to cotton IPM often overlook that adoption of a management

tactic or strategy depends on short-term financial returns to the producer, whose primary goal is an abundant crop. Management of pests is secondary to this goal, and is therefore subject to the forces driving the economics of overall production. For this reason, producers typically select varieties, adopt cultural practices and establish fertility levels independent of pest management considerations (Headley, 1975). Until very recently this was particularly true with respect to varietal selection because characters conferring resistance to one pest often increased susceptibility to other pests (Phillips *et al.*, 1980; Jenkins and Wilson, 1996). However, the recent release of genetically transformed pest- or herbicide-resistant cottons has increased the importance of pest management considerations in varietal selection in many production systems.

One area offering potential for improvement is simply the more efficient and timely production of the crop. Reynolds *et al.* (1975) point out that excessive irrigation and fertilization may increase susceptibility to pest problems. If the crop is produced as efficiently as possible with respect to agronomically appropriate planting dates, seeding rates, irrigation and fertility management, the pest manager often has increased latitude because the risks of pest damage resulting from slow fruiting rates, low fruit retention, poor plant vigour and uneven or untimely crop development are minimized. Production of a timely and uniform crop also minimizes problems with poor lint quality that are often associated with attempts to mature late-set bolls. Except for the influences of crop insurance programmes, the best economic strategy would normally involve practices that provide the highest net returns. In many instances these practices do not result in maximum production, but rather in marginally lower yields, reduced inputs and improvements in lint quality (Adkisson *et al.*, 1982; Chu *et al.*, 1996). Given the availability of inexpensive and effective conventional pesticides, these practices may not necessarily be those that most effectively avoid pest problems. Water availability is an emerging issue in some regions, and more efficient water usage will likely influence other aspects of the production system (Geng *et al.*, 1990).

Cotton is a highly plastic crop that responds rapidly to prevailing conditions (Huffaker et al., 1978). This plasticity allows cotton to be grown across a continuum of production season durations, irrigation regimes and fertility levels. Within specific systems, this plasticity offers the potential to manipulate the crop to the pest manager's advantage, depending on the timing and severity of pest problems and the costs of managing them. However, in the absence of severe problems such as pesticide resistance in a key pest, pest management issues seldom drive changes in production practices. Instead, these changes are motivated by economy and simplicity of production, government subsidies and insurance, and yield considerations. Under these conditions, it is necessary for the pest management researcher to react to changes in the system and be cognizant of these overriding incentives in the development and integration of new management tactics or strategies.

Putting pest biology in a cropping system context

Relationships among various crop and non-crop components of the agroecosystem influence the management of cotton pests (Reynolds et al., 1975; Phillips et al., 1980; Luttrell, 1994). Because the most important relationships vary with production region, the pest(s) of interest and the management approach, the development of successful ecologically or behaviourally based pest management tactics is decidedly context-specific. An implication of this specificity is that tactics or approaches developed within one production system may not be easily transferable to other systems.

A more detailed knowledge of the interactions among biological components of the production system, among these components and various management methods, and among different management tactics is needed (Quisenberry and Schotzko, 1994). Although these needs are clear, development of this information will not be straightforward. For example, Verkerk et al. (1998) stated that rigorous studies of tri-trophic interactions within a single crop are rare because of their complexity. Introduction of other cropping and non-cropping components adds an additional layer of complexity to the research problem. Quisenberry and Schotzko (1994) emphasized the need to focus on the basic biological interactions within production systems. Elucidation and application of the underlying mechanisms controlling these interactions would provide a more practical approach to complexity than direct experimental examination of an almost unlimited array of factors. However, the applicability of such research will be largely determined by the extent to which it reflects and addresses components of specific production systems.

Role of modelling in understanding and managing complex systems

The complexity associated with developing and integrating multiple control tactics into management strategies was apparent to early IPM workers. Modelling was seen as a solution to the problem of evaluating these systems. It was obvious that direct observation and analysis of the processes involved in crop production, the impacts of pest control tactics, the cost–benefit relationships associated with each management practice and the interactions among these factors were not possible. Therefore, it was hoped that simulation models would provide a means of evaluating the outcomes resulting from combinations of changes in crop management practices (Huffaker et al., 1978). Despite considerable effort to model agricultural systems, and significant contributions of those efforts to our understanding of certain components of production systems, those original goals have not been attained.

Among the uses of models, prediction has probably received the most attention despite the models doing this least well (Wormer, 1991). The typically poor performance of predictive models stems from several sources. First, even comparatively simple natural systems are highly complex. Corresponding models are abstractions or conceptualizations of reality that typically require assumptions of unknown validity, and the arbitrary grouping or organization of processes into manageable units (Wormer, 1991). Some theoretical assumptions, such as the homogenous and continuous nature of populations, are incorrect in most cases. Metapopulation models that account for patchiness and other sources of variation in populations are becoming more relevant to population modelling efforts (Hoy, 2000), but these methods still fail to accommodate some of the most important sources of model error. All components of the production system, particularly the insect components, are highly subject to environmental conditions. Even early workers recognized that accurate long-term population prediction was unlikely if weather was an important factor (Huffaker *et al.*, 1978). The problem of realistic incorporation of weather into predictive models becomes more difficult by orders of magnitude when the relevant scale is the microclimate because of the heterogeneous array of microclimatic conditions existing at a given time in even a relatively uniform crop. Regardless of the difficulties associated with incorporating environmental effects into predictive models, Nylin (2001) points out that the seasonal and life history plasticity displayed by many organisms in response to a given environment are additional sources of model error that are poorly understood.

One of the most serious sources of predictive error results from the relationship between model complexity and error in measurement of model parameters. Many ecological inputs are difficult to measure with precision. As the number of model parameters is increased, so is the accumulated prediction error (Wormer, 1991; Colbert, 1995). This limitation complicates efforts to select or validate models because quantitatively similar models may yield very different predictions, and sensitivity analysis is not always sufficient to indicate problems with parameter estimates (Hoy, 2000). Thus, even if a relatively simple model is specified exactly correctly, accurate long-term predictions may not be possible. The problem becomes even more difficult and complex when factors such as spatial dynamics (Colbert, 1995) and pest migration (Pedgley, 1993) are involved.

Another use of models is as decision support tools. Although the widespread adoption of models for management of pests in cotton has not occurred, some authors suggest that their use is likely to become commonplace (Atkinson and McKinlay, 1997). However, realization of the potential of model-based decision supports would require improvements in current ETs and sampling technologies. Cox (1996) outlined several causes for the meagre rates of development and adoption of model-driven decision support systems. Among them were the facts that decision support models are usually complex enough to require technical support for users, and they often require inputs that are impractical for producers to obtain. In addition, researchers often fail to specify models in terms that reflect the constraints and concerns of farmers, and when these tools are proposed for use in evaluating management options, they often fail to provide an understanding of associated benefits and costs. Cox (1996) further argues for simpler models if the output is to be interpreted by users other than those who built the models.

The point is not to argue that models are not appropriate or useful tools in pest management, but rather that their limitations should be recognized if they are to be used to full advantage. Virtually all researchers use mathematical or statistical models to evaluate and describe research results. Models assist in structuring information and providing insight into complex processes (Wormer, 1991). Pedigo (1995) suggested that a factor inhibiting the development of multiple-tactic management systems is the lack of a simple conceptual basis for integration, and that modelling has been useful in this regard but mainly as a research tool. Therefore, there seems a continuing need for improved modelling methods in understanding and manipulating complex agroecosystems, but these methods will likely be most useful in areas other than simulation and decision support.

Impacts of scale in cotton IPM

Except for a few 'areawide' or 'community-wide' management efforts, IPM has traditionally focused on a field-by-field approach.

This approach persists because the field is an arbitrary, but convenient, management unit for most aspects of crop production. Also, pesticide-based management tactics are not usually amenable to implementation on larger units because of the distinct differences in crop and pest conditions that occur between adjacent fields (and sometimes within fields). However, most pests have one or more mobile stages and this mobility creates the need to develop ecologically based IPM strategies operable on an appropriate ecological scale. Indeed, the term 'mobile' has no interpretable meaning except in reference to the scale of mobility. In most production systems cotton culture is characterized as a monoculture, despite being grown in the presence of other crops such as maize, sorghum or soybean. Therefore, vegetational diversity that is not apparent at the field level becomes obvious at the farm or community level. It is well known that relationships among the various crops and non-cropping areas within a production system have important implications to pest management (Luttrell, 1994; Kennedy and Storer, 2000). However, the seasonal distributions of many pests among production system components and the causes of periodically intense infestations are not well known (Pedgley, 1993).

Issues of habitat diversity and scale are difficult to address because they vary with context. Scales of movement differ greatly among different pests and natural enemies. In systems impacted by pests that are long-range migrants, the relevant scale of study may exceed the size of the production system of interest. Yet the question of scale may be critically important in integrating management tactics (Sheehan, 1986), and in determining how best to implement specific management techniques (Meerman et al., 1996). Atkinson and McKinlay (1997) suggest that only studies that include monitoring of whole systems will provide answers that adequately address interspecific and scale-related factors. However, the scales of such studies are necessarily limited by practical concerns. Relatively recent conceptual methods, such as metapopulation ecology, where landscapes are viewed in terms of

habitat patches (Hanski, 1998), may be helpful in the analysis of scale-related problems but their use to date has been limited.

System dynamics and the state of constant change

Changes in crop development and associated pest populations occur in seasonal patterns. These changes are predictable in a general sense, given knowledge of previous production cycles. However, more insidious changes to agricultural systems occur across production seasons. For example, the typical market lifespan of a cotton cultivar within a production region is a few years at best. Changes in cultural practices such as row width, seeding rates, tillage methods, irrigation methods and timing, and even the acreage devoted to cotton occur in response to new information, availability of new equipment and market forces. Changes in management practices are often associated with changes in pest complexes (Luttrell, 1994). The constantly changing nature of most production systems serves to make associated ecological problems moving targets, and addressing those targets a reactive and iterative process.

Examples of Ecologically Based Tactics

What constitutes an ecologically based tactic, and in particular a successful ecologically based tactic, is a matter of opinion. The following examples concentrate on instances in which some combination of qualities of the pest, natural enemy and/or environment have been manipulated or exploited to reduce or eliminate a pest problem.

Case studies

Cotton stalk destruction in Texas and Arizona

The importance of the timely destruction of crop residues has long been recognized in the management of cotton pests (Hinds, 1908; Newell and Paulsen, 1908). Early efforts to use this cultural practice were hampered by the lack of an efficient means of destroying harvested fields and the extended growing season necessitated by hand harvesting. In recent decades, with the availability of effective chemical and mechanical crop termination tools and mechanized harvest, this method has been effectively used to solve severe pest problems in the USA.

In the early 1950s the pink bollworm inflicted heavy losses to cotton in parts of southern Texas, and by the mid-1950s the insect was severely damaging cotton in Central Texas (Adkisson and Gaines, 1960). The magnitude of the pink bollworm problem in Texas, and the prospects for its spread to adjacent cotton production areas prompted an expanded research effort by federal and state scientists (Noble, 1969). Much of this research concentrated on maximizing postharvest mortality of pink bollworm larvae. Findings of this research indicated that the combined effects of flail-shredding the harvested cotton stalks and ploughing the shredded residue produced up to 95% mortality of pink bollworm larvae in infested bolls (Adkisson and Gaines, 1960). When chemical harvest aids were used to speed harvest so that stalks could be destroyed before short days induced diapause in the larvae, control was further enhanced (Adkisson, 1962). The stalk destruction effort was most effective when implemented in a community-wide manner, and such implementation was facilitated by state regulations establishing mandatory plough-down dates in most of the state. These regulations persist today, and the pink bollworm is not normally a pest of economic significance in Texas except in the western-most (Trans-Pecos) production region. The pink bollworm remains an economic problem in far western Texas, partly because larval mortality caused by stalk destruction is highest where winters are characterized by high rainfall and low temperatures (Adkisson and Gaines, 1960), and partly because the state-mandated plough-down date is too late (1 February) for maximal effectiveness.

Stalk destruction has also been used to reduce the severity of boll weevil infestations (Niles et al., 1978). An agronomic practice periodically used in production regions of the western USA is stub cotton. In the production of stub cotton, the current crop is produced from plants remaining from the previous cropping cycle. In some cases the above-ground portions of the plants are shredded or cut back after harvest, but the lower stalks and roots are left intact. More commonly entire plants, including immature fruit, are left standing in the field. This practice permits plants to begin growth and fruiting earlier than those produced from planted seed, but it also provides a year-round supply of food and shelter for cotton pests. Particularly in Arizona, stub cotton has been a feature of the production system except when increasing pest problems resulted in regulatory prohibition (Moore et al., 1996). In the early 1960s, during a period when stub cotton was allowed, localized infestations of the boll weevil were detected in Arizona cotton (Brazzel et al., 1996; Moore et al., 1996). The state responded in 1965 by instituting a ban on stub cotton and establishing mandatory plough-down dates (Bergman et al., 1982). By the following production season losses from the boll weevil were negligible where the mandated cultural controls were properly followed (Fye, 1968). The boll weevil remained of little concern until the state withdrew the ban on stub cotton in 1978. By August of that year boll weevil infestations were again documented (Bergman et al., 1982). By 1981 the boll weevil was widespread in Arizona. In 1983, the state of Arizona again banned stub cotton, and that same year a boll weevil eradication programme was initiated. That programme subsequently expanded to cover the southern California valleys, northwestern Mexico, and central and western Arizona, and few weevils have been trapped since the early 1990s (Moore et al., 1996). The precise role of stalk destruction in elimination of the boll weevil from Arizona in the 1980s is not clear, because the mandated cotton-free period was as short as 45 days during that period (Henneberry and Phillips, 1996). It is unlikely that the maximal benefit of a cotton-free period could be achieved by such a late plough-down.

Uniform delayed planting

Cotton production in the Texas Rolling Plains is typified by low inputs and relatively low yields (Masud *et al.*, 1985). The low production costs of the region kept cotton production profitable until the 1960s, when the boll weevil began to overwinter in large numbers (Walker and Smith, 1996). Subsequently, the traditional production system became threatened by increased control costs for the boll weevil and associated secondary pests (Masud *et al.*, 1985). White and Rummel (1978) observed that many boll weevils emerged from their overwintering habitat before squaring cotton was present, and that their host-free longevity was limited. Also, Slosser (1978) observed that colonization by overwintered weevils and subsequent infestation levels were lower in June plantings than in May plantings. These observations led to the recommendation for uniform delayed planting. In uniform delayed planting, recommended planting dates typically 2 weeks later than those previously used were established for different communities. Within communities efforts were made to complete planting in as short a period as possible (Walker and Smith, 1996). In addition to reduced early-season weevil pressure, delayed planting maximized the mortality of boll weevil larvae in fallen squares by virtue of exposure to high soil temperatures in July (Fuchs *et al.*, 1998). Agronomic advantages included improved plant stands, seedling vigour and reduced seeding rates because of the more favourable growing conditions at the time of plant emergence (Masud *et al.*, 1985). An economic evaluation of the practice indicated that uniform delayed planting generally resulted in higher lint yields, reduced insecticide inputs and increased net returns (Masud *et al.*, 1985). Despite the success of uniform delayed planting and its contribution to the Rolling Plains production system, emphasis on this tactic has recently declined because of progress in eradication of the boll weevil and efforts to boost yields using longer-season varieties.

Short-season production

The most widely adopted alternative management strategy in United States cotton systems has been pest avoidance by short-season production (Luttrell *et al.*, 1994). Several important pest problems in cotton have been addressed with some degree of success by shortening the production cycle. Practices leading to early and rapid fruiting were recognized as important means of escaping damage from the boll weevil soon after it became established in the USA (Bennett, 1908). The short-season production strategy has remained an important tool in many modern management programmes for the boll weevil (Bradley and Phillips, 1978; Henneberry and Phillips, 1996), pink bollworm (Adkisson and Gaines, 1960; Noble, 1969; Chu *et al.*, 1996) and the bollworm *Helicoverpa zea* (Boddie) (Kennedy and Storer, 2000).

In recent decades much production has shifted to earlier-maturing varieties, particularly in the Mid-south and South-west (Luttrell, 1994; Jenkins and Wilson, 1996). Historically, the determinate, short-season varieties have found greater acceptance in the drier production regions of the Southwest than in the higher rainfall areas of the cotton belt (Walker and Smith, 1996). This is because successful short-season production involves more than varietal selection. In addition, planting dates, plant populations, fertility levels (especially nitrogen and phosphorous) and the timing and numbers of irrigations must also be carefully controlled (Niles *et al.*, 1978). In particular, excessive or poorly timed irrigations may promote vegetative growth and delayed crop maturity (Noble, 1969; Niles *et al.*, 1978), and availability of soil moisture often cannot be regulated in areas typically receiving heavy late-season rains. Finally, the rapid production and maturation of fruit is of limited usefulness if the crop cannot be harvested in a timely manner. Most short-season production systems should feature use of chemical harvest aids to speed harvest and early stalk destruction efforts. These materials have been of considerable use in management of late-season boll weevil and pink bollworm populations (Niles *et al.*, 1978; Henneberry *et al.*, 1988; Chu *et al.*, 1996; Moore *et al.*, 1996).

The benefits of community-wide, short-season production for management of the boll weevil were demonstrated in the area surrounding Laveen, Arizona, in the mid-1980s (Henneberry and Phillips, 1996). This demonstration featured uniform planting dates, in-season insecticide treatments to keep boll weevil populations below treatment thresholds, irrigation termination by 1 September, use of plant growth regulators to remove non-harvestable fruit in mid- to late September and a 60- to 75-day increase in the host-free period compared with the normal practice. Although the numbers of insecticide treatments were not reduced compared with the year previous to implementation of the demonstration, yields were much higher. Similarly, Chu et al. (1996) reported the results of a mandatory cultural programme initiated in 1989 for management of the pink bollworm in the Imperial Valley of California. The programme changed cotton production in the region from a full- to a moderately short-season production system, featuring treatments with a defoliant or growth regulator by 1 September, and stalk destruction by 1 November. Adoption of the short-season production practices not only reduced pink bollworm population levels and damage, but also resulted in improved lint grade.

Heilman et al. (1979) evaluated a short-season production system in the Lower Rio Grande Valley of Texas, where the primary pests were the boll weevil and tobacco budworm, H. virescens. Their system concentrated on short-season cultivars and reduced irrigation and fertilizer inputs compared with normal practices. Under the short-season system the growing season was shortened by about 20 days with no loss of yield, and the number of insecticide applications was reduced to six compared with ten for the conventional system.

Despite the demonstrated value of short-season systems, use of this strategy has varied among locations and production years. Less emphasis is now placed on short-season practices in Arizona because of the elimination of the boll weevil. Although short-season production proved an effective means of managing the pink bollworm in the Imperial Valley, the introduction of transgenic, insecticidal cottons has facilitated the reversion to a more traditional system. A number of producers in the Lower Rio Grande Valley of Texas, especially dryland producers, still utilize many of the short-season production practices; yet they are not used on a community-wide basis because of the allure of potential yield increases from a late-season top crop.

Intercropping or strip-cropping and ecological diversity

The recent literature indicates considerable interest in managing pests through habitat modification or increased cropping system diversity. Although the idea that diversity itself confers system stability has proved inconsistent (Smith and McSorley, 2000), specific cases of reduced pest damage in cotton caused by increased cropping diversity have been demonstrated. An example involves strip-cropping or intercropping lucerne in association with cotton to reduce damage from lygus bugs (Lygus spp.).

In the 1960s lygus bugs, and secondary pest problems resulting from pesticide treatments for their control, threatened the viability of cotton production in the San Joaquin Valley of California (Stern, 1969). In addition to cotton, major components of the agricultural system included production of certified lucerne seed and safflower. Both crops produced large populations of lygus that moved to cotton when the lucerne was stressed for moisture or cut for hay, or when lucerne or safflower seed crops matured (Stern et al., 1967; Mueller and Stern, 1973; Sevacherian and Stern, 1974). To reduce the lygus problem, Stern et al. (1967) devised a method of strip-cutting lucerne in which the hay crop was harvested in alternate strips. By offsetting the harvest dates of alternating strips by 12–14 days, a stable environment was afforded adult lygus, while high mortality of eggs and nymphs occurred in the mowed strips. This harvest pattern discouraged the movement of adult lygus to cotton and limited their population build-up in lucerne. However, successful implementation of the strip-cutting technique required careful control of irrigation schedules. Stern et al. (1967) provide detailed recommendations for appropriate irrigation systems. It is evident from their descriptions that the strip-cutting technique required substantial investment in irrigation management. As a consequence, adoption of the tactic was localized and more extensive where irrigation from an underground pipe minimized problems with movement of equipment over irrigation ditches.

During this time, cotton acreage was regulated through a price support programme and producers typically had productive land available in excess of their allotments (Stern, 1981). This arrangement led to the production of 'skip-row' cotton (two rows of cotton alternating with two unplanted rows across the field), and facilitated the development of a technique in which lucerne was intercropped with cotton (Stern, 1969). In the intercropping tactic, strips of lucerne about 20 ft wide alternated with strips of cotton about 500 ft wide. Although the harvest of lucerne hay from alternating halves of the strips at 2- to 3-week intervals was anticipated, hay production proved too limited for continuation of this practice. The intercropping tactic also required special irrigation measures to maintain the lucerne strips in a lush condition. Stern (1969) indicated that the intercropping scheme essentially eliminated the need for lygus treatments in cotton. Despite its effectiveness, intercropping was not widely adopted because it required a higher level of planning and irrigation management than did the use of conventional insecticides (Reynolds *et al.*, 1975). Intercropping also interfered with the production of certified lucerne seed, and the elimination of acreage restrictions on cotton production led to the abandonment of both skip-row culture and the intercropping tactic (Stern, 1981).

Aphid pathogens in Arkansas

In the late 1970s through the 1980s the cotton aphid, *A. gossypii* Glover, became a widespread pest of United States cotton (King and Phillips, 1989; Slosser *et al.*, 1989; Leser *et al.*, 1992), whereas it had previously been considered a secondary pest. In addition to a change in the pest status of the aphid, resistance to organophosphate insecticides (Grafton-Cardwell, 1991; O'Brien

et al., 1992) complicated control efforts. In 1989, epizootics caused by *Neozygites fresenii* (Nowakowski) Batko were observed in Arkansas (Steinkraus *et al.*, 1991). These epizootics, which have recurred each year, were marked by rapid declines in aphid populations (Steinkraus *et al.*, 2003). Field surveys indicated that epizootics of the fungal pathogen generally occurred during July or early August in Arkansas (Steinkraus and Hollingsworth, 1994; Steinkraus *et al.*, 1995), and that they could be predicted based on the prevalence of fungus-killed aphids (Steinkraus and Hollingsworth, 1994). Additional studies defined sampling guidelines for detection of the fungus, and established that impending declines in aphid populations were associated with a fungal prevalence of more than 15% (Hollingsworth *et al.*, 1995). In 1993, a service was initiated in which participants collected and forwarded aphids from infested fields. In return they received information on the extent and stage of fungal infection, and advice regarding the need for chemical controls (Steinkraus *et al.*, 2003). The results of these examinations were also posted on the service's Internet web site and used in extension newsletters to inform producers and consultants of the occurrence of epizootics in their areas. The service has expanded to include Arkansas, Alabama, Florida, Georgia, Louisiana, Mississippi and South Carolina, processing samples from about 300–400 fields yearly. Although the aphid fungus diagnostic service has had substantial impact in reducing the use of unnecessary chemical controls (Steinkraus *et al.*, 2003), its future is contingent on the continued predictability of epizootics, funding and perhaps the availability of alternative tactics including new chemical pesticides.

Characteristics of successful tactics

Classification of an ecologically based pest management tactic as successful depends on whether 'success' means the tactic was effective, or whether such classification also requires adoption by producers. From a practical standpoint, any tactic that is not adopted, regardless of its effectiveness, cannot be considered successful. Given that a tactic is effective, a critical factor governing its success is the availability of effective and economical alternatives. Lack of alternatives obviously favours adoption. But as demonstrated by the cyclical adoption of ploughdown practices in Arizona, a high level of effectiveness may also favour complacency and subsequent abandonment of important control measures. A factor strongly favouring adoption is legislation, which makes the control tactic mandatory. While such action ensures at least temporary adoption, it carries with it the difficulties of enforcement. Also, government mandates are not necessarily the most satisfactory path to adoption, as they sometimes are ill-conceived, and therefore quickly abandoned (see Morrill, 1921). While the underlying ecological basis for a given tactic need not be simple, unless the tactic is mandated or no reasonable alternative is available, success often hinges on simplicity of implementation or ease of integration into the production system. Tactics that are difficult to implement correctly or that are perceived as being incompatible with one or more aspects of the production system tend to encourage producers to continue to search for alternatives. These alternatives frequently come in the form of new classes of synthetic insecticides.

Factors Limiting Durability or Adoption of Ecologically Based Tactics

Many ecologically based pest management tactics or strategies that are initially adopted are not durable. They are only used until a simpler, less expensive or more effective alternative becomes available. Some tactics, such as uniform delayed planting, are optimally effective only when implemented on a community-wide basis. Thus, abandonment of the tactic by only part of the producers in an area may lead to ineffectiveness of the tactic. In other cases such as the interplanting of lucerne and cotton, the realization

of the complexity of proper management or changes in government regulation may make the tactic less appealing compared to alternatives that were previously considered undesirable. In any case, durability of any successful ecologically based pest management tactic or strategy is likely to be temporary. The extent of a tactic's durability depends on how well it continues to compete with available alternatives in the face of changing management, economic and sociological conditions.

The need for risk avoidance and short-term profitability

Farming is a for-profit enterprise. This means that any management practice, including those intended to manage pests, must consistently deliver a financial return in excess of its cost. Headley (1975) listed three ways in which this return could be realized: an increase in production with no increase in input costs; maintenance of production with fewer inputs; and reduced variation in production (and therefore income). Reduced variability in production is a function of the consistency of performance of the pest management practice. The value of reduced variability is in reducing risk. Headley (1975) also argues against extrapolating the value of a pest management practice to a production region based on the returns to early adopters. If the practice results in higher outputs and the market is elastic (price adjusts to reflect production levels), benefits realized by early adopters may diminish after widespread adoption because of price adjustments. This may then lead to abandonment of the management practice.

Several authors have argued on philosophical grounds that changes must be made to the ways costs and profits are calculated in agricultural systems so that indirect and long-term environmental costs are included (Geng et al., 1990; Schaller, 1993). However, there are no means by which producers can recover these indirect costs. The fact that rising production costs and flat or declining commodity prices have made agriculture less profitable than in the past (Pedigo, 1995; Prokopy *et al.*, 2000) only serves to emphasize the need for a financial return from pest management practices. These financial conditions serve to place increased emphasis on short-term profitability at the expense of longer-term benefits, as farmers at financial risk are usually unable to implement practices that require long-term investment for realization of profit (Prokopy *et al.*, 2000). Therefore, adoption of a pest management tactic is heavily dependent on realization of its short-term benefits, whether they may be in the form of increased profit or decreased risk. Durability of a tactic is then dependent on the continuation of those benefits relative to the benefits offered by alternatives.

Sociological tendencies favouring simpler management

Both Grumbly (1978) and Huffaker *et al.* (1978) allude to the importance of simplicity in delivering and implementing IPM programmes. Part of the attractiveness of pesticides is their simplicity of use, whereas implementation of alternative, ecologically based methods often requires greater ecological knowledge and increased management effort (Ehler and Bottrell, 2000). Management of the cotton crop is already a complicated affair, and pest management tactics or strategies that substantially increase the required level of management are unlikely to be widely adopted. Higher levels of management generally translate into higher costs, whether from increased time and effort for planning or increased dollar outlays for implementation. Higher levels of management also increase the opportunity for management errors that decrease effectiveness of a tactic or increase production risks. The need for simplicity does not infer that the ecological basis for a tactic need be simple, only that the tactic should be simple to implement. Transparency of implementation to producers (compatibility with production goals and minimization of apparent management effort) is desirable and may be essential for successful adoption and long-term use of ecologically based pest management tactics.

Competition with the pesticide 'gold standard'

Conventional pesticides have an important role in United States agriculture (Ragsdale, 2000), and they will continue to be important management tools in cotton IPM for the foreseeable future. Goals of developing wholly non-chemical management strategies are not realistic (Gianessi, 1993). Even where ecologically based tactics are incorporated into an overall management strategy, conventional pesticides are usually required for their augmentation (Pedigo, 1995). Pesticides generally provide incremental benefits far in excess of their costs (Zilberman *et al.*, 1991), are simple to use, are rapid in their action and require the user to possess little ecological knowledge (Quisenberry and Schotzko, 1994; Ehler and Bottrell, 2000). In fact, conventional pesticides are typically used to rescue crops from pest infestations, and in this role may provide considerable ability to mask or minimize the deleterious effects of otherwise poor management. Because conventional pesticides are easy to use and familiar to producers, emphasis on more complex ecologically based management tactics tends to be reduced so long as effective pesticide-based methods are available (Luttrell, 1994; Pedigo, 1995). Even when goals of alternative tactics are met, such as target specificity and low environmental toxicity, these advantages are often insufficient to overcome disadvantages of cost and reduced impact on other pest species compared with conventional pesticides (Foster and Harris, 1997). Thus, widespread and durable replacement of chemical controls will require that the alternatives be consistently effective, economical and simple to implement.

Dynamics of an ever-changing production system

Changes in the economics of production, or in pest management strategies themselves, influence the applicability of a given management tactic (Foster and Harris, 1997).

Ecologically based management tactics, in particular, often involve explicit changes to the production system. These changes are intended, by definition, to change the dynamics of pest populations. As a result, management priorities identified before integration of the tactic may shift. The entomological literature contains numerous examples of successfully integrated management tactics that lacked long-term durability because of changes to production goals or methods (see examples of ecologically based tactics described earlier). In many instances limited durability is probably unavoidable, but in others technological durability and transferability to other systems may be improved through knowledge of the specific ecological mechanisms involved and verification of their roles within the agroecosystem.

Major Developments Offering New Opportunities and Challenges

Cotton production systems are constantly changing. Many of these changes are relatively subtle in their impacts on pest situations. However, key recent developments, such as progress in the eradication of the boll weevil or the relatively rapid and widespread adoption of transgenic varieties, have markedly changed the nature and objectives of pest management in some production systems. Additional changes, such as conservation tillage, narrow row spacings or prescription pest control, have the potential to influence cotton production and pest management if associated improvements in economics or technology increase their adoption.

Boll weevil eradication

In 1983, shortly after the conclusion of the boll weevil eradication trial in North Carolina and Virginia, the boll weevil eradication programme was initiated in 95,000 acres of North and South Carolina cotton. By 1987, the programme had expanded to approximately 400,000 acres in Georgia,

Florida and parts of Alabama (Brazzel et al., 1996). By 1998, approximately 4.5 million acres were engaged in some phase of eradication (Williams, 1999), and by 2002, for the first time the boll weevil was not reported as one of the top five pests of United States cotton in the Cotton Insect Losses report of the Beltwide Cotton Conferences (Williams, 2002). By 2003, the boll weevil was eradicated from nearly 6 million acres, and active programmes were under way in most of the remaining US cotton acreage (Grefenstette and El-Lissy, 2003). In the wake of this programme, production and management constraints previously determined by the presence of the boll weevil were lifted, new complexes of pests arose and production levels responded to the changed economics of the cotton system. These changes constitute both new challenges and new opportunities for cotton pest management.

Elimination of a dominant pest

Before the elimination of the boll weevil as an economic pest in much of the south-eastern USA it was impossible to accurately determine the impact of many minor or occasional pests. This was especially true for early-season pests because the overriding need to control the boll weevil resulted in treatments that affected both pests and non-target arthropod populations. Prior to suppression of boll weevil populations some minor pests were inadvertently controlled by treatments for the weevil, and potential impacts of other pests were likely masked by weevil-induced damage. The chemical controls that reduced populations of certain pests also reduced populations of beneficial arthropods (Brazzel et al., 1996). They are generally considered to have exacerbated early-season problems with the tobacco budworm and cotton bollworm (Lambert et al., 1996). As a result of the removal of the boll weevil as the dominant cotton pest in a substantial portion of the United States cotton belt, researchers' attention has shifted to improve management of heliothines (see Transgenic varieties section), and to understanding and managing species that were previously considered minor or occasional pests.

Release of 'second-tier' pests

Areas from which the boll weevil has been eradicated no longer feature chemical controls for the weevil except in instances of reinfestation. The omission of these chemical controls from the management system has made a major contribution towards the occurrence of what Leonard and Emfinger (2002) referred to as 'low spray environments'. An interesting consequence of such environments has been the elevation of populations of certain species formerly considered minor or occasional pests. Apparently sprays for the boll weevil previously suppressed these pests. Factors other than eradication likely contributed to the elevated status of some pests, such as the European maize borer in North Carolina (Ellsworth and Bradley, 1992). But the elimination of sprays for the boll weevil has been a major factor in the increased pest status of a cadre of species including a complex of stinkbug species, the cotton fleahopper, *Pseudatomoscelis seriatus* (Reuter), and plant bugs (*Lygus* spp.) (Lambert *et al.*, 1996; Leonard and Emfinger, 2002). As boll weevil eradication progresses, the pest status of United States cotton continues to change with generally increasing losses from assorted minor pests (Williams, 2002, 2003). Such changes present new challenges for the economic entomologist, but they also provide new opportunities for the study and development of alternative, ecologically based management strategies.

Production level responses

An important consequence of boll weevil eradication has been the tremendous increase in cotton acreage in some parts of the south-eastern cotton belt. These are areas of relatively high annual rainfall that promoted greater in-season boll weevil survival and population growth compared with the more arid western production regions. Before the boll weevil was eradicated, the combined cotton acreage of Virginia and North Carolina was less than 50,000 acres, of South Carolina was about 70,000 acres and of Georgia and Florida was less than 290,000

and 110,000 acres, respectively (Grefenstette and El-Lissy, 2003). By 2003, cotton acreage had increased to about 100,000 acres in Virginia, 200,000 acres in South Carolina and 120,000 acres in Florida, while acreage in North Carolina approached 1 million acres and in Georgia 1.4 million acres (Williams, 2003). These increases in production are making the marketplace more competitive, and may add incentives to develop and adopt ecologically based management methods that reduce production inputs.

Transgenic varieties

One of the most technically impressive and globally controversial events in cotton production has been the development and adoption of insect- and herbicide-resistant transgenic varieties. Insecticidal cotton expressing the gene for the Cry1Ac protein from *Bacillus thuringiensis* (Bt cotton; Bollgard, Monsanto, St. Louis, Missouri) became commercially available in 1996 when it was planted in approximately 1.8 million acres in the USA (Perlak *et al.*, 2001). By 2001, more than 75% of United States cotton acreage was planted with transgenic varieties expressing insect, herbicide or both types of resistances (May *et al.*, 2002). The level of adoption of transgenic varieties is still changing, and in 2002 the acreage planted with Bt cotton actually decreased by nearly 1 million acres (Williams, 2003). The ultimate level of adoption will vary regionally, as the value of transgenic varieties to producers depends on pest types and infestation levels, technology fees associated with seed purchases, and lint yield and quality attributes (Klotz-Ingram *et al.*, 1999; Bryant *et al.*, 2002, 2003).

Insecticidal transgenics

As boll weevil eradication has progressed across the south-eastern cotton belt, emphasis has shifted to the control of heliothines, especially the tobacco budworm. Concurrently, the tobacco budworm has become increasingly difficult to control. By the early 1990s significant levels of resistance to all available classes of insecticide were documented in this species (Elzen, 1997). Assays in the late 1990s confirmed that high frequencies of resistance to some materials persisted (Hardee *et al.*, 2001). This situation encouraged the rapid adoption of Bt varieties shortly after they became available. Since their introduction, associated control of tobacco budworm and pink bollworm has been very good (Agi *et al.*, 2001; Perlak *et al.*, 2001). However, control of the cotton bollworm has been less consistent (Hardee *et al.*, 2001; Perlak *et al.*, 2001). In fact, benefits of supplemental insecticidal controls for the bollworm on Bt cottons have been demonstrated (Agi *et al.*, 2001; Jackson *et al.*, 2003). Although some problems in controlling the bollworm may be related to lack of Cry1Ac expression in selected plant parts (Perlak *et al.*, 2001) or changing levels of gene expression with increasing plant maturity (Adamczyk *et al.*, 2001), susceptibility of the bollworm to the Cry1Ac toxin is generally lower and more variable than that of the tobacco budworm (Tabashnik, 1997; Luttrell *et al.*, 1999; Jackson *et al.*, 2003). Furthermore, transformed varieties expressing the Cry1Ac gene are not effective against the beet armyworm, *Spodoptera exigua* (Hübner), or fall armyworm, *Spodoptera frugiperda* (J.E. Smith) (Perlak *et al.*, 2001). As a consequence of the high adoption rate of the Bt varieties, the bollworm has become the predominant lepidopterous pest of United States cotton (Williams, 1999, 2002).

In areas traditionally experiencing high populations of the tobacco budworm or pink bollworm the Bt cottons have become an important production component, and substantial effort has been directed towards developing management strategies to preserve their effectiveness. Resistance to the Cry1Ac toxin in the tobacco budworm has been detected in laboratory experiments and field populations (Elzen, 1997; Gould *et al.*, 1997), but no trend towards resistance development has been observed in the field (Elzen and Hardee, 2003). Likewise, resistance is known from field populations of the pink bollworm but efficacy of Bt cotton has remained high (Tabashnik *et al.*, 2000).

Resistance management efforts have relied primarily on the strategy of a high dose of Cry-toxin combined with the production of susceptible insects in non-transgenic refuges (Elzen and Hardee, 2003). Refuge requirements vary in size and placement depending on whether they are protected by conventional (non-Bt) insecticides or left completely untreated. The refuge strategies are based on simulations that make assumptions regarding the genetics of resistance, local population movement or mixing and random mating of susceptible and resistant moths. In addition, the spatial arrangement and management of refuges, and contributions made to local populations by long-range migrants and individuals developing on non-crop plants, may play important roles in the development of resistance (Kennedy and Storer, 2000). Refuge strategies have not been rigorously tested (Tabashnik, 1997), and producers and some scientists have begun to doubt the need for refuges.

Recently, Bt varieties containing two Cry-toxins (Cry1Ac and Cry2Ab; Bollgard II, Monsanto, St. Louis, Missouri) became commercially available. These varieties have exhibited higher levels of bollworm control than those expressing the single Cry-toxin gene (Bacheler and Mott, 2003). Jackson *et al.* (2003) found that the addition of the second Cry-toxin negated the need for supplemental chemical controls for the bollworm. Incorporation of different or additional insecticidal toxins into commercial varieties may aid in long-term resistance management (Elzen and Hardee, 2003). However, the extent of adoption of these new varieties will likely depend on the levels of pest pressure and the cost of seed and technology fees in addition to efficacy. In areas or years of low pest pressure, the technology fees may be greater than the cost of conventional chemical controls (Edge *et al.*, 2001).

Adoption of the Bt varieties in areas of traditionally high tobacco budworm populations has substantially contributed to the occurrence of 'low spray environments' (see Boll weevil eradication section). Leonard and Emfinger (2002) and Bacheler and Mott (2003) suggested that increased adoption of

Bollgard II varieties would further reduce insecticide inputs for lepidopterous pests while increasing the potential for damage from other pests not affected by Bt toxins. Concern has also been expressed for potential environmental impacts on non-target species. Verkerk *et al.* (1998) suggested that release of previously minor pests, and consequential use of conventional chemical controls, could be ecologically disruptive by reducing populations of natural enemies. Hilbeck (2001) cites possible effects on food web structure and population regulation processes at multitrophic levels, but also indicates that such problems have not yet been reported despite large-scale use of transgenic varieties in the USA. Lewis *et al.* (1997) and Hilbeck (2001) imply that the insecticidal transgenic varieties represent a 'single solution strategy' or a 'silver bullet solution', respectively, at odds with the development of ecologically based IPM. While the philosophical aspects of development and utilization of insecticidal transgenic varieties can be debated, in practice this technology represents a tool not unlike highly effective conventionally bred insect resistance. At present, the transgenic varieties have neither solved all of the important problems facing cotton pest management nor been universally adopted. This technology can be used to simplify crop management where the species composition and severity of occurrence of lepidopterous pest problems warrant its use. In this role use of the transgenic varieties may facilitate study of the cotton agroecosystem through reduction or elimination of disruptive insecticidal controls for key lepidopterous pests.

Herbicide resistance

Transgenic cotton varieties expressing resistance to specific broad-spectrum herbicides have increased management options for controlling problem weeds, including cocklebur (*Xanthium strumarium* Linnaeus) and the morning glories (*Ipomea* spp.) (Bryant *et al.*, 2003). These varieties have been relatively widely adopted because they often offer efficient weed control at reduced costs (Baldwin, 2000). Few reports have addressed repercus-

sions of the use of herbicide-resistant varieties on the management of arthropod pests. Gurr *et al.* (2003) suggested that use of the herbicide-resistant cottons could effectively increase botanical and invertebrate diversity if herbicide applications were delayed until after crop emergence. Similarly, Leonard and Emfinger (2002) reported that delays in weed control allow insects to use weeds as refuges until cotton becomes available. Although the occurrence of specific problems caused by such delayed weed control has not been reported, in central Texas substantial populations of the cotton fleahopper have been observed on weeds growing among seedling cotton plants before the application of chemical weed controls (D.W.S. and C. P.-C. Suh, 2003 unpublished observation). In addition, the herbicide-resistant varieties have alleviated weed control problems that previously limited the adoption of conservation tillage (Leonard and Emfinger, 2002) and ultra-narrow row cotton (Vories *et al.*, 1999; Jones, 2001; Jost and Cothren, 2001). Therefore, the use of transgenic herbicide-resistant cottons has the potential to influence the management of arthropod pests indirectly through associated changes in cultural practices.

Complicating cultural shifts

Changes in cultural practices often have important direct and indirect effects on populations of pests and beneficial arthropods in cotton. Two practices in particular, conservation tillage and narrow row spacing, have received considerable attention in the past, but their practicality was limited by the lack of adequate weed control methods. Development of transgenic herbicide-resistant varieties has substantially alleviated this constraint. Consequently, these practices are receiving renewed attention in some areas of the cotton belt. At the same time, a few researchers have taken the philosophy associated with precision or 'site-specific' farming and extended it to the 'prescription management' of insect pests. While none of these practices currently enjoys a wide level of adoption in United

States cotton production systems, they each offer the potential for increased adoption, and therefore may impact on cotton pest management strategies of the near future.

Conservation tillage

Reported effects of conservation tillage practices on insect populations in cotton have not indicated clear trends (Ruberson et al., 1997). Gaylor et al. (1984) found no evidence that conservation tillage in cotton influenced pest populations except that cutworm populations were higher when a legume winter cover crop was used. Leonard and Emfinger (2002) indicated that, compared with conventional tillage, problems with cutworms, cotton aphids and the false chinch bug (Nysius raphanus Howard) were more common in systems featuring reduced tillage in conjunction with winter cover crops. Likewise, Stewart (2003) indicated an increased potential for problems with cutworms, grasshoppers, false chinch bugs and spider mites in conservation tillage systems. Reported effects of conservation tillage practices on beneficial arthropods have also been inconsistent. Gaylor et al. (1984) reported higher peak predator populations in conventional rather than conservation tillage treatments, while Ruberson et al. (1997) indicated typically higher beneficial population levels in row-tilled rather than conventionally tilled plots. McCutcheon (2000) found that populations of big-eyed bugs (Geocoris spp.) were reduced in a rye/no-till system because of increased population levels of predatory ants. Finally, Stapel et al. (1998) found that higher populations of beneficial insects were associated with conservation tillage and cover crops, but pest populations were also higher in the reduced-tillage cotton.

Because the term 'conservation tillage' encompasses a wide variety of tillage practices, inconsistencies among reports may be the result of several factors including variation in the practices used, the extent and types of associated cover crops, and local or regional differences in the types and population levels of pests and beneficial arthropods. In production systems where timely destruction of cotton stalks is an important pest management component, effective chemical control of weeds must be accompanied by effective chemical termination of the postharvest cotton crop. Especially where the ability to effectively terminate the cotton crop is impinged, consistent effects of conservation tillage on cotton pests may not become evident until a high level of adoption is achieved within a production region.

Ultra-narrow row culture

Ultra-narrow row production involves planting cotton in rows of 38 cm (15 inches) or less at plant populations that are slightly to considerably higher than in conventional plantings (Nichols et al., 2003). Ultra-narrow row production is most often viewed as an option on marginal land where plant growth is limited (Allen et al., 1998; Kerby, 1998). One advantage of ultra-narrow row cotton is in machinery costs. Ultra-narrow row cotton is stripper-harvested, and both equipment purchase and maintenance costs are lower for strippers than for spindle pickers (Jones, 2001; Shurley et al., 2002). Stripper harvesting requires that plant growth be carefully managed in a manner that is consistent with a low-input system.

Reports of the effects of ultra-narrow row production on cotton plant development, yield and production costs have been variable. In some cases, plants grown under narrow row, high plant population conditions display reduced plant height, fewer nodes and earlier maturity (Vories et al., 1999). However, effects of row spacing on these parameters are inconsistent (Gerik et al., 1998) and appear to depend on soil type and growing conditions (Jost and Cothren, 2001). Under marginal production conditions most results indicate some yield advantage of ultra-narrow row production over conventional row spacings (Shurley et al., 2002). Yield advantages furnished by narrow row production may be negated by reductions in gin turnout, fibre quality or trash content (Jones, 2001; Nichols et al., 2003). Although some production costs may be higher for ultra-narrow row cotton, such as seed and technology fees (Vories *et al.*, 1999; Shurley *et al.*, 2002), net returns may still be increased compared with conventional cotton (Parvin *et al.*, 2002).

Allen *et al.* (1998), Gerik *et al.* (1998) and Kerby (1998) all suggest that ultra-narrow row plantings managed for early maturity provide savings in insect control costs by virtue of a shortened fruiting period. However, such a conclusion may be premature without additional investigation. Such savings may not be expected for some pests such as the boll weevil. For example, Slosser *et al.* (1986) found that plantings on 51 cm row spacings were more heavily damaged by the boll weevil than were plantings on wider spacings. Also, Pierce *et al.* (2001) found that boll weevil survival and emergence was enhanced by narrow (17 cm) row spacings compared with conventional (96 cm) spacings. Although the boll weevil may not provide the most relevant example for most of the United States cotton belt, similar instances in which pest problems are exacerbated by narrow row spacings may yet become apparent. Whether insecticide savings materialize may depend on the specific production system and its corresponding pest complex, and on whether the pest avoidance provided by the shortened period of fruit susceptibility is negated by increased pest reproduction or survival in response to altered microclimate. Finally, ultra-narrow row plantings would appear to complicate efforts to estimate arthropod populations for the purposes of research or management. In these plantings, higher plant populations may necessitate increased sample sizes, and special methods may be required to physically traverse and sample such dense plantings.

Precision farming and prescription management

Scientists have begun to combine the use of recently available spatial tools such as global positioning system (GPS), multispectral aerial photography and satellite technologies to extend precision farming approaches to cotton pest control. These technologies provide a means to spatially delineate fields

based on any criteria, and to use these maps to address site-specific production problems (Leonard *et al.*, 2003a). Most recent efforts utilize remote-sensing methods that are not new, but are more accessible and less expensive than in the past. Objectives range from early pest detection or identification of high-risk areas of fields to 'prescription' pesticide application in which insecticides are restricted to selected areas of a field (Leonard *et al.*, 2003a). Accomplishment of these objectives would result in reduced pesticide cost and environmental contamination, as well as conservation of non-target and beneficial species in untreated field portions.

Results of efforts to delineate in-field pest populations using remote sensing have been mixed but some positive accomplishments have been reported. Fitzgerald *et al.* (1999) evaluated the use of aerial photography for early detection of mite populations in cotton so that producers could limit pesticide applications to areas of high infestation. Mite damage was successfully detected based on the reflectance and shape of the damaged areas, but whether this system could detect infestations early enough to allow spot treatment was not determined. In addition, image-processing difficulties caused by shadows necessitated limiting flights to within 15 min of solar noon, and conversion of the digital signal to percent reflectance required the use of radiometer data collected from ground panels. Goodell *et al.* (2002) used Landsat imagery to monitor spatial and temporal patterns of availability of western tarnished plant bug (*L. hesperus* Knight) host plants in the San Joaquin Valley of California. Potential lygus hosts were identified using the percent change in the normalized difference vegetation index (NDVI) between consecutive observations. Although ground checks confirmed the location and condition of identified vegetation, utility of the system was limited by the periodicity of satellite overflights (16 days) and interference by cloud cover on some days. Stewart *et al.* (2002) examined the utility of multispectral remote-sensing imagery for monitoring cotton crop development and arthropod populations. In early-season cotton the spectral signature was dominated by soil reflectance. Later spectral signatures were predictive of plant height, numbers of nodes and numbers of bolls, but the strengths of relationships varied among study sites. Although insect pressure was generally light, a weak relationship was observed between spectral signature and aphid populations. Relationships with other insect populations were generally poor.

In an effort to improve sampling efficiency for the tarnished plant bug Willers *et al.* (1999) developed a scheme that exploited an observed relationship between plant bug population levels and spatial differences in crop growth patterns indicated by remote sensing. These methods proved useful by focusing monitoring efforts for nymphs and teneral adults to specific portions of fields after initiation of squaring but before canopy closure. Sudbrink *et al.* (2002) and Leonard *et al.* (2003a, b) used a similar approach to restrict insecticide treatments to designated portions of fields. 'Prescription' maps for each field were developed based on NDVI measurements from multispectral images. Combined with field inspections, these maps were used to direct insecticide treatments for tarnished plant bugs (Sudbrink *et al.*, 2002; Leonard *et al.*, 2003a) and stink bugs (Sudbrink *et al.*, 2002) to indicated areas of the fields, thereby reducing the total insecticidal inputs. Leonard *et al.* (2003a, b) also propose basing prescriptions on historical yield maps so that production inputs can be focused on higher yielding portions of the fields.

Even where remote sensing has been successfully used to detect or define spatial patterns of pest infestations, additional research is needed to demonstrate or improve robustness and reliability of these methods. These methods presuppose distinguishable differences in crop growth and yield potential, and corresponding insect preferences, within fields. Their validation poses a difficult sampling problem because of the need for relatively precise estimates of insect population levels within individual portions of fields. Because producers will likely be unwilling or unable to invest in costly routine ground-truthing efforts,

adoption of these methods may hinge on the economy, robustness and reliability of remote-sensing information. In particular, producer aversion to risk may dictate that prescription management techniques rarely result in crop losses, while avoidance of risk through conservative treatment thresholds or prescription maps would reduce or eliminate cost savings.

Continuing Constraints and a Prognosis for Cotton IPM

The progression of boll weevil eradication programmes and the adoption of insecticidal transgenic cotton varieties have simplified management in some production systems traditionally dominated by the boll weevil, tobacco budworm and pink bollworm. In most of these systems, species previously considered occasional or secondary pests are achieving more elevated status. This has introduced considerable uncertainty into the system regarding the best approach to managing complexes of pests that are perhaps less predictable and less well understood than our traditional foes were. Our historical approach, referred to by Ehler and Bottrell (2000) as 'supervised control', served production needs reasonably well when we dealt primarily with a few, very serious pest species. Therefore, it should not be surprising if that model were extended to newer pests. In fact, removal of dominant pests from a production system may diminish producer incentive to move away from the traditional pesticide-based management schemes because novel or more complicated alternatives are most attractive when traditional approaches are inadequate. Still, reductions in the use of pesticides to control dominant pests provide the researcher opportunity to better understand the ecological foundations of the production systems. Improved understanding could lead to development and implementation of more ecologically sound management strategies if it is recognized that the other economic, ecological and sociological constraints to development and adoption of ecologically based IPM still remain.

Continuing constraints

Economics and simplicity of management

Adoption of cotton production practices, including methods of managing pests, will continue to be driven by efforts to improve profitability and simplify management. For the most part, pest management concerns have a large influence on the overall production strategy only when associated pest problems are perceived as being severe or limiting to production. Otherwise, the researcher or pest management practitioner must develop or alter management strategies in reaction to production system changes brought about by other forces. Thus, alternative pest management practices usually experience a high level of voluntary adoption only when they solve important economic problems or simplify management. For this reason, ecologically based management techniques that compete with effective pesticide-based controls should not only be compatible with production goals, but should offer a marginal improvement to profitability, risk reduction or ease of implementation compared with current management methods.

System dynamics and local specificity

Agroecosystems are both complex and dynamic, and the extent of these qualities increases with scale. There is little doubt that effective ecologically based pest management strategies must be tailored to local conditions because the underlying ecological relationships are often context-specific. Yet as pest managers we desire simple solutions that are durable, robust and practical to implement, at least within a given production system. Reconciliation of these disparate goals will require an astute knowledge of the ecological mechanisms regulating pest population dynamics within the crop or cropping system. This knowledge will be necessary not only to design and modify management tactics to fit specific contexts, but also to better identify appropriate vulnerabilities in pest ecology for exploitation.

Dwindling applied resources

A hiring boon in applied entomology during the 1960s and 1970s provided a large cadre of professionals who were responsible for many of the advancements in IPM. In recent years a major portion of that experienced brain trust and much of our 'institutional memory' has been lost to retirement. Furthermore, the combination of an apparent decrease in the perceived value of traditional applied ecology and tight departmental budgets has resulted in the loss of many applied faculty positions. The dwindling base of researchers and practitioners with actual field experience can only increase the challenge associated with development of new knowledge necessary for improvements in cotton pest management.

At the same time, changes in the way research is funded have increased dependence on extramural sources. This shift in funding mechanism influences both the process and focus of research. An inordinate dependence on extramural funding potentially impacts on the quality and diversity of research and instruction as faculty members are forced to spend more time writing proposals and reports, and less time conducting research in the first person. A resultant reliance on graduate students and support staff for data collection can deprive the researcher of the opportunity to observe – which may impact on the level of experience and insight that can be brought to bear on applied problems. Reliance on external funding also impacts on the type of research conducted because funding availability may not be consistent with the needs and opportunities identified by the researcher. As a result, customers of IPM may perceive a disconnect between research and the real world (e.g. Ohmart, 2000). Although technological advancements are often touted as the future of cotton IPM, the usefulness and durability of new technologies will be determined by how well they can be incorporated into agricultural systems. This incorporation will be dependent on the insight, experience and vision of qualified applied ecologists.

A prognosis

Context-specific research and technology transfer

The early research mandates of IPM were comparatively clear; we needed appropriate means of estimating pest populations, an understanding of pest biology and physiological ecology, and knowledge of the effects of pest populations on crop yields. In addition, we needed this information integrated into a systems framework. Many of the goals regarding sampling, pest biology and ETs were substantively addressed. However, a sophisticated level of information integration was mostly not accomplished. A factor perhaps contributing to this failure is the manner in which research goals have been identified and prioritized. Much of the basic biological and physiological information was developed independently of an applied context. Consequently, this valuable information was often made available in a form that did not facilitate direct application to pest management problems. On the other hand, much of the applied research focused on the bigger picture of pest control. Although it accurately documented phenomena observed in the field, it provided comparatively little concrete information about the mechanisms that regulated those phenomena. Expectations for timely solutions to pest problems have helped to maintain an emphasis on pesticide-based management methods because this approach has consistently yielded useable short-term solutions. These methods will continue to be an important and necessary component of most cotton pest management systems. However, in a few cases such as those cited herein (Examples of Ecologically Based Tactics section), some particular insight or opportunity prompted the development of basic biological information in a context that was directly applicable to a problem. This same approach can be leveraged against large, complex and intractable ecological problems by breaking them into solvable components. Rather than focusing on the larger and less well-defined problem, this approach would address the most important problems, or components of problems, that are likely to be solvable. The anticipated outcome would be a series of research products that addresses the larger problem in incremental and readily transferable changes to the production system.

An emphasis on mechanisms, simplicity

The many interrelations among components of agroecosystems are too dynamic and complex for direct manipulation in controlled experiments. Yet these are exactly the kinds of studies that are needed to accurately assess and understand the roles and functions of the various ecological components. An alternative to direct observation of entire systems that provides a higher probability of yielding useable results and insights is to study specific components of the ecological question with emphasis on uncovering underlying mechanisms. These insights then provide hypotheses that can be tested or validated in real-world systems. Such an astute understanding of production systems will be necessary to produce solutions that are simple enough to promote a high level of adoption.

In many instances, reasonable expectations for the successful evaluation or implementation of a management practice require a high level of adoption on a community-wide or regional basis. Even when this requirement is recognized it is usually not possible to accurately define the extent of area required. A research approach in which the system is dissected and its components intensively examined may offer the only practical approach to study. This may be the case not only because of the inordinate expense of conducting an area- or region-wide study, but also because statistically correct experimental designs and methods are usually not applicable due to the lack of sufficient numbers of equivalent experimental units. The most practical substitution for a lack of true replication may be sufficient documentation of enough facets of the system so that findings can be examined for consistency and compatibility.

In this respect, an understanding of the mechanisms involved can focus efforts on the most relevant measurements. Failure to adequately document system details in an areawide study forces the investigators to place inordinate reliance on assumptions and intuition when interpreting the results. This reliance offers little in the way of substantive information that may be necessary to explain the success or failure of a given management tactic or strategy, or identify modifications that may improve a tactic's effectiveness or economy.

Unprecedented opportunities in the face of unprecedented challenges

Regional elimination or abatement of problems associated with specific dominant pests has simplified pest management in some production systems. This simplification, along with an expanding repertoire of technically sophisticated equipment and methods, presents unique opportunities to study the ecology of cotton production systems with minimal interference from pesticides. However, identifying fruitful avenues and approaches for research will require a high level of insight and sophistication if the objective is to solve pest management problems rather than to simply publish scientific papers. Whether these opportunities result in a break from our traditional emphasis on control tactics for individual pests or pest complexes, towards the development of a more integrated and holistic management strategy, will be determined in a large part by the research and technology transfer approaches of an increasingly diminishing cadre of applied research and extension scientists.

References

Adamczyk, J.J. Jr, Hardee, D.D., Adams, L.C. and Sumerford, D.V. (2001) Correlating differences in larval survival and development of bollworm (Lepidoptera: Noctuidae) and fall armyworm (Lepidoptera: Noctuidae) to differential expression of Cry1A(c) δ-endotoxin in various plant parts among commercial cultivars of transgenic *Bacillus thuringiensis* cotton. *Journal of Economic Entomology* 94, 284–290.

Adkisson, P.L. (1962) Timing of defoliants and desiccants to reduce populations of the pink bollworm in diapause. *Journal of Economic Entomology* 55, 949–951.

Adkisson, P.L. and Gaines, J.C. (1960) Pink bollworm control as related to the total cotton insect control program of Central Texas. *Texas Agricultural Experiment Station Miscellaneous Publication* 444, 1–7.

Adkisson, P.L., Niles, G.A., Walker, J.K., Bird, L.S. and Scott, H.B. (1982) Controlling cotton's insect pests: a new system. *Science* 216, 19–22.

Agi, A.L., Burd, A., Bradley, J.R. Jr and Van Duyn, J.W. (2001) Planting date effects on heliothine larval numbers, fruit damage, and yield of transgenic *Bt* cotton in North Carolina. *Journal of Entomological Science* 36, 402–410.

Allen, C.T., Kennedy, C., Robertson, B., Kharboutli, M., Bryant, K., Capps, C. and Earnest, L. (1998) Potential of ultra narrow row cotton in southeast Arkansas. In: Duggar, P. and Richter, D.A. (eds) *Proceedings Beltwide Cotton Conferences.* National Cotton Council, Memphis, Tennessee, pp. 1403–1406.

Altieri, M.A. (1999) The ecological role of biodiversity in agroecosystems. *Agriculture, Ecosystems and Environment* 74, 19–31.

Atkinson, D. and McKinlay, R.G. (1997) Crop protection and its integration within sustainable farming systems. *Agriculture, Ecosystems and Environment* 64, 87–93.

Bacheler, J.S. and Mott, D.W. (2003) Efficacy of Bollgard II cotton under non-enhanced agronomic conditions in North Carolina, 1996–2002. In: Duggar, P. and Richter, D.A. (eds) *Proceedings Beltwide Cotton Conferences.* National Cotton Council, Memphis, Tennessee, pp. 1011–1014.

Baldwin, F.L. (2000) Transgenic crops: a view from the United States Extension Service. *Pest Management Science* 56, 584–585.

Benbrook, C. (2000) Forum: managing agricultural pests. *Issues in Science and Technology* 17, 10.

Bennett, R.L. (1908) A method of breeding early cotton to escape boll weevil damage. *USDA Farmers' Bulletin* 314, USDA, Washington DC, pp. 1–30.

Bergman, D., Henneberry, T.J. and Bariola, L.A. (1982) Distribution of the boll weevil in southwestern Arizona cultivated cotton from 1978–1981. In: Brown, J.M. (ed.) *Proceedings of the Beltwide Cotton Production Research Conferences*. National Cotton Council, Memphis, Tennessee, pp. 204–207.

Binns, M.R. and Nyrop, J.P. (1992) Sampling insect populations for the purpose of IPM decision making. *Annual Review of Entomology* 37, 427–453.

Bradley, J.R. Jr and Phillips, J.R. (1978) Biology and population dynamics. *Boll Weevil: Management Strategies, Southern Cooperative Service Bulletin* 228, 15–22.

Brazzel, J.R., Smith, J.W. and Knipling, E.F. (1996) Boll weevil eradication. In: King, E.G., Phillips, J.R. and Coleman, R.J. (eds) *Cotton Insects and Mites: Characterization and Management*. Cotton Foundation Reference Book Series, No. 3. The Cotton Foundation Publisher, Memphis, Tennessee, pp. 625–652.

Bryant, K.J., Robertson, W.C., Lorenz, G.M., Ihrig, R. and Hackman, G. (2002) Six years of transgenic cotton in Arkansas. In: Duggar, P. and Richter, D.A. (eds) *Proceedings Beltwide Cotton Conferences*. National Cotton Council, Memphis, Tennessee, CD-ROM.

Bryant, K.J., Nichols, R.L., Allen, C.T., Benson, N.R., Bourland, F.M., Earnest, L.D., Kharboutli, M.S., Smith, K.L. and Webster, E.P. (2003) Transgenic cotton cultivars: an economic comparison in Arkansas. *Journal of Cotton Science* 7, 194–204.

Burris, E., Pavloff, A.M., Leonard, B.R., Graves, J.B. and Church, G. (1990) Evaluation of two procedures for monitoring populations of early season pests (Thysanoptera: Thripidae and Homoptera: Aphidae) in cotton under selected management strategies. *Journal of Economic Entomology* 83, 1064–1068.

Byerly, K.F., Gutierrez, A.P., Jones, R.E. and Luck, R.F. (1978) A comparison of sampling methods for some arthropod populations in cotton. *Hilgardia* 46, 257–282.

Chu, C.-C., Henneberry, T.J., Weddle, R.C., Natwick, E.T., Carson, J.R., Valenzuela, C., Birdsall, S.L. and Staten, R.T. (1996) Reduction of pink bollworm (Lepidoptera: Gelechiidae) populations in the Imperial Valley, California, following mandatory short-season cotton management systems. *Journal of Economic Entomology* 89, 175–182.

Colbert, J.J. (1995) Introduction. In: Colbert, J.J. (ed.) Models as links between empiricism and theory in insect ecology. *Computers and Electronics in Agriculture* 13, 87–90.

Copping, L.G. and Menn, J.J. (2000) Biopesticides: a review of their action, applications and efficacy. *Pest Management Science* 56, 651–676.

Cox, P.G. (1996) Some issues in the design of agricultural decision support systems. *Agricultural Systems* 52, 355–381.

Cullen, E.M., Zalom, F.G., Flint, M.L. and Zilbert, E.E. (2000) Quantifying trade-offs between pest sampling time and precision in commercial IPM sampling programs. *Agricultural Systems* 66, 99–113.

Dillon, G.E. and Fitt, G.P. (1995) Reassessment of sampling relationships for *Helicoverpa* spp. (Lepidoptera: Noctuidae) in Australian cotton. *Bulletin of Entomological Research* 85, 321–329.

Edge, J.M., Benedict, J.H., Carroll, J.P. and Reding, H.K. (2001) Bollgard cotton: an assessment of global economic, environmental, and social benefits. *Journal of Cotton Science* 5, 121–136.

Ehler, L.E. and Bottrell, D.G. (2000) The illusion of integrated pest management. *Issues in Science and Technology* 16, 61–64.

Ellington, J. and Southward, M. (1999) Quadrat sample precision and cost with a high-vacuum insect sampling machine in cotton ecosystems. *Environmental Entomology* 28, 722–728.

Ellsworth, P.C. and Bradley, J.R. Jr (1992) Comparative damage potential and feeding dynamics of the European corn borer (Lepidoptera: Pyralidae) and cotton bollworm (Lepidoptera: Noctuidae) on cotton bolls. *Journal of Economic Entomology* 85, 402–410.

Elzen, G.W. (1997) Changes in resistance to insecticides in tobacco budworm populations in Mississippi, 1993–1995. *Southwestern Entomologist* 22, 61–72.

Elzen, G.W. and Hardee, D.D. (2003) United States Department of Agriculture–Agricultural Research Service research on managing insect resistance to insecticides. *Pest Management Science* 59, 770–776.

Fitzgerald, G.J., Maas, S.J. and DeTar, W.R. (1999) Early detection of spider mites in cotton using multispectral remote sensing. In: Duggar, P. and Richter, D.A. (eds) *Proceedings Beltwide Cotton Conferences*. National Cotton Council, Memphis, Tennessee, pp. 1022–1042.

Fleischer, S.J., Gaylor, M.J. and Edelson, J.V. (1985) Estimating absolute density from relative sampling of *Lygus lineolaris* (Heteroptera: Miridae) and selected predators in early to mid-season cotton. *Environmental Entomology* 14, 709–717.

Foster, S.P. and Harris, M.O. (1997) Behavioral manipulation methods for insect pest-management. *Annual Review of Entomology* 42, 123–146.

Fuchs, T.W., Rummel, D.R. and Boring, E.P. III (1998) Delayed uniform planting for areawide boll weevil suppression. *Southwestern Entomologist* 23, 325–333.

Fye, R.E. (1968) Populations of boll weevil in selected fields in Arizona in 1965 and 1966. *Journal of Economic Entomology* 61, 377–380.

Garcia, A., Gonzalez, D. and Leigh, T.F. (1982) Three methods for sampling arthropod numbers on California cotton. *Environmental Entomology* 11, 565–572.

Gaylor, M.J., Fleischer, S.J., Muehleisen, D.P. and Edelson, J.V. (1984) Insect populations in cotton produced under conservation tillage. *Journal of Soil and Water Conservation* 39, 61–64.

Geng, S., Hess, C.E. and Auburn, J. (1990) Sustainable agricultural systems: concepts and definitions. *Journal of Agronomy and Crop Science* 165, 73–85.

Gerik, T.J., Lemon, R.G., Faver, K.L., Hoelewyn, T.A. and Jungman, M. (1998) Performance of ultra-narrow row cotton in central Texas. In: Duggar, P. and Richter, D.A. (eds) *Proceedings Beltwide Cotton Conferences*. National Cotton Council, Memphis, Tennessee, pp. 1406–1408.

Gianessi, L. (1993) The quixotic quest for chemical-free farming. *Issues in Science and Technology* 10, 29–36.

Goodell, P.B., Lynn, K. and McFeeters, S.K. (2002) Using GIS approaches to study western tarnished plant bug in the San Joaquin Valley of CA. In: Duggar, P. and Richter, D.A. (eds) *Proceedings Beltwide Cotton Conferences*. National Cotton Council, Memphis, Tennessee, CD-ROM.

Gould, F., Anderson, A., Jones, A., Sumerford, D., Heckel, D.G., Lopez, J., Micinski, S., Leonard, R. and Laster, M. (1997) Initial frequency of alleles for resistance to *Bacillus thuringiensis* toxins in field populations of *Heliothis virescens*. *Proceedings of the National Academy of Sciences USA* 94, 3519–3523.

Grafton-Cardwell, E.E. (1991) Geographical and temporal variation in response to insecticides in various life stages of *Aphis gossypii* (Homoptera: Aphididae) infesting cotton in California. *Journal of Economic Entomology* 84, 741–749.

Grefenstette, B. and El-Lissy, O. (2003) Boll weevil eradication update. In: Duggar, P. and Richter, D.A. (eds) *Proceedings Beltwide Cotton Conferences*. National Cotton Council, Memphis, Tennessee, pp. 131–141.

Grumbly, T.P. (1978) Policy coherence through a redefinition of the pest control problem, or "If you can't beat 'em, join 'em". In: Smith, E.H. and Pimentel, D. (eds) *Pest Control Strategies*. Academic Press, New York, pp. 261–269.

Gurr, G.M., Wratten, S.D. and Luna, J.M. (2003) Multi-function agricultural biodiversity: pest management and other benefits. *Basic and Applied Ecology* 4, 107–116.

Hanski, I. (1998) Metapopulation dynamics. *Nature* 396, 41–49.

Hardee, D.D., Smith, M.T., Weathersbee, A.A. III. and Snodgrass, G.L. (1994) Sampling of the cotton aphid (Homoptera: Aphididae) in cotton. *Southwestern Entomologist* 19, 33–44.

Hardee, D.D., Adams, L.C. and Elzen, G.W. (2001) Monitoring for changes in tolerance and resistance to insecticides in bollworm/tobacco budworm in Mississippi, 1996–1999. *Southwestern Entomologist* 26, 365–372.

Headley, J.C. (1975) The economics of pest management. In: Metcalf, R.L. and Luckmann, W.H. (eds) *Introduction to Insect Pest Management*. John Wiley & Sons, New York, pp. 75–99.

Heilman, M.D., Namken, L.N., Norman, J.W. and Lukefahr, M.J. (1979) Evaluation of an integrated short-season management production system for cotton. *Journal of Economic Entomology* 72, 896–900.

Henneberry, T.J. and Phillips, J.R. (1996) Suppression and management of cotton insect populations on an areawide basis. In: King, E.G., Phillips, J.R. and Coleman, R.J. (eds) *Cotton Insects and Mites: Characterization and Management*. Cotton Foundation Reference Book Series, No. 3. The Cotton Foundation Publisher, Memphis, Tennessee, pp. 601–624.

Henneberry, T.J., Meng, T., Hutchinson, W.D., Bariola, L.A. and Deeter, B. (1988) Effects of ethephon on boll weevil (Coleoptera: Curculionidae) population development, cotton fruiting, and boll opening. *Journal of Economic Entomology* 81, 628–633.

Higley, L.G. and Wintersteen, W.K. (1992) A novel approach to environmental risk assessment of pesticides as a basis for incorporating environmental costs into economic injury levels. *American Entomologist* 38, 34–39.

Hilbeck, A. (2001) Implications of transgenic, insecticidal plants for insect and plant biodiversity. *Perspectives in Plant Ecology, Evolution and Systematics* 4, 43–61.

Hinds, W.E. (1908) The first and last essential step in combating the boll weevil. *Journal of Economic Entomology* 1, 233–243.

Hollingsworth, R.G., Steinkraus, D.C. and McNew, R.W. (1995) Sampling to predict fungal epizootics in cotton aphids (Homoptera: Aphididae). *Environmental Entomology* 24, 1414–1421.

Hoy, M.A. (2000) Transgenic arthropods for pest management programs: risks and realities. *Experimental and Applied Acarology* 24, 463–495.

Huffaker, C.B., Shoemaker, C.A. and Gutierrez, A.P. (1978) Current status, urgent needs, and future prospects of integrated pest management. In: Smith, E.H. and Pimentel, D. (eds) *Pest Control Strategies*. Academic Press, New York, pp. 237–259.

Hutchins, S.H. (1995) Free enterprise: the only sustainable solution to IPM implementation. *Journal of Agricultural Entomology* 12, 211–217.

Hutchins, S.H. and Gehring, P.J. (1993) Perspective on the value, regulation, and objective utilization of pest control technology. *American Entomologist* 39, 12–15.

Jackson, R.E., Bradley, J.R. Jr and Van Duyn, J.W. (2003) Field performance of transgenic cottons expressing one or two *Bacillus thuringiensis* endotoxins against bollworm, *Helicoverpa zea* (Boddie). *Journal of Cotton Science* 7, 57–64.

Jenkins, J.N. and Wilson, F.D. (1996) Host plant resistance. In: King, E.G., Phillips, J.R. and Coleman, R.J. (eds) *Cotton Insects and Mites: Characterization and Management*. Cotton Foundation Reference Book Series, No. 3. The Cotton Foundation Publisher, Memphis, Tennessee, pp. 563–597.

Jones, M.A. (2001) Evaluation of ultra-narrow row cotton in South Carolina. In: Duggar, P. and Richter, D.A. (eds) *Proceedings Beltwide Cotton Conferences*. National Cotton Council, Memphis, Tennessee, pp. 522–524.

Jost, P.H. and Cothren, J.T. (2001) Phenotypic alterations and crop maturity differences in ultra-narrow row and conventionally spaced cotton. *Crop Science* 41, 1150–1159.

Karandinos, M.G. (1976) Optimum sample size and comments on some published formulae. *Bulletin of the Entomological Society of America* 22, 417–421.

Kennedy, G.G. and Storer, N.P. (2000) Life systems of polyphagous arthropod pests in temporally unstable cropping systems. *Annual Review of Entomology* 45, 467–493.

Kerby, T. (1998) UNR cotton production system trial in the Mid-south. In: Duggar, P. and Richter, D.A. (eds) *Proceedings Beltwide Cotton Conferences*. National Cotton Council, Memphis, Tennessee, pp. 87–88.

King, E.G. Jr and Phillips, J.R. (1989) 42nd annual conference report on cotton insect research and control. In: Brown, J.M. and Richter, D.A. (eds) *Proceedings of the Beltwide Cotton Production Research Conferences*. National Cotton Council, Memphis, Tennessee, pp. 180–191.

Klotz-Ingram, C., Jans, S., Fernandez-Cornejo, J. and McBride, W. (1999) Farm-level production effects related to the adoption of genetically modified cotton for pest management. *AgBioForum* 2, 73–84.

Kogan, M. (1998) Integrated pest management: historical perspectives and contemporary developments. *Annual Review of Entomology* 43, 243–270.

Kogan, M. and Bajwa, W.I. (1999) Integrated pest management: a global reality? *Anais da Sociedade Entomologica do Brasil* 28, 1–25.

Lambert, W.R., Bacheler, J.S., Dickerson, W.A., Roof, M.E. and Smith, R.H. (1996) Insect and mite management in the Southeast. In: King, E.G., Phillips, J.R. and Coleman, R.J. (eds) *Cotton Insects and Mites: Characterization and Management*. Cotton Foundation Reference Book Series, No. 3. The Cotton Foundation Publisher, Memphis, Tennessee, pp. 655–672.

Landis, D.A., Wratten, S.D. and Gurr, G.M. (2000) Habitat management to conserve natural enemies of arthropod pests in agriculture. *Annual Review of Entomology* 45, 175–201.

Leonard, B.R. and Emfinger, K. (2002) Insects in low spray environments and modified cotton ecosystems. In: Duggar, P. and Richter, D.A. (eds) *Proceedings Beltwide Cotton Conferences*. National Cotton Council, Memphis, Tennessee, CD-ROM.

Leonard, B.R., Bagwell, R., Price, R., Downer, R., Magoun, D., Barham, E., Hardwick, J. and Lewis, D. (2003a) Prospects of insect management using remote sensing and variable rate technology. In: Duggar, P. and Richter, D.A. (eds) *Proceedings Beltwide Cotton Conferences*. National Cotton Council, Memphis, Tennessee, pp. 121–124.

Leonard, B.R., Bagwell, R.D., Temple, J., Culli, D., Magoun, D., Barham, E. and Hardwick, J. (2003b) Opportunities for prescription insecticide applications in Louisiana: development and evaluation of aerial SVI. In: Duggar, P. and Richter, D.A. (eds) *Proceedings Beltwide Cotton Conferences*. National Cotton Council, Memphis, Tennessee, pp. 1004–1010.

Leser, J.F., Allen, C.T. and Fuchs, T.W. (1992) Cotton aphid infestations in west Texas: a growing management problem. In: Herber, D.J. and Richter, D.A. (eds) *Proceedings Beltwide Cotton Conferences*. National Cotton Council, Memphis, Tennessee, pp. 823–827.

Lewis, W.J., Van Lenteren, J.C., Phatak, S.C. and Tumlinson, J.H. III (1997) A total system approach to sustainable pest management. *Proceedings of the National Academy of Sciences USA* 94, 12243–12248.

Luckmann, W.H. and Metcalf, R.L. (1975) The pest-management concept. In: Metcalf, R.L. and Luckmann, W.H. (eds) *Introduction to Insect Pest Management*. John Wiley & Sons, New York, pp. 3–35.

Luttrell, R.G. (1994) Cotton pest management. Part 2. A US perspective. *Annual Review of Entomology* 39, 527–542.

Luttrell, R.G., Fitt, G.P., Ramalho, F.S. and Sugonyaev, E.S. (1994) Cotton pest management. Part 1. A worldwide perspective. *Annual Review of Entomology* 39, 517–526.

Luttrell, R.G., Wan, L. and Knighten, K. (1999) Variation in susceptibility of noctuid (Lepidoptera) larvae attacking cotton and soybean to purified endotoxin proteins and commercial formulations of *Bacillus thuringiensis*. *Journal of Economic Entomology* 92, 21–32.

Masud, S.M., Lacewell, R.D., Boring, E.P. III and Fuchs, T.W. (1985) Economic implications of a regional uniform planting date cotton production system: Texas Rolling Plains. *Journal of Economic Entomology* 78, 535–541.

May, O.L., Culpepper, A.S., Shurley, D. and Roberts, P.M. (2002) Cultivar evaluation: evolution of systems trials to compare transgenic and non-transgenic cultivars. In: Duggar, P. and Richter, D.A. (eds) *Proceedings Beltwide Cotton Conferences*. National Cotton Council, Memphis, Tennessee, CD-ROM.

McCutcheon, G.S. (2000) Beneficial arthropods in conservation tillage cotton – a three-year study. In: Duggar, P. and Richter, D.A. (eds) *Proceedings Beltwide Cotton Conferences*. National Cotton Council, Memphis, Tennessee, pp.1303–1306.

Meerman, F., Van de Ven, G.W.J., Van Keulen, H. and Breman, H. (1996) Integrated crop management: an approach to sustainable agricultural development. *International Journal of Pest Management* 42, 13–24.

Micinski, S., Kirkpatrick, T.L. and Colyer, P.D. (1995) An improved plant washing procedure for monitoring early season insect pests in cotton. *Southwestern Entomologist* 20, 17–24.

Moore, L., Beasley, C.A., Leigh, T.F. and Henneberry, T.J. (1996) Insect and mite management in the West. In: King, E.G., Phillips, J.R. and Coleman, R.J. (eds) *Cotton Insects and Mites: Characterization and Management*. Cotton Foundation Reference Book Series, No. 3. The Cotton Foundation Publisher, Memphis, Tennessee, pp. 741–752.

Morrill, A.W. (1921) Arizona wild cotton or thurberia and its insect enemies in relation to the cotton industry of the Southwest. *Journal of Economic Entomology* 14, 472–478.

Mueller, A.J. and Stern, V.M. (1973) *Lygus* flight and dispersal behavior. *Environmental Entomology* 2, 361–364.

Myers, J.H., Savoie, A. and van Randen, E. (1998) Eradication and pest management. *Annual Review of Entomology* 43, 471–491.

Naranjo, S.E., Flint, H.M. and Henneberry, T.J. (1996) Binomial sampling plans for estimating and classifying population density of adult *Bemisia tabaci* in cotton. *Entomologia Experimentalis et Applicata* 80, 343–353.

Newell, W. and Paulsen, T.C. (1908) The possibility of reducing boll weevil damage by autumn spraying of cotton fields to destroy the foliage and squares. *Journal of Economic Entomology* 1, 113–116.

Nichols, S.P., Snipes, C.E. and Jones, M.A. (2003) Evaluation of row spacing and mepiquat chloride in cotton. *Journal of Cotton Science* 7, 148–155.

Niles, G.A., Harvey, L.H. and Walker, J.K. (1978) Cultural control of the boll weevil. *Boll Weevil: Management Strategies, Southern Cooperative Service Bulletin* 228, 23–38.

Noble, L.W. (1969) Fifty years of research on the pink bollworm in the United States. *USDA Agriculture Handbook* 357, 1–62.

Nylin, S. (2001) Life history perspectives on pest insects: what's the use? *Austral Ecology* 26, 507–517.

O'Brien, P.J., Abdel-aal, Y.A., Ottea, J.A. and Graves, J.B. (1992) Relationship of insecticide resistance to carboxylesterases in *Aphis gossypii* (Homoptera: Aphididae) from Mid-south cotton. *Journal of Economic Entomology* 85, 651–657.

Ohmart, C.P. (2000) Forum: Managing agricultural pests. *Issues in Science and Technology* 17, 11.

Parvin, D.W., Gentry, J.W., Cooke, F.T. and Martin, S.W. (2002) Three years experience with ultra-narrow-row cotton production in mississippi, 1999–2001. In: Duggar, P. and Richter, D.A. (eds) *Proceedings Beltwide Cotton Conferences*. National Cotton Council, Memphis, Tennessee, CD-ROM.

Pedgley, D.E. (1993) Managing migratory insects pests–a review. *International Journal of Pest Management* 39, 3–12.

Pedigo, L.P. (1995) Closing the gap between IPM theory and practice. *Journal of Agricultural Entomology* 12, 171–181.

Pedigo, L.P. and Higley, L.G. (1992) The economic injury level concept and environmental quality. *American Entomologist* 38, 12–21.

Perlak, F.J., Oppenhuizen, M., Gustafson, K., Voth, R., Sivasupramaniam, S., Heering, D., Carey, B., Ihrig, R.A. and Roberts, J.K. (2001) Development and commercial use of Bollgard cotton in the USA – early promises versus today's reality. *The Plant Journal* 27, 489–501.

Phillips, J.R., Gutierrez, A.P. and Adkisson, P.L. (1980) General accomplishments toward better insect control in cotton. In: Huffaker, C.B. (ed.) *New Technology of Pest Control*. John Wiley & Sons, New York, pp. 123–153.

Pierce, J.P. Breen, Yates, P.E. and Hair, C.J. (2001) Crop management and microclimate effects on immature boll weevil mortality in Chihuahuan desert cotton fields. *Southwestern Entomologist* 26, 87–93.

Pimentel, D. (1978) Socioeconomic and legal aspects of pest control. In: Smith, E.H. and Pimentel, D. (eds) *Pest Control Strategies*. Academic Press, New York, pp. 55–71.

Pimentel, D., McLaughlin, L., Zepp, A., Lakitan, B., Kraus, T., Kleinman, P., Vancini, F., Roach, W.J., Graap, E., Keeton, W.S. and Selig, G. (1993) Environmental and economic effects of reducing pesticide use in agriculture. *Agriculture Ecosystem and Environment* 46, 273–288.

Prokopy, R., Leskey, T., Pinero, J., Rull, J. and Wright, S. (2000) Forum: managing agricultural pests. *Issues in Science and Technology* 17, 11–12.

Quisenberry, S.S. and Schotzko, D.J. (1994) Integration of plant resistance with pest management methods in crop production systems. *Journal of Agricultural Entomology* 11, 279–290.

Ragsdale, N.N. (2000) The impact of the food quality protection act on the future of plant disease management. *Annual Review of Phytopathology* 38, 577–596.

Reynolds, H.T., Adkisson, P.L. and Smith, R.F. (1975) Cotton insect pest management. In: Metcalf, R.L. and Luckmann, W.H. (eds) *Introduction to Insect Pest Management*. John Wiley & Sons, New York, pp. 379–443.

Rothrock, M.A. and Sterling, W.L. (1982a) A comparison of three sequential sampling packages for arthropods in cotton. *Southwestern Entomologist* 7, 39–49.

Rothrock, M.A. and Sterling, W.L. (1982b) Sequential sampling for arthropods of cotton: its advantages over point sampling. *Southwestern Entomologist* 7, 70–81.

Ruberson, J.R., Phatak, S.C. and Lewis, W.J. (1997) Insect populations in a cover crop/strip tillage system. In: Duggar, P. and Richter, D.A. (eds) *Proceedings Beltwide Cotton Conferences*. National Cotton Council, Memphis, Tennessee, pp. 1121–1124.

Schaller, N. (1993) Sustainable agriculture and the environment: the concept of agricultural sustainability. *Agriculture, Ecosystems and Environment* 46, 89–97.

Sevacherian, V. and Stern, V.M. (1974) Host plant preferences of Lygus bugs in alfalfa-interplanted cotton fields. *Environmental Entomology* 3, 761–766.

Sheehan, W. (1986) Response by specialist and generalist natural enemies to agroecosystem diversification: a selective review. *Environmental Entomology* 15, 456–461.

Shurley, W.D., Bader, M.J., Bednarz, C.W., Brown, S.M., Harris, G. and Roberts, P.M. (2002) Economic assessment of ultra narrow row cotton production in Georgia. In: Duggar, P. and Richter, D.A. (eds) *Proceedings Beltwide Cotton Conferences*. National Cotton Council, Memphis, Tennessee, CD-ROM.

Slosser, J.R. (1978) The influence of planting date on boll weevil management. *Southwestern Entomologist* 3, 241–246.

Slosser, J.E., Puterka, G.J. and Price, J.R. (1986) Cultural control of the boll weevil (Coleoptera: Curculionidae): effects of narrow-row spacing and row direction. *Journal of Economic Entomology* 79, 378–383.

Slosser, J.E., Pinchak, W.E. and Rummel, D.R. (1989) A review of known and potential factors affecting the population dynamics of the cotton aphid. *Southwestern Entomologist* 14, 302–313.

Smith, H.A. and McSorley, R. (2000) Intercropping and pest management: a review of major concepts. *American Entomologist* 46, 154–161.

Smith, R.F. (1978) History and complexity of integrated pest management. In: Smith, E.H. and Pimentel, D. (eds) *Pest Control Strategies*. Academic Press, New York, pp. 41–53.

Snodgrass, G.L. (1993) Estimating absolute density of nymphs of *Lygus lineolaris* (Heteroptera: Miridae) in cotton using drop cloth and sweep-net sampling methods. *Journal of Economic Entomology* 86, 1116–1123.

Sparks, A.N. Jr, Norman, J.W., Spurgeon, D.W., Raulston, J.R. and Tanner, B.F. (1997) Effects of selected pesticides on populations of beneficial arthropods in Lower Rio Grande Valley cotton. In: Duggar, P. and Richter, D.A. (eds) *Proceedings Beltwide Cotton Conferences*. National Cotton Council, Memphis, Tennessee, pp. 1313–1316.

Stapel, J.O., Lewis, W.J., Phatak, S.C. and Ruberson, J.R. (1998) Insect pest management as a component of a sustainable cotton production system. In: Duggar, P. and Richter, D.A. (eds) *Proceedings Beltwide Cotton Conferences*. National Cotton Council, Memphis, Tennessee, pp. 1107–1111.

Steinkraus, D.C. and Hollingsworth, R.G. (1994) Predicting fungal epizootics on cotton aphids. *Arkansas Farm Research* 43, 10–11.

Steinkraus, D.C., Kring, T.J. and Tugwell, N.P. (1991) *Neozygites fresenii* in *Aphis gossypii* on cotton. *Southwestern Entomologist* 16, 118–122.

Steinkraus, D.C., Hollingsworth, R.G. and Slaymaker, P.H. (1995) Prevalence of *Neozygites fresenii* (Entomophthorales: Neozygitaceae) on cotton aphids (Homoptera: Aphididae) in Arkansas cotton. *Environmental Entomology* 24, 465–474.

Steinkraus, D., Boys, G., O'Leary, P., Bagwell, R., Freeman, B., Layton, B., Lorenz, G., Roberts, P. and Walker, T. (2003) Ten years of the cotton aphid fungus sampling service. In: Duggar, P. and Richter, D.A. (eds) *Proceedings Beltwide Cotton Conferences*. National Cotton Council, Memphis, Tennessee, pp. 1064–1070.

Stern, V.M. (1969) Interplanting alfalfa in cotton to control lygus bugs and other insect pests. *Proceedings Tall Timbers Conference on Ecological Animal Control by Habitat Management* 1, 55–69.

Stern, V.M. (1981) Environmental control of insects using trap crops, sanitation, prevention, and harvesting. In: Pimentel, D. (ed.) *Handbook of Pest Management in Agriculture*, Vol. 1. CRC Press, Boca Raton, Florida, pp. 199–209.

Stern, V.M., van den Bosch, R., Leigh, T.F., McCutcheon, O.D., Saller, W.R., Houston, C.E. and Garber, M.J. (1967) Lygus control by strip cutting alfalfa. *University of California Agricultural Extension Service* AXT-241, 1–13.

Stewart, S.D. (2003) Insect management in reduced tillage systems. In: Duggar, P. and Richter, D.A. (eds) *Proceedings Beltwide Cotton Conferences*. National Cotton Council, Memphis, Tennessee, pp. 110–112.

Stewart, S.D., Vilarroel, D. and Thomasson, A. (2002) Using remote sensing in insect pest management? In: Duggar, P. and Richter, D.A. (eds) *Proceedings Beltwide Cotton Conferences*. National Cotton Council, Memphis, Tennessee, CD-ROM.

Sudbrink, D.L. Jr, Harris, F.A., Robbins, J.T., English, P.J. and Hanks, J.A. (2002) Site specific management in Mississippi Delta cotton: experimental field studies and on-farm application. In: Duggar, P. and Richter, D.A. (eds) *Proceedings Beltwide Cotton Conferences*. National Cotton Council, Memphis, Tennessee, CD-ROM.

Tabashnik, B.E. (1997) Seeking the root of insect resistance to transgenic plants. *Proceedings of the National Academy of Sciences USA* 94, 3488–3490.

Tabashnik, B.E., Patin, A.L., Dennehy, T.J., Liu, Y., Carriere, Y., Sims, M.A. and Antilla, L. (2000) Frequency of resistance to *Bacillus thuringiensis* in field populations of pink bollworm. *Proceedings of the National Academy of Sciences USA* 97, 12980–12984.

Thomas, M.B. (1999) Ecological approaches and the development of "truly integrated" pest management. *Proceedings of the National Academy of Sciences USA* 96, 5944–5951.

Verkerk, R.H.J., Leather, S.R. and Wright, D.J. (1998) The potential for manipulating crop–pest–natural enemy interactions for improved insect pest management. *Bulletin of Entomological Research* 88, 493–501.

Vories, E.D., Glover, R.E., Bryant, K.J. and Valco, T.D. (1999) A three-year study of UNR cotton. In: Duggar, P. and Richter, D.A. (eds) *Proceedings Beltwide Cotton Conferences*. National Cotton Council, Memphis, Tennessee, pp. 1480–1482.

Walker, J.K. and Smith, C.W. (1996) Cultural control. In: King, E.G., Phillips, J.R. and Coleman, R.J. (eds) *Cotton Insects and Mites: Characterization and Management*. Cotton Foundation Reference Book Series, No. 3. The Cotton Foundation Publisher, Memphis, Tennessee, pp. 471–509.

White, J.R. and Rummel, D.R. (1978) Emergence profile of overwintering boll weevils and entry into cotton. *Environmental Entomology* 7, 7–14.

Willers, J.L., Seal, M.R. and Luttrell, R.G. (1999) Remote sensing, line-intercept sampling for tarnished plant bugs (Heteroptera: Miridae) in Mid-south cotton. *Journal of Cotton Science* 3, 160–170.

Williams, M.R. (1999) Cotton insect loss estimates – 1998. In: Duggar, P. and Richter, D.A. (eds) *Proceedings Beltwide Cotton Conferences*. National Cotton Council, Memphis, Tennessee, pp. 807–809.

Williams, M.R. (2002) Cotton insect losses – 2001. In: Duggar, P. and Richter, D.A. (eds) *Proceedings Beltwide Cotton Conferences*. National Cotton Council, Memphis, Tennessee, CD-ROM.

Williams, M.R. (2003) Cotton insect losses – 2002. In: Duggar, P. and Richter, D.A. (eds) *Proceedings Beltwide Cotton Conferences*. National Cotton Council, Memphis, Tennessee, pp. 101–109.

Williams, M.R., Wagner, T.L. and Willers, J.L. (1995) Revised protocol for scouting arthropod pests of cotton in the Midsouth. *Mississippi Agricultural and Forestry Experimental Station Technical Bulletin* 206, 1–37.

Wilson, L.T., Leigh, T.F. and Maggi, V. (1981) Presence—absence sampling of spider mite densities on cotton. *California Agriculture* 35, 10.

Wormer, S.P. (1991) Use of models in applied entomology: the need for perspective. *Environmental Entomology* 20, 768–773.

Zilberman, D., Schmitz, A., Casterline, G., Lichtenberg, E. and Siebert, J.B. (1991) The economics of pesticide use and regulation. *Science* 253, 518–522.

17 Ecological Implications for Post Harvest Integrated Pest Management of Grain and Grain-based Products

James F. Campbell and Frank H. Arthur

United States Department of Agriculture, Agricultural Research Service, Grain Marketing and Production Research Center, 1515 College Avenue, Manhattan, KS 66502, USA

Introduction

Flint and van den Bosch (1981) defined integrated pest management (IPM) as 'an ecologically based pest control strategy that relies heavily on natural mortality factors and seeks out control tactics that disrupt these factors as little as possible'. IPM also includes the integration of pest control tactics that emphasize prevention, avoidance, monitoring and suppression of pest problems, and these tactics should meet economic, public health and environmental goals. Underlying these and the many other proposed definitions of IPM is an understanding of the ecological interactions among pests, natural enemies and the environment within and around the system being managed. This is true in both pre- and postharvest applications, but the latter takes place in landscapes created and maintained by humans that are quite artificial, and thus presents unique challenges, but also some unique opportunities in terms of implementing an ecologically based IPM programme.

For durable commodities, such as grain and grain-based products, stored-product insect infestation can occur beginning at harvest and continuing through bulk storage; conversion into processed commodities in food-processing facilities; storage in warehouses; transportation in trucks, railcars and ships; presentation on retail shelves; and ultimately storage in consumer pantries. At each of the locations in the processing and distribution channel there are differences in the nature of the environment and management strategies and tactics used, but all share the same characteristic of being spatially and temporally patchy landscapes. This type of landscape structure has important implications for pest ecology and presents challenges in terms of evaluating pest populations and targeting pest management. As a result, application of IPM to these different stages of the postharvest process varies due to differences in the spatial and temporal scales of the patchiness, the ability to monitor the pest populations, the number and types of management tools available and the tolerance level for insect infestation. Because stored products are located in facilities created and maintained by humans there is considerable potential for the manipulation of environmental conditions and landscape structure to make the habitat less favourable to pest populations. As a result, postharvest ecologically based IPM focuses less on natural mortality factors maintaining populations below damaging levels, except perhaps in some bulk storage situations, and more on making the landscape less favourable to establishment and growth of infestations and controlling these infestations before they cause economic damage.

The Problem

Stored-product insects have an economic impact in three general ways: direct loss of biomass due to insects consuming stored commodities; loss of product 'quality' or 'value' due to the presence of insects; and costs associated with preventing or treating insect infestations. It has been estimated that 5–10% of stored grain in developed countries and 35% of stored grain in developing countries is lost to insect damage (Boxall, 1991). However, in many parts of the developing world, stored grain loss due to insect consumption and contamination can range up to 75% and directly threaten human health (Gorham, 1991). Estimates of the cost of grain loss due to insect, mould, and mycotoxin damage to the 15 billion bushels of grain stored in the USA each year have ranged from $500 million (Harein and Meronuck, 1995) to in excess of $1 billion (Cuperus and Krischik, 1995).

In bulk-stored grain in the USA, insect damage is only a direct cost if the grain is rejected due to high levels of insect damaged kernels (IDK) or assigned a lower grade, but managers can manipulate damage levels by blending grain (Hagstrum et al., 1999). Domestic flour millers have a low tolerance for live insects (Kenkel et al., 1993) and the presence of live insects or signs of insect feeding can result in rejection of grain. Rejection produces additional expenses associated with fumigation and transportation and can be as much as 10–20% of the value of the grain (Hagstrum et al., 1999).

As grain is processed into different foods it becomes a value-added commodity. Tolerance for insect infestation of processed food for human and animal consumption is typically very low and the direct and indirect costs to the food industry resulting from insect infestation of grain-based products are substantial. However, published economic estimates are hard to find due to a variety of reasons, including the difficulty in measuring economic impact and the proprietary nature of this information to the food industry. For processed cereal products it is the contamination of the food that is the major issue, rather than loss of food material due to consumption by insects. Stored-product insects and their fragments can indicate that the food is adulterated, cause a health hazard, and provoke allergic reactions (Olsen, 1998; Olsen et al., 2001), produce excretions that change the taste of food, and potentially carry disease-causing microorganisms (Foil and Gorham, 2000). Infestation of packaged commodities has a wide range of negative impacts including loss of customer goodwill, damage to commercial brand identity, failure to meet government regulations or pass plant inspections, and costs associated with handling of product returns and consumer complaints. If an insect is found in a packaged food, the consumers are not only hesitant to purchase the product in the future, but will also tell other consumers about their bad experience and thus compound the negative impact.

The food industry is sensitive to issues involving food quality and safety. However, there is considerable variation within the industry in terms of the quality of pest management programmes, with some continuing to rely primarily on calendar-based pesticide treatments. In addition, responsibility for pest management often changes hands as a processing product moves through distribution channels and passes through different locations that can have different pest infestation pressures. Despite the low tolerance for insects in food by consumers and the food industry, pest infestations are still a problem. Surveys of insects in food plants, warehouses and retail environments frequently document insect populations (Good, 1937; Evans and Porter, 1965; Zimmerman, 1990; Arbogast et al., 2000; Doud and Phillips, 2000; Campbell et al., 2002; Roesli et al., 2003b).

Stored-product Pest Complex

A diverse community of organisms (microflora, arthropods, rodents, birds) is associated with human structures and stored food. There are four major categories of pests that may contaminate processed food products (Zimmerman *et al.*, 2003):

1. Obligatory pests are associated with human environments, are attracted to human stored food and live and breed in the food product (e.g. stored-product insects).
2. Opportunistic pests are attracted to human food, are associated with human modified habitats and often inhabit human structures, but usually do not live in the products they contaminate (e.g. flies, cockroaches, ants, rats and mice).
3. Adventive pests are associated with human environments but are not particularly attracted to human food and not strongly associated with human structures (e.g. birds, bats, and some insect species).
4. Natural enemies are predators and pathogens of the other three groups and can occur in human structures, be attracted to food odours and hosts in or near human food and end up being contaminates of food.

Arthropods comprise a large portion of the biodiversity in food storage and processing facilities. Six hundred species of beetles in 34 families (Hinton, 1945) and 70 species of moths in primarily four families (Cox and Bell, 1991) have been reported to be associated with stored products, but only a portion of these typically become major pests. Many of these species are generalists and feed on a wide range of stored commodities, not just grain-based foods. The specific community of arthropods associated with food-processing, storage and retail facilities, and species that become pests, is influenced by a wide range of factors; geographic location, season, building construction and condition, food products available, management practices, etc. Because a suite of species can become pests in a given facility and species vary in their behaviour, ecology and the type of damage they cause, one of the

initial components of an ecologically based IPM programme is correct identification of the pests present.

Stored-product insects can be grouped in different ways based on taxonomy, feeding preference and life history traits (Fig. 17.1). Some species feed on whole seeds, typically with one portion of the life cycle occurring within the seed, and some feed on damaged (e.g. broken kernels, grain dust) and processed seeds. Insects in the second group tend to be secondary pests in bulk-stored grain, but become much more important as pests of processed food. Many of these species are opportunistic feeders and can feed on a wide range of non-whole-grain or processed grain-based products, including spices, dried fruit and vegetables, nuts, beans, oil seeds, fungi, other insects and animal products. Food facilities vary in terms of the types of potential food materials present at a site and thus often have to deal with multiple types of pests. For example, flour mills often have bulk grain storage as well as flour storage and retail stores may have a wide range of processed foods and also whole-seed products such as bird seed. This complicates both the monitoring of pests and the selection and application of pest management tactics. Some examples are listed below to provide an overview, but see Gorham (1987, 1991) for more detailed information.

Species that are pests of whole undamaged seeds are fewer in number and tend to be less general in their range of foods, than those that can exploit processed or damaged seeds. They are the primary pest species of stored whole grain and their presence can lead to contamination issues with processed commodities because larvae, pupae and sometimes adults can occur inside whole kernels. This makes them more difficult to detect and remove them prior to processing and thus they can be ground up along with the grain and contribute to insect fragment in processed products. Processed foods have limits on the number of insect fragments that they can contain. Some of the major pests of whole grain include the grain weevils in the family Curculionidae (*Sitophilus*

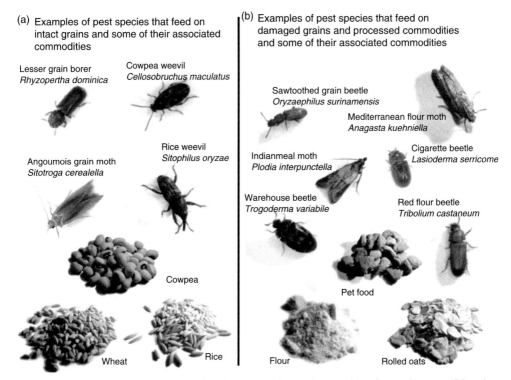

(a) Examples of pest species that feed on intact grains and some of their associated commodities

Lesser grain borer
Rhyzopertha dominica

Cowpea weevil
Cellosobruchus maculatus

Angoumois grain moth
Sitotroga cerealella

Rice weevil
Sitophilus oryzae

Cowpea

Wheat

Rice

(b) Examples of pest species that feed on damaged grains and processed commodities and some of their associated commodities

Sawtoothed grain beetle
Oryzaephilus surinamensis

Mediterranean flour moth
Anagasta kuehniella

Indianmeal moth
Plodia interpunctella

Cigarette beetle
Lasioderma serricome

Warehouse beetle
Trogoderma variabile

Red flour beetle
Tribolium castaneum

Pet food

Flour

Rolled oats

Fig. 17.1. Some common stored-product insect species and examples of stored commodities that they infest.

oryzae (L.) (rice weevil), *S. granarius* (L.) (granary weevil) and *S. zeamais* Motschulsky (maize weevil)), and grain borers in the family Bostrichidae (*Rhyzopertha dominica* (F.) (lesser grain borer) and *Prostephanus truncatus* (Horn) (larger grain borer)). For these species, both adults and larvae feed and tunnel through grain and feeding leads to fragmented kernels, powdery residues, and for some species a pungent odour. *Trogoderma granarium* Everts (khapra beetle) is dermestid species that prefers whole grain and cereal products to animal-based products (Lindgren *et al.*, 1955) and is the most destructive pest of grain in the parts of the world where it occurs. However, it is the only quarantine stored-product pest in the USA. Some moth species also attack intact seeds; for example, the Angoumois grain moth (*Sitotroga cerealella* (Olivier)) develops inside whole seeds.

A diverse group of beetles and moths share the characteristic that they are pri-marily pests' of processed foods. Many of these species have worldwide distributions, at least in association with environmentally controlled food facilities, and attack a wide range of commodities. Several examples are listed below: see *Oryzaephilus surinamensis* (L.) (sawtoothed grain beetle) and *O. mercator* (merchant grain beetle) (Silvanidae) are cosmopolitan pests of a wide range of foods and are good invaders of packaged foods. *Tribolium castaneum* (Herbst) (red flour beetle) and *T. confusum* (Jacquelin du Val) (confused flour beetle) (Tenebrionidae) are some of the most widespread and economically important pest species, particularly for the flour milling industry. A large number of dermestid (Dermestidae) beetles are associated with human structures, but only *Trogoderma variabile* Ballion (warehouse beetle) is considered a major processed food pest. In addition to damage caused by feeding, they can contaminate food with body parts, hairs, or cast larval cuticles that cause gastrointestinal irritation and allergic reactions for asthmatics and sensitized individuals (Olsen *et al.*, 2001). The cigarette beetle, *Lasioderma serricorne* (F.), despite its name, is a pest of more than just tobacco products, with immatures developing on a wide range of food products. Some of the beetle species are long lived, lay large numbers of eggs over their lifespan, and can persist on relatively small amounts of food, which contributes to their pest status.

A number of moth species are found associated with food storage structures. *Plodia interpunctella* (Hübner) (Indianmeal moth) is one of the most damaging stored-product moths for the food-processing industry, retail stores, and homeowners. Larvae feed on a wide range of foods and produce silk that also contributes to food contamination. Some other related moth species that attack a wide range of commodities and can be important pests in certain regions are *Pyralis farinalis* L. (meal moth), *Ephestia kuehniella* Zeller (Mediterranean flour moth) and *Cadra cautella* (almond moth). Moths tend to be short lived as adults, but can lay large numbers of eggs.

Stored-Product Environment

Landscape structure

Most landscapes created or modified by humans tend to be highly fragmented (Wiens, 1976). Fragmented landscapes are a mosaic of resource patches that are separated from each other by barriers to movement or by patches of less hospitable habitat. The structure and dynamics of the landscape mosaic influence ecological processes such as population dynamics and spatial distribution (Turner, 1989; Wiens *et al.*, 1993; Wiens, 1997). If the foundation of an effective pest management programme is an understanding of pest ecology and behaviour, this understanding must be at the relevant landscape conditions and the appropriate spatial and temporal scales for a given insect species. Stored-product pests of the food industry are pests in large part due to their effectiveness at exploiting the temporally and spatially fragmented landscapes within food facilities and within which food facilities are located. For postharvest pest management, we are typically starting with material that is considered to be free of live insect infestation (e.g. freshly harvested grain, milled flour, extruded food) and this material is stored in ways that spatially separate the food into patches. Most infestations of grain-based products, whether it is grain in a bin or a food package, result either from failure to adequately remove insects from where product is being stored or from stored-product insects finding and exploiting the new patch of resource.

The pattern of distribution of the food patches in a food facility ultimately influences the spatial distribution and abundance of the insects. Other resources important for stored-product insects may also be patchy and directly, or in combination with other factors, influence insect distribution (e.g. favourable environmental conditions, structural features such as harbourages and refugia). Food resources for stored-product insects are patchy at a range of spatial scales: individual pieces of food, packages of food material surrounded by packaging barriers, packages arranged on pallets, a warehouse, a processing plant in a landscape that includes other food storage and processing facilities. The spatial scale at which this patchiness is important to an organism depends on the organism being studied, and its perceptual abilities and movement patterns, and the questions asked (Wiens, 1989, 1997; Pearson *et al.*, 1996). A patch, for a whole grain feeding species such as the rice weevil, might be considered a single seed kernel (Campbell, 2002; Cope and Fox, 2003) or a whole bin (Hagstrum, 2001). In a warehouse, a patch may be an individual piece of spilled product, a crack filled with food material, or a whole package of commodity. At a larger spatial scale and with more mobile species, structures such as grain bins, food plants, or warehouses may be patches within a broader landscape. Barriers to movement (e.g. packaging and walls) and inhospitable habitat (e.g. exposed cement floors and walls) can separate all of these patch types from each other.

Landscape structure at a range of spatial scales probably influences stored-product insect populations, although our understanding of these processes is still very limited. Studies have found that stored-product insects are not evenly distributed at spatial scales ranging from patches of flour (Campbell and Hagstrum, 2002; Campbell and Runnion, 2003; Toews et al., 2005a), inside buildings (Arbogast et al., 1998; Arbogast et al., 2000; Campbell et al., 2002), within bulk grain (Nansen et al., 2004) and even around the outside of facilities (Campbell and Mullen, 2004). Targeting pest management to where pests are present can increase the probability of suppressing the pest population, and reduce the cost of management and risk of negative non-target affects (Brenner et al., 1998).

Spatial distribution of pest populations

Movement patterns in heterogeneous landscapes and how much time is spent in different patches together determine spatial distribution (With and Crist, 1995) and the degree to which resource patches are interconnected (Wiens et al., 1997). In food facilities, the distribution and movement of insects among different resource patches can be due to their own dispersal behaviour or through human intervention (e.g. mixing of infested grain with uninfested grain, bringing infested packages into a warehouse, moving the location of spillage through housekeeping activities). The extent of insect movement among patches of food will influence the probability that stored products will become infested, the persistence of populations within storage facilities and many aspects of pest management (e.g. the interpretation of trap catches, the effectiveness of insecticides and insect resistant packaging).

Although studies measuring stored-product insect dispersal ability are limited, they indicate that many stored-product insects are highly mobile and capable of moving among food patches: Chesnut (1972) demonstrated that *S. zeamais* flew up to 400 m; Hagstrum and Davis (1980) found that *C. cautella* flew 300 m during a 10-min

flight; Fadamiro (1997) estimated based on laboratory wind tunnel measurements that *P. truncatus* could disperse in still air up to 1620 m in an hour. Using self-mark recapture techniques in a commercial food facility, Campbell et al. (2002) found that male *Trogoderma variabile* were highly mobile, moving across multiple floors and from 7 to 216 m through a warehouse. These data suggest that there is considerable potential for these species to move between resource patches and colonize and exploit patchy resources throughout a food facility.

Stored-product pests are often trapped outside grain storage and processing structures (Throne and Cline, 1989, 1991; Fields et al., 1993; Dowdy and McGaughey, 1994; Doud and Phillips, 2000; Campbell and Arbogast, 2004; Campbell and Mullen, 2004). They are sometimes captured far away from anthropogenic structures (Strong, 1970; Cogburn and Vick, 1981; Sinclair and Haddrell, 1985; Vick et al., 1987) which suggests the capability for long distance flight, although these captures may also indicate feral populations (Khare and Agrawal, 1964; Howe, 1965a; Wright et al., 1990). Outside of a commercial food facility, male *T. variabile* were recaptured on average 75 m (range 21–508) and male *P. interpunctella* were recaptured on average 136 m (range 21–276) from where they were marked (Campbell and Mullen, 2004). Both of these species have also been demonstrated, using self-mark recapture, to find routes of entry and move from outside to the inside of food facilities (Campbell and Arbogast, 2004; Campbell and Mullen, 2004). Hagstrum (2001) also found considerable rates of immigration by a range of pest species into farm grain storage bins. These patterns suggest the potential for stored-product insects to move among food facilities in a landscape and the potential importance of movement into food facilities from outside sources in pest management.

Direct behavioural evidence of how stored-product insects move among resource patches is limited, but what is available shows that stored-product pests readily leave patches of food, can find and exploit multiple patches, and that these processes are influenced by a variety of endogenous and exogenous factors. The time *Cryptolestes ferrugineus* (Stephens), the rusty grain beetle, spent in patches has been shown to be influenced by strain, sex and age (Cox *et al.*, 1989, 1990; Cox and Parish, 1991). A variety of factors have been shown to influence the decision by red flour beetles, *T. castaneum* to leave food patches, including insect density (Naylor, 1961; Hagstrum and Gilbert, 1976; Ziegler, 1977b), age (Hagstrum and Gilbert, 1976; Ziegler, 1976) and patch quality (Ogden, 1970; Ziegler, 1977a). Campbell and Hagstrum (2002) found that red flour beetles were often observed outside of food patches and that females visited and laid eggs in multiple patches. Campbell and Runnion (2003) found that female red flour beetles adjusted the distribution of eggs among food patches in response to the amount of food in the patches. In sheds with small patches of flour, *T. castaneum* adults were found to move among food patches (Toews *et al.*, 2005b).

Importance of landscape patterns on pest populations and the importance in terms of pest management will vary depending on the type of storage facility. For large grain elevators, storing thousands of bushels of wheat, the movement of pests into the grain from outside sources such as spillage residue or movement from other grain elevators may be less significant from a pest management perspective than the dynamics of the populations within the bulk grain. However, for a food-processing facility or a retail store the impact of landscape structure and pest movement patterns is likely to be critical for both monitoring (e.g. it is individuals moving among resource patches that are captured in pheromone traps) and selecting and targeting pest management (e.g. applying pesticides where they will contact the infested resource patches or major routes of movement).

In a patchy landscape, typically insects occupy some patches and some are empty. Pest populations can be made up of subpopulations occupying the same patches; this type of population structure has been termed a metapopulation (Hanski and Simberloff, 1997). The degree of movement among patches is what defines the type of

metapopulation structure and the scale at which individuals can be considered to be from different populations. When there is considerable movement among patches and population trends within patches are correlated with each other this is considered a patchy population; but as the level of recruitment from within a patch increases relative to immigration from other patches, the population becomes more of a true metapopulation (Harrison and Taylor, 1997). Another scenario is a source–sink or a mainland–island metapopulation where large (mainland) or higher quality (source) patches persist and produce individuals that immigrate into smaller (island) or lower quality (sink) patches and enable subpopulations to persist in these locations. The presence of these types of populations for stored-product insects in and around food facilities is untested, but there are numerous lines of evidence based on the physical characteristics of the facilities, spatial distribution of populations, and how populations respond to treatments that indicate these are applicable scenarios for stored-product insects.

Evidence for pest population structure in food facilities

For stored-product insect IPM, the population structure is of relevance because of the implications in terms of how much of the population is actually being managed. At food facilities, it is likely that a range of population spatial distributions occur, which impacts the effectiveness of pest management tools, and the interpretation of monitoring programmes (Fig. 17.2). Spatial distribution of the population determines the proportion of individuals exposed to treatment and the potential for recolonization. This is especially important when dealing with treatments such as fumigation that do not have any residual effect. When pest subpopulations are interacting over spatial scales larger than an individual structure, rapid pest resurgence after treatments applied to that structure could occur. Given that pheromone

(a) Aerial view of a food-processing
plant and surrounding property

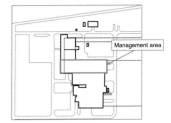

(b) Pest population located only
within a small portion of the
management area (response
example: targeted sanitation)

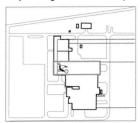

(c) Pest population located throughout
the management area, but primarily
contained within area (response
example: aerosol pesticide application,
heat treatment)

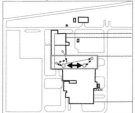

(d) Some of population located
inside management area, but
important sources located outside
(response example: identify and
eliminate major sources)

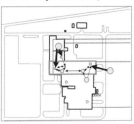

(e) Large portion of population located
offsite (sources), only a small portion
within management area (sink)
(response example: pest exclusion)

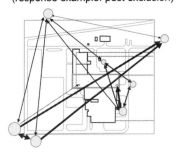

Fig. 17.2. Examples of how stored-product insect subpopulations (illustrated using circles, diameter proportional to subpopulation size) and movement patterns (illustrated using arrows, with thickness of the arrow proportional to amount of movement) could be distributed in and around a food-processing facility. A potential management area, such as a warehouse, is indicated. Differences in the percentage of the population being treated within this management area and the level of movement will impact the efficacy of different management tactics. These hypothetical distributions are likely to differ among species, locations and seasons.

traps capture dispersing individuals, the spatial scale over which subpopulations interact also impacts our ability to monitor efficacy. Thus it makes a difference if a population is contained within a single room, within a building or within a broader landscape containing multiple food storage and processing facilities. It also matters if an occupied patch is a source or mainland subpopulation that is contributing to infestation of other patches. For example, pests moving from one large infested patch (e.g. mothballed piece of uncleaned equipment) through the building and into the warehouse where they could infest packaged products (Campbell *et al.*, 2002).

A recent study by Campbell and Arbogast (2004), suggests that at the same flour mill location pest subpopulations exhibited one of two patterns, suggesting that they were interacting over two very different spatial scales. The first pattern was that source patches for the insects lay over a spatial scale greater than the mill itself and there was considerable movement of individuals across this larger spatial scale; effec-

tively linking activity inside and outside the mill (Fig. 17.3). This pattern applied to the *P. interpunctella* and the *T. variabile* at this site. Pheromone trap captures outside were higher than those inside the mill and inside and outside trap captures were correlated: cycling according to a seasonal pattern; declining in the fall even in the absence of treatment. These species, although readily captured in pheromone traps inside, were rarely, if ever, found, in product samples. Mark-recapture data demonstrated that *P. interpunctella* was capable of entering the building from outside. Fumigation treatments did not appear to impact trap captures of these species. The result of this

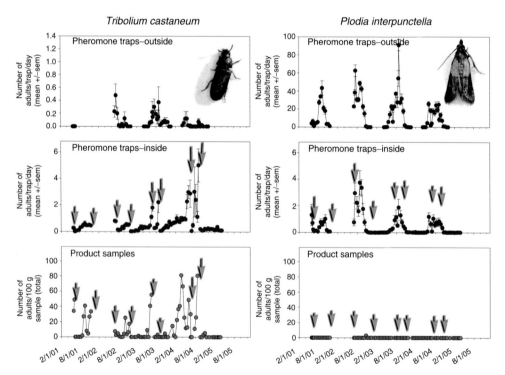

Fig. 17.3. Changes over time in the number of *Tribolium castaneum*, the red flour beetle, and *Plodia interpunctella*, the Indianmeal moth, associated with a flour mill. Pheromone traps – outside: average daily capture of beetles or moths in pheromone traps placed outside the flour mill. Pheromone traps – inside: average daily capture of beetles/trap or moths in pheromone traps baited with pheromone and food attractant placed within the flour mill. Product samples: total number of live adult beetles or immature moths recovered per 500 g product: combined product samples collected at four locations in the product stream (third, fourth, and fifth mids, and purifiers) and a trash bucket. Arrows indicate times when the mill was fumigated with either methyl bromide or sulphuryl fluoride. Differences in patterns with season and in response to fumigation are consistent with differences in population distribution. Portion of data originally presented in Campbell and Arbogast (2004).

pattern was that pheromone trap captures within the mill for these species indicated the potential for infestation and the capacity for insects to immigrate, but did not accurately reflect the current level of infestation within the mill. This pattern of pest spatial distribution and movement also explains why fumigations did not typically reduce pheromone trap captures of these species. This lack of impact is less likely to be due to the failure of the fumigation to suppress the population, and more from not treating all of the subpopulations or excluding individuals from untreated subpopulations outside the fumigated area from recolonizing patches. This pattern suggests a mainland–island or source–sink dynamic with the major sources of these species being outside the mill building.

The second observed pattern suggests source patches for the insects lay over a spatial scale contained primarily within the mill itself with pheromone traps capturing primarily insects moving among these internal patches. This pattern was seen with the major pest of this mill, the red flour beetle, where trap captures tended to be lower outside compared to inside and followed a pattern of sharp decline after fumigation treatment and then steady increase in numbers until the next fumigation (Fig. 17.3). This rebound, other than potentially the rate of increase, was not impacted by season and outside trap capture levels. Red flour beetle was the primary species recovered in product samples and pheromone trap captures were correlated with numbers in the product samples. Rebound after fumigation may result from persistence of individuals within some of the patches within the mill or limited movement of new individuals into the mill either actively or in infested products. This pattern suggests some type of metapopulation with either the mill itself being a subpopulation with limited immigration from outside sources or the subpopulations within the mill making up a metapopulation with some patches surviving treatment.

From an ecologically based pest management perspective, the focus of managing metapopulations needs to be on reducing immigration and establishment in the commodity patches being protected (e.g. grain bin, bagged product) and identifying and eliminating infested patches. Some resource patches represent the product(s) that we are trying to protect from infestation, some patches may be in or around the building and within the scope of a pest management programme, and other patches may be in areas outside of a pest manager's control. This patchy spatial distribution of resources is also temporally variable due to anthropogenic (e.g. bins are emptied and refilled, packages are moved, sanitation removes or moves patches of spillage) and non-anthropogenic reasons. Some food patches in a structure will tend to persist longer than others (e.g. food material in a wall void that is not accessible vs. a spillage in an aisle) and have a higher probability of becoming infested and contributing to the increase and persistence of pest populations. Identification of important source or mainland subpopulations and eliminating/reducing these sources, or, if they are outside sources that cannot be managed, reducing immigration and establishment is an important management strategy.

In bulk storages, it is likely that there is movement of insects into and out of the bins (Hagstrum, 2001), but, at least after initial establishment, internal dynamics appear to be a more dominant factor than immigration. The role these bulk storages may play as important source or mainland populations has not been investigated, but Doud and Phillips (2000) found Indianmeal moth captured at pheromone traps declined with distance from a mill/elevator. Pest populations within bulk grain also contributes to monitoring, are also spatially patchy, but given that management of bulk grain typically requires treating the whole mass, focusing pest management at pest 'hot spots' is rarely feasible.

Ecologically Based IPM

Common characteristics of the many proposed IPM definitions include the appropriate selection and integration of control methods with the aid of decision rules, consideration of multiple pests, enhancing the economic benefits to growers and society and environmental benefits (Kogan, 1998). In stored-product pest management, there are, in many cases, multiple tactics available to monitor and manage pest populations and their use is consistent with the IPM concept, but there has been limited monitoring-based decision making (Cuperus et al., 1990; White, 1992; Longstaff, 1994; Hagstrum et al., 1999; Phillips et al., 2000a). Pest monitoring focuses on either directly sampling for insects in patches or indirectly sampling individuals moving among patches, often using pheromone traps, and the relationship between the two measurements is not always clear. Postharvest pest management practices fall into one of three categories: (i) reducing/eliminating insects in food patches that are currently infested (e.g. fumigation, heat, sanitation, structural modification); (ii) putting up barriers to insect movement outside patches (e.g. screening on windows, fogging with pesticide); and (iii) reducing/preventing pest from entering new patches (e.g. insect resistant packages, repellents, sealing).

Monitoring

The foundation of a successful IPM programme is an effective monitoring system that supplies information on not only the number and type of pests present, but also detects changes in pest populations over time and locates foci of infestation and routes of entry (Burkholder, 1990). Insect monitoring can involve sampling of the commodity itself using visual inspection or traps to determine if the patch is infested or indirect sampling of the insects dispersing among resource patches using tools such as pheromone traps. Directly sampling the product is often destructive (e.g. packaged commodities) and can be difficult or prohibitively expensive (e.g. wall voids, disassembling equipment), while indirect sampling is often easier to perform the information obtained is more difficult to interpret and use for making pest management decisions. This is because we are sampling primarily dispersing individuals, and often the methods used to trap these individuals bias capture towards a particular sex and/or physiological state. In most situations, we do not know the relationship between indirect sampling methods (i.e. sampling dispersing individuals) and direct sampling methods (i.e. sampling individuals in infested material). Nansen et al. (2004) have shown that in a maize storage warehouse there was not a good spatial relationship between P. interpunctella adults captured on passive (no attractant) sticky traps above the surface of the grain and larval captures in the bulk maize. In contrast, in empty shed experiments a decline in trap catch with distance from insect release point has been shown for Lasioderma serricorne (Arbogast et al., 2003) and Plodia interpunctella (Arbogast et al., 2005).

Monitoring strategies and tactics differ between bulk-stored raw commodities and processed commodity facilities. Bulk-stored commodity monitoring relies primarily on direct sampling for insects in the product, but for processed commodity facilities a combination of direct sampling and indirect sampling is more widely used. The major difficulty with bulk grain sampling is that we need to estimate the number of insects present in a very large volume, but, because of the small volume of samples relative to the total volume of stored grain, low density and non-uniform distribution of insects, and the difficulty in taking samples from throughout the grain mass, extrapolating from the sample data is not highly accurate.

Various sampling tools (e.g. grain trier, pelican sampler, vacuum probe) are available for collecting grain samples depending on the volume of grain to sample and whether the grain is in a bin or being moved (Hagstrum, 1994; Subramanyam and Hagstrum, 1996). These samples are then sieved using either a hand sieve or an inclined sieve to remove external insects from the grain (White, 1983). However, many of the important bulk grain pests are internal feeders and are difficult to detect because only the adult stages that have emerged from the grain can be sieved out of samples. A range of techniques to detect internal feeding insects have been developed (e.g. staining, flotation, X-ray examination, sound detection, nuclear magnetic resonance, ELISA, near-infrared reflectance spectroscopy (NIR)), but most are still relatively labour and time intensive to perform on large amounts of grain (Pedersen, 1992; Dowell et al., 1998). An indirect approach commonly used by industry to assess insect density is to determine the number of IDK present in samples. Unfortunately, IDK and number of internally feeding insects are not always related (Perez-Mendoza et al., 2004).

Traps, which capture insects as they move through the grain, can be more efficient at detecting the presence of insects than direct sampling. However, species differences in mobility can lead to differences in trap capture and, as with direct sampling, only a relatively small portion of the grain mass is being sampled. Probe traps, which are a type of pit-fall trap, can be used to sample and detect insect populations in bulk grain (Hagstrum et al., 1998), but they provide an estimate of relative abundance, not absolute numbers. A version of the probe trap that automatically counts the insects that are captured has been developed and commercialized (Epsky and Shuman, 2001; Toews et al., 2003). In some cases, traps placed in the headspace above the raw commodity can provide a good prediction of future pest problems. For example, capture of C. ferruginens in sticky traps in the headspace of farm bins during the first 3 weeks of storage provided a good indication that the bin would become infested (Hagstrum et al., 1994).

Monitoring of insects in food-processing and warehouse structures involves either direct visual sampling or the use of traps. Visual inspection done on a regular basis is one of the primary means by which insect infestation is monitored in food facilities (Mills and Pedersen, 1990). The strength of this approach is that not only does it detect signs of insect infestation, but it can also identify potential problem areas

such as accumulations of spillage before they become infested. However, in many cases food patches are not detectable or access requires destructive sampling, which makes detection difficult until populations build to very high levels.

Use of traps to monitor insects is also common in food storage and processing environments and a range of trap types are available (e.g. pheromone traps, food attractant traps, sticky boards, light traps) and pheromone trap use is increasing in commercial facilities (Phillips et al., 2000b). Traps have the advantage that they sample continuously and with appropriate stimuli they can attract insects from a wide area. Thus, trapping can provide information more quickly and easily and in many cases earlier than visual inspections. Since most of these traps capture insects that are dispersing between resource patches, it can be difficult to make the connection between numbers captured in traps and actual levels of product infestation. The best use of this information may be to use the relative numbers captured and their spatial distribution to make targeted pest management decisions (i.e. indicative interpretation) rather than trying to estimate total abundance (Arbogast and Mankin, 1999).

Pheromones have been isolated and lures are commercially available for many stored-product insects (Chambers, 1990; Phillips et al., 2000b) and several traps designed specifically for stored-product insects are available commercially (Vick et al., 1990; Mullen, 1992; Mullen and Dowdy, 2001; Collins and Chambers, 2003). Food odours are also important as attractants for traps both alone and in combination with pheromone lures as synergists or additive attractants. Pheromones combined with food odour can be more attractive than either alone (Trematerra and Girgenti, 1989; Phillips et al., 1993; Landolt and Phillips, 1997). Food odours can also be used to increase the trap capture of species without commercially available pheromone lures, of females, and of immature stages.

Pheromone traps have been demonstrated to be effective at capturing stored-product pests in food-processing and storage environments (Vick *et al.*, 1986; Soderstrom *et al.*, 1987; Pierce, 1994; Bowditch and Madden, 1996). A number of studies have used pheromone trapping to address the temporal and spatial patterns to stored-product pest abundance in bulk grain storage containers (Arbogast *et al.*, 1998; Brenner *et al.*, 1998), flour mills (Doud and Phillips, 2000; Campbell and Arbogast, 2004), food-processing plants (Rees, 1999), warehouses (Campbell *et al.*, 2002) and retail stores (Arbogast *et al.*, 2000; Roesli *et al.*, 2003b). However, many questions remain about the use of these monitoring tools, from the very practical issues such as how many traps are needed and which types work best, to fundamental issues concerning the relationship between pheromone trap captures and actual pest population density, distribution and level of product infestation (Arbogast and Mankin, 1999). There has been limited research into the relationship between pheromone trap capture and the absolute number of insects ['representative' trap interpretation (Arbogast and Mankin, 1999)] present in a structure (Hagstrum and Stanley, 1979; Mankin *et al.*, 1983, 1999; Leos-Martinez *et al.*, 1986; Wileyto *et al.*, 1994; Rees, 1999).

The application of geostatistical techniques to pest monitoring data has been increasingly used to help determine the spatial distribution of the insects inside and outside of food facilities (Liebhold *et al.*, 1993; Arbogast *et al.*, 1998; Brenner *et al.*, 1998). As a practical tool, contour mapping helps plant managers visualize and incorporate spatial distribution information into pest management programmes. However, it is also important to take into account the assumptions behind this approach and to set up monitoring programmes that will generate data of sufficient quality to address the questions needed. For example, the number of traps and degree of spatial autocorrelation among traps strongly influences our ability to make accurate contour maps (Nansen *et al.*, 2003b). Alternative methods for analysing spatial distribution are also available (Nansen *et al.*, 2003a). In addition, insect movement behaviour

and the environment around each trap can influence trap capture in ways that do not reflect the true distribution of infestation (Campbell *et al.*, 2002).

Management Tactics

Sanitation is critical for the production, manufacture and distribution of food (Mills and Pedersen, 1990; Gould, 1994). A sanitation programme is much more than just housekeeping and in its usage in the food industry is similar to an IPM programme in that it involves the integration of multiple components. Additional components of a sanitation programme include: inspection, examination of raw materials, processes and processing equipment, facilities and finished products for infestation; physical manipulations, such as the use of barriers, sealing gaps, controlling temperatures and humidity; and mechanical methods such as traps and irradiation. Sanitation basically functions to prevent problems from occurring by removing food sources and favourable conditions for stored-product insects, and thorough continual vigilance responds quickly to eliminate pest problems when they arise.

Housekeeping (i.e. removing resource patches) and pest exclusion (i.e. preventing movement among resource patches) are arguably the most critical components of a food protection programme. Housekeeping involves cleaning spillage, removing or sealing damaged packages and removing out-of-condition products. Housekeeping is the process of maintaining cleanliness and orderliness and is important in reducing the ability of pests to become established. Degradation of food material and presence of moisture can lead to pest problems that would not occur under good housekeeping conditions. Pest exclusion involves two major components; reducing the number of external sources of insects and preventing their entry into the facility. Many aspects of food-processing facility design, modification and repair concern the prevention of pest entry (Imholte and Imholte-Tauscher, 1999): 'buffer' zones around a food plant; reducing the attraction of insects by the proper placement of lighting and using lights that are less attractive to insects; proper sealing of the building walls and roofs and openings such as windows, doors, loading docks, track-wells, intake and exhaust vents that can serve as pest entry points. To prevent human movement of insects into a facility, an effective monitoring programme is also needed to assure that products are pest-free when they arrive on the site.

Packages are designed to protect food products from the point of manufacture to consumption and can play an important role in pest exclusion. Infestation of packaged products depends on the package design, handling during manufacture, shipping and storage, and the time exposed to potential infestation. Almost all non-perishable food package designs have openings that can allow insects to enter due to manufacturing specifications, flaws in manufacturing or damage that occurs during shipment and storage (Mullen, 1994; Mowery et al., 2002). Insects can enter packages by chewing through the material or by invading through existing openings (Highland, 1984, 1991), but most infestations are the result of invasion through seams and closures rather than chewing (Mullen, 1997). Adult sawtoothed grain beetle, for example, can enter packaging through openings less than 1 mm in diameter and the adult red flour beetle can enter holes in packaging that are less than 1.35 mm in diameter (Highland, 1984). Stored-product pests can be attracted to the food odours coming from holes in packages and lay eggs in or near the holes which enable infestation through even smaller openings (Barrer and Jay, 1980; Mowery et al., 2002; 2004). Packages can serve as odour barriers and this can reduce infestation, but any hole in the package will negate its odour-proof qualities (Mowery et al., 2004).

Insecticides are widely used in the food industry as a pest management tool, and in many situations are considered the primary management tool. Ideally pesticides should be used in response to pest problems that escape the methods of management already described. Insecticides used in food-processing facilities can be classified into three general groups: fumigants, aerosols and surface treatments.

Fumigants are insecticides that are toxic in the gaseous phase (e.g. methyl bromide, phosphine and sulfuryl fluoride). Fumigants penetrate into bulk and bagged commodities, packaging, and can disperse throughout a structure to kill hidden insects and immature states. Fumigants offer no residual protection; therefore, reinfestation can be rapid. Modified or controlled atmospheres are also toxic as gases, and have been used with some success to control insects in stored bulk grains (Adler et al., 2000), but are not generally used by the milling and processing industry as whole-plant treatments (White and Leesch, 1996). However, modified atmospheres, vacuum sealing or low pressure treatments may be useful for small-scale or specialty applications (Mbata and Phillips, 2001). The fumigant methyl bromide has been widely used for structural treatments, but its use is being phased-out and the search is ongoing for viable replacements (Fields and White, 2002).

Aerosols are pesticides applied in small droplets as a mist or fog inside a structure. Aerosols and space sprays are targeted primarily at exposed insects that are flying or walking on a surface and do not penetrate well and, as a result, are not used against pests in bulk commodities. They are dispensed from an aerosol fogger, can have vapour and contact toxicity, and offer very little residual production. Synergized pyrethrins, a natural product, and the organophosphate dichlorvos are two insecticides that have historically been used for aerosol applications inside processing facilities and in food warehouses. More recently, new pyrethroid and insect growth regulators as surface treatments and aerosols have been registered for use against stored-product insects, but most documented reports of efficacy describe tests done at laboratory conditions (Arthur and Gillenwater, 1990; Arthur, 1993).

Pesticides are applied directly to flooring and wall surfaces either to cover surfaces or limited to specific areas, such as cracks, crevices and wall baseboards. These pesticides have contact toxicity and generally give some degree of residual control. Examples of registered products for surface and/or crack and crevice treatments include malathion, the pyrethroid insecticide cyfluthrin, the IGRs hydroprene and methoprene, and the inert dust diatomaceous earth (DE). Insecticide resistance has been reported for some of these compounds (Collins, 1990; Subramanyam and Hagstrum, 1996; Wright et al., 2002). Pheromones have also been investigated as pest suppression tools (e.g. mass trapping, mating disruption) (Trematerra, 1988), sometimes in conjunction with pesticides (e.g. lure and kill) (Nansen and Phillips, 2004).

Manipulating temperature can also be used as a pest management tool. Many milling and processing facilities have attached bulk storage bins that are used as part of normal operations and management of insects in bulk storage is important not only for the protection of the stored commodity but also to reducing bulk storage as a source of infestation for other portions of the facility. The use of aeration, or low-volume ambient air, to cool stored grains is an important component of most management programmes in temperate climates (Reed and Arthur, 2000). Aeration for insect pest management usually modifies the temperature of the bulk grain mass to limit insect population development, and since most stored-product insects cease development at 15°C (Howe, 1965b; Fields, 1992), this temperature is often used as a target for aeration management. Typical airflow rates used for aeration are in the range of 0.0013–0.0065 $m^3/s/m^3$ (0.1–0.5 ft^3/min/bushel, or CFM/bushel) (McCune et al., 1963; Noyes et al., 1987).

Simulation modelling studies using historical weather data have shown that controlled aeration for specified time periods can suppress S. zeamais, the maize weevil, in maize stored throughout the eastern USA (Arthur et al., 1998, 2001), and C. ferrugineus, the rusty grain beetle, in wheat stored in the southern plains of the USA (Flinn et al., 1997; Arthur and Flinn, 2000). These concepts have been expanded to predict the potential of aeration to control Sitophilus oryzae (L.), the rice weevil, in rough rice and milled rice stored in Japan (Arthur et al., 2003). Recent field research has also shown aeration to be effective in suppressing insect populations in stored rice in the south-central USA (Ranalli et al., 2003). In addition, modifications to existing strategies, such as cooling stored wheat during the initial months of summer storage, can also limit pest population growth (Arthur and Casada, 2005). Aeration is compatible with other integrated control strategies used for storage facilities.

Heating flour mills to temperatures necessary to kill insects was first documented as a control strategy in the early part of the 20th century (Dean, 1911). With the impending loss of the fumigant methyl bromide, heat treatments are receiving increased attention as a whole-plant structural treatment for insect control (Dowdy, 1999; Dowdy and Fields, 2002). Several recent publications document the use of heat treatments to kill insects infesting the flour and feed mills, model the response of different insect species, and describe temperature thresholds necssary to eliminate pest populations (Wright et al., 2002; Mahroof et al., 2003a, b; Roesli et al., 2003a; Arthur, 2005).

There is a long history of research into the biological control of stored-product pests (Burkholder and Faustini, 1991; Brower et al., 1995; Schöller and Flinn, 2000). Some experimental successes using biological control have been reported for both whole and processed commodity storage situations, but the use of biological control as a component of IPM in the food industry remains very limited. Although not well documented, natural enemies occur in food facilities and can impact pest populations, even if they are not dramatically suppressing populations due to either intrinsic factors or constraint by other management tools. A heavy reliance on fumigation, and other broad-spectrum insecticides, has probably contributed to limited pest suppression by natural enemies. The incorporation of biological control as a component of IPM may increase in the future with the reduction in the use of broad-spectrum insecticides and better monitoring to identify source patches of pests. However, many facilities are extremely sensitive to food contamination with any foreign substance and there is a reluctance to use biological control.

Stored-product insects have a number of pathogens (e.g. fungi, bacteria, protozoa, viruses) that have been investigated for their effectiveness in pest management, but this investigation has been limited primarily to basic research and rarely have they been used in IPM programmes (Moore et al., 2000). There is little evidence that consuming food with microbial insecticides presents a health hazard (Burges, 1981; Siegel and Shadduck, 1990), but the use of pathogens, even if specific for insects, around human food does present challenges in terms of public perspective. Some pathogen species have been registered for use on stored-products and some species occur naturally in stored-product pest populations and storage environments where they may impact pest population dynamics (Krieg, 1987; Morris et al., 1998; Oduor et al., 2000).

A suite of parasites and predators are associated with stored-product insects and they have received considerable experimental research, but relatively little field study. Schöller (1998) reports that 58 species of parasitoids and predators that attack 79 species of stored-product pests have been studied in at least 900 published articles. Most species are widely distributed geographically and are often found associated with human storage of food. Internal feeding grain pests are susceptible to some parasitoid species that are able to move through bulk grain, detect seeds that are infested and parasitize the immature stage inside the seed (e.g. species are Anisopteromalus calandrae (Howard), Lariophagus distinguendus Förster, Pteromalus cerealellae (Ashmead) and Theocolax elegans (Westwood)).

These species tend to be facultative and attack multiple internally feeding species. The parasitoid, Anisopteromalus calandrae, has been demonstrated to reduce rice weevil infestations in wheat spillage by 90% (Press et al., 1984) and to reduce infestation of bagged wheat (Cline et al., 1985).

Stored-product moths are susceptible to a wide range of parasitoid species. The eggs of several important stored-product moths are susceptible to attack by wasps in the genus Trichogramma (Brower, 1983a, b). Trichogramma spp. have been reported from groundnut storage environments (Brower, 1984) and releases have reduced moth populations in groundnuts (Brower, 1988). Trichogramma evanescens Westwood has been used commercially in Europe for management of stored-product moths in retail facilities (Schöller and Flinn, 2000). Habrobracon (Bracon) hebetor, a braconid wasp, and Venturia canescens, an ichneumonid wasp, parasitize pyralid moth larvae, including P. interpunctella, and have been recovered from flour mills and other food storage facilities. In simulated warehouses, the parasitoid wasps Habrobracon hebetor and Venturia canescens are capable of reducing Cadra cautella infestation in food spillage and subsequent infestation of packaged commodities (Cline et al., 1984, 1986; Cline and Press, 1990). Cline and Press (1990) proposed that the combination of packaging that reduces infestation and parasitoid wasps to reduce pest populations in the structure of the building may be an effective approach in certain situations.

The Hemipteran predator Xylocoris flavipes (Reuter), the warehouse pirate bug, is probably the most studied predator that persists in commodity storage facilities such as groundnut warehouses and grain bins and can effectively suppress pest populations (Press et al., 1975; Arbogast, 1976; LeCato et al., 1977; Brower and Mullen, 1990; Brower and Press, 1992). A variety of Coleoptera species are facultative predators of stored-product pests, but a number of these species (e.g. Tribolium castaneum) are also directly damaging to grain or processed commodities (LeCato, 1975).

Potential for Application of Ecologically Based IPM

There are a wide range of monitoring and management tools available for stored-product pest management in the food industry, but often the effectiveness of these approaches and how best to integrate them are not well understood. Even as some tactics are being lost, new ones are being developed and tested, and older approaches that have not been extensively used due to the reliance on tactics such as fumigation are being revived. The difficulty for the food industry from an IPM perspective has been how to integrate these various tools into a coherent and effective programme. Because of the artificial nature of these environments and the low tolerance in many situations for presence of insects, ecologically based IPM will rely less on promoting population regulation using natural enemies and more on modifying the environment to make it less favourable to pest establishment and persistence. The exception to this is bulk storage, where biological control shows more potential for success since some insects can be tolerated in many situations and natural enemies can be cleaned out of the material before processing.

In current practice, many locations still rely on calendar-based pesticide applications and have little understanding of the ecological basis of pest management. This attitude is changing, but for reasons already discussed pest management of stored-product insects in facilities storing and processing food has some unique challenges compared to pre-harvest IPM. Many of the components of an IPM programme are known and are available for use, but our understanding of how to optimally integrate and target these tactics as part of an ecologically based IPM is very limited. Adoption has also been hindered by a poor understanding of pest populations in the spatially and temporally complex landscapes in which food is processed and stored, the difficulty of evaluating pest populations, and finally the limited information on field efficacy and how to optimally select and combine management tools.

For bulk-stored commodities, it is often difficult to adequately monitor and sample large grain bulks, particularly in commercial elevator facilities, because of the need to directly sample the large volume of grain and detect relatively low densities of insects. Precise treatment thresholds and economic injury levels have not been developed, and standards and rejection criteria are inconsistent and difficult to apply. As a result, treatments based on an economic threshold are not typically performed and control strategies are often applied preventively, even when using tactics that do not have any residual effect. However, expert systems have been developed to help manage farm-stored grain in the USA, which includes interpretation of sampling data and predictive models for insect population growth using different management strategies (Flinn and Hagstrum, 1990). A similar system has also been developed for managing insects in grain elevators (Flinn *et al.*, 2003). This system uses vacuum probe sampling to determine pest densities and performs a risk benefit analysis to determine the level of risk of economic losses and the need for fumigation. Aeration systems to cool the temperature of stored grain can also be an effective management tool for bulk grain (Reed and Arthur, 2000) and is potentially more compatible with biological control than fumigation. For example, Flinn's (1998) study on the effect of grain temperature on parasitoid wasp *Theocolax elegans* (Westwood) suppression of *R. dominica* populations in wheat indicated that aeration of the grain bin could increase parasitoid effectiveness.

As grain products move from bulk storage to processing and milling facilities, then through distribution and marketing channels to consumers, the concept of economic injury level becomes more difficult to apply. Often there is 'zero tolerance' for insects, and controls become more preventative. There are no precise damage thresholds or injury levels, and it may be difficult to adequately determine pest levels or to estimate all of the direct and indirect costs. In the food industry, multiple tactics are used to manage pests, although often these tactics are not necessarily integrated optimally and there is still a heavy reliance on chemical insecticides. A recent survey of grocery stores in Oklahoma showed that management practices are still pesticide intensive, with little use of IPM alternatives (Platt *et al.*, 1998).

Combining and integrating different management tools and careful selection and timing of different approaches, combined with an understanding of pest behaviour and ecology, can result in improved effectiveness. Heat combined with DE effectively reduced the temperatures necessary to kill stored-product insects (Dowdy and Fields, 2002). In another test, high temperatures typically attained during a heat treatment had no deleterious effects on contact insecticides such as cyfluthrin WP and hydroprene, and may have even enhanced toxicity of cyfluthrin WP (Arthur and Dowdy, 2003). Synergistic interactions have been achieved by combining *B. bassiana* with desiccant dusts (DE) to control stored-product beetles (Lord, 2001). Autoinoculation releases using food and pheromone baits to attract insects that pick up the pathogen and then disseminate it through the environment have also been explored (Vail *et al.*, 1993). Many of these studies have been under laboratory conditions, so there is limited information on integration under field conditions. The Toews *et al.* (2005b) study in simulated warehouses illustrates that it is difficult to impact populations in hidden refugia with pesticide applications not applied directly to the food patches, and pheromone trap captures evaluating the impact of the treatment did not always reliably track changes in pest populations.

Often there is reluctance or lack of interest on the part of food industry to move away from calendar-based pesticide treatments to a more integrated approach. In large part this is due to a justifiable concern about making mistakes with pest control in an industry with an extremely low pest threshold. From the scientific perspective, there is also a shortage of experimental data from real-world situations with which to make recommendations. With the pending loss of major management tools such as methyl bromide and organophosphate insecticides due to government regulations and market demands directly from consumers, there will be increasing pressure to develop IPM programmes to keep our food supply safe from insect infestation and a need for the scientific community and the food industry to work together to find these solutions.

References

Adler, C., Corinth, H. and Reichmuth, C. (2000) Modified Atmospheres. In: Subramanyam, B. and Hagstrum, D.W. (eds) *Alternatives to Pesticides in Stored Product IPM*. Kluwer Academic Publishers, Boston, Massachusetts, pp. 105–146

Arbogast, R.T. (1976) Suppression of *Oryzaephilus surinamensis* (L.)(Coleoptera: Cucujidae) on shelled corn by the predator *Xylocoris flavipes* (Reuter) (Hemiptera: Anthocoridae). *Journal of Georgian Entomological Society* 11, 67–71.

Arbogast, R.T. and Mankin, R.W. (1999) The utility of spatial analysis in management of storage pests. In: Zuxun, J., Quan, L., Yongsheng, L., Xianchang, T. and Lianghua, G. (eds) *Stored Product Protection: Proceedings of the 7th International Working Conference on Stored-product Protection*. Sichuan Publishing House of Science and Technology, Chengdu, China, pp. 1519–1527.

Arbogast, R., Kendra, P. and Chini, S. (2003) *Lasioderma serricorne* (Coleoptera: Anobiidae): spatial relationship between trap catch and distance from a source of infestation. *Florida Entomologist* 86, 437–444.

Arbogast, R., Chini, S. and McGovern, J. (2005) *Plodia interpunctella* (Lepidoptera: Pyralidae): spatial relationship between trap catch and distance from a source of emerging adults. *Journal of Economic Entomology* 98, 326–333.

Arbogast, R.T., Weaver, D.K., Kendra, P.E. and Brenner, R.J. (1998) Implications of spatial distribution of insect populations in storage ecosystems. *Environmental Entomology* 27, 202–216.

Arbogast, R.T., Kendra, P.E., Mankin, R.W. and McGovern, J.E. (2000) Monitoring insect pests in retail stores by trapping and spatial analysis. *Journal of Economic Entomology* 93, 1531–1542.

Arthur, F.H. (1993) Evaluation of prallethrin aerosol to control stored product insect pests. *Journal of Stored Product Research* 29, 253–257.

Arthur, F.H. (2005) Initial and delayed mortality of late-instar larvae, pupae, and adults of *Tribolium castaneum* and *Tribolium confusum* (Coleoptera: Tenebrionidae) exposed at variable temperatures and time intervals. *Journal of Stored Product Research* 42, 1–7.

Arthur, F.H. and Casada, M. (2005) Feasibility of summer aeration for management of wheat stored in Kansas. *Applied Engineering in Agriculture* 21, 1027–1038.

Arthur, F.H. and Dowdy, A.K. (2003) Impact of high temperatures on efficacy of cyfluthrin and hydroprene applied to concrete to control *Tribolium castaneum* (Herbst). *Journal of Stored Products Research* 39, 193–204.

Arthur, F.H. and Flinn, P. (2000) Aeration management for stored hard red winter wheat: simulated impact on rusty grain beetle (Coleoptera: Cucujidae) populations. *Journal of Economic Entomology* 93, 1364–1372.

Arthur, F.H. and Gillenwater, H.B. (1990) Evaluation of esfenvalerate aerosol for control of stored product insect pests. *Journal of Entomological Science* 25, 261–267.

Arthur, F.H., Throne, J., Maier, D. and Montross, M. (1998) Feasibility of aeration for management of maize weevil populations in corn stored in the southern United States: model simulations based on recorded weather data. *American Entomologist* 44, 118–123.

Arthur, F.H., Throne, J., Maier, D. and Montross, M. (2001) Impact of aeration on maize weevil (Coleoptera: Curculionidae) populations in corn stored in the northern United States: simulation studies. *American Entomologist* 47, 104–110.

Arthur, F.H., Takahashi, K., Hoernemann, C.K. and Soto, N. (2003) Potential for autumn aeration of stored rough rice and the potential number of generations of *Sitophilus zeamais* Motschusky in milled rice in Japan. *Journal of Stored Product Research* 39, 471–487.

Barrer, P. and Jay, E.G. (1980) Laboratory observations on the ability of *Ephestia cautella* (Walker) (Lepidoptera: Phycitidae) to locate, and to oviposit in response to a source of grain odour. *Journal of Stored Product Research* 16, 1–7.

Bowditch, T.G. and Madden, J.L. (1996) Spatial and temporal distribution of *Ephestia cautella* (Walker) (Lepidoptera: Pyralidae) in a confectionery factory: causal factors and management implications. *Journal of Stored Products Research* 32, 123–130.

Boxall, R.A. (1991) Post-harvest lossess to insects – a world overview. In: Rossmoore, H.W. (ed.) *Biodeterioration and Biodegradation 8*. Elsevier Applied Science, London, pp. 160–175.

Brenner, R.J., Focks, D.A., Arbogast, R.T., Weaver, D.K. and Shuman, D. (1998) Practical use of spatial analysis in precision targeting for integrated pest management. *American Entomologist* 44, 79–101.

Brower, J.H. (1983a) Eggs of stored-product Lepidoptera as hosts for *Trichogramma evanescens* (Hym: Trichogrammatidae). *Entomophaga* 28, 355–362.

Brower, J.H. (1983b) Utilization of stored-product Lepidoptera eggs as hosts by *Trichogramma periosum* Riley (Hymenoptera: Trichogrammatidae). *Journal of the Kansas Entomological Society* 56, 50–54.

Brower, J.H. (1984) The natural occurrence of the egg parasite, *Trichogramma*, on almond moth eggs in peanut storages in Georgia. *Journal of Georgia Entomology* 19, 283–290.

Brower, J.H. (1988) Population suppression of the almond moth and the Indianmeal moth (Lepidoptera: Pyralidae) by release of *Trichogramma pretiosum* (Hymenoptera: Trichogrammatidae) into simulated peanut storages. *Journal of Economic Entomology* 81, 944–948.

Brower, J.H. and Mullen, M.A. (1990) Effects of *Xylocoris flavipes* (Hemiptera: Anthocoridae) releases on moth populations in experimental peanut storages. *Journal of Entomological Science* 25, 268–276.

Brower, J.H. and Press, J.W. (1992) Suppression of residual populations of stored-product pests in empty corn bins by releasing the predator *Xylocoris flavipes* (Reuter). *Biological Control* 2, 66–72.

Brower, J.H., Smith, L., Vail, P.V. and Flinn, P.W. (1995) Biological control. In: Subramanyam, B. and Hagstrum, D.W. (eds) *Integrated Management of Insects in Stored Products*. Marcel Dekker, New York, pp. 223–286.

Burges, H.D. (1981) Safety, safety testing and quality control of microbial pesticides. In: Burges, H.D. (ed.) *Microbial Control of Pests and Plant Disease 1970–1980*. Academic Press, London, pp. 737–767.

Burkholder, W.E. (1990) Practical use of pheromones and other attractants for stored-product insects. In: Ridgway, R.L., Silverstein, R.M. and Inscoe, M.N. (eds) *Behavior-modifying Chemicals for Insect Management: Applications of Pheromones and Other Attractants*. Marcel Dekker, New York, pp. 497–516.

Burkholder, W.E. and Faustini, D.L. (1991) Biological methods of survey and control. In: Gorham, J.R. (ed.) *Ecology and Management of Food Industry Pests*. Association of Official Analytical Chemists, Arlington, Virginia, pp. 361–372.

Campbell, J.F. (2002) Influence of seed size on exploitation by the rice weevil, *Sitophilus oryzae*. *Journal of Insect Behavior* 15, 429–445.

Campbell, J.F. and Arbogast, R.T. (2004) Stored-product insects in a flour mill: population dynamics and response to fumigation treatments. *Entomologia Experimentalis et Applicata* 112, 217–225.

Campbell, J.F. and Hagstrum, D.W. (2002) Patch exploitation by *Tribolium castaneum*: movement patterns, distribution, and oviposition. *Journal of Stored Products Research*, 38, 55–68.

Campbell, J.F. and Mullen, M.A. (2004) Distribution and dispersal behavior of *Trogoderma variabile* Ballion and *Plodia interpunctella* (Hubner) outside a food processing plant. *Journal of Economic Entomology* 97, 1455–1464.

Campbell, J.F. and Runnion, C. (2003) Patch exploitation by female red flour beetles, *Tribolium castaneum*. *Journal of Insect Science*, 3, 20. available online: insectscience.org/3.20.

Campbell, J.F., Mullen, M.A. and Dowdy, A.K. (2002) Monitoring stored-product pests in food processing plants: a case study using pheromone trapping, contour mapping, and mark-recapture. *Journal of Economic Entomology* 95, 1089–1101.

Chambers, J. (1990) Overview on stored-product insect pheromones and food attractants. *Journal of the Kansas Entomological Society* 63, 490–499.

Chestnut, T.L. (1972) Flight habits of the maize weevil as related to field infestation of corn. *Journal of Economic Entomology* 65, 434–435.

Cline, L.D. and Press, J.W. (1990) Reduction in almond moth (Lepidoptera: Pyralidae) infestations using commercial packaging of foods in combination with the parasitic wasp, *Bracon hebetor* (Hymenoptera: Braconidae). *Journal of Economic Entomology* 83, 1110–1113.

Cline, L.D., Press, J.W. and Flaherty, B.R. (1984) Preventing the spread of the almond moth (Lepidoptera: Pyralidae) from infested food debris to adjacent uninfested packages, using the parasite *Bracon hebetor* (Hymenoptera: Braconidae). *Journal of Economic Entomology* 77, 331–333.

Cline, L.D., Press, J.W. and Flaherty, B.R. (1985) Suppression of the rice weevil, *Sitophilus oryzae* (Coleoptera: Curculionidae), inside and outside of burlap, woven polypropylene, and cotton bags by the parasitic wasp, *Anisopteromalus calandrae* (Hymenoptera: Pteromalidae). *Journal of Economic Entomology* 78, 835–838.

Cline, L.D., Press, J.W. and Flaherty, B.R. (1986) Protecting uninfested packages from attack by *Cadra cautella* (Lepidoptera: Pyralidae) with the parasitic wasp *Venturia canescens* (Hymenoptera: Ichneumonidae). *Journal of Economic Entomology* 79, 418–420.

Cogburn, R.R. and Vick, K.W. (1981) Distribution of angoumois grain moth, Almond moth, and Indianmeal moth in rice fields and rice storage in Texas as indicated by pheromone-baited adhesive traps. *Environmental Entomology* 10, 1003–1007.

Collins, L.E. and Chambers, J. (2003) The I-SPy indicator: an effective trap for the detection of insect pests in empty stores and on flat surfaces in the cereal and food trades. *Journal of Stored Product Research* 39, 277–292.

Collins, P.J. (1990) A new resistance to pyrethroids in *Tribolium castaneum* (Herbst). *Pesticide Science* 28, 101–115.

Cope, J.M. and Fox, C.W. (2003) Oviposition decisions in the seed beetle, *Callosobruchus maculatus* (Coleoptera: Bruchidae): effects of seed size on superparasitism. *Journal of Stored Product Research* 39, 355–365.

Cox, P.D. and Bell, C.H. (1991) Biology and ecology of moth pests of stored foods. In: Gorham, J.R. (ed.) *Ecology and Management of Food Industry Pests*. FDA technical bulletin 4, Association of Official Analytical Chemists, Arlington, Virginia, pp. 181–193.

Cox, P. and Parish, W.E. (1991) Effects of refuge content and food availability on refuge-seeking behaviour in *Cryptolestes ferrugineus* (Stephens) (Coleoptera: Cucujidae). *Journal of Stored Product Research* 27, 135–139.

Cox, P., Parish, W.E. and Beirne, M.A. (1989) Variations in the refuge-seeking behaviour of four strains of *Cryptolestes ferrugineus* (Stephens) (Coleoptera: Cucujidae) at different temperatures. *Journal of Stored Product Research* 25, 239–242.

Cox, P., Parish, W.E. and Ledson, M. (1990) Factors affecting the refuge-seeking behaviour of *Cryptolestes ferrugineus* (Stephens) (Coleoptera: Cucujidae). *Journal of Stored Product Research* 26, 169–174.

Cuperus, G.W. and Krischik, V. (1995) Why stored product integrated pest management is needed. In: Krischik, V., Cuperus, G.W. and Galliart, D. (eds) *Stored Product Management*. Oklahoma Cooperative Extension Service Publication, E-912, p. 199.

Cuperus, G.W., Noyes, R.T., Fargo, W.S., Clary, B.L., Arnold, D.C. and Anderson, K. (1990) Management practices in a high risk stored wheat system in Oklahoma. *American Entomologist* 36, 129–134.

Dean, D.A. (1911) Heat as a means of controlling mill insects. *Journal of Economic Entomology* 4, 142–158.

Doud, C.W. and Phillips, T.W. (2000) Activity of *Plodia interpunctella* (Lepidoptera: Pyralidae) in and around flour mills. *Journal of Economic Entomology* 93, 1842–1847.

Dowdy, A.K. (1999) Mortality of red flour beetle, *Tribolium castaneum* (Coleoptera: Tenebrionidae) exposed to high temperature and diatomaceous earth combination. *Journal of Stored Product Research* 35, 175–182.

Dowdy, A.K. and Fields, P.G. (2002) Heat combined with diatomaceous earth to control the confused flour beetle (Coleoptera: Tenebrionidae) in a flour mill. *Journal of Stored Product Research* 38, 11–22.

Dowdy, A.K. and McGaughey, W.H. (1994) Seasonal activity of stored-product insects in and around farm-stored wheat. *Journal of Economic Entomology* 87, 1351–1358.

Dowell, F.E., Throne, J.E. and Baker, J.E. (1998) Automated nondestructive detection of internal insect infestation of wheat kernels by using near-infrared reflectance spectroscopy. *Journal of Economic Entomology* 91, 899–904.

Epsky, N.D. and Shuman, D. (2001) Laboratory evaluation of an improved electronic grain probe insect counter. *Journal of Stored Product Research* 37, 187–197.

Evans, B.R. and Porter, J.E. (1965) The incidence, importance, and control of insects found in stored food and food-handling areas of ships. *Journal of Economic Entomology* 19, 173–180.

Fadamiro, H.Y. (1997) Free flight capacity determination in a sustained flight tunnel: effects of age and sexual state on the flight duration of *Prostephanus truncatus*. *Physiological Entomology* 22, 29–36.

Fields, P.G. (1992) The control of stored-product insects and mites with extreme temperatures. *Journal of Stored Product Research* 28, 89–118.

Fields, P.G. and White, N.D.G. (2002) Alternatives to methyl bromide treatments for stored product and quarantine insects. *Annual Review of Entomology* 47, 331–359.

Fields, P.G., Van Loon, J., Dolinski, M.G., Harris, J.L. and Burkholder, W.E. (1993) The distribution of *Rhyzopertha dominica* (F.) in western Canada. *Canadian Entomologist* 125, 317–328.

Flinn, P.W. (1998) Temperature effects of efficacy of *Choetospila elegans* (Hymenoptera: Pteromalidae) to suppress *Rhyzopertha dominica* (Coleoptera: Bostrichidae) in stored wheat. *Journal of Economic Entomology* 91, 320–323.

Flinn, P.W. and Hagstrum, D.W. (1990) Stored grain advisor: a knowledge-based system for management of insect pests of stored grain. *AI Applications in Natural Resource Management* 4, 44–52.

Flinn, P.W., Hagstrum, D.W. and Muir, W.E. (1997) Effects of time of aeration, bin size, and latitude on insect populations in stored wheat: a simulation study. *Journal of Economic Entomology* 90, 646–651.

Flinn, P.W., Hagstrum, D., Reed, C. and Phillips, T.W. (2003) United States Department of Agriculture–Agricultural Research Service stored-grain areawide Integrated Pest Management program. *Pest Management Science* 59, 614–618.

Flint, M. and van den Bosch, R. (1981) *Introduction to Integrated Pest Management*. Plenum Press, New York.

Foil, L.D. and Gorham, J.R. (2000) Mechanical transmission of disease agents by arthropods. In: Eldridge, B.F. and Edman, J.F. (eds) *Medical Entomology*. Kluwer Academic Publishers, Dordrecht, The Netherlands, pp. 461–514.

Good, N.E. (1937) Insects found in the milling streams of flour mills in the southwestern milling area. *Journal of the Kansas Entomological Society* 10, 135–148.

Gorham, J.R. (1987) *Insect and Mite Pests in Food: An Illustrated Key*. Agriculture Handbook Number 655. US Department of Agriculture, Washington, DC.

Gorham, J.R. (1991) *Ecology and Management of Food Industry Pests*. Association of Official Analytical Chemists, Arlington, Virginia.

Gould, W.A. (1994) *CGMP's/Food Plant Sanitation*. CTI Publications, Timonium, Massachusetts.

Hagstrum, D.W. (1994) Field monitoring and prediction of stored-grain insect populations. *Postharvest News and Information* 5, 39N–45N.

Hagstrum, D.W. (2001) Immigration of insects into bins storing newly harvested wheat on 12 Kansas farms. *Journal of Stored Product Research* 37, 221–229.

Hagstrum, D.W. and Davis, L.R. Jr (1980) Mate-seeking behavior of *Ephestia cautella*. *Environmental Entomology* 9, 589–592.

Hagstrum, D.W. and Gilbert, E.E. (1976) Emigration rate and age structure dynamics of *Tribolium castaneum* populations during growth phase of a colonization episode. *Environmental Entomology* 5, 445–448.

Hagstrum, D.W. and Stanley, J.M. (1979) Release–recapture estimates of the population density of *Ephestia cautella* (Walker) in a commercial peanut warehouse. *Journal of Stored Product Research* 15, 117–122.

Hagstrum, D.W., Dowdy, A.K. and Lippert, G.E. (1994) Early detection of insects in stored wheat using sticky traps in bin headspace and prediction of infestation level. *Environmental Entomology* 23, 1241–1244.

Hagstrum, D.W., Flinn, P.W. and Subramanyam, B. (1998) Predicting insect density from probe trap catch in farm-stored wheat. *Journal of Stored Product Research* 34, 251–262.

Hagstrum, D.W., Reed, C. and Kenkel, P. (1999) Management of stored wheat insect pests in the USA. *Integrated Pest Management Reviews* 4, 127–142.

Hanski, I. and Simberloff, D. (1997) The metapopulation approach, its history, conceptual domain, and application to conservation. In: Hanski, I. and Gilpin, M. (eds) *Metapopulation Biology: Ecology, Genetics, and Evolution*. Academic Press, San Diego, California, pp. 1–26.

Harein, P. and Meronuck, R. (1995) Stored grain losses due to insects and molds and the importance of proper grain management. In: Krischik, V., Cuperus, G.W. and Galliart, D. (eds) *Stored Product Management*. Oklahoma Cooperative Extension Service Publication, E-912, Oklahoma, pp. 29–31.

Harrison, S. and Taylor, A. (1997) Empirical evidence for metapopulation dynamics. In: Hanski, I. and Gilpin, M. (eds) *Metapopulation Biology: Ecology, Genetics, and Evolution*. Academic Press, San Diego, California, pp. 27–42.

Highland, H.A. (1984) Insect infestation of packages. In: Baur, F.J. (ed.) *Insect Management for Food Storage and Processing*. American Association of Cereal Chemists, St Paul, Minnesota, pp. 309–320.

Highland, H.A. (1991) Protecting packages against insects. In: Gorham, J.R. (ed.) *Ecology and Management of Food Industry Pests*. Association of Official Analytical Chemists, Arlington, Virginia, pp. 345–350..

Hinton, H.E. (1945) *A Monograph of the Beetles Associated with Stored Products*, Vol. 1. British Museum (Natural History), London.

Howe, R.W. (1965a) *Sitophilus granarius* (L.) (Coleoptera: Curculionidae) breeding in acorns. *Journal of Stored Product Research* 1, 99–100.

Howe, R.W. (1965b) A summary of estimates of optimal and minimal conditions for population increase of some stored products insects. *Journal of Stored Product Research* 1, 177–184.

Imholte, T.J. and Imholte-Tauscher, T.K. (1999) *Engineering for Food Safety and Sanitation: A Guide to the Sanitary Design of Food Plants and Food Plant Equipment*. Technical Institute for Food Safety, Woodinville, Washington.

Kenkel, P., Criswell, J.T., Cuperus, G., Noyes, R.T., Anderson, K., Fargo, W.S., Shelton, K., Morrison, W.P. and Adams, B. (1993) Current management practices and impact of pesticide loss in the hard red wheat post-harvest system. Oklahoma Cooperative Extension Service Circular E-930, Oklahoma.

Khare, B.P. and Agrawal, N.S. (1964) Rodent and ant burrows as sources of insect inoculum in the threshing floors. *Indian Journal of Entomology* 26, 97–102.

Kogan, M. (1998) Integrated pest management: historical perspectives and contemporary developments. *Annual Review of Entomology* 42, 243–270.

Krieg, A. (1987) Diseases caused by bacteria and other prokaryotes. In: Fuxa, J.R. and Tanada, Y. (eds) *Epizootiology of Insect Diseases*. John Wiley & Sons, New York, pp. 323–355.

Landolt, P.J. and Phillips, T.W. (1997) Host plant influences on sex pheromone behavior of phytophagous insects. *Annual Review of Entomology* 42, 371–391.

LeCato, G.L. (1975) Red flour beetle: population growth on diets of corn, wheat, rice or shelled peanuts supplemented with eggs or adults of the Indianmeal moth. *Journal of Economic Entomology* 68, 763–765.

LeCato, G.L., Collins, J.M. and Arbogast, R.T. (1977) Reduction of residual populations of stored-product insects by *Xylocoris flavipes* (Hemiptera: Anthocoridae). *Journal of Kansas Entomological Society* 50, 84–88.

Leos-Martinez, J., Granovsky, T.A., Williams, H.J., Vinson, S.B. and Burkholder, W.E. (1986) Estimation of aerial density of the lesser grain borer (Coleoptera: Bostrichidae) in a warehouse using Dominicalure traps. *Journal of Economic Entomology* 79, 1134–1138.

Liebhold, A.M., Rossi, R.E. and Kemp, W.P. (1993) Geostatistics and geographic information systems in applied insect ecology. *Annual Review of Entomology* 38, 303–327.

Lindgren, D.L., Vincent, L.E. and Krohne, M.E. (1955) The khapra beetle, *Trogoderma granarium. Hilgardia* 24, 1–36.

Longstaff, B.C. (1994) The management of stored product pests by non-chemical means – an Australian perspective. *Journal of Stored Product Research* 30, 179–185.

Lord, J.C. (2001) Desiccant dusts synergize the effect of *Beauveria bassiana* (Hyphomycetes: Moniliales) on stored-grain beetles. *Journal of Economic Entomology* 94, 367–372.

Mahroof, R., Subramanyam, B. and Eustace, D. (2003a) Temperature and relative humidity profiles during heat treatment of mills and its efficacy against *Tribolium castaneum* (Herbst) life stages. *Journal of Stored Product Research* 39, 555–569.

Mahroof, R., Subramanyam, B., Throne, J. and Menon, A. (2003b) Time–mortality relationships of *Tribolium castaneum* (Coleoptera: Tenebrionidae) life stages exposed to elevated temperatures. *Journal of Economic Entomology* 96, 1345–1351.

Mankin, R.W., Vick, K.W., Coffelt, J.A. and Weaver, B.A. (1983) Pheromone-mediated flight by male *Plodia interpunctella* (Hubner) (Lepidoptera: Pyralidae). *Environmental Entomology* 12, 1218–1222.

Mankin, R.W., Arbogast, R.T., Kendra, P.E. and Weaver, D.K. (1999) Active spaces of pheromone traps for *Plodia interpunctella* (Lepidoptera: Pyralidae) in enclosed environments. *Environmental Entomology* 28, 557–565.

Mbata, G.N. and Phillips, T.W. (2001) Effects of temperature and exposure time on mortality of stored-product insects exposed to low pressure. *Journal of Economic Entomology* 94, 1302–1307.

McCune, W., Person, N. and Sorenson, W. (1963) Conditioned air storage of grain. *Transactions of American Society of Agricultural Engineering* 6, 186–189.

Mills, R. and Pedersen, J. (1990) *A Flour Mill Sanitation Manual*. Eagan Press, St Paul, Minnesota.

Moore, D., Lord, J.C. and Smith, S.M. (2000) Pathogens. In: Subramanyam, B. and Hagstrum, D.W. (eds) *Alternatives to Pesticides in Stored-product IPM*. Kluwer Academic Publishers, Boston, Massachusetts, pp. 193–228.

Morris, O.N., Converse, V., Kanagaratnam, P. and Cote, J.C. (1998) Isolation, characterization, and culture of *Bacillus thuringiensis* from soil and dust from grain storage bins and their toxicity for *Mamestra configurata* (Lepidoptera: Noctuidae). *Canadian Entomologist* 130, 515–537.

Mowery, S.V., Mullen, M.A., Campbell, J.F. and Broce, A.B. (2002) Mechanisms underlying sawtoothed grain beetle (*Oryzaephilus surinamensis* [L.]) (Coleoptera: Silvanidae) infestation of consumer food packaging materials. *Journal of Economic Entomology* 95, 1333–1336.

Mowery, S.V., Campbell, J.F., Mullen, M.A. and Broce, A.B. (2004) Response of *Oryzaephilus surinamensis* (Coleoptera: Silvanidae) to food odor emanating through consumer packaging films. *Environmental Entomology* 33, 75–80.

Mullen, M.A. (1992) Development of a pheromone trap for monitoring *Tribolium castaneum. Journal of Stored Product Research* 28, 245–249.

Mullen, M.A. (1994) Rapid determination of the effectiveness of insect resistant packaging. *Journal of Stored Product Research* 30, 95–97.

Mullen, M.A. (1997) Keeping bugs at bay. *Feed Management* 48, 29–33.

Mullen, M.A. and Dowdy, A.K. (2001) A pheromone-baited trap for monitoring the Indianmeal moth, *Plodia interpunctella* (Hubner) (Lepidoptera: Pyralidae). *Journal of Stored Product Research* 37, 231–235.

Nansen, C. and Phillips, T.W. (2004) Attractancy and toxicity of an attracticide for Indianmeal moth, *Plodia interpunctella* (Lepidoptera: Pyralidae). *Journal of Economic Entomology* 97, 703–710.

Nansen, C., Subramanyam, Bh. and Roesli, R. (2003a). Characterizing spatial distribution of trap captures of beetles in retail pet stores using SADIE® software. *Journal of Stored Products Research* 40, 471–483.

Nansen, C., Campbell, J.F., Phillips, T.W. and Mullen, M.M. (2003b) The impact of spatial structure on the accuracy of contour maps of small data sets. *Journal of Economic Entomology* 96, 1617–1625.

Nansen, C., Phillips, T.W., Parajulee, M.N. and Franqui, R.A. (2004) Comparison of direct and indirect sampling procedures for *Plodia interpunctella* in a maize storage facility. *Journal of Stored Products Research* 40, 151–168.

Naylor, A.F. (1961) Dispersal in the red flour beetle, *Tribolium castaneum* (Tenebrionidae). *Ecology* 42, 231–237.

Noyes, R., Clary, B. and Cuperus, G.W. (1987) Maintaining the quality of stored wheat by aeration. Oklahoma State University Cooperative Extension Service Fact Sheet 1100, Oklahoma.

Oduor, G.I., Smith, S.M., Chandi, E.A., Karanja, L.W., Agano, J.O. and Moore, D. (2000) Occurrence of *Beauveria bassiana* on insect pests of stored maize in Kenya. *Journal of Stored Product Research* 36, 177–185.

Ogden, J.C. (1970) Aspects of dispersal in *Tribolium* flour beetles. *Physiological Zoology* 43, 124–131.

Olsen, A.R. (1998) Regulatory action criteria for filth and other extraneous materials. II. Allergenic mites: an emerging food safety issue. *Regulatory Toxicology Pharmacology* 28, 190–198.

Olsen, A.R., Gecan, J.S., Ziobro, G.C. and Bryce, J.R. (2001) Regulatory action criteria for filth and other extraneous materials. V. Strategy for evaluating hazardous and nonhazardous filth. *Regulatory Toxicology Pharmacology* 33, 363–392.

Pearson, S.M., Turner, M.G., Gardner, R.H. and O'Neill, R.V. (1996) An organism-based perspective of habitat fragmentation. In: Szaro, R.C. (ed.) *Biodiversity in Managed Landscapes: Theory and Practice.* Oxford University Press, Oxford, pp. 77–95.

Pedersen, J.R. (1992) Insects: identification, damage, and detection. In: Sauer, D.B. (ed.) *Storage of Cereal Grains and Their Products*, 4th edn. American Association of Cereal Chemists, St Paul, Minnesota, pp. 435–489.

Perez-Mendoza, J., Flinn, P., Campbell, J.F., Hagstrum, D. and Throne, J. (2004) Detection of stored-grain insect infestation in wheat transported in railroad hopper cars. *Journal of Economic Entomology* 97, 1474–1483.

Phillips, T.W., Jiang, X.-L., Burkholder, W.E., Phillips, J.K. and Tran, H.Q. (1993) Behavioral responses to food volatiles by two species of stored-product Coleoptera, *Sitophilus oryzae* (Curculionidae) and *Tribolium castaneum* (Tenebrionidae). *Journal of Chemical Ecology* 19, 723–734.

Phillips, T.W., Berberet, R.C. and Cuperus, G.W. (2000a) Post-harvest integrated pest management. In: Francis, F.J. (ed.) *The Wiley Encyclopedia of Feed Science and Technology.* John Wiley & Sons, New York, pp. 2690–2701.

Phillips, T.W., Cogan, P.M. and Fadamiro, H.Y. (2000b) Pheromones. In: Subramanyam, B. and Hagstrum, D.W. (eds) *Alternatives to Pesticides in Stored-product IPM.* Kluwer Academic Publishers, Boston, Massachusetts, pp. 273–302.

Pierce, L.H. (1994) Using Pheromones for location and suppression of Phycitid moths and cigarette beetles in Hawaii – a five-year summary. In: Highley, E., Wright, E.J., Banks, H.J. and Champ, B.R. (eds) *Proceedings of the 6th International Working Conference on Stored-product Protection.* CAB International, Wallingford, UK, pp. 439–443.

Platt, R.R., Cuperus, G.W., Payton, M.E., Bonjour, E.L. and Pinkston, K.N. (1998) Integrated pest management perceptions and practices and insect populations in grocery stores in south-central United States. *Journal of Stored Product Research* 34, 1–10.

Press, J.W., Flaherty, B.R and Arbogast, R.T. (1975) Control of the red flour beetle, *Tribolium castaneum*, in a warehouse by a predaceous bug, *Xylocoris flavipes. Journal of Georgian Entomological Society* 10, 76–78.

Press, J.W., Cline, L.D. and Flaherty, B.R. (1984) Suppression of residual populations of the rice weevil, *Sitophilus oryzae*, by the parasitic wasp, *Anisopteromalus calandrae. Journal of the Georgia Entomological Society* 10, 76–78.

Ranalli, R., Howell, T., Arthur, F.H. and Gardisser, D. (2003) Controlled ambient aeration during rice storage for temperature and insect control. *Applied Engineering and Agriculture* 18, 485–490.

Reed, C. and Arthur, F.H. (2000) Aeration. In: Subramanyam, B. and Hagstrum, D. (eds) *Alternatives to Pesticides in Stored-product IPM.* Kluwer Academic Publishers, Boston, Massachusetts, pp. 51–72.

Rees, D.P. (1999) Estimation of the optimum number of pheromone baited flight traps needed to monitor phycitine moths (*Ephestia cautella* and *Plodia interpunctella*) at a breakfast cereal factory – a case study. In: Zuxun, J., Quan, L., Yongsheng, L., Xianchang, T. and Lianghua, G. (eds) *Stored Product Protection: Proceedings of the 7th International Working Conference on Stored-product Protection.* Sichuan Publishing House of Science and Technology, Chengdu, China, pp. 1464–1471.

Roesli, R., Subramanyam, B., Fairchild, F.J. and Behnke, K.C. (2003a) Trap catches of stored-product insects before and after heat treatment in a pilot feed mill. *Journal of Stored Product Research* 39, 521–540.

Roesli, R., Subramanyam, B., Campbell, J.F. and Kemp, K. (2003b) Stored-product insects in Kansas retail pet stores. *Journal of Economic Entomology* 96, 1958–1966.

Schöller, M. (1998) Integration of biological and non-biological methods to control arthropods infesting stored products. *Postharvest News and Information* 9, 15–20.

Schöller, M. and Flinn, P.W. (2000) Parasitoids and predators. In: Subramanyam, B. and Hagstrum, D.W. (eds) *Alternatives to Pesticides in Stored-product IPM.* Kluwer Academic Publishers, Boston, Massachusetts, pp. 229–272.

Siegel, J.P. and Shadduck, J.A. (1990) Safety of microbial insecticides to vertebrates–humans. In: Laird, M., Lacey, L.A. and Davidson, E.W. (eds) *Safety of Microbial Insecticides.* CRC Press, Boca Raton, Florida, pp. 101–113.

Sinclair, E.R. and Haddrell, R.L. (1985) Flight of stored products beetles over a grain farming area in southern Queensland. *Journal of the Australian Entomological Society* 24, 9–15.

Soderstrom, E.L., Hinsch, R.T., Bongers, A.J., Brandl, D.G. and Hoogendorn, H. (1987) Detecting adult Phycitinae (Lepidoptera: Pyralidae) infestations in a raisin-marketing channel. *Journal of Economic Entomology* 80, 1229–1232.

Strong, R.G. (1970) Distribution and relative abundance of stored-product insects in California: a method of obtaining sample populations. *Journal of Economic Entomology* 63, 591–596.

Subramanyam, B. and Hagstrum, D.W. (1995) Sampling. In: Subramanyam, B. and Hagstrum, D.W. (eds) *Integrated Management of Insects in Stored Products*. Marcel Dekker, New York, pp. 135–193.

Subramanyam, B. and Hagstrum, D.W. (1996) Resistance measurement and management. In: Subramanyam, B. and Hagstrum, D.W. (eds) *Integrated Management of Insects in Stored Products*. Marcel Dekker, New York, pp. 331–398.

Throne, J.E. and Cline, L.D. (1989) Seasonal flight activity of the maize weevil, *Sitophilus zeamais* Motschulsky (Coleoptera: Curculionidae), and the rice weevil, *S. oryzae* (L.), in South Carolina. *Journal of Agricultural Entomology* 6, 183–192.

Throne, J.E. and Cline, L.D. (1991) Seasonal abundance of maize and rice weevils (Coleoptera: Curculionidae) in South Carolina. *Journal of Agricultural Entomology* 8, 93–100.

Toews, M.D., Phillips, T.W. and Shuman, D. (2003) Electronic and manual monitoring of *Cryptolestes ferrugineus* (Stephens) (Coleoptera: Laemophloeidae) in stored wheat. *Journal of Stored Product Research* 34, 712–718.

Toews, M.D., Arthur, F.H. and Campbell, J.F. (2005a) Role of food and structural complexity on capture of *Tribolium castaneum* (Herbst) (Coleoptera: Tenebrionidae) in simulated warehouses. *Environmental Entomology* 34, 164–169.

Toews, M.D., Arthur, F.H. and Campbell, J.F. (2005b) Role of food and structural complexity on capture of *Tribolium castaneum* (Herbst) (Coleoptera: Tenebrionidae) in simulated warehouses. *Environmental Entomology* 34, 833–843.

Trematerra, P. (1988) Suppression of *Ephestia kuehniella* Zeller by using a mass-trapping method. *Tecnica Molitoria* 18, 865–869.

Trematerra, P. and Girgenti, P. (1989) Influence of pheromone and food attractants on trapping of *Sitophilus oryzae* (L.) (Col., Curculionidae): a new trap. *Journal of Applied Entomology* 108, 12–20.

Turner, M.G. (1989) Landscape ecology: the effect of pattern on process. *Annual Review of Ecology and Systematics* 20, 171–197.

Vail, P.V., Hoffmann, D.F. and Tebbets, J.S. (1993) Autodissemination of *Plodia interpunctella* (Hübner) (Lepidoptera: Pyralidae) Granulosis virus by healthy adults. *Journal of Stored Product Research* 29, 71–74.

Vick, K.W., Koehler, P.G. and Neal, J.J. (1986) Incidence of stored-product Phycitinae moths in food distribution warehouses as determined by sex pheromone-baited traps. *Journal of Economic Entomology* 79, 936–939.

Vick, K.W., Coffelt, J.A. and Weaver, W.A. (1987) Presence of four species of stored-product moths in storage and field situations in north-central Florida as determined with sex pheromone-baited traps. *Florida Entomologist* 70, 488–492.

Vick, K.W., Mankin, R.W., Cogburn, R.R., Mullen, M., Throne, J.E., Wright, V.F. and Cline, L.D. (1990) Review of pheromone-baited sticky traps for detection of stored-product insects. *Journal of the Kansas Entomological Society* 63, 526–532.

White, G.G. (1983) A modified inclined sieve for separation of insects from wheat. *Journal of Stored Product Research* 19, 89–91.

White, N.D.G. (1992) A multidisciplinary approach to stored-grain research. *Journal of Stored Product Research* 28, 127–137.

White, N.D.G. and Leesch, J.G. (1996) Chemical control. In: Subramanyam, B. and Hagstrum, D.W. (eds) *Integrated Management of Insects in Stored Products*. Marcel Dekker, New York, pp. 267–330.

Wiens, J.A. (1976) Population responses to patchy environments. *Annual Review of Ecology and Systematics* 7, 81–120.

Wiens, J.A. (1989) Spatial scaling in ecology. *Functional Ecology* 3, 385–397.

Wiens, J.A. (1997) Metapopulation dynamics and landscape ecology. In: Hanski, I. and Gilpin, M.E. (eds) *Metapopulation Biology*. Academic Press, San Diego, California, pp. 43–62.

Wiens, J.A., Stenseth, N.C., Van Horne, B. and Ims, R.A. (1993) Ecological mechanisms and landscape ecology. *Oikos* 66, 369–380.

Wiens, J.A., Schooley, R.L. and Weeks, R.D. Jr (1997) Patchy landscapes and animal movements: do beetles percolate? *Oikos* 78, 257–264.

Wileyto, E.P., Ewens, W.J. and Mullen, M.A. (1994) Markov-recapture population estimates: a tool for improving interpretation of trapping experiments. *Ecology* 75, 1109–1117.

With, K.A. and Crist, T.O. (1995) Critical thresholds in species' responses to landscape structure. *Ecology* 76, 2446–2459.

Wright, V.F., Fleming, E.E. and Post, D. (1990) Survival of *Rhyzopertha dominica* (Coleoptera, Bostrichidae) on fruits and seeds collected from Woodrat nests in Kansas. *Journal of Kansas Entomological Society* 63, 344–347.

Wright, E.J., Sinclair, E.A. and Annis, P.C. (2002) Laboratory determination of the requirements for control of *Trogoderma variabile* (Coleoptera: Dermestidae) by heat. *Journal of Stored Product Research* 38, 147–155.

Ziegler, J.R. (1976) Evolution of the migration response: emigration by *Tribolium* and the influence of age. *Evolution* 30, 579–592.

Ziegler, J.R. (1977a) Dispersal and reproduction in *Tribolium*: the influence of food level. *Insect Physiology* 23, 955–960.

Ziegler, J.R. (1977b) Dispersal and reproduction in *Tribolium*: the influence of initial density. *Environmental Entomology* 7, 149–156.

Zimmerman, M.L. (1990) Coleoptera found in imported stored-food products entering southern California and Arizona between December 1984 and December 1987. *Coleopterists Bulletin* 44, 235–240.

Zimmerman, M.L., Olsen, A.R. and Friedman, S.L. (2003) Filth and extraneous material in food. In: Hui, Y.H., Bruinsma, B.L., Gorham, J.R., Nip, W.-K., Tong, P.S. and Ventresca, P. (eds) *Food Plant Sanitation*. Marcel Dekker, New York, pp. 69–76.

18 Diffusion of IPM Programmes in Commercial Agriculture: Concepts and Constraints

Thomas W. Fuchs

*Extension IPM Coordinator, Texas Cooperative Extension
Center, San Angelo, TX 76901, USA*

Introduction

Integrated pest management (IPM) by definition is a challenging subject due to its complexity and its application to commercial agriculture, which requires active management. Proper implementation of IPM requires that those who wish to implement it have an understanding of ecological principles and interactions involved in crop management. Successful implementation generally occurs only when technology is developed through research, demonstrated in the location in which it will be used and an educational process has been conducted to make end-users aware of the technology, its benefits and how it fits into a production system. While research and education in agriculture are probably as old as agriculture itself and are intimately tied to incremental improvement in technology, agricultural innovations historically were relatively slow to develop (Zalom, 1999). Not until the late 18th century were special organizations and societies formed in the USA to discuss and promote agricultural progress and disseminate information. However, by the mid-19th century, hundreds of such societies were operating, some of which were supported by state agricultural societies or boards of agriculture. These early societies often sponsored speakers and travelling 'Farmers' Institutes' providing a structure to deliver information to farmers (Scheuring, 1988).

The US Congress recognized the need for adult education in agriculture when the land grant college system was established through the Morrill Act in 1862, which provided land to each state to be used for a state agricultural college where students could learn about advanced production practices. Later, in 1887, the Hatch Act established State Agricultural Experiment Stations where agricultural research was to be conducted to address the needs of farmers and rural inhabitants. While land grant colleges successfully addressed many agricultural issues and developed specialized courses and publications to deliver new information, the adoption of research-based practices was very slow to develop (Zalom, 1999). Some land grant colleges even sponsored Farmers' Institutes, held travelling schools for farmers, and conducted on-farm demonstrations. However, it was not until the Smith-Lever Act of 1914, which established an Agricultural Extension Service within the land grant college system that real progress in diffusing agricultural technology began to take place.

©CAB International 2007. *Ecologically Based Integrated Pest
Management* (eds O. Koul and G.W. Cuperus)

This act provided funding to place county agents in communities to work directly with farmers to demonstrate and transfer educational information and new technology at the farm level.

Cooperation of the land grant university system with the US Department of Agriculture (USDA) and various agricultural organizations and industries over the past century has resulted in the development and implementation of many incremental advances in agriculture. These advances led to the present technically advanced state of US production systems. Today many other countries throughout the world have developed systems somewhat parallel to the US system. The remainder of this chapter will address the process of diffusion of technology used in IPM systems in commercial agriculture and will address the process primarily in the USA.

Theoretical Basis of Diffusion of Technology

There is a strong theoretical connection between diffusion of technology and behavioural psychology/sociology. One of the theoretical models used to describe the connection is Bennett's KASA model, which states that people adopt new practices or technology after progressing through a systematic hierarchy of changes in knowledge, attitudes, skills and aspirations (Bennett, 1977). This model recognizes that development and availability of new technology does not guarantee that it will be used and that training, which adds to a client's knowledge and skills, is not sufficient to bring about adoption.

Much of the literature regarding innovation and technology transfer is based upon a broad social psychological/sociological theory called diffusion theory. Rogers (1995) defined diffusion in this context as the process by which innovation is communicated through channels over time among members of a social system. Rogers (1983) identified five attributes, which most affect the adoption on a new idea or new technology.

These include: *relative advantage* – how the innovation is better than the current way of doing things; *compatibility* – how the innovation fits in with other management practices; *complexity* – the level of difficulty to understand and use the innovation; *trialability* – the degree to which an innovation can be used on a trial or experimental basis; and *observability* – how visible the results of the innovation are to others. Copp *et al.* (1958) identified the following stages in the adoption process: *awareness*, where the client learns of the innovation; *interest*, where the client feels that the innovation would be useful on his or her farm; *trial*, where the farmer actually tries the innovation on his or her farm; and *adoption*, when the farmer uses the innovation as a part of his or her normal production practices. Nowak (1987) indicated that awareness may come about in two distinct ways: either the producer has a problem and seeks a solution or a third party may call the producer's attention to a problem not previously recognized by the producer. The producer then seeks information about potential solutions and evaluates potential costs and benefits.

Farmers may be classified into categories based upon the time of adoption of innovations (Fig. 18.1). The distribution of farmers among five categories of adopters of technology approaches a normal population distribution (Fliegel, 1993; Rogers, 1995). These authors describe farmers with two standard deviations earlier in time of adoption than the average as *innovators*, and those between one and two standard deviations earlier than average as *early adopters*. Those within one standard deviation earlier or later than average were labelled as *early* and *late majority*, respectively. Finally, those with more than one standard deviation later than average in adoption were labelled as *late adopters* or *laggards*. While this classification is useful, it ignores those who fail to adopt; and rarely do all farmers adopt any technology.

Each category of adopter exhibits certain characteristics or values. Innovators, for example, are often called venturesome, daring or risk takers. Innovators may not be highly respected in their communities because they often try things that do not work out satisfactorily but they do play an

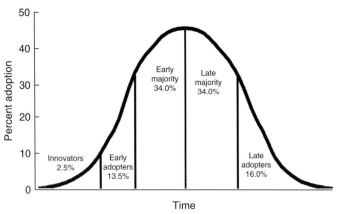

Fig. 18.1.   Distribution of farmers among five categories according to time of adoption of new technology. Adapted from North Central Rural Sociology Committee, Subcommittee for the Study of Diffusion of Farm Practices, North Central Regional Extension Publication, No. 13. East Lansing, Michigan, Michigan State University Cooperative Extension Service, 1961.

important role in diffusing innovations by acting as gatekeepers in the flow of new ideas into a production region. They will try almost any new technology and take pride in being the first to do so.

Early adopters tend to be the most highly respected opinion leaders in an area. They tend to have large operations, be well-educated and have more financial resources than later adopters. Others often seek the advice and leadership of this group of adopters because early adopters will generally make the technology work successfully in their farming operations.

Early majority adopters tend to be very deliberate and adopt just before the average members of the community while late majority adopters tend to be sceptical and adopt when it becomes an economic necessity for them or after considerable peer pressure to adopt.

Finally, late adopters or laggards tend to look back at the way things have been done in the past and are very reluctant to change. They tend to be suspicious of both innovations and the change agents attempting to implement innovations. They must be virtually certain that the new idea will not fail before they are willing to adopt.

While early adopters have much informal influence over the behaviours of others as opinion leaders in the adoption process,

another group, *change agents*; provide an equally important function in positively influencing innovation decisions by mediating between the change agency delivering the new technology and the relevant social system. The success of a change agent in bringing about adoption of innovations by clients is positively related to the extent that he or she works with opinion leaders or early adopters of technology (Rogers, 1995). The Cooperative Extension Service is currently and historically spending a great deal of time and effort in identifying opinion leaders and early adopters and has been very successful in the role of change agents. Rogers (1995) stated: 'The Agricultural Extension Service of the United States is reported to be the world's most successful change agency. Certainly it is the most admired and widely copied.' In recent years, agricultural consultants have played a very important role as change agents also (Post, 1988; Zalom, 1999).

Because adoption of an innovation follows a normal bell-shaped curve when plotted over time on a frequency basis, when the accumulative numbers of adopters is plotted over time, the result is an S-shaped (or sigmoid) curve as seen in Fig. 18.2 (Rogers, 1983, 1995; Dent, 1998). The S-shaped adopter distribution curve indicates that early adoption tends to proceed at a relatively slow rate.

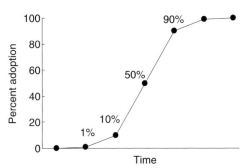

Fig. 18.2. Accumulative percent adoption of innovations through time. (From Dent, 1999.)

After approximately 10–20% have adopted, the rate accelerates as the new technology is recognized as attractive until 80–90% of the individuals who will adopt have done so. Then adoption tapers off as only a few laggards remain who might find the technology worthwhile. Most attempts to estimate adoption curves based upon empirical data have followed the procedures of Griliches (1957) who fitted a logistic function to time-series data on the adoption of hybrid maize in the USA.

Sorensen *et al.* (2000) recently challenged diffusion theory by hypothesizing that in the case of adoption of new IPM practices such as precision agriculture by producers who are already adopters of other IPM practices, producers are adapting their IPM systems to fit the new IPM technologies rather than vice versa as diffusion theory suggests. Further, Fernandez-Cornejo *et al.* (2002) introduced the concept of agricultural technology disadoption that explains how adoption levels can decline rather than increase. Disadoption can occur when a more effective technology becomes available or the technology adopted proves less successful than first thought or projected.

Practical Attributes of Technology Necessary for Successful Implementation

The reasons growers implement new technology, including IPM practices or any other technology have often been debated. It is critical to understand that the adoption is not a discrete, dichotomous event by which one moves from a non-adopter to an adopter by a single decision, but rather involves a process (Nowak *et al.*, 1966). One of the most basic reasons clientele adopt new technology is need, i.e. the grower recognizes a problem or need with which the new technology has potential to provide help. One of the most important considerations is that the technology has a cost advantage over current practice (Smith and Huffaker, 1973; Way, 1977; Frisbie and Adkisson, 1985; Wearing, 1988). A grower's bottom line profitability is highly motivating and the most likely avenue for change.

One need, historically important in growers implementing IPM, is the necessity to control a pest during a pest control crisis, particularly one involving resistance of the pest to pesticides (Corbet and Smith, 1976; Graebner *et al.*, 1984; Harris, 2001). When pesticides fail to provide economic control, growers are often anxious to seek other alternatives such as the use of IPM. The key event that led to the statewide Texas IPM programme, for example, was resistance of the tobacco budworm, *Heliothis virescens* (Fabricius), to all available insecticides in the 1960s. The IPM effort led growers from almost total dependence on insecticides to using a multi-tactic IPM approach for managing this key pest of Texas cotton (Frisbie and McWhorter, 1986). Resistance management continues to be an important component of the IPM programme today. Resistance management has historically focused on foliar-applied insecticides; but herbicide resistance, fungicide resistance, resistance to genetically altered plant toxins and even resistance to cultural practices such as crop rotation are of major concern today.

Compatibility of new technology with current practices and consistency with existing values and past experiences are important considerations in grower adoption of IPM (Masud *et al.*, 1985; Harris, 2001). New innovations or practices that vary little from existing practice are viewed as being less risky and easier to implement. The degree to which IPM is implemented on a specific crop can vary as growers may modify or adapt, as well as adopt, the technology to fit their individual needs. Grower modification of technology before adopting is referred to 'adaptive implementation' or 'reinvention' of the technology (Rogers and Shoemaker, 1971; Berman, 1980). An example of adoption vs. adaption occurred when monitoring techniques were implemented for processing tomatoes in California. Over a 5-year period after monitoring techniques were first available, 25% of all growers had adopted the IPM techniques, whereas 31% had adapted or reinvented the techniques to fit their operation. The degree to which the techniques were reinvented or adapted had little effect on the results but resulted in a better fit with other production practices (Grieshop et al., 1988; Zalom, 1999). The perceived complexity of a new IPM technology is important in its implementation. Growers are reluctant to implement IPM practices that appear to be very complex and complexity is often inversely related to the rate and degree of adoption (Nowak, 1987).

Processes Required for Successful Adoption of IPM

Three elements or processes are required for successful IPM diffusion and adoption. These are development of technology, delivery of technology and integration of the technology into the farming system (Goodell and Zalom, 2000). The development of new technology has historically been within the realm of researchers, much of it within the USDA and State Agricultural Experiment Stations within the land grant university system. From a historical perspective, several large government-funded projects were vitally important in the development of many of the concepts, tactics and knowledge that would become the foundation of IPM (Goodell and Zalom, 2000). The Huffaker Project (1972–1978) funded $13 million to 18 universities and enabled 300 researchers including multidisciplinary teams of agronomists, ecologists,

economists, entomologists, plant pathologists and system analysts, to develop ecological principles and basic concepts that improved our understanding of the dynamics of crop and pest development. The project, which focused on lucerne, pomes and stone fruits, citrus, cotton, pine forests and soybeans, resulted in advances in insect monitoring, use of insecticides, biological controls and evaluation of environmental and economic impacts of pest management (Frisbie, 1985). This project was the first major agricultural research project in the USA to use computer technology and system analysis to build models of crop ecosystems (Olsen et al., 2003).

The Consortium of IPM (CIPM), which operated between 1979 and 1985 built on the principles of the Huffaker Project and more fully incorporated encompassed pathogens, nematodes, weeds and vertebrate pests by bringing in 15 universities and a systems approach to total crop production (Goodell and Zalom, 2000). The USDA National IPM Programme initiated in 1972 and continuing to the present stresses the importance of multi-state and institutional collaboration through an interdisciplinary approach. It provides tools and practices for regional IPM projects and it has led to many advances in IPM knowledge. In 2000, the USDA created four regional IPM centres in the USA to continue and increase the emphasis on regional approaches and interstate collaborations. Also, in recent years, commercial companies, private consultants and Cooperative Extension specialists have all been more involved in development of new IPM technology and tactics. The development of transgenic crops is a prime example of commercial company involvement.

Delivery of technology, tactics and knowledge to end-users has long been a goal of Cooperative Extension. In recent years, with the reduction in numbers of Extension professionals due to budgetary constraints, many producers have increasingly relied on private consultants, certified crop advisers, seed and chemical companies or personnel employed by farm cooperatives (Wearing, 1988; Alston and Reding, 1998; Gray, 2001; Propst and Smolen, 2002).

Integration of technology, tactics and knowledge into farming systems occur largely through private consultants and farmers themselves although many states have Extension Agents, Extension specialists or farm advisers who facilitate integration of technology through demonstration and/or verification programmes. States such as Texas have Extension Agents-IPM who interact directly with growers on a one-to-one basis at the farm level to help integrate technology through educational and demonstration programmes. This approach has proven to be highly successful in implementing new IPM technology (Fuchs *et al.*, 1997; Teetes *et al.*, 1998). Annually, 23 Extension Agents-IPM in Texas working with agricultural producers conduct approximately 300 on-farm demonstrations to demonstrate and adapt new technology to local production systems.

Factors Affecting Adoption of IPM

In addition to the attributes of the technology, a number of social factors that involve the end-users affect diffusion of technology and adoption of IPM. These include farm ownership, farm size, grower demographics, educational outreach and many more. Growers who own their farm or whose family owns the farm are more likely to adopt technology than those who do not. Farmers who do not own their own farms often have to consider the attitudes and opinions of their landlords in making management decisions. Farms with corporate ownership are more likely to adopt than those owned by a partnership (Grieshop *et al.*, 1988). Larger and more specialized operations are more likely to adopt technology than smaller, more diversified operations (Nowak *et al.*, 1966; Alston and Reding, 1998; Lamboy, 2002).

Important grower characteristics include age, educational level, farm experience and gross farm sales (Nowak *et al.*, 1966; Ashby, 1982; Carlson and Dillman, 1988; Thomas *et al.*, 1990; Drost, 1996; Alston and Reding, 1998). Younger, better-educated farmers

tend to have more interactions with other farmers, information sources (including computers) and with change agents (Rogers, 1983; Nowak, 1987; Sorenson, 1993). They also tend to be less influenced by peer pressure or prevailing attitudes in the farming community.

The degree of educational outreach is another important factor in adoption of IPM. A number of studies documented that Extension outreach including technical information, publications, field meetings and demonstrations are important (Grieshop *et al.*, 1988; Thomas *et al.*, 1990; Alston and Reding, 1998). Ridgley and Brush (1992) documented that growers who place value on information obtained from Cooperative Extension were more likely to adopt IPM than those who place higher value on other information sources. However, private sector sources, producer cooperatives and publicly held companies are becoming increasingly important as a source of IPM information in many areas.

Other factors that impact adoption include availability of labour and management expertise, interaction with producer groups and commodity organizations, peer pressure and availability of financing. Growers tend to be more adverse to what some perceive as risks of using IPM if they have a large capital investment in the crop (Tait, 1978) and some simply fail to allocate sufficient management time to adopt and implement IPM effectively (Corbet, 1981).

Peer pressure can be an important factor in adoption of IPM practices and technology especially in programmes of countywide or regional scope. Implementation of a regional uniform planting date in the Texas Rolling Plains cotton production system as a cultural control technique to manage the boll weevil serves an example of where peer pressure was used to bring about essentially 100% conformity (Masud *et al.*, 1985). In this programme growers were urged to plant uniformly and late enough in the spring so that boll weevil emergence from overwintering habitat was essentially completed before cotton squares were available as a feeding and ovipositional site. The result was 'suicidal emergence' by a large percentage of overwintered weevils, thus delaying the need to use insecticides to control the population.

Barriers to Adoption of Technology

Farmers choose not to adopt technology for a number of reasons but reasons generally fit into two categories; either they are unwilling or unable to adopt (Nowak *et al.*, 1966). Sometimes they may be both unwilling and unable. Just because the grower chooses not to adopt a technology does not necessarily mean that the technology would not be beneficial nor does it mean that the grower does not like or want to adopt. Likewise, it does not mean the grower is not progressive or a 'good' grower. Often there are very good and logical reasons for non-adoption.

Being unable to adopt a technology generally means that some barrier or obstacle to adoption exists such that the decision not to adopt is rational and correct. The grower may be willing to adopt but for some reason is unable to make the decision. There are a number of reasons why a grower may make this decision.

A grower may be unable to adopt an IPM practice, for example, because he or she does not have all of the information needed to make a sound agronomic or economic decision or because the information is too difficult or expensive to obtain. Even in today's electronic age, obtaining information usually comes at some cost to the grower in terms of time and/or money. In other instances, there may be technical obstacles such as lack of specific knowledge of biology of pests, natural enemies and their interactions or lack of economic thresholds, guidelines or adequate sampling procedures.

The complexity of the practice or the expense of adoption in terms of investment in equipment may be too high for grower to afford. Or the grower may not have the labour resources, in terms of quantity or quality to properly implement the technology. Lack of managerial skills to implement the technology himself and limited access to support from Extension, crop consultants or other professionals might lead to non-adoption. Research studies indicate that adoption rates of technology are inversely related to complexity of the technology. The growers may also lack approval of a partner, a landlord, a financial lender or other party necessary to make the decision to adopt.

In addition to being unable to adopt a practice, a grower may be unwilling to adopt because he has not been fully convinced that the technology or practice will work, that it will increase net profits or is appropriate for his particular farming operation. There are also good and appropriate reasons for a grower to come to that conclusion (Nowak *et al.*, 1966). One reason could be that the information he has seen or heard regarding the practice is conflicting or inconsistent. Another reason might be that the grower is not convinced that the information is relevant to his particular farming operation, soil type, or environmental conditions or that it conflicts with other production practices on the farm. Research information that comes from another state or part of the state with very different growing conditions compared to the local situation which has not been confirmed by local experimentation or demonstration often leaves a grower unsure of how it applies to his farm or farming operation.

A grower might also believe that adopting the new practice will increase economic risks or result in some negative outcome that he is unwilling to accept. He might be satisfied and much more comfortable with traditional methods, which have supported him and his family while witnessing negative outcomes in others who attempted new techniques or technologies.

Agricultural policy also plays an important role in adoption of new technology including IPM. Farmers, like the rest of us, generally do that for which they are rewarded. Farm legislation and specific farm programmes for which growers may receive financial rewards may contain requirements that conflict with the goals of IPM. One example is where federal farm programmes provide incentives to growers that may be more advantageous for a specific crop compared with other crops, which the farmer might choose to grow. This perhaps results in the farmer growing the preferred crop in a certain field for several consecutive years rather than rotating to another crop that might reduce disease inoculum, insect pests or weed infestations. Another example occurs when federally subsidized crop insurance programmes contain provisions that treat one crop more favourably than others and that could serve as rotational crops or crops with more tolerance to prevailing insect or disease problems.

Methods of Delivery of IPM Information to Clientele

The land grant university model

Much of the information and technology on IPM developed and delivered in the 1970–1990 era was through land grant universities and the USDA. Land grant universities have three missions – teaching, research and outreach. Researchers (both university and USDA scientists) develop new technology, which is taught to students, interpreted and extended to clientele, generally by Cooperative Extension. Extension, historically, works primarily through County Extension Agents located at the county level who conduct educational programmes. County Extension Agents are trained and supported by Extension specialists often located at the university or at Research and Extension Centers located in strategic areas across the state. A number of states now have 'specialty' Extension Agents who are highly trained in specific areas such as entomology, weed science, agronomy, horticulture or IPM.

The land grant system model has been, and continues to be, highly successful in implementing IPM programmes. In Texas, for example, use of this system has resulted in a large percentage of growers of cotton, sorghum, pecans and vegetables practising IPM (Fuchs et al., 1997; Harris et al., 1998; Teetes et al., 1998; Harris, 2001; Smith et al., 2002).

Education of the user is generally regarded as the key to successful implementation of IPM (Wearing, 1988). Extension

delivers educational information by a number of different methods depending upon the audience and needs of the clientele. For example, mass media such as television, newspapers and now the Internet are useful for creating awareness. Many studies have confirmed that verbal communication is most effective and one-on-one consultation between Extension professionals and growers is the most effective form of verbal communication (Frisbie and Adkisson, 1985; Rojotte, 1985; Wearing, 1988). Field meetings in which an Extension educator, researcher or both meet with growers to provide hands-on training in scouting, problem diagnosis or pest management decision making have proven to be an excellent educational approach. Instances where field meetings are held at regular intervals throughout the production season have been particularly effective.

In addition to verbal communication, several other forms of communication with clientele are very useful. Written communication including fact sheets, brochures, bulletins and management guides are generally considered less effective than verbal communication but important. Newsletters, especially those containing up to date and timely information based upon field monitoring and offering management options to growers, are highly effective and highly rated by growers.

Computer-based information had little impact on crop protection decision making as late as 1988 (Wearing, 1988) but has gained in importance as the percentage of growers who use computerized decision aids and access the Internet increases. This source of educational information is becoming increasingly important as growers become more skilled at accessing electronic information.

Field demonstrations on grower's farms are perhaps the most powerful method of transferring technology to end-users. Dr Seaman A. Knapp, considered as the father of Cooperative Extension, was quoted as saying, 'What a man hears, he may doubt; what he sees, he may possibly doubt; but what he does himself, he cannot doubt.' The 'tried and true' concept of field

demonstrations is currently being emphasized in many successful Extension efforts
in the USA and it has become increasingly
important in developing countries (Anonymous, 2003). Growers not only learn of the
value of the new practice or technology on
their own farm and under their own production system and local conditions, but
often growers 'reinvent' or modify the technology to work even more effectively after
it is demonstrated.

One of the most powerful field demonstration techniques has the grower,
researcher, Extension specialists and Extension Agents interacting during the planning,
implementation and evaluation phases of
the demonstration. One pitfall that has often
occurred in the land grant system happened
when a researcher developed a new practice
or new technique, which had great potential, but was not developed adequately for
field implementation at the time the project
was terminated. This has been especially
prevalent in recent years when researchers have relied upon short-term grants to
fund their research and had to terminate
the project when the grant terminated and
often before the technology had been tested
under a range of environmental conditions
or production systems.

Two relatively recent changes to the
traditional land grant university model,
however, are helping in this regard. First,
Extension specialists and Extension Agents,
particularly specialty agents such as IPM
Agents are becoming increasingly involved
in adaptive research in cooperation with
research scientists. This allows the devel-
oper and the implementers of technology
to work together with growers to test and
demonstrate the technology over a wider
range of environmental conditions and production systems to test the robustness of the
technology and help to adapt or 'reinvent'
it for field use.

A second change is the increasing
emphasis on joint research/Extension
appointments whereby the development
and delivery of the technology to end-users
are both within the responsibility of the
same individual.

Role of private practitioners

Private practitioners of IPM whether
labelled private consultants, IPM consultants or PCAs (Pest Control Advisors) have
historically played a vital role in implementation of IPM and their role will be increasingly important role in the future. Many of
the early IPM consultants had direct ties to
university research and/or extension programmes, having been employed in the university system. Most are highly trained and
experienced in IPM. The economic benefits
of IPM demonstrated through university-
based programmes helped convince growers
of the benefits of hiring an IPM professional
to help manage their crops. As a result, the
number of private consultants practicing
in the USA increased dramatically, especially in the 1980s and 1990s. The National
Alliance of Independent Crop Consultants
expanded membership from 16 members
in 1979 to 180 in 1989 and to 498 in 1999.
As a result, the percentage of crops scouted
by professionals increased dramatically.
Cotton acreage scouted by professionals in
the 14 major producing states, for example,
increased from 41% in 1982 to 55% in 1989
(Ferguson and Yee, 1995). The use of private consultants has been shown to have a
positive influence on the adoption of IPM
practices (Thomas et al., 1990).

Private consultants work directly with
growers on a one-on-one basis and generally
help manage not only pests but also other
aspects of crop production including selection of the cultivar planted, tillage methods,
fertility, irrigation and other inputs. Many
work with the same producers year after
year and develop an intimate knowledge of
the farm and the management style of the
farmer. They are thus able to provide a very
high level of service to the grower.

The need for personnel to provide information and recommendations to producers is
exemplified by interest in certification programmes such as the Certified Crop Adviser
(CCA) programme offered by the American
Society of Agronomy. The CCA programme
is a voluntary programme, which provides a
base level of standards through testing and

the requirement for continuing education credits. To become certified, the applicant is required to pass two examinations and submit credentials detailing their education and crop advising experience. They are also required to provide two references and sign a code of ethics. The examinations cover four major competency areas including nutrient management, soil and water management, IPM and crop management. Almost 15,000 persons have been certified through this programme (Anonymous, 2004).

California, in 1972, became the first state to require that all individuals who recommend the use of any pest control method, including pesticides, be licensed by the state (Goodell and Zalom, 2000). Currently, many states have licensing requirements, most tied to making pesticide and/or fertilizer recommendations.

Chemical and seed suppliers have historically provided the necessary pesticides and seeds to growers. In recent years, increased competition has stimulated dealers to diversify the products they offer and, in many cases, provide information and even scouting services in addition to conventional products. Some suppliers/dealers charge a fee for scouting services while others provide scouting services in exchange for an 'assumed' agreement by the grower to purchase pesticides, seeds or other inputs from the dealer or supplier.

Chemical dealers face a number of challenges in providing information-based services to producers. For example, they must compete with independent and unaffiliated crop consultants and while contending with potential conflict of interest in terms of the products they sell (Sorensen et al., 2000). However, it appears that an increasing number of growers are using the services of dealers and suppliers for pest scouting and other services.

Industry involvement

Campbell Soup Company provides one of the best examples of industry developing and delivering IPM strategies and informa-tion to growers, which led to a very successful IPM programme. Each year, the Campbell Soup Company contracts with growers for thousands of hectares of processing tomatoes in California, Ohio and Mexico, carrots grown in California, Texas and New Jersey and celery grown in California, Michigan and Florida. In 1990, the company adopted a 'total system' approach to pesticide management with the corporate goal to reduce the number of synthetic pesticide applications per season by 50% by 1994 using IPM. After 4 years of developing and implementing IPM strategies and providing information and assistance to growers, the Campbell Soup Company attained and surpassed its corporate goal (Bolkan and Reinert, 1994).

More and more individual companies have taken on the role of delivering technical information relative to their products directly to end-users. Many conduct grower meetings or seminars and some have established databases of individual customers so that information can be delivered electronically directly to customers. Companies that sell transgenic technology, delivered in the seeds of crops, often are active in having technology agreements with individual growers.

Grower groups and commodity organizations

Many grower groups and commodity organizations have an educational outreach component to the services they provide. The National Cotton Council, for example, annually sponsors the Beltwide Cotton Conference where participants representing all aspects of the cotton industry come together to share information and hear the latest research results and industrial developments. Many other commodity organizations have similar events to inform and educate their membership.

National and state grower groups such as the Farm Bureau as well as groups formed for specific purposes such as the Texas Pest Management Association and the Lodi–Woodbridge Winegrape Commission in California take an active role in educating members and promoting the adoption of new technology (Sorensen, 1993; Zalom, 1999).

Farmer-to-farmer methods

Farmer-to-farmer communication and/or groups of farmers directing their own resources towards a coordinated effort have proven to be highly effective in diffusion of technology (Stark *et al.,* 1990; Thomas *et al.,* 1990; Alston and Reding, 1998; Zalom, 1999). Farmer involvement in field meet-

ings and field demonstrations where growers discuss their success in adopting new IPM technology has proven to be highly successful (Goodell and Zalom, 2000).

Because IPM systems are not static, there will always be a need to develop and deliver new knowledge and with IPM evolving towards more complex integrated systems, approaches to the delivery of information will continue to change. The increasing rate of development and the introduction of new technologies will challenge the agricultural community's ability to provide training and delivery systems to incorporate the new tools into existing IPM programmes.

References

Alston, D.G. and Reding, M.E. (1998) Factors influencing adoption and educational outreach of integrated pest management. Available at: http://www.joe/1998june/a3.html.

Anonymous (2003) *Bangladeshi Farmers Benefit from IPM.* IPM CRSP Progress Report No. 4. Office of International Research, Education and Development, Virginia Tech, Blacksburg, Virginia.

Anonymous (2004) American Society of Agronomy Certified Crop Adviser. Available at: http://www.agronomy. org./cca/faq.htm)

Ashby, J.A. (1982) Technology and ecology: implications for innovation research in peasant agriculture. *Rural Sociology* 47, 234–250.

Bennett, C.F. (1977) *Analyzing Impacts of Extension Programs.* Publication ESC-575, Extension Service, US Department of Agriculture, Washington, DC.

Berman, P. (1980) Thinking about programmed and adaptive implementation: matching strategies to situations. In: Ingram, H. and Mann, D. (eds) *Why Policies Succeed or Fail.* Sage, Beverly Hills, California, pp. 205–230.

Bolkan, H.A. and Reinert, W.R. (1994) Developing and implementing IPM strategies to assist farmers: an industry approach. *Plant Disease* 78, 454–550.

Carlson, J.F. and Dillman, D.A. (1988) The influence of farmers' mechanical skill on the development and adoption of a new agricultural practice. *Rural Sociology* 53, 235–245.

Copp, J.H., Sill, M.L. and Brown, E.J. (1958) The function of information sources in the farm practice adoption process. *Rural Sociology* 23, 146–157.

Corbet, P.S. (1981) Non-entomological impediments to the adoption of integrated pest management. *Population Ecology* 3, 183–202.

Corbet, P.S. and Smith, R.F. (1976) Integrated control: a realistic alternative to misuse of pesticides? In: Huffaker, C.B. and Messenger, P.S. (eds) *Theory and Practice of Biological Control.* Academic Press, New York, pp. 661–682.

Dent, H.S. Jr (1998) *The Roaring 2000s.* Simon & Schuster, New York.

Dent, H.S. Jr (1999) *The Great Boom Ahead.* Simon & Schuster, New York.

Drost, D. (1996) Barriers to adopting sustainable agricultural practices. Available at: http://www.joe. org.1996december/index.html

Ferguson, W. and Yee, J. (1995) A logit model of cotton producer participation in professional scout programs. *Journal of Sustainable Agriculture* 5, 87–96.

Fernandez-Cornejo, J., Alexander, C. and Goodhue, R.E. (2002) Dynamic diffusion with disadoption: the case of crop biotechnology in the U.S.A. *Agriculture Resource Economic Review* 31, 112–126.

Fliegel, F.C. (1993) *Diffusion Research in Rural Sociology: The Record and Prospects for the Future.* Greenwood Press, Westport, Connecticut.

Frisbie, R.E. (1985) Consortium for integrated pest management organization and administration. In: Frisbie, R.E. and Adkisson, P.L. (eds) *CIPM: Integrated Pest Management on Major Agricultural Crops.* Texas Agricultural Experiment Station, Texas, MP-1616, pp. 1–9.

Frisbie, R.E. and Adkisson, P.L. (1985) IPM: definitions and current status in US agriculture. In: Frisbie, R.E. and Adkisson, P.L. (eds) *Biological Control in Agricultural IPM Systems.* Academic Press, New York, pp. 41–51.

Frisbie, R.E. and McWhorter, G.M. (1986) Implementing a statewide pest management program for Texas, U.S.A. In: Palti, J. and Ausher, R. (eds) *Advisory Work in Crop Pest and Disease Management.* Springer-Verlag, Berlin, pp. 234–262.

Fuchs, T.W., Smith, D.T. and Holloway, R. (1997) Status of IPM and insecticide use in Texas cotton. In: Duggar, P. and Richter, D.A. (eds) *Proceedings Beltwide Cotton Conference.* National Cotton Council, Memphis, Tennessee, pp. 1140–1144.

Goodell, P.B. and Zalom, F.G. (2000) Delivering IPM: progress and challenges. In: Kennedy, G.C. and Sutton, T.B. (eds) *Emerging Technologies for Integrated Pest Management.* APS Press, St Paul, Minnesota, pp. 483–496.

Graebner, L., Moreno, D.S. and Baritelle, J.L. (1984) The Fillmore Citrus Protective District: a success story in integrated pest management. *Bulletin Entomological Society of America* 30, 27–33.

Gray, M.E. (2001) The role of extension in promoting IPM programs. *American Entomologist* 47, 134–137.

Grieshop, J.I., Zalom, F.G. and Miyao, G. (1988) Adoption and diffusion of integrated pest management innovations in agriculture. *Bulletin Entomological Society of America* 34, 72–78.

Griliches, Z. (1957) Hybrid corn: an exploration in the economics of technological change. *Econometrica* 25, 501–522.

Harris, M.K. (2001) IPM, what has it delivered? *Plant Disease* 8, 112–121.

Harris, M.K., Ree, B., Cooper, J., Jackman, J., Young, J., Lacewell, R. and Knutson, A. (1998) Economic impact of pecan integrated pest management in Texas. *Journal of Economic Entomology* 91, 1011–1020.

Lamboy, J.S. (2002) *Level of Adoption of IPM in New York Greenhouses.* New York State Pest Management Program, IPM Publication Number 417, Ithaca, New York.

Masud, S.M., Lacewell, R.D., Boring III, E.P. and Fuchs, T.W. (1985) Economic implications of a regional uniform planting date cotton production system: Texas Rolling Plains. *Journal of Economic Entomology* 78, 535–541.

Nowak, P. (1987) The adoption of agricultural conservation technologies: economic and diffusion explanations. *Rural Sociology* 52, 208–220.

Nowak, P., Padgett, S. and Hoban, T.J. (1966) Practical considerations in assessing barriers to IPM adoption. *Proceedings of the Third National IPM Symposium/Workshop.* US Dept of Agricultural Economic Research Services, Washington, DC, MP 1542.

Olsen, L., Zalom, F. and Adkisson, P. (2003) Integrated pest management in the USA. In: Maredia, K.M., Dakouo, D. and Mota-Sanchez, D. (eds) *Integrated Pest Management in the Global Arena.* CAB International, Wallingford, UK, pp.249–271.

Post, G. (1988) The private consultant: benefit or burden? *HortScience* 23, 490–492.

Propst, T.L. and Smolen, M.D. (2002) Demonstration of best management practices in the Salt Fork Watershed. Final Report CWA Section 319 (h) FY1997. Oklahoma Conservation Task #96. Oklahoma Cooperative Extension Service. Available at: http://biosystems.okstate.edu/waterquality

Ridgley, A.-M. and Brush, S.B. (1992) Social factors and selective technology adoption: the case of integrated pest management. *Human Organization* 51, 367–378.

Rogers, E.M. (1983) *Diffusion of Innovations,* 3rd edn. Free Press, New York.

Rogers, E.M. (1995) *Diffusion of Innovations,* 4th edn. Free Press, New York.

Rogers, E.M. and Shoemaker, F.F. (1971) *Communication Innovations: A cross-Cultural Approach.* Free Press, New York.

Rojotte, E.G. (1985) IMPACTS – The newsletter of the national evaluation of extension integrated pest management programs. Virginia Cooperative Extension Service, USDA, Blacksburg, Virginia.

Scheuring, A.F. (1988) *A Sustaining Comradship.* University of California, Division of Agriculture and Natural Resources, Oakland, California.

Smith, D.T., Harris, M.K. and Liu, T.X. (2002) Adoption of pest management by vegetable growers: a case study. *American Entomologist* 48, 236–242.

Smith, R.F. and Huffaker, C.B. (1973) Integrated control strategy in the USA and its practical implementation. *EPPO Bulletin* 3, 31–49.

Sorensen, A. (1993) IPM and growers: an evolution of thinking. In: Leslie, A.R. and Cuperus, G.W. (eds) *Successful Implementation of Integrated Pest Management for Agricultural Crops*. Lewis Publishers, Boca Raton, Florida, pp.129–151.

Sorensen, A., Day, E. and Steward, P.A. (2000) Factors affecting the adoption of new technologies. In: Kennedy, G.G. and Sutton, T.B. (eds) *Emerging Technologies for Integrated Pest Management*. APS Press, St Paul, Minnesota, pp. 12–31.

Stark, J.A., Cuperus, G.W., Ward, C., Huhnke, R., Stritzke, J. and Berberet, R. (1990) *Alfalfa Integrated Management: A Case Study*. Oklahoma State University Cooperative Extension Service, Oklahoma, Circular E-899.

Tait, E.J. (1978) Factors affecting the usage of insecticides and fungicides on fruit and vegetables crops in Great Britain. II. Farmer-specific factors. *Journal of Environment Management* 6, 127–142.

Teetes, G.L., Pendleton, B.B. and Parker, R.D. (1998) *Quantifying the Use of IPM by Sorghum Growers*. Project Termination Report, Sorghum Insect Laboratory Entomology, Texas A&M University, College Station, Texas.

Thomas, J.K., Ladewig, H. and McIntosh, W.A. (1990) The adoption of integrated pest management practices among Texas cotton growers. *Rural Sociology* 55, 395–410.

Way, M.J. (1977) Integrated control – practical realities. *Outlook Agriculture* 9, 127–135.

Wearing, C.H. (1988) Evaluating the IPM implementation process. *Annual Review of Entomology* 33, 17–38.

Zalom, F. (1999) Professional training and technology transfer. In: Ruberson, R. (ed.) *Handbook of Pest Management*. Marcel Dekker, New York, pp. 765–787.

Index